Calculus for Engineering and the Sciences

PRELIMINARY VERSION ▪ VOLUME 1

ELGIN H. JOHNSTON
Iowa State University

JEROLD MATHEWS
Iowa State University

HarperCollins*CollegePublishers*

To Gail, for her patience, support, and love
To Ellie, with gratitude

Sponsoring Editor: Kevin Connors
Senior Development Editor: Kathy Richmond
Production Administrator: Randee Wire
Project Coordination and Text Design: Publication Services, Inc.
Photo Researcher: Nina Page
Cover Design: Lesiak/Crampton Design: Lucy Lesiak
Printer and Binder: R. R. Donnelley & Sons Company
Cover Printer: The Lehigh Press, Inc.

Cover Photo: The gravitational force ($\mathbf{F}_g$) and the normal force ($\mathbf{N}$) perpendicular to the track govern the motion of this loop-the-loop roller coaster. The magnitude and direction of the velocity ($\mathbf{V}$) and the normal force are continually changing. Calculus is the branch of mathematics that deals with such changes; here engineers apply calculus and physics in designing a safe but thrilling ride.

Photo and Permission Credits: **Chapter 1:** Section 1.3 exercise #24, p. 33: "Portion of a Topographical Map." Copyright ©1988 by Trails Illustrated. Reprinted by permission of Trails Illustrated, a division of Ponderosa Publishing Company. Fig. 1.35, p. 30: "Plot by Kamerlingh Onnes" from *Introduction to Solid State Physics*, 3/e by C. Kittle, 1966. Reprinted by permission of John Wiley & Sons. Photo, p. 89: Courtesy of IBM. **Chapter 2:** Fig. 2.5, p. 102: "Chain Block Hoist" from *Van Nostrand's Scientific Encyclopedia*, 5/e edited by D.M. Considine, p. 482. Reprinted by permission of Van Nostrand Reinhold. Photo, p. 193: Scott Foresman. **Chapter 3:** Photo, p. 251: The Bettman Archive. Photo, p. 266: Scott Foresman. **Chapter 4:** Photo, p. 346: Neuhart Donges Neuhart. **Chapter 5:** Photo, p. 425: The MIT Museum. **Chapter 6:** Photo, p. 502: ITAR-TASS/SOVFOTO.

Calculus for Engineering and the Sciences Preliminary Version   Volume 1

Library of Congress Cataloging-in-Publication Data

Johnston, Elgin.
    Calculus for engineering and the sciences / Elgin Johnston, Jerold
  Mathews. — Prelim. version.
       p.   cm.
    Includes index.
    ISBN 0-06-501577-0 (v. 1)
    1. Calculus.   I. Mathews, Jerold C.   II. Title.
QA303.J595   1995
515'.0245–dc20   95-35571

95 96 97 98 9 8 7 6 5 4 3 2 1

Preface

Approach

Since 1970, the content, teaching, and learning of calculus have been affected by the increasing importance of mathematics in the engineering, mathematical, and physical sciences, by the opportunities presented by student-accessible computing technology, and by the needs of the increasingly diverse students who are its clientele. These developments make calculus teaching an exciting challenge. This book, written for students majoring in science, engineering, or the mathematical sciences, is our response to this challenge.

With the help of grants from the National Science Foundation, we have written a text intended to help these students learn calculus more effectively, so that students are better prepared for subsequent technical courses and possess more experience in solving realistic, multistep problems. Our intent/constraint was to do this within a fairly traditional syllabus.

We have provided for more effective learning and problem-solving experience in several ways. Among the skills of good problem solvers is flexibility, including the use of numerical, graphical, and symbolic representations of a problem or its solution. We often use such representations and ask students to model our practice in their work.

In the first three chapters we discuss approximation and error, trigonometric and exponential functions, and vectors and parametric equations, all of which are important in the engineering, mathematical, and physical sciences. We return to these ideas in subsequent chapters, thus giving students repeated practice with important concepts. We have included as many realistic problems as possible, both in section exercises and in the student projects at the end of each chapter.

Pedagogical Features

To better prepare for advanced courses in mathematics, physics, or other disciplines, students need to work actively at solving problems, trying examples, and investigating troublesome points through graphical, numerical, and analytic representations. This is asking a lot of students, but effective learning of the subject demands it. Toward these ends, we provide a number of features to help students explore, learn, and understand calculus.

Minimum Technology We assume that each student has convenient access to at least a graphics calculator for reading the text and working on exercises. This is reflected in the importance we give to the rule of three (discussed below) in presentations and exercises and in our de-emphasis of graphing by hand.

Investigations New concepts often are introduced through an investigation of a physical or mathematical problem. For example, rate of change is introduced by investigating the rate of change of pressure in a piston chamber as the piston moves, vectors and parametric equations emerge from a discussion of position and velocity, a formula for areas of regions bounded by closed curves is inferred from a formula used by surveyors to calculate the area of a polygonal region (this, together with parts of the section on work, provides an early look at line integrals), and Taylor polynomials are developed in an effort to find better and better approximations to the exponential function.

Examples   Every idea developed in the text is demonstrated in one or more examples. We use the examples to model good problem-solving techniques and especially to demonstrate how numerical, graphical, and analytic techniques can be used in investigating a problem. Many examples are solved in more than one way to remind students that there is no one right way to solve a problem.

Graphs and Tables   When every student has access to a graphics calculator or computer algebra system, the figures and tables in a text can play a larger role. To encourage more active reading, we often urge students to verify table entries or to reproduce graphs that accompany many investigations and examples. In most cases, students can obtain qualitatively similar graphs with a graphics calculator.

Exercises   Each exercise set consists of three sections: **Basic, Growth,** and **Review/Preview.** The Basic exercises are routine applications of the formulas and techniques illustrated in the examples. By working through these exercises students get practice in the skills necessary to solve problems using calculus. The Growth exercises provide opportunities for students to build upon the ideas in the text discussion and examples. Some are more challenging versions of Basic exercises, some require the solution of intermediate problems, some are open-ended, and some require the student to experiment before making and justifying a conjecture. Many Growth exercises ask the student to respond with a brief, written explanation. Through a careful choice of Growth exercises instructors can give assignments appropriate for their students. Every exercise section ends with several Review/Preview problems. These problems review important topics covered earlier in the text or in previous courses, especially when these topics will be needed in future sections, or preview ideas that will be introduced later.

Chapter-End Material   Each chapter ends with review materials, beginning with a Review of Key Concepts section that briefly summarizes the chapter highlights. Next comes a Chapter Summary review grid that includes many of these ideas with a graph, a formula, or a short example. Following the review grid are Chapter Review Exercises for the entire chapter.

Student Projects   One of the goals of our NSF-sponsored activities was to produce a collection of projects to give students significant problem-solving experience. Several of these projects are included at the end of each chapter, and more will appear in a supplementary collection. A sample of projects is given in "To the Student."

Since 1987 we have assigned at least three student projects each semester. Working in teams of two or three, students solve these projects, using computers or calculators as necessary, and write up their results. We have adapted our course outline and the content of this book to provide for both the in- and out-of-class time needed for these projects.

We have appended the student projects to the chapters instead of integrating them into the discussion since we presume that some instructors will not assign student projects or will assign different materials.

History of Measuring and Calculating Tools   For thousands of years, men and women have been using mechanical devices to help them calculate and measure. Many of these ingenious tools were designed to solve problems that are still of interest to calculus students. Some of these devices were important to the development of present-day calculators and computers.

Major Features and Content

Introduction to the Rule of Three
Chapter 1 reviews functions and introduces students to the rule of three, which is used to study functions numerically, graphically, and analytically. Using the familiar idea of slope of a line, we discuss rate of change by zooming in on the graph of the nonlinear equation relating the pressure and volume of a gas being compressed by a piston, thus linking the idea of rate of change to a concrete, physical interpretation and to the geometric idea of slope. We continue the graphical, numerical, and analytic approach in a discussion of limits through ideas of approximation and error. At the end of the chapter we again discuss rate of change and formally define the derivative.

Quick Start to Calculus
In Chapter 2 we continue a quick start to calculus. We review the elementary functions, calculate their derivatives, and discuss the mechanics of differentiation. In Chapter 3 we discuss vectors and parametric equations in the context of describing the motion of an object.

We believe the early study of vectors and parametric equations has several advantages. The study of motion was important in the development of calculus and remains one of its richest sources of applications. The early introduction of vectors allows us to use motion in motivating several important topics, and it broadens the range of applications-based problems. The early introduction of vectors gives students more time to become familiar with vectors, for use in physics (and other courses) as well as mathematics. The rich connections between the position vector and the time parameter of motion in two dimensions leads in a natural way to parametric equations. The study of the length and direction of a vector leads to the concept of polar coordinates. The entire discussion of vectors and parametric equations provides a review of portions of trigonometry.

Modeling
We tell students that calculus is the language of science and engineering because it can be used to describe the world around us. But how do scientists, engineers, and mathematicians arrive at a mathematical description of a phenomenon? In Chapter 2 and Chapter 3 we discuss a variety of mathematical models. We describe several real-life situations, discuss some of the simplifying assumptions that modelers might make, and then show how these assumptions are translated into equations. The wide range of examples, from the spread of a disease to Newton's laws, show that calculus is a powerful and versatile tool for describing the world around us.

Differential Equations
We discuss topics from differential equations when it serves our purpose of helping students effectively learn calculus. After using motion to motivate the ideas of position and velocity vectors and asking students to solve several "forward problems," in which they calculate the velocity and acceleration of an object from its position vector, we ask them to solve several inverse problems. Using simple models, mostly familiar from solving related forward problems, students determine the position vector of an object from its velocity and initial position. These problems provide good applications of calculus and background for the Fundamental Theorem of Calculus. Additional background is given in Chapter 4, where the tangent line approximation and Euler's method are discussed. Using vector notation, Euler's method is extended to the second-order differential equation describing the motion of a simple pendulum. Finally, in

Chapter 7 power series are used in solving several first- and second-order differential equations.

Problem Solving and Applications

Referring particularly to the sections on related rates, optimization, and volumes by cross-section, we attempt to model and explain good problem-solving techniques, including graphical and numerical exploration to suggest a conjecture, the use of analytical techniques to verify or disprove conjectures suggested by graphs or numerical evidence, interpretation of a solution in the context of the problem setting, and reflection on a solution to make sure it is consistent with common sense.

Integration

From the discussion in Chapter 4 of the forward and inverse problems in the context of motion and the connection with area that is motivated by Euler's method, the definition of the integral is straightforward. The Fundamental Theorem of Calculus is completed by the end of Section 5.2. Apart from a section on numerical integration, in the balance of Chapter 5 we discuss integration techniques. Emphasis is placed on three integration techniques and the use of tables or a CAS. We present the techniques of substitution, integration by parts, and partial fractions. Trigonometric substitutions are mentioned but given less emphasis. Their use in partial fractions is avoided by giving reduction formulas.

Approximations and Power Series

The main goal of the last chapter is to develop Taylor series, Taylor polynomials, and the tools to study them. We begin by seeking simple polynomial approximations to the exponential function. This leads to the definition of Taylor polynomials and a study of the error that arises when they are used in approximations. To better understand this error we move on to studying sequences and series of constants. This leads to Taylor series, and the behavior of these series brings us back to the Taylor polynomial approximation question. We conclude the chapter by discussing how Taylor series and polynomials can be used to find solutions or approximations to solutions of differential equations.

The Supplement Package to Accompany the Text

Calculus for Engineering and the Sciences is supported by an extensive supplement package available for purchase by your students.

The **Student Solution Manual,** written by Kirk Jones of Eastern Kentucky University and Tim Ray of Southeast Missouri State University, contains solutions to every odd-numbered exercise from the text. To order, use ISBN 0-06-502359-5.

An HP 48G® Calculus Companion by Jerold C. Mathews and Jack Eidswick is co-authored by one of the authors of **Calculus for Engineering and the Sciences.** This versatile book approaches calculus topics using the graphical, symbolic, and numerical power of the HP 48G® graphics calculator. To order, use ISBN 0-06-500165-6.

The DERIVE® Calculus Workbook by Lisa Townsley Kulich and Barbara Victor, both of Illinois Benedictine College, is a supplementary laboratory manual that encourages students to explore the concepts of calculus and improve their ability to communicate mathematics while developing their skills using DERIVE® software. To order, use ISBN 0-673-99455-4.

Calculus Labs Using Maple® by James Braselton, Arthur Sparks, and John Davenport, all of Georgia Southern University (ISBN 0-673-99814-2), and

Calculus Labs Using Mathematica® by Arthur Sparks, John Davenport, and James Braselton (ISBN 0-06-501196-1) are textbook-independent manuals to guide students through investigations of calculus concepts using a computer algebra system.

Laboratory Explorations in Calculus with Applications to Physics by Joan Hundhausen and F. Richard Yeatts, both of Colorado School of Mines, is a lab manual consisting of over 30 self-contained textbook-independent projects, divided into pencil-and-paper exercises, numerical exercises, and sample problems and can be used to support a course in either calculus or physics. To order, use ISBN 0-06-501719-6.

Acknowledgments

Acknowledgments   Many friends, students, and colleagues were of great help to us in writing this textbook. We were encouraged and guided by our own calculus students at Iowa State University as we tested these materials in our classrooms. The students were always willing to tell us what was working, what was not, and how things might be improved. We have lost track of the number of errors, both typographical and more serious, that our students have, sometimes with great pride, pointed out. Many of our colleagues at Iowa State University also contributed by test-teaching some materials and refining many of our student projects. These include Roger Alexander, Clifford Bergman, Alan Heckenbach, Irvin Hentzel, Justin Peters, Richard Tondra, and Bruce Wagner. The name of the Candidate Theorem is due to A. M. Fink. The idea for the Chapter Summary review grids came from Keith Stroyan. Our sincere thanks to our colleagues and students.

During the summers of 1992 and 1993 we offered two workshops in calculus reform. The 1992 workshop was at the University of Wisconsin at Lacrosse, and the 1993 workshop was given at Idaho State University in Pocatello. Although we went there to present our ideas and to demonstrate some of our student projects, we learned a lot by listening to participants' ideas on the directions calculus should take. We thank all of the workshop participants for their insights and ideas:

The University of Wisconsin at Lacrosse workshop: Carrie Ash-Mott, John Bruha, Robert Coffman, Wesley Day, Ronald Dettmers, G. S. Gill, David Hardy, Marian Harty, Linda H. Host, Erna Jensen, Clement Jeske, Charles Kolsrud, Larry Krajewski, Robert Kreczner, Don Leake, Steve Leth, Franklyn Lightfoot, Rich Maresh, Dan Nicol, Gerald W. Niedfeldt, David Oakland, Marlene Pinzka, Kay Strangman, Gordon Sundberg, Jack Unbehaun, Calvin Van Niewaal, Paul Williams, Randy Wills, J. D. Wine, and Elizabeth Wood.

The Idaho State University at Pocatello workshop: Sam Berney, Janet Burgoyne, John C. Eilers, Chaitan Gupta, Joseph Hwang, Jim Brennan, Bob Davis, Bob Firman, Roger Higdem, Ken Meerdink, Tom Misseldine, Eric Rowley, Madeline Schaal, Peter Wildman, Mike Prophet, Dan Schaal, Don Shimamoto, Suzanne Wisner, and Larry Ford.

We are grateful to the instructors who reviewed drafts of this book. Comments from these reviewers were invaluable in suggesting ways to improve our manuscript. Our sincere thanks to each of them:

Joe Albree, Auburn University–Montgomery

Daniel D. Anderson, University of Iowa

Alfred D. Andrew, Georgia Institute of Technology

William J. Barnier, Sonoma State University

William Bauldry, Appalachian State University

Maurino Bautista, Rochester Institute of Technology

Frank Beatrous, University of Pittsburgh

Mark Bridger, Northeastern University

Janet Burgoyne, South Dakota School of Mines and Technology

George Cain, Georgia Institute of Technology

Sandra Dawson, Glenbrook South High School

Audrey Douthit, Pennsylvania State University

Deborah D. Faust, William Rainey Harper College

Robert P. Flagg, University of South Maine

Donald Hartig, California Polytechnic State University–San Luis Obispo

L. Carl Leinbach, Gettysburg College

Steven C. Leth, University of Northern Colorado

Joan H. McCarter, Arizona State University

Steven J. Merrill, Marquette University

Joseph D. Myers, U.S. Military Academy

James Osterburg, University of Cincinnati

Justin Peters, Iowa State University

Richard D. Porter, Northeastern University

Rod Smart, University of Wisconsin–Madison

Kirby C. Smith, Texas A&M University

Craig Steenberg, Lewis-Clark State College

Patrick Sullivan, Valparaiso University

Ken E. Thomas, Andrews University

Jan Vandever, South Dakota State University

Many other people have been very helpful in finding materials on which to base problems and discussions. These include Susan North of the Ames Public Library, the reference staff at the Iowa State University Library, the reference staff at the Alaska State Library, the U.S. Geological Survey, and Michael Graff of the Iowa State University Computation Center. We also thank Rich Wolfson, Middlebury College, for his counsel about the cover photograph.

We invite suggestions from readers—students, professors, and others—for improvements to our text. We will try to incorporate appropriate suggestions in later printings. Please write us at HarperCollins or contact us by electronic mail.

Elgin H. Johnston
Iowa State University
johnston@pollux.math.iastate.edu

Jerold Mathews
Iowa State University
mathews@iastate.edu

To the Student: A Guide for Solving Calculus Problems

He bought Descartes's Geometry & read it by himself . . . when he was got over 3 or 4 pages he could understand no farther than he began again & got 3 or 4 pages farther till he came to another difficult place, than he began again & advanced farther & continued so doing till he made himself Master of the whole. . . .

This short quote was taken from *Never at Rest*, a biography of Isaac Newton, written by Richard S. Westfall (Cambridge University Press, 1983). Newton's struggles with Descartes's book occurred in 1664, one year before he calculated the area beneath a hyperbola to "two & fifty figures."

In his notebooks Newton left us clues to his method of working and solving problems. He read and reread until he understood, solved special cases, sketched diagrams, worked out numerical details to gain insight into a problem, and repeatedly returned to problems he could not solve.

Your job is not to invent calculus, but to come to understand and apply its main ideas. Both the understanding and the application of calculus are based on problem solving. This book features several kinds of problems. First come the examples, which are worked out in detail. The Basic exercises are problems similar in content and difficulty to the easier examples. Growth exercises include problems that are variations on examples, problems whose solutions depend on the recognition and solution of a subproblem, or problems in which the main ideas of the section are discussed further. At the end of each chapter we include one or more student projects.

We mentioned above that Newton calculated the area beneath a hyperbola to 52 figures. One of the two student projects in Chapter 5 reproduces Newton's drawing, analysis, and a sample of his numerical calculations. The solution of this problem requires careful reading, an interpretation of a graph, symbolic and arithmetic calculation, and an overall understanding of Newton's analysis. While the answer to many Basic and Growth exercises may be a graph or a single number, the answer to a student project is a description of your solution, sufficiently well planned and written that another person can follow your reasoning.

Although working on a student project takes effort and time, we believe the short- and long-term payoffs toward understanding and applying calculus are worth this investment. Many projects require some kind of computing technology, sometimes nothing more than a graphing calculator. Learning how, when, and to what extent computing technology should be used in problem solving is part of what you can learn from the student projects (and from some of the Basic and Growth exercises as well).

Here are some examples of the student projects.

Chapter 1:   Exploring Chaos. Iterates of the function $f(x) = cx(1 - x)$ for various values of c between 2 and 4, graphed in several ways.

Chapter 2:   Porsche 911RS America 1993 Test Results. Calculation of acceleration and position from published velocity data. Interpretation of results.

Chapter 3:   Timing a Rifle Bullet. Justification of a rule of thumb using Newton's laws of motion.

Chapter 4:   Retrograde Motion of Mars. Vector/graphical analysis of the retrograde motion of a planet.

Chapter 5: Improved Calculation of "Hyperbola-Areas." Interpreting and verifying Newton's calculation of a logarithm.

Chapter 6: Tsar-Kolokol (Tsar Bell). Vectors, numerical data, and the trapezoid rule are used in calculating the volume of a solid of revolution.

Chapter 7: Signal Processing and Series. A brief introduction to Fourier series and a discussion of how these series are applied in signal processing.

Solving Calculus Problems   Almost certainly, you will get stuck on some of the problems in the Basic exercises, Growth exercises, or student projects. Getting stuck is normal. A guide for getting unstuck follows.

Step 1   As you read and *re-read* the problem, determine as clearly as possible what is being asked and what a proper solution would involve. If necessary, focus on a subproblem whose solution may help in understanding the original problem. Try to capture the essentials of the problem or subproblem in a phrase, a sketch, or other diagram.

Step 2   If a solution strategy is not suggested by your understanding of the problem, minimize inactive waiting for an insight and maximize active investigation. Ask yourself: Is the problem similar to one I've solved before? Can I identify a special case? Solving a special case may help you choose a solution strategy for the general case.

Step 3   Apply your solution strategy. If you encounter difficulty, modify your strategy or try a different one. If you are seeking a symbolic solution to the problem, it may be useful to first solve it numerically or graphically.

Step 4   Decide if your solution is reasonable. This important step can help you avoid absurd results and alert you to possible mistakes.

The title of Westfall's biography of Newton—*Never at Rest*—suggests the importance of such qualities as intensity, energy, and persistence in Newton's life. They are important qualities of your approach to problem solving and, more broadly, to understanding and applying calculus.

CONTENTS

Chapter 1

Rates of Change, Limits, and the Derivative

What Is Calculus?

Change is all around us. There are many obvious examples of change that we experience every day: the change in the position of a car moving down the street, the change in the temperature of tap water when the hot water faucet is turned on, or the change in volume as we turn the knob on a stereo. Other manifestations of change are less obvious: the stresses on a beam in a skyscraper, the shape of a loop in a roller coaster, or the marketing strategy for a new product. Diverse as these phenomena may seem, they can all be described with one powerful tool . . . the calculus.

Credit for the invention of calculus usually goes to two men: Sir Isaac Newton (1642–1727), an English scientist and mathematician, and Gottfried Wilhelm Leibniz (1646–1716), a German philosopher and mathematician. However, as is the case with almost every great scientific discovery, the work of Leibniz and Newton rested on the work of many other great scientists, philosophers, and mathematicians, each of whom contributed knowledge that is now part of calculus. These include Archimedes, who devised processes that foreshadowed integral calculus; Galileo, who discovered a famous law of falling bodies; and Kepler, who formulated important laws of planetary motion.

In this chapter we lay the groundwork for our study of calculus and some of its applications. The ideas described in this chapter—functions, limits, and rates of change—are the foundation for the rest of the text. All of these will be seen again in almost every section of the text, though often in new situations and applied to new problems.

In this chapter (and the rest of the text) much of our approach is driven by three realizations.

- The people who use calculus (physicists, chemists, biologists, sociologists, engineers, economists, mathematicians, etc.) work not only with formulas, but also with graphs and tables of data. Thus we begin the chapter by studying functions and how they can be presented or interpreted in formula, graphical, and tabular form. We stick with this trichotomy throughout the chapter as we study limits and rates of change.

- The people who use calculus also use not only pencil and paper, but also graphing calculators and computer algebra systems (CASs). Thus we too use these tools to study functions, limits, and rates of change. You, as a student of calculus, should use these tools at every opportunity to verify examples, to experiment and gather data about a problem, and to check the answers to your own work.

- The people who use calculus need to explain their answers to other people. A problem is not necessarily "finished" when an answer is produced or a graph is drawn. The answer may have to be explained and interpreted in the context of the problem.

So do you have a pencil and some paper? Is your graphing calculator or CAS close at hand? If so, we're ready to start.

1.1   FUNCTIONS

We deal with functions every day. Students in science and engineering are accustomed to using functions that are defined by formulas. For example, the function that returns the volume V of a sphere with radius r is described by

$$(1) \qquad V = V(r) = \frac{4}{3}\pi r^3.$$

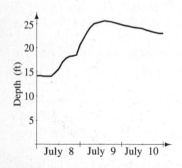

Figure 1.1. Skunk River levels during the flood of 1993. (Data from U.S. Department of the Interior Geological Survey.)

But functions can be described in other ways too. Graphs and tables that appear in newspapers and magazines may also represent functions, though we usually don't think "function" when we see such things. As an example, look at the graph in Fig. 1.1. This graph shows the level of the Skunk River as measured in Ames, Iowa, on July 8, 9, and 10. This graph describes a function that tells the level of the river at a given time during these three days. We do not have an equation for this function, but the graph is nonetheless the graph of a function. So what exactly is a function? A function is not a formula, nor is it a graph, though it can be represented by either one. In this section we find out what is a function and what is not.

The next time you use a word processor, take a moment to think about what is happening. You press the key marked "D," and the letter "d" appears on the computer screen. You press "<shift> W," and the letter "W" appears. Of course, you are seeing a miraculous piece of technology at work. You are also confident that what you type is what you will see on the screen. Why? Because the computer and word-processing software are designed to evaluate the function that assigns a particular "output" on the screen to a given "input" on the keyboard.

A function must have a **domain,** or collection of inputs. For the word processor function, the domain is a set of keystrokes. Some elements of this domain are single keystrokes. For example, hit the key marked "K" to obtain "k." Some of the elements of the domain are two keys hit at the same time, as when you depress

<shift> to get a capital. We use $\mathcal{D}_W$ to denote the set of keystrokes that result in some output on the screen. A function must also have a **range,** or collection of outputs. In this case, the range is a set of letters, numbers, and other symbols that can appear on the computer screen. We will denote this set by $\mathcal{R}_W$. The function we are discussing is the rule that matches each keystroke input with the corresponding output on the screen. We will call this rule W (for *word processor*). You should check that W satisfies the definition of function:

DEFINITION   function

A function with domain $\mathcal{D}$ and range in $\mathcal{R}$ is a rule that assigns to each element in the set $\mathcal{D}$, one and only one element in the set $\mathcal{R}$. The range of the function is the set of elements in $\mathcal{R}$ that are assigned to an element of $\mathcal{D}$.

What makes W a function? Given that the domain and a set containing the range are specified, W is a function because it is **well defined.** This means that the rule W assigns to each keystroke input in the domain one and only one screen output in the range. In plainer language, this means that every time you strike the key labeled "V" (and no other key) you will get "v" on the screen (assuming your computer is working properly!). You won't get "v" sometimes, with an occasional "a" or "b"—you will always get "v." That's reassuring.

In the definition of function, we introduced a set $\mathcal{R}$ that *contains the range* of the function. It is not necessary to know the range of a function to define the function, but it is important to know some set $\mathcal{R}$ that contains the range. For many of the functions that we will work with, the range is very difficult to determine, though it is usually no trouble to give a set that contains the range.

Sometimes it is helpful to think of a function as a machine that takes some item as input and returns some output, as illustrated in Fig. 1.2. For a function, the input is an element of the domain. The output is the corresponding element of the range. For example, if the domain item keystroke "R" is input into the word processor function W, the output is "r" on the computer screen. As shorthand for this input-output relation, we write

$$W(\text{keystroke R}) = \text{r on computer screen.}$$

In general, if f is a function and x is something in the domain of f, then we write $f(x)$ to denote the range element that corresponds to x. The input is x and the output is $f(x)$. This output is often called the value of $f(x)$ or the value of f at x.

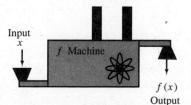

Figure 1.2. The f machine returns output $f(x)$ for input x.

Sometimes we will want to give this output a name, such as y. In this case we will write

$$y = f(x)$$

and say that y **is a function of** x.

Representing Functions   According to the previous discussion, a function is a rule for linking elements of one set (the domain) to elements of another set (the range). Thus a function is a rule, but not a formula, or a graph, or a table. Nonetheless, formulas, graphs, and tables are very useful ways to describe functions. When a function is described in one of these ways, explicit mention of the domain and range of the function might be omitted. In these cases it is usually assumed that the person working with the function can determine its domain and range from experience. For most of the functions that we work with, the domain and range are subsets of the real numbers. We will also work with functions that have sets of ordered pairs or ordered triples of real numbers as their domain or range. In these cases, if the domain is not given, we can try to determine the "likely" domain by using knowledge of algebra and trigonometry.

■ **EXAMPLE 1**   Discuss the domain and range of the function g defined by the equation $g(x) = \sqrt{x}$.

Solution.   We've all heard that "you can't take the square root of a negative number." Thus it seems natural that the domain of g is the set $[0, \infty) = \{x : x \geq 0\}$ of nonnegative real numbers. To determine the range of g, we appeal to tradition. Before negative numbers were known, the side of a square with area x was regarded as a (positive) number $\sqrt{x}$. Once negative numbers came into use, for any $x > 0$ there were two numbers whose squares were x. For example, $2^2 = 4$ and $(-2)^2 = 4$. Over the years the common resolution of this ambiguity has come to be that $\sqrt{x}$ shall denote the nonnegative number whose square is x. This fits tradition and common sense. Calculators and computer algebra systems use this convention.

Sometimes a CAS or graphing calculator can help in determining the domain of a function represented by an equation. The graph in Fig. 1.3 was produced by having a CAS graph $y = \sqrt{x}$ for $-5 \leq x \leq 5$. The CAS responded that $\sqrt{x}$ is not a real number for $x < 0$, but still produced the graph for those x values for which $\sqrt{x}$ is real. The graph produced suggests that the domain of g is $x \geq 0$.

A word of warning is appropriate here. Be careful when you use graphics packages to help determine the domain of a function. Sometimes the results are deceptive. Always think about the calculator or CAS result and interpret it in light of the problem that you are trying to solve. (See Example 9 and Exercises 44–47.)

■

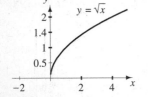

Figure 1.3. A CAS-produced graph of $y = \sqrt{x}$ for $-5 \leq x \leq 5$.

■ **EXAMPLE 2**   The Iditarod Trail sled dog race is an annual event to commemorate the courage of Leonhard Seppala and other dog mushers who delivered badly needed diphtheria serum to Nome, Alaska, in 1925. The 1100-mile course of the race runs from Anchorage to Nome. The first race was run in 1973. The table below gives the times of the winners for the races through 1994. Time is given as days:hours:minutes:seconds. The table represents a function mapping race year to winning time. Discuss this function.

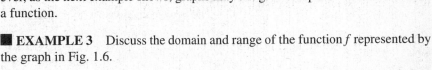

Year	Name	Time	Year	Name	Time
1973	Dick Wilmarth	20:00:49:41	1984	Dean Osmar	12:15:07:33
1974	Carl Huntington	20:15:02:07	1985	Libby Riddles	18:00:20:17
1975	Emmitt Peters	14:14:43:45	1986	Susan Butcher	11:15:06:00
1976	Jerry Riley	18:22:58:17	1987	Susan Butcher	11:02:05:13
1977	Rick Swenson	16:27:13:00	1988	Susan Butcher	11:11:41:40
1978	Dick Mackey	14:18:52:24	1989	Joe Runyan	11:05:24:34
1979	Rick Swenson	15:10:37:47	1990	Susan Butcher	11:01:53:23
1980	Joe May	14:07:11:51	1991	Rick Swenson	12:16:35:39
1981	Rick Swenson	12:08:45:02	1992	Martin Buser	10:19:17:15
1982	Rick Swenson	16:04:40:10	1993	Jeff King	10:15:38:15
1983	Dick Mackey	12:14:10:44	1994	Martin Buser	10:13:02:39

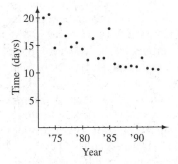

Figure 1.4. Winning times for the Iditarod Trail sled dog race.

Solution. Let I (for Iditarod) be the function that takes as input the year of the race and, as output, gives the winning time for that year. The domain of I is the set of years that the race has been run:

$$\{1973, 1974, 1975, \ldots, 1994\}.$$

The range of I is the set of winning times. A graph representing I is shown in Fig. 1.4. Note that the graph is not a smooth curve, but consists of several individual points. The graph of I is simply the collection of points with coordinates of the form $(t, I(t))$, for t in the domain of I. Because the domain of I has 22 elements, there are only 22 such points. These points make up the graph of I. What are some other functions suggested by the table? ∎

When a function f is represented graphically, we assume, unless told otherwise, that the domain is represented by some part of the horizontal axis and the range by some part of the vertical axis. The graph of f consists of the points with coordinates $(x, f(x))$, where x is in the domain of f. The domain of the function contains the set of points on the horizontal axis that lie above or below the graph. The range of the function contains the set of points on the vertical axis that lie to the left or right of the graph. Given an a in the domain (on the horizontal axis), we find the corresponding element in the range by extending a dotted line from a, either up or down, until it meets the graph at the point $(a, f(a))$. From this point extend the dotted line horizontally until it meets the vertical axis at $f(a)$. See Fig. 1.5. Graphs are a very useful way of studying and representing functions. However, as the next example shows, graphs may not give complete information about a function.

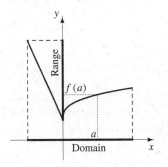

Figure 1.5. Graphical representation of a function.

■ **EXAMPLE 3** Discuss the domain and range of the function f represented by the graph in Fig. 1.6.

Solution. The graph appears to lie above or below the interval $[-2, 3]$ on the horizontal axis. Thus the domain of f contains the interval $[-2, 3]$. We cannot be certain that this is the entire domain of f since the portion of the graph shown does not tell us whether or not f is defined for real numbers outside of $[-2, 3]$; that is, the picture may show the graph for only part of the domain of f. The graph appears to lie to the left and right of the interval $[-0.7, 3.8]$ (approximately) on the vertical axis, so we conclude that the range of f contains this interval. ∎

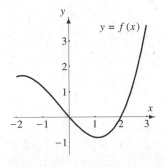

Figure 1.6. Graphical representation of f.

When doing experimental work, scientists often collect data to get a table of values for a function. While a table of values alone cannot tell us the domain and range of a function, we can make reasonable guesses about the domain and range through knowledge of the experiment and extrapolation of the tabulated data.

■ **EXAMPLE 4** A laboratory worker measures the temperature of a cup of coffee, taking a reading every 2 minutes. The temperature of the lab is maintained at 21°C. When the experiment starts, at time 0, the coffee is at 95°C, and a chunk of dry ice at temperature −79°C is dropped into the cup. The data recorded is collected in the accompanying table. Discuss the domain and range of the temperature function T.

Time (min)	Temperature (°C)
0	95
2	80
4	69
6	60
8	53
10	48

Solution. There is not much information to go on here, but some thought about the nature of the experiment can help us decide on reasonable possibilities for the domain and range of T. The experiment starts at time 0 and presumably can be continued indefinitely. Thus the domain of T can be taken to be all times $t \geq 0$, where t is measured in minutes. The range is less certain. At the last entry in the table the temperature was probably still falling. Was there enough dry ice to cause the temperature to drop below room temperature? If so, how far below? To be certain, we would have to continue the experiment until the temperature dropped below room temperature or we were convinced that it would not do so. However, the temperature of the coffee cannot drop below the temperature of the dry ice, which is −79°C. Hence the range of T is contained in the interval $[-79, 95]$.

Graphing the data and drawing a "smooth" curve through the plotted points can help us make a better guess about the range. Extending the graph in a way that seems natural (the dashed curve in Fig. 1.7), we see that the graph appears to be leveling out. Thus it looks like the temperature of the coffee will not go below room temperature, so we may be correct in stating that the range of T is $[21, 95]$. ■

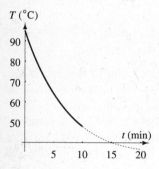

Figure 1.7. Graph of temperature.

Combining Functions

Usually we work with functions that are defined by equations, as in Example 1. Although such functions may be complicated, they usually constructed from simple pieces with which we are already familiar. These pieces are often referred to as the **elementary functions** and include

- The constant functions given by $f(x) = c$, where c is a real number.
- The power functions: $x, x^2, x^3, \ldots$.
- The reciprocal function defined by $f(x) = 1/x$.
- The root functions: $\sqrt{x}, \sqrt[3]{x}, \sqrt[4]{x}, \ldots$.
- The trigonometric functions: $\sin x, \cos x, \tan x, \sec x$, and so on.
- The inverse trigonometric functions: $\arcsin x, \arccos x, \arctan x$, and so on.
- The logarithmic functions: $\ln x, \log_{10} x, \log_2 x$.
- The exponential functions: $e^x, 10^x, 2^x$.

Though students of calculus will be familiar with most or all of these functions from previous courses, we will briefly review many of them in later sections. In the remainder of this section we review a few methods for combining functions to build new functions.

■ **EXAMPLE 5** The table on the left below shows the weight in pounds (lb) of six recreational runners. The table on the right shows the height in inches (in.) of seven runners. (Note that some of the people measured appear in both tables.) Discuss the functions represented by these tables. Discuss the function defined by dividing the weight function by the height function.

Runner	Weight (lb)	Runner	Height (in.)
Chuck	146	Bob	73
Dick	150	Chuck	70
Elgin	173	Elgin	72
Gary	124	Gary	67
Jerry K.	153	Gene	68
Jerry M.	140	George	69
		Jerry K.	68

Solution. Let w be the function represented by the weight table and h the function represented by the height table. Each of these functions has as its domain the set of runners listed in the corresponding table. The domain of the quotient function w/h is the set of four people common to the two lists. Dividing the weight by height for each of these runners, we can construct a table for the quotient. The range of w/h is the set of four lb/in. measurements appearing in the column on the right.

Runner	Weight/Height (lb/in.)
Chuck	$146/70 \approx 2.09$
Elgin	$173/72 \approx 2.40$
Gary	$124/67 \approx 1.85$
Jerry K.	$153/68 \approx 2.25$

■

■ **EXAMPLE 6** Discuss the sum and product of the functions f and g defined by $f(x) = \sqrt{x-1}$ and $g(x) = \sqrt{x+2}$.

Solution. The sum of f and g is denoted $f+g$ and is defined by

$$(2) \qquad (f+g)(x) = f(x) + g(x)$$

for those x in the domain of $f+g$. The product function is denoted $f \cdot g$ and defined by

$$(3) \qquad (f \cdot g)(x) = f(x)g(x)$$

for those x in the domain of $f \cdot g$.

The function f has domain $\mathcal{D}_f = \{x : x \geq 1\}$, since these are the x values for which $x-1$ is nonnegative. The function g has domain $\mathcal{D}_g = \{x : x \geq -2\}$. For $f(x) + g(x)$ or $f(x)g(x)$ to be defined, x must be in the domain of both f and g. In addition, (2) and (3) are defined for any x for which $f(x)$ and $g(x)$ are defined.

Hence the domain of $f + g$ and the domain of $f \cdot g$ are the same and are equal to the intersection of the domains of f and g,

(4) $$\mathscr{D} = \mathscr{D}_f \cap \mathscr{D}_g = \{x : x \geq 1\}.$$

An equation representing $f \cdot g$ can be written as

$$(f \cdot g)(x) = f(x)g(x) = \sqrt{x - 1}\sqrt{x + 2}.$$

We might be tempted to simplify this last expression by saying

$$\sqrt{x - 1}\sqrt{x + 2} = \sqrt{(x - 1)(x + 2)} = \sqrt{x^2 + x - 2}$$

and then writing

(5) $$(f \cdot g)(x) = \sqrt{x^2 + x - 2}.$$

But representing $f \cdot g$ by (5) is deceiving. Note that $x^2 + x - 2$ is nonnegative for $x \leq -2$, so $\sqrt{x^2 + x - 2}$ is defined for such x. However, these x values are not in the domain of $f \cdot g$, as seen from (4). Thus the function seemingly represented by (5) is different from $f \cdot g$, since it has a different domain. This shows that we have to be careful when combining functions. ∎

■ **EXAMPLE 7**   Figure 1.8 shows the graphs of two functions, f and g. Sketch the graph of the difference function, $g - f$.

Solution.   The domains of f and g both contain the portion of the horizontal axis shown in the figure. Hence the domain of $g - f$ also contains this portion of the horizontal axis. We graph $g - f$ over this same interval.

The difference (or sum) of functions given graphically can be found with the help of a straight-edge. At a point a in the domain of $g - f$, simply mark on the straight-edge the vertical distance d between $f(a)$ and $g(a)$. See Fig. 1.9. On the axes on which the difference is to be graphed, find a on the horizontal axis. If $g(a) \geq f(a)$ (i.e., g is above f at a), then

$$(g - f)(a) = g(a) - f(a) \geq 0,$$

and we mark a point at distance d above a. If $g(a) < f(a)$, then

$$(g - f)(a) = g(a) - f(a) < 0,$$

so we mark a point at distance d below a. Do this for as many points in the domain as necessary to get a feel for the graph of $g - f$. Then sketch the appropriate curve through the plotted points. This is illustrated in Fig. 1.10. ∎

Compositions   Use your calculator to find the approximate value of $\sqrt{\sin 0.67}$. You probably did this in two steps. For the first step you calculated $\sin 0.67 \approx 0.620986$, and for the second step you found $\sqrt{0.620986} \approx 0.788027$. In performing these two steps you were evaluating two functions. The first of these functions was the sine function, and the second was the square root function.

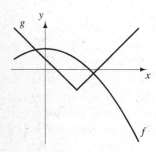

Figure 1.8. The graphs of f and g.

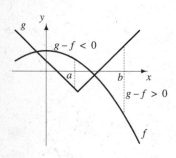

Figure 1.9. On a straight-edge mark the vertical distance between $f(a)$ and $g(a)$.

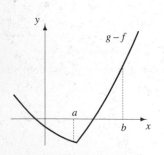

Figure 1.10. The graph of $g - f$.

The sine function was evaluated for the input value 0.67 radians. Then the output from the sine function was used as input for the square root function. The process of taking the output from one function as the input for another function is called function composition. What happens when you try to find the value of $\sqrt{\sin 3.8}$?

DEFINITION   composition of functions

Let f and g be functions. The composition of f with g is denoted by $f \circ g$. For x in the domain of $f \circ g$, the value of $(f \circ g)(x)$ is

(6) $$(f \circ g)(x) = f(g(x)).$$

Expression (6) makes sense only if x is in the domain of g and $g(x)$ is in the domain of f. Hence the domain of $f \circ g$ is

$$\mathcal{D}_{f \circ g} = \{x : x \text{ is in } \mathcal{D}_g \text{ and } g(x) \text{ is in } \mathcal{D}_f\}.$$

In working with functions it is important not only to understand how to find the composition of given functions, but also to write a given function as a composition of simpler ones.

■ **EXAMPLE 8**   Find $f \circ g$ and $g \circ f$ for the functions defined by the equations

$$f(x) = (x - 1)^2 \quad \text{and} \quad g(x) = \sqrt{x - 4}.$$

Give the domain of each composition.

Solution.   The domain of f is $(-\infty, \infty)$, the set of all real numbers. The domain of g is $[4, \infty) = \{x : x \geq 4\}$.

First consider $f \circ g$. This function is given by

$$(f \circ g)(x) = f(g(x)) = f\left(\sqrt{x - 4}\right) = \left(\sqrt{x - 4} - 1\right)^2 = x - 3 - 2\sqrt{x - 4}$$

and is defined when $\sqrt{x - 4}$ makes sense. Hence $f \circ g$ has domain $[4, \infty)$.

The function $g \circ f$ is represented by

$$(g \circ f)(x) = g(f(x)) = g\left((x - 1)^2\right) = \sqrt{(x - 1)^2 - 4} = \sqrt{x^2 - 2x - 3}$$

and is defined only when $f(x) = (x - 1)^2$ is in the domain of g. Thus we must have

$$(x - 1)^2 \geq 4, \quad \text{that is} \quad \begin{cases} x - 1 \geq 2 \\ \text{or} \\ x - 1 \leq -2 \end{cases}$$

Solving these inequalities, we find

$$x \geq 3 \quad \text{or} \quad x \leq -1.$$

Hence the domain of $g \circ f$ is $(-\infty, -1] \cup [3, \infty)$.   ■

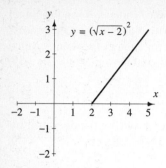

Figure 1.11. The function $f \circ g$ has domain $x \geq 2$.

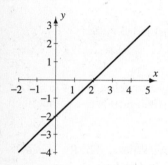

Figure 1.12. Sometimes a CAS may not produce the correct graph of $f \circ g$.

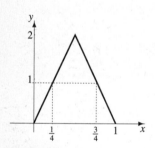

Figure 1.13. Graph of f.

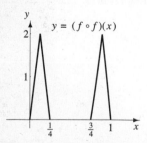

Figure 1.14. Graph of $f \circ f$.

■ **EXAMPLE 9**   Let f be given by $f(x) = x^2$ and g by $g(x) = \sqrt{x - 2}$. Sketch the graph of $f \circ g$.

Solution.   The function g is defined for $x \geq 2$. Since f is defined for all real numbers, the domain of $f \circ g$ is $\{x : x \geq 2\}$. For x in this domain, we have

$$(f \circ g)(x) = \left(\sqrt{x - 2}\right)^2 = x - 2, \quad x \geq 2.$$

The graph of $y = (f \circ g)(x)$, shown in Fig. 1.11, is the $x \geq 2$ portion of the line of slope 1 and y-intercept -2.

Be careful if you do this composition with a CAS or graphing calculator. Sometimes these devices perform unwanted simplifications and as a result may not give a true picture of the graph. The graph in Fig. 1.12 was produced when a CAS was used to graph $y = f(g(x))$ for f and g as defined above. Anyone looking at such a graph without thinking might be fooled into saying the domain for $f \circ g$ is all real numbers. There are two important lessons to be learned from this example:

1. Keep on thinking when using a calculator or computer!

2. Get to know the limitations and quirks of your calculator and CAS!   ■

■ **EXAMPLE 10**   The graph of a function f with domain $[0, 1]$ is shown in Fig. 1.13. Find the domain of $f \circ f$ and sketch the graph of the composition.

Solution.   Since

(7) $$(f \circ f)(x) = f(f(x)),$$

the composition is defined only for those x for which the inner $f(x)$ is between 0 and 1. Reading from the graph, we see this is the case when

$$0 \leq x \leq 1/4 \quad \text{or} \quad 3/4 \leq x \leq 1.$$

Thus the domain of $f \circ f$ is $[0, 1/4] \cup [3/4, 1]$.

To find the graph, let y denote the inner $f(x)$ in (7). As we let x run from 0 to $\frac{1}{4}$, $y = f(x)$ runs from 0 to 1. This is seen by looking at the portion of the graph above $[0, 1/4]$. As y runs from 0 to 1, $f(y)$ grows from 0 to 2 then drops back to 0, as indicated in Fig. 1.13. This means that as x runs from 0 to $1/4$, the graph of $f \circ f(x)$ is a sharp spike of height 2 centered over the interval $[0, 1/4]$. Similar reasoning shows that there is a second such spike over the interval $[3/4, 1]$. The graph is shown in Fig. 1.14.

The graph of $y = (f \circ f)(x)$ can also be found by finding a formula for $f(x)$ and then graphing the composition with a graphing calculator or computer algebra system. (See Exercise 38.)   ■

When two functions are given graphically, the graph of their composition can be constructed.

■ **EXAMPLE 11**   Figure 1.15 shows the graphs of two functions f and g, both of which have domain $[0, 1]$. Sketch the graph of $f \circ g$.

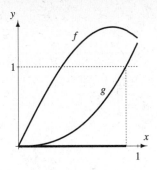

Figure 1.15. The domain of $f \circ g$ includes those x for which $0 \leq g(x) \leq 1$.

Solution. First determine the domain of $f \circ g$. Since the domain of f is $[0, 1]$, the expression $f(g(x))$ will make sense if and only if $0 \leq g(x) \leq 1$. Thus any x values for which $g(x) > 1$ or $g(x) < 0$ are not in the domain of $f \circ g$. From the graph of g we can see which horizontal-axis values correspond to elements of the domain of $f \circ g$. See Fig. 1.15.

Now suppose that a is in the domain of $f \circ g$. Use the following four steps to find the point $(a, (f \circ g)(a))$ on the graph of the composition.

> **1.** From a on the horizontal axis draw a dashed line vertically to the graph of g. This line hits the graph at the point $(a, g(a))$.
> **2.** Think of the y-coordinate $g(a)$ as the "output" from the function g. To find $f(g(a))$, this output must become "input" for the function f. This is done by drawing a horizontal line from $(a, g(a))$ to the point $(g(a), g(a))$ on the line $y = x$.
> **3.** To treat $g(a)$ as input for f, move vertically from this point to meet the graph of $y = f(x)$ at the point $(g(a), f(g(a)))$.
> **4.** From this point, extend the line horizontally until it hits the vertical line $x = a$. This final point has coordinates $(a, f(g(a)))$, and so is a point on the graph of $f \circ g$.

Repeat this process for several points in the domain of the composition, and then connect them with a likely curve. This is illustrated in Fig. 1.16. ■

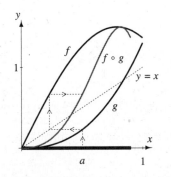

Figure 1.16. Plotting points of $f \circ g$.

■ **EXAMPLE 12**   A function f is given by

$$f(x) = \tan \sqrt{x^3 - 3x^2 + 7}.$$

Express f as a composition of simpler functions.

Solution. Given an x in the domain of f, the first step in evaluating $f(x)$ is to evaluate the polynomial $x^3 - 3x^2 + 7$. The result of this calculation is then put into the square root function, and this square root is then put into the tangent function. This suggests that f can be written as a composition of three functions. The first (inner) function is the polynomial function p given by

$$p(x) = x^3 - 3x^2 + 7.$$

The second (middle) function is the square root function s, represented by

$$s(x) = \sqrt{x},$$

and the last (outer) function is the tangent function t given by

$$t(x) = \tan x.$$

Thus $f = t \circ s \circ p$. For x in the domain of f, we have

$$f(x) = (t \circ s \circ p)(x) = t(s(p(x))).$$ ■

Exercises 1.1

Basic

Exercises 1–6:   Evaluate $f(a)$ for the given value of a.

1. $f(x) = -3x^2 + 2x + 7, \quad a = -2$

2. $f(s) = 2s^2 + 4s + 8, \quad a = 0.45$

3. $f(t) = -3t^2 - t + 1, \quad a = 1 + \sqrt{3}$

4. $f(x) = x^2 + x - 3, \quad a = t + h$

5. $f(\tau) = (2\tau + 1)^3, \quad a = x + h$

6. $f(z) = \dfrac{2z + 1}{5z - 3}, \quad a = \dfrac{x}{y}$

Exercises 7–12:   State the domain of the function defined by the given equation.

7. $f(x) = (|x - 3| + 1)^2$

8. $g(t) = \dfrac{t + 3}{t + 4}$

9. $y = \sqrt{x(x + 1)}$

10. $h(r) = \dfrac{1}{\sqrt{(r - a)(r - b)}}, \quad$ where $b > a$

11. $H(r) = \dfrac{1}{\sqrt{r - a}} \dfrac{1}{\sqrt{r - b}}, \quad$ where $b > a$

12. $s = \dfrac{1}{\sin t}$

Exercises 13–17:   State the domain and range of each function. A graph may be helpful in determining the range.

13. $f(x) = 4$

14. $y = -3t + 2$

15. $h(t) = \dfrac{1}{t^2 + 1}$

16. $g(x) = \dfrac{x^2 - 3x + 2}{x - 2}$

17. $y = \dfrac{|x|}{x}$

18. Sketch a graph for a function with domain $[-1, 3] = \{x : -1 \le x \le 3\}$ and range $[0, 2]$.

19. Write an equation for a function f whose assumed domain is $[-1, 3]$.

20. Write an equation for a function whose assumed domain is $(-\infty, -1) \cup (1, \infty)$. That is, the domain should be all real numbers *except* those in $[-1, 1]$.

Exercises 21 and 22:   The graphs of two functions, f and g, are shown. Sketch the graph of $f + g$ and $f - g$.

21.

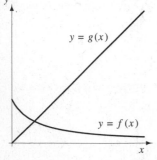

22.

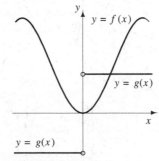

Exercises 23–26:   Find an equation for the composition and state the domain of the composition.

23. $(f \circ g)(x)$ if $f(x) = \dfrac{1}{x}$ and $g(x) = x^2 - 6x + 8$

24. $(r \circ r)(u)$ if $r(u) = \sqrt{1 - u}$

25. $(f \circ g)(t)$ if $f(x) = \dfrac{|x|}{x}$ and $g(t) = \sin t$

26. $(h \circ k)(x)$ if $h(x) = \sqrt{x}$ and $k(x) = (x - 1)(x - 2)(x - 3)^2(x - 4)$

Exercises 27 and 28: Sketch the graph of $(f \circ g)(x)$. The scales on the x- and y-axes are the same.

27.

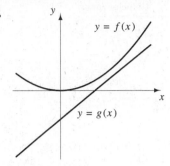

28.

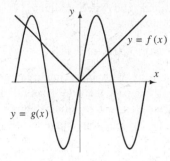

Exercises 29–32: Decompose each function into a composition of simpler functions.

29. $f(x) = \dfrac{2}{x^3 - 3x^2 + 1}$

30. $H(x) = \sin(\sqrt{x + 5})$

31. $h(x) = f\left(\dfrac{1}{g(x)}\right)$

32. $r(x) = \left(\dfrac{1 - 2\sin x}{1 + 2\sin x}\right)^{1/2}$

Growth

33. Plot $y = \cos x$, $y = \cos(\cos x)$, and $y = \cos(\cos(\cos(\cos x)))$. Describe what appears to be happening to the graphs as you graph "deeper and deeper" compositions of the cosine function. Try even longer cosine compositions to check your answer.

34. Plot $y = \sin x$, $y = \sin(\sin x)$, and $y = \sin(\sin(\sin(\sin x)))$. Describe what appears to be happening to the graphs as you graph "deeper and deeper" compositions of the sine function. Try even longer sine compositions to check your answer.

35. Suppose that the domain of f is $[-1, 3]$ and that $g(x) = 2x - 1$. What is the domain of $f \circ g$?

36. Suppose that the domain of f is $[-1, 4]$ and that $g(x) = x^2$. What is the domain of $f \circ g$?

37. Suppose that the domain of h is $[0, \infty)$ and that $k(x) = 1/(x + 1)$. What can be said about the domain of $k \circ h$?

38. Verify that the function f whose graph is given in Fig. 1.13 is described by the equation

$$f(x) = 2 - 4\left|x - \frac{1}{2}\right|, \quad 0 \le x \le 1.$$

Use a graphing calculator or CAS to graph $y = (f \circ f)(x)$ and discuss how the resulting graph leads to Fig. 1.14.

39. The *vertical line test* is a method for deciding if a graph represents a function. Let G be a graph in the (x, y)-plane. If every vertical line intersects G at

most once, then the graph represents y as a function of x. Explain why this test works.

40. Every year the owners of the Negative Outlook photography studio seek to enlarge their profits through advertising. Records of dollars spent on advertising and corresponding profits for the last six years are summarized in the table below. What function(s) are represented by this table? Does it appear that profits can be substantially increased by spending more on advertising?

Year	Advertising Costs ($)	Profit ($)
1	1000	163.93
2	2000	2105.26
3	3000	5744.68
4	4000	8101.26
5	5000	9124.08
6	6000	9557.52

41. In some modern supercomputers addition and multiplication "cost" the same amount, but exponentiations (powers) cost twice as much. (Cost here is directly proportional to computation time.) Hence for large computations it can be profitable to replace exponentiations by multiplications. For example, computing

(8) $ax^3 + bx^2 + cx + d$

in the normal way involves two exponentiations, three multiplications, and three additions. If we rewrite the polynomial as

$$(9) \qquad x(x(ax + b) + c) + d,$$

we see how to get by with three multiplications and three additions. Suppose that additions and multiplications cost one unit and exponentiations cost two units. Then the computation suggested by (8) costs

$$2 \cdot 2 + 3 + 3 = 10 \text{ units},$$

while the computation suggested by (9) costs

$$3 + 3 = 6 \text{ units}.$$

Make a table showing the costs of each computation method for polynomials of degree 1, 2, 3, 4, and 5. As the degree of the polynomial grows, what appears to happen to the ratio of the costs of the two methods?

42. At public swimming pools the concentration of chlorine is checked regularly. The following table shows the chlorine levels (in parts per million) in a pool for a 7-day period during which no chlorine was added. What are the domain and range of the function described by the table? If no chlorine is added, what do you predict the level will be in 10 days? Give reasons for your answer.

Day	Chlorine (ppm)
1	3
2	2.55
3	2.16
4	1.84
5	1.56
6	1.33

43. The two tables shown below represent two functions.

Table 1		Table 2	
x	—	x	$(f \circ g)(x)$
1	3	1	2
2	5	2	7
3	8	4	7
4	8	5	5
5	7	6	5
6	2		

a. If Table 1 represents the function f and Table 2 represents the function $f \circ g$, construct a table to represent g.

b. If Table 1 represents g and Table 2 represents the function $f \circ g$, construct a table to represent f.

Exercises 44–47: An equation for a function f is given. The accompanying graph of $y = f(x)$ for $-5 \le x \le 5$ was produced by a CAS. Decide if the graph shown actually is the graph of the function f given. If not, state what is wrong with the graph. Try graphing f on your own calculator or CAS. Is the graph produced correct or not?

44. $f(x) = \dfrac{x^2 - 9}{x + 3}$

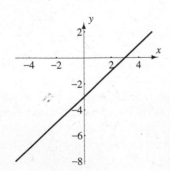

45. $f(x) = (2 - \sqrt{x + 1})^2$

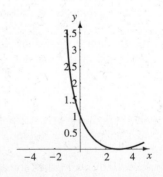

46. $f(x) = g(h(x))$ where $g(x) = x^2 + 1$ and

$$h(x) = \dfrac{1}{\sqrt{x-3}}$$

47. $f(x) = g(h(x))$ where $g(x) = \cos x$ and

$h(x) = \sqrt{x}$

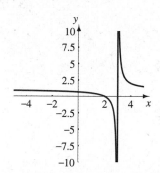

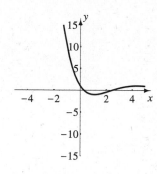

Exercises 48 and 49:   The graph of a function f is shown along with the dotted line $y = x$. Graphically trace the progress of $f(a)$, $f(f(a))$, $f(f(f(a)))$,... for the a shown on the x-axis. Describe what happens as the composition "deepens."

48.

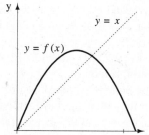

49.

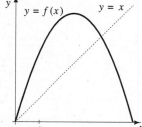

Review/Preview

The Review/Preview exercises at the end of each section give you a chance to brush up on concepts and skills that will be needed in subsequent sections.

50. What are the slope and y-intercept of the line with equation $y = 3x - 4$? Sketch a graph of the line.

51. Find the slope and y-intercept of the line with equation $4x - 7y + 5 = 0$. Sketch a graph of the line.

52. Find an equation for the line through the points $(-2, 5)$ and $(4, 1)$. Sketch a graph of the line.

53. Find an equation for the line through the points $(8, 0)$ and $(8, \sqrt{2})$. Sketch a graph of the line.

54. A cylindrical pail contains water to a depth of 2 inches. At 1:00 P.M. it starts raining, and the water in the pail rises at a steady rate of $1/2$ inch per hour for three hours. At 4:00 it stops raining. Write a formula that gives the depth of water in the pail t hours after the rain starts. Graph the water depth as a function of time.

55. A parachutist is 5000 feet above the ground and falling at a steady rate of 12 feet per second. When will she reach the ground? Find a formula for her height above ground t seconds after the time she is 5000 feet above the ground.

56. A bicyclist rides a distance of 54 kilometers in four hours. What is the bicyclist's average speed during the ride?

57. A bicyclist rides from 10:00 A.M. until 3:00 P.M. His average speed for the ride is 12 km per hour, but during the first hour he rides at a steady speed of 16 km per hour. What was his average speed during the last three hours of his ride?

58. Let $f(x) = x^2$. Simplify the expression

$$\dfrac{f(x+h) - f(x)}{h}.$$

59. Let $g(t) = (t + 1)^2$. Simplify the expression

$$\dfrac{g(t) - g(x)}{t - x}.$$

60. Find an equation for the line whose graph is shown below.

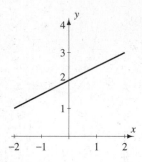

61. Find an equation for the line whose graph is shown below.

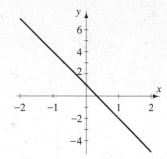

1.2 SLOPE AS A RATE OF CHANGE

Equations of lines are the most important and most often seen equations in science, engineering, and mathematics. Many real-life phenomena are described (or modeled) by equations of lines. The speed of an object under constant acceleration, the work you do in climbing a ladder, and the profit a theater owner makes by showing a movie are all examples of things that can be described by an equation for a line. Even situations that cannot be described exactly by lines can often be approximated very accurately by equations of lines. In fact, approximation by lines is one of the fundamental ideas behind the differential calculus.

In this section we begin our study of change by noting that the slope of a line can be interpreted as the rate of change of one quantity with respect to another. We use this idea to motivate a definition of rate of change for quantities that are related by a nonlinear function.

Lines and Slope

In algebra we learned that a nonvertical line in the Cartesian plane (or (x, y)-plane) is described by an equation of the form

$$y = mx + b.$$

The constant m is the *slope* of the line and tells us that when x increases by 1, y changes by m. This change in y will be an increase if $m > 0$, a decrease if $m < 0$, and no change if $m = 0$. See Fig. 1.17. The number b is called the y-intercept of the line and tells us that the point $(0, b)$ is on the graph of the line.

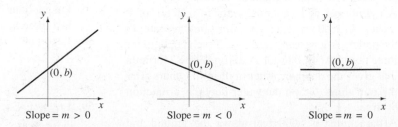

Figure 1.17. Slopes of lines.

■ **EXAMPLE 1**   The equation

(1) $$-3x + 4y - 8 = 0$$

is the equation of a line. Find the slope and y-intercept of the line and sketch a graph of the line.

Solution.   Solving (1) for y we have

(2) $$y = \frac{3}{4}x + 2.$$

From this form of the equation we see that the line has slope $3/4$ and y-intercept 2. To sketch the graph we first plot the y-intercept, $(0, 2)$. To find a second point on the line we use the slope. Since the slope is $3/4$, we can start at $(0, 2)$, move horizontally to the right 1 unit (i.e., increase x by 1), and then move up $3/4$ units (i.e., increase y by $3/4$). We are now at the point $(1, 11/4)$, which is another point on the line. To complete the graph, draw the line determined by the two points that we have plotted. See Fig. 1.18.   ■

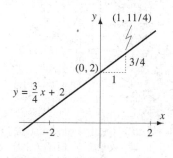

Figure 1.18. The line of slope $3/4$ and y-intercept 2.

Almost every problem in science, mathematics, and engineering can be solved in more than one way. One way of interpreting this statement is: "There is no *one* right way to solve a problem. Almost every problem has many equally valid solutions." Get into the habit of thinking about different ways to solve a problem. Even if you have already solved a problem in one way, looking at alternative solutions is a good way to check your solution and almost always adds to your understanding and insight. We can even apply this philosophy to the previous example.

■ **EXAMPLE 2**   Do Example 1 in another way.

Solution.   Since two points determine a line, we can use (1) to find two points on the line. One point can be the y-intercept, which is found by letting $x = 0$ in (1):

$$-3 \cdot 0 + 4y - 8 = 0.$$

Solving this equation, we find $y = 2$. Hence 2 is the y-intercept and the point $(0, 2)$ is on the graph of the line. Next, substitute $x = -4$ in (1) to get

$$(-3)(-4) + 4y = 8.$$

The solution to this equation is $y = -1$, so the point $(-4, -1)$ is also on the line. Plot the points $(0, 2)$ and $(-4, -1)$ and draw the line determined by these two points to get the graph, as shown in Fig. 1.19. Since we know two points on the line, we can determine the slope, m, by

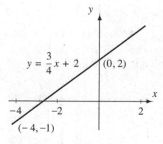

Figure 1.19. Two points determine a line.

$$m = \frac{\text{change in } y}{\text{change in } x} = \frac{2 - (-1)}{0 - (-4)} = \frac{3}{4}.$$ ■

Try using your graphing calculator or CAS to solve this example a third way. First produce the graph of the line on your calculator or CAS. Some graphics packages may take the equation as given in (1), but for most others you have to enter the equation in some form like (2). Once the graph is drawn, use the cursor to read the coordinates of the y-intercept and the coordinates of some other points on the line. Use this information to compute the slope. How do your answers compare with the results of Examples 1 and 2?

Units for Slope

In science and engineering, equations are usually used to represent real-life situations, so the variables often have associated units (e.g., feet, minutes, dollars). If the equation is the equation of a line, these units can be used to give more meaning to the slope.

■ **EXAMPLE 3**   During heavy rains the water level in a local river must be closely monitored. At 1:00 P.M. a county agent measures the depth of a river and finds it to be 12 feet. Three hours later she takes another measurement and finds the depth is 14 feet. She assumes that the river is "rising steadily," so the depth as a function of time can be described by a line. She also knows that the river is at flood stage when its depth is 22 feet. The agent must prepare a report for the 6:00 news. The report is to include the rate at which the river is rising and the time at which the river will reach flood stage. Help the agent prepare her report.

Solution.   Let t (hours) be time, with $t = 0$ corresponding to 12:00 noon. Let $d = d(t)$ (feet) be the depth of the river at time t. (We write $d = d(t)$ to remind us that d is a function of t.) Since we are assuming that $d(t)$ can be described by a line, we have

$$d = mt + b.$$

The agent's data gives us two points on this line: $(t, d) = (1 \text{ h}, 12 \text{ ft})$ and $(t, d) = (4 \text{ h}, 14 \text{ ft})$. Hence the slope of the line is

$$m = \frac{14 \text{ ft} - 12 \text{ ft}}{4 \text{ h} - 1 \text{ h}} = \frac{2}{3} \text{ ft/h}.$$

The slope tells us that the river is rising at a rate of $2/3$ ft/h. This is one of the pieces of information needed by the agent. Since the point $(1, 12)$ is on the line and we know the slope, we may use the point-slope form for the equation of a line to write

$$(3) \qquad (d - 12) = \frac{2}{3}(t - 1) \quad \text{from which} \quad d = \frac{2}{3}t + \frac{34}{3}.$$

To predict the time when the river will reach flood level, we need to know when $d = 22$. We can estimate this time from the graph in Fig. 1.20, or use the cursor and zoom-in feature on a calculator to see what t value corresponds to the point on the graph with a d value of 22. (Try it. What do you get?) This would certainly be accurate enough for broadcast purposes. We can also use the depth equation (3).

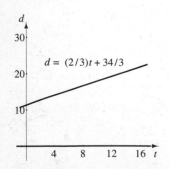

Figure 1.20. The river is rising!

Set $d = 22$ in the equation to get

$$22 = \frac{2}{3}t + \frac{34}{3}$$

and solve for t. The result is $t = 16$. Since the units are hours, and $t = 0$ corresponds to 12:00 noon, $t = 16$ corresponds to 4:00 A.M. the next morning.   ■

Slope as a Rate of Change   In Example 3 we found that the slope of the line $d = mt + b$ could be interpreted as the rate at which the river was rising. In other words, m was the rate at which the depth of the river was changing with respect to time. We can interpret the slope of any nonvertical line in this way.

DEFINITION   **rate of change for a linear function**

If y is related to x by the equation for a line,

$$y = mx + b,$$

then the **rate of change of y with respect to** x is m. If x and y have associated units, then the units of m are

$$\text{units of } m = \frac{\text{units of } y}{\text{units of } x},$$

which is read as "units of y per unit of x."

Thus for lines, slope and rate of change are the same. Knowing the units for the rate of change is important in giving the slope the correct physical interpretation in real-life situations.

■ **EXAMPLE 4**   The federal income tax regulations for 1993 state that if you are single and your 1993 income was equal to or greater than $250,000, then your federal income tax is $79,772 plus 39.6 percent of the amount over $250,000. Write a linear equation that gives the income tax for those who earned at least $250,000 in 1993. Interpret the slope of this equation as a rate of change and explain what it means.

Solution.   Let I be income, in dollars, and $T = T(I)$ the corresponding taxes, again in dollars. Then if $I \geq \$250,000$ we have

$$(4) \qquad T = 79,772 + 0.396(I - 250,000) = 0.396I - 19,228.$$

The slope of this equation is 0.396, so the rate of change of T with respect to I is 0.396. This rate of change has units

$$\frac{\text{tax dollars}}{\text{income dollar}}.$$

This means that if you earn more than $250,000 and your income goes up $1, then your taxes go up 39.6¢. In other words, 39.6¢ of each dollar earned above $250,000 goes to pay taxes. A graph of (4) is shown in Fig. 1.21.   ■

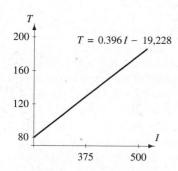

Figure 1.21. The I and T axis labels represent thousands of dollars.

Rate of Change

In the previous examples we saw that interpreting the slope of a line as a rate of change can provide important information. We use this idea to develop a definition for the rate of change (or slope) for a nonlinear function. The notion of rate of change for nonlinear functions is one of the most important ideas in calculus.

Investigation

In the investigations we motivate, explore, and develop new concepts. These concepts are then studied further in the examples.

A piston chamber holds 10^{23} molecules of an ideal gas at a temperature of $0°C$. As the piston moves, the volume V of the piston chamber varies. According to Boyle's law, the pressure P of the gas is

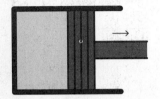

Figure 1.22. As the piston moves out, the available volume for the gas increases.

$$(5) \qquad P = P(V) = \frac{4}{V},$$

where V is measured in liters and P is measured in units of atmospheric pressure (atm). When the piston is moved outward, the volume V increases and the pressure changes. See Fig. 1.22. At what rate is the pressure changing with respect to volume when $V = 3$ liters?

The graph of (5) is shown in Fig. 1.23. The units on the horizontal axis represent liters, and those on the vertical axis represent atmospheres of pressure. This means that the rate of change has units

$$\frac{P \text{ units}}{V \text{ units}} = \text{atm/liter.}$$

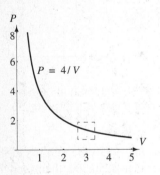

Figure 1.23. The graph of $P = 4/V$.

These are appropriate units because we are looking for the rate of change of pressure with respect to volume. If the graph of this function were a line, we could find the rate of change by finding the slope of the line. Even though the graph is not a line, we will see that a small piece of the graph looks very much like a line, and if we take a small enough piece, it is almost impossible to tell it from a line. With the aid of a graphing calculator or computer, it is easy to "zoom in" on a small piece of the graph.

Since we are interested in the rate of change when

$$V = 3 \quad \text{and} \quad P = P(3) = 4/3,$$

we look at a piece of the graph containing the point $(3, 4/3)$. In Fig. 1.23 there is a square of side 1 centered at $(3, 4/3)$. Enlarge this piece of the graph by using the zoom feature on your calculator or CAS or by replotting (5) for $2.5 \le V \le 3.5$. The result should be something like Fig. 1.24. This portion of the graph looks straighter than the part shown in Fig. 1.23, but it still doesn't look like a line. We can take a closer look by zooming in on the small rectangle shown in Fig. 1.24. When you do this yourself, you should see something like the graph in Fig. 1.25.

This piece of the graph looks very much like a line. (But it isn't really. Hold your book up so the page is flat and level with your eye. Look in the direction of

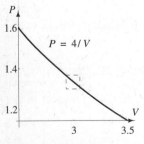

Figure 1.24. Zooming in on the graph near $(3, 4/3)$.

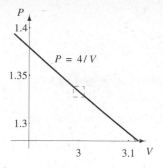

Figure 1.25. The graph of $P = 4/V$ for $2.9 \leq V \leq 3.1$.

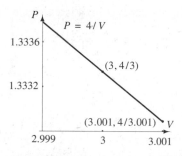

Figure 1.26. Graph of $P = 4/V$ for $2.999 \leq V \leq 3.001$.

the graph in Fig. 1.25 and you will see a slight curve.) Reading from the vertical and horizontal scales of this graph, the slope appears to be about

$$\frac{1.38 \text{ atm} - 1.28 \text{ atm}}{2.9 \text{ liters} - 3.1 \text{ liters}} = -0.5 \text{ atm/liter.}$$

This means that when the volume of the piston chamber is 3 liters and the piston is moved out to make the volume bigger, the pressure appears to be changing at a rate of about -0.5 atm/liter. The negative sign in front of this rate of change tells us that the pressure decreases as we make the piston chamber bigger. Is this consistent with common sense?

How accurate is the rate of change? Can we do better? Let's try zooming in even closer; that is, let's enlarge the portion of the graph in the square shown in Fig. 1.25. The result is shown in Fig. 1.26. The graph cannot be distinguished from the graph of a line, at least in this figure.

Again using the vertical and horizontal coordinates shown on the axes, we estimate the slope as

$$\frac{1.3338 \text{ atm} - 1.3329 \text{ atm}}{2.999 \text{ liters} - 3.001 \text{ liters}} = -0.45 \text{ atm/liter.}$$

Based on these estimates from graphs, we conclude that when the piston chamber has volume 3 liters and the piston is pulled out to make the volume larger, the pressure of the gas in the chamber changes at a rate of about -0.45 atm/liter.

Looking closely at the graph like this can give us a good feel for the rate of change of P with respect to V when $V = 3$. However, because it is difficult to get precise coordinate readings from a graph, our slope calculations may not be as accurate as we would like. We can do better by using equation (5) to actually find the coordinates of two points on the "line" of Fig. 1.26. Two such points are

$$(3, P(3)) = (3, 4/3) \quad \text{and} \quad (3.001, P(3.001)) = (3.001, 4/3.001).$$

With these two points used to compute the slope of the "line," our estimate for the rate of change is

$$\frac{\dfrac{4}{3} - \dfrac{4}{3.001}}{3 - 3.001} \text{ atm/liter} \approx -0.44430 \text{ atm/liter.}$$

Since we actually used coordinates of points on the graph for this last calculation, we have more confidence that this number really represents the slope of the "line" in Fig. 1.26. This also suggests an easier way to approximate the rate of change at $V = 3$. Find the coordinates of a point R on the graph very near the point $Q = (3, 4/3)$. The portion of the graph between Q and R is almost a line. The slope of the segment $\overline{QR}$ is an approximation to the slope of the line. Since the graph appears more "line-like" as we zoom in closer and closer, our calculations give better approximations to the rate of change by taking R closer to Q.

As an example, take $R = (2.99999, P(2.99999))$, a point on the graph very close to Q. As seen in Fig. 1.27, the graph between Q and R is again almost a line.

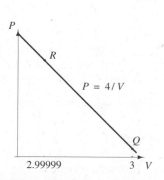

Figure 1.27. Computing slope using points $Q = (3, 4/3)$ and $R = (2.99999, 4/2.99999)$.

The slope of the segment joining Q and R is

(6)

$$\frac{P(3) - P(2.99999)}{3 - 2.99999} \text{ atm/liter} = \frac{\frac{4}{3} - \frac{4}{2.99999}}{3 - 2.99999} \text{ atm/liter} \approx -0.44445 \text{ atm/liter}.$$

In the next section we streamline the process of finding the rate of change.

Exercises 1.2

Basic

1. Find the slope and y-intercept of the line determined by the points $(3, -2)$ and $(-4, 1)$. Sketch a graph of the line.

2. A car accelerates down the highway in such a way that at time t seconds its velocity $v(t)$, in feet/second, is given by the linear function

$$v = v(t) = 6t + 15.$$

What are the units for the constants 6 and 15 in the above formula? What is the rate of change of v with respect to t? (Don't forget the units.) This rate of change of velocity with respect to time has a familiar name. What is it?

3. The county agent featured in Example 3 decides she would like to give the rate at which the water level of the river is rising in inches/minute. What is the rate of rise in this case? In these units, what is the linear equation that gives the height of the river at time t?

4. Referring to the Investigation, suppose that the piston chamber has a volume of 5 liters and that the piston is being pulled outward. Approximate the rate of change of P with respect to V:
   a. Graphically, by estimating the slope for small sections of the graph of $P(V)$ near $V = 5$.
   b. Numerically, by computing the slope of the segment determined by the point $(5, P(5))$ and some nearby point of the graph.

5. A empty, cylindrical can, open at the top, has radius 8 cm and height 20 cm. Water is poured into the can at a rate of 10 cm³/s.
   a. Find a formula for the height $h(t)$ of water in the can after t seconds.
   b. What is the rate of change of the height of water with respect to time?
   c. When will the can be full?

6. The equation used to convert degrees Fahrenheit (F) to degrees Celsius (C) is the equation of a line. A temperature of 32°F corresponds to 0°C, and a temperature of 212°F corresponds to 100°C.
   a. Use these data in finding a formula for converting a Fahrenheit temperature to Celsius.
   b. Sketch a graph of the equation found in *a*. Label your axes.
   c. What is the rate of change of degrees Celsius with respect to degrees Fahrenheit?
   d. Find a formula for converting a Celsius temperature to Fahrenheit.
   e. Sketch a graph of the equation found in *d*. Label your axes.
   f. What is the rate of change of degrees Fahrenheit with respect to degrees Celsius?

Growth

7. In Example 4 we found that for people earning more than $250,000, the rate of change of tax dollars, T, with respect to income dollars, I, is 0.396. For people in this tax bracket, what is the rate of change of *income dollars* with respect to *tax dollars*? Explain what this rate of change means.

8. If the temperature of the gas in the Investigation is raised to 273°C and held at that temperature, the pressure of the gas in the piston chamber will double:

$$P(V) = \frac{8}{V}.$$

How will the rate of change of P with respect to V be affected when $V = 3$ liters? Explain your answer.

9. Sketch the graph of $y = |x|$ and consider a small piece of the graph near the point $(0, 0)$. Will a small

enough piece of this graph ever look like a straight line? Why?

10. Sketch the graph of the function $y = x^{1/3}$. Use graphical and numerical techniques to investigate the rate of change of y with respect to x:
  a. at $x = 8$.
  b. at $x = 0$.

11. Fill in the blanks in the following sentence:
  The speedometer in a car measures the rate of change of _____ with respect to _____.

12. A calculus student might think of a speedometer as a "slope-meter." Explain why this terminology is appropriate.

13. As seen in many of the figures in this section, computers and graphing calculators often produce graphs in which the horizontal and vertical scales are different. Care is needed when working with such graphs since the appearance of the graph of a function may change as the horizontal and vertical scales change. The figure to the left shows the graph of a line $y = mx$ for $0 \le x \le 1$. Fill in the appropriate y-axis labels at the tick marks if:
  a. $m = 1$ **b.** $m = 3$ **c.** $m = 1/4$

14. The line $y = \frac{1}{2}x$ and the line $y = mx + 2$ were graphed on the same set of axes with a computer graphics package. In the resulting graph (shown below) the unit in the vertical direction is three times as long as the unit in the horizontal direction, and the two lines cross at right angles. Find m.

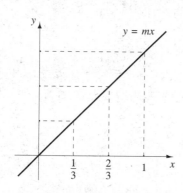

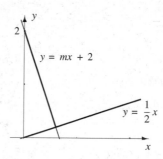

Exercises 15–18: A picture of a container is shown. Water is poured into the container at a constant rate (say 1 liter/s). As the water is poured in, the height changes. Draw a graph of the height of water as a function of time. Discuss your graph, telling which features in the container relate to which features on the graph.

15.

16.

17.

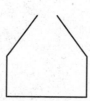

18.

Exercises 19–22:   Water is poured into a container at a constant rate (say 1 liter/s). The graph shown is the graph of the height of water as a function of time. Draw a possible picture of the container that resulted in the given graph.

19.

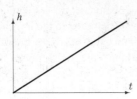

21.

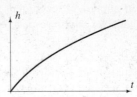

20.

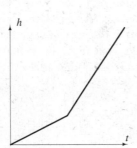

22.

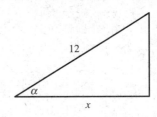

Review/Preview

23. Consider the polynomial $p(x) = x^2 - 3x + 5$.
  a. What number should be added to $p(x)$ to "complete the square"?
  b. For what values of b is $p(x) + bx$ a perfect square?
  c. For what values of a is $p(x) + ax^2$ a perfect square?

24. Find all values of x for which

$$\sqrt{x^2 + 4} = 9.$$

25. Find an equation for the line that passes through the point $(-1, 3)$ and makes an angle of $60°$ with the x-axis.

26. Sketch the graph of the line $y = -2x + 5$. At what angle does the line cross the x-axis?

27. Let $f(x) = x^2 + 4x$. Express $f(x^2 + 2)$ in simplest form.

28. On what intervals is $g(x) = x(x - 2)(x + 3)$ positive?

29. Find the coordinates of the point(s) where the line $y = 4x + 3$ intersects the parabola $y = 2x^2 - 3x - 2$.

30. If $\pi \le \theta \le 2\pi$ and $\tan \theta = -\sqrt{3}$, what is θ? What are the values of $\sin \theta$ and $\cos \theta$?

31. In the figure below, if $\sin \alpha = \frac{3}{5}$, what is x?

1.3   CALCULATING THE RATE OF CHANGE

In the previous section we worked with the formula

(1) $$P = P(V) = \frac{4}{V}$$

and looked at two ways to estimate the rate of change of P with respect to V when $V = 3$. The first method was a graphical approach. We saw that as we zoomed in near the point $(3, P(3))$ the graph of (1) looked more and more like a straight line. We estimated the "slope" of this "almost line" by reading the horizontal and vertical changes from the coordinate axes. The second method refined this process by using numerical calculation. We found the coordinates of two points on the graph and then found the slope of the segment joining them. One point was $Q = (3, P(3))$ and the second, R, was taken close to Q. When R was close enough to Q, segment $\overline{QR}$ was almost indistinguishable from the portion of the graph between Q and R. We used the slope of $\overline{QR}$ as an estimate of the rate of change of P with respect to V when $V = 3$.

The processes described above can become tedious if we need to continually refine our estimate of the rate of change. Each new estimate requires either that we zoom in closer or that we calculate the slope of $\overline{QR}$ for a new value of R. Furthermore, we can always get a better estimate by zooming in closer or by taking R closer to Q. When are we done? How can this process lead to a definite number that we can call *the* rate of change? In this section we investigate a symbolic approach to finding the rate of change and see that this approach does lead to a number that we can call the rate of change.

Refining the Rate-of-Change Calculations

Investigation

As in the Investigation of Section 1.2, let $P = 4/V$. Refine the rate-of-change calculations of the previous section and show that

$$-\frac{4}{9} \text{ atm/liter}$$

is a reasonable value for the rate of change of P with respect to V when $V = 3$.

Let $Q = (3, P(3)) = (3, 4/3)$, and let R be another point on the graph of

$$P = P(V) = \frac{4}{V}.$$

If R is close to Q, then we can express R as

$$R = (3 + h, P(3 + h)) = \left(3 + h, \frac{4}{3 + h}\right),$$

where h is a small negative number if R is to the left of Q, and h is a small positive number if R is to the right of Q. See Fig. 1.28.

The slope of segment $\overline{QR}$ is

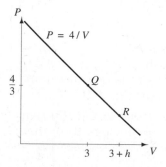

$P = 4/V$

Figure 1.28. The graph of $P = 4/V$ near $Q = (3, 4/3)$. Note $h > 0$ since R is to the right of Q.

$$(2) \qquad \frac{P(3 + h) - P(3)}{(3 + h) - 3} = \frac{\dfrac{4}{3 + h} - \dfrac{4}{3}}{h} = \frac{4 \cdot 3 - 4(3 + h)}{(3 + h)3h}$$

$$= -\frac{4h}{(3 + h)3h} = -\frac{4}{3(3 + h)}.$$

Because (2) makes sense for any small, nonzero h, we can use this expression to streamline the calculations done in the last section. For example, in equation (6) of Section 1.2 we calculated the slope of $\overline{QR}$ with $R = (2.99999, P(2.99999))$. This R has the form $(3 + h, P(3 + h))$ with $h = -0.00001$. Thus the result in (6) of Section 1.2 can be found by substituting -0.00001 for h in (2):

$$-\frac{4}{3(3+h)} = -\frac{4}{3(3 - 0.00001)} \approx -0.44445.$$

More important, we can use (2) to see what happens to our rate-of-change calculations as we zoom in closer and closer on the graph near $Q = (3, 4/3)$, that is, as we let R get closer and closer to Q. As $R = (3 + h, P(3 + h))$ is taken closer and closer to Q, it must be true that h gets closer and closer to 0. But then the rate-of-change estimate (2) gets closer and closer to

$$-\frac{4}{3(3 + 0)} = -\frac{4}{9} = -0.44444\ldots.$$

We take $-(4/9)$ atm/liter to be *the* rate of change of P with respect to V when $V = 3$. We interpret this to mean that if the piston is moving to expand the chamber, then at the instant when the chamber volume is 3 liters the pressure P is changing at a rate of $-(4/9)$ atm/liter. The minus sign in front of the rate of change tells us that as V increases, P decreases.

The procedure described above can be used to find rates of change in many other situations.

Calculating a Rate of Change

Let $y = f(x)$, where f is a function, and let $x = a$ be in the domain of f. To find the rate of change of y with respect to x at $x = a$:

1. Form and simplify the expression

(3)
$$\frac{f(a + h) - f(a)}{h}, \quad h \neq 0.$$

This is the slope of the segment determined by the points $(a, f(a))$ and $(a + h, f(a + h))$ on the graph of $y = f(x)$. When h is small, (3) can be used as an estimate of the rate of change of y (or f) with respect to x when $x = a$. See Fig. 1.29.

2. Investigate the behavior of (3) as h gets close to 0. If (3) gets close to one and only one number as h gets close to 0, then this number is the *rate of change of f (or y) with respect to x at x = a.*

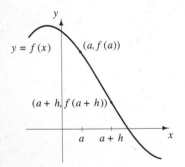

Figure 1.29. Find the slope of the segment determined by $(a, f(a))$ and $(a + h, f(a + h))$.

We refer to the process described above as the rate-of-change algorithm. In the remaining sections of this chapter we will study the rate-of-change algorithm and introduce new ideas and terminology that allow us to express the algorithm in a more efficient form and in more precise language. This will add to our understanding of the rate of change and its role in describing the world around us.

Nonetheless, we can use the rate-of-change algorithm as stated here to find rates of change for some simple functions.

■ **EXAMPLE 1**   Let $y = f(x) = x^2$.

a. Find the rate of change of y with respect to x at $x = -1$.

b. Find the rate of change of y with respect to x at the arbitrary point $x = a$.

Solution

a. Before using the rate-of-change algorithm to find the answer, we try zooming in to get an estimate for the rate of change. (You should play along on your own calculator or CAS.) Figure 1.30 shows the graph of $y = x^2$. The small square in the figure is centered at the point of interest, $(-1, (-1)^2) = (-1, 1)$. When we zoom in on the piece of the graph in the square we see something like Fig. 1.31.

The graph in Fig. 1.31 looks like the graph of a line. Using the tick marks on the axes we see that the y-coordinate of the "line" drops from 1.02 to 0.98 as the x-coordinate grows from -1.01 to -0.99. Hence the slope is about

$$\frac{1.02 - 0.98}{-1.01 - (-0.99)} = -2.$$

Thus we guess that when $x = -1$ the rate of change of y with respect to x is close to -2.

To apply the rate-of-change algorithm, use (3) with $f(x) = x^2$ and $a = -1$. We have

$$(4) \quad \frac{f(a + h) - f(a)}{h} = \frac{(-1 + h)^2 - (-1)^2}{2} = \frac{-2h + h^2}{h} = -2 + h.$$

This last expression is the slope of the segment determined by

$$Q = (-1, 1) \quad \text{and} \quad R = (-1 + h, (-1 + h)^2)$$

as seen in Fig. 1.32. For small h, (4) is an approximation of the rate of change of y with respect to x at $x = -1$. As we move R closer and closer to Q, we see that h becomes closer and closer to 0, and (4) gets closer and closer to -2:

$$-2 + h \xrightarrow[h \to 0]{} -2.$$

Thus when $x = -1$, the rate of change of y with respect to x is -2. Remember that the answer is a rate of change and so has units of y units$/x$ units.

b. Using (3), the rate of change of y with respect to x when $x = a$ is approximated by

$$\frac{f(a + h) - f(a)}{h} = \frac{(a + h)^2 - a^2}{h} = \frac{2ah + h^2}{h} = 2a + h$$

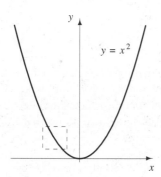

Figure 1.30. The graph of $y = x^2$. The small square is centered at $(-1, 1)$.

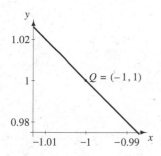

Figure 1.31. Zooming in on $y = x^2$ near $Q = (-1, 1)$.

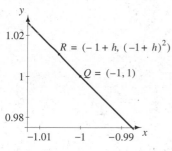

Figure 1.32. Take a point R close to Q.

when h is small. As h gets closer and closer to 0, this last expression tends to $2a$:

$$2a + h \xrightarrow[h \to 0]{} 2a.$$

Thus when $x = a$ the rate of change of y with respect to x is

$$2a \frac{y \text{ units}}{x \text{ units}}.$$

If we set $a = -1$, we get a rate of change of -2 as found in part a. ■

The rate-of-change algorithm worked well in finding rates of change for $P = 4/V$ and $y = x^2$ because in these cases (3) could be simplified to a point where we could determine the behavior as h got close to 0. Sometimes such simplification is beyond our capabilities. In such cases we can use graphical or numerical techniques to gain insight to the behavior of (3) as h gets close to 0.

■ **EXAMPLE 2**   Let $r = r(\theta) = \sin \theta$, where θ is measured in radians. Investigate the rate of change of r with respect to θ when $\theta = \pi/3$.

Solution.   Figure 1.33 shows the graph of $r = \sin \theta$ for $-2\pi < \theta < 2\pi$. In Fig. 1.34 we have zoomed in on the graph near the point

$$Q = \left(\pi/3, \sin(\pi/3)\right) = \left(\pi/3, \sqrt{3}/2\right) \approx (1.0472, 0.8660).$$

The graph in Fig. 1.34 is almost a line, and the slope appears to be about 0.5. Thus at $\theta = \pi/3$ the rate of change of r with respect to θ is approximately 0.5.

We can also estimate the rate of change by using (3). We have

$$(5) \quad \frac{r\left(\dfrac{\pi}{3} + h\right) - r\left(\dfrac{\pi}{3}\right)}{h} = \frac{\sin\left(\dfrac{\pi}{3} + h\right) - \sin\left(\dfrac{\pi}{3}\right)}{h} = \frac{\sin\left(\dfrac{\pi}{3} + h\right) - \dfrac{\sqrt{3}}{2}}{h}.$$

We need to know what happens to this expression as h gets close to 0. In previous examples we were able to simplify the rate-of-change expressions to the point where we could answer this question by simply substituting $h = 0$. This is not the

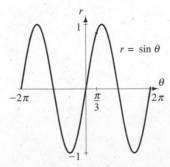

Figure 1.33. The graph of $r = \sin \theta$ for $-2\pi < \theta < 2\pi$.

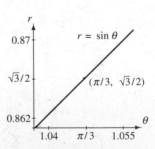

Figure 1.34. Zooming in near the point $(\pi/3, \sqrt{3}/2)$.

case with (5). There are no obvious simplifications, and simply substituting $h = 0$ into (5) results in $0/0$, which is undefined. Another approach is to gather numerical data by trying several values of h close to 0. With the aid of a calculator or computer we can generate the table below. You should verify some of the numbers in this table with your own calculator, and then try some additional h values even closer to 0.

h	$\dfrac{\sin(\pi/3 + h) - \sqrt{3}/2}{h}$
0.1	0.455902
−0.1	0.542432
0.05	0.478146
−0.05	0.521438
0.001	0.499567
−0.001	0.500433
0.000005	0.499998
−0.000005	0.500002

Using data from the table we guess that the rate-of-change estimate in (5) is getting close to 0.5 as h gets close to 0:

$$\frac{\sin\left(\dfrac{\pi}{3} + h\right) - \dfrac{\sqrt{3}}{2}}{h} \xrightarrow[h \to 0]{} 0.5.$$

We can also investigate the behavior of (5) by graphing this expression for h near 0. The graph is shown in Fig. 1.35. Keep in mind that this graph is the graph of the quotient shown in (5) (which is a function of h) and *not* the graph of $r = \sin\theta$. Although (5) (and hence the graph in Fig. 1.35) is not defined for $h = 0$, we can see from the graph that the value of the expression is closing in on 0.5 as h approaches 0.

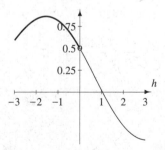

Figure 1.35. What appears to happen to $(\sin(\pi/3 + h) - \sqrt{3}/2)/h$ as h gets close to 0?

The results of the above investigations suggest that when $\theta = \pi/3$, the rate of change of $r = \sin\theta$ with respect to θ is close to (or perhaps equal to) 0.5. ■

Functions Without Formulas

Scientists do not always have a formula to express a relationship between two quantities. Sometimes these relationships can be given only graphically or as numerical data. In these cases we can still estimate rates of change.

■ **EXAMPLE 3**   The graph below shows the temperature in degrees Celsius inside an autoclave (used for sterilization of laboratory equipment) between 11:00 A.M. and 1:00 P.M. What is the rate of change of temperature with respect to time at 11:45 A.M.? At 12:15 P.M.?

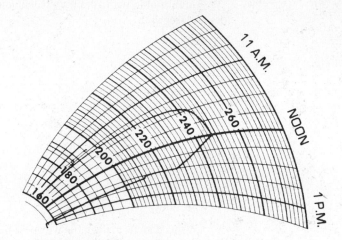

Solution.   At 11:45 A.M. the temperature in the autoclave was about 247°C. At 12:00 noon the temperature was 248°C. Since the temperature graph appears to be almost a line between these two points, the rate of change of the temperature is approximately the slope of this "line,"

$$\frac{\text{change in temperature}}{\text{change in time}} = \frac{1°C}{15 \text{ min}} \approx 0.067°C/\text{min}.$$

As the graph shows, and the positive sign on our answer indicates, the temperature was rising at 11:45 A.M.

At 12:15 P.M. the temperature in the autoclave was 237°. Since the graph seems almost linear between 12:00 noon and 12:15 P.M., the rate of change of the temperature at 12:15 P.M. should be close to the slope

$$\frac{\text{change in temperature}}{\text{change in time}} = \frac{237° - 248°}{15 \text{ min}} \approx -0.73°C/\text{min}.$$

The negative sign tells us that the temperature was decreasing at 12:15 P.M. This can be seen by looking at the graph.   ■

Functions Without a Rate of Change

There are some functions that do not have a rate of change at some points. The rate-of-change algorithm can be used to identify some of these points for a given function.

■ **EXAMPLE 4**   Let

(6) $$y = f(x) = |x - 2| + 1.$$

Investigate the rate of change of y with respect to x when $x = 2$.

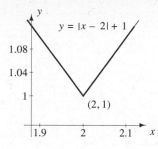

Figure 1.36. Zooming-in near the point (2, 1) on the graph of $y = |x - 2| + 1$.

Solution. Try zooming in on the graph of (6) near the point (2, 1) (see Fig. 1.36). We quickly find that no matter how close we zoom in, the portion of the graph we see is never going to look like a straight line because the corner at (2, 1) will not straighten out at any magnification. Thus we cannot approximate a rate of change at $x = 2$ by estimating the slope of an "almost line."

Next, try the rate-of-change algorithm. We investigate the expression

h	$\dfrac{|h|}{h}$
0.01	1
-0.01	-1
0.001	1
-0.001	-1
0.0005	1
-0.0005	-1

$$(7) \qquad \frac{f(2 + h) - f(2)}{h} = \frac{(|(2 + h) - 2| + 1) - 1}{h} = \frac{|h|}{h}$$

for h close to 0. A table of values for (7) is shown. The data in the table indicate that the value (7) does not settle down to one and only one number as h gets close to 0. This means that the rate of change of $y = |x - 2| + 1$ with respect to x is undefined at $x = 2$. ∎

Exercises 1.3

Basic

Exercises 1–10: Use the rate-of-change algorithm to find the rate of change for the given value of x.

1. $f(x) = 2x^2 - 3, \quad x = 1$

2. $f(x) = -x^2 + 4, \quad x = -2$

3. $y = 5x - 7, \quad x = 4$

4. $s = -2t + 8, \quad t = a$

5. $g(x) = x^3, \quad x = 2$

6. $g(x) = x^3, \quad x = a$

7. $F(x) = 2/x, \quad x = 2$

8. $h(t) = 3/t^2, \quad t = 1$

9. $f(x) = \sqrt{x}, \quad x = 9$

10. $y = \sqrt{x}, \quad x = a, \quad a > 0$

11. Let $y = f(x) = mx + b$ where m and b are constants. Use the rate-of-change algorithm to show that for any value of x, the rate of change of y with respect to x is m.

12. Let $y = \sin x$. Use both graphical and numerical methods, as in Example 2, to investigate the rate of change of y with respect to x when $x = 0$ and when $x = 2\pi/3$.

13. Let $y = 2^x$. Use both graphical and numerical methods, as in Example 2, to investigate the rate of change of y with respect to x when $x = 0$, $x = 1$, $x = 2$, and $x = 4$.

14. Let $h = h(t) = |2t - 1|$. For what t values is the rate of change of h with respect to t undefined? What is the rate of change at those t values for which the rate of change is defined?

15. Let $g = g(x) = |3 + x| - 2|4 - 3x|$. For what x values is the rate of change of g with respect to x undefined? What is the rate of change at those x values for which the rate of change is defined?

16. When a heavy metal ball is dropped from the top of a 500-foot building, the height (in feet) of the ball above the ground after t seconds is

$$(8) \qquad h(t) = 500 - 16t^2.$$

a. Given the conditions described in the problem, for what t values does (8) make physical sense?

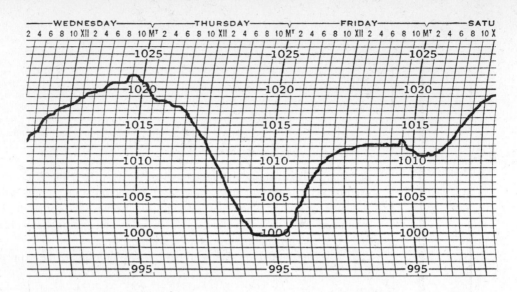

b. What are the units for the rate of change of h with respect to t? What does this rate of change tell us about the falling ball?

c. What is the rate of change of h with respect to t at any given time during the fall?

d. When does the ball hit the ground? How fast is the ball going when it hits the ground?

17. The graph above shows air pressure (in millibars) as a function of time in Ames, Iowa, from 4:00 A.M. Wednesday, December 11, 1991, to 2:00 P.M. Saturday, December 14, 1991. What was the rate of change of air pressure at 2:00 A.M. Thursday? Was the pressure rising or falling? At what time did the air pressure appear to be rising most rapidly? Falling most rapidly?

18. Let g be a function. The graph of $s = g(t)$ is shown at the right.

a. Indicate with an asterisk those points on the graph where the rate of change of s with respect to t is undefined.

b. Indicate with a 0 those points where the rate of change is 0.

c. Indicate with a 1 those points where the rate of change is 1.

Give reasons for all answers.

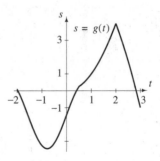

Growth

19. Let f be a function with $f(-2) = 4$, and let $y = f(x)$. Suppose the rate of change of y with respect to x at $x = -2$ is 0. If we zoom in very close to the graph of $y = f(x)$ near the point $(-2, 4)$, what will we see? Explain your answer.

20. Let f and g be two functions, with $f(1) = 3$ and $g(1) = 3$. Suppose that at $x = 1$ the rate of change of f with respect to x is 2 and the rate of change of g with respect to x is 1. If $y = f(x)$ and $y = g(x)$ are

graphed on the same set of axes and we then zoom in on the point $(1, 3)$, what will we see? Draw a sketch and explain.

21. Consider the functions f and g defined by $f(x) = x^2$ and $g(x) = x^2 + 5$. At a given $x = a$ how do the rates of change of the two functions compare?

22. Let h be a function and c a constant. How are the rates of change of $y = h(x)$ and $y = h(x) + c$ related? Explain your answer.

23. The table below gives U.S. census data for the years 1790–1990. Estimate the rate of change of U.S. population with respect to time in 1800, 1900, and 1980 (choose appropriate units). Explain how you got your estimates, and explain why someone else might arrive at different estimates.

Year	Population (millions)	Year	Population (millions)
1790	4	1900	76
1800	5	1910	92
1810	7	1920	106
1820	10	1930	123
1830	13	1940	132
1840	17	1950	151
1850	23	1960	178
1860	31	1970	203
1870	40	1980	227
1880	50	1990	247
1890	63		

24. The figure below is a topographical map of the region surrounding Jackstraw Mountain on the west side of Rocky Mountain National Park. The contour lines on the map indicate altitudes above sea level. The change in elevation between adjacent lines is 80 feet. A backpacker hiking off-trail passes through the point labeled P on the map. Estimate the rate of change of the backpacker's elevation as he passes through P for each of the following directions of travel. (Your answers should be in units of vertical feet/horizontal mile. You will need the scale of miles accompanying the map.)

a. Along the path labeled N and in the direction of the arrow.

b. Along the path labeled N and in the direction opposite the arrow.

c. Along the path labeled W and in the direction of the arrow.

25. A car on a freeway is moving at 60 miles per hour as it passes a white mark on the road. Estimate how far the car will be from the mark in 1 hour, in 1 minute, in 2 seconds, and in 0.01 second. Which of your four answers is likely to be the most accurate? Least accurate? Explain.

26. Let f be a function and $x = a$ a point in the domain of f. Explain why the rate of change of f with respect to x at $x = a$ can be found by considering the behavior of the expression

$$\frac{f(x) - f(a)}{x - a}$$

as x gets close to a. In particular, tell why this method always gives the same results as the rate-of-change algorithm.

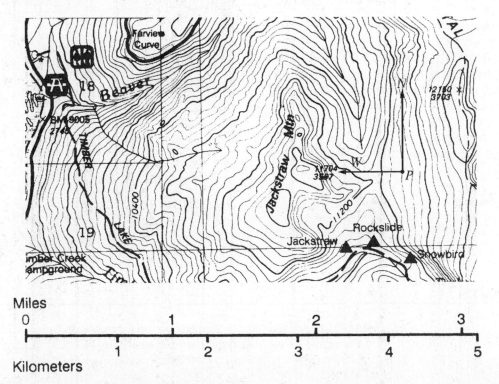

Miles
0   1   2   3

Kilometers
1   2   3   4   5

Review/Preview

27. For each $f(x)$ given below, simplify the expression

$$\frac{f(x) - f(a)}{x - a}.$$

a. $f(x) = 3x - 6$
b. $f(x) = x^2 + 2x - 7$
c. $f(x) = \dfrac{-3}{x + 2}$
d. $f(x) = \dfrac{4x}{x^2 + 3}$

28. For each $f(x)$ given in Exercise 27, simplify the expression

$$\frac{f(x + h) - f(x)}{h}.$$

29. For what values of b does the equation

$$x^2 + bx + 7 = 0$$

have two different real solutions? Exactly one real solution? No real solutions?

30. Solve the following inequalities. Check your answers by graphing.
a. $-2x + 6 < 0$
b. $x^2 - 4x + 5 \geq 0$
c. $\dfrac{x - 1}{x + 2} > 1$

31. a. For what θ between 0 and 2π are $\sin\theta$ and $\cos\theta$ both positive?
b. For what θ between 0 and 2π are $\sin\theta$ and $\cos\theta$ of opposite sign?
c. For what θ between 0 and 2π is $\sin\theta \geq \cos\theta$?

32. Sketch the graphs of $y = \sin x$ and $y = \cos x$ on the same set of axes. How are the two graphs different? How are they similar?

33. Find real numbers a and b with $-0.01 \leq a, b \leq 0.01$ and:
a. $a/b = 3$.
b. $a/b = -500$.
c. $a/b = 10^9$.
d. $a/b = -10^{-9}$.

34. Let a and b be two real numbers. Find the largest and smallest possible values of a/b if:
a. $3.9 \leq a \leq 4.1$ and $5 \leq b \leq 5.2$.
b. $-3.1 \leq a \leq -2.9$ and $5 \leq b \leq 5.2$.
c. $-0.1 \leq a \leq 0.1$ and $9.9 \leq b \leq 10.2$.

35. The graph of a function $s = r(t)$ is shown below. Sketch graphs of each of the following. Use a different set of axes for each.
a. $s = 2r(t)$
b. $s = -r(t)$
c. $s = r(-t)$
d. $s = |r(t)|$
e. $s = r(|t|)$
f. $s = r(t - 2)$

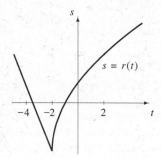

36. The graphs of two functions, $y = f(x)$ and $y = g(x)$ are shown below. On the same set of axes sketch the graphs of $y = f(x) + g(x)$ and $y = f(x) - g(x)$.

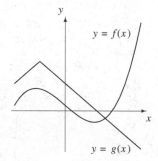

1.4   LIMITS

In 1911 Dutch physicist H. Kamerlingh Onnes measured the electrical resistance of a mercury sample at various temperatures. His plot of resistance versus temperature is shown in Fig. 1.37. At the *critical temperature* of 4.2 kelvins (K) the

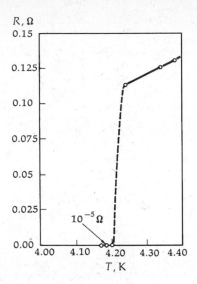

Figure 1.37. Resistance (R) of a sample of mercury as a function of temperature (T).

resistance showed a sudden decrease. Onnes had discovered the phenomenon of superconductivity. Since that time physicists have studied the electrical properties of matter at temperatures close to 0 K, the temperature at which all molecular motion ceases. Physicists know that a temperature of 0 K (or absolute zero) can never be attained, but they have been able to achieve temperatures just slightly above absolute zero. By studying electrical resistance at these temperatures, physicists hope to infer what will happen to the resistance as temperature falls toward 0 K. In mathematical language we might say that physicists are investigating "the limit of electrical resistance as temperature approaches absolute zero."

When we compute the rate of change of a function f at a point $x = a$ in its domain, we are in a similar situation to that of the physicist studying properties at low temperature. The rate of change is found by investigating the behavior of the quotient

$$(1) \qquad \frac{f(a + h) - f(a)}{h}$$

when h is close to 0. We cannot simply set $h = 0$ in (1) because the expression is undefined for such h. Instead, we investigate the value of (1) for real numbers h close to 0. We might do so numerically (with a table of values), graphically, algebraically, or by combining two or more of these approaches. In doing this our goal is to obtain a number or expression that we will call the rate of change of f with respect to x at $x = a$. This number or expression is *the limit of (1) as h approaches 0.*

The definition of limit is motivated by the need to study the behavior of expressions like (1) near a value of h where the expression may not be defined.

DEFINITION   limit of a function

Let g be a function and let a and L be real numbers. If we can make $g(x)$ as close to L as we like by taking x close to a, but not equal to a, then we say

$g(x)$ has **limit** L as x approaches a.

We denote this by

$$(2) \qquad \lim_{x \to a} g(x) = L.$$

Equation (2) is also read "the limit of $g(x)$ as x approaches a equals L."

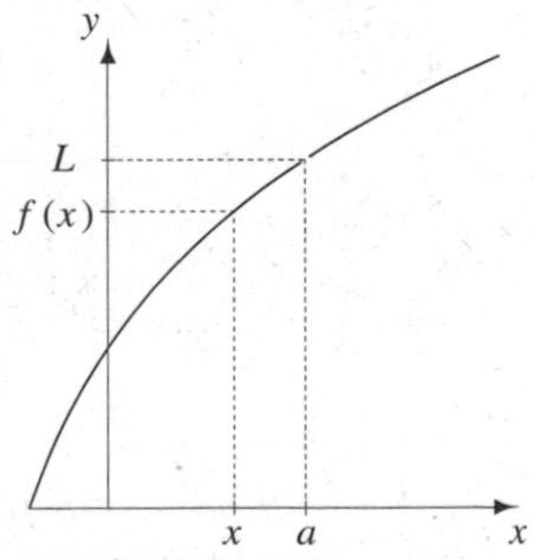

This definition of limit is meant to be a "working definition." By thinking of $\lim_{x \to a} f(x)$ as the number that $f(x)$ is close to when x is close to a, as illustrated in Fig. 1.38, we can evaluate most limits, including limits of expressions like (1). However, we will need to do more with limits than simply evaluate limits of functions. We will use limits to study rates of change, approximations, areas, and many other things. For this we will need a more precise definition than the one given above. In the next section we will see that mathematicians mean something very special when they say "close to." Once "close to" is properly defined, we will have a definition of limit that is good for more than just evaluation of limits.

Figure 1.38. When x is close to a, $f(x)$ is close to L. We write $\lim_{x \to a} f(x) = L$.

Nonetheless, knowing how to evaluate limits is important. Therefore we will look at several examples.

Many limits can be evaluated using little more than common sense and knowledge about some familiar functions.

■ **EXAMPLE 1**   Evaluate $\lim_{x \to 5} 2x^2 + 4$.

Solution.   When x is close to 5, $2x^2 + 4$ is close to $2 \cdot 5^2 + 4 = 54$. Hence

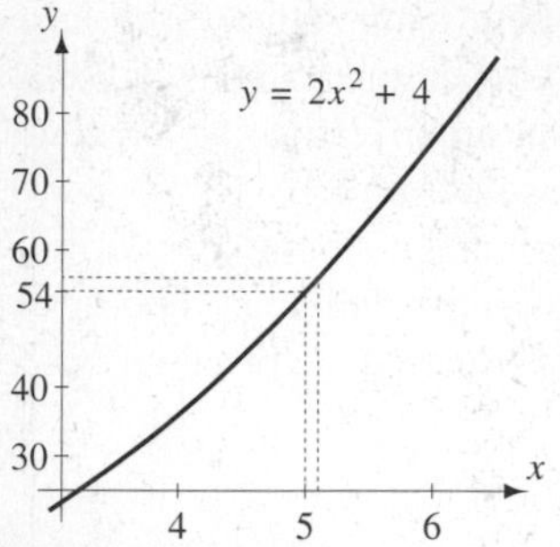

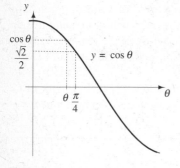

Figure 1.39. When x is close to 5, $2x^2 + 4$ is close to 54.

$$(3) \qquad\qquad \lim_{x \to 5} 2x^2 + 4 = 54.$$

The graph of $y = 2x^2 + 4$ near $x = 5$ is shown in Fig. 1.39. From the graph we see that when x is close to 5, the value of $y = 2x^2 + 4$ is close to 54. This is consistent with (3).   ■

■ **EXAMPLE 2**   Find the value of $\lim_{\theta \to \pi/4} \cos \theta$.

Solution.   When θ is close to $\pi/4$, $\cos \theta$ is close to $\cos \pi/4 = \sqrt{2}/2$. Therefore

$$\lim_{\theta \to \pi/4} \cos \theta = \frac{\sqrt{2}}{2}.$$

This is illustrated by looking at the graph of $y = \cos \theta$ near $\theta = \pi/4$. In Fig. 1.40 we see that if θ is close to $\pi/4$, then $\cos \theta$ is close to $\sqrt{2}/2$.   ■

Figure 1.40. When θ is close to $\pi/4$, $\cos \theta$ is close to $\sqrt{2}/2$.

More care is needed in evaluating a limit near a point where the function is not defined. In these cases the limit cannot be determined by a simple substitution as in the previous examples.

■ **EXAMPLE 3**   Evaluate

$$(4) \qquad\qquad \lim_{t \to 3} \frac{2t^2 - 4t - 6}{t^2 - 9}.$$

Solution.   We demonstrate three techniques for exploring limits: numerical, graphical, and analytic. First, however, let's see why the "common sense" approach used in the two previous examples fails here. If t is close to but not equal to 3, then

$$2t^2 - 4t - 6 \approx 2(3)^2 - 4(3) - 6 = 0$$

and

$$t^2 - 9 \approx 3^2 - 9 = 0.$$

Hence the numerator and denominator in (4) are both small, nonzero numbers. This means that when t is close to 3, the quotient in (4) is a ratio of two small, nonzero numbers. Since the ratio of two small numbers can be anything (see Exercises 33 and 34 in Section 1.3), we do not have enough information to decide what number, if any, is close to the ratio.

t	$\dfrac{2t^2 - 4t - 6}{t^2 - 9}$
2.5	1.27273
3.5	1.38462
2.9	1.32203
3.1	1.34426
2.99	1.33222
3.01	1.33444
2.9999	1.33332
3.0001	1.33334

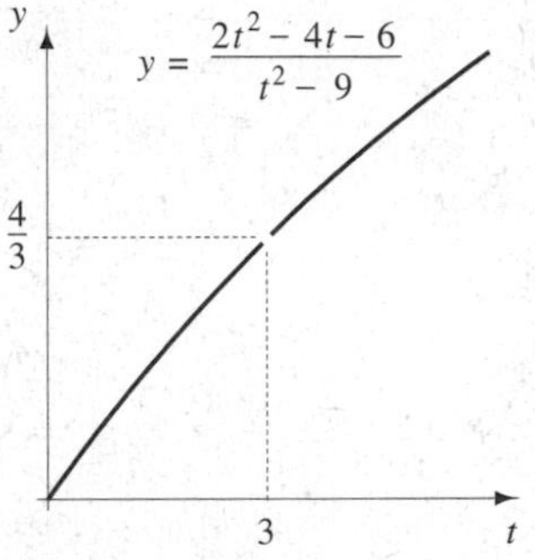

Figure 1.41. The graph indicates that $\lim_{t\to 3}(2t^2 - 4t - 6)/(t^2 - 9) \approx 1.3$.

Numerical approach. With a calculator or computer we can generate a list of values of

$$(5) \qquad \frac{2t^2 - 4t - 6}{t^2 - 9}$$

for t values close to 3. A table of such values is shown. (You should verify some of the examples in this table, and perhaps add one or two new lines to the table.) The numerical data seem to indicate that (5) gets close to $4/3 \approx 1.3333$ as t nears 3. Thus we have evidence that

$$\lim_{t\to 3} \frac{2t^2 - 4t - 6}{t^2 - 9} = \frac{4}{3} \approx 1.333.$$

Graphical approach. Figure 1.41 shows the graph of

$$y = \frac{2t^2 - 4t - 6}{t^2 - 9}$$

for $2 < t < 3$ and $3 < t < 4$. (You should produce a similar graph on your computer or calculator.) Let t be close to 3 on the horizontal axis. Move vertically from t to the graph, and from the graph over to the vertical axis. We meet the vertical axis at a point close to 1.33. This means that when t is close to 3, the value of (5) is close to 1.33. This suggests that

$$\lim_{t\to 3} \frac{2t^2 - 4t - 6}{t^2 - 9} \approx 1.33.$$

To get a better approximation to the limit, zoom in on the graph and investigate the graph for t values even closer to 3.

Analytic approach. After simplifying (5) algebraically, we can get very precise information about the limit. If t is close to 3 but not equal to 3, then factoring the numerator and denominator gives

$$(6) \qquad \frac{2t^2 - 4t - 6}{t^2 - 9} = \frac{2(t - 3)(t + 1)}{(t - 3)(t + 3)} = \frac{2(t + 1)}{t + 3}.$$

By canceling the factor of $t - 3$ we have eliminated the factor that causes the numerator and denominator of (5) to be 0 when $t = 3$. When t is close to 3, the numerator of the last expression is close to $2(3 + 1) = 8$, and the denominator is close to $(3 + 3) = 6$. Thus when t is close to 3, (6) is close to

$$\frac{2(3 + 1)}{(3 + 3)} = \frac{4}{3}.$$

This means that

$$(7) \qquad \lim_{t\to 3} \frac{2t^2 - 4t - 6}{t^2 - 9} = \lim_{t\to 3} \frac{2(t + 1)}{(t + 3)} = \frac{4}{3} \approx 1.33333.$$

Limits and the Rate of Change

In the previous section we stated the rate-of-change algorithm. This algorithm gives guidelines for determining the rate of change of a function f at a point a, providing this rate of change exists. We saw that the rate-of-change algorithm can also be used to identify some instances in which the rate of change does not exist. According to the algorithm, the rate of change of f at a is determined by the behavior of

$$(8) \qquad \frac{f(a + h) - f(a)}{h}$$

for values of h close to 0. Hence the rate of change, if it exists, is the limit of (8) as h tends to 0. We restate the rate-of-change algorithm in this more compact form.

The Rate-of-Change Algorithm

Let f be a function and a a real number in the domain of f. If the rate of change of f at a exists, then it is equal to

$$\lim_{h \to 0} \frac{f(a + h) - f(a)}{h}.$$

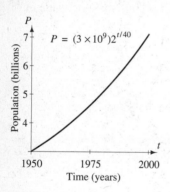

Figure 1.42. The graph of the population function.

■ **EXAMPLE 4** In 1950 the world population was 3 billion. Since that time the population has grown at a rate that results in a doubling of the population every 40 years. See Fig. 1.42.

a. Let $P = P(t)$ be the population in year t, with $t = 0$ corresponding to 1950. Show that the formula

$$(9) \qquad P = P(t) = (3 \times 10^9)2^{t/40}$$

gives the correct population in 1950 and models the 40-year doubling time.

b. Assuming that (9) is a correct model for population growth, how fast will the population be growing in 1997? What will be the percent per year rate of growth at this time?

Solution

a. To check that (9) gives the correct 1950 population, we substitute $t = 0$ into the formula. The result is

$$P(0) = (3 \times 10^9)2^{0/40} = 3 \times 10^9,$$

which is the correct figure for 1950. To verify that (9) also models a 40-year doubling time, we show that for any time t, $P(t + 40)$ is twice $P(t)$. We have

$$P(t + 40) = (3 \times 10^9)2^{(t+40)/40} = (3 \times 10^9)2^{1+(t/40)}$$
$$= (3 \times 10^9)2^1 \cdot 2^{t/40} = 2(3 \times 10^9)2^{t/40} = 2P(t).$$

This shows that the population modeled by (9) doubles during any 40-year period.

b. The year 1997 corresponds to $t = 47$. If (9) is correct, then the rate of change of population in 1997 will be

$$\lim_{h \to 0} \frac{P(47 + h) - P(47)}{h}$$

$$= \lim_{h \to 0} \frac{(3 \times 10^9)2^{(47+h)/40} - (3 \times 10^9)2^{47/40}}{h}$$

$$= \lim_{h \to 0} (3 \times 10^9) \frac{2^{47/40}2^{h/40} - 2^{47/40}}{h}$$

$$= \lim_{h \to 0} (3 \times 10^9)2^{47/40} \frac{2^{h/40} - 1}{h}$$

$$= \lim_{h \to 0} P(47) \frac{2^{h/40} - 1}{h}.$$

We approximate this last limit numerically by computing the value of

$$P(47) \frac{2^{h/40} - 1}{h}$$

for several values of h close to 0. These calculations are shown in the table. The values in the right column seem to be close to

h	$P(47)\dfrac{2^{h/40} - 1}{h}$
0.1	$0.017344P(47)$
-0.1	$0.017314P(47)$
0.005	$0.017329P(47)$
-0.005	$0.017328P(47)$
0.0001	$0.017329P(47)$
-0.0001	$0.017329P(47)$

(10) $P(47) \cdot 0.017329 \approx 117{,}383{,}000.$

Because the units for t are years and the units for P are people, the rate of change has units of people/year. Hence in 1997 the rate of change of population can be approximated as

117,383,000 people/year.

From (10) we see that the yearly rate of growth is about 1.7 percent of the 1997 population. ■

Working with Limits

Sums, Products, and Quotients   Many of the limits that we work with can be evaluated by algebraically combining simpler limits. For example, in (7) we evaluated

$$\lim_{t \to 3} \frac{2(t + 1)}{(t + 3)}$$

by noting that the numerator has limit 8 and the denominator has limit 6, and then reasoning that the quotient must have limit 8/6. This is an application of (14) in the following list of rules for limits.

Rules for Combining Limits

Let g and h be two functions with

$$\lim_{x \to a} g(x) = L \quad \text{and} \quad \lim_{x \to a} h(x) = M,$$

and let c be a real number. Then,

(11) $\qquad \lim_{x \to a}(c \cdot g(x)) = c\left(\lim_{x \to a} g(x)\right) = cL,$

(12) $\qquad \lim_{x \to a}(g(x) + h(x)) = \left(\lim_{x \to a} g(x)\right) + \left(\lim_{x \to a} h(x)\right) = L + M,$

(13) $\qquad \lim_{x \to a}(g(x)h(x)) = \left(\lim_{x \to a} g(x)\right)\left(\lim_{x \to a} h(x)\right) = LM.$

If $M \neq 0$, then

(14) $\qquad \lim_{x \to a} \dfrac{g(x)}{h(x)} = \dfrac{\lim_{x \to a} g(x)}{\lim_{x \to a} h(x)} = \dfrac{L}{M}.$

Although the results given above can all be "rigorously proven," we will accept these statements without proof because they seem to follow from common sense. For example, to explain (13) above we can reason as follows:

> When x is close to a, $f(x)$ is close to L and $g(x)$ is close to M. Thus when x is close to a, $f(x) \cdot g(x)$ is close to $L \cdot M$. This suggests that
>
> $$\lim_{x \to a}(g(x)h(x)) = LM.$$

Those who are interested in rigorous proofs for the limit rules should see Exercises 27 and 28 in Section 1.5.

Polynomials and Trigonometric Functions   To use the limit rules effectively, we must be able to break complicated limits into simple ones and then evaluate the simple limits. Here are some of the more common "simple limits."

Limits of Polynomials, Sines, and Cosines

Let a be a real number. Then for any polynomial

$$p(x) = c_n x^n + c_{n-1} x^{n-1} + \cdots + c_1 x + c_0,$$

(15) $\qquad \lim_{x \to a} p(x) = p(a).$

Also, for any real a,

(16) $\qquad \lim_{x \to a} \sin x = \sin a \quad \text{and} \quad \lim_{x \to a} \cos x = \cos a.$

We also accept these rules without proof because they seem "natural" given what we know about polynomials and trigonometric functions. These rules really say that for these functions, "what you expect is what you get." In fact, these rules seem so sensible that we have already applied them in working Examples 1 and 2. Can you see where? When we apply the rules for combining limits and for finding limits of polynomials, sines, and cosines, we will usually not mention the application. However, we give one example to illustrate how we might document uses of these rules.

■ **EXAMPLE 5**   Evaluate

(17) $$\lim_{t \to -1} (t^2 - 2t + 7) \tan t.$$

Solution.   First recall that $\tan t = \sin t / \cos t$. By (16),

$$\lim_{t \to -1} \sin t = \sin(-1) \quad \text{and} \quad \lim_{t \to -1} \cos t = \cos(-1).$$

Because $\cos(-1) \approx 0.540302 \neq 0$, it follows from (14) that

$$\lim_{t \to -1} \tan t = \lim_{t \to -1} \frac{\sin t}{\cos t} = \frac{\lim\limits_{t \to -1} \sin t}{\lim\limits_{t \to -1} \cos t} = \frac{\sin(-1)}{\cos(-1)} = \tan(-1).$$

Next, because $t^2 - 2t + 7$ is a polynomial, we use (15) to see

$$\lim_{t \to -1} t^2 - 2t + 7 = (-1)^2 - 2(-1) + 7 = 10.$$

We now know that the limit of each factor of (17) exists, so we apply (13) to get

(18) $$\lim_{t \to -1} (t^2 - 2t + 7) \tan t = \left(\lim_{t \to -1} (t^2 - 2t + 7) \right)\left(\lim_{t \to -1} \tan t \right)$$
$$= 10 \tan(-1) \approx -15.5741$$

The graph of $y = (t^2 - 2t + 7) \tan t$ for t near -1 is shown in Fig. 1.43. From the graph we see that when t is close to -1, $(t^2 - 2t + 7) \tan t$ is close to -15. This is consistent with (18). ■

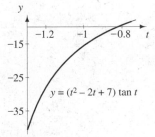

Figure 1.43. The graph of $y = (t^2 - 2t + 7) \tan t$ for t near -1.

Compositions   With (11), (12), (13), and (14) we can find the limits of functions that are sums, products, or quotients of other functions whose limits we know. Composition is another important way of combining functions. Hence it is also important to know how limits are affected by composition. The rule for limits of compositions seems to be a natural consequence of the definition of limit, so we will accept it without proof. However, there are some interesting subtleties behind the statement. These subtleties are discussed in Exercise 24 of Section 1.5.

Limits and Compositions

Let $h(x)$ and $g(x)$ be functions. Suppose that

$$\lim_{x \to a} g(x) = L$$

and that $h(x)$ is defined at L with

$$\lim_{x \to L} h(x) = h(L).$$

Then

(19) $$\lim_{x \to a} h(g(x)) = h\left(\lim_{x \to a} g(x)\right) = h(L).$$

Most uses of (19) seem consistent with common sense, so we usually do not mention applications of the result. However, here is one example in which we do point out the use of (19).

■ **EXAMPLE 6** Find the value of

$$\lim_{\theta \to 1/2} \sin(-\theta^2 + 2\theta + 4).$$

Solution. Let

$$p(\theta) = -\theta^2 + 2\theta + 4.$$

Because $p(\theta)$ is a polynomial, we can apply (15) to obtain

$$\lim_{\theta \to 1/2} p(\theta) = p(1/2) = -(1/2)^2 + 2(1/2) + 4 = \frac{19}{4}.$$

Next note that $\sin t$ is defined for $t = 19/4$ and that by (16)

$$\lim_{t \to 19/4} \sin t = \sin\left(\frac{19}{4}\right).$$

Now apply (19) to get

$$\lim_{\theta \to 1/2} \sin(-\theta^2 + 2\theta + 4) = \lim_{\theta \to 1/2} \sin(p(\theta))$$

(20)
$$= \sin\left(\lim_{\theta \to 1/2} p(\theta)\right) = \sin\left(\frac{19}{4}\right) \approx -0.999293.$$

The graph of $y = \sin(-\theta^2 + 2\theta + 4)$ for θ near $1/2$ is shown in Fig. 1.44. Is the answer in (20) consistent with the graph? ■

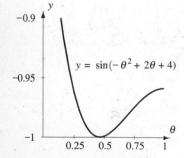

Figure 1.44. The graph of $y = \sin(-\theta^2 + 2\theta + 4)$ for θ near $1/2$.

A Trigonometric Limit

In Chapter 2 we will calculate the rates of change of many different functions. When we find the rates of change of the sine and cosine functions, we will need to know the value of

$$\lim_{h \to 0} \frac{\sin h}{h}.$$

We evaluate this limit in the next example. In doing so, we not only see more techniques for working with limits, but we also get to review some trigonometry.

■ **EXAMPLE 7**   Evaluate $\lim_{h \to 0} \sin h / h$, where h is in radians.

Solution.   First, let's get an estimate for the limit by graphing

(21) $$y = \frac{\sin h}{h}$$

for h values close to 0. The graph is shown in Fig. 1.45 and illustrates that (21) gets close to 1 as h gets close to 0. Hence we guess that

$$\lim_{h \to 0} \frac{\sin h}{h} \approx 1.$$

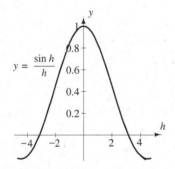

$y = \dfrac{\sin h}{h}$

Figure 1.45. The graph of $y = (\sin h)/h$ suggests that $\lim_{h \to 0}(\sin h)/h = 1$.

We could also have found this estimate by gathering numerical data. See Exercise 14.

To verify analytically that the value of the limit is 1, we first show that for h close to 0,

(22) $$\cos h \le \frac{\sin h}{h} \le \frac{1}{\cos h}.$$

We start by establishing (22) for $h > 0$. Consider the circle of center $O = (0, 0)$ and radius 1. Let $S = (1, 0)$, and label P on the upper half of the circle so that $\angle POS$ has measure h. Then

$$P = (\cos h, \sin h).$$

From P draw a segment perpendicular to the x-axis. This segment intersects the axis at $R = (\cos h, 0)$. Let the line perpendicular to the x-axis at S meet $\overrightarrow{OP}$ at Q. Since triangle OSQ is a right triangle with right angle at S and $OS = 1$, it follows that

$$QS = \frac{QS}{1} = \frac{QS}{OS} = \frac{\text{opposite}}{\text{adjacent}} = \tan h.$$

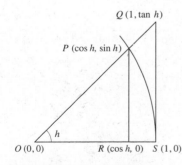

Figure 1.46. A right triangle in the circle of center $(0, 0)$ and radius 1.

As seen in Fig. 1.46,

(23) $$\text{area}(\triangle ORP) \le \text{area}(\text{sector } OSP) \le \text{area}(\triangle OSQ).$$

Because sector OSP has central angle h radians and is inscribed in a circle of radius 1, it follows that

$$(24) \qquad \text{area(sector } OSP) = \frac{1}{2}h \cdot 1^2 = \frac{1}{2}h.$$

In addition,

$$(25) \qquad \text{area}(\triangle ORP) = \frac{1}{2}\cos h \sin h$$

and

$$(26) \qquad \text{area}(\triangle OSQ) = \frac{1}{2}\tan h.$$

Combining (23), (24), (25), and (26), we have

$$\frac{1}{2}\cos h \sin h \le \frac{1}{2}h \le \frac{1}{2}\frac{\sin h}{\cos h}$$

We are assuming that h is a small positive number, so $\sin h > 0$. Multiplying the last expression by $2/\sin h$ gives

$$\cos h \le \frac{h}{\sin h} \le \frac{1}{\cos h},$$

from which

$$(27) \qquad \cos h \le \frac{\sin h}{h} \le \frac{1}{\cos h}.$$

This establishes (22) for h positive and close to 0.

Now suppose that h is small and negative. Then $-h$ is small and positive, so by (27) we have

$$(28) \qquad \cos(-h) \le \frac{\sin(-h)}{-h} \le \frac{1}{\cos(-h)}.$$

The cosine function is an even function, so

$$\cos(-h) = \cos h.$$

The sine function is an odd function, which means that

$$\sin(-h) = -\sin h.$$

Incorporating these identities into (28), we again get (27). This shows that (22) also holds for small negative h.

We now use (22) to show that

$$(29) \qquad \lim_{h \to 0} \frac{\sin h}{h} = 1.$$

By (16) we know that

$$\lim_{h \to 0} \cos h = \cos 0 = 1,$$

and by (14) we have

$$\lim_{h \to 0} \frac{1}{\cos h} = \frac{1}{\displaystyle\lim_{h \to 0} \cos h} = 1.$$

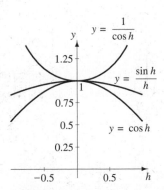

Figure 1.47. When h is small, $(\sin h)/h$ is between $\cos h$ and $1/\cos h$.

By (22) the expression $(\sin h)/h$ is "trapped" between $\cos h$ and $1/\cos h$. This is illustrated in Fig. 1.47. Since $\cos h$ and $1/\cos h$ both tend to 1 as h approaches 0, it must be true that $(\sin h)/h$ also approaches 1 as h tends to 0. Hence (29) is correct. ∎

Existence of Limits

Figure 1.48 shows the graphs of three functions, f, g, and h, near $x = 1$. The graphs of these three functions look very different near $x = 1$.

  The graph of f shows that if $x < 1$ and close to 1, then $f(x)$ is close to 3, but if $x > 1$ and close to 1, then $f(x)$ is close to 2.

  The graph of g shows that as x gets close to 1, the values of $g(x)$ get bigger and bigger. They seem to be growing to infinity, so $x = 1$ may be an asymptote to the graph of g.

  The graph of h oscillates wildly between the horizontal lines $y = 0$ and $y = 2$ as x gets close to 1.

Though these graphs look very different, they do have one thing in common. Each graph indicates that *as x gets close to 1, the function values do not approach one and only one finite real number.* More specifically, as x approaches 1, $f(x)$ gets close to two different numbers, $g(x)$ does not get close to any finite number, and $h(x)$, as it oscillates, gets close to every number between 0 and 2. For each of these three functions the limit as x approaches 1 does not exist.

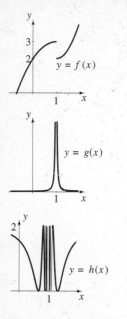

Figure 1.48. Sometimes a limit does not exist.

θ	$\tan \theta$
2.0708	-1.83047
1.0708	1.83050
1.6708	-9.96628
1.4708	9.96701
1.5718	-996.3
1.5698	1003.7

When Limits Do Not Exist

Let F be a function and a a real number. If there is no number L such that $F(x)$ gets close to L (and only L) as x gets close to a, then

$$\lim_{x \to a} F(x) \quad \text{does not exist.}$$

■ **EXAMPLE 8**   Show that

$$\lim_{\theta \to \pi/2} \tan \theta$$

does not exist.

Solution.   When evaluating a limit of a function, it is almost always a good idea to look at a graph of the function. In Fig. 1.49 we see the graph of $y = \tan \theta$ for θ near $\pi/2$. The graph suggests that the line $\theta = \pi/2$ is a vertical asymptote to the graph. When $\theta > \pi/2$ and close to $\pi/2$, the values of $\tan \theta$ are large and negative, so it appears that the function values are heading toward $-\infty$ as θ closes in on $\pi/2$ from the right. When $\theta < \pi/2$ and close to $\pi/2$, the values of $\tan \theta$ are large and positive, and it appears that the function values are approaching $+\infty$ as θ approaches $\pi/2$ from the left. Since $\tan \theta$ is not getting close to one and only one finite number as θ gets close to $\pi/2$, we conclude that

$$\lim_{\theta \to \pi/2} \tan \theta \quad \text{does not exist.}$$

We can also investigate the behavior of the tangent function near $\theta = \pi/2$ by computing $\tan \theta$ for several θ values near $\pi/2 \approx 1.5708$. A table of such values is shown. The data in the table reflect the behavior of the graph and so also suggest that

$$\lim_{\theta \to \pi/2} \tan \theta \quad \text{does not exist.} \qquad \blacksquare$$

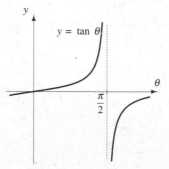

Figure 1.49. The graph indicates that $\lim_{\theta \to \pi/2} \tan \theta$ does not exist.

Exercises 1.4

Basic

Exercises 1–12: Determine the limit, if it exists. Use graphical, numerical, or analytic methods.

1. $\lim\limits_{x \to 2}(x^2 - 3x)/(x + 1)$

2. $\lim\limits_{t \to -1/2} \sin \pi t$

3. $\lim\limits_{h \to 0}(3h^3 - 7h^2 + h)/h$

4. $\lim\limits_{x \to -4}(2x^2 + 7x - 4)/(x + 4)$

5. $\lim\limits_{t \to 1/2} \dfrac{2t + 1}{2t - 1}$

6. $\lim\limits_{r \to \sqrt{2}} \dfrac{r^2 - 2}{r^3 + 2r^2 - 2r - 4}$

7. $\lim\limits_{h \to 0} \dfrac{g(4 + h) - g(4)}{h}$ where $g(t) = t^3 - 3t$

8. $\lim\limits_{h \to 0} \dfrac{G(-8 + h) - G(-8)}{h}$ where $G(s) = 2/(s + 1)$

9. $\lim\limits_{x \to -1/3} \dfrac{\sqrt{6x + 3} - 1}{6x + 2}$

10. $\lim\limits_{x \to -1} f(x),$ where

$$f(x) = \begin{cases} x + 1 & (x < -1) \\ 2 - 2x^2 & (-1 \le x \le 2) \\ 0 & (x > 2) \end{cases}$$

11. $\lim\limits_{x \to 2} f(x)$ for $f(x)$ as given in Exercise 10.

12. $\lim\limits_{h \to 0} \dfrac{f(-1 + h) - f(-1)}{h}$

for $f(x)$ as given in Exercise 10.

13. Investigate

$$\lim_{\theta \to 0} \frac{1 - \cos \theta}{\theta^2}.$$

Use graphical and numerical techniques.

14. Investigate

$$\lim_{\theta \to 0} \frac{\sin \theta}{\theta}.$$

Use graphical and numerical techniques.

15. Investigate

$$\lim_{\theta \to 0} \cos\left(\frac{1}{\theta}\right).$$

Use graphical and numerical techniques.

16. Investigate

$$\lim_{\theta \to \pi/4} \frac{\tan \theta - 1}{\theta - \pi/4}.$$

Use graphical and numerical techniques.

17. The figure below shows the graph of a function f.
    a. For what values of c does $\lim\limits_{x \to c} f(x) = 1$?
    b. For what values of c does $\lim\limits_{x \to c} f(x)$ not exist?
    c. Evaluate $\lim\limits_{x \to -1} f(x)$.
    d. Evaluate $\lim\limits_{x \to -1} f(x^2)$.
    e. Evaluate $\lim\limits_{x \to -1} f(\sqrt{x})$.

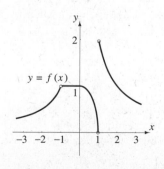

18. The figure below shows the graph of a function g.
    a. For what values of c does $\lim\limits_{x \to c} g(x)$ fail to exist?
    b. Evaluate $\lim\limits_{x \to 2} g(x)$.
    c. Evaluate $\lim\limits_{x \to 2}(g(x))^2$.
    d. Evaluate $\lim\limits_{x \to 2}(g(x))^3$.

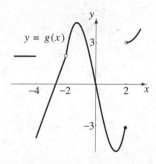

Growth

19. Investigate the following limits both graphically and numerically:

a. $\lim\limits_{h \to 0} \dfrac{2^h - 1}{h}$

b. $\lim\limits_{h \to 0} \dfrac{3^h - 1}{h}$

c. $\lim\limits_{h \to 0} \dfrac{(1/2)^h - 1}{h}$

20. Let a be a fixed positive number. Consider the limit

$$\lim_{h \to 0} \frac{a^h - 1}{h}.$$

By using numerical methods, find an approximate value of a for which the above limit is 1.

21. Find functions f and g for which

$$\lim_{x \to 3} f(x) = \lim_{x \to 3} g(x) = 0$$

and:

a. $\lim\limits_{x \to 3} \dfrac{f(x)}{g(x)} = 5$.

b. $\lim\limits_{x \to 3} \dfrac{f(x)}{g(x)} = -\sqrt{3}$.

c. $\lim\limits_{x \to 3} \dfrac{f(x)}{g(x)}$ does not exist.

22. Let $0 < c < \pi/2$ and $0 < d < \pi/2$. If $\lim_{\theta \to c} \sin\theta = 0.3$ and $\lim_{\theta \to d} \sin\theta = 0.9$, find:

a. $\lim\limits_{\theta \to c+d} \sin\theta$.

b. $\lim\limits_{\theta \to c-d} \cos\theta$.

c. $\lim\limits_{\theta \to 2d} \cos\theta$.

23. The graph of a function f is shown below. On the same set of axes draw the graph of a function g so that $\lim_{x \to -2}(f(x) + g(x))$ does not exist, $\lim_{x \to 0}(f(x) + g(x)) = 1$, and $\lim_{x \to 1}(f(x) + g(x)) = 0$.

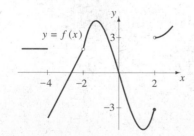

24. From the graph in Fig 1.37, what is the limit of electrical resistance of mercury as the temperature approaches 0 K?

25. Let f be a function and assume that $\lim_{x \to a} f(x) = L$. Is it then true that

$$\lim_{x \to a} |f(x) - L| = 0?$$

Give reasons for your answer.

26. Let g be a function and assume that $\lim_{x \to a} |g(x) - M| = 0$. What is the value of

$$\lim_{x \to a} g(x)?$$

Give reasons for your answer.

27. Let $f(x) = x^2 + 2x + 2$.

a. Verify that $\lim_{x \to 1} f(x) = 5$.

b. Answer the following question by looking at a graph: If x is within 0.05 of 1, then what is the largest $|f(x) - 5|$ can be?

c. By looking at a graph, find a positive number r that satisfies the following: If x is within r of 1, then $|f(x) - 5| < 0.1$.

28. Let $g(t) = t^3 - 4t - 4$.

a. Verify that $\lim_{t \to -2} g(t) = -4$.

b. Answer the following question by looking at a graph: If t is within 0.01 of -2, then what is the largest $|g(t) - (-4)|$ can be?

c. By looking at a graph, find a positive number r that satisfies the following: If t is within r of -2, then $|g(t) - (-4)| < 0.05$.

29. Let $r(\theta) = (\sin 2\theta)/\theta$.

a. Verify that $\lim_{\theta \to 0} r(\theta) = 2$.

b. Answer the following question by looking at a graph: If θ is within 0.1 of 0, then what is the largest $|r(\theta) - 2|$ can be?

c. By looking at a graph, find a positive number r that satisfies the following: If θ is within r of 0, then $|r(\theta) - 2| < 0.05$.

30. Let $f(x) = (\sqrt{x} - 2)/(x - 4)$.

a. Verify that $\lim_{x \to 4} f(x) = 1/4$.

b. Answer the following question by looking at a graph: If x is within 0.05 of 4, then what is the largest $|f(x) - 1/4|$ can be?

c. By looking at a graph, find a positive number r that satisfies the following: If x is within r of 4, then $|f(x) - 1/4| < 0.1$.

31. The following table shows the history of the world record for the 1-mile run since 1885. These data are plotted in the graph following the table.

The world record mile

Year	Name (Country)	Time	Year	Name (Country)	Time
1875	Walter Slade (Britain)	4:24.5	1942	Gunder Haegg (Sweden)	4:04.6
1880	Walter George (Britain)	4:23.2	1943	Arne Andersson (Sweden)	4:02.6
1882	Walter George (Britain)	4:21.4	1944	Arne Andersson (Sweden)	4:01.6
1882	Walter George (Britain)	4:19.4	1945	Gunder Haegg (Sweden)	4:01.4
1884	Walter George (Britain)	4:18.4	1954	Roger Bannister (Britain)	3:59.4
1894	Fred Bacon (Scotland)	4:18.2	1954	John Landy (Australia)	3:58.0
1895	Fred Bacon (Scotland)	4:17.0	1957	Derek Ibbotson (Britain)	3:57.2
1911	Thomas Connett (U.S.)	4:15.6	1958	Herb Elliot (Australia)	3:54.5
1911	John Paul Jones (U.S.)	4:15.4	1962	Peter Snell (New Zealand)	3:54.4
1913	John Paul Jones (U.S.)	4:14.6	1964	Peter Snell (New Zealand)	3:54.1
1915	Norman Tauber (U.S.)	4:12.6	1965	Michel Jazy (France)	3:53.6
1923	Paavo Nurmi (Finland)	4:10.4	1966	Jim Ryun (U.S.)	3:51.3
1931	Jules Ladoumegue (France)	4:09.2	1967	Jim Ryun (U.S.)	3:51.1
1933	Jack Lovelock (New Zealand)	4:07.6	1975	Filbert Bayi (Tanzania)	3:51.0
1934	Glenn Cunningham (U.S.)	4:06.8	1975	John Walker (New Zealand)	3:49.4
1937	Sydney Wooderson (Britain)	4:06.4	1981	Sebastian Coe (Britain)	3:47.33
1942	Gunder Haegg (Sweden)	4:06.2	1985	Steve Cram (Britain)	3:46.30
1942	Arne Andersson (Sweden)	4:06.2			

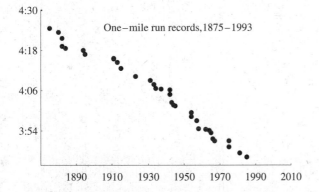

One–mile run records, 1875–1993

a. Use any method to extrapolate the mile run records to the year 2010 A.D.

b. Calculate the average of the times and the average of the dates using the table. Add the point for the average year and average time to the graph. (Such a point is sometimes called a *centroid*.) Put a nail through the centroid. Place a ruler against the nail and try to position the ruler so that if you were to draw a line with the ruler in this position, the line would "best fit" the data. Use this best fit line to estimate what the record will be in 2010 A.D.

c. From these data, does it appear that there is a "limit" to how fast humans can run the mile? Give reasons for your answer and discuss what you mean by "limit" in this situation.

32. The following table shows how the volume of a gas varies with temperature when the pressure of the gas is kept constant. Plot these data, and then use the method given in Exercise 31 to find a line approximating the data. With the aid of this line, find the limit of the temperature as the volume approaches 0. (These data were generated by computer as part of a chemistry lab to investigate methods of finding the Celsius temperature of absolute zero.)

Volume (cm^3)	Temperature ($^\circ$C)
230	9.4
233.9	14.3
238.1	19.4
242.9	25.3
246.7	30.0
250.9	35.1

33. Use Fig. 1.46 to give a geometric argument that $|\sin h| \leq |h|$.

34. In Example 7 we found the limit of $(\sin h)/h$ as h approaches 0 by trapping $(\sin h)/h$ between two functions with known limits. In doing so, we illustrated an application of the Squeeze Theorem:

Let f, g, and h be three functions, each defined for x values close to a but possibly not defined at $x = a$. Assume that

$$\lim_{x \to a} f(x) = L = \lim_{x \to a} h(x)$$

and that for x near a,

$$f(x) \le g(x) \le h(x).$$

Then

$$\lim_{h \to a} g(x) = L.$$

a. Draw a good picture illustrating the Squeeze Theorem.

b. Write a paragraph explaining why the Squeeze Theorem is true.

c. Explain how the Squeeze Theorem was used in Example 7.

35. With the aid of the figure below, give a geometric argument that:
a. $\left| \sin \theta - \sin a \right| \le \left| \theta - a \right|$.
b. $\left| \cos \theta - \cos a \right| \le \left| \theta - a \right|$.

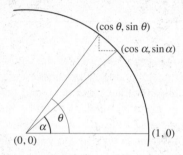

36. Use the Squeeze Theorem of Exercise 34 and the inequalities developed in Exercise 35 to show that:
a. $\lim_{\theta \to a} \sin \theta = \sin a$.
b. $\lim_{\theta \to a} \cos \theta = \cos a$.
How does this result relate to (16)?

37. Using Fig. 1.46, show that when $0 < |h| < \pi/2$ we have

$$\frac{\sin h}{h} - 1 \le \frac{1 - \cos h}{h} \le 1 - \frac{\sin h}{h}.$$

Then use the Squeeze Theorem of Exercise 34 to evaluate

$$\lim_{h \to 0} \frac{1 - \cos h}{h}.$$

Review/Preview

38. Find a cubic polynomial $p(x)$ that takes the value 0 at $x = -2$, $x = 1$, and $x = 3$.

39. Find a cubic polynomial $p(x)$ that takes the value 0 at $x = -1$, $x = 1/2$, and $x = 3$ and takes the value 5 at $x = 0$.

40. Find a cubic polynomial $p(x)$ that takes the value 0 at $x = 2$ and $x = 3$ but takes the value 0 at no other values of x.

41. Find a cubic polynomial $p(x)$ that takes the value 0 at $x = -3/2$ and $x = 5$, takes the value 0 at no other values of x, and takes the value -6 at $x = 1$.

42. Find θ if $\sin \theta = -1/2$ and $\pi \le \theta \le 3\pi/2$.

43. Find t if $\cos t = 1$ and $5\pi \le t \le 6\pi$.

44. Find $\sin \theta$ and $\cos \theta$ given that $\tan \theta = 5/12$ and $0 \le \theta \le \pi$.

45. Find $\sin \theta$ and $\cos \theta$ given that $\tan \theta = -12/5$ and $0 \le \theta \le \pi$.

46. Find the volume of a cube C if
a. One of the diagonals of a face has length 8.
b. One of the body diagonals of the cube has length of 8.

47. With the help of the accompanying figure and the distance formula, verify the law of cosines: *If a triangle has sides of length a, b, and c, and the angle between the sides of length a and b has measure θ, then*

$$c^2 = a^2 + b^2 - 2ab \cos \theta.$$

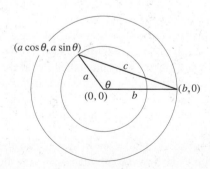

48. A light bulb is suspended 8 feet above the floor of a room. The bulb is covered by a conical shade. Angle AVR has measure $\pi/8$, where A is a point on the altitude of the cone, V is the vertex of the cone, and R is a point on the rim of the cone. The bulb illuminates a circular region on the floor. Find the area of this circle. (See the accompanying figure.)

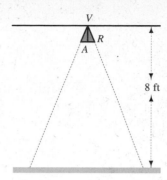

1.5 MORE WORK WITH LIMITS

For a real number x, let

(1) $$s(x) = -\frac{1}{216}x^2 + \frac{1}{4}x + \frac{9}{8}.$$

Use your calculator to check the following:

$$s(9) = 3 \quad \text{and} \quad \sqrt{9} = 3$$
$$s(8.9) \approx 2.98329 \quad \text{and} \quad \sqrt{8.9} \approx 2.98329$$
$$s(8.75) \approx 2.95804 \quad \text{and} \quad \sqrt{8.75} \approx 2.95804$$
$$s(9.37) \approx 3.06103 \quad \text{and} \quad \sqrt{9.37} \approx 3.06105.$$

Does this mean that we can use (1) to calculate square roots? Before answering, let's collect some more data. So far all of the $s(x)$ values were calculated for x values close to 9. If we try positive x values that are not close to 9, $s(x)$ does not give a good approximation to $\sqrt{x}$:

$$s(2) \approx 1.60648 \quad \text{and} \quad \sqrt{2} \approx 1.41421$$
$$s(53) \approx 1.3704 \quad \text{and} \quad \sqrt{53} \approx 7.28011.$$

Based on these examples, it appears that $s(x)$ is a good approximation to $\sqrt{x}$ for x close to 9, but not for x values far from 9. Why isn't $s(x)$ close to $\sqrt{x}$ for all x? Or, alternatively, why should $s(x)$ be close to $\sqrt{x}$ for any values of x? Suppose that we want $\sqrt{x}$ accurate to two decimal places. For what values of x can we use $s(x)$? When must we use another method to calculate $\sqrt{x}$? We answer these questions by taking another look at the definition of limit.

Another Look at Limits

In the previous section we saw the following definition of the limit.

DEFINITION **limit of a function**

Let g be a function and let a and L be real numbers. If we can make $g(x)$ as close to L as we like by taking x close to a, but not equal to a, then we say

$g(x)$ has limit L as x approaches a.

We denote this by

(2)
$$\lim_{x \to a} g(x) = L.$$

Equation (2) is read "the limit of $g(x)$ as x approaches a equals L."

For some work with limits we need to be a bit more precise about the phrase "close to" in this definition. When we talk about the number $g(x)$ being close to the number L, it is natural to ask, "how close?" To answer this question, we look at

(3)
$$|g(x) - L|.$$

If $g(x)$ is close to L, then $|g(x) - L|$ should be small. To describe how close $g(x)$ is to L, we tell how small (3) is. We do this with a number. For example, if $g(x)$ is within 0.01 of L, we know that

$$|g(x) - L| \le 0.01.$$

The definition of limit says: We can make $g(x)$ as close to L as we like by taking x close to a. This means that we can make (3) as small as we like by making $|x - a|$ small. We illustrate this with some examples.

■ **EXAMPLE 1**   Let $g(x) = x^2 - x + 5$, and consider the statement

$$\lim_{x \to 2} g(x) = 7.$$

a. Show that if $|x - 2| < 0.03$, then $g(x)$ is within 0.1 of 7.

b. Suppose we want

(4)
$$|g(x) - 7| < 0.001.$$

Find a positive number d so that (4) is true if $|x - 2| < d$.

Solution

a. We verify this graphically. First note that

$$|x - 2| < 0.03 \quad \text{is equivalent to} \quad 1.97 < x < 2.03.$$

The graph of $y = g(x)$ on this interval is shown in Fig. 1.50. We see that for $1.97 < x < 2.03$, the points on the graph lie between the lines $y = 6.9$ and $y = 7.1$. This means that if

$$1.97 < x < 2.03 \quad \text{then} \quad 6.9 < g(x) < 7.1.$$

Hence if $|x - 2| < 0.03$, then $g(x) = x^2 - x + 5$ is within 0.1 of 7.

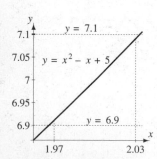

Figure 1.50. When $1.97 < x < 2.03$, we have $6.9 < x^2 - x + 5 < 7.1$.

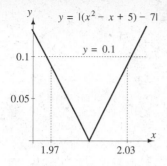

Figure 1.51. If $|x - 2| < 0.03$, then $|(x^2 - x + 5) - 7| < 0.1$.

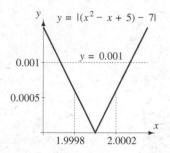

Figure 1.52. When $|x - 2| < 0.0002$, then $|(x^2 - x + 5) - 7| < 0.001$.

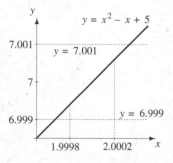

Figure 1.53. When $1.9998 < x < 2.0002$, we have $6.999 < g(x) < 7.001$.

For a different approach, graph $y = |g(x) - 7|$ for $1.97 < x < 2.03$. See Fig. 1.51. For x in this interval, the points on the graph lie below the line $y = 0.1$. This means that

$$|g(x) - 7| < 0.1 \quad \text{if} \quad 1.97 < x < 2.03.$$

Thus $g(x) = x^2 - x + 5$ is within 0.1 of 7 when x is within 0.03 of 2.

b. Since we want $|g(x) - 7| < 0.001$ for x close to 2, draw the graph of $y = |g(x) - 7|$ on some small interval centered at 2. In Fig. 1.52 we see the graph for $1.9995 < x < 2.0005$. Next draw the line $y = 0.001$ on the graph. We then see that

$$y = |g(x) - 7| < 0.001 \quad \text{for} \quad 1.9998 < x < 2.0002,$$

that is, when

$$|x - 2| < 0.0002.$$

Hence we may take $d = 0.0002$.

In Fig. 1.53 we show the graph of $y = g(x)$ for $|x - 2| < 0.0002$. Note that for x values in this interval the points of the graph lie between the lines $y = 6.999$ and $y = 7.001$. This again shows that $d = 0.0002$ works. Are there other acceptable values of d? How many other values? ∎

■ **EXAMPLE 2** In the previous section we showed that

$$\lim_{\theta \to 0} \frac{\sin \theta}{\theta} = 1.$$

Find $d > 0$ so that

$$(5) \qquad \text{if} \quad 0 < |\theta - 0| < d \quad \text{then} \quad \left|\frac{\sin \theta}{\theta} - 1\right| < 0.05.$$

Solution. Figure 1.54 shows the graph of $y = (\sin \theta)/\theta$ for $-1 < \theta < 1, \theta \neq 0$. Since we want

$$0.95 < \frac{\sin \theta}{\theta} < 1.05,$$

we include the lines $y = 0.95$ and $y = 1.05$ in the figure. The graph of $y = (\sin \theta)/\theta$ lies between these two horizontal lines when $-0.5 < \theta < 0$ or $0 < \theta < 0.5$. Thus we can take $d = 0.5$. For this value of d equation (5) is true. Explain why any positive $d < 0.5$ would also work.

We could also find an acceptable value of d by looking at numerical data. The table below gives the value of $(\sin \theta)/\theta$ for many θ values close to 0.

θ	−0.8	−0.6	−0.4	−0.2	0.2	0.4	0.6	0.8
$\dfrac{\sin \theta}{\theta}$	0.897	0.941	0.974	0.993	0.993	0.974	0.941	0.897

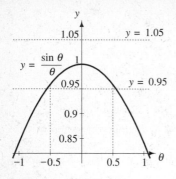

Figure 1.54. When $-0.5 < x < 0.5$, we have $0.95 < (\sin \theta)/\theta < 1.05$.

The data in the table suggest that

$$\left| \frac{\sin \theta}{\theta} - 1 \right| < 0.05 \quad \text{when} \quad 0 < |x| < 0.4.$$

(It would be a good idea to verify this graphically. Why?) Based on the data, we can take $d = 0.4$. ■

The two previous examples illustrate the meaning of "close to" in the definition of limit. In this definition the phrase "we can make $g(x)$ as close to L as we like" means that given any small number ϵ (pronounced ep'-si-lon), we can make $g(x)$ within ϵ of L. That is, we can make

(6) $$|g(x) - L| < \epsilon.$$

We make $g(x)$ close to L by taking x close to a but not equal to a. This means that once ϵ is known, we can produce another number δ (pronounced del'-ta) so that (6) is true when $0 < |x - a| < \delta$. That is, δ is found so that

$$0 < |x - a| < \delta \quad \text{implies} \quad |g(x) - L| < \epsilon.$$

In Example 1b we were given $\epsilon = 0.001$ and produced $\delta = d = 0.0002$. In Example 2 we were given $\epsilon = 0.05$ and produced $\delta = d = 0.5$. We also showed that $\delta = d = 0.4$ was acceptable.

With this new understanding of "close to," we restate the definition of limit.

DEFINITION **limit of a function**

Let a and L be real numbers, and let g be a function whose domain includes the intervals

(7) $$a - r < x < a \quad \text{and} \quad a < x < a + r \quad \text{for some } r > 0.$$

We say that $g(x)$ has limit L as x approaches a if for each $\epsilon > 0$ there is a corresponding $\delta > 0$ such that

(8) $$\text{if} \quad 0 < |x - z| < \delta, \quad \text{then} \quad |g(x) - L| < \epsilon.$$

See Fig. 1.55. We denote this by

$$\lim_{x \to a} g(x) = L.$$

If there is no real number L for which the above is true, then we say $g(x)$ has no limit as x approaches a, or that $\lim_{x \to a} g(x)$ is undefined.

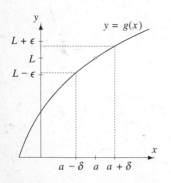

Figure 1.55. First we are given $\epsilon > 0$. There is a $\delta > 0$ so that when $0 < |x - a| < \delta$, we have $|g(x) - L| < \epsilon$.

According to (7), g must be defined in some interval $a - r < x < a + r$ except, possibly, at $x = a$. Thus $g(a)$ may or may not be defined. From (8) we see that we are only concerned with $g(x)$ values for x close to a but not equal to a. Hence whether $g(a)$ is defined or not has nothing to do with the value of $\lim_{x \to a} g(x)$.

■ **EXAMPLE 3**

a. Evaluate

(9)
$$\lim_{t \to 3} \frac{\sqrt{3t - 5} - 2}{t - 3}.$$

b. Let L be the value of the limit in part a and let $\epsilon = 0.035$. Find a number $\delta > 0$ so that

$$0 < |t - 3| < \delta \quad \text{implies} \quad \left| \frac{\sqrt{3t - 5} - 2}{t - 3} - L \right| < \epsilon.$$

Use graphical methods.

Solution

a. Notice that when $t = 3$ the numerator and denominator of

(10)
$$\frac{\sqrt{3t - 5} - 2}{t - 3}$$

are both 0. Thus we cannot hope to evaluate (9) by substituting $t = 3$ in (10). To simplify (10), rationalize the numerator by multiplying the numerator and denominator of the expression by $\sqrt{3t - 5} + 2$:

$$\frac{\sqrt{3t - 5} - 2}{t - 3} = \frac{\sqrt{3t - 5} - 2}{t - 3} \cdot \frac{\sqrt{3t - 5} + 2}{\sqrt{3t - 5} + 2}$$

$$= \frac{(\sqrt{3t - 5})^2 - 2^2}{(t - 3)(\sqrt{3t - 5} + 2)}$$

$$= \frac{3(t - 3)}{(t - 3)(\sqrt{3t - 5} + 2)}$$

$$= \frac{3}{\sqrt{3t - 5} + 2}.$$

Thus

$$\lim_{t \to 3} \frac{\sqrt{3t - 5} - 2}{t - 3} = \lim_{t \to 3} \frac{3}{\sqrt{3t - 5} + 2} = \frac{3}{\sqrt{3 \cdot 3 - 5} + 2} = \frac{3}{4}.$$

b. With a calculator or CAS, sketch the graph of

$$y = \left| \frac{\sqrt{3t - 5} - 2}{t - 3} - \frac{3}{4} \right|$$

for t values close to 4. See Fig. 1.56. We see for $x \neq 3$ and $2.8 < x < 3.2$, the graph lies below the line $y = \epsilon = 0.035$. Thus, if $\delta = 0.2$

$$0 < |t - 3| < 0.2 = \delta$$

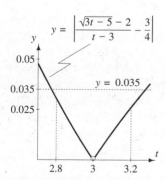

$$y = \left| \frac{\sqrt{3t - 5} - 2}{t - 3} - \frac{3}{4} \right|$$

Figure 1.56. If $0 < |t - 3| < 0.2$, then $|(\sqrt{3t - 5} - 2)/(t - 3) - 3/4| < 0.035$.

implies

$$\left| \frac{\sqrt{3t - 5} - 2}{t - 3} - L \right| < 0.035 = \epsilon.$$

Note that any positive number smaller than 0.2 would also be an acceptable value for δ. ∎

Limits and Approximation

When we use a calculator to find the value of sin 1 (where the 1 is in radians), we see the answer 0.841471 on the display. This decimal expression is not the *exact* value of sin 1, but we assume that the answer we see is correct to the number of digits shown. How does the calculator obtain this answer?

In calculators and computers the computations of numerical values of square roots, trigonometric functions, logarithms, exponentials, and so on are based on algorithms and formulas that approximate these functions. To guarantee that the calculator answer is correct to the number of digits shown, the people designing the calculator must know how accurate the approximation is. Limits are important in finding good approximations and in studying the accuracy of an approximation.

The statement

$$\lim_{x \to a} f(x) = L$$

means that $f(x)$ is close to L when x is close to a. Suppose that for a given x value we need an approximation of $f(x)$. Would it be acceptable to say

(11) $$f(x) \approx L?$$

We need a lot more information before we can answer this question. For our value of x, how good is the approximation given in (11)? How good an approximation do we need? The answer to the second question depends on what we want to do with the approximation. To answer the first question, we need to be able to say something about the *error in the approximation,*

(12) $$E(x) = f(x) - L.$$

We use the error, $E(x)$, to measure how good the approximation (11) is. If $E(x)$ is small enough, the approximation is a good one and may serve our needs. If $E(x)$ is large, the approximation may not be of any use. It is usually impossible or impractical to calculate $E(x)$ exactly, but we can often find a number larger than $|E(x)|$. If this larger number is small, then the error is small. In Examples 2 and 3 we learned techniques for finding δ given ϵ. These same techniques can be used to analyze $E(x)$. In fact, the E in $E(x)$ not only stands for error, but also reminds us of ϵ.

∎ **EXAMPLE 4**   Because $\lim_{x \to 3} 2^x = 8$, we consider using the approximation

$$2^x \approx 8$$

when x is close to 3. For what values of x is this approximation within 0.01 of the actual value of 2^x?

Solution.   We need to find x values such that

$$|E(x)| = |2^x - 8| < 0.01.$$

In the language of limits, we are given $\epsilon = 0.01$ and need to find $\delta > 0$ so that

$$|x - 3| < \delta \quad \text{implies} \quad |E(x)| = |2^x - 8| < \epsilon = 0.01.$$

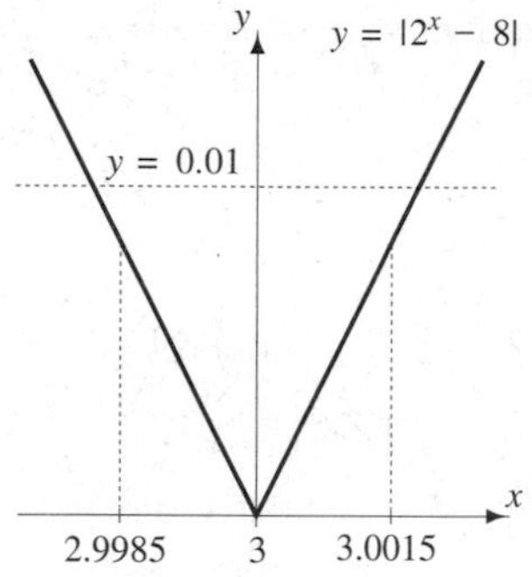

On your calculator or CAS graph $y = |2^x - 8|$ for x near 3. (Remember, the first graph you produce may not be what you want. The authors used their CAS to plot six different graphs of $y = |2^x - 8|$ before settling on the one in Fig. 1.57.) The graph of $y = |2^x - 8|$ lies below the line $y = 0.01$ for $2.9985 < x < 3.0015$. Hence for x in this interval we can say

$$2^x \approx 8$$

with an error of at most 0.01   ■

Figure 1.57. When $2.9985 < x < 3.0015$, we have 2^x within 0.01 of 8.

Sometimes we can manipulate a limit result to get a useful approximation. For example, in the previous section we showed that

$$(13) \qquad \lim_{x \to 0} \frac{\sin x}{x} = 1.$$

We use this result to find a good approximation to $\sin x$ for small x. Let

$$(14) \qquad E(x) = \frac{\sin x}{x} - 1.$$

Because

$$\lim_{x \to 0} E(x) = \lim_{x \to 0} \left(\frac{\sin x}{x} - 1 \right) = 1 - 1 = 0,$$

we know that $E(x)$ can be made small by taking x close to 0. Rewrite (14) as

$$\frac{\sin x}{x} = 1 + E(x).$$

Then multiply by x to get

$$\sin x = x + x \cdot E(x).$$

If x is small, then $E(x)$ is small and

$$\sin x - x = x \cdot E(x)$$

is small. Hence

(15) $$\sin x \approx x$$

and the error, $x \cdot E(x)$, is small when x is small. We say more about this error in the next example.

■ **EXAMPLE 5**   Discuss the error in the approximation

$$\sin x \approx x$$

for $-0.5 < x < 0.5$.

Solution.   In Example 2 we showed that

$$|E(x)| = \left| \frac{\sin x}{x} - 1 \right| < 0.05$$

if $-0.5 < x < 0.5$. Hence for x in this interval the error in the approximation $\sin x \approx x$ is

(16) $|\sin x - x| = |x \cdot E(x)| = |x| \cdot |E(x)| \le 0.05|x| \le 0.05 \cdot 0.5 = 0.025.$

From (16) we see that

$$|\sin x - x| \le 0.05|x|.$$

This means that when $-0.5 < x < 0.5$, the error in (15) is never more than 5 percent of $|x|$. Thus the error gets smaller as $|x|$ gets smaller, and the error is small compared to $|x|$.

To further understand the error, we graph the equations $y = \sin x$ and $y = x$ for $-0.5 < x < 0.5$ on the same set of coordinate axes. The graph is shown in Fig. 1.58. The two graphs are close, which means that the numbers $\sin x$ and x are close for $-0.5 < x < 0.5$.

For another look at the error, graph the equation $y = |E(x)| = |\sin x - x|$ on $-0.5 < x < 0.5$. The graph is shown in Fig. 1.59. This graph shows plainly that the error in the approximation gets small as $|x|$ gets small.   ■

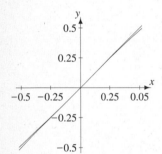

Figure 1.58. The graphs of $y = \sin x$ and $y = x$ are close for $-0.5 < x < 0.5$.

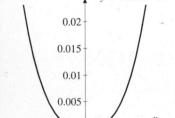

Figure 1.59. The graphs of $y = |E(x)| = |\sin x - x|$ shows that the error is small for $-0.5 < x < 0.5$.

■ **EXAMPLE 6**   Evaluate

(17) $$\lim_{t \to 9} \frac{\sqrt{t} - 3}{t - 9}$$

and use the result to find a good approximation to $\sqrt{t}$ for t close to 9. Discuss the error in the approximation for t in the interval $8.8 < t < 9.2$.

Solution.   For $t \ne 9$ we have

$$\frac{\sqrt{t} - 3}{t - 9} = \frac{\sqrt{t} - 3}{(\sqrt{t})^2 - 3^2} = \frac{\sqrt{t} - 3}{(\sqrt{t} - 3)(\sqrt{t} + 3)} = \frac{1}{\sqrt{t} + 3}.$$

Therefore

$$\lim_{t \to 9} \frac{\sqrt{t} - 3}{t - 9} = \lim_{t \to 9} \frac{1}{\sqrt{t} + 3} = \frac{1}{\sqrt{9} + 3} = \frac{1}{6}.$$

Now let

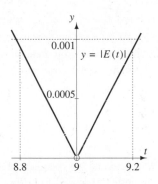

Figure 1.60. The graph shows that $y = |E(x)| = |(\sqrt{t} - 3)/(t - 9) - 1/6| < 0.001$ when $8.8 < t < 9.2$.

(18)
$$\frac{\sqrt{t} - 3}{t - 9} - \frac{1}{6} = E(t).$$

Solve this equation for $\sqrt{t}$. To do this, add $\frac{1}{6}$ to both sides of (18), then multiply both sides of the result by $t - 9$, and then add 3 to both sides. We have

(19)
$$\sqrt{t} = 3 + \frac{1}{6}(t - 9) + E(t)(t - 9).$$

When t is close to 9, we know that $E(t)$ is close to 0 and that $(t - 9)$ is small. Hence for t close to 9, $E(t)(t - 9)$ is small. If we drop this small term from (19), we get an approximation for $\sqrt{t}$:

(20)
$$\sqrt{t} \approx 3 + \frac{1}{6}(t - 9).$$

From (19), the absolute value of the error in this approximation is

(21)
$$\left| \sqrt{t} - \left(3 + \frac{1}{6}(t - 9) \right) \right| = |(t - 9)E(t)|.$$

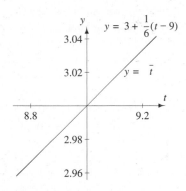

Figure 1.61. The graphs of $y = \sqrt{t}$ and $y = 3 + \frac{1}{6}(t - 9)$ are so close we cannot see the difference.

To estimate this error for $8.8 < t < 9.2$, first graph

$$y = |E(t)| = \left| \frac{\sqrt{t} - 3}{t - 9} - \frac{1}{6} \right|$$

on this interval. The graph appears in Fig. 1.60 and shows that $|E(t)| < 0.001$ when $8.8 < t < 9.2$. Use this upper estimate for $|E(t)|$ in (21) to see that

$$\left| \sqrt{t} - \left(3 + \frac{1}{6}(t - 9) \right) \right| \le 0.001|t - 9|.$$

Hence the error in the approximation (20) is at most 0.1% of the value of $|t - 9|$. Furthermore, for $8.8 < t < 9.2$, we have

$$|(t - 9)E(t)| \le (0.001)(0.2) = 0.0002,$$

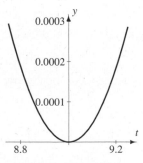

Figure 1.62. The graph of $y = |\sqrt{t} - (3 + \frac{1}{6}(t - 9))|$ shows that the error is very small for $8.8 < t < 9.2$.

so the error in (20) is no more than 0.0002. In Fig. 1.61 we show the graphs of $y = \sqrt{t}$ and $y = 3 + \frac{1}{6}(t - 9)$ on the interval $8.8 < t < 9.2$. At this resolution we see no difference in the graphs. To better see the error, we show the graph of $y = |(t - 9)E(t)|$ in Fig. 1.62. For a more practical test, use the approximation we've developed to approximate $\sqrt{8.85}$. Putting $t = 8.85$ into (20), we have

$$\sqrt{8.85} \approx 3 + \frac{1}{6}(8.85 - 9) = 2.975$$

while a calculator gives a value of 2.97489. Not bad! In Exercise 40 we do more work with this example to obtain the approximation presented at the beginning of this section.   ■

One-Sided Limits

According to Einstein's special theory of relativity, the length of a rod measured at rest is different from the length measured when the rod is in motion. If you measure an arrow on your desk and find it has length L_0, and then measure the same arrow as it flies by at speed v, the length will be

$$(22) \qquad L = L_0 \sqrt{1 - \left(\frac{v}{c}\right)^2},$$

where $c = 2.998 \times 10^{10}$ cm/s is the speed of light. At the everyday speeds v of cars and planes, this effect is not noticeable, but if v is large compared to c, the difference between L_0 and L will be significant.

Although an arrow can never travel at the speed of light, physicists are interesting in studying objects at speeds close to c. This suggests taking the limit of (22) as v approaches c. However, because (22) is defined for $0 < v < c$ but not for $v > c$, we really want to study (22) only for speeds v close to but less than c. In this case, we say that we are evaluating the limit of (22) as v approaches c from the left.

DEFINITION   **left-hand limit**

Let $g(x)$ be defined in an interval $a - r < x < a$ for some $r > 0$. We say that $g(x)$ has **left-hand limit** L as x approaches a if we can make $g(x)$ as close to L as we like by taking $x < a$ and close to a. We write

$$(23) \qquad \lim_{x \to a-} g(x) = L.$$

This is read "the limit of $g(x)$ as x approaches a from the left is L."

We illustrate the idea of the left-hand limit in Fig. 1.63. We can also define a right-hand limit, $\lim_{x \to b+} g(x)$. See Exercise 26.

■ **EXAMPLE 7**   Find the limit of the length of the flying arrow as v approaches c from the left.

Solution.   When $v < c$ and close to c, expression (22) is defined. For such values of v

$$L = L_0 \sqrt{1 - \left(\frac{v}{c}\right)^2} \quad \text{is close to} \quad L_0 \sqrt{1 - \left(\frac{c}{c}\right)^2} = 0.$$

Hence

$$\lim_{v \to c-} L_0 \sqrt{1 - \left(\frac{v}{c}\right)^2} = 0.$$

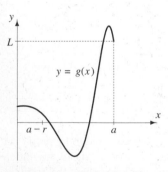

Figure 1.63. When $x < a$ and x is close to a, $g(x)$ is close to L. We write $\lim\limits_{x \to a-} g(x) = L$.

This means that as the speed of the arrow approaches the speed of light, the length of the arrow approaches 0. ∎

■ **EXAMPLE 8**   The function f is defined by

$$f(x) = \begin{cases} -2x + 1 & x < 1 \\ 3 & x = 1 \\ 2x^2 & x > 1. \end{cases}$$

Evaluate

$$\lim_{x \to 1-} f(x), \quad \lim_{x \to 1+} f(x), \quad \text{and} \quad \lim_{x \to 1} f(x).$$

Solution.   The graph of $y = f(x)$ is shown in Fig. 1.64. From the graph we see that if x is close to 1 and less than 1, $f(x)$ is close to -1. In fact, when $x < 1$, we have $f(x) = -2x + 1$, so

$$\lim_{x \to 1-} f(x) = \lim_{x \to 1-} (-2x + 1) = -2 \cdot 1 + 1 = -1.$$

When $x > 1$, we have $f(x) = 2x^2$. Hence

$$\lim_{x \to 1+} f(x) = \lim_{x \to 1+} 2x^2 = 2 \cdot 1^2 = 2.$$

This can also be seen by looking at the graph for $x > 1$ and close to 1. The above discussions also show that when x is close to 1, then $f(x)$ might be close to -1 or close to 2, depending on whether $x < 1$ or $x > 1$. Because the values of $f(x)$ do not get close to one and only one number L as x approaches 1, we conclude that

$$\lim_{x \to 1} f(x) \quad \text{docs not exist.} \qquad ∎$$

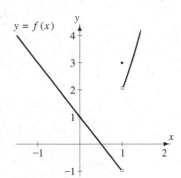

$y = f(x)$

Figure 1.64. The graph of $y = f(x)$.

Continuous Functions

In Section 1.4 we saw that if $p(x)$ is a polynomial and c is a real number, then

$$\lim_{x \to c} p(x) = p(c).$$

In fact, we have seen many examples with

$$\lim_{x \to c} f(x) = f(c).$$

Functions with such "well-behaved" limits are very important.

DEFINITION   **continuous function**

The function f is **continuous at c** if

(24) $$\lim_{x \to c} f(x) = f(c).$$

See Fig. 1.65. If f is continuous at every point in an interval $a < x < b$, we say that f is **continuous on the interval** $a < x < b$. If f is continuous at every point in its domain, then we say that f is **continuous**.

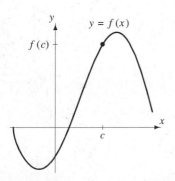

$y = f(x)$

Figure 1.65. If $\lim_{x \to c} f(x) = f(c)$, then f is continuous at c.

When we look closely at this definition, we see that three things must be true for a function f to be continuous at c:

a. $f(x)$ must be defined at $x = c$.

b. $\lim_{x \to c} f(x)$ must exist.

c. If a and b are both true, then the value of the limit in b must be $f(c)$.

In Example 1 of Section 1.4 we showed that

$$\lim_{x \to 5} 2x^2 + 4 = 2 \cdot 5^2 + 4.$$

Hence the function defined by $2x^2 + 4$ is continuous at $x = 5$.

In Example 2 of Section 1.4 we showed that

$$\lim_{\theta \to \pi/4} \cos \theta = \cos(\pi/4).$$

This means that the cosine function is continuous at $\theta = \pi/4$.

If one or more of the three conditions listed above is not satisfied, then $f(x)$ is **discontinuous** (or not continuous) at $x = c$. See Figs. 1.66 and 1.67. We have already seen many examples of discontinuity. For example, we have seen that

$$\lim_{x \to 0} \frac{\sin x}{x} = 1.$$

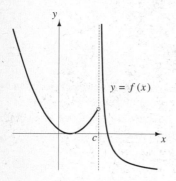

Figure 1.66. If f is not defined at c or $\lim_{x \to c} f(x)$ does not exist, then f is discontinuous at c.

However, the function defined by $(\sin x)/x$ is not continuous at $x = 0$ because the function is not defined for this value of x.

The function f of Example 8 is defined at $x = 1$ but is not continuous at this point because

$$\lim_{x \to 1} f(x) \quad \text{does not exist.}$$

As mentioned earlier, if p is any polynomial function

$$p(x) = a_n x^n + a_{n-1} x^{n-1} + \cdots + a_1 x + a_0,$$

then for any real number c,

$$\lim_{x \to c} p(x) = p(c).$$

This means that a polynomial is continuous at every point in its domain. Hence any polynomial is a continuous function.

In Section 1.4 we also saw that for any real number c

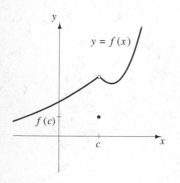

Figure 1.67. If $\lim_{x \to c} f(x)$ exists but is not equal to $f(c)$, then f is discontinuous at c.

$$\lim_{x \to c} \sin x = \sin c \quad \text{and} \quad \lim_{x \to c} \cos x = \cos c.$$

Thus the sine function and the cosine function are continuous.

In addition, the rules for combining limits given in the previous section imply that most combinations of continuous functions are continuous. Likewise, the result on limits and compositions shows that compositions of continuous functions are also continuous.

Continuous functions have many important and interesting properties. In one sense, the graph of a continuous function is well behaved.

> Let f be continuous on an interval $a < x < b$. Then the graph of $y = f(x)$ on $a < x < b$ is an unbroken curve.

According to this statement, the graph of a function continuous on an interval cannot have jumps or breaks like those seen in Fig. 1.66, Fig. 1.67, or Fig. 1.48. The converse of this result is "almost" true. A precise statement of the converse would take some terminology not usually encountered in calculus courses. However, for our purposes the following statement will be enough.

> For a function f suppose that the graph $y = f(x)$ for $a < x < b$ is an unbroken curve and the graph exhibits no oscillatory behavior like that seen the third graph in Fig. 1.48. Then f is continuous on the interval $a < x < b$.

Thus the graph in Fig. 1.65 is the graph of a continuous function.

Exercises 1.5

Basic

1. Let $f(x) = 3x^2 - 2x + 1$.
   a. Find the value of $\lim_{x \to -1} f(x)$.
   b. Let L be the value of the limit found in part a and let $\epsilon = 0.1$. Use graphical methods to find a $\delta > 0$ so that for $0 < |x - (-1)| < \delta$ we have $|f(x) - L| < \epsilon$.

2. Let $f(x) = x + 2\cos x$.
   a. Find the value of $\lim_{x \to \pi} f(x)$.
   b. Let L be the value of the limit found in part a and let $\epsilon = 0.01$. Use graphical methods to find a $\delta > 0$ so that for $0 < |x - \pi| < \delta$ we have $|f(x) - L| < \epsilon$.

3. Let $r(\theta) = 6$.
   a. Find the value of $\lim_{\theta \to \sqrt{5}} r(\theta)$.
   b. Let L be the value of the limit found in part a and let $\epsilon = 0.1$. Use graphical methods to find a $\delta > 0$ so that for $0 < |\theta - \sqrt{5}| < \delta$ we have $|r(\theta) - L| < \epsilon$.

4. Let $g(t) = 2^t$.
   a. Find the value of $\lim_{t \to 1} g(t)$.
   b. Let L be the value of the limit found in part a and let $\epsilon = 0.05$. Use graphical methods to find a $\delta > 0$ so that for $0 < |t - 1| < \delta$ we have $|g(t) - L| < \epsilon$.

5. Let $H(r) = \dfrac{r^3 + 8}{r^2 - 4}$.
   a. Find the value of $\lim_{r \to -2} H(r)$.
   b. Let L be the value of the limit found in part a and let $\epsilon = 0.1$. Use graphical methods to find a $\delta > 0$ so that for $0 < |r - (-2)| < \delta$ we have $|H(r) - L| < \epsilon$.

6. Let $t(\theta) = \dfrac{\cos \theta - 1}{2\theta^2}$.
   a. Find the value of $\lim_{\theta \to 0} t(\theta)$.
   b. Let L be the value of the limit found in part a and let $\epsilon = 0.005$. Use graphical methods to find a $\delta > 0$ so that for $0 < |\theta - 0| < \delta$ we have $|t(\theta) - L| < \epsilon$.

Exercises 7–13: A function f and a number a are given. (*a*) Find the value of $\lim_{x \to a} f(x)$. Use analytic, numerical, or graphical means. (*b*) Let L be the value of the limit found in part *a*. Find an interval containing the point $x = a$ on which $f(x) \approx L$ with an error of at most 0.01.

7. $f(x) = -4x^2 + 2x - 1, \quad a = 2$

8. $f(x) = \dfrac{1}{\sqrt{x+1}}, \quad a = 1$

9. $f(x) = \cos(2x), \quad a = 0$

10. $f(x) = \dfrac{10^x - 1}{x}, \quad a = 0$

11. $f(x) = \dfrac{x - 5}{\sqrt{x} - \sqrt{5}}, \quad a = 5$

12. $f(x) = \dfrac{\tan x - 1}{x - \pi/4}, \quad a = \pi/4$

13. $f(x) = \dfrac{1/(x+1) - \frac{1}{2}}{x - 1}, \quad a = 1$

14. a. Evaluate

$$\lim_{x \to 2} \frac{3x^2 - 2x - 8}{x - 2}.$$

b. Use the result of *a* to find an approximation to $3x^2 - 2x - 8$ for x values near 2.

c. Find an interval containing 2 on which the error in this approximation is less than 0.001.

d. For $x = 1.98$, compare the actual value of $3x^2 - 2x - 8$ with the value given by the approximation.

15. a. Evaluate (or estimate)

$$\lim_{r \to -2} \frac{\sqrt{2r^2 + 1} - 3}{r + 2}.$$

b. Use the result of *a* to find an approximation to $\sqrt{2r^2 + 1}$ for r values near -2.

c. Find an interval containing -2 on which the error in this approximation is less than 0.001.

d. For $r = -1.98$, compare the actual (calculator) value of $\sqrt{2r^2 + 1}$ with the value given by the approximation.

16. a. Use graphical or numerical means to find the value of

$$\lim_{t \to 2} \frac{3^t - 3^2}{t - 2}.$$

b. Use the result of *a* to find an approximation to 3^t for t values near 2.

c. Find an interval containing 2 on which the error in this approximation is less than 0.01.

d. For $t = 2.015$, compare the actual (calculator) value of 2^t with the value given by the approximation.

17. a. Use graphical or numerical means to find the value of

$$\lim_{x \to \pi/4} \frac{\tan x - 1}{x - \pi/4}.$$

b. Use the result of *a* to find an approximation to $\tan x$ for x values near $\pi/4$.

c. Find an interval containing $\pi/4$ on which the error in this approximation is less than 0.001.

d. For $x = 0.75$, compare the actual value of $\tan x$ with the value given by the approximation.

18. In Fig. 1.58, which is the graph of $y = x$ and which is the graph of $y = \sin x$? Give reasons for your answer.

19. Let $h(t) = t/|t|$.

a. Evaluate $\lim_{t \to 0+} h(t)$ and $\lim_{t \to 0-} h(t)$.

b. What can be said about $\lim_{t \to 0} h(t)$?

c. Is $h(t)$ continuous at $t = 0$? Why or why not?

20. Let $P(w)$ be the cost of first-class postage for a letter that weighs w ounces. If $0 < w < 1$, then $P(w) = 32¢$. If $w > 1$, the cost is $32¢$ plus $23¢$ for each ounce or fraction of an ounce above 1.

a. Sketch a graph of the function $C = P(w)$ for $0 < w < 5.5$.

b. Evaluate $\lim_{w \to 2-} P(w)$ and $\lim_{w \to 2+} P(w)$.

c. Tell why $P(w)$ is discontinuous for $w = 1, 2, 3, 4, \ldots$.

21. The graph of a function f is shown below.
   a. Evaluate $\lim_{x \to -1+} f(x)$ and $\lim_{x \to -1-} f(x)$.
   b. Evaluate $\lim_{x \to 0+} f(x)$ and $\lim_{x \to 0-} f(x)$.
   c. Evaluate $\lim_{x \to 1+} f(x)$ and $\lim_{x \to 1-} f(x)$.
   d. For what values of x between -3 and 3 is f discontinuous?

22. The graph of a function g is shown below.
   a. Evaluate $\lim_{x \to -4+} g(x)$ and $\lim_{x \to -4-} g(x)$.
   b. Evaluate $\lim_{x \to -2+} g(x)$ and $\lim_{x \to -2-} g(x)$.
   c. Evaluate $\lim_{x \to 2+} g(x)$ and $\lim_{x \to 2-} g(x)$.
   d. For what values of x between -5 and 3 is g discontinuous?

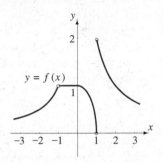

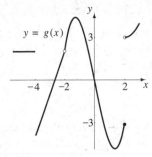

Growth

23. Restate the rule for limits and compositions in Section 1.4 in such a way that the conditions on the function h are expressed in terms of the continuity of h at L.

24. a. For $x \neq 0$ let

$$g(x) = x \sin\left(\frac{1}{x}\right).$$

   Show that $\lim_{x \to 0} g(x) = 0$.
   b. For $x \neq 0$ let

$$h(x) = \frac{\sin x}{x}.$$

   Note that $h(x)$ is not defined for $x = 0$ but that $\lim_{x \to 0} h(x) = 1$. Show, however, that $\lim_{x \to 0} h(g(x))$ does not exist.
   c. Carefully tell why this example shows that it is necessary to have h defined at L in the rule for limits and compositions in Section 1.4.

25. Give an ϵ, δ definition for the left-hand limit. Model it on the ϵ, δ definition for the limit given in this section.

26. Write a definition for the right-hand limit of a function. Model your definition on the definition of the left-hand limit given in this section.

27. In this exercise we "verify" equation (12) from the rules for combining limits in Section 1.4. Let g and h be functions and suppose that

$$\lim_{x \to a} g(x) = 3, \qquad \lim_{x \to a} h(x) = 5.$$

Take $\epsilon = 0.1$. Use the following two steps as a guide to argue why there must be a $\delta > 0$ so that $0 < |x - a| < \delta$ implies

$$|(g(x) + h(x)) - (3 + 5)| < \epsilon.$$

   a. First tell why there must be a $\delta > 0$ so that when $0 < |x - a| < \delta$ we have

$$|g(x) - 3| < 0.05$$

   and

$$|h(x) - 5| < 0.05.$$

   b. Let δ have the value discussed in part a. Show that $0 < |x - a| < \delta$ implies

$$|(g(x) + h(x)) - 8| < 0.1 = \epsilon.$$

28. In this exercise we "verify" equation (14) from the rules for combining limits in Section 1.4. Let g and h be functions and suppose that

$$\lim_{x \to a} g(x) = 3, \qquad \lim_{x \to a} h(x) = 5.$$

Take $\epsilon = 0.1$ use the following two steps as a guide to argue why there must be a $\delta > 0$ so that $0 < |x - a| < \delta$ implies

$$\left| \frac{g(x)}{h(x)} - \frac{3}{5} \right| < \epsilon.$$

   a. First tell why there must be a $\delta > 0$ so that when $0 < |x - a| < \delta$ we have

$$|g(x) - 3| < 0.25 \quad \text{and} \quad |h(x) - 5| < 0.25.$$

b. Let δ have the value discussed in part a. Show that $0 < |x - a| < \delta$ implies

$$\left| \frac{g(x)}{h(x)} - \frac{3}{5} \right| < 0.1 = \epsilon.$$

29. Let

$$p(x) = x^3 - 4x + 6.$$

The following steps outline a method for using the squeeze theorem (See Exercise 34 of Section 1.4) to show that $\lim_{x \to 1} p(x) = p(1) = 3$.

a. Note that

$$p(x) - p(1)$$

takes the value 0 when $x = 1$. Hence the polynomial $p(x) - p(1)$ has a factor of $x - 1$. Verify that

$$p(x) - p(1) = (x - 1)(x^2 + x - 3).$$

b. For $|x - 1| < 1$ (i.e., $0 < x < 2$) show that

$$|x^2 + x - 3| < 9.$$

c. Show that for $|x - 1| < 1$,

$$|p(x) - 3| \leq 9|x - 1|.$$

d. Use the squeeze theorem and part **c** to show that

$$\lim_{x \to 1} p(x) = p(1) = 3$$

Relate this result to rule for limits of polynomials, sines, and cosines in Section 1.4.

30. Use the technique outlined in the previous problem to show that

$$\lim_{x \to 2} (-2x^3 + 3x^2 + x - 3) = -5.$$

31. Let $p(x)$ be a polynomial and a a real number. Based on the outline given in Exercise 29, discuss a method for showing that $\lim_{x \to a} p(x) = p(a)$. Note that this gives a method of proving equation (15) of the rule for limits of polynomials, sines, and cosines in Section 1.4.

32. Define $r = r(c)$ to be the number of distinct real zeros of the polynomial

$$x^2 - 3x + c.$$

For example, $r(-4) = 2$ because $x^2 - 3x - 4 = 0$ has two different roots, $x = 4$ and $x = -1$. However, $r(5) = 0$ because the equation $x^2 - 3x + 5 = 0$ has no real solutions. For what values of c is $r(c) = 2$? When is $r(c) = 1$? When is $r(c) = 0$? Graph the function $r = r(c)$ and list the values of c where the function r is not continuous.

33. Define $r = r(b)$ to be the number of distinct real zeros of the polynomial

$$x^2 + bx + 4.$$

For example, $r(3) = 0$ because $x^2 + 3x + 4 = 0$ has no real solutions, and $r(5) = 2$ because the equation $x^2 + 5x + 4 = 0$ has two real solutions. For what values of b is $r(b) = 2$? When is $r(b) = 1$? When is $r(b) = 0$? Graph the function $r = r(b)$ and list the values of b where the function r is not continuous.

34. Define $r(d)$ to be the number of distinct real zeros of the polynomial

$$2x^3 + 3x^2 - 12x + d.$$

For example, $r(0) = 3$ because $2x^3 + 3x^2 - 12x + 0 = 0$ has three real solutions, as shown by the graph below. For what values of d is $r(d) = 3$? When is $r(d) = 2$? When is $r(d) = 1$? When is $r(d) = 0$? Graph the function $r = r(d)$ and list the values of d for which r is not continuous.

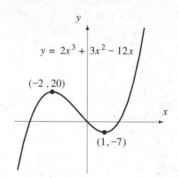

35. Find the value of

$$\lim_{x \to 8} \frac{\sqrt[3]{x} - 2}{x - 8}.$$

Use this result to develop an approximation to $\sqrt[3]{x}$ for x values close to 8. Find an interval containing 8 on which the error in the approximation is less than 1 percent of the value of $|x - 8|$.

36. Use graphical or numerical means to find the value of

$$\lim_{x \to 3} \frac{2^x - 8}{x - 3}.$$

Use this result to develop an approximation to 2^x for x values close to 3. Find an interval containing 3 on which the error in the approximation is less than 1 percent of the value of $|x - 3|$.

37. Find the value of

$$\lim_{x \to -2} \frac{1/\sqrt{x^2 + 5} - 1/3}{x + 2}.$$

Use this result to develop an approximation to $1/\sqrt{x^2 + 5}$ for x values close to -2. Find an interval containing -2 on which the error in the approximation is less than 1 percent of the value of $|x - (-2)|$.

38. Use graphical or numerical means to find the value of

$$\lim_{\theta \to 0} \frac{\cos \theta - 1}{\theta^2}$$

(or see Exercise 13 in Section 1.4). Use this result to find an approximation to $\cos \theta$ for θ values close to 0. Find an interval containing 0 on which the error in the approximation is less than 1 percent of the value of θ^2. On this interval, compare $\cos \theta$ and the approximation graphically.

39. Use graphical or numerical means to find the value of

$$\lim_{\theta \to 0} \frac{\sin \theta - \theta}{\theta^3}.$$

Use this result to find an approximation to $\sin \theta$ for θ values close to 0. Find an interval containing 0

on which the error in the approximation is less than 1 percent of the value of θ^3. Compare $\sin \theta$ and the approximation graphically.

40. In Example 6 we showed that

$$\sqrt{t} \approx 3 + \frac{1}{6}(t - 9).$$

We use this result to derive the approximation discussed at the beginning of this section.

a. Verify that

$$\lim_{t \to 9} \frac{\sqrt{t} - \left(3 + \frac{1}{6}(t - 9)\right)}{(t - 9)^2} = -\frac{1}{216}.$$

b. Use the result of *a* to derive an approximation for $\sqrt{t}$ and show that this approximation is the one given at the beginning of the section.

c. Discuss the error in this approximation on the interval $8.5 \le t \le 9.5$.

41. Suppose that $\lim_{x \to c} f(x) = L$. What can be said about

$$\lim_{x \to c-} f(x) \quad \text{and} \quad \lim_{x \to c+} f(x)?$$

42. Suppose that

$$\lim_{x \to c-} f(x) = L \quad \text{and} \quad \lim_{x \to c+} f(x) = M.$$

Under what conditions does $\lim_{x \to c} f(x)$ exist?

Review/Preview

43. Let $g(x) = x^2 - 3x + 1$ and $h(x) = (x - 1)^2$. Find and simplify:
   a. $g(h(x))$.
   b. $h(g(x))$.
   c. $h(h(x))$.
   d. $h(h(h(x)))$.

44. Let $f(x) = \sqrt{x + 1}$ and $g(x) = 2(x - 1)^2$. Find and simplify:
   a. $g(f(x))$.
   b. $f(g(x))$.
   c. $f(f(x))$.
   d. $g(f(g(x)))$.

45. Let $p(x)$ be a polynomial that takes the value 0 only at $x = -3, 2, 5$ and the value 2 only at $x = -5, 3, 6$. If $q(x) = 2x - 4$:
   a. Find the solutions to $p(q(x)) = 0$.
   b. Find the solutions to $q(p(x)) = 0$.

46. Let $p(x)$ be a polynomial that takes the value 0 only at $x = -3, 2, 5$ and the value -3 only at $x = -5, 3, 6$. If $q(x) = 3x + 9$:
   a. Find the solutions to $p(q(x)) = 0$.
   b. Find the solutions to $q(p(x)) = 0$.

47. Find x if:
   a. $2^x = 8$.
   b. $5^x = 1/125$.
   c. $3^x = 9\sqrt{3}$.
   d. $x^{-2} = 9/4$.

48. Find y if:
   a. $2^y = 64$.
   b. $(1/3)^y = 9\sqrt{3}$.
   c. $y^4 = 32$.
   d. $6^{(1-y)} = 0$.

49. For $0 \le \theta \le 2\pi$, find the number of distinct solutions for each of the following equations.
   a. $\sin \theta = 1/2$
   b. $\cos \theta = -2\sqrt{3}$
   c. $\sin 3\theta = -0.1$
   d. $\tan 2\theta = 100$

50. For $0 \le \theta \le 2\pi$, find the number of distinct solutions for each of the following equations.
   a. $\cos \theta = \sqrt{3}/2$
   b. $\cos 4\theta = 0$
   c. $\sin 10\theta = 1/\pi$
   d. $\tan 3\theta = -1$

51. The lengths of the diagonals of the faces of a rectangular box are 10 cm, 18 cm, and 20 cm. Find the volume of the box.

52. A wall of a house contains a window that is 6 feet wide. Twenty feet away a sidewalk runs parallel to the wall of the house. A man inside the house is 8 feet from the center of the window. How long is the stretch of sidewalk that the man can see?

53. In finding the height of a mountain, a geologist makes two measurements. Several miles from the foot of the mountain she measures the angle of elevation of the summit and finds it to be 13°. She then moves 4 miles further from the foot of the mountain and measures the angle of elevation of the summit from her new position. The second measurement is 8°. See the accompanying figure. What is the height of the mountain?

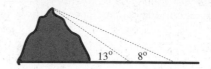

1.6   THE DERIVATIVE

In Sections 1.2, 1.3, and 1.4 we answered many questions by using techniques we developed for finding the rate of change of a function. For example, we investigated the rate at which pressure changes with respect to volume in a piston chamber, the rate at which temperature rises or falls in an autoclave, and the rate at which the world population is growing. When something is used in many ways to answer questions in many different areas it is not unusual for it to have many different names. This is the case with the rate of change. The rate of change is also called the *derivative*.

Because the concept of derivative is so important, we repeat the definition of rate of change given in Section 1.4, but use it to define the derivative.

DEFINITION   **derivative**

Let f be a function and suppose that f is defined at the point a (i.e., $f(a)$ is defined). The **derivative of f at a** is

(1) $$\lim_{h \to 0} \frac{f(a + h) - f(a)}{h}.$$

If the limit in (1) exists, it is denoted $f'(a)$. If this limit does not exist, we say f has no derivative at $x = a$.

The derivative, $f'(a)$, has at least one other name. It is sometimes called the *slope of f at a.* Usually the terminology used (derivative, rate of change, or slope) depends on the problem to be solved.

When we call the derivative a rate of change, we are usually concerned with two variables, say y and x, and a function f relating the two variables by $y = f(x)$. Suppose we set $x = a$ and are interested in how the y value changes when x changes from the value a. In Sections 1.2 and 1.3 we saw that when h is small, the quotient

(2) $$\frac{f(a + h) - f(a)}{h}$$

gives us a measure of the rate of change of y (or f) in response to a change in x. Taking the limit of (2) as $h \to 0$ results in the derivative $f'(a)$. In such a case this derivative may be referred to as the rate of change of f (or y) with respect to x when $x = a$.

When we call the derivative a slope, we often have a geometric interpretation in mind. As we saw in Sections 1.2 and 1.3, expression (2) gives the slope of the "line" we see when we zoom in on the graph of $y = f(x)$ near $x = a$. See Fig. 1.68. When a derivative arises in this way, it might be more meaningful to call it the *slope of f at x = a*.

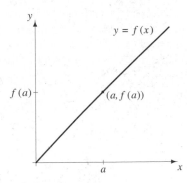

Figure 1.68. We see a "line" of slope $f'(a)$ when we zoom in on the graph of $y = f(x)$ near $(a, f(a))$.

Notation for the Derivative

There are also many different notations for the derivative (or rate of change). The notation used often depends on how the function f is presented or on how the derivative is to be used. When a function is defined by a formula $y = f(x)$, the following are all notations for the derivative of f (or y) with respect to x at $x = a$:

$$f'(a) \qquad \frac{dy}{dx}\Big|_{x=a} \qquad \frac{df}{dx}\Big|_{x=a} \qquad D_x f(a).$$

We shall generally use the first three of these notations.

The Derivative as a Function

Given a numerical value $x = a$, the derivative of f at a is a number denoted by $f'(a)$. If we allow a to be a variable, we can think of the derivative as a function, f'.

The Derivative as a Function

Let f be a function. The derivative function f' is defined by the equation

(3)
$$f'(x) = \lim_{h \to 0} \frac{f(x + h) - f(x)}{h}.$$

The function f' is defined for all those x for which the limit in (3) exists.

The derivative of f at a, $f'(a)$, is the value of the function f' at the value $x = a$. When $y = f(x)$, we will also use

$$\frac{dy}{dx} \quad \text{and} \quad \frac{df}{dx}$$

to denote the function f'.

Another Formula for the Derivative

In Sections 1.2 and 1.3 we estimated the rate of change of a function f at a point $x = a$ by calculating the slope of the line determined by the point $P = (a, f(a))$ and a nearby point $Q = (a + h, f(a + h))$. Since Q was close to P, we knew that h had to be a small number. (See Fig. 1.69.) The rate of change (or slope) was found by first calculating

$$\Delta f = \text{change in } f = f(a + h) - f(a)$$

and

$$\Delta x = \text{change in } x = (a + h) - a = h$$

and then forming the quotient

$$\frac{\text{change in } f}{\text{change in } x} = \frac{\Delta f}{\Delta x} = \frac{f(a + h) - f(a)}{h}.$$

In view of the definition of derivative, the rate of change of f with respect to x at $x = a$ is then

$$f'(a) = \lim_{\Delta x \to 0} \frac{\Delta f}{\Delta x} = \lim_{h \to 0} \frac{f(a + h) - f(a)}{h}.$$

Notations such as Δf and Δx are often used to represent small changes in quantities. This notation reminds us that the rate of change (or derivative) can be estimated by calculating the quotient

$$\frac{\Delta f}{\Delta x}$$

for small Δx.

In the above discussions we used $Q = (a + h, f(a + h))$ to denote a point close to $P = (a, f(a))$. We could just as easily have denoted this nearby point $(x, f(x))$ with the understanding that x is close to a. This is illustrated in Fig. 1.70. Substituting x for $a + h$ and noting that $h \to 0$ is equivalent to $x \to a$ gives us another form for the definition of the derivative of f at a,

(4) $$f'(a) = \lim_{x \to a} \frac{f(x) - f(a)}{x - a}.$$

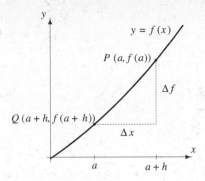

Figure 1.69. $Q(a+h, f(a+h))$ is near $P(a, f(a))$ on the graph of $y = f(x)$.

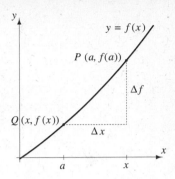

Figure 1.70. $Q(x, f(x))$ is near $P(a, f(a))$ on the graph of $y = f(x)$.

Feel free to use (1) or (4) when computing a derivative. Depending on the function f, calculations with the two formulas can involve different algebra.

Interpreting the Derivative

To effectively use the derivative as a tool, it is important to know how to interpret it in different contexts. Remember that the derivative (or rate of change) of $y = f(x)$ carries the units

$$\frac{\text{units of } y}{\text{units of } x}.$$

Knowing these units often helps in determining the meaning of the derivative.

■ **EXAMPLE 1**   A high diver jumps from a tower into a pool of water 100 feet below. After she has fallen for t seconds, her height (in feet) above the pool is $H(t) = 100 - 16t^2$. Find the derivative of H and discuss the meaning of the derivative. How long after she jumps does the diver reach the pool? What is her speed at this time?

Solution.   By (3) the derivative is

$$H'(t) = \lim_{h \to 0} \frac{H(t + h) - H(t)}{h}$$
$$= \lim_{h \to 0} \frac{(100 - 16(t + h)^2) - (100 - 16t^2)}{h}.$$

We cannot evaluate this limit by simply substituting $h = 0$ because this leads to $0/0$. Thus we do some algebra to find an equivalent expression whose limit we can determine. After simplifying the numerator in the last expression above, we can factor and cancel an h from the numerator and denominator.

$$H'(t) = \lim_{h \to 0} \frac{-32th - 16h^2}{h} = \lim_{h \to 0} (-32t - 16h) = -32t.$$

Since the height $H(t)$ is in feet and time t is in seconds, the units for the derivative are feet/second. Thus at time t the rate of change of the diver's height with

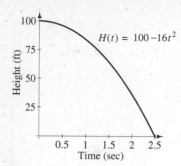

Figure 1.71. The graph illustrates the height of the diver as a function of time.

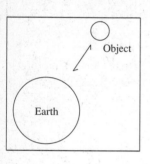

respect to time is $-32t$ ft/s. The units are units of speed, so the derivative tells us how fast the diver is moving at time t. The minus sign tells us that when we zoom in on the graph of $H = H(t)$ near a point $(t, H(t))$, we will see a "line" of negative slope. Hence $H(t)$ decreases as time t increases. (This makes sense because as the diver falls, her height above the water is decreasing.)

When the diver reaches the water, her height $H(t)$ above the pool will be 0. Solving

$$H(t) = 100 - 16t^2 = 0 \quad \text{we find} \quad t = \pm 2.5 \text{ seconds.}$$

Since the t we seek must be positive, we conclude that the diver hits the water 2.5 seconds after jumping off the tower. See Fig. 1.71. The derivative at this time is

$$H'(2.5) = -32(2.5) = -80 \text{ ft/s.}$$

Thus the diver is moving downward at 80 ft/s when she hits the water.   ■

■ **EXAMPLE 2**   According to Newton's law of gravitation, a 100 kg object r meters above the surface of the earth is subject to a gravitational force of

$$F(r) = \frac{100GM}{(R_0 + r)^2} \text{ newtons (N),}$$

where R_0 is the radius of the earth ($\approx 6.37 \times 10^6$ m), M is the mass of the earth ($\approx 5.98 \times 10^{24}$ kg), and G is the gravitational constant ($\approx 6.67 \times 10^{-11}$ Nm^2/kg^2). Find the rate of change of F with respect to r as the object moves away from the surface of the earth along a ray from the center of the earth. What is the rate of change when the object is 10^7 m above the surface of the earth?

Solution.   The rate of change is the derivative. Using (4), we have

$$F'(r) = \lim_{t \to r} \frac{F(t) - F(r)}{t - r} = \lim_{t \to r} \frac{\dfrac{100GM}{(R_0 + t)^2} - \dfrac{100GM}{(R_0 + r)^2}}{t - r}.$$

Adding the fractions in the numerator of this expression gives

$$F'(r) = \lim_{t \to r} \frac{100GM\left((R_0 + r)^2 - (R_0 + t)^2\right)}{(R_0 + r)^2(R_0 + t)^2(t - r)}.$$

Next factor $(R_0 + r)^2 - (R_0 + t)^2$ as a difference of squares to obtain

$$F'(r) = \lim_{t \to r} \frac{100GM(r - t)(2R_0 + r + t)}{(R_0 + r)^2(R_0 + t)^2(t - r)}$$

$$= \lim_{t \to r} \frac{-100GM(2R_0 + r + t)}{(R_0 + r)^2(R_0 + t)^2}$$

$$= \frac{-100GM(2R_0 + 2r)}{(R_0 + r)^4}$$

$$= \frac{-200GM}{(R_0 + r)^3}.$$

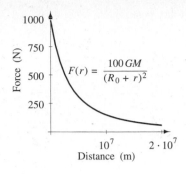

Figure 1.72. The graph illustrates the force on the object as a function of distance from the Earth's surface.

The units for the derivative are newtons/meter. When $r = 10^7$ m, the derivative has value

$$F'(10^7) = \frac{-200GM}{(R_0 + 10^7)^3} \approx -1.82 \times 10^{-5} \text{ N/m}.$$

When the object is 10^7 m above the earth's surface, the gravitational force on the object is $F(10^7) \approx 148.8$ N. See Fig. 1.72. The derivative tells us that the force on the object decreases by about 1.82×10^{-5} N if the object moves 1 m further from the surface of the earth. ∎

■ **EXAMPLE 3**   The owner of the Good Lookin' Glass Company asks a group of engineers and accountants to reflect on the cost of manufacturing several one-way mirrors. They report that the cost of producing $N > 0$ mirrors will be about

$$(5) \qquad\qquad C(N) = 3000 + 5N + 2\sqrt{N}$$

dollars. Compute the derivative of the function C and discuss the significance of the derivative.

Solution.   We first discuss the meaning of (5). The glass company cannot manufacture $100\frac{1}{2}$ or $\sqrt{2}$ mirrors, but only a positive whole number of mirrors. Thus $C(N)$ is only defined for $N = 1, 2, 3, \ldots$. The graph of $C(N)$ just shows the individual points $(1, C(1)), (2, C(2)), (3, C(3)), \ldots$ (see Fig. 1.73). Such a function does not have a derivative. When we zoom in on this graph, we will never see a straight line. This suggests that the rate of change (or derivative) of $C(N)$ with respect to N does not exist. Indeed, if the derivative of C did exist, it would be given by

$$(6) \qquad\qquad \lim_{t \to N} \frac{C(t) - C(N)}{t - N}.$$

However, the quotient $(C(t) - C(N))/(t - N)$ is not defined for t in intervals on either side of N, since any such interval contains noninteger values of t. This means that the conditions for the existence of the limit (6) are not satisfied. Hence the limit (6) does not exist.

Even though (5) may not be meaningful from a manufacturing point of view for nonintegers N, useful information can be obtained by assuming $C(N)$ is defined

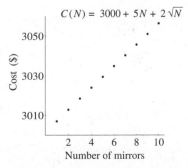

Figure 1.73. The cost function is defined only for nonnegative integers N.

for all positive N. We can then compute $C'(N)$ for any positive N:

$$\begin{aligned}
C'(N) &= \lim_{t \to N} \frac{C(t) - C(N)}{t - N} \\
&= \lim_{t \to N} \frac{(3000 + 5t + 2\sqrt{t}) - (3000 + 5N + 2\sqrt{N})}{t - N} \\
&= \lim_{t \to N} \left(\frac{5(t - N)}{t - N} + \frac{2(\sqrt{t} - \sqrt{N})}{(\sqrt{t} - \sqrt{N})(\sqrt{t} + \sqrt{N})} \right) \\
&= \lim_{t \to N} \left(5 + \frac{2}{\sqrt{t} + \sqrt{N}} \right) \\
&= 5 + \frac{1}{\sqrt{N}}.
\end{aligned}$$

When $C(N)$ is the cost of manufacturing N units of a product, $C'(N)$ has units of dollars/product and is called the **marginal cost**. The marginal cost can be used as an estimate of the cost of producing a unit of the product after N units have already been manufactured. That is, $C'(N)$ is used as an estimate of $C(N + 1) - C(N)$. This is justified by noting that when h is small,

$$\frac{C(N + h) - C(N)}{h} \approx C'(N).$$

Taking $h = 1$ to be "small" in this context gives our approximation,

$$C(N + 1) - C(N) \approx C'(N).$$

In manufacturing processes where the average cost per product goes down as more products are produced, this approximation is often a good one. To check this, assume that 1000 mirrors have been produced. The extra cost to produce mirror 1001 is

$$\begin{aligned}
C(1001) - C(1000) &= (3000 + 5 \cdot 1001 + 2\sqrt{1001}) \\
&\quad - (3000 + 5 \cdot 1000 + 2\sqrt{1000}) \\
&\approx \$5.031615
\end{aligned}$$

The approximation to this figure given by the marginal cost is

$$C(1001) - C(1000) \approx C'(1000) = 5 + 1/\sqrt{1000} \approx \$5.031623.$$

Pretty close! ∎

The Tangent Line

Let f be a function with derivative $f'(a)$ at $x = a$. When we zoom in on the graph of $y = f(x)$ near the point $(a, f(a))$, the graph appears to be a straight line with slope $f'(a)$. This motivates the following definition.

DEFINITION   **tangent line**

Let $f(x)$ have derivative $f'(a)$ at $x = a$. The line with slope $f'(a)$ through $(a, f(a))$ is the **tangent line** to the graph of $y = f(x)$ at the point $(a, f(a))$.

Since the line tangent to $y = f(x)$ at the point $(a, f(a))$ has slope $f'(a)$, we can easily write down an equation for the tangent line:

$$y - f(a) = f'(a)(x - a)$$

or

(7) $$y = f(a) + f'(a)(x - a)$$

The Tangent Line as an Approximation

If we draw the line tangent to the graph of $y = f(x)$ at the point $(a, f(a))$ then zoom in on the graph near this point, the graph and the tangent line will be almost indistinguishable, as illustrated in Fig. 1.74.

Since the graphs of the functions

$$y = f(x) \quad \text{and} \quad y = f(a) + f'(a)(x - a)$$

are close, the values of the two functions must be close. Thus for x close to a,

(8) $$f(x) \approx f(a) + f'(a)(x - a).$$

Expression (8) says what when x is close to a, the function value $f(x)$ is approximated by the corresponding y-value for the tangent line.

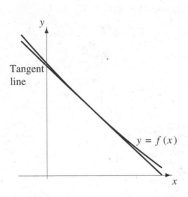

Figure 1.74. The graph of the tangent line is close to the graph of the function near the point of tangency.

■ **EXAMPLE 4**   Find the equation of the line tangent to $f(x) = 1/(x + 1)$ at $(2, 1/3)$. Use the equation of the tangent line to approximate $1/(1.93 + 1)$.

Solution.   The tangent line passes through the point of tangency, $(2, 1/3)$. The slope of the tangent line is the derivative

$$f'(2) = \lim_{x \to 2} \frac{f(x) - f(2)}{x - 2} = \lim_{x \to 2} \frac{\dfrac{1}{x + 1} - \dfrac{1}{3}}{x - 2}.$$

Add the fractions in the numerator and simplify the resulting complex fraction. We then have

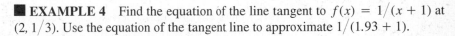

$$f'(2) = \lim_{x \to 2} \frac{2 - x}{3(x + 1)(x - 2)} = \lim_{x \to 2} \frac{-1}{3(x + 1)} = -\frac{1}{9}.$$

The equation of the line tangent to $y = 1/(x + 1)$ at $(2, 1/3)$ is

$$y = f(2) + f'(2)(x - 2) = \frac{1}{3} - \frac{1}{9}(x - 2).$$

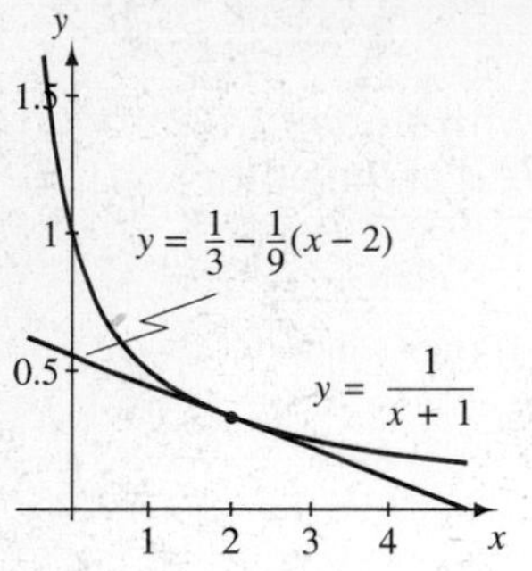

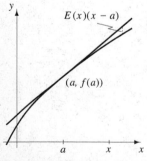

Figure 1.75. The line tangent to $y = 1/(x + 1)$ at $(2, 1/3)$.

When x is close to 2, the y coordinates for the graph of $y = 1/(x + 1)$ and the tangent line should be close. See Fig. 1.75. Hence for x near 2,

$$\frac{1}{x + 1} \approx \frac{1}{3} - \frac{1}{9}(x - 2).$$

Using this last expression, we have

$$(1.93 + 1)^{-1} = \frac{1}{1.93 + 1} \approx \frac{1}{3} - (1/9)(1.93 - 2) \approx 0.341111.$$

How does this compare with the value of $(1.93 + 1)^{-1}$ given by your calculator? ∎

Error in the Tangent Line Approximation   We can use the techniques developed in Section 1.5 to estimate the error in the tangent line approximation. Let

(9) $$\frac{f(x) - f(a)}{x - a} - f'(a) = E(x).$$

Because

$$\lim_{x \to a} E(x) = \lim_{x \to a} \left(\frac{f(x) - f(a)}{x - a} - f'(a) \right) = f'(a) - f'(a) = 0,$$

we can make $E(x)$ small by taking x close to a. Solving (9) for $f(x)$, we find

(10) $$f(x) = f(a) + f'(a)(x - a) + E(x)(x - a).$$

Hence the difference between the function value $f(x)$ and the value $f(a) + f'(a)(x - a)$ of the tangent line expression is

(11) $$f(x) - (f(a) + f'(a)(x - a)) = E(x)(x - a).$$

From (11) we see that the difference is small compared to $|x - a|$ when $E(x)$ is small, that is, when x is close to a. The error is illustrated in Fig. 1.76.

   In Section 1.5 we actually discussed the error in the tangent line in two examples. In Example 5 we investigated the error in approximation

$$\sin x \approx x$$

for x near 0. Note that $y = x$ is the equation for the line tangent to the graph of $y = \sin x$ at $(0, 0)$. In Example 6 we looked at the error in the approximation

$$\sqrt{x} \approx 3 + \frac{1}{6}(x - 9)$$

for x near 3. The line with equation $y = 3 + \frac{1}{6}(x - 9)$ is the line tangent to $y = \sqrt{x}$ at $(9, 3)$.

Figure 1.76. The difference between $f(x)$ and the y-coordinate of the tangent line is $E(x)(x - a)$.

Derivatives and Continuous Functions As another consequence of (10) we obtain a necessary condition for a function f to have a derivative at a point $x = a$.

Continuity of Differentiable Functions

If the function f has a derivative at $x = a$, then f is continuous at $x = a$, that is,

$$\lim_{x \to a} f(x) = f(a).$$

To justify this statement, take the limit as $x \to a$ of both sides of (10):

$$\lim_{x \to a} f(x) = \lim_{x \to a}(f(a) + f'(a)(x - a) + E(x)(x - a)).$$

Applying results from Section 1.4 for limits of sums and products of functions, we obtain

$$\lim_{x \to a} f(x) = \lim_{x \to a} f(a) + \lim_{x \to a}\left(f'(a)(x - a)\right) + \lim_{x \to a}(E(x)(x - a))$$

$$= f(a) + f'(a) \lim_{x \to a}(x - a) + \left(\lim_{x \to a} E(x)\right)\left(\lim_{x \to a}(x - a)\right)$$

$$= f(a) + f'(a) \cdot 0 + 0 \cdot 0$$

$$= f(a).$$

Since $\lim_{x \to a} f(x) = f(a)$, the function f is continuous at $x = a$.

Exercises 1.6

Basic

Exercises 1–10: Determine the derivative in each case. Use (1) for some of the problems and (4) for others.

1. $f(x) = x^2 - 3$
2. $y = 2x^2 - 6$
3. $h(t) = -3t^2 + 4t - \sqrt{2}$
4. $g(u) = \dfrac{2}{u + 2}$
5. $y = \dfrac{4}{2 - x}$

6. $s(t) = t - 2\sqrt{t}$
7. $y = 2t + 3\sqrt{t + 9}$
8. $r(s) = as^2 + bs + c$ where a, b, c are constants.
9. $y = \dfrac{1}{ax + b}$ where a and b are constants.
10. $h(t) = \sqrt{at + b}$ where a and b are constants.

Exercises 11–18: Find an equation for the line tangent to the graph of the function at the specified point.

11. $y = x^2 - 3x$ at $(1, -2)$
12. $s(t) = -3t^2 - 2t + 1$ at $(-2, -7)$
13. $h(r) = 4r + 2$ at $(1, 6)$
14. $y = \dfrac{1}{x}$ at $(a, 1/a)$, $a \neq 0$

15. $g(u) = \dfrac{2}{u^2 - 1}$ at $(2, 2/3)$
16. $w = \sqrt{2v - 1}$ at $(b, \sqrt{2b - 1})$, $b > 1/2$
17. $f(x) = 2|x|$ at $(2, 4)$

18. Use the equation for the line tangent to $y = x^2$ at $(4, 16)$ to approximate $(4.14)^2$ and $(3.91)^2$. In each case compute the error in the approximation.

19. Use the equation for the line tangent to $y = 1/x$ at $(3, 1/3)$ to approximate $1/(3.15)$ and $1/(2.85)$. In each case compute the error in the approximation.

20. In Chapter 2 we will see that the derivative of $\sin x$ is $\cos x$.
    a. Find an equation for the line tangent to the graph of $y = \sin x$ at $(\pi/3, \sqrt{3}/2)$.
    b. Use the result of part a to find an approximation to $\sin(\pi/3 + 0.1)$. (All angle measures are in radians.) Use your calculator to check the accuracy of the approximation.

21. In Chapter 2 we will see that the derivative of $\tan x$ is $\sec^2 x$.
    a. Find an equation for the line tangent to the graph of $y = \tan x$ at $(\pi/4, 1)$.
    b. Use the result of part a to find an approximation to $\tan(\pi/4 - 0.05)$. (All angle measures are in radians.) Use your calculator to check the accuracy of the approximation.

22. The planet Jupiter has mass $M \approx 1.9 \times 10^{27}$ kg and radius $R \approx 6.98 \times 10^7$ m. Suppose the 100 kg mass of Example 2 is 10^7 m above the surface of Jupiter. Find the rate of change of the gravitational force as the object moves away from Jupiter along a ray through the center of the planet. Discuss the meaning of this rate of change. (See Example 2.)

23. Repeat Exercise 22 for the moon. The moon has mass $M \approx 7.34 \times 10^{22}$ kg and radius $R \approx 1.74 \times 10^6$ m.

24. The accountants at the Seed & Sod Turf Company estimate that the cost of a grassroots advertising campaign that will reach $N \times 10^5$ consumers is roughly $C(N) = 5000 + 1000\sqrt{N} + 0.05(N - 1)^2$ dollars.
    a. Find the derivative of $C(N)$ and discuss the meaning of this derivative as a marginal cost.
    b. Use the derivative to estimate the extra cost of a campaign to reach a total of 1,100,000 consumers over that of a campaign to reach 1,000,000 consumers.

Growth

25. A function f has derivative
$$f'(x) = \frac{1}{x^2 + 1}.$$
    Given that $f(0) = 2$:
    a. Find an equation for the line tangent to the graph of $y = f(x)$ at the point $(0, 2)$.
    b. Find an approximation to $f(0.05)$. To $f(-0.1)$.

26. A function h has derivative
$$h'(u) = u\sin(2u).$$
    Given that $h(3\pi/4) = -3$, find an approximation to
$$h\left(\frac{3\pi}{4} + 0.15\right).$$

27. The line $y = 5x - 4$ is tangent to the graph of $y = g(x)$ at the point $(-1, -9)$. Find $g'(-1)$ and an approximation to $g(-0.88)$.

28. The line $y = -2t + 3$ is tangent to the graph of $y = h(t)$ at the point $(3, -3)$. Find $h'(3)$ and an approximation to $h(3.1)$.

29. Let $f(x) = x^3$.
    a. Find an equation for the line tangent to the graph of $y = f(x)$ at the point $(-2, -8)$.
    b. Let $y = t(x)$ be an equation for the tangent line of part a. Suppose we wish to use the approximation
$$f(x) \approx t(x)$$
    for x values near -2. Find r so that when $|x - (-2)| < r$, the error in the approximation is
$$|f(x) - t(x)| < 0.01|x - (-2)|.$$
    (See Example 6 in Section 1.5.)

30. Let $f(x) = 1/\sqrt{x + 1}$.
    a. Find an equation for the line tangent to the graph of $y = f(x)$ at the point $(3, 1/2)$.
    b. Let $y = t(x)$ be an equation for the tangent line of part a. Suppose we wish to use the approximation
$$f(x) \approx t(x)$$
    for x near 3. Find r so that when $|x - 3| < r$, the error in the approximation is
$$|f(x) - t(x)| < 0.01|x - 3|.$$
    (See Example 6 in Section 1.5.)

31. Let f be a function with $f(4) = 3$ and $f'(4) = 7$. Define $h(x) = f(-x)$.
   a. Tell why the point $(-4, 3)$ is on the graph of $y = h(x)$.
   b. What is the value of $h'(-4)$? Justify your answer by comparing the graphs of $y = f(x)$ and $y = h(x) = f(-x)$.
   c. By arguing graphically, tell why if the derivative of $h(x)$ at $x = a$ is defined, then $h'(a) = -f'(-a)$.

32. Let f be a function with $f(4) = 3$ and $f'(4) = 7$. Define $h(x) = -f(x)$.
   a. Tell why the point $(4, -3)$ is on the graph of $y = h(x)$.
   b. What is the value of $h'(4)$? Justify your answer by comparing the graphs of $y = f(x)$ and $y = h(x) = -f(x)$.
   c. By arguing graphically, tell why if the derivative of $h(x)$ at $x = a$ is defined, then $h'(a) = -f'(a)$.

33. Let f be a function with $f(4) = 3$ and $f'(4) = 7$. Define $h(x) = 8f(x)$.
   a. Tell why the point $(4, 24)$ is on the graph of $y = h(x)$.
   b. What is the value of $h'(4)$? Justify your answer by comparing the graphs of $y = f(x)$ and $y = h(x) = 8f(x)$.
   c. Express $h'(a)$ in terms of the derivative of f. Justify your answer with the aid of graphs.

34. Let f be a function with $f(4) = 3$ and $f'(4) = 7$. Define $h(x) = f(x + 7)$.
   a. Tell why the point $(-3, 3)$ is on the graph of $y = h(x)$.
   b. What is the value of $h'(-3)$? Justify your answer by comparing the graphs of $y = f(x)$ and $y = h(x) = f(x + 7)$.
   c. Express $h'(a)$ in terms of the derivative of f. Justify your answer with the aid of graphs.

35. The limit statement
$$\lim_{x \to 2} \frac{(x^9 - 4x^7 + 3x - 2) - 4}{x - 2} = 515$$
is a statement about the derivative of some function f at some value $x = a$. What are f, a, and $f'(a)$?

36. The limit statement
$$\lim_{h \to 0} \frac{\sec(\pi/4 + h) - \sqrt{2}}{h} = \sqrt{2}$$
is a statement about the derivative of some function g at some $t = a$. What are g, a, and $g'(a)$?

37. A function f satisfies
$$f(1) = f(2) = f(3) = f(4) = f(5) = 0$$
$$f'(1) = f'(3) = f'(5) = 1$$
and
$$f'(2) = f'(4) = -1.$$
Use this information to sketch a possible graph for $y = f(x)$.

Review/Preview

38. Reduce each of the following expressions.
   a. $\dfrac{x^2 - a^2}{x - a}$
   b. $\dfrac{x^3 - a^3}{x - a}$
   c. $\dfrac{x^4 - a^4}{x - a}$
   d. $\dfrac{x^5 - a^5}{x - a}$

39. Using the pattern developed in the previous exercise, reduce the expression
$$\frac{x^n - a^n}{x - a}.$$

40. Factor $p(x) = x^3 + 7x^2 + 14x + 6$, given that $p(-3) = 0$.

41. Factor $q(x) = 6x^3 + 55x^2 + 86x - 35$, given that $q(1/3) = 0$.

42. Factor $r(x) = 2x^4 - x^3 - 96x^2 + 62x - 7$ given that $r(1/2) = 0$ and $r(-7) = 0$.

43. Let $p(x) = x^3 + 2x^2 - 3x - 5$. Factor $p(x) - p(3)$.

44. Let $p(x) = -2x^3 + x^2 + 8x + 1$. Factor $p(x) - p(-2)$.

45. Let $p(x)$ be any polynomial of degree at least 1 and a a real number. Why must the polynomial $p(x) - p(a)$ have a factor of $x - a$?

46. If you enter the number 2 in your calculator and then press the square root button 10 times, the result is 2^x for some x. What is x?

47. If you enter the number 5 in your calculator and then press the square root button 15 times and the squaring button 20 times, the result is 5^y for some y. What is y?

48. The figure below shows angles of measure $-\alpha$ and β inscribed in a unit circle. The terminal ray angle of measure $-\alpha$ intersects the unit circle in $P = (\cos(-\alpha), \sin(-\alpha))$, while the terminal ray of the angle of measure β intersects the circle in $Q = (\cos\beta, \sin\beta)$. Hence angle POQ has measure $\alpha + \beta$. Find PQ^2 in two different ways: first using the distance formula, then using the law of cosines. Equate the two expressions and show that they lead to the identity

$$\cos(\alpha + \beta) = \cos\alpha \cos\beta - \sin\alpha \sin\beta$$

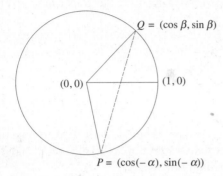

49. Find the value of:
  a. $\log_{10} 100$.
  b. $\log_{10}(1/100)$.
  c. $\log_5 5\sqrt{5}$.
  d. $\log_7 1$.

REVIEW OF KEY CONCEPTS

We began this chapter with a review of functions. We saw that functions can be represented in a number of ways: symbolically (with a formula), graphically, or numerically (with a table of values). Functions represented in these ways arise regularly in science, engineering, and mathematics. Therefore it is important to know how to work with functions in all of these forms, not only with paper and pencil, but also with the aid of a calculator or computer algebra system.

In the following sections we interpreted the slope of a line as a rate of change and used this to motivate a definition for the rate of change of an arbitrary function. We defined the rate of change to be the "slope" of the "line" we see when we zoom in on a function. From this graphical definition we developed a numerical understanding of the rate of change and an analytic (or symbolic) definition for the rate of change. We found that a graphing calculator can be a valuable tool in helping us understand the rate of change.

The rate of change definition led naturally to the idea of limit. Once limit was defined, we reformulated our definition of rate of change in a more precise form. We also saw that limits are very closely related to the notions of approximation and error. Approximation is a very important idea in modern science, engineering, and mathematics. Hence we spent some time discussing approximations and errors and will continue to do so in subsequent chapters.

We concluded the chapter by defining the derivative of a function and noted that the derivative is really just another name for the rate of change. As one geometric application of the derivative, we defined the tangent line and noted that it is really just the "line" we see when we zoom in on a point of a graph. We saw that tangent lines can be used to obtain good approximations to functions.

On the next page, we summarize many of the ideas mentioned above in table form. For each concept we present a general definition and a specific example, and support both with a graph.

Chapter Summary

Representing Functions

A function can be represented by a graph: 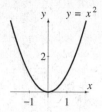	A function f can be represented by an equation: $$f(x) = x^2$$	A function f can be represented by a table of values:

Table for the third column:

x	$f(x)$
0	0
1	1
2	4
3	9

Rate of Change

The rate of change at $x = a$ is the slope of the "line" we see when we zoom in on the point $(a, f(a))$. 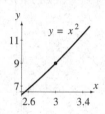	The rate of change of $y = f(x)$ with respect to x at $x = a$ is $$\lim_{h \to 0} \frac{f(a + h) - f(a)}{h}.$$	If $y = f(x) = x^2$, the rate of change on y with respect to x at $x = 3$ is $$\lim_{h \to 0} \frac{(3 + h)^2 - 3^2}{h} = 6.$$

Limit of *f*(*x*) at *a*

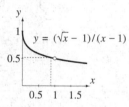	If we can make $f(x)$ as close to L as we like by taking x close to (but not equal to) a, then $$\lim_{x \to a} f(x) = L.$$	$$\lim_{x \to 1} \frac{\sqrt{x} - 1}{x - 1} = \frac{1}{2}$$ As x gets close to 1, $(\sqrt{x}-1)/(x-1)$ gets close to $1/2$:

Table for the third column:

x	$(\sqrt{x} - 1)/(x - 1)$
1.1	0.488088
0.9	0.513167
1.005	0.499377

Limit of *f*(*x*) at *a*

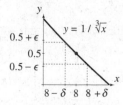	$$\lim_{x \to a} f(x) = L$$ means that if $\epsilon > 0$ is given, then there is a number $\delta > 0$ such that $$|f(x) - L| < \epsilon$$ whenever $$0 < |x - a| < \delta.$$	$$\lim_{x \to 8} \left(1/\sqrt[3]{x}\right) = 1/2$$ Let $\epsilon = 0.05$. Note (by graphing) that when $$0 < |x - 8| < 0.2,$$ we have $$|1/\sqrt[3]{x} - 1/2| < 0.05 = \epsilon.$$ Thus for $\epsilon = 0.05$, we can take $\delta = 0.2$.

Continuous Function

A continuous function has an unbroken graph.

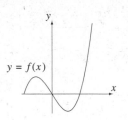

$y = f(x)$

A function f is continuous at $x = a$ if

$$\lim_{x \to a} f(x) = f(a).$$

If a function is continuous at every point in an interval $a \le x \le b$, then on this interval the graph of $y = f(x)$ is an unbroken curve.

Let c be a real number and $p(x)$ a polynomial. Then

$$\lim_{x \to c} p(x) = p(c).$$

Hence every polynomial is continuous on the real line. The sine and cosine functions are also continuous on the real line because for every real number c,

$$\lim_{\theta \to c} \sin \theta = \sin c$$

and

$$\lim_{\theta \to c} \cos \theta = \cos c.$$

The Derivative

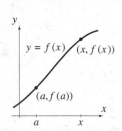

$y = f(x)$, $(x, f(x))$

$(a, f(a))$

The derivative of f at a is

$$f'(a) = \lim_{h \to 0} \frac{f(a + h) - f(a)}{h}$$

provided this limit exists. The derivative at a is also given by

$$f'(a) = \lim_{x \to a} \frac{f(x) - f(a)}{x - a}.$$

Let $f(x) = x^2$. The derivative of f is the function f' defined by

$$f'(x) = \lim_{t \to x} \frac{t^2 - x^2}{t - x} = 2x.$$

The Line Tangent to a Graph

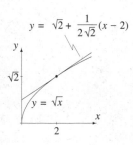

$y = \sqrt{2} + \dfrac{1}{2\sqrt{2}}(x - 2)$

$y = \sqrt{x}$

Suppose that the function f has derivative $f'(a)$ at $x = a$. The line tangent to the graph of $y = f(x)$ at the point $(a, f(a))$ is the line of slope $f'(a)$ that passes through $(a, f(a))$. This line has equation

$$y = f(a) + f'(a)(x - a).$$

The derivative of $f(x) = \sqrt{x}$ at $x = 2$ is

$$f'(2) = \frac{1}{2\sqrt{2}}.$$

The line tangent to the graph of $y = \sqrt{x}$ at the point $(2, \sqrt{2})$ has equation

$$y = \sqrt{2} + \frac{1}{2\sqrt{2}}(x - 2).$$

The Tangent Line Approximation

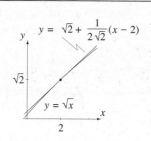

$$y = \sqrt{2} + \frac{1}{2\sqrt{2}}(x - 2)$$

$y = \sqrt{x}$

When x is close to a, the approximation

$$f(x) \approx f(a) + f'(a)(x - a)$$

can be used. The error is

$$|f(x) - (f(a) + f'(a)(x - a))|$$
$$= |E(x)(x - a)|$$

where

$$E(x) = \frac{f(x) - f(a)}{x - a} - f'(a).$$

Note that $E(x)$ is small when x is close to a.

The error in the approximation

$$\sqrt{x} \approx \sqrt{2} + \frac{1}{2\sqrt{2}}(x - 2)$$

is no more than 1 percent of $|x - 2|$ when x is within 0.2 of 2. This is because when $0 < |x - 2| < 0.2$, we have

$$|E(x)| = \left| \frac{\sqrt{x} - \sqrt{2}}{x - 2} - \frac{1}{2\sqrt{2}} \right|$$
$$< 0.01.$$

CHAPTER REVIEW EXERCISES

Exercises 1–6:   The two functions f and g are defined by the tables given below.

x	$f(x)$	x	$g(x)$
0	-3	-3	0
1	π	-0.5	4.5
2	$\sin 1$	0	4
3	0	2	$-\pi$
4	3	3	3
5	12.4		

1. State the domain of f and the domain of g.
2. What is the range of f? Of g?
3. What is the domain of the function $f + g$? Construct a table for $f + g$.
4. What is the domain of the function $\sqrt{g}$? Construct a table for $\sqrt{g}$.

5. What is the domain of the function f/g? Construct a table for f/g.
6. What is the domain of the function $f \circ g$? Construct a table for $f \circ g$.

Exercises 7–14:   The two functions f and g are defined by the graphs given below.

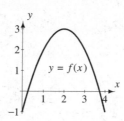

$y = f(x)$

$y = g(x)$

7. State the domain of f and the domain of g.
8. What is the range of f? Of g?

9. What is the domain of the function $f + g$? Draw a graph of $f + g$.

10. What is the domain of the function $\sqrt{f}$? Draw a graph of $\sqrt{f}$.

11. What is the domain of the function g/f?

12. What is the domain of the function $f \circ g$? Estimate the value of $(f \circ g)(1)$. Of $(f \circ g)(2)$.

15. In Section 1.5 we discussed the formula

$$L = L_0 \sqrt{1 - \left(\frac{v}{c}\right)^2},$$

which gives the length L of an object moving at speed v if its length at rest is L_0. Assume that L is measured in meters and v in centimeters/second. What are the units of the rate of change of L with respect to v? Explain carefully the physical interpretation of this rate of change.

16. The speed of sound in air at 1 atm pressure and $0°C$ is approximately 1090 ft/s. A hiker yells and waits to hear her echo from a mountain d feet across a valley.
   a. Find a formula for the length of time T it takes for the echo to return to the hiker.
   b. What is the rate of change of T with respect to d? What are the units of this rate of change?
   c. Explain carefully the physical interpretation of this rate of change.

13. Estimate the rate of change of f with respect to x at $x = 1$. Explain how you arrived at your answer.

14. Estimate the rate of change of g with respect to x at $x = 3$. Explain how you arrived at your answer.

17. The inflation rate tells how much the average cost of goods changes in a year. For example, if the rate of inflation is 3 percent for a year, then a product that cost \$1.00 at the beginning of the year would cost, on average, \$1.03 at the end of the year. At the end of every month the United States government publishes the *annual* inflation rate for that month. But these annual rates often change from month to month! For example, in March 1994 the annual inflation rate was 3.5 percent, in April it was 3.2 percent, and in May it was 4.1 percent. Interpret these figures as rates of change and explain how the *annual* inflation rate can change every month.

Exercises 18–20:   Investigate the limit (*a*) numerically, (*b*) graphically, and (*c*) analytically.

18. $\lim\limits_{x \to 2} x^2 - 3x + 2$

19. $\lim\limits_{x \to -1} \dfrac{x^2 - 1}{x + 1}$

20. $\lim\limits_{t \to 3} \dfrac{\sqrt{t + 3} - \sqrt{6}}{t - 3}$

Exercises 21–23:   Investigate the limit (*a*) numerically and (*b*) graphically.

21. $\lim\limits_{\theta \to \pi/3} \dfrac{\cos \theta - 1/2}{\theta - \pi/3}$

22. $\lim\limits_{r \to 3} \dfrac{10^r - 1000}{r - 3}$

23. $\lim\limits_{x \to 10} \dfrac{\log_{10} x - 1}{x - 10}$

24. Let L be the value of the limit in Exercise 19. Find a $\delta > 0$ so that when

$$0 < |x - (-1)| < \delta$$

we have $\left| \dfrac{x^2 - 1}{x + 1} - L \right| < 0.01.$

25. Let L be the value of the limit in Exercise 21. Find a $\delta > 0$ so that when

$$0 < |\theta - \pi/3| < \delta$$

we have $\left| \dfrac{\cos \theta - 1/2}{\theta - \pi/3} - L \right| < 0.01.$

26. Let L be the value of the limit in Exercise 23. Find a $\delta > 0$ so that when

$$0 < |x - 10| < \delta$$

we have $\left| \dfrac{\log_{10} x - 1}{x - 10} - L \right| < 0.01.$

Exercises 27–30: Find $f'(x)$ for the given f.

27. $f(x) = 2x - 5$
28. $f(x) = -3x^3 + 4x$

29. $f(x) = x + \dfrac{1}{x}$

30. $f(x) = (x + a)^2$, a constant.

31. Find the equation for the line tangent to $y = \sqrt{x + 2} - 1$ at the point $(7, 2)$. Find $r > 0$ so that for $7 - r < x < 7 + r$ the error in the tangent line approximation to $\sqrt{x + 2} - 1$ is no more than 0.001 in absolute value.

32. Use graphical or numerical methods to find the derivative of $f(x) = \cos x$ at $x = \pi/2$. Find an equation for the line tangent to $y = \cos x$ at the point $(\pi/2, 1)$. Find $r > 0$ so that for $\pi/2 - r < x < \pi/2 + r$ the error in the tangent line approximation to $\cos x$ is no more than 0.005 in absolute value.

33. Write a short paragraph that clearly tells what is meant by the statement

$$\lim_{x \to a} f(x) \quad \text{does not exist.}$$

Illustrate your explanation with some graphs.

34. In Section 1.6 we showed that if a function f has a derivative at $x = a$, then it is continuous at $x = a$. Is the converse true? That is, if a function f is continuous at $x = a$, must it have a derivative at $x = a$? Justify your answer.

35. In Section 1.6 we showed that if a function f has a derivative at $x = a$, then it is continuous at $x = a$. Explain why if f is not continuous at $x = a$, then we will never see a "straight line" when we zoom in on the graph of $y = f(x)$ at the point $(a, f(a))$.

Student Project

Exploring Chaos

What Is Chaos?   Meteorologists say that "a butterfly flapping its wings in China can cause a tornado in Kansas several weeks later." Although it is very unlikely that the truth of this statement will ever actually be demonstrated, the statement does illustrate a problem with long-range weather prediction. Very small disturbances in the atmosphere can, over time, lead to large, noticeable phenomena. To accurately predict the weather a month from today, meteorologists would need precise information (pressure, humidity, velocity, temperature, etc.) now, and they would need this information about every cubic foot of the atmosphere. And even if it were possible to get information on this scale, small errors in some of these measurements might be enough to make weather patterns differ drastically from predictions.

The atmosphere is an example of a *chaotic system*. A chaotic system is one in which very small changes made in one part of the system can lead to very big changes in another part. If the butterfly in China does not flap its wings at 2:00 P.M. on April 21, 1996, then it will be a nice day in Kansas on May 30. But if the butterfly does flap its wings, this small change in the atmosphere will propagate and eventually lead to a tornado on May 30. In recent years mathematicians, scientists, and engineers have come to realize that many natural phenomena once thought to be well understood are really chaotic systems.

Compositions and Chaos.   In this project we will look at a fascinating chaotic system that arises by composing the polynomial

$$p(x) = cx(1 - x)$$

repeatedly with itself. Pick a number c with $0 < c < 4$. The system we examine will be the list of numbers we get by setting $x = 0.3$ and computing

(1) $$p(0.3), \ p(p(0.3)), \ p(p(p(0.3))), \ p(p(p(p(0.3)))), \ \ldots.$$

(The choice $x = 0.3$ is arbitrary. Any x with $0 < x < 1$ would work as well.) In constructing (1) we use $x = 0.3$ as input for the polynomial p. We get output $p(0.3)$. We then use this output as input for p, and so get as a next output $p(p(0.3))$. Now we use this output as input for p, and so on. Repeat this process several times. What happens to the list (1)? We will see that the answer depends on the value of c, and that very small changes in c can cause substantial changes in the list (1).

Problem 1. Let $c = 2.1$, so

$$p(x) = 2.1x(1 - x).$$

Use a calculator or CAS to compute the first 10 terms of the list (1). Describe what happens to the numbers in the list. Now let $c = 2.5$, so

$$p(x) = 2.5x(1 - x).$$

Compute the first 15 terms of the list (1). Describe what happens to the number in the new list. How is the new list similar to the previous list? How is it different?

Problem 2. The number 0.52381 should be familiar from Problem 1. Again setting $c = 2.1$, compute $p(0.52381)$. The number 0.52381 is called a *fixed point* for $p(x)$. Explain why this terminology is appropriate. The exact value of the fixed point can be found by solving the equation

$$p(x) = x.$$

Explain why a solution of this equation is a fixed point of p. Find the exact value of the fixed point that is approximated by 0.52381. Repeat for the number 0.6 with $c = 2.5$.

Problem 3. Now let $c = 3.4$ so

$$p(x) = 3.4x(1 - x).$$

Use a calculator or CAS to compute the first 100 terms of the list (1). Describe what happens to the numbers in the list. Repeat with $c = 3.5$, $c = 3.83$, $c = 3.845$, and $c = 3.862$. How do the lists produced differ?

Problem 4. In the previous problem you found that the values in the list eventually start to repeat. For $c = 3.4$ the list (1) eventually settled down to alternate between the two values 0.451963 and 0.842154. Explain why these two numbers are fixed points of the polynomial $p(p(x))$. Explain why these two numbers must be (approximate) solutions to the equation

$$p(p(x)) = x.$$

For $c = 3.5$, $c = 3.83$, and $c = 3.845$ list (1) also settled down to cycle repeatedly through a few values. What are these values? For which polynomial are these values fixed points?

Where's the Chaos?   At the time of this writing there does not seem to be a universally accepted definition of chaos. However, most of the proposed definitions state that if there is chaos, then it must be true that small changes in the input to a system can result in large changes in the output. We saw hints of such behavior in the previous problems. In working these problems we found that the behavior of list (1) changes as c changes. Sometimes the list settles down to one number; sometimes it cycles repeatedly between 2, 3, 4, or 6 numbers; and sometimes there appears to be no eventual pattern. In addition, we saw that in some cases a very small change in c substantially changed the behavior of the list. Thus the list (1) exhibits chaos. In particular, very small changes in the number c can change the list from one that cycles through a few numbers to one with no apparent pattern. Indeed, it can be shown that for any positive integer n there is a value of c between 0 and 4 such that the list (1) eventually settles down to cycle through n different values. In addition, as c gets closer and closer to 4 (but remains less than 4), it takes smaller and smaller changes in c to effect substantial changes in (1).

Graphing the Composition List.   By graphing, we can obtain a different view of how list (1) differs for different values of c. Since (1) consists of compositions, we can graph the compositions as we did in Section 1.1. For example, take $c = 2.5$, and then draw the graph of $y = p(x)$ and the line $y = x$. See Fig. 1.77. Starting from 0.3 on the x-axis, move up to the graph of p. We meet the graph of p at the point $(0.3, p(0.3))$. Change the output $p(0.3)$ to input by moving horizontally to the line $y = x$. We meet the line in the point $(p(0.3), p(0.3))$. Next move vertically to the graph of p again. We meet the graph in a point with y-coordinate $p(p(0.3))$. Continue the process just described, moving horizontally to the line $y = x$, and then vertically to the graph of p. Each time we meet p, the y-coordinate is the next element in the list (1). A programmable graphing calculator or CAS can be programmed to produce the resulting composition diagram very quickly. See Figs. 1.77 and 1.78.

For another way to investigate the behavior of (1), form the collection of ordered pairs

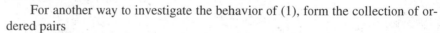

$$(0, 0.3), (1, p(0.3)), (2, p(p(0.3))), (3, p(p(p(0.3)))), \ldots,$$

where the first coordinate of an ordered pair tells how many times p was used in obtaining the second coordinate. When we plot several of these points (say, the first 200), we can get an idea of whether or not (1) settles down to cycle through some values. See Figs. 1.79 and 1.80.

Problem 5.   Generate graphs like those shown in Figs. 1.77–80 for three other values of c.

A Picture of Chaos.   To get a better picture of the chaotic nature of (1), we collect data on the behavior of the list for many values of c. For a given value of c, compute the list (1) until it seems to start cycling. For each number b in a cycle, construct the ordered pair (c, b). For example, when $c = 3.5$, list (1) eventually cycles through

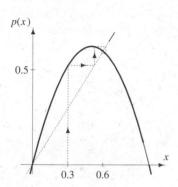

Figure 1.77. When $c = 2.5$, the list of compositions quickly closes in on 0.6.

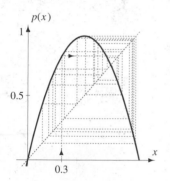

Figure 1.78. When $c = 3.862$, the list of compositions does not seem to enter a cycle.

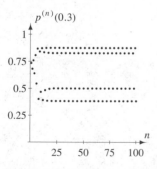

Figure 1.79. When $c = 3.5$, the list of compositions soon enters a cycle of length 4.

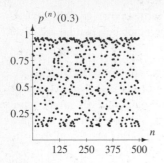

Figure 1.80. When $c = 3.862$, the list of compositions does not appear to enter a cycle.

the four numbers 0.38282, 0.826941, 0.500884, and 0.874997. Thus we collect the four ordered pairs

$$(3.5, 0.38282), (3.5, 0.826941), (3.5, 0.500884), \text{ and } (3.5, 0.874997).$$

Do this for hundreds of values of c between 2 and 4 and then plot all of the points collected. Of course there are many values of c for which (1) does not cycle, and others for which the cycle values become apparent only after many, many iterations. However, if we carry the list (1) to 1000 places for each of several hundred values of c, collect the last 100 entries in each list, and then use these 100 entries as the second coordinate in ordered pairs with first coordinate c we obtain the graph shown in Fig. 1.81.

In this graph the c values run along the horizontal axis. For a given c value the point or points above indicate entries 901 through 1000 in (1). In this picture we can see evidence of c values that result in cycles of length 2, 4, 8, and 3 as well as other c values that suggest more erratic behavior. This intriguing picture is also a fractal. Draw any small square in the graph with its right edge on the line $c = 4$. If we were to magnify the small portion of the graph inside the square we would see a picture very similar to the original graph.

For more about the interesting behavior of (1) and about the dynamics of iterations of simple maps, see *Chaos and Fractals*, edited by Robert Devaney and Linda Keen and published by the American Mathematical Society.

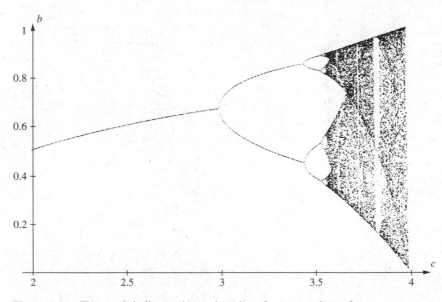

Figure 1.81. The graph indicates the cycle values for some values of c.

A History of Measuring and Calculating Tools
Napier's Bones

5	6	9	Index
1/0	1/2	1/8	2
1/5	1/8	2/7	3
2/0	2/4	3/6	4
2/5	3/0	4/5	5
3/0	3/6	5/4	6
3/5	4/2	6/3	7
4/0	4/8	7/2	8
4/5	5/4	8/1	9

Napier's Bones (also called Napier's Rods) were invented by John Napier in 1617. In the early 1600s scholars of arithmetic were looking for mechanical devices to facilitate calculation. Napier's Bones were an easy-to-use and easy-to-construct device for helping with multiplication, and remained popular for over a half century after their invention. (See the photo below.) Napier's Bones consisted of a collection of long rectangular rods, each divided into nine cells. The nine cells of a rod contained the first nine multiples of the number in the first cell. Each cell was divided in half by a diagonal that ran from the upper left to the lower right of the cell and separated the tens digit from the units digit. (See the accompanying figure.)

To multiply two numbers, one of the numbers is formed by placing rods with the proper top digits side by side, and an "index" with the numbers 2–9 is laid to the right of these rods. The figure shows the index laid to the right of rods that display the number 569. To multiply 569 by, say, 384, start by finding 4 on the index. The result of multiplying 569 by 4 is found by recording the numbers on the rods to the left of 4 after adding adjacent digits on adjacent rods, to give 2276. (The 7 is results from adding the 4 and 3 on the "6" and "9" rods, and the 2 in the hundreds place results from adding the 0 and 2 on the "5" and "6" rods.) Next we use the rods to multiply 569 by 8, to get 4552. (The 5 in the hundreds place results from adding the 0 and 4 on the "5" and "6" rods and the 1 carried from the addition that gives the tens digit.) Next the product of 569 and 3 is read as 1707. The results of these products read from the rods are recorded as shown below and added to get the product.

$$
\begin{array}{ccccccc}
 & &   & 5 & 6 & 9 \\
 & &   & 3 & 8 & 4 \\
\hline
 & & 2 & 2 & 7 & 6 \\
 & 4 & 5 & 5 & 2 & \\
1 & 7 & 0 & 7 & & \\
\hline
2 & 1 & 8 & 4 & 9 & 6 \\
\end{array}
$$

Chapter 2

Finding the Derivative

The Derivatives of Elementary Functions

In Chapter 1 we looked at many different ways to find or estimate the rate of change of a function. We saw that the derivative (or rate of change) of a function f can be estimated by zooming in on the graph of $y = f(x)$ and reading off the slope of the "line" we see. From this we learned that a differentiable function looks like a line when we get close enough, and that we can approximate such a function with a linear function. We also saw that the derivative of f can be estimated numerically by calculating a ratio of the form

$$(1) \qquad \frac{f(x) - f(a)}{x - a}$$

for values of x close to a. This approach strongly suggests the derivative as a measure of "the rate of change in f with respect to x." The units of (1) are

$$\frac{f \text{ units}}{x \text{ units}}.$$

Knowing these units helps us interpret the derivative in the context of the problem in which it arises. The analytical approach

$$f'(x) = \lim_{t \to x} \frac{f(t) - f(x)}{t - x}$$

puts the other two approaches on a sound mathematical basis and gives us a way to calculate the derivative as a function. All three of these approaches give important insights into the meaning and interpretation of the derivative and hence into the understanding of physical situations that can be described by a function.

In spite of the important ideas conveyed by each of these approaches, in this chapter we will study a more mechanical approach to differentiation. The derivative is a widely used tool in science, engineering, and mathematics. Often, applications of the derivatives are as routine as, say, performing a multiplication. For this reason, it is important that we be able to calculate derivatives quickly and efficiently. That is the goal of this chapter.

In Sections 2.1, 2.2, and 2.3 we learn how to find the derivative of a function that is made from other functions whose derivatives we do know. If we know the derivatives of f and g, what is the derivative of $f + g$? of $f \cdot g$? of $f \circ g$?

In Sections 2.4, 2.5, 2.6, and 2.7 we find the derivatives of the building blocks for almost all of the functions that we will encounter. These building blocks are the so-called elementary functions: the power functions, the sine and cosine functions, the exponential and logarithm functions, and the inverse trigonometric functions. Once we know (i.e., memorize!) the derivatives of these few functions and know how to find the derivatives of functions built from these pieces, we can find the derivative of almost any function we encounter.

In Section 2.8 we take a look at several mathematical models. These models illustrate the many and diverse situations that can be described by derivatives. In addition, they remind us that the derivative is much more than just an algorithm to manipulate formulas. It is an important and useful tool for describing the world around us.

Computer algebra systems and many calculators have routines for finding derivatives. But with practice, you will find the calculation of derivatives to be a straightforward mechanical process that in many cases is done more quickly by hand than by machine. Nonetheless, get to know the differentiation routines on your computer or calculator. These routines will be very useful when finding a derivative is just a minor procedure in solving a much bigger problem that is being done with the aid of technology. In the chapter we will concentrate on finding derivatives by hand. However, as part of getting to know your calculator or computer, get in the habit of checking some of your differentiation answers by machine.

2.1   DERIVATIVES OF POLYNOMIALS

Polynomial functions are among the most often seen functions in science, engineering, and mathematics. Every nonvertical line can be described by a first-degree polynomial. The force of gravity and the intensity of a light source are described by inverse square laws, and so both involve second-degree polynomials. When we study Taylor series, we will see that many nonpolynomial functions, such as the sine function and the logarithm function, can be approximated to any desired accuracy by polynomial functions.

Polynomials

A polynomial function is a function p that can be described by

$$(1) \qquad p(x) = a_n x^n + a_{n-1} x^{n-1} + a_{n-2} x^{n-2} + \cdots + a_1 x + a_0,$$

where n is a positive integer and the coefficients $a_n, a_{n-1}, a_{n-2}, \ldots, a_1, a_0$ are real numbers. The degree of p is the largest exponent on x. Hence, if $a_n \neq 0$, then p is a polynomial of degree n. Some explicit examples are

$$p(x) = -4x^5 + 3x^4 - 7x^2 + x - 2 \quad \text{and} \quad q(x) = -x^3 + \sqrt{2}x^2 + \pi x - 13.$$

Polynomial p has degree 5, and polynomial q has degree 3.

The basic building blocks of polynomials are the power functions, described by

$$1, x, x^2, x^3, \ldots.$$

Polynomials are formed by multiplying some of these power functions by constants and then adding the resulting products together. Thus we can find the derivative of any polynomial if we know three things:

1. The derivatives of the power functions.
2. How to find the derivative of a constant multiple of a function given the derivative of the function.
3. How to find the derivative of a sum of functions given the derivative of each function in the sum.

The Derivative of x^n

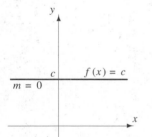

Figure 2.1. The graph of a constant function is a line of slope 0.

In Chapter 1 we saw that the derivative of a function f at a point $x = a$ is the slope of the "line" that we see as we zoom in on the graph of f near the point $(a, f(a))$. If f is a constant function,

$$(2) \qquad\qquad f(x) = c$$

for some real number c, then we always see a horizontal line (of slope 0) when we zoom in on the graph of $y = f(x)$. See Fig. 2.1. This means that the constant function (2) has derivative 0:

$$\frac{dy}{dx} = f'(x) = (c)' = 0.$$

If g is the function defined by

$$g(x) = x,$$

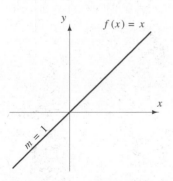

Figure 2.2. The graph $y = x$ is a line of slope 1.

then the graph of $y = g(x)$ is a line of slope 1. See Fig. 2.2. When we zoom in on the graph at any point, we see a line of slope 1. Hence

$$g'(x) = (x)' = 1.$$

We summarize these results below.

> **Derivatives of c and x**
>
> The constant function f defined by $y = f(x) = c$ has derivative
>
> $$\frac{dy}{dx} = f'(x) = 0$$
>
> for any real number x.
>   The function g given by $y = g(x) = x$ has derivative
>
> $$\frac{dy}{dx} = g'(x) = 1$$
>
> for any real number x.

Now let n be a positive intger with $n \geq 2$. When we zoom in on the graph of $y = x^n$ at different points, we see "lines" of different slope. Thus zooming in alone is not enough to determine the derivative of x^n. We use the definition of derivative given in Section 1.6.

DEFINITION   derivative

Let f be a function and let x be in the domain of f. The derivative of f at the point x is

$$(3) \qquad f'(x) = \lim_{t \to x} \frac{f(t) - f(x)}{t - x}$$

provided that this limit exists.

Applying (3) with $f(x) = x^n$, we find

$$(4) \qquad (x^n)' = f'(x) = \lim_{t \to x} \frac{t^n - x^n}{t - x}$$

As t approaches x, both the numerator and the denominator in (4) approach 0. Hence we simplify the quotient in (4) before evaluating the limit. To do this, we recall some factorization formulas. We are familiar with the difference-of-squares factorization

$$(5) \qquad t^2 - x^2 = (t - x)(t + x)$$

and the difference-of-cubes factorization

$$(6) \qquad t^3 - x^3 = (t - x)(t^2 + tx + x^2).$$

It is not hard to verify a difference-of-fourth-powers factorization

$$(7) \qquad t^4 - x^4 = (t - x)(t^3 + t^2x + tx^2 + x^3).$$

If n is a positive integer, then the polynomial $t^n - x^n$ is 0 when $t = x$. This means that the polynomial has a factor of $t - x$. From the pattern established in (5), (6), and (7), we can make a good guess at the other factor:

$$(8) \qquad t^n - x^n = (t - x)\underbrace{(t^{n-1} + t^{n-2}x + t^{n-3}x^2 + \cdots + tx^{n-2} + x^{n-1})}_{n \text{ terms}}.$$

This is the difference-of-nth-powers factorization formula. This formula can be verified by multiplication of the terms on the right. See Exercise 20.

Now use (8) to factor the numerator in (4). This leads to

$$(9) \qquad \begin{aligned} f'(x) = (x^n)' &= \lim_{t \to x} \frac{(t - x)(t^{n-1} + t^{n-2}x + t^{n-3}x^2 + \cdots + tx^{n-2} + x^{n-1})}{t - x} \\ &= \lim_{t \to x}(t^{n-1} + t^{n-2}x + t^{n-3}x^2 + \cdots + tx^{n-2} + x^{n-1}). \end{aligned}$$

If we think of x as a constant, then

$$t^{n-1} + t^{n-2}x + \cdots + tx^{n-2} + x^{n-1}$$

is a polynomial in variable t. Hence we can evaluate this last limit by substituting x for t to get

$$\begin{aligned} f'(x) = (x^n)' &= \underbrace{(x^{n-1} + x^{n-2}x + x^{n-3}x^2 + \cdots + xx^{n-2} + x^{n-1})}_{n \text{ terms}} \\ &= nx^{n-1}. \end{aligned}$$

We have proved the following important result.

Derivative of the Power Functions

If n is a positive integer, and f is defined by $y = f(x) = x^n$, then

$$\frac{dy}{dx} = f'(x) = (x^n)' = nx^{n-1}.$$

Multiplication by a Constant

When a function f is multiplied by a constant c, the derivative of the product, cf, is the product of c and the derivative of f.

Derivative of a Constant Multiple of a Function

If f is a function and c a real number, then

$$(10) \qquad \frac{d}{dx}\big(cf(x)\big) = \big(cf(x)\big)' = cf'(x).$$

Though this result is not hard to prove using the definition of derivative (see Exercise 22), we give here a graphical argument. Suppose that $f'(a) = m$. Then when we zoom in on the graph of f near the point $P = (a, f(a))$, we see a "line" of slope m. This means that if $Q = (a + h, f(a + h))$ is a nearby point, then

$$\frac{f(a + h) - f(a)}{h} \approx m.$$

Hence the vertical difference between P and Q is

$$f(a + h) - f(a) \approx mh.$$

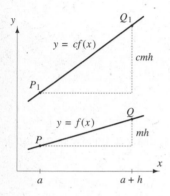

See Fig. 2.3. Now consider the graph of cf, and zoom in on the graph near the point $P_1 = (a, cf(a))$. Again we see a "line," and the slope of this line is approximated by the slope of $\overline{P_1 Q_1}$, where $Q_1 = (a + h, cf(a + h))$. This slope is

$$\frac{cf(a + h) - cf(a)}{h} = \frac{c(f(a + h) - f(a))}{h} \approx \frac{cmh}{h} = cm.$$

This slope is the derivative of cf at $x = a$. Therefore,

$$(cf(x))'|_{x=a} = cm = cf'(a).$$

Figure 2.3. Zooming in on graphs of f and cf.

Now we can quickly write down the derivative of any function of the form $f(x) = cx^n$ when n is a positive integer.

■ **EXAMPLE 1**   Find the derivative of $f(x) = -7x^{20}$.

Solution.   The function given by $-7x^{20}$ is the power function x^{20} multiplied by the constant -7. Thus by (10),

$$(-7x^{20})' = (-7)(x^{20})' = -7(20x^{19}) = -140x^{19}. \qquad ■$$

■ **EXAMPLE 2**   The point $(-0.2, 0.35)$ is on the graph of $y = f(x)$, and the slope of the tangent to the graph at this point is 1.6. Find the equation for the line tangent to the graph of $y = 3f(x)$ at the point where $x = -0.2$.

Solution.   Since $(-0.2, 0.35)$ is on the graph of f, we have $f(-0.2) = 0.35$. Then

$$(3f)(-0.2) = 3f(-0.2) = 1.05,$$

so $(-0.2, 1.05)$ is on the graph of $y = 3f(x)$. The slope of the line tangent to the graph at this point is

$$(3f)'(-0.2) = 3f'(-0.2) = 3(1.6) = 4.8.$$

Since the tangent line contains the point of tangency, $(-0.2, 1.05)$, an equation for the line is

$$y - 1.05 = 4.8(x + 0.2). \qquad ■$$

The Derivative of a Sum

When two functions, f and g, are added, the derivative of the sum $f + g$ is the sum of the derivative of f and the derivative of g.

Derivative of a Sum

Let f and g be differentiable functions. Then

(11) $$(f + g)'(x) = f'(x) + g'(x).$$

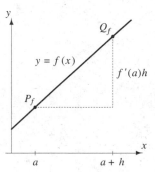

Graph of f. Here $f'(a) > 0$.

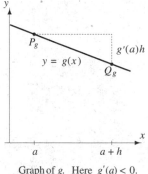

Graph of g. Here $g'(a) < 0$.

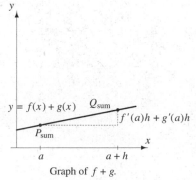

Graph of $f + g$.

Figure 2.4. The derivative of the sum of two functions is the sum of the derivatives of the functions.

We will look at this result in several ways. We do this to demonstrate that a problem can have several solutions and to illustrate different ways of interpreting the derivative. First we relate the formula to a physical situation.

Two hoses, a big one and a little one, are used to fill a swimming pool. The water is turned on at the time $t = 0$. Let $H(t)$, in ft^3, be the volume of water that flows from the big hose during the first t hours, and let $h(t)$ be the volume of water that flows from the little hose in the first t hours. Then

$$H'(t) \text{ has units of ft}^3/\text{h}$$

and is the rate at which water is flowing from the big hose at time t. A similar statement holds for $h'(t)$.

If the big hose alone is used to fill the pool, then at the time t the pool has $H(t)$ ft^3 of water and the amount of water in the pool is changing at a rate of $H'(t)$ ft^3/h. If the little hose alone is used, then at the time t the pool has $h(t)$ ft^3 of water and the volume of water in the pool is changing at a rate of $h'(t)$ ft^3/h.

If both hoses are used simultaneously, then at time t the volume of water in the pool is

(12) $$(H + h)(t) = H(t) + h(t) \text{ ft}^3.$$

The derivative of (12),

(13) $$(H + h)'(t) \text{ ft}^3/\text{h},$$

is the rate at which the volume of water in the pool is changing at time t. On the other hand, common sense tells us that the rate of change of water in the pool is simply the sum of the rate at which water enters from each hose,

(14) $$H'(t) + h'(t) \text{ ft}^3/\text{h}.$$

Since (13) and (14) represent the same rate of change, the two expressions are equal. Hence

$$(H + h)'(t) = H'(t) + h'(t).$$

Next we illustrate (11) by zooming in. If we zoom in on the graph of $y = f(x)$ near the point $P_f = (a, f(a))$, we see a "line" of slope $f'(a)$. See Fig. 2.4. If

$Q_f = (a + h, f(a + h))$ is a nearby point on the graph, then

$$\frac{f(a + h) - f(a)}{h} \approx f'(a).$$

Hence the vertical separation between P_f and Q_f is

$$f(a + h) - f(a) \approx f'(a)h.$$

If we zoom in on the graph of $y = g(x)$ near $P_g = (a, g(a))$, we see that the vertical separation between this point and the nearby point $Q_g = (a + h, g(a + h))$ is

$$g(a + h) - g(a) \approx g'(a)h.$$

When we zoom in on the graph of $y = (f + g)(x)$ near $P_{sum} = (a, (f + g)(a))$, we then see that the vertical separation between

$$P_{sum} = (a, (f + g)(a)) \quad \text{and} \quad Q_{sum} = (a + h, (f + g)(a + h))$$

is

$$(f + g)(a + h) - (f + g)(a) = (f(a + h) - f(a)) + (g(a + h) - g(a)).$$
$$\approx f'(a)h + g'(a)h.$$

Thus when we zoom we see a "line" with the following slope:

$$(15) \qquad \text{slope of } \overline{P_{sum}Q_{sum}} = \frac{\text{rise}}{\text{run}} \approx \frac{f'(a)h + g'(a)h}{h} = f'(a) + g'(a).$$

The slope of this line is the derivative of $f + g$ at $x = a$. Hence we also know that

$$(16) \qquad \text{slope of } \overline{P_{sum}Q_{sum}} \approx (f + g)'(a).$$

Equating the expressions in (15) and (16), we find

$$(f + g)'(a) = f'(a) + g'(a).$$

This again verifies (11).

For a third way of justifying (11), we use the definition of derivative. Because f and g are differentiable, we have

$$(17) \qquad f'(x) = \lim_{t \to x} \frac{f(t) - f(x)}{t - x} \quad \text{and} \quad g'(x) = \lim_{t \to x} \frac{g(t) - g(x)}{t - x}.$$

The derivative of $f + g$ is given by

$$(18) \qquad (f + g)'(x) = \lim_{t \to x} \frac{(f + g)(t) - (f + g)(x)}{t - x}$$

provided this limit exists. To evaluate this expression, we use the results for manipulating limits given in Section 1.4. Using these results, we obtain

$$\lim_{t \to x} \frac{(f + g)(t) - (f + g)(x)}{t - x} = \lim_{t \to x} \frac{(f(t) + g(t)) - (f(x) + g(x))}{t - x}$$

$$= \lim_{t \to x} \left(\frac{f(t) - f(x)}{t - x} + \frac{g(t) - g(x)}{t - x} \right)$$

$$= \lim_{t \to x} \frac{f(t) - f(x)}{t - x} + \lim_{t \to x} \frac{g(t) - g(x)}{t - x}$$

$$= f'(x) + g'(x),$$

where the last equality follows from (17). This argument shows that the limit in (18) exists and is equal to $f'(x) + g'(x)$. This again verifies (11). This last argument, the hose argument, or the graphical argument can all be extended to sums of three or more functions. Hence

$$(f(x) + g(x) + h(x))' = f'(x) + g'(x) + h'(x),$$

and analogous formulas are true for sums of four, five, or more functions.

The results of this section may seem obvious, but they are nonetheless important and will be used over and over. As an immediate application we can now differentiate any polynomial.

■ **EXAMPLE 3**   Find the derivative of $p(x) = 4x^5 + 7x^3 - 3x^2 + 6$.

Solution.   First write the derivative as a sum of derivatives:

$$p'(x) = (4x^5 + 7x^3 - 3x^2 + 6)' = (4x^5)' + (7x^3)' + (-3x^2)' + (6)'.$$

Next factor the constant multiples out of the derivatives:

$$p'(x) = 4(x^5)' + 7(x^3)' - 3(x^2)' + (6)'.$$

Next use the formulas for the derivative of a power function and the fact that a constant function has derivative 0:

$$p'(x) = 4(5x^4) + 7(3x^2) - 3(2x^1) + 0 = 20x^4 + 21x^2 - 6x. \qquad ■$$

■ **EXAMPLE 4**   True or false: The rate of change of the average of a finite collection of functions is equal to the average of the rate of change of the functions.

Solution.   This statement is true for all domain values where the derivatives of all of the functions are defined. We demonstrate for the case of three functions f, g, and h. Let $\mathcal{D}_{f'}$ be the domain of f', that is, the set of all x values for which $f'(x)$ is defined. Likewise, let $\mathcal{D}_{g'}$ and $\mathcal{D}_{h'}$ denote the domains of g' and h', respectively. The average of the three functions is given by

$$A(x) = \frac{f(x) + g(x) + h(x)}{3}$$

and has a derivative at all points in $\mathcal{D}_{f'} \cap \mathcal{D}_{g'} \cap \mathcal{D}_{h'}$. Furthermore, the rate of change of the average is

$$A'(x) = \left(\frac{1}{3}f(x) + \frac{1}{3}g(x) + \frac{1}{3}h(x)\right)'.$$

Since the derivative of a sum of functions is the sum of the derivatives of the functions, this becomes

$$= \left(\frac{1}{3}f(x)\right)' + \left(\frac{1}{3}g(x)\right)' + \left(\frac{1}{3}h(x)\right)'.$$

Next note that $((1/3)f(x))' = (1/3)f'(x)$, with analogous results for the other two terms above. Hence we have

$$A'(x) = \frac{1}{3}f'(x) + \frac{1}{3}g'(x) + \frac{1}{3}h'(x)$$

$$= \frac{f'(x) + g'(x) + h'(x)}{3},$$

which is the average of the rates of change. ∎

In the solutions to Examples 3 and 4 we took great care to indicate where we were applying the results about derivatives of sums and derivatives of constant multiples. Though it is important to recognize when we are applying these results, in the future we will usually omit the details of these applications.

Exercises 2.1

Basic

Exercises 1–8: Find the derivative.

1. $f(x) = 4x - 7$

2. $g(t) = 3t^3 - 2t + \frac{1}{2}$

3. $y = \frac{1}{4}x^4 + \frac{1}{3}x^3 + \frac{1}{2}x^2 + x + 1$

4. $h(r) = ar^3 + br^2 + cr + d$, where a, b, c, d are constants.

5. $H(s) = \pi^2$

6. $s = t^{n+3}$, where n is a nonnegative integer.

7. $f(x) = (x + 1)(2x + 1)$

8. $y = 16(t + 2)^3$

9. Find the derivative of f where $f(x) = \frac{1}{3}x^3$. Then find a different function with the same derivative as f.

10. Find the derivative of g where $g(x) = -3x + 8$. Then find a different function with the same derivative as g.

11. Let $y = x^4$. Find the derivative of y with respect to x. Then write down a function that has derivative $\frac{1}{3}x^3$.

12. Let n be a positive integer. Find the derivative of f where $f(x) = x^{n+1}$. Then write down two different functions that have derivative x^n.

Exercises 13–15: Let f, g, and h be functions with $f'(3) = -4$, $g'(3) = 7$, and $h'(3) = 1$. Find $s'(3)$ if:

13. $s(x) = 3f(x) - 2g(x) + \frac{1}{2}h(x)$.

14. $f(x) = 2(g(x) + h(x)) - 3s(x)$.

15. $f(x) + g(x) + h(x) + s(x) = 20\sqrt{3}$.

Exercises 16–19:   Write an equation for the line tangent to the graph of the function at the given point.

16. $f(x) = 2x^2 - 3x + 4$ at $x = 2$

17. $h(t) = 2\sqrt{2}$ at $t = -\sqrt{5}$

18. $g(x) = ax^2 + bx + c$ at $t = d$ $(a, b, c$ constants)

19. $f(x) = x^n$ at $x = a$ (n a positive integer)

Growth

20. Verify the factorizations given in (5), (6), and (7).

21. Verify the formula for the factorization of the difference of nth powers given in Equation (8).

22. Let f be a differentiable function and c a real number. Use the definition of derivative and the rules for manipulation of limits developed in Chapter 1 to show that

$$(cf)' = cf'$$

23. Is it true that

$$(p(x)q(x))' = p'(x)q'(x)$$

for every pair of polynomials p and q? Try some examples before answering.

24. Let $f(x) = 1/x^n$, where n is a positive integer. Use the definition of derivative and the difference-of-nth-powers factorization formula given in (8) to show that

$$f'(x) = \frac{d}{dx}\frac{1}{x^n} = \frac{d}{dx}x^{-n} = -nx^{-n-1}.$$

Exercises 25–28:   Use the result of Exercise 24 to find the derivative.

25. $f(x) = \dfrac{2}{x^3}$

26. $r(\theta) = \theta + 6 + \dfrac{1}{\theta}$

27. $s = 3t^3 - 4t + 7 - 6t^{-1} + \sqrt{2}t^{-5}$

28. $h(r) = \left(1 + \dfrac{2}{r^2}\right)^2$

29. Two balloonists, Sir Bass and Madam Alto, leave the ground in their balloons at the same instant. The height of Sir Bass's balloon at time t minutes is $h(t)$ meters, while the height of Madam Alto's balloon at time t is $2h(t)$ meters. At any given time, what can be said about the speeds at which the two balloons are ascending or descending? Give reasons for your answer.

30. Two gymnasts, Matt and Bart, climb a rope. At time $t = 0$ Matt starts climbing from floor level while Bart starts from 10 feet above ground. It is observed that at all times during the climb, Matt is ascending twice as fast as Bart. At time $t = 20$ minutes Bart is 2000 feet above the ground. How high is Matt at this time? Give reasons for your answer and interpret the problem in terms of derivatives.

Review/Preview

31. Find x if $\dfrac{3 - 2x}{4 + x} = 7$.

32. For what values of b does the equation

$$x^2 + bx - 7 = 0$$

have one or more real solutions?

33. Let

$$f(x) = (x - 1)(x + 7)$$

and

$$g(x) = (2x + 3)(x + 4).$$

If $f(a) > 0$ and $g(a) < 0$, what can be said about a?

34. Write $h(t) = \sqrt{2 - \sin(4t - 1)}$ as a composition of three simple functions.

35. Explain why $\log_{10} x$ is not defined for $x \le 0$.

36. If $\log_{10} x = a$ and $\log_{10} y = a + 1$, what is the value of x/y?

37. In a right triangle, one angle has measure θ and the side adjacent to this angle has length 1. What is the length of the other side? What is the length of the hypotenuse? When the Pythagorean theorem for this triangle is written down, what familiar trigonometric identity do you see?

38. Explain why $\sin \theta$ cannot be greater than 1 and cannot be less than -1.

39. Evaluate

$$\lim_{x \to 2} \frac{x^2 - 6x + 8}{2x - 4}$$

numerically, graphically, and analytically.

40. Find a function $f(x)$ that is not defined at $x = -2$ but for which $\lim_{x \to -2} f(x) = 5$.

2.2   DERIVATIVES OF PRODUCTS AND QUOTIENTS

The chain block, shown in Fig. 2.5, is an ingenious mechanism that allows people to hoist loads of 3 or 4 tons by hand. When the chain is pulled at point P to rotate wheel A through one revolution, wheel B is raised $\pi(R - r)$ units. To lift an object of weight W, one has to pull at P with a force of

(1)
$$F = W \frac{R - r}{2Re}$$

where e is the mechanical efficiency of the mechanism (usually about 0.30 because of friction). Note that when R and r are close, it takes very little force to raise the weight. However, when this is the case, the weight is not lifted very far. To understand the fascinating physics behind mechanisms like the chain block, we need to be able to study the derivative of expressions such as (1). Such expressions often involve products and quotients of functions we already know how to differentiate. In this section we see how to differentiate functions that can be described

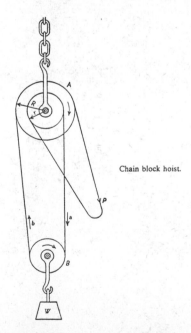

Chain block hoist.

Figure 2.5. The chain block. Pull at P to raise the object below wheel B.

in the form

$$u(x)v(x) \quad \text{or} \quad \frac{u(x)}{v(x)}$$

given that we already know the derivatives of u and v.

The Product Rule

Suppose that we know the rates of change of the functions u and v. What can we say about the rates of change of the product uv? The product rule tells us how the derivative of the product is related to the derivatives of the factors.

The Product Rule

Let u and v be differentiable functions and let $w = uv$ be the product of these two functions. Then

(2) $\qquad w'(x) = (u(x)v(x))' = u'(x)v(x) + u(x)v'(x).$

As (2) shows, the formula for differentiating a product is more complicated than the rule for differentiating a sum. The product rule reflects the way that the rates of change of the factors affect the rate of change of the product. This is illustrated by a very familiar process: that of changing the size of a window on a computer screen.

We change the size of a computer window by "grabbing" a corner and then "dragging" the mouse. Depending on how we move the mouse, the length and height of the window change, and may change quickly or slowly. As a result, the area of the window changes. The rate of change of the area of the window depends on the rates of change of the height and length and provides an illustration of the product rule.

Let $l(t)$, in centimeters, be the length of the window at time t seconds. The rate of change of the length at time t is

$$l'(t) \text{ cm/s.}$$

A short time later, at time $t + \Delta t$, the length of the window is $l(t + \Delta t)$. Since Δt is small, we know that

$$l'(t) \approx \frac{l(t + \Delta t) - l(t)}{\Delta t}.$$

Solving for $l(t + \Delta t)$, we find

$$l(t + \Delta t) \approx l(t) + l'(t)\Delta t.$$

Thus the length of the window has changed by approximately

$$l'(t)\Delta t \text{ cm}$$

during the time interval from t to $t + \Delta t$. Similarly, if $h(t)$ is the height of the window at time t, then the height changes by approximately

$$h'(t)\Delta t \text{ cm}$$

during this time interval.

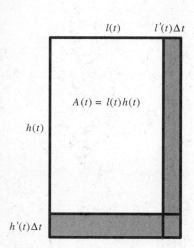

The area of the window at time t is $A(t) = l(t)h(t)$. Now compare the windows at times t and $t + \Delta t$. (See Fig. 2.6.) The shaded region in the figure shows how the rates of change of the factors affect different parts of the area. The change in the area is the area of this shaded region,

$$A(t + \Delta t) - A(t) \approx \bigl(l'(t)\Delta t\bigr)h(t) + l(t)\bigl(h'(t)\Delta t\bigr) + \bigl(l'(t)\Delta t\bigr)\bigl(h'(t)\Delta t\bigr).$$

Dividing by Δt, we find that the rate of change of the area is

$$A'(t) = (l(t)h(t))' \approx \frac{A(t + \Delta t) - A(t)}{\Delta t}$$
$$\approx l'(t)h(t) + l(t)h'(t) + l'(t)h'(t)\Delta t$$
$$\approx l'(t)h(t) + l(t)h'(t)$$

where in the last line we have dropped the small term $l'(t)h'(t)\Delta t$.

The product rule can also be derived directly from the definition of derivative. Though this argument gives less insight into the product rule than the mouse argument, it gives us a chance to use some of the limit results from Chapter 1.

Figure 2.6. The shaded region indicates the change in the window size from time t to time $t + \Delta t$.

The Product Rule by Definition

Let $w = uv$ and assume that u and v are both differentiable at $x = a$. Then

$$w'(a) = \lim_{x \to a} \frac{w(x) - w(a)}{x - a} = \lim_{x \to a} \frac{u(x)v(x) - u(a)v(a)}{x - a}.$$

Add and subtract $u(a)v(x)$ in the numerator to obtain

$$w'(a) = \lim_{x \to a} \frac{(u(x)v(x) - u(a)v(x)) + (u(a)v(x) - u(a)v(a))}{x - a}$$

(3)
$$= \lim_{x \to a} \frac{u(x)v(x) - u(a)v(x)}{x - a} + \lim_{x \to a} \frac{u(a)v(x) - u(a)v(a)}{x - a}$$

$$= \lim_{x \to a} \frac{u(x) - u(a)}{x - a} \lim_{x \to a} v(x) + u(a) \lim_{x \to a} \frac{v(x) - v(a)}{x - a}.$$

By the definition of derivative, we know

$$\lim_{x \to a} \frac{u(x) - u(a)}{x - a} = u'(a) \quad \text{and} \quad \lim_{x \to a} \frac{v(x) - v(a)}{x - a} = v'(a).$$

In addition, since $v(x)$ has a derivative at a, it is also continuous at a. See Section 1.6. Hence

$$\lim_{x \to a} v(x) = v(a).$$

Substituting these results in the last expression in (3), we have

$$w'(a) = (uv)'(a) = u'(a)v(a) + u(a)v'(a).$$

■ **EXAMPLE 1**   If $y = (2x^3 - 4x + 2)(x^2 - x - 1)$, find $\dfrac{dy}{dx}$:

a. Using the product rule.

b. By first expanding the product and then differentiating the resulting polynomial.

Solution

a. Using the template given in (2), we have

$$\frac{dy}{dx} = \left(\underbrace{(2x^3 - 4x + 2)}_{u(x)} \underbrace{(x^2 - x - 1)}_{v(x)} \right)'$$

$$= \underbrace{(2x^3 - 4x + 2)'}_{u'(x)} \underbrace{(x^2 - x - 1)}_{v(x)} + \underbrace{(2x^3 - 4x + 2)}_{u(x)} \underbrace{(x^2 - x - 1)'}_{v'(x)}$$

$$= (6x^2 - 4)(x^2 - x - 1) + (2x^3 - 4x + 2)(2x - 1).$$

b. Expanding the given product, we find

$$y = 2x^5 - 2x^4 - 6x^3 + 6x^2 + 2x - 2.$$

We can differentiate any polynomial in this form by using the techniques developed in Section 2.1. This gives

$$\frac{dy}{dx} = 10x^4 - 8x^3 - 18x^2 + 12x + 2.$$

Are these two answers the same?   ■

■ **EXAMPLE 2**   Let $u(x) \geq 0$ and let

(4) $$w(x) = \sqrt{u(x)}.$$

Assuming that $u(x)$ and $w(x)$ are both differentiable, find a formula for $w'(x)$.

Solution.   Squaring both sides of (4), we have

$$w(x)w(x) = u(x).$$

Differentiate both sides of this equation, using the product rule on the left side. We obtain

$$w'(x)w(x) + w(x)w'(x) = u'(x).$$

Solving this equation for $w'(x)$ and then substituting $w(x) = \sqrt{u(x)}$ gives

$$w'(x) = \frac{u'(x)}{2w(x)} = \frac{1}{2}\frac{u'(x)}{\sqrt{u(x)}}.$$ ■

There is no need to assume that w is differentiable in Example 2. If $u(x) > 0$ and $u'(x)$ exists, then $w'(x)$ must also exist. See Exercise 24.

Differentiating a Quotient

A formula for differentiating a quotient of two functions can be found by using the product rule.

■ **EXAMPLE 3**   Derive the "quotient rule." Assuming that $u(x)$, $v(x)$, and $w(x)$ are differentiable and that

$$(5) \qquad\qquad w(x) = \frac{u(x)}{v(x)},$$

find a formula for $w'(x)$.

Solution.   Multiply both sides of (5) by $v(x)$ to get

$$w(x)v(x) = u(x).$$

Differentiate both sides of this equation, using the product rule on the left side. We obtain

$$w'(x)v(x) + w(x)v'(x) = u'(x).$$

Solve this equation for $w'(x)$ and substitute $\dfrac{u(x)}{v(x)}$ for $w(x)$ to get

$$w'(x) = \frac{u'(x) - w(x)v'(x)}{v(x)}$$

$$= \frac{u'(x) - \left(\dfrac{u(x)}{v(x)}\right)v'(x)}{v(x)}$$

$$= \frac{u'(x)v(x) - u(x)v'(x)}{(v(x))^2}. \qquad\qquad ■$$

There is no need to assume that w is differentiable in Example 3. If u and v are differentiable at x and $v(x) \neq 0$, then $w = u/v$ is differentiable at x. (See Exercise 23.) We will often use the result established in Example 3 since it gives an easy, direct way to differentiate a quotient. As a special case, we can use the quotient rule to differentiate $1/x^n$, where n is a positive integer. (See Exercise 24 in Section 2.1.)

$$\left(\frac{1}{x^n}\right)' = \frac{(1)'x^n - 1(x^n)'}{(x^n)^2} = \frac{0 \cdot x^n - nx^{n-1}}{x^{2n}} = -nx^{-n-1}$$

Combining this with the formula for the derivative of the power function found in the previous section, we have:

The Derivative of x^n

If n is an integer and f is defined by $y = f(x) = x^n$, then

$$\frac{dy}{dx} = f'(x) = nx^{n-1}.$$

■ **EXAMPLE 4**   Good weather in Iowa often results in large corn harvests called "bin busters" because there is not enough room in grain elevators for all of the corn harvested. At such times it is not uncommon to see large conical piles of grain on the ground near the elevators. One such pile at a local elevator was in the shape of a right circular cone of height 15 ft. (The height was kept at 15 ft since corn was dumped from a chute 20 ft above ground level. Machines moved the grain in the pile around to keep the height at 15 ft during dumping.) Late in the harvest season this pile had a base radius of 100 ft and corn was being dumped at a rate of 2000 ft^3/day. How fast was the radius of the base of the pile changing at this time?

Solution.   Let $r(t)$ be the radius of the pile at time t, where t is measured in days. (See Fig. 2.7.) We need to find $r'(t)$ since this derivative is the rate of change of r. Let $V(t)$ be the corresponding volume of corn in the pile at time t. At the time in question, the volume is increasing at a rate of 2000 ft^3/day, so

$$V'(t) = 2000$$

at this time. Since the corn is piled in a cone of height 15 and base radius $r(t)$, we have

$$V(t) = \frac{1}{3}\pi \cdot 15(r(t))^2.$$

Differentiate both sides of this equation, using the product rule on the right to differentiate $(r(t))^2 = r(t)r(t)$. We obtain

(6) $$V'(t) = \frac{\pi}{3}15(r'(t)r(t) + r(t)r'(t)) = 10\pi r(t)r'(t).$$

At the time in question we have $V'(t) = 2000$ ft^3/day and $r(t) = 100$ ft. Putting these values, with units, into (6), we have

$$2000 \text{ ft}^3/\text{day} = 10\pi(100 \text{ ft})r'(t).$$

Solving for $r'(t)$, we find that the radius of the pile is increasing at a rate of

$$r'(t) = \frac{2}{\pi} \text{ ft/day} \approx 0.63662 \text{ ft/day}. \qquad ■$$

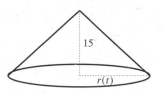

Figure 2.7. A bin buster!

Exercises 2.2

Basic

Exercises 1–14:   Find the derivative of the function represented by the given equation. Use the product, quotient, or square root rules developed in this section.

1. $y = (x^2 - 3x + 1)(2x + 4)$

2. $y = 2x(3x^3 - 3x + \pi)$

3. $H(t) = (-9t^5 + 6t^4 - 21t^2 + 8t - 1)^2$

4. $q(t) = \dfrac{2t - 1}{2t + 1}$

5. $f(x) = \dfrac{2x^2 - 3x + 10}{x^2 - 7}$

6. $p(x) = -3x^2 + 2x - 7 + \dfrac{4}{x} + \dfrac{21}{x^{21}}$

7. $s(r) = \left(1 - \dfrac{1}{r} + \dfrac{2}{r^2}\right)^2$

8. $g(u) = \dfrac{(u^2 - 2u + 3)(4u^2 - u + 1)}{u^3 + 8}$

9. $r = (\theta^2 - 2)(\theta^3 - 3)(\theta^4 - 4)$

10. $f(z) = z(z - z^{-1})(2z^2 - 4 + 2z^{-2})$

11. $p(r) = \sqrt{r^2 + 2r + 7}$

12. $w(v) = \sqrt{\dfrac{v^2 - 3v + 1}{2v + 5}}$

13. $F(s) = \sqrt{2s - 1}\sqrt{6s + 7}$

14. $y = \sqrt{\sqrt{x} + 1}$

Exercises 15–18:   Find $f'(2)$, given that $u(2) = 1$, $v(2) = 2$, $w(2) = 3$, $u'(2) = 4$, $v'(2) = 5$, and $w'(2) = 6$.

15. $f(x) = u(x)v(x)$

16. $f(x) = \dfrac{u(x)v(x)}{w(x)}$

17. $f(x) = \sqrt{2w(x) - 3u(x)}$

18. $v(x) = \sqrt{2f(x) + 5}$

19. If $y = (2x^2 - 1)(3x + 7)$, find the coordinates (x, y) of all points where $f'(x) = 0$.

20. If $y = \sqrt{3x^2 + 2x + 7}$, find the coordinates (x, y) of all points where $dy/dx = 0$.

21. Find an equation for the line tangent to

$$y = \dfrac{2x^2 - 7}{7x + 1}$$

at the point where $x = 2$.

22. Find an equation for the line tangent to

$$g(t) = (t + 1)(t + 2)(t + 3)$$

at the point where $t = -2$.

Growth

23. Derive the quotient rule (see Example 3) using the definition of derivative. Assume that $u(x)$ and $v(x)$ are differentiable at all points in the domain of $w(x) = u(x)/v(x)$. First show that

$$w'(x) = \lim_{t \to x} \frac{(u(t)/v(t)) - (u(x)/v(x))}{t - x}$$

$$= \lim_{t \to x} \frac{u(t)v(x) - u(x)v(t)}{t - x} \cdot \frac{1}{v(t)v(x)}.$$

Now use manipulations similar to those used in deriving the product rule from the definition of derivative. Use the final form of the quotient rule to guide you in these manipulations.

24. Use the definition of derivative to derive the rule for differentiating functions of the form $w(x) = \sqrt{u(x)}$. (See Example 2.) Assume that $u(x)$ is differentiable at all points where $u(x) > 0$. Start by showing that

$$w'(x) = \lim_{t \to x} \frac{\sqrt{u(t)} - \sqrt{u(x)}}{t - x}$$

$$= \lim_{t \to x} \left(\frac{u(t) - u(x)}{t - x} \cdot \frac{1}{\sqrt{u(t)} + \sqrt{u(x)}}\right).$$

Now use the definition of the derivative of u and the continuity of u to obtain the final formula.

25. A function w is expressed as a product of two functions and differentiated using the product rule. The result is

$$w'(x) = (2x - 3)(4x^3 + 3x - 1)$$
$$+ (x^2 - 3x + 8)(12x^2 + 3).$$

What is $w(x)$?

26. A function w is expressed as the square root of another function and differentiated using the "square root" rule of Example 2. The result is

$$w'(x) = \frac{3x^2 - 2x + 1}{\sqrt{x^3 - x^2 + x - 32}}.$$

What is $w(x)$?

27. Let $u(x)$ be a differentiable function. Find and simplify the derivative of each of the following:

a. $(u(x))^2$ **c.** $(u(x))^4$

b. $(u(x))^3$ **d.** $(u(x))^5$

You may wish to use the results of a and b in doing c and d. Based on your answers, what would you expect for the derivative of $(u(x))^n$?

28. Let n be a positive integer and $u(x)$ a differentiable function. Show that

$$u(t)^n - u(x)^n =$$
$$(u(t) - u(x))(u(t)^{n-1} + u(t)^{n-2}u(x) + \cdots$$
$$+ u(t)u(x)^{n-2} + u(x)^{n-1})$$

and then use this factorization to find a formula for the derivative of $(u(x))^n$. Compare with your answers to the previous exercise.

29. Differentiate the identity $|x| = \sqrt{x^2}$ to find a formula for the derivative of $|x|$. For what values of x is this derivative defined? Use a graph of $y = |x|$ to check that your answer is correct.

30. Use the idea introduced in the previous exercise to find the derivative of:

a. $|2x - 1|$.   **b.** $|x^2 - 4|$.

31. Oil spills from a ruptured tanker, forming a circular oil slick on the surface of the ocean. Concerned observers in a helicopter note that the oil slick is 2 km in radius and that the radius seems to be increasing at 10 m per hour. How fast is the area of the slick increasing at this time? If the oil slick is 0.5 cm thick, how fast is oil spilling from the tanker? (Be sure to include units with your answers.)

32. During a recent newspaper drive, volunteers collected papers and stacked them in a rectangular pile. At mid-morning the pile of papers measured 10 ft on the west side, 15 ft on the north side, and 5 ft high. At this time some people were stacking papers on the west side, causing this side to grow westward at 2 ft/h. Others were stacking papers on the north side, causing this side to creep northward at $1/2$ ft/h, and a third group were stacking papers on top, causing the height to increase by 3 ft/h. How fast was the volume of the pile increasing at this time? How fast should papers be hauled away to keep the volume of the pile constant?

Review/Preview

33. Write

$$f(x) = \frac{1}{\sin \sqrt{x - 3}}$$

as a composition of simple functions.

34. Write

$$h(t) = \sqrt[3]{1 + \sqrt[4]{t^2 - 1}}$$

as a composition of simple functions.

35. Find two different functions that have derivative $4x^3 + 1$. How do the graphs of these two functions differ?

36. Find two different functions that have derivative 6. How do the graphs of these functions differ?

37. A right triangle has an acute angle of measure θ, adjacent side of length a, and hypotenuse of length c. A second right triangle has an acute angle of measure 2θ and also has an adjacent side of length a. What is the length of the hypotenuse of this second triangle in terms of a and c?

38. A surveyor looks through her scope. Her partner is exactly 100 m away holding a 3 m stick perpendicular to the ground and with one end on the ground. In order to center the top of the stick in the cross hairs of the scope, she must incline the scope at $2°$ above horizontal. If the scope is 1.5 m above ground, what is the net change in elevation between the surveyor's position and her partner's position?

39. For real numbers a and b, discuss the difference, if any, between $(2^a)^b$, 2^{a^b}, and 2^{ab}.

40. Given a numerical value for x, explain to a friend the steps needed in using a calculator to find the value of

$$\tan\left(\frac{1 + \sqrt{x}}{2 - \sqrt{x}}\right).$$

41. Use a graph or calculator to determine

$$\lim_{x \to 3} \frac{2}{(x - 3)^2}.$$

42. Use a graph or calculator to investigate $\dfrac{2x - 4}{(x - 3)^3}$ near $x = 3$, and then discuss

$$\lim_{x \to 3} \frac{2x - 4}{(x - 3)^3}.$$

43. Consider the chain block shown in Fig. 2.5. Show that if you pull at P so that wheel A turns through one rotation, then wheel B will rise a distance of $\pi(R - r)$.

2.3   DIFFERENTIATING COMPOSITIONS

Many important quantities can be expressed as a composition of simple functions. For example, if you put $1000 in a savings account that pays r percent annual interest compounded quarterly, then at the end of 20 years your account will contain

$$A(r) = 1000(1 + 0.25r)^{80} \text{ dollars.}$$

The function A is a composition of two simple functions. One of these is the linear function

$$l(r) = 1 + 0.25r,$$

and the other is the power function

$$p(x) = 1000x^{80}.$$

In Section 2.1 we learned how to find the derivatives of l and p. In this section we see how we can use the derivatives of l and p to find the derivative of

$$A(r) = p(l(r)).$$

We will find that the technique for computing such derivatives is not only a valuable practical tool, but also provides a useful means for studying rates of change of inverse functions and of functions defined implicitly.

Differentiation of Compositions

The formula for differentiating compositions is often called the *chain rule*.

The Chain Rule

Let f and g be differentiable functions and let $h = f \circ g$. Then

(1) $h'(x) = (f \circ g)'(x) = \big(f(g(x))\big)' = f'\big(g(x)\big)g'(x).$

To see why the chain rule has the form given in (1), we work with the approximation results developed in Sections 1.5 and 1.6. Let c be a number such that $f'(c)$

exists, and let t be close to c. Then

$$(2) \qquad f(t) = f(c) + f'(c)(t - c) + E(t)(t - c)$$

where

$$E(t) = \frac{f(t) - f(c)}{t - c} - f'(c).$$

Because

$$\lim_{t \to c} \frac{f(t) - f(c)}{t - c} = f'(c),$$

we know that $E(t)$ approaches 0 as t approaches c.

Now assume that $g'(a)$ exists and that $f'(g(a))$ exists (i.e., that $g(a)$ is in the domain of f'). In (2) subsitute $g(a)$ for c and $g(x)$ for t. We then obtain

$$(3) \quad f(g(x)) = f(g(a)) + f'(g(a))(g(x) - g(a)) + E(g(x))(g(x) - g(a))$$

where $E(g(x))$ approaches 0 as $g(x)$ approaches $g(a)$. But note that as

$$x \to a \quad \text{we have} \quad g(x) \to g(a).$$

(Why?) It follows that as

$$x \to a \quad \text{we have} \quad E(g(x)) \to 0.$$

Now compute the derivative of $f \circ g$ at $x = a$ using the definition of derivative.

$$(f \circ g)'(a) = \lim_{x \to a} \frac{(f \circ g)(x) - (f \circ g)(a)}{x - a}$$

$$= \lim_{x \to a} \frac{(f(g(x)) - f(g(a)))}{x - a}$$

Substituting the expression for $f(g(x))$ given in (3), we have

$(f \circ g)'(a)$

$$(4) \quad \begin{aligned} &= \lim_{x \to a} \frac{\left(f(g(a)) + f'(g(a))(g(x) - g(a)) + E(g(x))(x - a) \right) - f(g(a))}{x - a} \\ &= \lim_{x \to a} \left(f'(g(a)) \frac{g(x) - g(a)}{x - a} + E(g(x)) \right) \\ &= f'(g(a)) \lim_{x \to a} \frac{g(x) - g(a)}{x - a} + \lim_{x \to a} E(g(x)) \end{aligned}$$

By the definition of derivative,

$$\lim_{x \to a} \frac{g(x) - g(a)}{x - a} = g'(a).$$

Also recall that

$$\lim_{x \to a} E(g(x)) = 0.$$

Using these results in (4), we obtain the chain rule,

$$(f \circ g)'(a) = f'(g(a))g'(a).$$

In applying the chain rule, it is important to identify the simple functions that make up a more complicated function. It is also important to know how these simple functions are combined, that is, when they are added, multiplied, composed, and so on.

■ **EXAMPLE 1**   Find $\dfrac{dy}{dx}$ if $y = (-6x^8 + \sqrt{3x^4} + x - 7)^{12}$.

Solution.   We first note that $y = (f \circ g)(x)$, where g and f are given by

$$g(x) = -6x^8 + \sqrt{3x^4} + x - 7 \quad \text{and} \quad f(x) = x^{12}.$$

We next find

$$g'(x) = -48x^7 + 4\sqrt{3x^3} + 1 \quad \text{and} \quad f'(x) = 12x^{11}.$$

Thus,

$$\begin{aligned}
\frac{dy}{dx} = (f \circ g)'(x) &= f'(g(x))g'(x) \\
&= 12(g(x))^{11}(-48x^7 + 4\sqrt{3x^3} + 1) \\
&= 12(-6x^8 + \sqrt{3x^4} + x - 7)^{11}(-48x^7 + 4\sqrt{3x^3} + 1).
\end{aligned}$$

How can the derivative be found by the methods of Section 2.1?   ■

We will often work with functions that are compositions of three or more functions. The chain rule also applies in these cases.

■ **EXAMPLE 2**   Find a formula for $\dfrac{dy}{dx}$ if $y = f(g(h(x)))$.

Solution.   Let $c(x) = g(h(x))$. Then $y = f(c(x))$, so by the chain rule,

$$(5) \qquad\qquad \frac{dy}{dx} = (f \circ c)'(x) = f'(c(x))c'(x).$$

Applying the chain rule to $c(x) = g(h(x))$, we have

$$c'(x) = g'(h(x))h'(x).$$

Substituting this result in (5) gives

$$\frac{dy}{dx} = (f \circ g \circ h)'(x) = f'(g(h(x)))g'(h(x))h'(x). \qquad\qquad ■$$

Differentiation of Power Functions

With the chain rule we can now extend our rule for differentiation of x^n (see Section 2.2) to the case where the exponent n is a rational number.

■ **EXAMPLE 3** Let $h(x) = x^{p/q}$, where p and q are integers and $q \neq 0$. Assuming $h(x)$ is differentiable for $x > 0$, find $h'(x)$.

Solution. Raise both sides of the equation $h(x) = x^{p/q}$ to the qth power to get

$$(6) \qquad (h(x))^q = (x^{p/q})^q = x^p.$$

Now differentiate both sides of this equation. We need to use the chain rule to differentiate the left side because $(h(x))^q$ is the composition of the functions h and g, where $g(t) = t^q$. When we differentiate (6), we obtain

$$q(h(x)^{q-1})h'(x) = px^{p-1}.$$

Solve this equation for $h'(x)$ and substitute $h(x) = x^{p/q}$ to find

$$h'(x) = \frac{px^{p-1}}{q(h(x)^{q-1})} = \frac{p}{q}\frac{x^{p-1}}{(x^{p/q})^{q-1}} = \frac{p}{q}x^{(p/q)-1}.$$

Actually, there is no need to assume $x^{p/q}$ is differentiable to get this result. See Exercise 34 in Section 2.6. ■

Combining the result of Example 3 with the rules for differentiating power functions from Section 2.2, we have:

The Derivative of the Power Function

Let $y = f(x) = x^r$, where r is any rational number. Then

$$(7) \qquad \frac{dy}{dx} = rx^{r-1}.$$

The Chain Rule Reformulated

To calculate the derivative of a composition, it may be easier to work with a different form of the chain rule.

The Chain Rule

Suppose that $y = (f \circ g)(x)$. We can express this composition as a *composition chain*:

$$y = f(u)$$
$$u = g(x)$$

Then

$$(8) \qquad \frac{dy}{dx} = (f \circ g)'(x) = \frac{dy}{du}\frac{du}{dx}.$$

Since $dy/du = f'(u) = f'(g(x))$ and $du/dx = g'(x)$, we see that (8) is equivalent to (1). Hence (8) is another way of stating the chain rule.

■ **EXAMPLE 4** Find $\dfrac{dy}{dx}$ if

$$(9) \qquad\qquad y = \sqrt{3x^2 - 4x + 1/2}$$

Solution. Begin by writing a composition chain that displays the given function as a composition of simple pieces:

$$y = \sqrt{u}$$
$$u = 3x^2 - 4x + \frac{1}{2}.$$

Since $y = u^{1/2}$, we apply (7) to get

$$(10) \qquad\qquad \frac{dy}{du} = \frac{1}{2}u^{-1/2}.$$

Because u is a polynomial, we can immediately write down its derivative:

$$(11) \qquad\qquad \frac{du}{dx} = 6x - 4.$$

Substituting (10) and (11) into (8), we have

$$\frac{dy}{dx} = \frac{dy}{du}\frac{du}{dx} = \frac{1}{2}u^{-1/2}(6x - 4).$$

Since (9) was given in terms of x, our answer should be in terms of x only. Thus we replace u by $3x^2 - 4x + \frac{1}{2}$. This leads to the answer

$$\frac{dy}{dx} = \frac{1}{2}(3x^2 - 4x + 1/2)^{-1/2}(6x - 4). \qquad\blacksquare$$

The chain rule formulation given in (8) extends readily to compositions of three or more functions (see Exercise 31).

■ **EXAMPLE 5** Find $\dfrac{dy}{dx}$ if

$$y = \frac{3}{(\sqrt{x^2 + 2} + 2)^2}.$$

Solution. In this case the composition chain consists of three simple functions. We have

$$y = \frac{3}{u^2} = 3u^{-2}$$
$$u = \sqrt{v} + 2 = v^{1/2} + 2$$
$$v = x^2 + 2.$$

By (7) we have,

$$\frac{dy}{du} = -6u^{-3}, \quad \frac{du}{dv} = \frac{1}{2}v^{-1/2}, \quad \text{and} \quad \frac{dv}{dx} = 2x.$$

By the chain rule,

$$\frac{dy}{dx} = \frac{dy}{du}\frac{du}{dv}\frac{dv}{dx}$$

$$= (-6u^{-3})\left(\frac{1}{2}v^{-1/2}\right)(2x).$$

Now substitute the expressions for u and v in (12).

$$\frac{dy}{dx} = (-6(\sqrt{v}+2)^{-3})\left(\frac{1}{2}v^{-1/2}\right)(2x)$$

$$= (-6(\sqrt{x^2+2}+2)^{-3})\left(\frac{1}{2}(x^2+2)^{-1/2}\right)(2x)$$

$$= \frac{-6x}{(\sqrt{x^2+2}+2)^3\sqrt{x^2+2}} \qquad \blacksquare$$

Implicitly Defined Functions

Investigation 1

In Section 1.1 we remarked that when an equation is used to define a function, we often rely on our experience to supply a natural domain and range for the function if none are given. However, there are many equations for which the domain and range of the function are not at all obvious. For example, consider the equation

$$(12) \qquad y^3 - 2xy^2 - x^2y + 3x - 1 = 0.$$

We can view this equation as defining y in terms of x, because if we put an x value into the equation, we can solve for the corresponding y value. For example, when $x = 1$, we find that $y = -1$ solves (12). So a function defined by this equation might map the domain element 1 to the range element -1. However, we also note that when $x = 1$, the y values 1 and 2 also work in the equation. Which should we choose as the range element corresponding to $x = 1$? If we are to use (12) to define a function, we need to choose a domain and then decide which y value is to correspond to a given x in the domain. One way to gain insight into an equation such as (12) is by graphing, that is, plotting the set of all ordered pairs (x, y) that satisfy (12). Graphs of such equations can be very difficult to draw by hand. However, many computer algebra systems and plotting packages have commands to quickly draw graphs of the solutions to such equations.

The graph of (12) is shown in Fig. 2.8. From the graph we see that given a value x on the horizontal axis, there may be one, two, or three y values that correspond to x. To define a function using (12), we need to specify a domain for the function and a rule that tells us which y to choose for each x in the domain. Suppose that we want (12) to define a function f so that $y = -1$ corresponds to the domain value $x = 1$, that is, so that $f(1) = -1$. We express this by saying:

Equation (12) defines y as a function of x near $(1, -1)$.

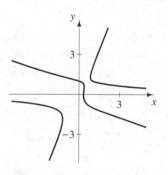

Figure 2.8. A graph of the equation $y^3 - 2xy^2 - x^2y + 3x - 1 = 0$.

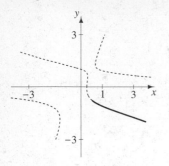

Figure 2.9. The highlighted portion gives the graph of a function f for which $f(1) = -1$.

We can find the graph of such an f by using the graph in Fig. 2.8. To do this, we take a piece of the graph of the equation, a piece that contains $(1, -1)$, and identify this piece as the graph of a function f. We need only be sure that the piece of the graph chosen contains $(1, -1)$ and that for each x value that corresponds to a point on the graph, there is only one corresponding y value; that is, the piece of graph we choose must satisfy the "vertical line test" (see Exercise 39 in Section 1.1). We have indicated such a portion of the graph in Fig. 2.9. This solid piece of the graph defines a function with domain something like $\{x : 0.5 \le x \le 4\}$ and range something like $\{y : -2 \le y \le -0.8\}$.

To evaluate $f(x)$ for an x in the domain, we can put the x value into (12) and solve the resulting equation for y. To check that we have the right y value for the given x, we can make sure that the point (x, y) does lie on the solid portion of the graph in Fig. 2.9. Can we find a formula for $f(x)$? Yes, but not easily. Furthermore, if (12) were more complicated, it might be impossible to find such a formula. When a function f as just described is "indirectly" defined through an equation such as (12), we say that f is defined **implicitly.**

Differentiating Implicitly Defined Functions

Investigation 2

As in Investigation 1, suppose that

$$(13) \qquad y^3 - 2xy^2 - x^2y + 3x - 1 = 0$$

defines y as a function of x near $(1, -1)$. Call the function f. As mentioned above, a formula for $y = f(x)$ is very difficult to find. Surprisingly, however, it is not hard to find an expression for f'. The process for finding this is called **implicit differentiation.**

For a given x in the domain of f, we know that the pair of numbers

$$(x, y) = (x, f(x))$$

satisfies (13). Substituting this pair into the equation, we have

$$(14) \qquad (f(x))^3 - 2x(f(x))^2 - x^2 f(x) + 3x - 1 = 0.$$

Differentiate both sides of (14) with respect to x. The derivative of the right side is 0. Why? The expression on the left side involves some products and compositions, so we will use the product rule and the chain rule. First break the derivative of the left side into the sum of several derivatives:

$$\left((f(x))^3\right)' - 2\left(x(f(x))^2\right)' - (x^2 f(x))' + 3x' - 1' = 0'.$$

Apply the product rule to the second and third terms of this expression to get

$$(15) \qquad \left((f(x))^3\right)' - 2\left[x'(f(x))^2 + x\left((f(x))^2\right)'\right]$$
$$- [(x^2)' f(x) + x^2(f(x))'] + 3 = 0.$$

Next differentiate $(f(x))^2$ and $(f(x))^3$ by the chain rule. Because $(f(x))^2$ is the function f composed with the squaring function, we have

(16) $$\left((f(x))^2\right)' = 2f(x)f'(x).$$

Similarly,

(17) $$\left((f(x))^3\right)' = 3(f(x))^2 f'(x).$$

Substituting (16) and (17) into (15), we have

$$3(f(x))^2 f'(x) - 2[1 \cdot (f(x))^2 + x(2f(x)f'(x))]$$
$$- [2xf(x) + x^2 f'(x)] + 3 = 0.$$

Expanding and regrouping this last equation, we see it is a linear expression in $f'(x)$:

$$f'(x)[3(f(x))^2 - 4xf(x) - x^2] + [-2(f(x))^2 - 2xf(x) + 3] = 0.$$

Solve for $f'(x)$ to get

$$f'(x) = \frac{2(f(x))^2 + 2xf(x) - 3}{3(f(x))^2 - 4xf(x) - x^2}.$$

Replacing $f(x)$ by y, we have an expression for the desired derivative:

(18) $$\frac{dy}{dx} = f'(x) = \frac{2y^2 + 2xy - 3}{3y^2 - 4xy - x^2}.$$

When $x = 1$, we have $y = -1$. The rate of change of y with respect to x at $(1, -1)$ is found by substituting these x and y values into (18):

$$\left.\frac{dy}{dx}\right|_{x=1, y=-1} = \frac{2(-1)^2 + 2 \cdot 1(-1) - 3}{3(-1)^2 - 4 \cdot 1(-1) - 1^2} = -\frac{1}{2}.$$

In Fig. 2.10 we zoom in on the graph of $y^3 - 2xy^2 - x^2 y + 3x - 1 = 0$ near the point $(1, -1)$ and include the graph of the line tangent to the graph at this point. What is the equation for this line?

Note that we can evaluate (18) for a given x value only if we know the corresponding y value. So even though dy/dx is not hard to find, it may be very difficult to evaluate the derivative for a given x value!

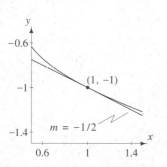

Figure 2.10. The line tangent to $y^3 - 2xy^2 - x^2 y + 3x - 1 = 0$ at the point $(1, -1)$ has slope $-1/2$.

In practice, it is not really necessary to replace y by $f(x)$ before performing the differentiation. This substitution was just an aid to help us remember to think of y as a function of x.

■ **EXAMPLE 6** Find the coordinates of the points on the graph of

(19) $$(x - 2y - 1)^2 + (x + y)^2 = 16$$

where the tangent line is horizontal.

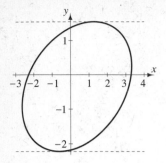

Figure 2.11. Graph of
$(x - 2y - 1)^2 + (x + y)^2 = 16$.

Solution.   To get an idea of what we are looking for, use a computer or calculator to graph this relation. As seen in Fig. 2.11, there appear to be two places where the tangent line is horizontal. The coordinates of the points of tangency appear to be about $(1, 1.5)$ and $(-0.5, -2.2)$.

The tangent to the graph is horizontal at points where the derivative

$$\frac{dy}{dx} = 0.$$

We find this derivative by differentiating (19) implicitly, keeping in mind that the equation defines y as a function of x. Differentiate both sides of (19) with respect to x and apply the chain rule to get

$$2(x - 2y - 1)(x - 2y - 1)' + 2(x + y)(x + y)' = 0.$$

Since the derivative of y with respect to x is $\dfrac{dy}{dx}$, this last equation becomes

$$2(x - 2y - 1)\left(1 - 2\frac{dy}{dx}\right) + 2(x + y)\left(1 + \frac{dy}{dx}\right) = 0.$$

Expand, and group the terms involving $\dfrac{dy}{dx}$ to get

$$\frac{dy}{dx}(-2x + 10y + 4) + (4x - 2y - 2) = 0.$$

Solve for $\dfrac{dy}{dx}$ to find

$$\frac{dy}{dx} = \frac{-2x + y + 1}{-x + 5y + 2}.$$

At a point where the tangent is horizontal (has slope 0) this derivative is 0. Hence at the points (x, y) where the tangent line is horizontal,

(20)
$$\frac{-2x + y + 1}{-x + 5y + 2} = 0.$$

A fraction is equal to 0 when its numerator is 0. Hence from (20) we get

(21)
$$-2x + y + 1 = 0.$$

If (x, y) is a point on the graph where the tangent has slope 0, then this point must also be on the graph of the equation and hence must satisfy (19). From (19) and (21) we get a system of two equations and two unknowns,

(22)
$$(x - 2y - 1)^2 + (x + y)^2 = 16$$
$$-2x + y + 1 = 0.$$

Solve the second equation for y,

(23)
$$y = 2x - 1,$$

and substitute the result into the first equation in (22). We then have

$$(x - 2(2x - 1) - 1)^2 + (x + (2x - 1))^2 = 16.$$

This equation can be solved using a calculator, computer, or the quadratic formula, or by completing the square. The solutions are

$$x = \frac{1 + 2\sqrt{2}}{3} \approx 1.27 \quad \text{and} \quad x = \frac{1 - 2\sqrt{2}}{3} \approx -0.609.$$

Substituting these values into (23), we find corresponding y values of

$$y = \frac{-1 + 4\sqrt{2}}{3} \approx 1.55 \quad \text{and} \quad y = \frac{-1 - 4\sqrt{2}}{3} \approx -2.22.$$

Hence the coordinates of the two points on the graph where the tangent is horizontal are

$$\left(\frac{1 + 2\sqrt{2}}{3}, \frac{-1 + 4\sqrt{2}}{3}\right) \approx (1.27, 1.55)$$

and

$$\left(\frac{1 - 2\sqrt{2}}{3}, \frac{-1 - 4\sqrt{2}}{3}\right) \approx (-0.609, -2.22).$$

How do these answers compare with the estimates read from the graph in Fig. 2.11? ■

Exercises 2.3

Basic

Exercises 1–12:   Find the derivative of the function represented by the given equation.

1. $f(x) = 2x^{13/3} + 2$

2. $s = \sqrt[5]{u - 7}$

3. $y = (2x^2 - 3x + 1)^{23}$

4. $f(t) = \sqrt{2t - 7}$

5. $g(x) = \left(x^3 - 5x + \frac{1}{2}\right)^{5/4}$

6. $r = (2\theta - 1)^6(4\theta + \pi)^8$

7. $h(u) = \sqrt[3]{\dfrac{4u - 1}{3u + 7}}$

8. $y = \sqrt{1 + \sqrt{1 + \sqrt{x + 1}}}$

9. $H(z) = (3z^2 - 1)^4(4z^{-3} + 2z^{-2} + 4)\sqrt{1 - \dfrac{1}{z}}$

10. $l(x) = \dfrac{1}{(1 + x + x^2)^{10}}$

11. $y = f(f(x))$ where $f(x) = x^2 - 3x + 1$

12. $h(t) = (g \circ g \circ g)(t)$ where $g(t) = \dfrac{1}{t^2} + 1$

Exercises 13–16:   Find an expression for dy/dx.

13. $y = u^2 + 1, u = x^{-3/2} + 1$

14. $y = \sqrt{w}, w = \dfrac{1}{v}, v = 2x - 1$

15. $y = f(u), u = -2x^3 + x + 1$, where $f'(u) = 2\sin u$.

16. $y = H(t^2 + 1), t = (3x - 1)$, where $H'(v) = \dfrac{1}{v}$.

Exercises 17–20:   Assume that the equation defines y as a function of x. Use implicit differentiation to find an expression for dy/dx.

17. $x^3 + y^3 - x + 2y = 3$

18. $2x^2 - 3xy + y^2 = 2$

19. $(x - 2y)^2 - xy^2 = -2xy + 4$

20. $\sqrt{x^2 + y^2} - x^2 y^3 = x + y$

Exercises 21 and 22:   Assume that the equation defines y as a function of x near the given point P. Find the value of dy/dx at P.

21. $4y^2 - xy + 2x - 3y = 3$, $P = (2, 1)$

22. $\dfrac{1}{x + y} + 3 - 4(x + y)^2 = -2$, $P = (1, -2)$

Growth

Exercises 23 and 24:   In Exercise 29 of Section 2.2 we showed that $(d/dx)|x| = x/|x|$ for $x \neq 0$. Use this formula as needed to find the derivative of the given function.

23. $f(x) = |(x - 2)(x + 1)|$

24. $h(t) = \dfrac{2t^2 - 3}{|t^2 - 6|}$

Exercises 25–29:   The figures below show the graph of $y = f(x)$ and the graph of $y = f'(x)$. Sketch the graphs of $y = g(x)$ and $y = g'(x)$.

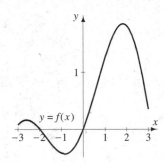

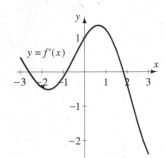

25. $g(x) = f(x - 2)$

26. $g(x) = f(|x|)$

27. $g(x) = |f(x)|$

28. $g(x) = f(-x)$

29. $g(x) = \dfrac{|f(x)|}{f(x)}$

30. In Example 3 we used the chain rule to say

$$((h(x))^q)' = q(h(x))^{q-1} h'(x).$$

Work out the details of this chain rule application.

31. Suppose that a function can be decomposed to a composition chain of length 3:

$$y = f(u)$$
$$u = g(v)$$
$$v = h(x).$$

Show that

$$\frac{dy}{dx} = \frac{dy}{du}\frac{du}{dv}\frac{dv}{dx}.$$

32. Consider the equation

$$2xy^3 + x^2 y^2 - 3y + x = 1.$$

a. Assume that this equation defines y as a function of x. Find an expression for dy/dx.

b. Assume that the equation defines x as a function of y. Find an expression for dx/dy.

c. Show that $(dy/dx)(dx/dy) = 1$. Explain this graphically by interpreting the derivative as the slope of a tangent.

33. Let the equation $y = f(x)$ define x as a function of y. Show that $dx/dy = 1/(f'(x))$. Interpret this result graphically by interpreting the derivative as the slope of a tangent.

34. The accompanying figure shows the graph of the ellipse with equation

$$x^2 - xy + \frac{3}{4}y^2 = 7.$$

The ellipse is centered at the origin but rotated from "standard position."

a. Find the coordinates of all points where the ellipse crosses the coordinate axes.

b. Find the coordinates of all points where the tangent line is horizontal.

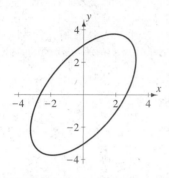

c. Find the coordinates of all points where the tangent line is vertical.

d. At a vertex of the ellipse, the tangent line is perpendicular to the segment from the origin to the vertex. Use this information to find the vertices of the ellipse. What are the lengths of the major and minor axes?

35. Fill in reasons and complete the following argument to give another proof of the chain rule. Let $h(x) = f(g(x))$. Then

$$h'(x) = \lim_{t \to x} \frac{f(g(t)) - f(g(x))}{t - x}$$

$$= \lim_{t \to x} \frac{f(g(t)) - f(g(x))}{g(t) - g(x)} \frac{g(t) - g(x)}{t - x}.$$

Set $u = g(t)$ and note that as $t \to x$, $u = g(t) \to g(x)$. Hence

$$h'(x) = \lim_{u \to g(x)} \frac{f(u) - f(g(x))}{u - g(x)} \lim_{t \to x} \frac{g(t) - g(x)}{t - x}.$$

Review/Preview

36. Find two different functions that have derivative $5(x + 2)^4$. Check by taking the derivative of your answers.

37. Find two different functions that have derivative $-3(x+2)^{-4}$. Check by taking the derivative of your answers.

38. Describe what happens to the value of $\sin \theta$ as θ increases from 0 to 2π.

39. Describe what happens to the value of $\cos \theta$ as θ increases from 0 to 2π.

40. Describe what happens to the value of $\tan \theta$ as θ increases from 0 to 2π.

41. What are the period and amplitude of the functions given by $s(t) = 5 \sin(2t)$?

42. What are the period and amplitude of the functions given by $s(t) = -2 \cos\left(\frac{1}{4}t\right)$?

43. On the same set of axes sketch the graphs of $y = \sin x$ and $y = \cos x$. How many times do the graphs cross in the interval $[0, 2\pi]$?

44. Given that $\sin 1 \approx 0.841471$, find a different number x between 0 and 2π with $\sin x \approx 0.841471$. Find a number y between 0 and 2π with $\sin y \approx -0.841471$. (Do this without using a calculator or computer.)

45. Given that $\tan 1 \approx 1.557408$, find a different number x between 0 and 2π with $\tan x \approx 1.557408$. Find a number y between 0 and 2π with $\tan y \approx -1.557408$. (Do this without using a calculator or computer.)

2.4 TRIGONOMETRIC FUNCTIONS

If you are ever at a carnival near noon, go look at the ferris wheel. When the sun is directly overhead, the shadow of the ferris wheel will be on the ground below the wheel. Watch the shadow of one of the cars on the ferris wheel. The shadow will move back and forth across the ground as the wheel turns. The moving shadow illustrates two very important ideas. First, the moving shadow is an example of *periodic motion.* The back-and-forth motion of the shadow repeats regularly: the shadow moves from left to right under the ferris wheel, then from right to left, then

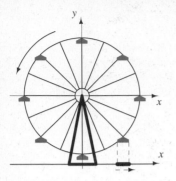

Figure 2.12. The shadow of a car moves back and forth along the ground.

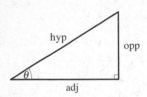

Figure 2.13. The side opposite angle θ is labeled "opp." The side adjacent to the angle is labeled "adj," and the hypotenuse is labeled "hyp."

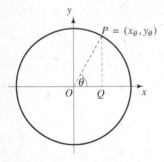

Figure 2.14. The terminal side of the angle of measure θ meets the unit circle in the point with coordinates $(\cos\theta, \sin\theta)$.

from left to right, and so on. Periodic phenomena are everywhere in our world. They are seen in the tides, the orbit of the earth about the sun, the waving of tree branches in the wind, and the motion of a piston in an internal combustion engine.

The moving shadow also illustrates the cosine function. Let the ferris wheel have radius 1, and put a coordinate system in the plane of the wheel with the axle at the origin and the vertical (y) axis perpendicular to the ground. Let θ be the angle the ray from the origin through the car makes with the positive horizontal (x) axis. The x-coordinate of the car is $\cos\theta$. Since the position of the car's shadow on the ground is just this x-coordinate, you are seeing an animation of the cosine function.

This example shows us that there is a close relationship between periodic phenomena and trigonometric functions.

Now look more closely at the motion of the shadow. You will notice that the shadow moves very slowly near the ends of its back-and-forth motion, and it moves faster when it is under the center of the wheel. Since velocity is the rate of change of position with respect to time, the speed of the shadow illustrates the derivative of the cosine function.

Usually students first see the trigonometric functions when studying right triangle trigonometry. Given a right triangle with one acute angle of measure θ, let "opp" denote the length of the side opposite angle θ, "adj" denote the length of the other side, and "hyp" the length of the hypotenuse. (See Fig. 2.13.) Then we have

$$\sin\theta = \frac{\text{opp}}{\text{hyp}} \qquad \cos\theta = \frac{\text{adj}}{\text{hyp}} \qquad \tan\theta = \frac{\text{opp}}{\text{adj}}.$$

These ratios are the same for any right triangle with an acute angle θ because any two such triangles are similar. It is because the values of these functions depend only on the angles in the triangle and not on the size of the triangle that trigonometry is such a useful and powerful method for analyzing geometric figures.

The trigonometric functions can be defined for angles that are not acute angles by referring to the circle of center $O = (0, 0)$ and radius 1. Starting from the positive x-axis, measure an acute angle of measure θ and let $P = (x_\theta, y_\theta)$ be the point in the first quadrant where the terminal side of the angle intersects the circle. Drop a perpendicular from P to meet the x-axis at Q. Then triangle OPQ is a right triangle with acute angle θ. (See Fig. 2.14.) Side PQ, which is opposite θ, has length y_θ, side OQ has length x_θ, and the hypotenuse has length 1. Thus

$$\cos\theta = \frac{\text{adj}}{\text{hyp}} = x_\theta \quad \text{and} \quad \sin\theta = \frac{\text{opp}}{\text{hyp}} = y_\theta.$$

Motivated by this matching of the values of $\cos\theta$ and $\sin\theta$ with the x- and y-coordinates of the point on the circle, we extend the domain of these functions from positive acute angles to all angles. (See Fig. 2.15.)

DEFINITION sine and cosine

Let θ be any real number. From the positive x-axis, measure an angle of θ radians. (Measure counterclockwise for $\theta > 0$, clockwise for $\theta < 0$.) Let (x_θ, y_θ) be the coordinates of the point where the terminal side of the angle meets the circle of center $(0, 0)$ and radius 1. Then

$$\cos\theta = x_\theta \quad \text{and} \quad \sin\theta = y_\theta.$$

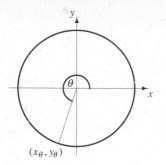

Figure 2.15. With the unit circle we can define sine and cosine for all angles.

As mentioned in Chapter 1, we will assume, unless told otherwise, that all angle measures are given in radians. The reason for this is simply that the calculus of trigonometric functions works out better in radians than it does in degrees. We will say more about this later in the section.

As illustrated repeatedly in Chapter 1, many of the trigonometric identities that you learned (or memorized) in high school can be quickly found in the geometry of the unit circle. Get in the habit of using the circle to work with easy trigonometric identities. This will not only increase your understanding of the trigonometric functions, but also shorten the list of facts you might feel you need to memorize.

■ **EXAMPLE 1**   Use the unit circle to show that

$$\cos\theta = \sin\left(\frac{\pi}{2} - \theta\right) \quad \text{and} \quad \sin\theta = \cos\left(\frac{\pi}{2} - \theta\right).$$

Solution.   First observe that the average of the two angles involved is

$$\frac{\theta + \left(\frac{\pi}{2} - \theta\right)}{2} = \frac{\pi}{4}.$$

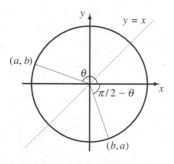

Figure 2.16. The angles of measure θ and $\pi/2 - \theta$ are symmetric with respect to the line $y = x$.

Thus the terminal sides of θ and $(\pi/2) - \theta$ are symmetrically located with respect to the angle $\pi/4$. Hence one angle is the reflection of the other across the line $y = x$. (See Fig. 2.16.) Thus if the terminal side of the angle of measure θ intersects the circle at point

$$P = (a, b) = (\cos\theta, \sin\theta),$$

then the terminal side of the angle of measure $\left(\dfrac{\pi}{2} - \theta\right)$ intersects the circle at point

$$Q = (b, a) = \left(\cos\left(\frac{\pi}{2} - \theta\right), \sin\left(\frac{\pi}{2} - \theta\right)\right).$$

We can then read off

$$\cos\left(\frac{\pi}{2} - \theta\right) = b = \sin\theta \quad \text{and} \quad \sin\left(\frac{\pi}{2} - \theta\right) = a = \cos\theta. \quad ■$$

Once the sine and cosine are found, it is easy to evaluate the other four trigonometric functions.

DEFINITION   **tangent, cotangent, secant, and cosecant**

If θ is an angle for which $\cos\theta \neq 0$, then the tangent of θ is defined by

$$\tan\theta = \frac{\sin\theta}{\cos\theta}$$

and the secant of θ by

$$\sec\theta = \frac{1}{\cos\theta}.$$

If θ is an angle for which $\sin \theta \neq 0$, then the cotangent of θ is defined by

$$\cot \theta = \frac{\cos \theta}{\sin \theta}$$

and the cosecant of θ by

$$\csc \theta = \frac{1}{\sin \theta}.$$

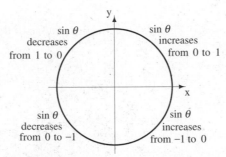

Figure 2.17. The rise and fall of the sine function.

The graph of $w = \sin \theta$ is not hard to sketch if we take note of the behavior of the y-coordinate of points on the circle as we move around the circle. When $\theta = 0$, this y-coordinate has value $\sin 0 = 0$. As angle θ increases from 0 to $\pi/2$, this y-coordinate increases from 0 to $\sin(\pi/2) = 1$. (See Fig. 2.17.) The value of $\sin \theta$ then decreases from 1 to 0 as θ grows from $\pi/2$ to π, and then continues to decrease to -1 as θ increases to $3\pi/2$. As θ increases from $3\pi/2$ to 2π, the value of $\sin \theta$ increases from -1 to 0. We have now moved once around the circle, and as θ grows from 2π to 4π, the $\sin \theta$ values will repeat the behavior just discussed. This behavior is reflected in the familiar graph shown in Fig. 2.18.

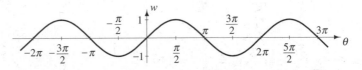

Figure 2.18. Graph of $w = \sin \theta$.

A similar analysis leads to the graph of $w = \cos \theta$, shown in Fig. 2.19.

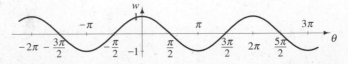

Figure 2.19. Graph of $w = \cos \theta$.

Note that the graph of the cosine appears to be the graph of the sine function moved $\pi/2$ units to the left. This suggests the identity

$$\cos\theta = \sin\left(\theta + \frac{\pi}{2}\right)$$

Try proving this identity using the geometry of the unit circle. (See Exercise 27.)

The Derivatives of Sine and Cosine

We begin by using the geometry of the unit circle to find the derivative of $\sin\theta$. While the argument is not "mathematically rigorous," it does give some insight into why the derivative of the sine function is the cosine function. For a more formal argument using the definition of derivative and some limit results derived in Chapter 1, see Exercise 32.

Investigation

For a given value of θ, the derivative of the sine function is the slope of the "line" we see as we zoom in on the graph of $y = \sin t$ near the point $(\theta, \sin\theta)$. We estimate this slope by considering a nearby point $(\theta + h, \sin(\theta + h))$ on the graph and finding the slope of the segment determined by these two points. See Fig. 2.20. This slope is

(1)
$$\frac{\sin(\theta + h) - \sin\theta}{h}.$$

To estimate the numerator of (1), look at the unit circle. Let the ray at angle θ intersect the circle at point $P = (\cos\theta, \sin\theta)$ and the ray of angle $\theta + h$ intersect the circle at point $Q = (\cos(\theta + h), \sin(\theta + h))$. Then the numerator of (1) is the difference Δy in the y-coordinates of these two points. Join P and Q with a segment and make this the hypotenuse of a right triangle with one side Δy. (See Fig. 2.21.) Since h is small,

$$\text{length of segment } PQ \approx \text{length of arc } PQ = h \cdot 1 = h.$$

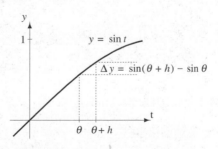

Figure 2.20. Finding the "slope" of the sine graph.

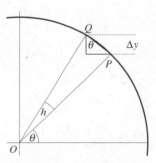

Figure 2.21. Finding Δy on the unit circle.

In addition, segment PQ is almost tangent to the circle at P, so PQ is almost perpendicular to segment OP. It follows that the small triangle with hypotenuse PQ has angle of measure approximately θ at Q. Hence

$$\sin(\theta + h) - \sin\theta = \Delta y \approx h\cos\theta.$$

Substituting this into (1), we find that when we zoom in on the graph of $y = \sin t$ near $(\theta, \sin\theta)$, we see a segment of slope approximately $\cos\theta$. Since this slope should be a good estimate for the derivative, we conclude that

$$\frac{d}{dt}\sin t = (\sin t)' = \cos t.$$

Once we have the derivative of the sine function, the derivative of the other trigonometric functions can be obtained in many ways.

■ **EXAMPLE 2**   Find $\dfrac{d}{dt}\cos t$.

Solution.   In Example 1 we showed that

$$w = \cos t = \sin\left(\frac{\pi}{2} - t\right).$$

This displays w as a composition of the sine function and the function given by $(\pi/2) - t$. As a composition chain, this is

$$w = \sin u$$

$$u = \frac{\pi}{2} - t.$$

By the chain rule,

$$\frac{dw}{dt} = \frac{dw}{du}\frac{du}{dt} = (\cos u)(-1) = -\cos\left(\frac{\pi}{2} - t\right) = -\sin t. \qquad ■$$

Summarizing the last two results, we have:

Derivatives of the Sine and Cosine

$$\frac{d}{d\theta}\sin\theta = (\sin\theta)' = \cos\theta$$

$$\frac{d}{d\theta}\cos\theta = (\cos\theta)' = -\sin\theta$$

■ **EXAMPLE 3**   Find $\dfrac{d}{d\theta}\tan\theta$.

Solution.   Since

$$\tan\theta = \frac{\sin\theta}{\cos\theta}$$

we can use the quotient rule developed in Example 3 of Section 2.2. We have

$$(\tan \theta)' = \left(\frac{\sin \theta}{\cos \theta}\right)'$$

$$= \frac{(\sin \theta)'(\cos \theta) - (\sin \theta)(\cos \theta)'}{(\cos \theta)^2}$$

$$= \frac{(\cos \theta)(\cos \theta) - (\sin \theta)(-\sin \theta)}{(\cos \theta)^2}$$

$$= \frac{\cos^2 \theta + \sin^2 \theta}{\cos^2 \theta}$$

$$= \frac{1}{\cos^2 \theta}$$

$$= \sec^2 \theta$$

All of the differentiation techniques studied in earlier sections may be needed to find the derivatives of functions that involve trigonometric functions.

■ **EXAMPLE 4**   Find $\dfrac{dy}{dx}$ if $y = \cos(\sqrt{x \cos x})$.

Solution.   First decompose the given function into a composition chain:

$$y = \cos u$$

$$u = \sqrt{v}$$

$$v = x \cos x.$$

We then have

$$\frac{dy}{du} = -\sin u$$

$$\frac{du}{dv} = \frac{1}{2\sqrt{v}}$$

$$\frac{dv}{dx} = 1 \cdot \cos x + x(-\sin x).$$

By the chain rule,

$$\frac{dy}{dx} = \frac{dy}{du}\frac{du}{dv}\frac{dv}{dx}$$

$$= (-\sin u)\left(\frac{1}{2\sqrt{v}}\right)(\cos x - x \sin x).$$

Substitute $u = \sqrt{v}$ and then $v = x \cos x$ in the last expression to get

$$\frac{dy}{dx} = \frac{-\sin(\sqrt{x \cos x})(\cos x - x \sin x)}{2\sqrt{x \cos x}}.$$

■ **EXAMPLE 5**   Suppose that the equation

(2) $x = \tan y$

implicitly defines y as a function of x. Find a formula for $\dfrac{dy}{dx}$.

Solution.   Display (2) as a composition chain:

$$x = \tan y$$
$$y = y(x).$$

By the chain rule,

(3) $$\dfrac{dx}{dx} = \dfrac{dx}{dy}\dfrac{dy}{dx}.$$

Since $\dfrac{dx}{dx} = 1$ and $\dfrac{dx}{dy} = \dfrac{d}{dy}\tan y = \sec^2 y$, equation (3) becomes

$$1 = \sec^2 y \dfrac{dy}{dx}.$$

Solving for $\dfrac{dy}{dx}$, we find

(4) $$\dfrac{dy}{dx} = \dfrac{1}{\sec^2 y}.$$

There is nothing wrong with this answer, but we can put it in a form that might be easier to work with. Noting that

$$\sec^2 y = 1 + \tan^2 y = 1 + x^2,$$

we see that (4) can be written as

$$\dfrac{dy}{dx} = \dfrac{1}{1 + x^2}.$$ ■

Why We Work in Radians

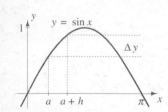

Figure 2.22. Graph of $y = \sin x$ with x in radians.

As mentioned earlier in this section, the calculus of trigonometric functions is a little easier to manage if angles are measured in radians. We can illustrate this graphically by noting how the rate of change of $y = \sin x$ differs when x is measured in radians and in degrees. Figure 2.22 shows the graph of $y = \sin x$. We assume x is in radians and have marked the x-axis accordingly. Consider the point $x = a$ and the nearby point $x = a + h$ on the x-axis. Let

$$\Delta y = \sin(a + h) - \sin a$$

be the change in the value of the sine function as x changes from a to $a + h$. Then we know that

(5) $$\dfrac{\Delta y}{h} \approx \dfrac{d}{dx}\sin x \bigg|_{x=a} = \cos a.$$

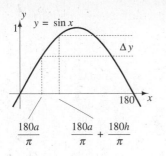

Figure 2.23. Graph of $y = \sin x$ with x in degrees.

The units of this rate of change are y units/radian. Now suppose that x is given in degrees and recalibrate the x-axis accordingly. (See Fig. 2.23.) Note that nothing about the graph changes except the units on the x-axis. Since

$$2\pi \text{ radians} = 360°,$$

the a radian point of Fig. 2.22 corresponds to $180a/\pi$ degrees, and the $a + h$ point is $(180a/\pi) + (180h/\pi)$ degrees. The change Δy in the value of the sine function for these two values is unchanged because the graph is unchanged and the units on the vertical axis are unchanged. However, the rate of change is now given by

$$(6) \qquad \frac{\Delta y}{180h/\pi} = \frac{\pi}{180}\frac{\Delta y}{h}.$$

From (5) we have $\Delta y/h \approx \cos a$, where a is in radians. Thus when x is given in degrees, the rate of change in (6) is

$$(7) \qquad \frac{\pi}{180}\cos a.$$

Expressions (5) and (7) illustrate an important reason for working in radians. In radians, the derivative of the sine function is "cleaner." In degrees, the derivative carries an inconvenient constant multiplier of $\pi/180$. As the above demonstration shows, one can think of radian measure as simply scaling the domain variable x to make the derivative work out as nicely as possible. Of course, radian measure has many other nice features, such as providing a natural way of relating angle measure to the circle.

Exercises 2.4

Basic

Exercises 1–6: Write down or describe all solutions to the equation. In some cases you may want to give decimal approximations.

1. $\sin\theta = \dfrac{1}{2}$

2. $\tan x = 1$

3. $\cos t = \dfrac{3}{5}$

4. $\sin 3\theta = -\dfrac{\sqrt{3}}{2}$

5. $\cos^2 x - 4\cos x - 5 = 0$

6. $\tan^2 x - 2\tan x - 4 = 0$

Exercises 7–18: Find the derivative.

7. $y = \sin(2x + 3)$

8. $q(t) = \sec t$

9. $f(x) = -\dfrac{\cos x}{x}$

10. $w = \dfrac{1 - \sin 2u}{1 + \sin 2u}$

11. $T(\theta) = \tan\left(\dfrac{1 + \theta}{1 - \theta}\right)$

12. $h(t) = \cot 3t$

13. $s(v) = \csc 4v$

14. $z = \sin r \cos r$

15. $y = \sin(x\cos x)$

16. $F(w) = (2 - \sqrt{\sin w})^5$

17. $f(x) = 1 + \tan^2(2x)$

18. $g(t) = p(\sin t)$, where $p(x) = 3x^4 - 2x^3 + x - 7$

Exercises 19–22: The given equation defines y as a function of x. Find the value of $\dfrac{dy}{dx}$ at point P.

19. $\sin y = x$ at $P = \left(\dfrac{\sqrt{2}}{2}, \dfrac{3\pi}{4}\right)$

20. $\sin^2 x + \cos^2 y = 1$ at $P = (0.5, 0.5)$

21. $\tan^2 x - \sec^2 y = -1$ at $P = \left(-\dfrac{\pi}{12}, -\dfrac{\pi}{12}\right)$

22. $\tan(x + y) = x + y + 1$ at $P = (a, b)$

23. Describe the values of θ for which $\sin \theta > 0$ but $\sin \theta/2 < 0$.

24. Describe the values of θ for which both $\cos \theta > 0$ and $\cos \theta/2 > 0$.

25. Describe the values of θ for which both $\tan \theta < 0$ and $\tan \theta/3 < 0$.

26. Describe the values of θ for which $\sin \theta = 0$ but $\sin \theta/2 < 0$.

27. Use the unit circle to show that
$$\sin(\theta + \pi/2) = \cos \theta.$$

28. Use the unit circle to show that
$$\sin(\pi - \theta) = \sin \theta.$$

29. Use the unit circle to show that
$$\cos(\pi - \theta) = -\cos \theta.$$

Growth

30. Take a stick of length $\pi/2$ and fasten a pencil to each end. Holding the stick parallel to the x-axis, use the point of the left pencil to draw the graph of $y = \sin x$. What graph is drawn by the pencil on the right?

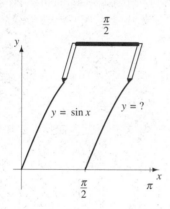

31. If the stick in the previous problem is held at an angle of $\pi/4$ with respect to the x-axis, and the left pencil again draws the graph of $y = \sin x$, what graph does the right pencil draw?

32. In Section 1.3 we showed that
$$\lim_{h \to 0} \frac{\sin h}{h} = 1$$
and
$$\lim_{h \to 0} \frac{\cos h - 1}{h} = 0.$$

The following is the beginning of an argument that $(\sin \theta)' = \cos \theta$. Fill in the justification for each step and complete the argument.

$$(\sin \theta)' = \lim_{h \to 0} \frac{\sin(\theta + h) - \sin \theta}{h}$$

$$= \lim_{h \to 0} \frac{(\sin \theta \cos h + \cos \theta \sin h) - \sin \theta}{h}$$

$$= \lim_{h \to 0} \left(\sin \theta \frac{\cos h - 1}{h}\right)$$

$$+ \lim_{h \to 0} \left(\cos \theta \frac{\sin h}{h}\right)$$

33. The figure below shows a metal rod of length 1 attached to the rim of a wheel of radius 1. The rod is attached to the wheel by a pin that allows the rod to pivot freely about the point of attachment. As the wheel rotates, the free end of the rod moves back and forth in a horizontal channel centered at $x = 0$. The wheel turns counterclockwise at a constant rate of two revolutions per second. Assume at time $t = 0$ the free end of the rod was at position $x = 2$.

a. On a set of axes with the horizontal axis representing time and the vertical axis representing position on the x-axis, sketch a graph showing the position of the end of the rod at any time t.

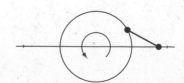

b. Find an equation that gives the position of the end of the rod at any time t.

c. Find the derivative with respect to t of the function found in part b. Interpret the derivative as a rate of change. What information does the derivative give us?

34. A cork floating in the ocean is lifted up and down by the waves. The height difference between the low point of the cork and the high point is 10 m, and the cork rises from its low point to its high point and then falls back to its low point six times per minute.

   a. On a set of axes with the horizontal axis representing time and the vertical axis representing height, sketch a graph showing the height of the cork at time t. What does height 0 mean in your graph? What is the significance of time $t = 0$ on your graph?

   b. Write an equation (using a trigonometric function) to describe the height of the cork at any time t given in minutes. Include a discussion of all assumptions and decisions you made in coming up with your equation.

   c. Using your answer to **b**, find the rate of change of height with respect to time. What does this rate of change tell you?

35. Let $A = (\cos a, \sin a)$ and $B = (\cos b, \sin b)$ be distinct points on the unit circle, with neither one equal to the point $P = (1, 0)$ and so that the segment $\overline{AB}$ is not vertical. See the accompanying figure.

   a. Find the slope of segment AB.

   b. Write down an equation for the line through P and parallel to $\overline{AB}$.

   c. Find the other point, Q, where this line intersects the unit circle.

   d. Show by geometry that

$$Q = (\cos(a + b), \sin(a + b)).$$

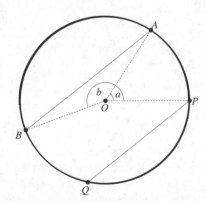

e. Combine **c** and **d** to obtain an angle addition formula for the sine and cosine. (Warning: These formulas will look different from the ones you are used to seeing!)

36. This exercise outlines a geometric argument that the derivative of $\tan \theta$ is $\sec^2 \theta$ (at least for first-quadrant angles θ). In the accompanying diagram, $\angle DOA = \theta$, $\angle DOB = \theta + h$, and $\overline{AC}$ is perpendicular to $\overline{OA}$.

   a. Show that $\overline{AB}$ has length

$$\tan(\theta + h) - \tan \theta$$

   and that $\overline{OA}$ has length $\sec \theta$.

   b. Show that $\angle BAC = \theta$.

   c. Justify the statement $AC \approx h \sec \theta$.

   d. Combine a, b, and c to conclude that

$$\frac{\tan(\theta + h) - \tan \theta}{h} \approx \sec^2 \theta.$$

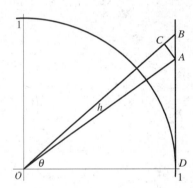

37. In the figure used in the preceding exercise, show that

$$BC = \sec(\theta + h) - \sec \theta.$$

Use this to show that

$$\frac{\sec(\theta + h) - \sec \theta}{h} \approx \tan \theta \sec \theta.$$

38. At the beginning of this section we remarked that the speed of the shadow of the car of a ferris wheel illustrates the rate of change of the cosine function. Check that the statements made about the shadow's motion are consistent with the fact that

$$(\cos \theta)' = \sin \theta.$$

Review/Preview

39. Solve for x:

$$y = \frac{3x + 2}{2x - 7}$$

40. If y is a real number, for how many real x might

$$y = x^3 - 3x?$$

41. The function

$$f(t) = (2t^2 - 3t, (2 - t)^3)$$

takes a real number t and returns an ordered pair of real numbers. What is $f(-1)$? $f(1/2)$?

42. The function

$$h(u) = (\cos 2u, u, u^2)$$

takes a real number u and returns an ordered triple of real numbers. What is $h(\pi/4)$? $h(0)$?

43. The function

$$g(x, y) = 6x^2 - 3xy + 4y^2$$

takes an ordered pair (x, y) of real numbers and returns a real number. What is $g(-3, 5)$? $g(0, 2)$? $g(a, a)$?

44. Write down a formula for a function that takes an ordered pair (x, y) of real numbers and returns an ordered triple of real numbers.

45. Let $f(x) = x^3 - 3x$. Graph the function on the interval $[-2, 3] = \{x : -2 \le x \le 3\}$, and then use the graph to help answer the following questions:
 a. For what x does $f(x) = 0$?
 b. On what interval(s) does the graph rise from left to right?
 c. On what interval(s) does the graph fall from left to right?
 d. At which x values does the graph change from rising to falling? From falling to rising? What is special about the graph at these points?

46. Let $T(\theta) = \sin \theta + 2 \cos \theta$. Graph the function on the interval $[0, \pi] = \{x : 0 \le \theta \le \pi\}$, and then use the graph to help answer the following questions:
 a. For what θ does $T(\theta) = 0$?
 b. On what interval(s) does the graph rise from left to right?
 c. On what interval(s) does the graph fall from left to right?
 d. At which θ values does the graph change from rising to falling? From falling to rising? What is special about the graph at these points?

2.5 EXPONENTIAL FUNCTIONS

The world's population doubles every 25 years. A star of the first magnitude is 100 times brighter than a star of the sixth magnitude. Nuclear waste must be stored safely for 25,000 years. A dollar invested at 4 percent interest per year compounded daily will provide your descendants 1000 years later with over ten thousand trillion dollars. We have all heard facts like these quoted in the news or in books of "Amazing Facts." Diverse as they may seem, these facts all have one thing in common. All are described by the exponential function. The exponential function is also used in describing more subtle aspects of our daily lives, from the warming of a can of soda to the swaying of tall skyscrapers in the wind.

Investigation 1

In the March 1992 issue of *Applied and Environmental Microbiology*, Derek Lovely and Elizabeth Phillips of the U.S. Geological Survey presented data on the role of bacterium *Desulfovibrio desulfuricans* in the reduction of uranium (VI) to uranium (IV). Uranium (VI) in solution was added to a solution containing the bacteria. The amount of uranium (VI) (in micromoles, or mM) was measured every hour for four hours. The purpose of the research was to study how bacteria contribute to the reduction of uranium (VI) in sedimentary environments. (One possible benefit of such research is a way to use microorganisms to help recover uranium from contaminated waters and waste streams.) Data from the experiment are summarized in Table 1 and Fig. 2.24. Find the rate of change of the amount of uranium with respect to time and discuss the significance of the result.

The graph in Fig. 2.24 shows that the amount of uranium (VI) decreases with time. The graph is very steep near $t = 0$ and very shallow near $t = 4$. This suggests a very rapid decrease in the amount of uranium (VI) at the beginning of the experiment, but a more gradual decrease later. To better understand how the amount of uranium changes with time, we investigate the rate of change. Let $U(t)$ be the amount of uranium at time t. Then

$$U'(0) \approx \frac{U(1) - U(0)}{1 - 0} = \frac{0.4 - 0.8}{1} = -0.4 \text{ mM/h}$$

$$U'(1) \approx \frac{U(2) - U(1)}{2 - 1} = \frac{0.21 - 0.4}{1} = -0.19 \text{ mM/h}$$

$$U'(2) \approx \frac{U(3) - U(2)}{3 - 2} = \frac{0.21 - 0.11}{1} = -0.1 \text{ mM/h}$$

$$U'(3) \approx \frac{U(4) - U(3)}{4 - 3} = \frac{0.11 - 0.06}{1} = -0.05 \text{ mM/h}.$$

These results are summarized in Table 2. In this table we have added a third column, showing the value of $U'(t)/U(t)$. The values in this third column are very close and suggest that

$$(1) \qquad \frac{U'(t)}{U(t)} \approx -0.48 \quad \text{or} \quad U'(t) \approx -0.48U(t).$$

We interpret (1) as saying

The rate of change in the amount of uranium is proportional to the amount of uranium.

This makes sense. When there is a lot of uranium in solution, there is a lot to be removed, so the decrease is large. When there is not very much uranium, only a little bit is removed, so the decrease is small.

We can use (1) to say even more about $U(t)$. From

$$\frac{U(t + 1) - U(t)}{(t + 1) - t} \approx U'(t) \approx -0.48U(t),$$

we solve for $U(t + 1)$ to get

$$(2) \qquad U(t + 1) \approx U(t) - 0.48U(t) = 0.52U(t).$$

Put $t = 0$ into (2) and we have

$$(3) \qquad U(1) \approx 0.52U(0) = 0.8(0.52).$$

Next set $t = 1$ in (2) and use (3) to get

$$(4) \qquad U(2) \approx 0.52U(1) \approx 0.8(0.52)^2.$$

Table 1

t (h)	$U(t)$ (mM)
0	0.8
1	0.4
2	0.21
3	0.11
4	0.06

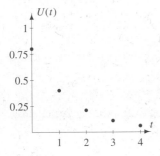

Figure 2.24. Uranium (VI) in solution.

Table 2

t (h)	$U'(t)$ (mM/h)	$U'(t)/U(t)$
0	−0.4	−0.5
1	−0.19	−0.475
2	−0.1	−0.476
3	−0.05	−0.456
4	—	—

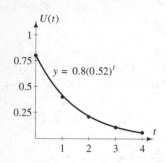

Figure 2.25. The graph of $U(t) = 0.8(0.52)^t$ fits the uranium (VI) data.

Continue this process by taking $t = 2$, and then $t = 3$ in (2). We then find

$$(5) \qquad\qquad U(3) \approx 0.52U(2) \approx 0.8(0.52)^3$$

and

$$(6) \qquad\qquad U(4) \approx 0.52U(3) = 0.8(0.52)^4.$$

From (3), (4) , (5), and (6) it appears reasonable to guess that

$$(7) \qquad\qquad U(t) \approx 0.8(0.52)^t.$$

How good is this guess? In Fig. 2.25 we show the graph of (7) and the data points from Fig. 2.24. Pretty good agreement.

The function described by the right side of (7) is an example of an exponential function.

DEFINITION   exponential function

Let b be a positive real number with $b \neq 1$. The exponential function of base b is the function f given by

$$(8) \qquad\qquad f(x) = b^x.$$

The domain of the exponential function is the set of real numbers.

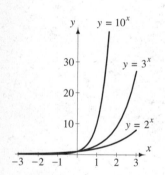

Figure 2.26. Graphs of $y = b^x$ for some $b > 1$.

Use a calculator or computer to graph $y = b^x$ for various values of the base b. The graphs illustrate some of the important features of these functions. First graph $y = b^x$ for several values of $b > 1$. See Fig. 2.26. The graphs all rise as we move from left to right, and appear to rise more rapidly for larger values of b. As we look far to the left, where x is large and negative, we see that the graph is very close to the x-axis. When $b > 1$, the graph of $y = b^x$ is asymptotic to the negative x-axis.

Now graph $y = b^x$ for several values of b with $0 < b < 1$ (see Fig. 2.27). These graphs all fall as we move from left to right, and appear to fall more rapidly for smaller values of b. The graphs also suggest that for $0 < b < 1$, the graph of $y = b^x$ is asymptotic to the positive x-axis.

The graphs suggest another very important feature of the exponential functions.

For all real x, $b^x > 0$. Thus, an exponential function is never equal to 0.

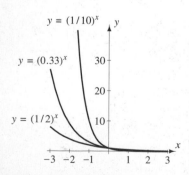

Figure 2.27. Graphs of $y = b^x$ for some $b < 1$.

Remember that the exponential functions b^x are different from the power functions x^n is many ways. Even though 2^x and x^2 look very similar, the two functions described by these expressions are really very different. For example, the graph of the exponential function $f(x) = 2^x$ is always rising as we move from left to right, whereas the graph of the power function $g(x) = x^2$ is falling for $x < 0$ but

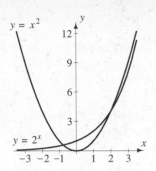

Figure 2.28. Graphs of $y = 2^x$ and $y = x^2$.

rising for $x > 0$. (See Fig. 2.28.) The exponential function has an asymptote, but the power function does not. The exponential function is never 0, but the power function does take on the value 0. The differences in the two functions are further illustrated by their derivatives.

Investigation 2

Find $f'(x)$ if $f(x) = b^x$.

By the definition of derivative, we have

$$(9) \qquad f'(x) = \lim_{h \to 0} \frac{f(x + h) - f(x)}{h} = \lim_{h \to 0} \frac{b^{x+h} - b^x}{h}.$$

From rules for working with exponents, we have

$$b^{x+h} = b^x b^h.$$

Substitute this into (9) and factor b^x out of the numerator to get

$$(10) \qquad f'(x) = \lim_{h \to 0} \frac{b^x b^h - b^x}{h} = \lim_{h \to 0} b^x \frac{b^h - 1}{h} = b^x \lim_{h \to 0} \frac{b^h - 1}{h}.$$

Since b^x is independent of h, we were able to factor it outside the limit to obtain the last expression in (10). Now let

$$(11) \qquad C_b = \lim_{h \to 0} \frac{b^h - 1}{h}.$$

Combining (10) and (11) we see that the derivative of $f(x) = b^x$ is

$$f'(x) = C_b b^x.$$

This value of C_b is independent of x but takes on different values for different values of the base b of the exponential function. For example, when $b = 2$, we find

$$\frac{d}{dx} 2^x = C_2 2^x$$

where

$$(12) \qquad C_2 = \lim_{h \to 0} \frac{2^h - 1}{h}.$$

To get an idea of the value of the limit in (12) we can graph $y = (2^h - 1)/h$ for values of h close to 0 (see Fig. 2.29) or calculate the value of this quotient for values of h close to 0 (see Table 3).

Reading from the table or the graph, we see that $C_2 \approx 0.693$. We could get more accuracy by zooming closer to the graph in Fig. 2.29 or by extending Table 3 with smaller values of h. In any case, this shows us that

$$\frac{d}{dx} 2^x \approx (0.693) 2^x.$$

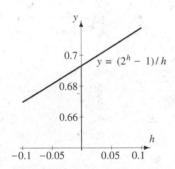

Figure 2.29. Graph of $y = (2^h - 1)/h$.

Table 3

h	$\dfrac{(2^h - 1)}{h}$
−0.1	0.66967
0.1	0.717735
−0.01	0.69075
0.01	0.695555
−0.001	0.692907
0.001	0.693387
−0.0001	0.693123
0.0001	0.693171

Thus, the derivative of the exponential function 2^x is a constant times the function itself, and the constant is approximately 0.693.

We can repeat the above procedures for any positive b and find an estimate for C_b. In Exercise 24 you are asked to verify some of the C_b values given in Table 4. Using the values in this table, we see that

$$\frac{d}{dx}3^x = C_3 3^x \approx (1.0986)3^x \quad \text{and} \quad \frac{d}{dx}10^x = C_{10}10^x \approx (2.3026)10^x.$$

We summarize the results of the previous discussion in the following statement.

Table 4. Some values for C_b

b	$C_b \approx$
2.0	0.693147
2.2	0.788457
2.4	0.875469
2.6	0.955511
2.7	0.993252
2.8	1.02962
3.0	1.09861
4.0	1.38629
10.0	2.30259

The Derivative of an Exponential Function

Let f be the function defined by $f(x) = b^x$. Then

(13) $$f'(x) = \frac{d}{dx}b^x = C_b b^x$$

where the constant C_b is given by

$$C_b = \lim_{h \to 0} \frac{b^h - 1}{h}.$$

Equation (13) reveals one of the most important facts about the exponential function:

If f is an exponential function, then the rate of change of $f(x)$ with respect to x is proportional to $f(x)$.

This property of the exponential function was illustrated by the function U in Investigation 1. Many familiar quantities change at a rate proportional to the size of the quantity itself. Some examples are:

Population: The larger a population is, the faster it grows, since more organisms means more births.

Invested money: A large sum of money earns more interest than a small sum.

Radioactive decay: Radium-227 has a half-life of 6.7 years. A 100-gram sample will decrease to 50 grams in 6.7 years, while a 10-gram sample will decrease to 5 grams over the same time period.

Cooling: A very hot copper wire will lose more heat in 10 minutes than a copper wire just slightly above room temperature.

Exponential functions are used in describing all of these phenomena, and many more.

The Number *e*

The values of $C_2 \approx 0.693147$, $C_{2.7} \approx 0.993252$, $C_{2.8} \approx 1.02962$, and $C_3 \approx 1.09861$ in Table 4 suggest that C_b increases in value as b increases. Hence there should be a value of b between 2.7 and 2.8 with $C_b = 1$. This special value of b is one of the most important numbers in mathematics. We use e to denote this special number. (The letter e honors the great mathematician Leonhard Euler.)

The Number e

There is a number e for which $C_e = 1$. Hence

$$\frac{d}{dx}e^x = C_e e^x = e^x.$$

Since the number e is the special value of b for which

(14)
$$C_b = \lim_{h \to 0} \frac{b^h - 1}{h} = 1,$$

we can approximate e by trying different values of b in (14) and numerically or graphically checking the limit to see if it is equal to 1. (See Exercise 20 in Section 1.4 and Exercise 36 at the end of this section.)

Here we will take an alternative approach. Set h to a small value (say $h = 0.00001$) and graph

(15)
$$C_b = \frac{b^{0.00001} - 1}{0.00001}$$

for b values between 2 and 3. In effect, this allows us to estimate the limit in (14) for all b between 2 and 3. The graph is shown in Fig. 2.30. We are interested in the b value for which (15) is equal to 1. The graph shows that this happens for some value of b between 2.7 and 2.8, so $2.7 < e < 2.8$.

Now take h a little smaller, say $h = 0.000001$, and graph

(16)
$$C_b = \frac{b^{0.000001} - 1}{0.000001}$$

for $2.7 \leq b \leq 2.8$. From the graph in Fig. 2.31 we see that $C_b = 1$ when b is slightly smaller than 2.72. Hence $e \approx 2.72$ and $2.71 < e < 2.72$.

By continuing this zooming-in process or trying different values of b in (14), we can get better and better approximations to e. With enough patience and computing power, we can find that

$$e \approx 2.718281828459045.\ldots$$

Like the constant π, the number e is irrational; that is, it cannot be written as a fraction in which the numerator and denominator are integers. Hence the decimal representation for e is nonterminating and nonrepeating. Like π, the number e is also transcendental. This means that e is not a zero of any polynomial with integer coefficients. Just as mathematicians and scientists prefer to use radians when working with angles and trigonometric functions, they prefer to use e as the base for exponential functions. Calculus with trigonometric functions is "simpler" when done in radians. The calculus of exponential functions is "simpler" when the base for the exponential is e. Every scientific calculator has a button for evaluating the exponential function. This button is usually labeled e^x or "exp."

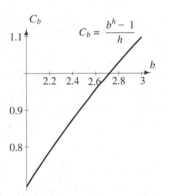

Figure 2.30. Graph of $(b^h - 1)/h$ for $2 \leq b \leq 3$ and $h = 0.00001$.

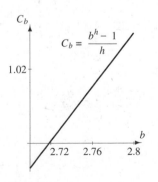

Figure 2.31. Graph of $(b^h - 1)/h$ for $2.7 \leq b \leq 2.8$ and $h = 0.000001$.

The Exponential Function e^x

The exponential function with base e is usually referred to as *the* exponential function.

The Exponential Function

The exponential function, sometimes denoted by "exp," is the function given by

$$\exp(x) = e^x.$$

The derivative of the exponential function is

$$\frac{d}{dx} \exp(x) = e^x.$$

The graph of the exponential function is shown in Fig. 2.32. We end this section with a few examples involving exponential functions.

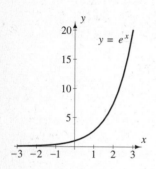

Figure 2.32. Graph of $y = e^x$.

■ **EXAMPLE 1** In the early stages of growth, the population P of a colony of bacteria can often be described by an exponential function. Assuming time t is in hours, estimate how long it takes a population to double if at time t the population is:

    a. $P(t) = 2^t$.   **b.** $P(t) = e^t$.   **c.** $P(t) = 10^t$.

Solution

a. At time t_0 the population is $P(t_0) = 2^{t_0}$. Note that one hour later the population is

$$P(t_0 + 1) = 2^{t_0+1} = 2 \cdot 2^{t_0} = 2P(t_0).$$

Hence the population doubles in one hour. Notice that the doubling time is always one hour, regardless of the starting time t_0.

b. Now let $P(t) = e^t$ and suppose that at time t_1 the population is double what it was at time t_0. Then the doubling time t_d is $t_d = t_1 - t_0$ and

(17) $$P(t_1) = 2P(t_0).$$

This means that

$$e^{t_1} = 2e^{t_0}.$$

Divide both sides of the last equation by e^{t_0} to get

(18) $$2 = \frac{e^{t_1}}{e^{t_0}} = e^{t_1 - t_0} = e^{t_d}.$$

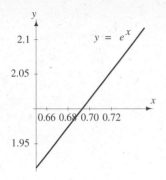

Figure 2.33. Graph of $y = e^x$ and $y = 2$ near the intersection point of the two graphs.

Table 5

t	10^t
0.0	1.00000
0.1	1.25893
0.2	1.58489
0.3	1.99526
0.4	2.51189

Table 6

t	10^t
0.300	1.99526
0.301	1.99986
0.302	2.00447

Thus the doubling time is t_d hours, where t_d is the number for which $e^{t_d} = 2$. There are several ways to estimate t_d. We do this graphically by zooming in on the graph of $y = e^x$ at the point where the graph crosses the line $y = 2$. See Fig. 2.33. From the graph we see that $t_d \approx 0.693$. Hence the time for the population to double is approximately 0.693 hours, or about 41.5 minutes.

We could estimate for t_d by using a calculator to try different possibilities for t_d in (18) and refining our estimate with each try. We might also recall from high school algebra that $2 = e^{t_d}$ is solved by

$$t_d = \log_e 2 = \ln 2 \approx 0.693147.$$

We've seen this number earlier in this section. Where?

c. Now let $P(t) = 10^t$ and let t_1 be the time when the population is double what it was at time t_0. The doubling time is again given by $t_d = t_1 - t_0$. Equation (17) now leads to

$$2 = 10^{t_d}.$$

We can estimate t_d by trying possibilities with our calculator and refining. Since $10^0 = 1$ and $10^1 = 10$, we know that t_d is between 0 and 1. To refine this estimate, calculate the value of 10^t for $t = 0.1, 0.2, 0.3 \ldots$. The results of these calculations are shown in Table 5.

From the table we see that t_d is between 0.3 and 0.4. Next we see that $10^{0.31} \approx 2.04174$. Hence t_d is between 0.30 and 0.31. Now try values between these two numbers. Starting from 0.300 and increasing by 0.001 at each step, we get the values shown in Table 6. We find that t_d is between 0.301 and 0.302. So with exponential growth given by 10^t, the doubling time is approximately 0.301 hours, or about 18 minutes. ∎

■ **EXAMPLE 2** Find $h'(t)$ if

$$h(t) = \exp(2t \sin 2t) = e^{2t \sin 2t}.$$

Solution. The function given by $h(t)$ is a composition. Let $y = h(t)$ and write a composition chain:

$$y = e^u$$
$$u = 2t \sin 2t.$$

We then have

$$\frac{dy}{du} = \frac{d}{du} e^u = e^u.$$

By the product rule,

$$\frac{du}{dt} = (2t \sin 2t)'$$
$$= (2t)'(\sin 2t) + 2t(\sin 2t)'$$
$$= 2 \sin 2t + 2t\big((\cos 2t)(2t)'\big)$$
$$= 2 \sin 2t + 4t \cos 2t,$$

where we have also used the chain rule to differentiate $\sin 2t$. Combine the above derivatives to get

$$h'(t) = \frac{dy}{dt} = \frac{dy}{du}\frac{du}{dt} = e^u(2\sin 2t + 4t\cos 2t)$$

$$= (2\sin 2t + 4t\cos 2t)e^{2t\sin 2t}.$$ ∎

■ **EXAMPLE 3**   The equation

(19) $$x = e^y$$

defines y as a function of x. Find an expression for $\dfrac{dy}{dx}$.

Solution.   We write $y(x)$ for y to help us remember that y is a function of x. Then (19) becomes

$$x = \exp(y(x)).$$

Differentiate both sides of this equation with respect to x, using the chain rule on the right. We obtain

$$1 = \exp'(y(x))y'(x) = \exp(y(x))y'(x) = e^y\frac{dy}{dx}.$$

Solve for $\dfrac{dy}{dx}$ to get

$$\frac{dy}{dx} = \frac{1}{e^y}.$$

Since (19) tells us that $e^y = x$, we can also express this derivative as

$$\frac{dy}{dx} = \frac{1}{x}.$$ ∎

Exercises 2.5

Basic

Exercises 1–7:   Find all solutions to the equation. In some cases decimal approximations are appropriate.

1. $2^{-3x} = 16$
2. $5^{2t} - 2 \cdot 5^t = 0$
3. $e^\theta = e$
4. $e^r = \frac{1}{2}$
5. $e^{2x} - 3e^x + 2 = 0$

6. $\dfrac{e^{3t} - 1}{e^{3t} + 1} = 4$

7. $\dfrac{e^u + e^{-u}}{2} = 5$

Exercises 8–19: Find the derivative.

8. $y = e^{2x}$

9. $f(x) = \sin(2e^{2x})$

10. $g(t) = 4^t$ $(C_4 \approx 1.38629)$

11. $z = e^{e^t}$

12. $w = \sqrt{(e^x)^2 - e^x + 3}$

13. $h(r) = r^4 e^{4r}$

14. $q(s) = 2^s e^{4s}$

15. $s(t) = \tan\left(\dfrac{1 + e^t}{1 - e^t}\right)$

16. $R(\tau) = \tau^2 e^{\tau^2}$

17. $v = \dfrac{(e^u + 2)^2}{\sin 3u}$

18. $c(t) = \dfrac{e^t + e^{-t}}{2}$

19. $s(t) = \dfrac{e^t - e^{-t}}{2}$

Exercises 20–23: Find an expression for $\dfrac{dy}{dx}$ if the equation defines y as a function of x.

20. $(x + y)^2 = ye^x$

21. $ye^{x+y} = 1$

22. $e^x \tan(xy^2) - 2y = x + y$

23. $e^{y^2} = xy$

24. Verify any three of the C_b values in Table 4.

25. Suppose that in investigating the derivative of the exponential function, we decide that we want a base b so that

$$(b^x)' = 2 \cdot b^x.$$

What is the value of b? Find the answer by numerical experimentation, and then by considering the equation $(e^{bx})' = b \cdot e^{bx}$.

26. When the sources of pollution are eliminated, a large polluted body of water eventually "cleanses" itself through natural processes. Observations have shown that during this cleansing period the concentration of pollutants can be described by an exponential function of the form

$$C(t) = C_0 e^{kt}.$$

In this expression, $C(t)$ has units of parts per million (ppm) and is the concentration of pollutants at time t, C_0 is the concentration at time $t = 0$, and k is a constant. An environmentalist measures the pollutants in the lake at time $t = 0$ and finds a concentration of 150 ppm. One year later the concentration is 100 ppm.

  a. Find C_0 and k.

  b. What will the concentration of pollutants be 5 years after the initial measurement?

  c. When will the concentration of pollutants be less than 10 ppm?

27. When money in a savings account earns interest at a rate of r percent per year, compounded daily, then the balance of the account T years after the money is deposited is well approximated by

$$A(t) = A_0 e^{0.01r \cdot T}$$

where A_0 is the amount of the initial deposit. In the introduction to this section we stated that "a dollar invested at 4 percent interest per year compounded daily will provide your descendants 1000 years later with over ten thousand trillion dollars." Verify this statement.

28. Two thousand years ago Greek astronomers Hipparchus and Ptolemy created a system to describe the apparent brightness of stars. The 20 brightest visible stars were denoted magnitude 1 stars, and the faintest visible stars were declared to be of magnitude 6. Other stars were assigned magnitudes based on their apparent brightness compared with those of magnitude 1 or 6. In 1850 Pogson refined the system. He noticed that magnitude 1 stars were about 100 times brighter than magnitude 6 stars. Since there are five steps of magnitude from 1 to 6, Pogson proposed that a star of magnitude m should be $\sqrt[5]{100} \approx 2.512$ times brighter than a star of magnitude $m + 1$.

  a. Explain how Pogson came up with the number $\sqrt[5]{100}$ and why this number is reasonable.

  b. Antares is a magnitude 1.2 star, and Marsik is a magnitude 5.3 star. Let b_1 be the brightness of Antares and b_2 be the brightness of Marsik. Find the numerical value of b_1/b_2.

Growth

29. Give a convincing argument that the equation $e^{-2x} = k$ has a solution for every $k > 0$.

30. Give a convincing argument that the equation $e^x + 3e^{3x} = k$ has a solution for every $k > 0$.

31. Consider the equation $e^x + 3e^{-3x} = k$. Find two positive k values for which this equation does *not* have a solution.

32. Let $b > 0$ with $b \neq 1$ and let $r > 0$. Give a convincing argument that the equation

$$b^x = r$$

has exactly one solution. Support your argument with graphs.

33. Let a, b, k be real numbers. For what values of a and b does the equation

$$e^{ax} + 2e^{bx} = k$$

have a solution for every $k > 0$?

34. Compute e^{C_2}, e^{C_3}, and $e^{C_{10}}$. Make a conjecture about the value of e^{C_b}.

35. **a.** Differentiate both sides of the equation $(ab)^x = a^x b^x$, and use the result to show that $C_a + C_b = C_{ab}$.
    b. Differentiate both sides of the equation $(a/b)^x = a^x/b^x$, and use the result to show that $C_a - C_b = C_{a/b}$.
    c. Differentiate both sides of the equation $(a^b)^x = a^{bx}$, and use the result to show that $bC_a = C_{a^b}$.
    d. What familiar function satisfies the identities proved in a, b, and c?

36. In finding the value of e, we looked for a number b for which

$$\frac{b^h - 1}{h} \approx 1$$

for small h. Solve this expression for b to obtain

$$b \approx (1 + h)^{1/h}.$$

Substitute small values of h in the expression for b and obtain an approximation to e. Argue that

$$\lim_{h \to 0}(1 + h)^{1/h} = e.$$

37. Find two different functions that have derivative e^x.

38. Find two different functions that have derivative e^{3x}. Check by finding the derivative of your answers.

39. **a.** Find the equation of the line that passes through the origin and is tangent to the graph of $y = e^x$.

b. Consider the equation

$$e^x = ax.$$

How many solutions does this equation have if $a < 0$? If $a = 0$? If $a > 0$?

40. **a.** Investigate the behavior of x/e^x, x^2/e^x, x^{10}/e^x, and x^{50}/e^x for very large values of x. (You may do this numerically, graphically, or both.)
    b. Make a conjecture about the behavior of x^n/e^x as x grows toward infinity.

41. Simple population models often describe population by an exponential formula

$$P(t) = P_0 e^{kt}$$

where P_0 is the population at some starting time $t = 0$ and k is some positive constant. For such populations, the rate of change is

$$P'(t) = kP_0 e^{kt}.$$

Write a short paragraph explaining the meaning of this rate of change. In particular, explain why larger populations increase faster than smaller ones.

42. If amount A_0 of a radioactive material is present at time $t = 0$, then the amount $A(t)$ present at a later time $t > 0$ is given by

$$A(t) = A_0 e^{-kt}$$

where k is a positive constant. Find the rate of change of the amount of radioactive material with respect to time, and write a short paragraph explaining the meaning of this rate of change. When radioactive material decays, it gives off radiation. Use the rate-of-change equation to explain why it is more dangerous to be near a large amount of radioactive material than near a small amount.

43. The *hyperbolic functions* are defined by the equations

$$\cosh x = \frac{e^x + e^{-x}}{2} \qquad \sinh x = \frac{e^x - e^{-x}}{2}.$$

The function $\cosh x$ is the *hyperbolic cosine* of x, and $\sinh x$ is the *hyperbolic sine* of x. These particular combinations of exponential functions arise often in science and engineering and have been assigned the notations $\cosh x$ and $\sinh x$ for convenience. (See also Exercises 18 and 19.)

a. Show that

$$(\cosh x)' = \sinh x$$

and

$$(\sinh x)' = \cosh x.$$

b. Show that

$$(\cosh x)^2 - (\sinh x)^2 = 1.$$

(This identity is the reason for the word "hyperbolic" in the names of these functions. If $x = \cosh t$ and $y = \sinh t$, then the point (x, y) lies on the hyperbola $x^2 - y^2 = 1$.)

c. Show that

$$\cosh 2x = (\cosh x)^2 + (\sinh x)^2$$

and

$$\sinh 2x = 2 \sinh x \cosh x.$$

d. The *hyperbolic tangent* and *hyperbolic secant* are defined by

$$\tanh x = \frac{\sinh x}{\cosh x}$$

and

$$\operatorname{sech} x = \frac{1}{\cosh x},$$

respectively. Show that

$$(\tanh x)' = (\operatorname{sech} x)^2$$

and

$$(\operatorname{sech} x)' = -\operatorname{sech} x \tanh x.$$

44. When a drug is used in treatment, it is important to know the rate at which the drug is removed from the body. The following data were obtained by monitoring the levels of a drug in a patient's blood. Blood samples were taken every hour, and the concentration of the drug was determined and reported in milligrams per liter.

t (h)	Concentration (mg/liter)
0	10.0
1	7.0
2	5.0
3	3.5
4	2.5
5	2.0
6	1.5
7	1.0
8	0.7
9	0.5

Analyze the data as in Investigation 1.

a. Estimate the rate of change of concentration with respect to time.

b. Show that the rate of change is (roughly) proportional to concentration.

c. Come up with an exponential function that seems to describe the data. Plot your function and the data points on the same set of axes.

Review/Preview

45. Solve by graphing: $(x + 2)(x - 3) > 0$.

46. Use graphs to estimate the solution to

$$x^3 - 3x^2 + 4x - 5 < 0.$$

47. Find y' if $y = -3x^3 + 2x - (3/x^4)$.

48. Find $dr/d\theta$ if $r = \theta(2\theta + 1)(3\theta + 2)$.

49. How many solutions (x, y) are there to the following system of equations?

$$y = x^2 + 4x - 6$$

$$x = -y^2 - 3y + 1$$

50. Find an equation for the line through the points $(0, b)$ and $(a, 0)$.

51. If a nonvertical line has nonzero slope, then the equation for the line can be written in the form

$$\frac{x}{a} + \frac{y}{b} = 1.$$

This equation is called the *two-intercept* form for the line. Explain the reason for this terminology.

52. Write an equation that is satisfied precisely by the set of points (x, y) that are three units from $(2, 3)$.

53. Write an inequality that is satisfied precisely by the set of points (x, y) that are less than three units from $(2, 3)$.

54. Find an equation for the line that bisects the first-quadrant angle determined by the lines $y = 0$ and $y = x$.

2.6 LOGARITHMS

In April 1994 Derek Atkins, Michael Graff, Paul Leyland, and Arjen Lenstra announced that the 129-digit number

114381625757888867669235779976146612010218296721242362562523-5639-
587056184293570693524573389783059710589890751475992900268795435-41

$$= 3490529510847650949147849619903-8-
9813341776463849338784399082057-7$$

$$\times 3276913299326670954996198819083-4-
461413177642967992942539798288533.$$

The 129-digit number, known as RSA 129, appeared in a 1977 paper by Ronald Rivest, Adi Shamir, and Leonard Adelman. In the paper they presented a coding system now known as "public key encryption." This encoding system requires a large number, like RSA 129, that is the product of two large prime numbers. Anyone who has the large number and a publicly known encoding number can use these to encode a message. However, these messages can be decoded only by someone who knows the prime factors of the large number. Thus the security of a public key encryption system depends on the feasibility of factoring large numbers. When RSA 129 was presented as an example of a number that could be used for this encryption scheme, the authors estimated that given the current state of knowledge about factoring and the speed of 1977 computers, it would take about 40 quadrillion (4×10^{16}) years to factor the number. Atkins, Graff, Leyland, and Lenstra did it in about eight months.

Of course, computers in 1994 were much faster than the computers of 1977. In addition, the problem was cut into pieces and farmed out to over 600 volunteers worldwide. Each volunteer did a part of the problem and relayed the results to a central location for the final analysis. But the use of faster computers and 600 volunteers is nowhere near enough to cut 40 quadrillion years down to less than 1. The search for factors used a technique called quadratic sieving, developed by Carl Pomerance and announced in a 1982 paper. With this technique the amount of time required to factor a positive integer N is proportional to

$$(1) \qquad\qquad L(N) = e^{\sqrt{(\ln N)(\ln \ln N)}}$$

where $\ln N = \log_e N$ is the natural logarithm of N.

Today many businesses and governments use public key encryption to send confidential messages. Hence the speed with which large numbers can be factored is of tremendous importance. As (1) shows, logarithms are essential in assessing the security of public key systems because these codes can be broken only by factoring a large number. This illustrates one way in which logarithms play an important part in today's world. Logarithms are also used in describing the intensity of an earthquake, in measuring the noise of a Rolling Stones concert, and in establishing a schedule for the administration of medication.

The Logarithm Function

Find the number s that solves the equation

$$(2) \qquad\qquad 7^s = 34.$$

With a little thought we see that s is between 1 and 2. With some trial and error and the help of a calculator, we could find s accurate to one or two decimal places. Try it! But no matter how hard we work, we will never be able to write the exact value of s. However, equations similar to (2) show up often, and it is important to be able to work with the solutions to such equations. Because we cannot write down the solution to such equations, we simply introduce new notation to represent the solution. Thus we denote the solution to (2) by

$$s = \log_7 34$$

and say "s is the logarithm to the base 7 of 34." This terminology may sound confusing at first, but just remember that it is just a convenient means for talking about the solution to an equation such as (2).

DEFINITION **logarithm to the base b**

Let b be a positive real number with $b \neq 1$, and let $r > 0$. The solution to the equation

$$b^s = r$$

is called the logarithm to the base b of r and is denoted

$$s = \log_b r.$$

In other words, $\log_b r$ is the number to which b must be raised to get an answer of r. That is,

(3) $$b^{\log_b r} = r.$$

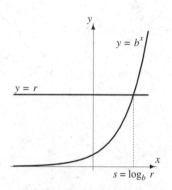

y = b^x

y = r

s = log_b r

Figure 2.34. When $r > 0$, the line $y = r$ meets the graph of $y = b^x$ in exactly one point.

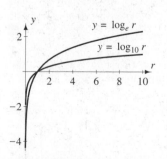

y = log_e r

y = log_10 r

Figure 2.35. Graphs of two logarithm functions.

If $b > 0$ and $b \neq 1$, then for each $r > 0$ the equation

(4) $$b^x = r$$

has a unique solution $x = s$. To see this, first observe that the horizontal line $y = r$ intersects the graph of $y = b^x$ in exactly one point (see Fig. 2.34). Denote the co-ordinates of this intersection point by (s, r). Then $b^s = r$, so $s = \log_b r$. Furthermore, there is no other solution to (4). Thus for each $r > 0$, there is one and only one number $\log_b r$ that satisfies (3). This means that the equation

$$f(r) = \log_b r$$

describes a (well-defined) function f whose domain is the set of positive real numbers.

Most scientific or graphing calculators have a button for $\log_{10}$ (usually labeled "log") and a button for $\log_e$ (usually labeled "ln"). The graphs of $y = \log_{10} r$ and $y = \log_e r$ are shown in Fig. 2.35. The graph reminds us of some facts about logarithms to a base $b > 1$. The graphs rise as we move from left to right. As x gets close to 0, the graph falls toward $-\infty$; thus the negative y-axis is an asymptote for each of the graphs. The graphs of $y = \log_e r$ and $y = \log_{10} r$ are unbroken, so these functions are continuous for $r > 0$.

■ **EXAMPLE 1**

a. Find the value of:

    i. $\log_5 125$.   **ii.** $\log_5 \dfrac{1}{\sqrt{5}}$.

b. Estimate the value of $\log_{1/2} 5$.

Solution

a. i. Let $c = \log_5 125$. According to the definition of logarithm, we must have

$$5^c = 125.$$

Hence $c = 3$, so

$$\log_5 125 = 3.$$

ii. Now let $d = \log_5 \dfrac{1}{\sqrt{5}}$. Then

$$5^d = \frac{1}{\sqrt{5}} = 5^{-1/2},$$

which means that $d = -1/2$. Hence

$$\log_5 \frac{1}{\sqrt{5}} = -1/2.$$

b. Set $c = \log_{1/2} 5$. Then

$$(1/2)^c = 5.$$

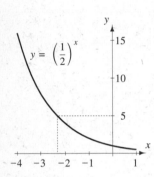

Figure 2.36. Graph of $y = (1/2)^x$.

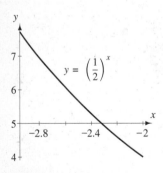

Figure 2.37. Graphically estimating the solution to $(1/2)^x = 5$.

The graph of $y = \left(\frac{1}{2}\right)^x$ is shown in Fig. 2.36. From the graph we see that there is only one value of x for which $(1/2)^x = 5$, and this occurs for some x between -2 and -3. Zoom in on the graph between these values. See Fig. 2.37. We see that the function takes the value 5 when x is approximately -2.3. This tells us that $\log_{1/2} 5 \approx -2.3$. Check this answer with your calculator. Is $\left(\frac{1}{2}\right)^{-2.3} \approx 5$? ■

Before the invention of cheap, accurate, easy-to-use calculators, logarithms were used to simplify arithmetic operations. Logarithms could be used to convert multiplication problems to addition problems, divisions to subtractions, and exponentiations to multiplications. (These techniques are not taught much nowadays, but you may be able to get your teacher to talk about the "old days.") Logarithms can be used to convert problems to different forms because of several identities. These identities are still useful for manipulating and simplifying complicated expressions involving logarithms.

Identities for Logarithms

Let b be positive with $b \neq 1$, let $x > 0$ and $y > 0$, and let r be any real number. Then

a. $\log_b 1 = 0$

b. $\log_b xy = \log_b x + \log_b y$

c. $\log_b \dfrac{x}{y} = \log_b x - \log_b y$

d. $\log_b x^r = r \log_b x$

e. If $a > 0$ and $a \neq 1$, then $\log_a x = \dfrac{\log_b x}{\log_b a}$.

We verify two of these identities here and leave the others for the exercises. See Exercises 29 and 30.

■ **EXAMPLE 2**   Verify the logarithm identity

$$\log_b \frac{x}{y} = \log_b x - \log_b y.$$

Solution.   Using the definition of logarithm, we can write

$$b^{\log_b(x/y)} = \frac{x}{y} = \frac{b^{\log_b x}}{b^{\log_b y}} = b^{\log_b x - \log_b y}.$$

The first and last expressions in the above display are equal and both are expressed as exponents to base b. Hence the exponents must be equal. (Why?) Equating these exponents gives our result:

$$\log_b \frac{x}{y} = \log_b x - \log_b y.$$ ■

■ **EXAMPLE 3**   Show that the logarithm identity

$$\log_a x = \frac{\log_b x}{\log_b a}$$

is true.

Solution.   By the definition of logarithm,

(5) $$b^{\log_b x} = x = a^{\log_a x}.$$

Now we also have

(6) $$a = b^{\log_b a}.$$

Substitute (6) into (5) to get

$$b^{\log_b x} = (b^{\log_b a})^{\log_a x} = b^{(\log_b a)(\log_a x)}.$$

The first and last expressions are both exponentials with base b. Hence the exponents must be equal. Thus

(7) $$\log_b x = (\log_b a)(\log_a x).$$

Since $a \neq 1$, we know $\log_b a \neq 0$. Dividing both sides of (7) by $\log_b a$ to get

$$\log_a x = \frac{\log_b x}{\log_b a}.$$

This property of logarithms shows us that there is, in a sense, just one logarithm function. That is, if $a, b > 0$ with neither equal to 1, then $\log_b x$ is a constant multiple of $\log_a x$.

The Derivative of the Logarithm Function

Using the definition of the logarithm and results from Section 2.5 about differentiation of the exponential functions, we can find a formula for the derivative of the logarithm to any base.

■ **EXAMPLE 4**   Find $\dfrac{d}{dx} \log_b x$.

Solution.   By the definition of logarithm,

(8) $$x = b^{\log_b x}.$$

The right side is the composition of two functions. If

$$f(x) = b^x \quad \text{and} \quad g(x) = \log_b x,$$

then

$$b^{\log_b x} = f(g(x)).$$

In Section 2.6 we saw that

$$f'(x) = (b^x)' = C_b b^x$$

where

$$C_b = \lim_{h \to 0} \frac{b^h - 1}{h}.$$

Now differentiate both sides of (8) with respect to x. Applying the chain rule for the derivative of the expression on the right, we have

$$1 = (f(g(x)))' = f'(g(x))g'(x)$$
$$= C_b b^{g(x)} g'(x) = C_b b^{\log_b x} (\log_b x)' = C_b x (\log_b x)'.$$

Dividing the last result by $C_b x$, we find

$$(\log_b x)' = \frac{d}{dx} \log_b x = \frac{1}{C_b x}.$$

With logarithm notation we can say more about the constants C_b.

■ **EXAMPLE 5** Let $y = b^x$ where $b > 0$ and $b \neq 1$. Show that

$$\frac{dy}{dx} = (b^x)' = (\log_e b)b^x.$$

From this conclude that

$$C_b = \lim_{h \to 0} \frac{b^h - 1}{h} = \log_e b.$$

Solution. Because $b = e^{\log_e b}$, we can write

$$y = b^x = e^{(\log_e b)x}.$$

Thus y can be written as a composition chain:

(9)
$$y = e^u$$
$$u = (\log_e b)x.$$

Because $\log_e b$ is a constant, we know

$$\frac{du}{dx} = \frac{d}{dx}(\log_e b)x = \log_e b.$$

Hence

$$\frac{dy}{dx} = \frac{dy}{du}\frac{du}{dx} = e^u(\log_e b) = (\log_e b)e^{(\log_e b)x} = (\log_e b)b^x.$$

In the previous section we showed that

$$\frac{dy}{dx} = (b^x)' = C_b b^x$$

for some constant C_b. Equate these two expression for $\frac{dy}{dx}$ to get

$$C_b b^x = \frac{dy}{dx} = (\log_e b)b^x.$$

Divide both sides of the equation by the nonzero number b^x to get

$$C_b = \log_e b.$$

The Natural Logarithm

In the previous section we argued that for calculus, the "best base" for the exponential function is the base e. This is because $C_e = 1$ and hence

$$(e^x)' = e^x.$$

In Example 4 we showed that

$$(\log_b x)' = \frac{1}{C_b x}.$$

This derivative is simplest when $C_b = 1$. Hence for those who do calculus, the "logarithm of choice" is the logarithm to base $b = e$. This logarithm is called the *natural logarithm*.

DEFINITION   natural logarithm

The natural logarithm is the logarithm to the base e. The natural logarithm of x is denoted by

$$\ln x$$

and is defined for all $x > 0$.

The graph of the natural logarithm is shown in Fig. 2.38.

From the definition of logarithm we see that the natural logarithm and the exponential function are closely related.

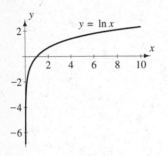

Figure 2.38. Graph of the natural logarithm function.

The Natural Logarithm and the Exponential

If x is a real number, then

(10) $\ln e^x = x.$

If x is a positive real number, then

(11) $e^{\ln x} = x.$

Equations (10) and (11) say that the exponential function is the inverse of the natural logarithm function and that the natural logarithm function is the inverse of the exponential function. We will study inverse functions in the next section.

We choose to work with the natural logarithm because its derivative is the "simplest" of the logarithm derivatives.

The Derivative of ln x

The derivative of the natural logarithm function is

$$\frac{d}{dx} \ln x = (\ln x)' = \frac{1}{x}.$$

■ **EXAMPLE 6**   Find $f'(t)$ if $f(t) = \ln(2t^2 - 3)$.

Solution.   Let $y = \ln(2t^2 - 3)$ and write the function as a composition chain:

$$y = \ln u$$
$$u = 2t^2 - 3.$$

Then

$$\frac{dy}{du} = \frac{1}{u} \quad \text{and} \quad \frac{du}{dt} = 4t.$$

By the chain rule, we have

$$f'(t) = \frac{dy}{dt} = \frac{dy}{du}\frac{du}{dt} = \frac{1}{u}4t = \frac{4t}{2t^2 - 3}.$$ ■

The preceding example illustrates a general "rule" for differentiating expressions of the form $y = \ln u(x)$. With the equation expressed as a composition chain, we have

$$y = \ln u$$
$$u = u(x).$$

By the chain rule,

(12) $$(\ln u(x))' = \frac{dy}{dx} = \frac{dy}{du}\frac{du}{dx} = \frac{1}{u}u'(x) = \frac{u'(x)}{u(x)}.$$

The rules of logarithms can be used to simplify computations by converting products to sums, quotients to differences, and exponentials to products. These rules can also be used to simplify some differentiations that might otherwise involve the product rule, the quotient rule, or the chain rule. The process is called **logarithmic differentiation.** We illustrate the process with an example.

■ **EXAMPLE 7**   Find $\dfrac{dy}{dx}$ if

(13) $$y = \frac{(2 + x^2)\sqrt{\sin 2x}}{(3x^3 - 2x - 4)^5}.$$

Solution.   First note that this derivative could be found using the techniques developed in earlier sections. Since the function is a quotient, we could use the quotient differentiation formula developed in Example 3 of Section 2.2. We would also need the product rule to differentiate the numerator, and the chain rule to differentiate the denominator and the $\sqrt{\sin 2x}$ expression in the numerator.

Instead, start by taking logarithms of both sides, and then use logarithm identities to write the logarithm of the quotient as a difference of logarithms:

(14) $$\ln y = \ln\left(\frac{(2 + x^2)\sqrt{\sin 2x}}{(3x^3 - 2x - 4)^5}\right)$$

$$= \ln\left((2 + x^2)(\sin 2x)^{1/2}\right) - \ln\left((3x^3 - 2x - 4)^5\right)$$

Next convert the logarithm of the product in the last expression to a sum of logs, and then drag the exponents out in front as multipliers to get

$$(15) \qquad \ln y = \ln(2 + x^2) + \frac{1}{2}\ln(\sin 2x) - 5\ln(3x^3 - 2x - 4).$$

Now differentiate both sides of (15), remembering that y is a function of x. Using (12), we obtain

$$\frac{y'}{y} = \frac{(2 + 2x^2)'}{2x^2 + 2} + \frac{1}{2}\frac{(\sin 2x)'}{\sin 2x} - 5\frac{(3x^3 - 2x - 4)'}{(3x^3 - 2x - 4)}$$

$$= \frac{4x}{2x^2 + 2} + \frac{1}{2}\frac{2\cos 2x}{\sin 2x} - 5\frac{9x^2 - 2}{(3x^3 - 2x - 4)}.$$

Multiply both sides of this last expression by y and replace y by the expression on the right of (13). We then have an expression for y':

$$y' = \left(\frac{(2 + x^2)\sqrt{\sin 2x}}{(3x^3 - 2x - 4)^5}\right)\left(\frac{4x}{2x^2 + 2} + \cot 2x - 5\frac{9x^2 - 2}{(3x^3 - 2x - 4)}\right). \qquad \blacksquare$$

Analyzing Exponential Data

The logarithm is a valuable tool in the analysis of data and functions. Assume that we gather some data from an experiment and that the data come in ordered pairs:

$$(16) \qquad (x_1, y_1), (x_2, y_2), (x_3, y_3), (x_4, y_4), (x_5, y_5)$$

where each x-coordinate represents some input in the experiment and the y-coordinate is the corresponding output. Suppose we suspect that the data should be described by an exponential function. That is, there are real numbers b and c such that

$$(17) \qquad\qquad\qquad y = cb^x.$$

How can we find b and c? One way is to take the logarithm of both sides of (17) and expand the right side using rules for logarithms. This gives

$$(18) \qquad\qquad \ln y = \ln(cb^x) = \ln c + (\ln b)x.$$

The expression on the right describes a line of slope $\ln b$ and y-intercept $\ln c$. If the points (16) satisfy (17), then the points

$$(x_1, \ln y_1), (x_2, \ln y_2), (x_3, \ln y_3), (x_4, \ln y_4), (x_5, \ln y_5)$$

lie on this line. We can use these points to find the slope and y-intercept of the line and then calculate the values of b and c.

■ **EXAMPLE 8**   At various times during the day a microbiologist counts the number of microorganisms in a culture. (This is done by diluting a small sample of the culture in solution, counting the number of organisms in the dilution, and

Table 1

t (h)	Population $P(t)$
1	108
3	2,387
4	11,223
6	247,922
8	5.477×10^6
9	2.574×10^7

Table 2

t (h)	$\ln P(t)$
1	4.683
3	7.778
4	9.326
6	12.421
8	15.516
9	17.064

then using this to infer the microbe population.) The data obtained are given in Table 1. What is the exponential function that describes the population?

Solution. Take the natural logarithm of population for each time. The resulting numbers are given in Table 2.

When we plot the data in this second table, we see that the points appear to lie on a line, as seen in Fig. 2.39. We can find an equation for this line by using the coordinates of two of the points, say $(1, 4.683)$ and $(9, 17.064)$. The line determined by these two points has slope

$$\frac{17.064 - 4.683}{9 - 1} \approx 1.548.$$

Hence an equation for this line is given by (approximately)

$$y - 4.683 = 1.548(t - 1) \quad \text{which leads to} \quad y = 1.548t + 3.135.$$

The graph of this line is shown in Fig. 2.39. Because this line describes the data in the previous table, we have

$$\ln P(t) = 1.548t + 3.135.$$

Exponentiating both sides of this equation, we obtain an exponential that may do a good job of describing the population growth:

$$\begin{aligned} P(t) &= e^{\ln P(t)} \\ &= e^{(1.548t + 3.135)} \\ &= e^{1.548t} e^{3.135} \\ &= e^{3.135}(e^{1.548})^t \\ &\approx 22.989(4.702^t). \end{aligned}$$

The graph of this exponential with some of the original data points is shown in Fig. 2.40. Not all of the data could fit on the graph because of the scale. ■

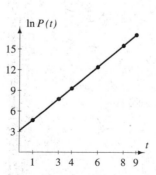

Figure 2.39. Graph of the logarithm of population data and a line that seems to fit the data.

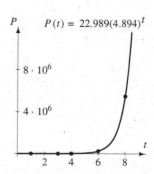

Figure 2.40. Graph of population data points and a possible exponential curve describing the population growth.

Exercises 2.6

Basic

Exercises 1–6:   Find the exact value of the expression.

1. $\log_3 81$
2. $\log_2(1/\sqrt[3]{4})$
3. $\log_{0.23} 0.012167$
4. $\log_e e^{1995}$
5. $\log_{\sqrt{6}} 36\sqrt{6}$
6. $\log_{4.567} 1$

Exercises 7–14:   Solve for x.

7. $2^{3x} = 5$
8. $e^x = \dfrac{1}{2}$
9. $2e^{6x} - 5e^{3x} + 6 = 0$
10. $2e^{6x} + 5e^{3x} - 6 = 0$
11. $\log_2 3x = 5$
12. $\log_4(x + 2) + \log_4(x - 2) = 1$
13. $(\log_2 x) + (\log_5 x) = \log_{10} x$
14. $(\log_2 x) + (\log_5 x) = \log_{10} x + 1$

Exercises 15–24:   Find the derivative.

15. $y = \ln(x + 2)$
16. $h(t) = t \ln t$
17. $r = \ln(\sin \theta)$
18. $G(u) = \ln(\ln u)$
19. $r(t) = (1 + \ln 2t)^{10}$
20. $f(x) = \dfrac{1 + \ln x}{2 - \ln x}$
21. $q(z) = \ln\left(\dfrac{1 - 2z + z^3}{z \sin z}\right)$
22. $y = \dfrac{1}{\sqrt{\ln x}}$
23. $r = e^{(\ln \theta)^2}$
24. $y = e^{x \ln x}$

Exercises 25–28:   Find $\dfrac{dy}{dx}$ by using logarithmic differentiation.

25. $y = \dfrac{x^2}{2x + 1}$
26. $y = x^{x+1}$
27. $y = \sqrt{x \tan x}$
28. $y = \left(\dfrac{1 + e^x}{1 - e^x}\right)^{-10}$

29. Use the definition of logarithm and rules for working with exponentials to prove the identity

$$\log_b xy = \log_b x + \log_b y.$$

30. Use the definition of logarithm and rules for working with exponentials to prove the identity

$$\log_b x^r = r \log_b x.$$

31. Why don't we ever talk about logarithms to base 1?
32. The pairs (x, y) given in the following table are data points from an exponential function. Find an equation for the function.

x	y
1	9.02501
2	81.4509
3	735.095
4	6634.24
5	59874.1

33. The pairs (x, y) given in the following table are data points from an exponential function. Find an equation for the function.

x	y
-3	8103.08
-1	20.0855
0	1.0000
4	6.14421×10^{-6}
5	3.05902×10^{-7}
6	1.523×10^{-8}

34. In Section 2.3 we saw that if r is a rational number, then

$$\frac{d}{dx} x^r = r x^{r-1}.$$

With the help of logarithms we can find the derivative of x^a for any real number a. Define $f(x) = x^a$ for $x > 0$, and write

$$f(x) = x^a = e^{a \ln x}.$$

Differentiate this expression and simplify the result to show that

$$f'(x) = ax^{a-1}.$$

Why is the restriction $x > 0$ necessary in defining f?

Growth

35. Show that the function $y = \ln |x|$ has derivative

$$\frac{dy}{dx} = \frac{1}{x} \quad (x \neq 0).$$

    (Hint: First consider the case $x > 0$, and then the case $x < 0$.)

36. Find the derivative of each of the following functions.
    a. $y = \ln |\sin x|$
    b. $r = \ln |4\theta^2 - 10|$
    c. $f(t) = \ln \left| \dfrac{1-t}{1+t} \right|$

37. Find two different functions with derivative $1/(x-4)$ for $x > 4$.

38. Find two different functions with derivative $2x/(x^2 + 1)$.

39. Tell why the point (a, b) is on the graph of $y = e^x$ exactly when the point (b, a) is on the graph of $y = \ln x$. What does this say about the relationship between the graphs of the two functions?

40. By looking at the graph of

$$y = \frac{\ln x}{x}$$

    determine the behavior of $\dfrac{\ln x}{x}$ as x gets large and positive.

41. By looking at the graph of

$$y = \frac{\ln x}{\sqrt[4]{x}}$$

    determine the behavior of $\dfrac{\ln x}{\sqrt[4]{x}}$ as x gets large and positive.

42. By looking at the graph of

$$y = x \ln x$$

    determine the behavior of $x \ln x$ as $x \to 0$ through positive values.

43. Let f be a function. What is the domain of

$$\ln(f(x)) + 4 \ln(-f(x))?$$

    Give reasons for your answer.

44. For what real numbers k does the equation $\ln x = k$ have a solution?

45. Determine (approximately) the positive values of x for which 2^x is larger than x^8.

46. Determine (approximately) the values of x for which $x^{1/10}$ is larger than $\ln x$.

47. Consider the function described by

$$y = \sin 2x - \cos 4x + (2x^3 - 2x^2 + 4)^{1/2}.$$

    Explain why logarithmic differentiation would not be very useful in computing dy/dx.

48. In Investigation 1 of Section 2.5 we worked with data obtained from studies of the ability of bacteria to remove uranium from a solution. Use the techniques of Example 8 to find an exponential function

$$U(t) = cb^t$$

    that describes the data fairly well. Plot your exponential function and the data on the same set of axes.

49. Exercise 44 of Section 2.5 presented data on drug levels in a patient's body as a function of time. Use the techniques of Example 8 to find an exponential function

$$A(t) = cb^t$$

    that describes the data fairly well. Plot your exponential function and the data on the same set of axes.

50. A modern-day Rip van Winkle decides to take a very long nap. Having no faith in modern timepieces, he gets a 10-gram sample of radium-226. After t years have passed, the amount of radium-226 that remains will be

$$A = 10e^{-k \cdot t}$$

    where $k \approx 0.000427869$.
    a. Mr. van Winkle solves this equation for t so he can tell the time as a function of the amount A of radium that remains. What is Mr. van Winkle's formula for the time?
    b. Mr. van Winkle finally dozes off on January 1, 2000. When he awakens, 1 gram of radium remains. What is the (approximate) date?

51. The figure below shows the graph of an exponential function $y = e^{kx}$ (solid line) and the graph of $y = \ln x$ (dotted line). Use the method for composing graphs discussed in Section 1.1 to find the graph of $y = \ln(e^{kx})$, and from this graph determine (approximately) the value of k.

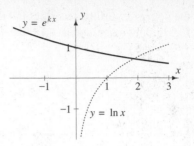

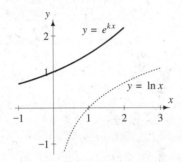

52. The following figure shows the graph of an exponential function $y = e^{kx}$ (solid line) and the graph of $y = \ln x$ (dotted line). Use the method for composing graphs discussed in Section 1.1 to find the graph

of $y = \ln(e^{kx})$. From this graph determine (approximately) the value of k.

53. Given that it took about eight months to factor RSA 129 and that the factoring time is proportional to $L(N)$ as given in (1), estimate the amount of time needed to factor a 150-digit number and a 200-digit number.

54. Suppose that you wanted a public key number that would take one million years to factor using the methods used on RSA 129. Estimate the number of digits needed. What if you wanted your number to be unfactored for one trillion (10^{12}) years?

Review/Preview

Exercises 55–58:   Solve for x in terms of y.

55. $y = -3x + 4$

56. $y = x^2 - 3x$

57. $y = -3x^2 + 4x - 8$

58. $y = (x + 2)^5 - 2000$

59. A function whose range consists of ordered pairs of real numbers is defined by

$$f(t) = (t^2 - 2t + 1, 4t + 8).$$

Find $f(-1)$ and $f(3)$.

60. A function whose range consists of ordered triples of real numbers is defined by

$$G(t) = (t \sin t, t \cos t, t).$$

Find $G(\pi/2)$ and $G(0.3)$.

61. Find the distance between the points $(-1, 2)$ and $(4, 7)$.

62. Find the perimeter of the triangle with vertices $(0, 0)$, $(-3, 1)$, and $(4, 5)$.

63. Find the first-quadrant angle determined by the positive x-axis and the line $y = 2x$.

64. A line l passes through the origin and makes an angle of $\pi/3$ with the positive x-axis. Find the equation of the line.

2.7   INVERSE FUNCTIONS

In Bemidji, Minnesota (47.5° north latitude, 95° west longitude), the sun only climbs 10° above the horizon on December 22 but gets as high as 57° above the horizon on June 21. One day between December 22, 1995, and December 22, 1996, you visit Bemidji and notice that the sun is 30° above the horizon at its highest point in the sky. Is this enough to determine the date of your visit? This question is not as strange as it may seem. Pre-twentieth century navigators determined their positions in the ocean using only the date and the angle of the sun above the horizon at noon. Moreover, given any two of these three pieces of information, they could calculate the third.

Nonetheless, the answer to our question is no. Let $A(t)$ be the angle of the sun above the horizon when the sun is at its highest point in the sky on day t. Between December 22, 1995, and June 21, 1996, $A(t)$ increases from $10°$ to $57°$, so a measurement of $30°$ is possible on some day before June 21. From June 21, 1996, to December 22, 1996, $A(t)$ decreases from $57°$ to $10°$, so a measurement of $30°$ is also possible on some day after June 21. The angle $a = A(t)$ (in degrees) on day t is pretty well approximated by

$$(1) \qquad a = A(t) = -13.5 + \frac{70.5}{1 + 0.00006(t - 182.5)^2} \qquad 0 \le t \le 365,$$

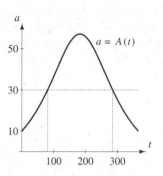

Figure 2.41. The angle of the sun in Bemidji, Minnesota.

where $t = 0$ corresponds to December 22. (In Chapter 8 we will use vector methods to obtain a more accurate expression for $A(t)$. For now we will take (1) as exact.) The graph of $a = A(t)$ is shown in Fig. 2.41. Notice that the line $a = 30$ hits the graph twice, indicating that there are two days for which $a = 30°$. So given *only* information about the angle of the sun, it is not possible to determine the date.

By solving (1) for t, we can find the date(s) that correspond to angle a. Add 13.5 to both sides, and then multiply by $1 + 0.00006(t - 182.5)^2$. We then have

$$(a + 13.5) + 0.00006(a + 13.5)(t - 182.5)^2 = 70.5.$$

This leads to

$$(t - 182.5)^2 = \frac{57 - a}{0.00006(13.5 + a)}.$$

Solve for t to get

$$(2) \qquad t = 182.5 - \sqrt{\frac{57 - a}{0.00006(13.5 + a)}} \quad \text{or} \quad t = 182.5 + \sqrt{\frac{57 - a}{0.00006(13.5 + a)}}.$$

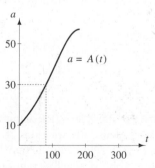

Figure 2.42. Each angle corresponds to just one t between 0 and 182.5.

These two formulas for t mean that for a value of a between $10°$ and $57°$, there might be two days t when this angle is realized. This can also be seen from the graph in Fig. 2.41.

The graph in Fig. 2.41 has its high point at $t = 182.5$. (This corresponds to June 21, the summer solstice.) If we know whether the date is before or after June 21, we can decide which value of t gives the correct date. For example, suppose that we know that the date is before June 21. Then we are concerned only with the portion of the graph between $t = 0$ and $t = 182.5$. As seen in Fig. 2.42, no horizontal line intersects this portion of the graph more than once. Hence, for a value of a between $10°$ and $57°$, there is only one corresponding value of t. Because $0 \le t \le 182.5$, we know the date is given by the first of the two expressions for t in (2):

$$t = 182.5 - \sqrt{\frac{57 - a}{0.00006(13.5 + a)}}.$$

When we put $a = 30$ into this formula, we get

$$t \approx 80.8,$$

which is March 13.

The Inverse of a Function

In the Bemidji problem we were given a function A. The input for A was a number t corresponding to a day of the year, and the output was the angle $a = A(t)$ of the sun above the horizon at its highest point on day t. The angle of $30°$ was treated as "output" from the function A. We were asked to find the input value that produced this output. The process of determining the input for a function given the output of the function is called *finding the inverse of a function*.

In the discussion of this problem we saw that there may be more than one input that gives a desired output. If we want each output value to correspond to exactly one input value, extra information may be needed to help us choose among candidates for the input. The information that the date was before June 21 told us that $t \leq 182.5$. This allowed us to conclude that the day t was given by

$$(3) \qquad t = T(a) = 182.5 - \sqrt{\frac{57 - a}{0.00006(13.5 + a)}}.$$

The function T is the inverse of the function A for $0 \leq t \leq 182.5$. Suppose that some t value with $0 \leq t \leq 182.5$ is used as input in (1), and that the output is a. If we use this output as input in (3), we get back the original t as output. Thus if $0 \leq t \leq 182.5$, then

$$t = T(A(t)) = (T \circ A)(t).$$

We use this relation to define the inverse of a function.

DEFINITION   inverse of a function

Let f be a function with domain $\mathcal{D}$. A function g is the **inverse of** f if the domain of g is equal to the range of f and

$$(4) \qquad\qquad (g \circ f)(x) = g(f(x)) = x$$

for every x in $\mathcal{D}$.

Equation (4) says that if some value $f(x)$ of f is used as input for g, the output for the function g will be x. Thus g just "undoes" f. We can also illustrate this by thinking of functions as machines. When x is put into the f machine, the output is $f(x)$. This $f(x)$ is then used as input for the g machine, and the resulting output is x. See Fig. 2.43.

The inverse of a function f is sometimes denoted by f^{-1}. Hence in the above definition we have

$$g = f^{-1}.$$

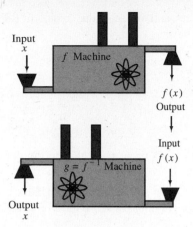

Figure 2.43. The function $g = f^{-1}$ takes output $f(x)$ from f and returns x.

■ EXAMPLE 1

a. Let $f(x) = \ln x$ and $g(x) = e^x$. Show that g is the inverse of f and that f is the inverse of g.

b. Show that $g(x) = (x - 1)^3 + 1$ is the inverse of $f(x) = \sqrt[3]{x - 1} + 1$.

c. Let $f(x) = x^2$ for all real numbers x. Show that $g(x) = \sqrt{x}$ is *not* the inverse of f.

d. Let $f(x) = x^2$ for domain $\mathcal{D} = [0, \infty)$. Show that $g(x) = \sqrt{x}$ is the inverse of f.

Solution

a. The range of the natural logarithm is the set of all real numbers. This set is also the domain of the exponential function. Hence the range of g is equal to the domain of f. The natural logarithm function is defined by the relation

$$e^{\ln x} = x \quad (x > 0).$$

This shows that

$$g(f(x)) = x$$

for every x in the domain of the logarithm function. Thus the exponential function g is the inverse of the natural logarithm function f.

 The range of the exponential function g is the set of positive real numbers, and this set is the domain of the logarithm function f. In addition,

$$(f \circ g)(x) = \ln(e^x) = x(\ln e) = x \cdot 1 = x$$

for all real x. Thus the natural logarithm is the inverse of the exponential function.

b. Since $f(x) = \sqrt[3]{x - 1} + 1$ is defined for all real numbers, we take the domain of f to be the set of all real numbers. The range of f is also the set of all real numbers, and this set is the domain of the polynomial g. For any real number x, we have

$$(g \circ f)(x) = g(f(x))$$
$$= (f(x) - 1)^3 + 1$$
$$= \left(\left(\sqrt[3]{x - 1} + 1\right) - 1\right)^3 + 1$$
$$= x.$$

Since $(g \circ f)(x) = x$ for all x in the domain of f, we conclude that g is the inverse of f.

c. If $f(x) = x^2$ and $g(x) = \sqrt{x}$, then the range of f is equal to the domain of g. Thus $g(f(x))$ makes sense for all for real numbers x. For real x we have

(5) $$(g \circ f)(x) = g(x^2) = \sqrt{x^2} = |x|.$$

Thus $(g \circ f)(x) \neq x$ if x is negative, so g is not the inverse of f.

d. Now let $f(x) = x^2$ and take the domain of f to be $[0, \infty)$, the set of nonnegative real numbers. As in part c, the range of f is equal to the domain of g. For any nonnegative x, we have

$$(g \circ f)(x) = g(x^2) = \sqrt{x^2} = |x| = x.$$

Hence g is the inverse of f. This example shows that the existence of an inverse function f can depend on the domain of f as well as the definition of f on its domain. ■

In part a of the preceding example we showed that the natural logarithm function is the inverse of the exponential function and that the exponential function is the inverse of the natural logarithm function. In fact, all inverse functions occur in such pairs.

■ **EXAMPLE 2**   Let g be the inverse of f. Show that f is also the inverse of g.

Solution.   Because g is the inverse of f, we know that the range of f is equal to the domain of g and that

$$(g \circ f)(x) = g(f(x)) = x$$

for all x in the domain of f. Since the output from g is back in the domain of f, it follows that the range of g is equal to the domain of f. Now let y be in the domain of g. Then y is also in the range of f, so there is an x in the domain of f with $f(x) = y$. But then we have

$$(f \circ g)(y) = f(g(y))$$
$$= f(g(f(x))).$$

Because $g(f(x)) = x$,

$$f(g(f(x))) = f(x)$$
$$= y.$$

Because $f(g(y)) = y$ for every y in the domain of g, we know that f is also the inverse of g. ■

Existence of the Inverse

The Bemidji problem shows that not every function has an inverse. Suppose that x_1 and x_2 are two different domain (or input) values for a function f and

$$f(x_1) = b = f(x_2).$$

Then f has no inverse function. If the function g were the inverse of f, we would have

$$g(b) = g(f(x_1)) = x_1 \quad \text{and} \quad g(b) = g(f(x_2)) = x_2.$$

Because g is a function, it cannot give two different values for the input b. Thus f can have no inverse function.

This argument shows that if a function f with domain $\mathcal{D}$ has an inverse, then any two different inputs (domain values) for f must result in different outputs. Functions with this property are said to be *one-to-one* on their domain.

DEFINITION   **one-to-one function**

Let f be a function with domain $\mathcal{D}$. We say f is **one-to-one** if whenever $x_1, x_2 \in \mathcal{D}$ with

$$x_1 \neq x_2 \quad \text{then} \quad f(x_1) \neq f(x_2).$$

We can quickly check whether or not a function f is one-to-one by looking at the graph of f.

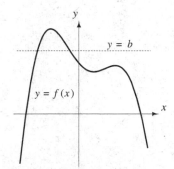

The Horizontal Line Test
A function f is one-to-one on its domain if and only if each horizontal line intersects the graph of f in at most one point.

Figure 2.44. There is a horizontal line that intersects the graph of f more than once. Hence f is not one-to-one.

Indeed, if a horizontal line $y = b$ intersects the graph of f at two different points (x_1, b) and (x_2, b), then we have

$$x_1 \neq x_2 \quad \text{but} \quad f(x_1) = b = f(x_2).$$

This shows that f is not one-to-one.

On the other hand, if f is not one-to-one, then there are domain values $x_1 \neq x_2$ for which $f(x_1) = f(x_2)$. Call this common value b. Then the horizontal line $y = b$ intersects the graph of f at points (x_1, b) and (x_2, b). See Figs. 2.44 and 2.45.

Checking whether or not a function is one-to-one is equivalent to checking whether or not the function has an inverse.

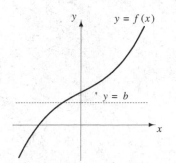

Figure 2.45. Each horizontal line intersects the graph of f at most once. Hence f is one-to-one.

Existence of an Inverse
A function has an inverse if and only if the function is one-to-one on its domain.

■ **EXAMPLE 3**   Let f be defined by

(6)
$$f(x) = \frac{2x - 3}{x + 1}$$

for domain $\mathcal{D} = [0, \infty) = \{x : 0 \le x < \infty\}$. Show that f is one-to-one on its domain and find the inverse of f.

Solution.   The graph of f is shown in Fig. 2.46. We cannot see the graph of f for the whole domain $\mathcal{D}$, but it appears that any horizontal line intersects the graph of f in at most one point. Thus f is one-to-one. We can also check this directly using (6). Suppose that u, v are in $\mathcal{D}$ and that $f(u) = f(v)$. Then

$$\frac{2u - 3}{u + 1} = \frac{2v - 3}{v + 1}.$$

Cross-multiplying leads to

$$(2u - 3)(v + 1) = (2v - 3)(u + 1).$$

Expanding and canceling like terms on each side gives

$$2u - 3v = 3u - 2v,$$

from which we see $u = v$. Hence if $u \ne v$, we must have $f(u) \ne f(v)$, so f is one-to-one.

Since f is one-to-one on its domain, f has an inverse. Let g be this inverse function. Then

(7)
$$x = (f \circ g)(x) = f(g(x)) = \frac{2g(x) - 3}{g(x) + 1}.$$

We solve (7) for $g(x)$. To keep notation simple, replace $g(x)$ by y in (7). This gives

$$x = \frac{2y - 3}{y + 1}.$$

Multiply both sides of this equation by $y + 1$ to get

$$x(y + 1) = 2y - 3.$$

Rearrange this last expression to get

$$xy - 2y = -x - 3.$$

Divide by $x - 2$, and we have

$$y = g(x) = \frac{x + 3}{-x + 2}.$$

How could you check that this g is the inverse of f?

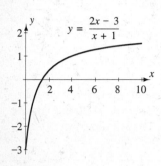

Figure 2.46. The graph of f appears to pass the horizontal line test.

In the Bemidji problem we worked with the function A defined by

$$A(t) = -13.5 + \frac{70.5}{1 + 0.00006(t - 182.5)^2} \qquad 0 \le t \le 365.$$

This function does not have an inverse because it is not one-to-one on its domain. We saw that A is one-to-one on the interval $0 \le t \le 182.5$ and that if we restrict A to this part of its domain, the resulting function does have an inverse.

This idea is an important one. If a function f is not one-to-one on its domain, then we can find a subset of the domain on which f is one-on-one and restrict f to this subset. We can then find an inverse function to f defined on this subset. We saw this demonstrated in parts c and d of Example 1. The function f described by

$$f(x) = x^2, \quad -\infty < x < \infty$$

does not have an inverse. However, the function F defined by

$$F(x) = x^2, \quad 0 \le x < \infty$$

does have an inverse.

■ **EXAMPLE 4**   Let f be defined by

$$f(x) = -x^2 + 2x - 7$$

with domain $\mathcal{D}$ the set of all real numbers. Show that f is not one-to-one on $\mathcal{D}$. Find a subset of $\mathcal{D}$ on which f is one-to-one, and find a formula for the inverse of the restricted version of f.

Solution.   The graph of $y = f(x)$ is shown in Fig. 2.47. With a quick glance at the graph, we see that there are many horizontal lines that intersect the graph in more than one point. Hence f is not one-to-one on $\mathcal{D}$.

We can also show f is not one-to-one by finding two different real numbers u, v such that $f(u) = f(v)$. For example, let $u = 0$ and $v = 2$. Then

$$f(u) = f(0) = -7 = f(2) = f(v).$$

When a function gives the same output for two different inputs, the function is not one-to-one.

There are many (in fact, infinitely many) subsets of $\mathcal{D}$ on which f is one-to-one. To find such a subset, erase a portion of the graph of f so that the remaining piece of the graph intersects each horizontal line in at most one point. Then f is one-to-one on the part of the domain corresponding to this piece of the graph. For example, the portion of the graph for $x \ge 1$ appears to satisfy the horizontal line condition. (See Fig. 2.48.) Thus f is one-to-one on $[1, \infty) = \{x : x \ge 1\}$.

Now let g be the inverse to the restricted version of f. Note that

(8)   domain of g = range of the restricted f = $(-\infty, -6]$

and

(9)   range of g = domain of the restricted f = $[1, \infty)$.

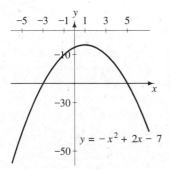

Figure 2.47. A horizontal line intersects the graph of f in more than one point.

$y = -x^2 + 2x - 7$

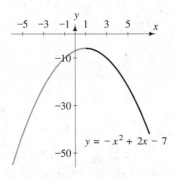

Figure 2.48. f is one-to-one on $[1, \infty)$.

$y = -x^2 + 2x - 7$

For any x in $(-\infty, -6]$, we have

$$x = f(g(x)) = -(g(x))^2 + 2g(x) - 7.$$

Rewrite this as

$$(g(x))^2 - 2g(x) + (x + 7) = 0.$$

Use the quadratic equation to solve this expression for $g(x)$:

(10) $$g(x) = 1 \pm \sqrt{-x - 6}.$$

The $\pm$ sign in (10) means that there are two possible choices for g. Because we want the range of g to be $[1, \infty)$, we choose the situation that corresponds to the $+$ sign. Hence if we restrict f to the interval $[1, \infty)$, this restricted version of f has inverse

$$g(x) = 1 + \sqrt{-x - 6}.$$ ∎

Let f be a function and $g = f^{-1}$ be its inverse function. If (a, b) is on the graph of $y = f(x)$, then a is in the domain of f and

$$f(a) = b.$$

It follows that

$$a = g(b),$$

so b is in the domain of g and (b, a) is on the graph of $y = g(x)$. Similarly, because f is the inverse of g, we find that if (b, a) is on the graph of $y = g(x)$, then (a, b) is on the graph of $y = f(x)$. Combining these facts, we see that a point (a, b) is on the graph of $y = f(x)$ exactly when the point (b, a) is on the graph of $y = g(x) = f^{-1}(x)$. In the (x, y)-plane, the point (b, a) is the reflection about the line $y = x$ of the point (a, b). This implies the following important relation between the graph of a function and the graph of its inverse. (See Fig. 2.49.)

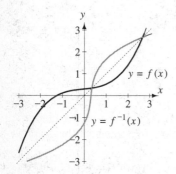

Figure 2.49. The graph of f^{-1} (in gray) is the reflection of the graph of f about $y = x$.

The Graph of an Inverse Function

Let g be the inverse of f. The graph of $y = g(x)$ is the reflection about the line $y = x$ of the graph of $y = f(x)$.

■ **EXAMPLE 5**   In Example 4 we found g, the inverse to

$$f(x) = -x^2 + 2x - 7 \quad x \geq 1.$$

Verify that the graph of $y = g(x)$ is the reflection about $y = x$ of the graph of $y = f(x)$.

Solution.   The graph of

$$y = g(x) = 1 + \sqrt{-x - 6}$$

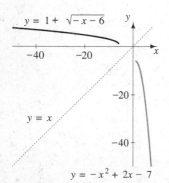

Figure 2.50. The graphs of $y = g(x) = f^{-1}(x)$ and $y = f(x)$.

for $x \leq -6$ is shown in Fig. 2.50. On the graph we have included (in gray) the graph of $y = f(x)$ for $x \geq 1$. The graph of $y = g(x)$ appears to be the reflection about the line $y = x$ of the graph of $y = f(x)$. ∎

The Derivative of an Inverse

The relationship between the graph of a function and the graph of its inverse function leads to a relationship between the derivative of a function and the derivative of its inverse. First note that if a line l of slope $m \neq 0$ is reflected about the line $y = x$, the result is a line of slope $1/m$. To see this, suppose that (a, b) and (c, d) are points on l. By the slope formula,

$$m = \text{slope of } l = \frac{d - b}{c - a}.$$

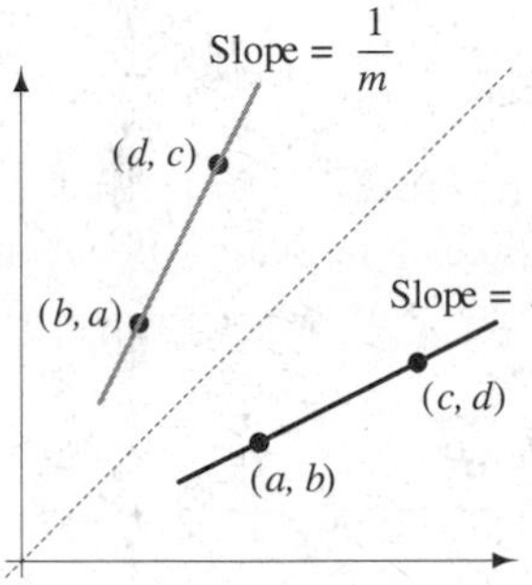

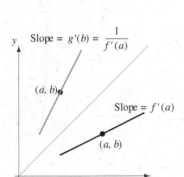

Figure 2.51. When a line of slope m is reflected about the line $y = x$, we get a line of slope $1/m$.

The reflection of l about $y = x$ is another line. This reflection contains the points (b, a) and (d, c). See Fig. 2.51. The slope of the reflection is

$$\frac{c - a}{d - b} = \frac{1}{m}.$$

Now suppose that a one-to-one function f is differentiable at a point a in its domain and that $f(a) = b$. Then the point (a, b) is on the graph of f. If we zoom in on the graph of $y = f(x)$ near (a, b), we see a "line" of slope $f'(a)$. The point (b, a) is on the graph of $y = g(x) = f^{-1}(x)$. When we zoom in on the graph of $y = g(x)$ near (b, a), we again see a "line." This line is the reflection about $y = x$ of the "line" of slope $f'(a)$ that we see when zooming in on the graph of $y = f(x)$ near (a, b). Hence near (b, a), the graph of $y = g(x) = f^{-1}(x)$ looks like a line of slope $1/(f'(a))$. (See Fig. 2.52.) However, when we zoom in on the graph of g near (b, a), the slope of the "line" we see is the derivative of g at b. Hence

$$(f^{-1})'(b) = g'(b) = \frac{1}{f'(a)} = \frac{1}{f'(g(b))}.$$

Figure 2.52. Zooming in on f and $g = f^{-1}$.

The Derivative of an Inverse

Let $g = f^{-1}$ be the inverse of f. If $f(a) = b$ and $f'(a) \neq 0$, then

$$(11) \qquad (f^{-1})'(b) = g'(b) = \frac{1}{f'(a)} = \frac{1}{f'(g(b))}.$$

The previous result can also be derived using implicit differentiation. (See Exercise 17.) Equation (11) is of interest because it provides geometric insight to the relation between a function and its inverse. We will rarely apply (11) directly. Results of this nature for particular functions can be readily derived using the differentiation techniques studied earlier in this chapter. We illustrate this in our investigation of the inverse trigonometric functions.

Inverse Trigonometric Functions

The inverse of the sine function takes a number x, with $-1 \leq x \leq 1$, and returns a number y for which $\sin y = x$. Every trigonometry student who has "solved a triangle" has worked with the inverse sine function and with the inverses of other trigonometric functions. The inverse trigonometric functions are important in calculus because their derivatives are simple, and these derivatives show up in many real-world applications. Before we discuss the the derivatives of these functions, we need to recall their definitions.

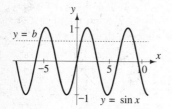

Figure 2.53. The sine function is not one-to-one on the real line.

The Inverse Sine Function

The sine function has domain $(-\infty, \infty)$ but is not one-to-one on this set. For each real b with $-1 \leq b \leq 1$, the horizontal line $y = b$ intersects the graph of $y = \sin x$ in infinitely many points. See Fig. 2.53. Before we can define an inverse for the sine function, we need to find a subset of the domain on which the function is one-to-one. The domain most often selected is

$$\left[-\frac{\pi}{2}, \frac{\pi}{2}\right] = \{x : -\pi/2 \leq x \leq \pi/2\}.$$

On this domain $\sin x$ is one-to-one and takes on every value from -1 to 1. See Fig. 2.54. The inverse of the sine function on this domain takes a number b between -1 and 1 and returns the number a between $-\pi/2$ and $\pi/2$ for which $\sin a = b$. This inverse function is called the *arcsine function*.

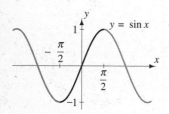

Figure 2.54. The sine function is one-to-one on $-\pi/2 \leq x \leq \pi/2$.

DEFINITION   **inverse sine function**

The inverse sine function (or arcsine function) has domain

$$[-1, 1] = \{x : -1 \leq x \leq 1\}.$$

For x in this domain,

$$\arcsin x = y$$

where y is the number between $-\dfrac{\pi}{2}$ and $\dfrac{\pi}{2}$ with

$$\sin y = x.$$

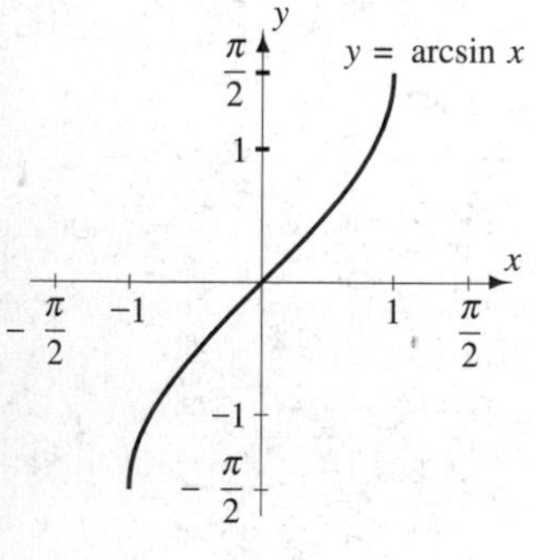

Figure 2.55. The graph of $y = \arcsin x$.

Since sine and arcsine are inverse functions, we have

(12) $$\arcsin(\sin x) = x \quad \text{for} \quad -\frac{\pi}{2} \leq x \leq \frac{\pi}{2}$$

and

(13) $$\sin(\arcsin x) = x \quad \text{for} \quad -1 \leq x \leq 1.$$

The graph of the inverse sine function is found by reflecting the graph of $y = \sin x \, (-\pi/2 \leq x \leq \pi/2)$ about the line $y = x$. See Fig. 2.55.

■ **EXAMPLE 6**   Find the value of:

a. $\arcsin\left(-\dfrac{\sqrt{2}}{2}\right)$.   **b.** $\sin\left(\arcsin\left(\frac{3}{5}\right)\right)$.   **c.** $\arcsin(\sin 10)$.

Solution

a. Since $\sin(-\pi/4) = -\sqrt{2}/2$ and $-\pi/4$ is between $-\pi/2$ and $\pi/2$, we have

$$\arcsin\left(-\frac{\sqrt{2}}{2}\right) = -\frac{\pi}{4}.$$

b. Since $-1 \le 3/5 \le 1$, we can apply (13) to obtain

$$\sin\left(\arcsin\left(\frac{3}{5}\right)\right) = \frac{3}{5}.$$

c. In this case (12) does not apply because 10 is not between $-\pi/2$ and $\pi/2$. However, 10 radians is a third-quadrant angle. (It is measured by wrapping once around the origin and continuing into the third quadrant. See Fig. 2.56.) Reflect the terminal side of the 10-radian angle about the y-axis to get the terminal side of a fourth-quadrant angle. These two angles meet the unit circle in points with the same y-coordinate, so these two angles have the same sine. Measuring from the positive x-axis down to the fourth-quadrant angle shows that the angle has measure $3\pi - 10$. Now $3\pi - 10$ is between $-\pi/2$ and $\pi/2$, so by (12),

$$(14) \qquad \arcsin(\sin 10) = \arcsin(\sin(3\pi - 10)) = 3\pi - 10.$$

The answer to this problem can be easily estimated with a calculator. All scientific calculators have an inverse sine button (usually labeled "arcsin" or "$\sin^{-1}$.") Use the calculator to find $\sin 10$, and then take the arcsin of this answer. This gives

$$\arcsin(\sin 10) \approx \arcsin(-0.544021) \approx -0.575222.$$

How does this compare with the answer in (14)? ■

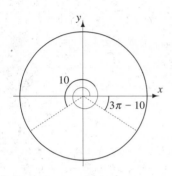

Figure 2.56. An angle of 10 radians has the same sine as an angle of $3\pi - 10$ radians.

■ **EXAMPLE 7** Let $y = \arcsin x$. Find $\dfrac{dy}{dx}$.

Solution. We could apply (11), but there is really no need to look up a formula. The derivative can quickly be found using (13) and the chain rule. By (13),

$$(15) \qquad \sin y = \sin(\arcsin x) = x.$$

Differentiate the left and right sides of this equation with respect to x, keeping in mind that y is a function of x. By the chain rule,

$$\cos y \frac{dy}{dx} = 1.$$

This leads to

$$(16) \qquad \frac{dy}{dx} = \frac{1}{\cos y}.$$

This derivative is defined when $\cos y \neq 0$. Since $-\dfrac{\pi}{2} \leq y \leq \dfrac{\pi}{2}$, we see the derivative is defined for

(17)
$$-\frac{\pi}{2} < y < \frac{\pi}{2}.$$

There is nothing wrong with the answer in (16), but it can be put into a more useful form. For y satisfying (17), we know $\cos y > 0$. By (15) we see that for such y,

$$\cos y = \sqrt{1 - \sin^2 y} = \sqrt{1 - x^2}.$$

Hence

$$\frac{dy}{dx} = (\arcsin x)' = \frac{1}{\sqrt{1 - x^2}} \quad \text{for } -1 < x < 1. \qquad \blacksquare$$

The Inverse Tangent Function   Inverses for the other trigonometric functions can be defined in much the same way that we defined the inverse for the sine function. We will briefly discuss the inverse of the tangent function here and discuss some of the others in the exercises. (See Exercises 27 and 30.)

The tangent function is defined for all real numbers except those for which the cosine is 0. Thus

$$\tan x = \frac{\sin x}{\cos x} \quad \text{for } x \neq \pm\frac{\pi}{2}, \pm\frac{3\pi}{2}, \pm\frac{5\pi}{2}, \dots.$$

The graph of $y = \tan x$ is shown in Fig. 2.57. Note that every horizontal line intersects the graph in infinitely many points. However, the tangent function is one-to-one when restricted to the domain

$$-\frac{\pi}{2} < x < \frac{\pi}{2},$$

and on this interval the tangent function takes on every real value. See Fig. 2.58. We define the inverse tangent (or arctangent) function to be the inverse of the tangent function when restricted to this interval.

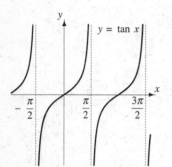

Figure 2.57. The tangent function is not one-to-one on the real line.

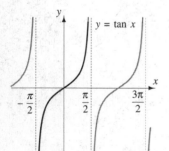

Figure 2.58. The tangent function is one-to-one on a suitable domain.

DEFINITION   inverse tangent function

The inverse tangent function (or arctangent function) has as its domain the set of all real numbers. For real x,

$$\arctan x = y$$

where y is the number between $-\dfrac{\pi}{2}$ and $\dfrac{\pi}{2}$ with

$$\tan y = x.$$

Since tangent and arctangent are inverse functions, we have

$$\arctan(\tan x) = x \quad \text{for} \quad -\frac{\pi}{2} < x < \frac{\pi}{2}$$

and

(18) $$\tan(\arctan x) = x \quad \text{for all real } x.$$

The graph of the inverse tangent function is found by reflecting the graph of $y = \tan x \, (-\pi/2 < x < \pi/2)$ about the line $y = x$. The graph of $y = \tan x$ has vertical asymptotes at $x = \pm\pi/2$. Under reflection about the line $y = x$, these correspond to horizontal asymptotes $y = \pm\pi/2$ for the graph of $y = \arctan x$. This means that as x gets large and positive, arctan x approaches $\pi/2$, and as x gets large and negative, arctan x approaches $-\pi/2$. The graph of the arctangent function is shown in Fig. 2.59.

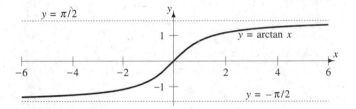

Figure 2.59. The graph of $y = \arctan x$ has two horizontal asymptotes.

We can find the derivative of the arctangent function by using (18) and the chain rule. Let

$$y = \arctan x.$$

Then

(19) $$\tan y = x.$$

Differentiate both sides of this equation with respect to x to obtain

$$\sec^2 y \frac{dy}{dx} = 1.$$

Use the identity $\sec^2 y = 1 + \tan^2 y$ and (19) to get

$$\frac{dy}{dx} = \frac{1}{\sec^2 y} = \frac{1}{1 + \tan^2 y} = \frac{1}{1 + x^2}.$$

Hence

$$\frac{dy}{dx} = (\arctan x)' = \frac{1}{1 + x^2}.$$

Exercises 2.7

Basic

Exercises 1–6: Find the exact value of each of the following.

1. $\arcsin(\sqrt{3}/2)$
2. $\arctan(-1)$
3. $\arcsin(-1)$

4. $\arctan(\tan 6)$
5. $\sin(\arcsin(-\pi/4))$
6. $\sin(\arctan(3/4))$

Exercises 7–12: Find the derivative.

7. $y = \arcsin(2x)$
8. $h(t) = t \arctan t$
9. $r = \arctan\left(\dfrac{1}{\theta}\right)$

10. $y = \arcsin\left(\dfrac{1+x}{1-x}\right)$
11. $r(t) = \arcsin t + \arcsin(\sqrt{1-t^2})$
12. $f(x) = e^{\arctan(x^2)}$

13. Show that

$$f(x) = \frac{2-x}{3+2x}$$

is one-to-one on its domain and find a formula for the inverse. What is the domain of the inverse?

14. Show that

$$g(x) = \frac{5x^3}{x^2+1}$$

is one-to-one on the real line. What is the domain of the inverse of g?

15. Show that

$$G(t) = \frac{1}{2}\ln(7t)$$

is the inverse of

$$F(t) = \frac{1}{7}e^{2t}.$$

16. Show that the linear function l defined by $l(x) = mx + b$ is one-to-one if $m \neq 0$. Write down a formula for the inverse of l.

17. Suppose that g is the inverse of f. Differentiate the expression

$$f(g(x)) = x$$

and from the result arrive at formula (11) for the derivative of the inverse.

18. Sometimes a function f is its own inverse. That is,

$$(f \circ f)(x) = x$$

for all x in the domain of f.
a. Write a short paragraph describing what the graph of such a function f must look like.
b. Draw three graphs for functions that are their own inverses. (No more than one of these graphs should be a line.)
c. Write a formula for a function that is its own inverse. Give two examples, at least one of which is not a line.

19. Estimate the angle of the sun above the horizon as seen from Bemidji, Minnesota, when the sun is at its highest point in the sky on January 31. On October 31. On your birthday. Use (1).

20. What are the possible dates when the sun's angle $A(t)$ in Bemidji measures $20°$? $40°$?

Growth

Exercises 21–26: Show that the function f is not one-to-one on its domain $\mathcal{D}$. Find a subset of the domain on which f is one-to-one, and then find a formula for the inverse of the restricted version of f.

21. $f(x) = x^2 - 2x + 1, \quad \mathcal{D} = (-\infty, \infty)$
22. $f(x) = -3x^2 + 7x, \quad \mathcal{D} = (-\infty, \infty)$
23. $f(x) = \dfrac{4x}{x^2+4}, \quad \mathcal{D} = (-\infty, \infty)$

24. $f(x) = \dfrac{-3}{x^2-9}, \quad \mathcal{D} = \{x : x \neq -3, 3\}$
25. $f(x) = \ln(x^2+1), \quad \mathcal{D} = (-\infty, \infty)$
26. $f(x) = e^{-x^2/4}, \quad \mathcal{D} = (-\infty, \infty)$

27. In this problem we define and discuss an inverse for the cosine function.
  a. Show that $\cos x$ is one-to-one on the interval $0 \le x \le \pi$.
  b. The inverse of the cosine function on this interval is called the arccosine function and is denoted arccos. Write a good definition of the arccosine function.
  c. Sketch the graph of $y = \arccos x$.
  d. Show that

$$(\arccos x)' = -\frac{1}{\sqrt{1 - x^2}}$$

  for $-1 < x < 1$.

28. The right triangle shown has one side of length x and hypotenuse of length 1.
  a. Show that angle B has measure $\arcsin x$ and angle A has measure $\arccos x$.
  b. Explain why part a suggests the identity

$$\arcsin x + \arccos x = \frac{\pi}{2},$$

  and then show that this identity holds for all x with $-1 \le x \le 1$.

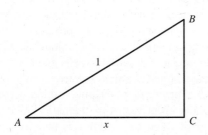

29. Use the figure for the previous problem to show that

$$\arcsin x = \arccos \sqrt{1 - x^2}$$

  for $0 \le x \le 1$.

30. In this problem we define and discuss an inverse for the secant function.
  a. Show that $\sec x$ is one-to-one on the domain

$$\mathcal{D} = \{x : 0 \le x \le \pi, x \ne \pi/2\}.$$

  b. The inverse of the secant function on this interval is called the arcsecant function and is denoted arcsec. Write a good definition of the arcsecant function.

c. Sketch the graph of $y = \operatorname{arcsec} x$.
d. Show that

$$(\operatorname{arcsec} x)' = -\frac{1}{|x| \sqrt{x^2 - 1}}$$

  for $|x| > 1$.

31. Use the diagram accompanying Exercise 28 to show that

$$\arccos x = \operatorname{arcsec} \frac{1}{x}.$$

32. Let f be one-to-one on its domain and suppose that g is the inverse of f. Let b be a fixed real number. The function F defined by

$$F(x) = f(x) + b$$

is also one-to-one and so also has an inverse. Find a formula for the inverse of F in terms of g.

33. Let f be one-to-one on its domain and suppose that g is the inverse of f. Let a be a fixed nonzero real number. The function F defined by

$$F(x) = af(x)$$

is also one-to-one and so also has an inverse. Find a formula for the inverse of F in terms of g.

34. Let f be one-to-one on its domain and suppose that g is the inverse of f. Let c be a fixed nonzero real number. The function F defined by

$$F(x) = f(cx)$$

is also one-to-one (but perhaps on a different domain) and so also has an inverse. Find a formula for the inverse of F in terms of g.

35. Let f be one-to-one on its domain. Suppose that we reflect the graph of f about the line $y = -x$.
  a. Show that the resulting graph is still the graph of a function. That is, show that this new graph also satisfies the vertical line condition.
  b. Let g be the function represented by this new graph. What can you say about $(f \circ g)(x)$?

36. Sometimes the graph of a function and its inverse intersect in one or more points. Explain why these points of intersection either fall on the line $y = x$ or else occur in pairs of the form (a, b), (b, a).

Review/Preview

37. Find the center and radius of the circle with equation

$$x^2 + y^2 - 4x + 8y - 2 = 0.$$

38. Find the center and radius of the circle with equation

$$3x^2 + 3y^2 + 8x = 0.$$

39. The graphs of $y = x^2$ and $y = 18 - x^2$ intersect at $(3, 9)$. Draw an accurate picture of what we would see if we zoomed in on the graphs at this point.

40. The graphs of $y = \sin t$ and $y = \cos t$ intersect at $(\pi/4, \sqrt{2}/2)$. Draw an accurate picture of what

we would see if we zoomed in on the graphs at this point.

41. Consider the three points $(-1, 0)$, $(0, -1)$, and $(2, 1)$.

   a. Find a, b, c so the parabola with equation $y = ax^2 + bx + c$ passes through all three points.

   b. Find d, f, g so the parabola with equation $x = dy^2 + fy + g$ passes through all three points.

   c. Find all points of intersection of the parabolas whose equations were found in parts a and b.

Exercises 42–46:   Find the derivative.

42. $y = (3x^4 - 6x^3 - 2x + 1)^{21}$

43. $h(s) = \ln(4s^2 + 3)$

44. $F(u) = \sqrt{1 + \sin^2 2u}$

45. $r = \dfrac{1}{1 + e^{\theta}}$

46. $y = \ln \sqrt{x(x + 1)(x + 2)(x + 3)(x + 4)}$

2.8   MODELING: TRANSLATING THE WORLD INTO MATHEMATICS

Mathematical models are used in almost every scientific and engineering field. A mathematical model is not a sculpture made of bronze or clay, nor is it made by gluing together pieces of plastic. A mathematical model is an equation or a set of equations. Mathematical models are used to help scientists and engineers better understand physical phenomena, to help economists make predictions about next year's economy, and to help sociologists learn more about human behavior.

A mathematical model usually does not describe exactly the phenomenon it is intended to model. This is because physical situations are usually very complex. There may be factors that play a major role in shaping a situation, as well as many minor factors that have a smaller influence. A mathematical model that attempts to account for all aspects of a situation may require so many equations and pieces of data that it becomes intractable. Thus, when mathematicians, scientists, and engineers create models, they try to take into account those factors that have a major influence on a situation but may ignore many of the minor ones. In other words, they make a trade-off between having a model that accounts for everything and having a model that they can work with. Because most mathematical models are not perfect, there is always room for improvement. As new mathematics is developed and faster computers become available, models can be changed to accommodate more of the factors that influence a situation.

In this section we look at some mathematical models that involve derivatives. Such models are not unusual because, perhaps surprisingly, it is often easier to describe the rate of change of a quantity than it is to describe the quantity directly. In this section we do not attempt to solve the equations that make up our models. We have chosen these examples to show the diverse sorts of phenomena that can be modeled mathematically, to illustrate the kinds of assumptions made by mathematical modelers, and to demonstrate how real-world situations are translated into mathematics.

Population Models

Year	Population (millions)
1790	4
1800	5
1810	7
1820	10
1830	13
1840	17
1850	23
1860	31
1870	40
1880	50
1890	63

The table shows U.S. census data for the years 1790–1890. Based on these data, can we predict the population for the year 2000? This is an example of the kind of question that mathematical modelers attempt to answer.

A Simple Model   Two major factors that influence the size of a population are births and deaths. The simplest population model is based on natural assumptions about the birth rate and the death rate.

1. There are more babies born in a large population than in a small population. Moreover, the number of babies born in a given year is proportional to the population for that year. Hence there is a positive constant β such that

$$(1) \qquad\qquad B(t) = \beta P(t),$$

where $B(t)$ is the number of births in year t and $P(t)$ is the population in year t.

2. There are more deaths in a large population than in a small population. Moreover, the number of deaths in a given year is proportional to the population for that year. Thus, there is a positive constant δ such that

$$(2) \qquad\qquad D(t) = \delta P(t),$$

where $D(t)$ is the number of deaths in year t.

The change in the population is the number of births minus the number of deaths. In year t this change is

$$\Delta P = B(t) - D(t) = \beta P(t) - \delta P(t) = k P(t),$$

where $k = \beta - \delta$. Because ΔP is the change in the number of people in a one-year period, the units of ΔP are people per year. Thus ΔP can be interpreted as the rate of change of $P(t)$ with respect to time. Because the rate of change is the derivative, we can describe the change in population by a *differential equation,*

$$(3) \qquad\qquad \frac{dP}{dt} = kP.$$

Equation (3) is a simple model for population growth. To see how well such a model can describe and predict the U.S. population, we need to solve (3) for P and determine the value of k.

In Section 2.5 we saw that one function that satisfies (3) is the exponential function. We can check that if C is a constant, then

$$P(t) = C e^{kt}$$

satisfies (3). To determine C and k, we use the data from the table. For convenience, we take $t = 0$ as 1790, so $t = 10$ is 1800, and so on. To determine C, use the 1790 figure ($t = 0$):

$$4 \times 10^6 = P(0) = C e^{k \cdot 0} = C \cdot 1 = C.$$

Next use, say, the 1840 figure ($t = 50$) to determine k. We have

$$17 \times 10^6 = P(50) = 4 \times 10^6 e^{k \cdot 50}.$$

Divide both sides of this equation by 4×10^6 and then take the logarithm of both sides to get

$$\ln(17/4) = \ln(e^{k \cdot 50}) = 50k.$$

Thus,

$$k = \frac{\ln(17/4)}{50} \approx 0.0289384,$$

and we have

(4) $$P(t) = 4 \times 10^6 e^{0.0289384t}.$$

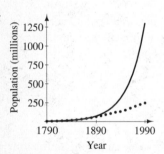

Figure 2.60. The simple model does not fit the data well after 1890.

To test this model, we compare it with the U.S. census data. In Fig. 2.60 the points represent the U.S. census data from 1790 through 1990, and the curve is the graph of $P = P(t)$. We see that the curve fits the data very well for the first 100 years or so, but it overestimates the population after 1900. In fact, for $t = 1990$, the P value for the point on the curve is more than five times the actual 1990 population. The year 2000 corresponds to $t = 210$. Putting this value for t into (4), we obtain

$$P(210) = 4 \times 10^6 e^{0.0289384 \cdot 210} \approx 1.7 \text{ billion.}$$

Judging by Fig. 2.60, this number is not a good prediction of the population for the year 2000.

A Better Model It is doubtful that the U.S. population will ever be 1.7 billion. Long before the population could reach this level the country would face shortages of food, living space, energy, and so on. All of these factors would slow the rate of population growth. With only simple assumptions about birth and death rates, (4) does a good job of predicting population growth under uncrowded conditions (e.g., before 1890). However, nothing in the model accounts for the fact that the United States, or any other nation, cannot accommodate an unlimited number of people. In an effort to improve the predictive power of our model, we modify it to take overcrowding into account.

Because (3) does a good job of describing the change in $P(t)$ when $P(t)$ is small, we keep the kP term in our model. We include a second term to account for the effects of overcrowding. In a population of size P, a particular individual might encounter any of the $P - 1$ other individuals. Because this is true for each of P individuals, there are $P(P - 1)$ different encounters possible. To keep the model simple, we assume that there are P^2 encounters possible. This assumption is reasonable because when P is large, $P(P - 1) \approx P^2$. Of course not all of these encounters will occur. However, it is reasonable to assume that the number of encounters is proportional to the number of possible encounters. Thus, when P is large, several such encounters take place and people feel crowded. Under such conditions people may be competing for resources such as food, space, and jobs. This will tend to slow the rate of growth of the population. Thus we modify (3) by adding a term $-cP^2$, so that

(5)
$$\frac{dP}{dt} = kP - cP^2.$$

In (5) the constant c is much smaller than k because the effects of overcrowding are not felt until P is large.

In Chapter 5 we will learn how to solve (5) using integration. Using the population data for times $t = 0, 50,$ and 100 to determine values for the constants in the solution, it can be shown that

(6)
$$P(t) = \frac{292 \times 10^6}{1 + 72e^{-0.03t}}.$$

In Fig. 2.61 we have plotted the census data and the graph of (6). The figure shows that (6) agrees pretty well with the data and thus might give a good prediction for the population in 2000. The predicted figure is

$$P(210) = \frac{292 \times 10^6}{1 + 72e^{-0.03 \cdot 210}} \approx 258,000,000.$$

In a few years we can see how close this prediction is.

In studying animal populations it is not unusual to run across interrelated species. For example, in a wilderness area populated by rabbits and foxes, the foxes eat the rabbits. This means that the rate of change of the rabbit population depends on the number of foxes (more foxes eat more rabbits). On the other hand, the rate of change of the fox population also depends on the size of the rabbit population (more rabbits mean more food and this leads to more foxes). Thus it is impossible to model the fox population without modeling the rabbit population at the same time. (See Exercise 14.) Models that must account for several things at once usually involve several interrelated equations. Our next four examples present models of such situations.

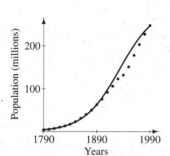

Figure 2.61. The model that takes overcrowding into account fits the data pretty well.

A Romantic Model

In the February 1988 issue of *Mathematics Magazine,* Steven Strogatz suggested a model for the following situation:

> Allen and Betty have known each other for years. Sometimes they are in love, sometimes not. Their feelings for each other change according to the state of their relationship. Allen is somewhat fickle in that the more Betty loves him, the more he dislikes her, but when she loses interest in him, his love for her grows. Betty, on the other hand, finds that her love for Allen grows when he loves her, but wanes when Allen loses interest in her.

Let $A(t)$ be the intensity of Allen's feelings for Betty at time t and $B(t)$ the strength of Betty's feelings for Allen at time t. Positive values of $A(t)$ and $B(t)$ denote affection, and negative values denote dislike. We assume that:

1. The rate of change of Allen's feelings for Betty is directly proportional to Betty's feelings for Allen.

2. The rate of change of Betty's feelings for Allen is directly proportional to Allen's feelings for Betty.

Because Allen's love for Betty increases when she dislikes him and decreases when she loves him, we know that $A'(t) > 0$ when $B(t) < 0$ and $A'(t) < 0$ when $B(t) > 0$. Combining this with assumption 1, we see that there is a positive constant α such that

$$(7) \qquad \frac{dA}{dt} = -\alpha B(t).$$

By similar reasoning, based on assumption 2, there is a positive constant β such that

$$(8) \qquad \frac{dB}{dt} = \beta A(t).$$

To solve (3) for $P(t)$ we needed to know the value of P for some values of t. To solve (7) and (8) for $A(t)$ and $B(t)$ we need to know the intensities of Allen's and Betty's feelings for each other at some time. Thus we assume that at time $t = 0$ we know $A(0) = A_0$ and $B(0) = B_0$. Combining these values at time $t = 0$ with (7) and (8), we have the model for Allen and Betty's relationship:

$$\frac{dA}{dt} = -\alpha B(t)$$

$$(9) \qquad \frac{dB}{dt} = \beta A(t)$$

$$A(0) = A_0$$

$$B(0) = B_0$$

To solve (9), we would need to determine numerical values for A_0, B_0, α, and β. These values might be determined by interviewing Betty and Allen or by having them fill out questionnaires.

Modeling the Spread of Disease

Almost everyone has had the flu at one time or another. During the flu season, all of us are in one of three stages, which we label S, I, and R. The individuals in stage S (for *susceptible*) are those who have not yet contracted the disease but are at risk of becoming infected. Those in stage I (for *infected*) have the disease and can pass it on to individuals in stage S. The people in stage R (for *recovered*) have had the flu and have recovered. By modeling the flow of people into and out of these three stages, we can model the spread of the disease.

Let $S(t)$, $I(t)$, and $R(t)$ be the number of individuals in stages S, I, and R, respectively, at time t. To create our model, we make some assumptions about the movement of people into and out of the three stages:

1. The total population remains constant at P_0. Thus, for all times t,

$$S(t) + I(t) + R(t) = P_0.$$

In particular, we assume there are no births or deaths in the population.

2. All susceptible individuals are equally at risk of infection. (This is likely to be true for such diseases as influenza and rubella. On the other hand, for diseases such as hepatitis, AIDS, and tuberculosis, some segments of the population are at greater risk of infection than others.)

3. Individuals leave stage S only by moving to stage I. The rate at which individuals in stage S become infected is proportional to the number of encounters between individuals in stage S and individuals in stage I.

4. After a fixed period of illness, individuals in stage I are no longer infectious. At this time we put them in stage R. (On entering this stage, individuals may still feel ill, but they can no longer pass the disease on to susceptibles.) The rate at which individuals leave stage I is proportional to the number of individuals in stage I.

5. Individuals in stage R are immune and cannot again contract the disease. Hence individuals never leave stage R.

Because of the nature of these assumptions, our model is often referred to as a *three-compartment model.* We can picture individuals as moving from one compartment to another as illustrated in Fig. 2.62.

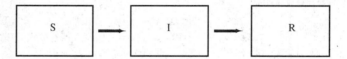

Figure 2.62. A three-compartment model for the spread of a disease.

If there are $S(t)$ individuals in stage S and $I(t)$ individuals in stage I, then there are $S(t)I(t)$ possible ways that an infected individual can encounter a susceptible individual. Not all of these encounters will occur, and not every such encounter will lead to a susceptible individual becoming infected. However, some portion of these encounters will result in susceptible individuals becoming ill. Thus assumption 3 seems reasonable, so the rate at which susceptible individuals leave stage S is $\alpha S(t)I(t)$, where α is some positive constant of proportionality. Because $S(t)$ decreases over time, dS/dt is negative. Thus

$$(10) \qquad \frac{dS}{dt} = -\alpha S(t)I(t).$$

Because all individuals who leave stage S move into stage I, new individuals become infected at the rate $\alpha S(t)I(t)$. If infected individuals recover, on the average, after δ days, then we may assume that each day $1/\delta$ infected people recover. Thus, infected people recover at a rate of about $(1/\delta)I(t)$ individuals per day. This leads to

$$(11) \qquad \frac{dI}{dt} = \alpha S(t)I(t) - \frac{1}{\delta}I(t).$$

Because the total population is constant, $R(t)$ can be determined once $S(t)$ and $I(t)$ are known; that is,

$$(12) \qquad R(t) = P_0 - S(t) - I(t).$$

To implement this model and predict the course of a disease, we would need to know the values of α and δ and the number of individuals in each stage at some time t_0. If S_0 and I_0 are the numbers of susceptibles and infecteds at time t_0, then our model for the spread of disease is found by combining these conditions with (10), (11), and (12):

$$\frac{dS}{dt} = -\alpha S(t)I(t)$$

$$(13) \qquad \frac{dI}{dt} = \alpha S(t)I(t) - \frac{1}{\delta}I(t)$$

$$R(t) = P_0 - S(t) - I(t)$$

$$S(t_0) = S_0$$

$$I(t_0) = I_0$$

Values for α and δ have been estimated for some diseases. For the 1988 Hong Kong Flu epidemic in New York City, these values are $P_0 \approx 7.9 \times 10^6$, $\alpha \approx 0.47/P_0$, and $\delta \approx 3$.

Modeling Combat

In a 1916 paper entitled "Aircraft in Warfare: The Dawn of the Fourth Arm," F. W. Lanchester developed a mathematical model for air combat. Over the years, Lanchester's model has been modified to describe many different kinds of battles, including combat between two ground forces. As an example, we model a simple battle in a field between two armies, the good guys and the bad guys. Let $G(t)$ be the number of good guys at time t and $B(t)$ the number of bad guys at time t. Because battles often last from one to several days, we measure t in hours. During a battle, members of each army are killed by members of the other army. Armies also lose troops through noncombat losses (accidents, desertion, etc.). In addition, the size of an army can increase if reinforcements are available. We make some assumptions to help us describe the rate at which each army gains and loses troops:

1. Both armies move with equal ease about the battlefield. In particular, neither army has a positional advantage over the other (however, see Exercise 11).

2. For each army, combat losses are at a rate proportional to the size of the opposing force. The constant of proportionality depends on the efficiency of the opposing army.

3. An army suffers noncombat losses at a rate proportional to the size of the army.

4. At time t, reinforcements join the good guys at a rate $g(t)$ troops per hour, and join the bad guys at a rate of $b(t)$ troops per hour. (Since reinforcements usually come in sporadically, $b(t)$ and $g(t)$ are usually 0 for most times t.)

According to assumption 2, the good guys suffer combat losses at a rate $-kB(t)$, where the positive constant k is indicative of the efficiency of the bad guys. According to assumption 3, the good guys suffer noncombat losses at a rate $-\ell G(t)$, for some positive constant ℓ. Combining these rates of change with the reinforcement function in assumption 4, we have

$$(14) \qquad \frac{dG}{dt} = -kB(t) - \ell G(t) + g(t).$$

Similarly, for the bad guys we have

(15) $$\frac{dB}{dt} = -mG(t) - nB(t) + b(t).$$

To complete the model, we need to know the initial (time $t = 0$) size of each army. Assume that the good guys start with $G(0) = G_0$ troops and the bad guys with $B(0) = B_0$ troops. Combining these initial conditions with (14) and (15), we have our combat model:

$$\frac{dG}{dt} = -kB(t) - \ell G(t) + g(t)$$

$$\frac{dB}{dt} = -mG(t) - nB(t) + b(t)$$

$$G(0) = G_0$$

$$B(0) = B_0$$

Models of this type have been used to analyze ongoing battles and wars. Special adaptations of this model to guerrilla warfare have been used to model the Vietnam War. Other applications have included studies of the Battle of the Bulge, the Battle of Iwo Jima, and the Battle of the Alamo.

A Model of Bilingualism

It has been estimated that there are between 2500 and 4000 languages spoken worldwide. Because there are between 150 and 200 countries, many countries have subpopulations that speak different languages. In a paper entitled "Can the Speakers of a Dominated Language Survive as Unilinguals?: A Mathematical Model of Bilingualism" (*Mathematical and Computer Modelling,* vol. 18, no. 6, 1993, pp. 9–18), I. Baggs and H. I. Freedman proposed a model to study the change in the number of unilingual and bilingual people in a society with two language groups. There are several current examples of such societies, for example, Quebec, where French and English are spoken; Miami, with large English- and Spanish-speaking populations; and Brussels, with both French and Flemish subcultures. Knowing something about the evolution of the different groups in such societies can help governments plan for the future needs of their citizens.

In constructing their model, Baggs and Freedman make the following assumptions (some of which we have simplified):

1. Everyone in the population speaks one or both of the two languages (say, English and French). Thus the population is broken into three "compartments": those who speak only English, those who speak only French, and those who speak both French and English.

2. For each of the three segments of the population, the birth rate and death rate are proportional to the size of the population segments.

3. Children of monolingual parents enter the population speaking the same language as the parents. Some children of bilingual parents enter the population as bilingual, and others enter as monolingual.

4. Monolingual individuals may become bilingual during their lifetimes. The rate at which members of a monolingual population become bilingual is

proportional to the number of contacts between members of the population and members of each of the other two populations.

5. Once an individual is bilingual, he or she remains bilingual.

Let $E(t)$ be the number of people who speak only English at time t, $F(t)$ the number who speak only French at time t, and $B(t)$ the number of bilingual individuals at time t. We first consider the rate of change of the segment of the population that speaks only English. According to the above assumptions, the size of this population is affected by three things. First is the birth/death rate within this segment. According to assumption 2, we can represent this factor by

$$(16) \qquad \alpha_E E(t),$$

where α_E is a constant. (We assume that the birth rate is higher than the death rate, so $\alpha_E > 0$.) Next is the rate at which bilingual parents produce children who only speak English. We assume that this rate is proportional to $B(t)$, so we represent it by

$$(17) \qquad \beta_E B(t),$$

where β_E is a nonnegative constant. According to assumption 4, members of the English population become bilingual at a rate proportional to the number of contacts with people who speak French. We account for the attrition due to encounters between English and French speakers by

$$(18) \qquad \gamma_{EF} E(t) \cdot B(t)$$

and the losses attributable to encounters between the English and bilingual population by

$$(19) \qquad \gamma_{EB} E(t) \cdot B(t).$$

We assume that γ_{EF} and γ_{EB} are positive, but we need to keep in mind that (18) and (19) represent losses to the English-speaking population. Thus we subtract (18) and (19) from the sum of (16) and (17) to get the rate of change of $E(t)$:

$$(20) \qquad \frac{dE}{dt} = \alpha_E E(t) + \beta_E B(t) - \gamma_{EF} E(t)F(t) - \gamma_{EB} E(t)B(t).$$

The factors that account for growth and attrition in the French-speaking population are similar. Thus, we also have

$$(21) \qquad \frac{dF}{dt} = \alpha_F F(t) + \beta_F B(t) - \gamma_{FE} F(t)E(t) - \gamma_{FB} F(t)B(t).$$

The bilingual population grows in two major ways. The first is from bilingual couples that produce bilingual children. We model this by

$$\frac{dB}{dt} = \beta_B B(t).$$

The bilingual population also grows as individuals from the English population learn French and people from the French population learn English. These factors are already accounted for as losses in (20) and (21). These losses correspond to gains for the bilingual population and so are accounted for by similar terms. Thus we have

$$(22) \qquad \frac{dB}{dt} = \beta_B B(t) + \gamma_{EF} E(t) F(t) + \gamma_{FE} F(t) E(t)$$
$$+ \gamma_{EB} E(t) B(t) + \gamma_{FB} F(t) B(t).$$

Equations (20), (21), and (22), together with data about the three population sizes at some time $t = t_0$, give us our model for the evolution of the language groups.

Actually, Baggs and Freedman's model was even more ambitious than the one outlined above. For example, they assumed that as the sizes of the English, French, and bilingual populations changed, so would the pressure to become bilingual or to speak, say, English. They took this into account by replacing the constants β_E, β_F, β_B, γ_{EF}, γ_{FE}, γ_{EB}, and γ_{FB} by expressions that depend on $B(t)$, $E(t)$, and $F(t)$.

Exercises 2.8

Basic

1. In our first population model we concluded that the U.S. population at time t might be described by

   $$(23) \qquad P(t) = Ce^{kt}.$$

   We then determined the values of C and k by using the population figures for 1790 and 1840, as given in the table on page 173. However, data from other years can be used to determine the values for C and k.
   a. Use the population figures for 1790 and 1880 to determine the values of C and k in (23).
   b. Plot the data from the table and the function found in part a on the same set of axes. Does the function fit the data well?
   c. Use the population figures for 1830 and 1890 to determine the values of C and k in (23).
   d. Plot the data from the table and the function found in part c on the same set of axes. Does the function fit the data well?

2. Money invested by a broker grows at a rate proportional to the value of the investment. Let $A(t)$ be the value of the investment at time t.
   a. Write a differential equation describing A.
   b. Suppose that the initial investment was $10,000 and that two years later the investment was worth $12,500. Find a formula for $A(t)$.
   c. Suppose that one year after the money was invested, the investment was worth $8000 and that three years later it was worth $11,000. Find a formula for $A(t)$.

3. Suppose that in the population model

   $$\frac{dP}{dt} = kP - cP^2$$

   we have $k = 0.025$ and $c = 8.33 \times 10^{-11}$.
   a. Is the population increasing or decreasing when $P = 250$ million? How about when $P = 400$ million?
   b. For what value of P is the population unchanging? Explain what this means in terms of the birth and death rates.

4. In experiments with fruit flies of the species *Drosophila willistoni*, M. E. Gilpin, F. J. Ayala, and J. G. Ehrenfeld found that the number $P(t)$ of flies in a colony at time t is well described by the equation

   $$\frac{dP}{dt} = 1.496P - 0.121P^{1.35}.$$

   a. Describe how the terms in the equation might represent aspects of the growth of the fruit fly population.
   b. For what values of P is the population increasing? Decreasing? Unchanging?

5. The radioactive isotope carbon-14 decays at a rate proportional to the amount of C^{14} present. Let $A(t)$ be the amount of C^{14} present in a sample at time t.

a. Write a paragraph explaining why $A(t)$ can be described by the model

(24) $$\frac{dA}{dt} = -kA.$$

Explain the significance of the positive constant k.

b. Verify that $A(t) = Ce^{-kt}$ (with C a constant) is a solution to (24).

c. Suppose that a sample of C^{14} initially has mass 10 grams and that the sample decays to 5 grams in 5568 years. Use these data to find the values of C and k.

6. The concentration of a drug in the bloodstream of an animal decreases at a rate proportional to the concentration of the drug. Let $C(t)$ be the concentration

of a drug in the body at time t. Write a differential equation that describes this function.

7. Once the source of pollution is stopped, a polluted body of water will eventually "purify itself." The concentration of pollutants decreases at a rate proportional to the concentration of pollutants. Let $P(t)$ be the concentration of pollutants in a lake at time t. Write a differential equation that describes P.

8. According to Newton's law of cooling, the temperature of a hot object decreases at a rate proportional to the difference between the temperature of the object and the temperature of the surroundings. Let $T(t)$ be the temperature of an object at time t and let T_0 be the temperature of the surroundings. Write a differential equation that describes T.

Growth

9. Assume that for our "romantic model" (9), feelings of affection are rated on a numerical scale with positive numbers denoting affection, negative numbers denoting dislike, and 0 corresponding to indifference. In addition, suppose that $\alpha = 1/2, \beta = 2/9, A_0 = 4$, and $B_0 = -1$.

a. Show that under these conditions,

$$A(t) = 4\cos(t/3) + \frac{3}{2}\sin(t/3)$$

$$B(t) = -\cos(t/3) + \frac{8}{3}\sin(t/3)$$

is a solution to (9). That is, verify that $A'(t) = -\alpha B(t), B'(t) = \beta B(t), A(0) = 4$, and $B(0) = -1$.

b. Graph the two solution functions on the same set of axes. Assuming that t is measured in months, write a brief synopsis of the first year of Allen's and Betty's romantic relationship.

c. Which of Allen and Betty seem subject to more intense feelings?

10. Our model for the spread of a disease is shown in (13). Suppose that after 30 days, individuals who have recovered from the disease again become susceptible. Modify the model in (13) to take this new aspect into account.

11. In his text *Modeling with Ordinary Differential Equations*, T. P. Dreyer gives a model for the Battle of the Alamo that considers two phases of the battle. At the start of the battle, there were 188 Texans and 3000 Mexicans. In the first phase of the battle

the Texan soldiers were inside the Alamo and were protected by the walls of the fort. The Mexican army was outside in the open attempting to gain entry. This first phase is modeled by

(25) $$\frac{dT}{dt} = -0.0007TM$$
$$\frac{dM}{dt} = -aT$$

where $T = T(t)$ is the number of Texans alive at time t and $M = M(t)$ is the number of Mexicans alive at time t, with t measured in hours. The positive constant a is a measure of the "efficiency" of the Texan forces. At the end of the first phase of the battle, 100 Texans and 1800 Mexicans were still alive, but the Mexicans had gained entrance to the Alamo. At this time the battle entered its second phase, involving hand-to-hand combat, with neither force having a positional advantage over the other. This second phase is modeled by

(26) $$\frac{dT}{dt} = -bM$$
$$\frac{dM}{dt} = -cT$$

a. What assumptions do you think were made about the first phase of the battle in coming up with (25) as a model for that phase?

b. How did the assumptions change as the battle entered its second phase? How do these new assumptions lead to (26)?

12. Imagine a pond that contains two species of fish, say A and B, both of which depend on the same food supply. Let $A(t)$ and $B(t)$ be, respectively, the number of fish of species A and B at time t. Because both species eat the same food, a large population for one species has an adverse effect on the other species. Suppose that:

- The number of offspring produced by species A is proportional to $A(t)$, and a similar statement holds for species B.

- Competition for scarce resources tends to limit the growth of each population. For each species, this effect is proportional to the number of encounters between members of different species.

a. Based on the above assumptions, write equations for dA/dt and dB/dt. Describe how each term in your equations is reflected in the assumptions.
b. Are there any circumstances under which both dA/dt and dB/dt are 0? What do you think happens to the populations in this case?

13. (Continuation of Exercise 12) Every member of a species is in competition for resources with every other member of the species. Thus, too many members of species A will have a detrimental effect on the growth of species A.
a. Alter the model in Exercise 12 to obtain a model that takes account of this intraspecies competition.

b. With this new model, are there any circumstances under which both dA/dt and dB/dt are 0? What do you suppose happens to the populations in this case?

14. Consider an environment with two animal populations, rabbits and foxes. Assume that the rabbits are the only food source for the foxes, but that food for the rabbits is plentiful. Let $F(t)$ and $R(t)$ be, respectively, the number of foxes and rabbits at time t. Assume that:

- New rabbits are born at a rate proportional to the rabbit population.

- Rabbits are killed and eaten by foxes at a rate proportional to the number of encounters between rabbits and foxes.

- Foxes can reproduce only if they have plenty of food. Thus, new foxes are born at a rate proportional to the number of encounters between rabbits and foxes.

- The fox population is very vulnerable to population pressures. This effect on the growth rate is proportional to the number of foxes.

a. Based on these assumptions, write expressions for dR/dt and dF/dt. Tell how each term in your equations is reflected in the assumptions.
b. Under what conditions are dR/dt and dF/dt both 0? What happens to the populations in this case?

Review/Preview

15. Find dy/dx if:
a. $y = (4x - 3)^2$
b. $y = 2x/\sqrt{1 + x}$
c. $y = e^{2x} \sin 3x$
16. Find $f'(t)$ if:
a. $f(t) = \dfrac{1 + t^2}{4 - 3t}$
b. $f(t) = \ln(1 + t + 4t^3)$
c. $f(t) = \tan(2e^{2t})$
17. An object moves at constant speed in the plane along the segment from $(-2, 3)$ to $(4, 5)$. If distance is measured in centimeters and the trip takes 6 seconds, at what speed is the object moving?
18. An object moves at constant speed in the plane along the segment from $(4, 0)$ to $(-9, -3)$. If distance is

measured in miles and the trip takes 25 minutes, at what speed is the object moving?
19. At time $t = 0$ seconds an object leaves $(0, 0)$ traveling along a line at a constant speed of 5 cm/s. During the motion, the object passes through the point $(3, 4)$. At what time is the object at $(3, 4)$? Where is the object at time $t = 11$ seconds?
20. At time $t = 0$ seconds an object leaves $(-1, 4)$ traveling along a line at a constant speed of 7 mph. During the motion, the object passes through the point $(3, 4)$. At what time is the object at $(3, 4)$? Where is the object at time $t = 11$ hours?

REVIEW OF KEY CONCEPTS

In this chapter we derived techniques and formulas for finding derivatives quickly and efficiently. We first developed a collection of "general rules" for finding the derivative of a function that is built from one or more simple functions. Among these were the product rule, the quotient rule, and the chain rule.

Next we collected the so-called elementary functions out of which many other functions are built. These include the power functions, the trigonometric functions, the exponential function, the natural logarithm, and the inverse trigonometric functions. We used the definition of the derivative to find the derivatives of many of these functions, and the "general rules" developed earlier to find others.

If we remember the derivatives of the elementary functions and know how to apply rules such as the product rule and the chain rule, then we can, with practice, quickly write down the derivatives of many functions that we see. These rules and formulas are valuable because with them we can avoid using the definition of derivative every time we need to perform a differentiation. However, it is important to always keep the definition of derivative in mind. The definition suggests many interpretations of the derivative: as a rate of change, as a slope, and so on. These interpretations are important in helping us understand what the derivative means in the context of a problem.

We summarize the derivatives of the elementary functions and the general rules in table form. We also provide a review grid for the elementary functions introduced and reviewed in this chapter.

General Rules for Differentiation

Let f and g be differentiable functions.

The derivative of a constant times a function
Let c be a constant.

$$(cf(x))' = cf'(x)$$

The derivative of a sum of functions

$$(f(x) + g(x))' = f'(x) + g'(x)$$

The product rule

$$(f(x)g(x))' = f'(x)g(x) + f(x)g'(x)$$

The quotient rule

$$\left(\frac{f(x)}{g(x)}\right)' = \frac{f'(x)g(x) - f(x)g'(x)}{(g(x))^2}, \quad g(x) \neq 0$$

The chain rule

$$(f \circ g)'(x) = (f(g(x)))' = f'(g(x))g'(x)$$

The Derivatives of the Elementary Functions

Chapter Summary

Function	Derivative	If u is a function of x:
c (a constant)	0	—
x^n (n a constant)	nx^{n-1}	$\dfrac{d}{dx}u^n = nu^{n-1}\dfrac{du}{dx}$
$\sin x$	$\cos x$	$\dfrac{d}{dx}\sin u = \cos u\dfrac{du}{dx}$
$\cos x$	$-\sin x$	$\dfrac{d}{dx}\cos u = -\sin u\dfrac{du}{dx}$
$\tan x$	$\sec^2 x$	$\dfrac{d}{dx}\tan u = \sec^2 u\dfrac{du}{dx}$
$\sec x$	$\sec x \tan x$	$\dfrac{d}{dx}\sec u = \sec u \tan u\dfrac{du}{dx}$
e^x	e^x	$\dfrac{d}{dx}e^u = e^u\dfrac{du}{dx}$
$\ln x$	$\dfrac{1}{x}$	$\dfrac{d}{dx}\ln u = \dfrac{1}{u}\dfrac{du}{dx}$
a^x ($a > 0, a \neq 1$)	$(\ln a)a^x$	$\dfrac{d}{dx}a^u = (\ln a)a^u\dfrac{du}{dx}$
$\log_a x$ ($a > 0, a \neq 1$)	$\dfrac{1}{x \ln a}$	$\dfrac{d}{dx}\log_a u = \dfrac{1}{u \ln a}\dfrac{du}{dx}$
$\arcsin x$	$\dfrac{1}{\sqrt{1 - x^2}}$	$\dfrac{d}{dx}\arcsin u = \dfrac{1}{\sqrt{1 - u^2}}\dfrac{du}{dx}$
$\arctan x$	$\dfrac{1}{1 + x^2}$	$\dfrac{d}{dx}\arctan u = \dfrac{1}{1 + u^2}\dfrac{du}{dx}$

Chapter Summary

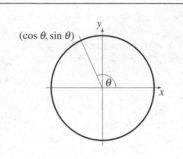

Measure angle θ from the positive x-axis. The terminal side of the angle intersects the circle $x^2 + y^2 = 1$ in the point $(\cos\theta, \sin\theta)$.

Because $(\cos\theta, \sin\theta)$ is a point on the circle $x^2 + y^2 = 1$, many identities can be obtained by using properties of the circle. For example,

$$\cos^2\theta + \sin^2\theta = x^2 + y^2 = 1.$$

By reflection about $y = x$,

$$\cos\left(\frac{\pi}{2} - \theta\right) = \sin\theta$$

$$\sin\left(\frac{\pi}{2} - \theta\right) = \cos\theta.$$

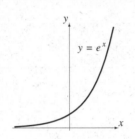

There is a number denoted by e such that

$$\lim_{h\to 0}\frac{e^h - 1}{h} = 1.$$

By exploring with a calculator, we find

$$e \approx 2.718281828459045\ldots.$$

The function exp defined by

$$\exp(x) = e^x$$

is called the exponential function.

Let $f(x) = ce^{kx}$. Then

$$f'(x) = kce^{kx} = kf(x).$$

Hence the rate of change of f is proportional to f. Many quantities change at a rate proportional to the quantity itself: population, amount of radioactive material, drug level in a body, and so on. Exponentials are essential in modeling such phenomena.

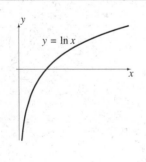

The logarithm of x to the base b is the number r for which $b^r = x$. We use the notation

$$r = \log_b x$$

to describe r. Hence

$$b^{\log_b x} = x.$$

The logarithm to the base e is called the natural logarithm and is denoted ln. Hence for $x > 0$,

$$\ln x = \log_e x.$$

Logarithms satisfy many useful identities. We present some of these for the natural logarithm:

$$\ln(xy) = \ln x + \ln y$$

$$\ln\left(\frac{x}{y}\right) = \ln x - \ln y$$

$$\ln x^r = r\ln x.$$

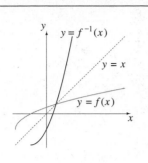	Let f be one-to-one on its domain $\mathcal{D}$. A function g is the inverse of f if the domain of g is equal to the range of f and $$(g \circ f)(x) = g(f(x)) = x$$ for all x in $\mathcal{D}$. We write $g = f^{-1}$. If g is the inverse of f, then f is also the inverse of g.	If g is the inverse of f, then the graph of $y = g(x)$ is the reflection about the line $y = x$ of the graph of $y = f(x)$. From this it also follows that $$g'(x) = (f^{-1})'(x) = \frac{1}{f'(g(x))}.$$
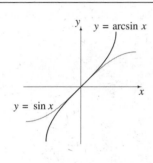	Let $-1 \le x \le 1$. The inverse sine of x is the number θ satisfying $$-\frac{\pi}{2} \le \theta \le \frac{\pi}{2}$$ and $$\sin \theta = x.$$ We denote the inverse sine function by arcsin and write $$\theta = \arcsin x.$$	If $-1 \le x \le 1$, then $$\sin(\arcsin x) = x.$$ If $-\pi/2 \le \theta \le \pi/2$, then $$\arcsin(\sin \theta) = \theta.$$
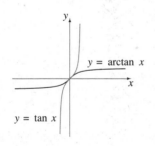	Let x be a real number. The inverse tangent of x is the number θ satisfying $$-\frac{\pi}{2} < \theta < \frac{\pi}{2}$$ and $$\tan \theta = x.$$ We denote the inverse tangent function by arctan and write $$\theta = \arctan x.$$	For any real number x, $$\tan(\arctan x) = x.$$ If $-\pi/2 < \theta < \pi/2$, then $$\arctan(\tan \theta) = \theta.$$

CHAPTER REVIEW EXERCISES

Exercises 1–20:   Find the derivative.

1. $y = -4x^5 + 7x^4 - \sqrt{2}x^2 + 3x - \pi$

2. $f(x) = (x^4 + x^2 + 1)^4$

3. $r = \left(\theta^2 - \dfrac{3}{\theta}\right)\sin \theta$

4. $Q(t) = \dfrac{3t^2 - 4t + 7}{-t^2 + 4}$

5. $y = \ln(4x^3 - 2x + 3)$

6. $b(s) = se^{s^2}$

7. $k(p) = \dfrac{p}{\sqrt{p^2 + 3}}$

8. $f(x) = \sqrt{\dfrac{2x - 3}{4x + 5}}$

9. $g(x) = \ln\left(\dfrac{2x - 3}{4x + 5}\right)$

10. $r(s) = 2^{2s}3^{3s}$

11. $y = \sqrt{t}\arcsin t^2$

12. $F(x) = (\sqrt{x} + 1)(\sqrt{x} + 2)(\sqrt{x} + 3)(\sqrt{x} + 4)$

13. $s(t) = \ln(\tan 2t)$

14. $z = \sqrt[3]{\dfrac{e^{2w} + 2w}{e^{-2w} - 2w}}$

15. $y = \dfrac{1}{\arctan\left(1 + \dfrac{1}{x}\right)}$

16. $H(z) = \dfrac{(z^2 - z)\sin z}{(1 + \tan z)}$

17. $r = \sin(\cos(1 + \tan 2\theta))$

18. $y = (ax^2 + bx + c)^d$, $\quad a, b, c, d$ constants

19. $H(s) = e^{as^2 + bs + c}$, $\quad a, b, c$ constants

20. $A(t) = \arcsin(at^2 + bt + c)$, $\quad a, b, c$ constants

Exercises 21–24: Use implicit differentiation to find $\dfrac{dy}{dx}$.

21. $y^3 + xy - 3x^2 = 0$

22. $\dfrac{2x - y}{3x + 2y} = xy$

23. $\sin(x + y) = x - y$

24. $e^{x-3y} = 2x + y^2 - 1$

Exercises 25 and 26: Four graphs are shown on one set of axes. Some of the graphs are mislabeled. Rearrange the labels so the graphs are properly labeled.

25.

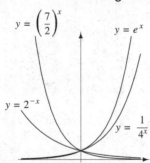

$y = \left(\dfrac{7}{2}\right)^x$ $y = e^x$ $y = 2^{-x}$ $y = \dfrac{1}{4^x}$

26.

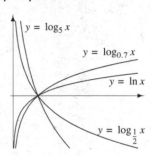

$y = \log_5 x$ $y = \log_{0.7} x$ $y = \ln x$ $y = \log_{\frac{1}{2}} x$

27. The graph of a trigonometric function is shown below. Label points where the graph intersects the x-axis and the tick marks on the y-axis if this is the graph of:

a. $y = \sin 2x$.

b. $y = 4\sin(2\pi x)$.

c. $y = \dfrac{1}{3}\sin\left(\dfrac{4}{\pi}x\right)$.

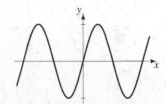

28. Let $f(x) = e^x$.

a. Find an equation for the line tangent to the graph of $y = f(x)$ at the point $(0, 1)$.

b. Estimate the maximum possible absolute error if we use the tangent line expression to approximate e^x for $-0.25 \le x \le 0.25$.

c. Use the methods of Sections 1.5 and 1.6 to find $r > 0$ so that on the interval $-r \le x \le r$, the error in the tangent line approximation is no more than $0.1|x|$ in absolute value.

29. Let $g(x) = \ln x$.

a. Find an equation for the line tangent to the graph of $y = g(x)$ at the point $(1, 0)$.

b. Estimate the maximum possible absolute error if we use the tangent line expression to approximate $\ln x$ for $0.5 \le x \le 1.5$.

c. Use the methods of Sections 1.5 and 1.6 to find $r > 0$ so that on the interval $1 - r \le x \le 1 + r$, the error in the tangent line approximation is no more than $0.1|x|$ in absolute value.

30. Let $h(x) = \arctan x$.

    a. Find an equation for the line tangent to the graph of $y = h(x)$ at the point $(0, 0)$.

    b. Estimate the maximum possible absolute error if we use the tangent line expression to approximate $\arctan x$ for $-0.5 \le x \le 0.5$.

    c. Use the methods of Sections 1.5 and 1.6 to find $r > 0$ so that on the interval $-r \le x \le r$, the error in the tangent line approximation is no more than $0.1|x|$ in absolute value.

31. A ground camera tracks a 100-foot rocket during a launch. The camera is fixed at a point 300 feet from the launch point and pivots so that it always points at the midpoint of the rocket. Let h be the height of the midpoint of the rocket above the ground.

    a. Find a formula for the angle θ determined by the ground and the line of sight of the camera. Your answer should be in terms of h and should involve inverse trigonometric functions.

    b. For what values of h is your formula valid? What are the possible values for θ?

    c. Find $d\theta/dh$.

    d. What are the units for the derivative? Explain what the derivative means in this situation.

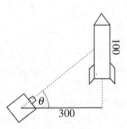

32. A tall cylindrical tower has a base radius of 50 feet. A surveyor stands x feet from the tower, as shown in the accompanying figure.

    a. Find a formula for the angle θ subtended by the tower. Your answer should be in terms of x and should involve inverse trigonometric functions.

    b. For what values of x is your formula valid? What are the possible values for θ?

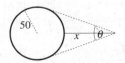

c. Find $d\theta/dx$.

d. What are the units for the derivative? Explain what the derivative means in this situation.

33. An airplane that is 150 feet long flies at a rate of 400 feet per second at a height of 500 feet off the ground. The plane passes directly above a boy on the ground, and the boy watches the plane as it passes over. Let y be the distance from the boy to the tail of the plane.

    a. Find a formula for the angle of vision ϕ subtended by the plane. Your formula should involve y and inverse trigonometric functions.

    b. For what values of y is your formula valid? What are the possible values for ϕ?

    c. Find $d\phi/dy$ and $d\phi/dt$.

    d. What are the units for the derivatives? Explain what the derivatives mean in this situation.

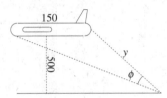

34. A wheel of radius 1 foot spins counterclockwise at one revolution per minute. A 6-foot rod is attached to the edge of the wheel and pivots as the wheel turns. The free end of the rod moves back and forth in a horizontal groove that is in line with the center of the wheel. Let α be the angle through which the wheel has rotated, and let θ be the angle the rod makes with the horizontal. See the accompanying diagram.

    a. Find a formula for θ. Your formula should involve α and inverse trigonometric functions.

    b. For what values of α is your formula valid? What are the possible values of θ?

    c. Find $d\theta/d\alpha$ and $d\theta/dt$.

    d. What are the units for the derivatives? Explain what the derivatives mean in this situation.

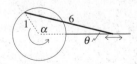

35. In Section 2.7 we gave an approximation for the maximum angle, a, of the sun above the horizon on day t in Bemidji, Minnesota. Another good approximation for this angle is given by

$$a = a(t) = -57.3 \arcsin(-0.53+0.33\cos(0.017t))$$

where $t = 0$ corresponds to December 22.

a. Estimate the measure of the angle $a(t)$ on January 31. On October 31.

b. Estimate the date if $a = 30°$ and it is after June 21.

c. Estimate the possible dates if $a = 45°$.

d. Find da/dt. What does the derivative mean in this context?

36. In Section 2.6 we remarked that the maximum length of time required to factor a large integer N is proportional to

$$L(N) = e^{\sqrt{(\ln N)(\ln \ln N)}}$$

Find dL/dN and explain what the derivative means in this context.

37. A ferris wheel has radius 60 feet and is mounted so that the bottom car is 5 feet off the ground. The wheel turns at a rate of one revolution every 40 seconds. You ride the ferris wheel at a time when the sun is directly overhead.

a. Write a formula that gives the position on the ground of the shadow of your car at any time t. Include all assumptions that you make in coming up with your formula.

b. Find the rate of change of the position of the shadow with respect to time. What does the rate of change mean in this context?

c. When is the rate of change the largest? When is it 0? How is the shadow moving at these times?

38. Consider the line $y = \frac{1}{2}t$. For $x \geq 0$, define $A(x)$ to be the area of the triangular region bounded by the t-axis, the line $y = \frac{1}{2}t$, and the vertical line $t = x$. See the following figure.

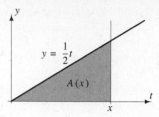

a. Find a formula for $A(x)$.

b. Find $A'(x)$. What does the derivative mean in this context?

39. Consider the line $y = 4t+3$. For $x \geq 0$, define $A(x)$ to be the area of the trapezoid bounded by the t-axis, the y-axis, the line $y = 4t + 3$, and the vertical line $t = x$. See the figure below.

a. Find a formula for $A(x)$.

b. Find $A'(x)$. What does the derivative mean in this context?

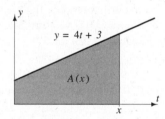

40. Let $m \geq 0$ and $b \geq 0$. Consider the line $y = mt + b$. For $x \geq 0$, define $A(x)$ to be the area of the region bounded by the t-axis, the y-axis, the line $y = mt + b$, and the vertical line $t = x$.

a. Find a formula for $A(x)$.

b. Find $A'(x)$. What does the derivative mean in this context?

Student Project

Porsche 911RS America Test Results

Porsche Acceleration Data.   The graph in Fig. 2.63 represents data from an acceleration test of a 1993 Porsche 911RS America. From a standing start, the Porsche accelerated for several seconds. The times (in seconds) at which the Porsche reached certain speeds (in miles per hour) were recorded. The data collected were presented in *Car and Driver* (November 1992, p. 80) and are summarized in the accompanying table. To obtain Fig. 2.63, the data were graphed and consecutive data points joined by a line segment.

In this project we will first estimate the distance $d(t)$ traveled by the Porsche during the first t seconds of the test, for $t = 0, 1.7, 2.4, \ldots, 35.4$. We will do so by

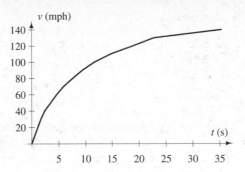

Figure 2.63. Porsche 911RS America 1993 model test results.

Time	Speed
0.0	0
1.7	30
2.4	40
3.5	50
4.6	60
5.9	70
7.6	80
9.45	90
11.6	100
14.8	110
18.9	120
22.8	130
35.4	140

estimating the distance traveled by the car in the time between two data points and keeping a running total of the distance traveled. We will also compute the average acceleration for the Porsche over the time intervals between consecutive velocity readings.

To get started, let's estimate the distance the Porsche traveled during the time interval from 5.9 to 7.6 seconds. During this 1.7 seconds the car accelerated from $v(5.9) = 70$ mph to $v(7.6) = 80$ mph. It is reasonable to assume that the average speed during this time was

$$\frac{70 + 80}{2} = 75 \text{ mph.}$$

Now, 1.7 seconds is the same as $1.7/3600$ hours. Thus during this 1.7 seconds of the test from $t = 5.9$ to $t = 7.6$, the Porsche travels approximately

$$(1) \qquad (75 \text{ miles}/\text{hour})(1.7/3600 \text{ hours}) \approx 0.0354 \text{ miles.}$$

In a similar way we can approximate the distance traveled by the Porsche during the time between any two consecutive data points. By adding these distances we can find the total distance covered by the car during the test.

Problem 1. Using the ideas outlined above, estimate the position $x(t)$ of the Porsche at times $0, 1.7, 2.4, 3.5, \ldots, 35.4$. How far did the Porsche travel up to the time that its velocity was 140 miles per hour? Using the results of these calculations, prepare a graph of the position $x(t)$ against time t.

Problem 2. Give a geometric interpretation of the computations used in Problem 1. First take another look at (1). We can rewrite the left side of this equation as

$$\frac{v(5.9) + v(7.6)}{2} \cdot (7.6 - 5.9)/3600.$$

This expression may be interpreted as the area of a certain trapezoid that can be drawn on the velocity graph. (For this it is helpful to think of the units on the t-axis as hours, so all of the t-axis labels should be divided by 3600.) Draw the appropriate trapezoid on the graph and carefully explain the connection between the area of this trapezoid and the distance traveled. Finally, relate the numerical value of $x(35.4)$ to an area associated with the velocity graph.

Finding the Acceleration.   In Problem 3 we will approximate the Porsche's acceleration as a function of time. In physics the average acceleration on a time interval $[t_1, t_2]$ is defined as

$$\frac{\text{change in speed}}{\text{change in time}} = \frac{v(t_2) - v(t_1)}{t_2 - t_1}.$$

Problem 3.   Compute the average acceleration of the Porsche on the time intervals $[0, 1.7]$, $[1.7, 2.4]$, $\ldots$, $[22.8, 35.4]$. Using the velocity graph, give a geometric interpretation of these computed average accelerations. Explain your interpretation carefully.

Problem 4.   Find the midpoint of each of the intervals $[0, 1.7]$, $[1.7, 2.4]$, $\ldots$, $[22.8, 35.4]$. Let t_m denote a typical midpoint. Explain why it is reasonable to use the average accelerations computed in Problem 3 as estimates of the actual acceleration at times $t = t_m$. Using these estimates, sketch a graph of the acceleration $a(t)$ against t. Could you have predicted the rough shape of this acceleration graph by looking at the velocity graph?

A History of Measuring and Calculating Tools
The Slide Rule

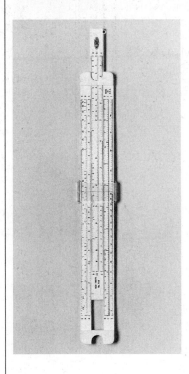

Before the invention of cheap, easy-to-use handheld calculators, engineers and scientists did many routine calculations with slide rules. The slide rule was an ingenious device that could be used to rapidly multiply, divide, and take square roots. The slide rule worked by mechanically implementing the logarithm identities

$$\log_{10}(xy) = \log_{10} x + \log_{10} y$$

(1) $$\log_{10}(x/y) = \log_{10} x - \log_{10} y$$

$$\log_{10}(\sqrt{x}) = \frac{1}{2}\log_{10} x$$

A simple slide rule consisted of two long pieces of wood or plastic constructed so that one could slide back and forth on the other. The pieces of wood were each thought of as part of the number line, with the left end of each piece corresponding to 0 and the right end to 1. On each of the pieces marks were drawn at distances of

$$\log_{10} 1 = 0$$

$$\log_{10} 2 \approx 0.301030$$

$$\log_{10} 3 \approx 0.477121$$

$$\vdots$$

$$\log_{10} 9 \approx 0.954243$$

$$\log_{10} 10 = 1$$

from the left end. However, the marks were not labeled $\log_{10} 1$, $\log_{10} 2, \ldots,$ $\log_{10} 9$, $\log_{10} 10$, but instead were labeled 1, 2, 3, $\ldots,$ 9, 10, respectively. For example, on each piece of the slide rule the mark labeled 3 was positioned $\log_{10} 3 \approx 0.477121$ units from the left end. See Fig. 2.64.

To demonstrate how the slide rule worked, we use it to find 2×3. First slide the upper piece so that the left end of this piece is above the 2 mark on the lower piece. Next find the number 3 on the upper piece. Notice that on the lower piece below this 3 is the answer, 6. See Fig. 2.65. How does it work? When we moved the upper piece so that the left end was above the 2 on the lower piece, we marked off a length of $\log_{10} 2$. From this mark we moved to 3 on the upper slide, so we moved an additional length of $\log_{10} 3$. Hence we are now at a distance $\log_{10} 2 + \log_{10} 3$ from the left end of the lower piece. By (1),

$$\log_{10} 2 + \log_{10} 3 = \log_{10}(2 \cdot 3)$$
$$= \log_{10} 6.$$

At this point the lower slide is marked 6, which is our answer.

Slide rules were usually calibrated more finely than shown in Figs. 2.64 and 2.65. There were always marks at positions $\log_{10} 1.1$, $\log_{10} 1.2, \ldots,$ $\log_{10} 9.9$, and larger models may have had even finer markings. To multiply numbers larger than

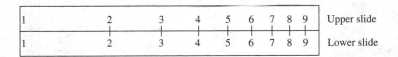

Figure 2.64. A simple slide rule.

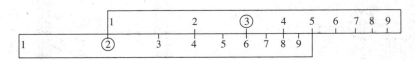

Figure 2.65. Using a slide rule to calculate 2×3.

10 or less than 1, the numbers were first converted to scientific notation. For example, to multiply 121 by 0.000398, first write

$$121 = 1.21 \times 10^2$$

$$0.000398 = 3.98 \times 10^{-4}.$$

Then

$$(121)(0.000398)$$

$$= (1.21 \times 10^2)(3.98 \times 10^{-4})$$

$$= (1.21 \times 3.98) \times 10^{-2}$$

The multiplication 1.21×3.98 can be done on the slide rule. The result of the slide rule computation is then multiplied by 10^{-2}.

See if you can figure out how a slide rule is used to divide two numbers. Can you think of a way to use a slide rule to find square roots? (Hint: Slide rules had a third scale marked 1 on the left end and 2 on the right. This scale sat above the lower scale and had marks at positions $\log_{10} 2$, $\log_{10} 3$, ..., $\log_{10} 100$.)

Chapter 3

Motion, Vectors, and Parametric Equations

Describing Motion

Among the most important sources and applications of calculus are the description and analysis of motion. Why? Because velocity and acceleration—key ideas in describing or analyzing motion—are rates of change. In this chapter we build on your knowledge of calculus and your experience with simple motion to motivate the ideas of velocity, acceleration, vector, and parametric equations. We begin by describing motion without regard to its cause, which is called *kinematics*. The words "kinematics" and "cinema" come from the Greek word "kinema," meaning motion. In Section 3.1 we study motion on a line, including average velocity, velocity, and speed. To study motion in more than one dimension, we use vectors to describe or calculate the position, velocity, and acceleration of an object. We also discuss the closely related idea of parametric equations. We discuss motion in a plane in Sections 3.2–3.5. Motion in space is deferred to Chapter 8. We take up the causes of motion, called *dynamics,* in Section 3.8.

3.1 MOTION ON A LINE

To motivate the idea of velocity we use Galileo's formula

$$s = kt^2,$$

which gives the distance s that an object falls (starting from rest and ignoring air resistance) in time t. In this formula k is a constant depending on the units of measurement. The equivalent of this formula is contained in Galileo's book *Dialogue Concerning Two New Sciences*. Galileo Galilei (1564–1642) argued against

Aristotle's ideas of motion, which claimed that heavier bodies fall faster than lighter ones, and founded dynamics. He died in the year in which Newton was born.

Investigation

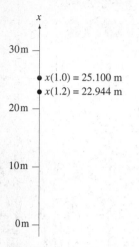

Figure 3.1. The position of the cannonball at two times.

A cannonball is dropped in a vacuum from a height of 30 meters above the earth's surface. Relative to the vertical line shown in Fig. 3.1, the position x of the ball at time t is

$$(1) \qquad x = x(t) = 30 - 4.9t^2, \quad 0 \le t \le \sqrt{30/4.9}.$$

Although this equation is based directly on Galileo's formula $s = kt^2$, it looks different. The constant k is $\frac{1}{2}9.8 = 4.9$. You may recognize this from the formula $s = \frac{1}{2}gt^2$, which is often included in a first physics course. We have chosen to measure lengths in meters and time in seconds. The constant 9.8 is the acceleration of gravity in SI units. We discuss these conventions in Section 3.8. The minus sign arises from the fact that we chose the positive direction of the x-axis as upward. The constant 30 is present because we chose to put the origin at the earth's surface. Note that when $t = 0$ in (1), $x = 30$. Finally, we are using x instead of s, where x measures not distance fallen in time t, which would be $30 - x$, but the coordinate position of the cannonball. The notation $x = x(t)$ is used to remind us that the symbol x is being used in two ways, as a dependent variable and as the name of a function with independent variable t. We obtained the upper limit $\sqrt{30/4.9} \approx 2.5$ seconds on t by setting $x = 0$ to calculate the time at which the ball hits the surface.

The black dots in the figure mark the positions of the cannonball at times $t = 1.0$ and $t = 1.2$. The positions $x(1.0)$ and $x(1.2)$ may be calculated from (1).

$$x(1.0) = 30 - 4.9(1.0)^2 = 25.100 \text{ m}$$
$$x(1.2) = 30 - 4.9(1.2)^2 = 22.944 \text{ m}.$$

From these positions we may calculate the distance D the ball falls during the time interval [1.0, 1.2], whose duration is 0.2 seconds. We have

$$(2) \qquad D = |x(1.2) - x(1.0)| = |22.944 - 25.100| = 2.156 \text{ m}.$$

To define and calculate the cannonball's velocity, which is the principal goal of this investigation, we recall the familiar formula $D = RT$.

For a Constant Rate, Distance Equals Rate Times Time

The formula "distance equals rate times time," usually written $D = RT$, gives the distance D traveled by an object moving at a *constant* rate R during time T. We use the term "speed" instead of "rate." For example, if between 10:00 A.M. and 11:06 A.M. your car were moving at a constant speed and the odometer changed from 20,000 km to 20,132 km, then the speed R of the car was

$$R = \frac{D}{T} = \frac{20{,}132 - 20{,}000}{11.1 - 10.0} = \frac{132}{1.1} = 120 \text{ km/h.}$$

We use this formula to calculate the velocity of the cannonball, whose speed is not constant. The bridging idea from constant speed to variable speed is that of *average velocity*.

Average Velocity

The average velocity of the ball on the time interval [1.0, 1.2] is

$$\frac{x(1.2) - x(1.0)}{1.2 - 1.0} = \frac{(30 - 4.9(1.2)^2) - (30 - 4.9(1.0)^2)}{1.2 - 1.0}$$

$$= \frac{-4.9(1.2^2 - 1.0^2)}{0.2} = -10.78 \text{ m/s.}$$

The average velocity is negative since the cannonball is moving downward and we chose the positive direction as upward. The meaning of the average velocity may be understood through the $D = RT$ formula. If the ball were traveling at the constant speed $|-10.78|$ m/s during the time interval [1.0, 1.2], it would have traveled a distance of $D = RT = (10.78)(0.2) = 2.156$ m. This is the same distance traveled by the "variable-rate ball," as calculated in (2). The "constant-rate ball" has the same positions as the variable-rate ball at times 1.0 and 1.2, and its change in position is the same. On smaller and smaller time intervals the behaviors of the constant-rate ball and the variable-rate ball become less and less distinguishable. We use the limiting value of the average velocity of the variable-rate ball—whose behavior we understand through the $D = RT$ formula—to give meaning to the velocity of the variable-rate ball at $t_1 = 1.0$. The average velocities given in Table 1 and shown in Fig. 3.2 give a sense of the limiting value of the average velocity on $[t_1, t_2]$ as $t_2 \to t_1$. Calculate the last three entries of the table. In the table and figure we have used the briefer notation $v(t_1, t_2)$ for the average velocity on $[t_1, t_2]$.

From Fig. 3.2 and Table 1, it appears that the velocity of the ball at $t = 1.0$ is close to -9.8 m/s.

This investigation suggests that the velocity of the ball is the derivative of its position function, that is, $v = dx/dt$. We now give formal definitions of average velocity, velocity, and speed of an object moving on a line.

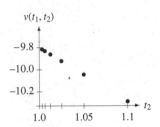

Figure 3.2. Average velocities as $t_2 \to t_1$.

Table 1. Average velocities as $t_2 \to t_1$

t_1	t_2	$v(t_1, t_2)$
1.0	1.1	-10.29
1.0	1.05	-10.045
1.0	1.025	-9.9225
1.0	1.0125	-9.86125
1.0	1.00625	$-9.8306\ldots$
1.0	1.003125	$-9.8153\ldots$

DEFINITION   average velocity, velocity, and speed

Let $x = x(t)$ be the position of an object moving on a line, where x is a function defined on an interval I. The **average velocity** of the object on a subinterval $[t_1, t_2]$ of I is

$$(3) \qquad\qquad v(t_1, t_2) = \frac{x(t_2) - x(t_1)}{t_2 - t_1}$$

If the function $x = x(t)$ is differentiable on I, then from (3) the limit of the average velocity function $v(t_1, t_2)$ as $t_2 \to t_1$ is the rate of change $x'(t_1)$ of x with respect to t at the point t_1. We define the **velocity** $v(t_1)$ of the object at t_1 as

$$v(t_1) = x'(t_1) = \lim_{t_2 \to t_1} \frac{x(t_2) - x(t_1)}{t_2 - t_1}$$

The **speed** of the object at time t is $|v(t)|$.

We give two more examples of motion on a line: the periodic motion of a mass attached to a spring and the motion of a projectile fired vertically upward.

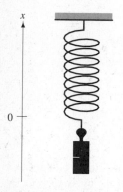

Figure 3.3. Spring and mass.

■ **EXAMPLE 1**   The motion of a mass attached to a spring (see Fig. 3.3) can be closely modeled by an equation of the form

(4) $$x = x(t) = A\sin(Bt + C), \quad t \geq 0.$$

To relate this equation and the spring-mass system shown in the figure, we place a coordinate system next to the spring-mass system and a reference mark on the mass. We agree that the coordinate position of the reference mark at time t is $x(t)$. To model a specific spring-mass system, that is, to find values of A, B, and C, we need some data. For example, suppose that the initial position of the mass (i.e., the reference mark) is at the -1.5 cm coordinate (i.e., $x(0) = -1.5$), that its initial velocity is -4.0 cm/s, and that it completes one oscillation in 2π seconds. From these data,

a. Calculate A, B, and C to two significant figures.
b. Calculate the least t for which $x = 0$.
c. Calculate the first time t^* at which the mass reaches its lowest point.

Solution

a. In our calculations we regard the given data as exact, but we round the final values of A, B, and C to two significant figures. Since $x(0) = -1.5$ and the period of the motion is 2π, we have

(5) $$-1.5 = A\sin C$$
(6) $$B = 1.$$

To use the fact that $v(0) = x'(0) = -4.0$, we differentiate (4), obtaining

(7) $$v(t) = x'(t) = AB\cos(Bt + C).$$

Since $B = 1$ and $v(0) = -4.0$, we have

(8) $$-4.0 = A\cos C.$$

We may calculate A and C using trigonometry. From (5) and (8), we have

$$A^2\sin^2 C + A^2\cos^2 C = A^2 = 2.25 + 16.00.$$

If we choose $A = \sqrt{18.25} \approx 4.3$, it follows from (5) and (8) that C is a third-quadrant angle and $\tan C = 0.375$. Hence,

$$C = \pi + \arctan 0.375 = 3.5003\ldots.$$

We take $C = 3.5$. From these calculations, we have

(9) $$x(t) = 4.3\sin(t + 3.5), \quad t \geq 0.$$

b. Note that for $t = 0$, the number $t + 3.5$ is in the third quadrant. Hence, the least t for which $x = 0$ satisfies $t + 3.5 = 2\pi$. This gives $t = 2.8$ s.

c. This question can be answered in either of two ways. Using the graph of x shown in Fig. 3.4, t^* must satisfy

$$x(t^*) = -4.3 = 4.3 \sin(t^* + 3.5).$$

Hence $t^* + 3.5 = 3\pi/2$. This gives $t' = 1.2$ s.

A more interesting answer is to note that at its lowest point the speed of the mass must be zero. Hence $|v(t^*)| = 0$; that is, we may calculate t^* by differentiating the position function and locating the least t for which $v = 0$. From (9) we have

$$v = x'(t) = 4.3 \cos(t + 3.5), \quad t \geq 0.$$

We calculate t^* by setting $t + 3.5 = 3\pi/2$. This gives $t^* = 1.2$ s.

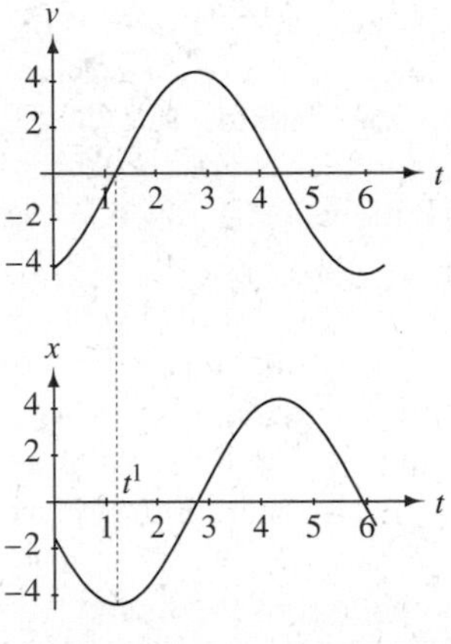

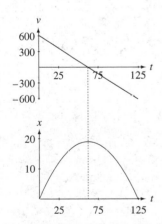

Figure 3.4. Connections between the graphs of x and v.

At time t^* the position function is at its lowest or minimum value and the velocity function is 0. Note that the graph of the position function has a horizontal tangent at $(t^*, x(t^*))$. Its slope there is 0. ∎

In the next example we give an equation for the position of a bullet as a function of time t. This equation and equation (1) in the investigation of the motion of a cannonball are variations on Galileo's equation. We are using these equations to introduce the mathematical ideas and procedures that make it possible for you to infer these same equations directly from Newton's laws.

■ **EXAMPLE 2** A rifle is fired vertically upward so that at $t = 0$ the bullet is 1.5 m above ground level and has speed 610 m/s ($\approx$ 2000 ft/s). If air resistance is neglected, then relative to a coordinate line with origin at ground level and with positive direction upward, the coordinate position x of the bullet is given by

$$x = x(t) = -4.9t^2 + 610t + 1.5, \quad t \in [0, b]$$

where b is the time when the bullet strikes the earth. We show a sketch of x against t in the graph at the bottom of Fig. 3.5. The units on the vertical scale are kilometers. Note that the bullet moves vertically, not on the parabola in the figure. Calculate b and the maximum height of the bullet.

Solution. The bullet undergoes a gradual decrease in velocity until the maximum height is reached, when the velocity is zero. The velocity of the bullet is

$$(10) \quad v = v(t) = x'(t) = \frac{d}{dt}(-4.9t^2 + 610t + 1.5) = -9.8t + 610.$$

Figure 3.5. Bullet's position and velocity versus time.

The velocity is graphed against time in the upper graph in Fig. 3.5. Both graphs show that the change in x per second decreases until zero is reached at the peak of the lower graph, which corresponds to the t-intercept of the upper graph. To find t_{max} so that $v(t_{max}) = 0$, we solve the equation $-9.8t + 610 = 0$. This gives $t_{max} \approx 62$ s. To find the maximum height, we calculate $x(t_{max}) \approx 1.9 \times 10^4$ m.

The value of b may be found by setting $x(t)$ equal to 0. The positive root of the resulting equation is 120 s, approximately. ∎

Inverse Problem

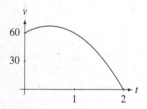

Figure 3.6. Forward and inverse problems.

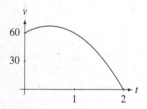

Figure 3.7. Plot of velocity versus time for broken odometer problem.

In the spring-mass and vertical bullet examples the position function $x = x(t)$ was given. If the position is known as a function of time, the motion is in some sense completely described. For example, the velocity can be calculated and questions concerning the motion can be answered by solving various equations. But, there's an *if* in our sentence: "If the position is known...." How can we describe motion if we're given not the position function but instead the velocity function? To oversimplify, that's an inverse problem.

Here's another way of thinking about the inverse problem. Instead of cannonballs, spring-mass systems, or bullets, suppose we discuss the motion of an automobile. In these terms, we've been solving the broken speedometer problem, as at the top of Fig. 3.6. We know the position of the automobile at any time t, that is, the odometer reading, and we urgently need the speedometer reading. Easy! Just differentiate the odometer.

A graphic for the inverse problem is shown at the bottom of the figure. The odometer broke two hours ago and we (less) urgently need to know its reading now. What do we do?

Inverse problems are not quite as easy as "forward problems." But they are so important that large parts of calculus were created just to solve them. To end the section, we solve a few easy inverse problems.

■ **EXAMPLE 3** Two hours ago your odometer broke. It read 314,159 miles at the time. Fortunately, a recording of the speedometer was made. A graph of the velocity (which, for motion in one direction, may be taken to be the same as the speed) against time is shown in Fig. 3.7. The units are hours and miles per hour. The time $t = 0$ is the time at which the odometer broke. Calculate the present odometer reading.

Solution. This is a reasonable problem as stated since speedometer readings would probably come in graphical form. However, to make the problem easier (but less realistic), we assume that the velocity is given by the equation

(11) $$v = v(t) = 30t - 30t^2 + 60.$$

Besides knowing the velocity, we know that the odometer read 314,159 two hours ago. We state this as

(12) $$x(0) = 314,159.$$

We must find a function x satisfying (11) and (12). The key thing to notice is that the given velocity function is the derivative of the unknown position function. If we think about how differentiation goes, surely $x = 15t^2 - 10t^3 + 60t$. For if you differentiate this expression, you will get the formula for v in (11). However, it's not quite this simple, for if we differentiate the slightly different function $x = 15t^2 - 10t^3 + 60t + 1000$, we also obtain $v = 30t - 30t^2 + 60$. In fact, if we differentiate any function of the form $x = 15t^2 - 10t^3 + 60t + c$, where c is a constant, we obtain $v = 30t - 30t^2 + 60$. We show in Chapter 4 that the only functions x having the property that $x' = v$ have the form $x = 15t^2 - 10t^3 + 60t + c$.

So from (11) we have inferred that

(13) $$x = 15t^2 - 10t^3 + 60t + c.$$

We may determine the constant c by using equation (12), which gives the odometer reading at the same time it broke. Substituting into (13), we have

$$314{,}159 = 15 \cdot 0^2 - 10 \cdot 0^3 + 60 \cdot 0 + c, \quad \text{which gives } c = 314{,}159.$$

Hence the position of your car—or the odometer reading—at any time t is

$$x(t) = 15t^2 - 10t^3 + 60t + 314{,}159.$$

To calculate the odometer reading at $t = 2$ hours, we replace t by 2 in this equation. This gives $x(2) = 314{,}259$. ∎

We solve two more inverse problems, using the spring-mass and bullet contexts. In each case we give the velocity and the initial position.

∎ **EXAMPLE 4** Suppose that the velocity and initial position of a spring-mass system are

$$x'(t) = v(t) = 4.2 \sin(0.57t + 1.8) \text{ m/s}$$
$$x(0) = 2.9 \text{ m}.$$

Calculate the position of the mass at $t = 1.5$.

Solution. It is reasonable to suppose that the function $x(t)$ must have the form

$$x = A \cos(0.57t + 1.8) + B.$$

Why is this reasonable? Because whatever x really is, differentiation must give $x' = v = 4.2 \sin(0.57t + 1.8)$. Since the derivative of $A \cos(0.57t + 1.8) + B$ is $v = -0.57A \sin(0.57t + 1.8)$, we see that $A = -4.2/0.57$. Hence

$$x(t) = \frac{-4.2}{0.57} \cos(0.57t + 1.8) + B.$$

We may evaluate B by using the initial condition $x(0) = 2.9$. Putting all of this together, we find

$$x(t) = \frac{-4.2}{0.57} \cos(0.57t + 1.8) + 2.9 + \frac{4.2}{0.57} \cos(1.8)$$
$$\approx 1.2 - 7.4 \cos(0.57t + 1.8).$$

From this result we may calculate $x(1.5)$. We have

$$x(1.5) \approx 7.7 \text{ m}. \qquad ∎$$

∎ **EXAMPLE 5** Suppose a rifle is fired vertically upward so that the bullet has velocity and initial position

$$v(t) = -9.8t + 615 \text{ m/s} \quad \text{and} \quad x(0) = 1.3 \text{ m}.$$

Find the maximum height of the bullet and the time for it to return to its initial position.

Solution.   Since $v = dx/dt = -9.8t + 615$, the function $x(t)$ must have the form $-\frac{1}{2}(9.8)t^2 + 615t + c$. To evaluate c, we use the initial condition $x(0) = 1.3$. We have

$$1.3 = -\frac{1}{2}(9.8) \cdot 0^2 + 615 \cdot 0 + c, \quad \text{so that} \quad c = 1.3.$$

The position function is therefore given by

(14) $$x = x(t) = -\frac{1}{2}(9.8)t^2 + 615t + 1.3.$$

To find the time t_{max} when the bullet attains its maximum height, we set v equal to 0 and solve for t. We find $t_{max} = 615/9.8$. The maximum height is $x(t_{max})$. From (14), we have

$$x(t_{max}) \approx 1.9 \times 10^4 \text{ m}.$$

The time it takes for the bullet to return to its starting position is $2t_{max}$. Depending on your audience, you may assert that this answer (*a*) is obvious (time up equals time down), (*b*) is geometrically clear (plot x against t and use symmetry—see Fig. 3.8), or (*c*) may be found by solving the quadratic equation $(-9.8/2)t^2 + 615t + 1.3 = 1.3$. ∎

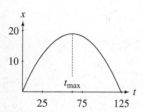

Figure 3.8. Bullet's position versus time.

Exercises 3.1

Basic

1. A cannonball is dropped from the floor of a bridge 100 m above the water, so that $x = x(t) = 100 - 4.9t^2$. Calculate its average velocity $v(1.5, 1.6)$. Using Table 1 and Fig. 3.2 as models, make a table and graph using average velocities $v(1.5, 1.5 + h)$, $h = 0.02, 0.01, 0.001$. Estimate $v(1.5)$ from your work. When does the ball hit the water?

2. A cannonball is dropped from the floor of a bridge 75 m above the water, so that $x = x(t) = 75 - 4.9t^2$. Calculate its average velocity $v(1.25, 1.35)$. Using Table 1 and Fig. 3.2 as models, make a table and graph using average velocities $v(1.25, 1.25 + h)$, $h = 0.02, 0.01, 0.001$. Estimate $v(1.25)$ from your work. When does the ball hit the water?

3. Answer questions *a–c* of Example 1 for the spring-mass system with $x(0) = 3.7$, $v(0) = -0.68$, and period 2π. Graph x and v together, as in Fig. 3.4.

4. Answer questions *a–c* of Example 1 for the spring-mass system with $x(0) = 3.8$, $v(0) = -3.1$, and period 2π. Graph x and v together, as in Fig. 3.4.

Exercises 5 and 6: Equation (1) has the form $x = x_0 - \frac{1}{2}gt^2$, where g is the acceleration of gravity and x_0 is the coordinate position of an object when $t = 0$. The units of g are m/s^2 if lengths are in meters and time is in seconds. The acceleration of gravity g varies with the source of the gravity. Assume that the coordinate system is the same as in Fig. 3.1. For earth, $g \approx 9.8$ m/s^2.

5. Find the velocity on impact of an object falling from 100 m on Jupiter, where $g \approx 26.5$ m/s^2.

6. Find the velocity on impact of an object falling from 100 m on the moon, where $g \approx 1.62$ m/s^2.

7. Repeat Example 2 for the case where the bullet is 2.0 m above ground level at $t = 0$ and has speed 630 m/s.

8. Repeat Example 2 for the case where the bullet is 0.7 m above ground level at $t = 0$ and has speed 600 m/s.

Exercises 9–16: Solve the inverse problems using ideas similar to those used in Examples 3–5.

9. Three hours ago your odometer broke. It read 271,828 miles at the time. During this time the speedometer was recorded. Your velocity was $v(t) = (-40/3)t^2 + (55/3)t + 65$. Calculate the present odometer reading. What was your highest speed during these three hours?

10. Two hours ago your odometer broke. It read 48,130 miles at the time. During this time the speedometer was recorded. Your velocity was $v(t) = (-110/3)t^2 + (145/3)t + 50$. Calculate the present odometer reading. What was your highest speed during these two hours?

11. Calculate $x(2.1)$ if $v(t) = 25 - 9.8t$ and $x(0) = 5.7$.

12. Calculate $x(3.9)$ if $v(t) = 48 - 9.8t$ and $x(0) = 2.1$.

13. Calculate $x(1.2)$ if $v(t) = 3.2\sin(1.3t + 2.7)$ and $x(0) = 1.0$.

14. Calculate $x(3.6)$ if $v(t) = 4.7\sin(2.6t + 1.3)$ and $x(0) = 0.0$.

15. Calculate the maximum height reached by an object for which $v(t) = -9.8t + 120$ and $x(0) = 2.5$.

16. Calculate the maximum height reached by an object for which $v(t) = -9.8t + 130$ and $x(0) = -3.5$.

Growth

17. This exercise refers to Example 1 and Fig. 3.4. Estimate the times $t \in [0, 2\pi]$ at which the speed of the mass is either a minimum or maximum.

18. An object was seen moving on a line at 20 m/s for 60 s and then at 30 m/s for 120 s. Use the definition of average velocity in calculating the object's average velocity for the 180 s trip.

19. An object was seen moving on a line at 25 m/s for 240 s and then at 32 m/s for 120 s. Use the definition of average velocity in calculating the object's average velocity for the 360 s trip.

20. If an object leaves from the point $x = 4.5$ at $t = 2.5$ and returns at $t = 5.6$, what is its average velocity on the interval $[2.5, 2.6]$?

21. If an object leaves from the point $x = 3.9$ at $t = 5.5$ and returns at $t = 8.1$, what is its average velocity on the interval $[5.5, 8.1]$?

22. Cities A and B are 150 km apart. If a motorist travels from A to B at 100 km/h and then goes back to A at 80 km/h, what meaning do you give to the "average" $(100 + 80)/2$?

23. Cities C and D are 170 km apart. If a motorist travels from A to B at 80 km/h and then back to A at 60 km/h, in what terms would you describe the "average" $(80 + 60)/2$?

24. Find the average velocity during ascent of the bullet whose position is given by $x = -4.9t^2 + 610t + 5.4$.

25. Find the average velocity during ascent of the bullet whose position is given by $x = -4.9t^2 + 630t + 3.9$.

26. Find the average velocity during the time interval $[t_1, t_2]$ for the bullet whose position is given by $x = -\frac{1}{2}gt^2 + v_0t + x_0$, where g is the acceleration of gravity, v_0 is the bullet's initial velocity (we assume $v_0 > 0$), and x_0 is its initial position. Use this result in finding the average velocity during ascent of this bullet.

27. Find the average velocity during the time interval $[-c/b, -c/b + \pi/(2b)]$ for a mass with position given by $x(t) = a\sin(bt + c)$, where a, b, and c are constants.

28. In Exercise 26 the average velocity $v(t_1, t_2)$ was calculated. Calculate $\lim_{t_2 \to t_1} v(t_1, t_2)$ and give a meaning to the result.

29. In Exercise 27 the average velocity $v(t_1, t_2)$ was calculated for specific values of t_1 and t_2. Write out the general case, calculate $\lim_{t_2 \to v_1} v(t_1, t_2)$, and give a meaning to the result. You may wish to use the identity

$$\sin A - \sin B = 2\sin\frac{A - B}{2}\cos\frac{A + B}{2}.$$

30. Your car has a broken speedometer. Plot a velocity graph from the following odometer data. Each data point is of the form (t, x), where t is the time in hours and x is the odometer reading in kilometers.

(0.0, 11084), (0.08, 11088), (0.16, 11092),

(0.24, 11095), (0.32, 11097), (0.40, 11099),

(0.48, 11100), (0.56, 11100)

31. Your car has a broken speedometer. Plot a velocity graph from the graph of the odometer reading shown in the following figure.

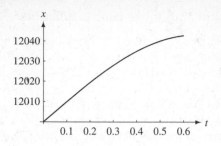

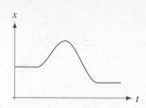

32. Does the bullet in Example 5 follow the path shown in Fig. 3.8?

33. An object is moving on a line. A qualitative graph of its position versus time is given in the following figure. Sketch a qualitative graph of its velocity versus time.

34. An object is moving on a line. A qualitative graph of its position versus time is given in the following figure. Sketch a qualitative graph of its velocity versus time.

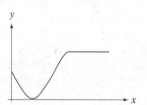

Exercises 35–52:   Determine $x = x(t)$.

35. $x'(t) = t^2 + 3t + 7$ and $x(0) = 1$

36. $x'(t) = 5t^2 - 3t$ and $x(0) = 5$

37. $x'(t) = \sqrt{2t + 3}$ and $x(0) = 11$

38. $x'(t) = \sqrt{5 - 7t}$ and $x(0) = 1$

39. $x'(t) = e^{3t}$ and $x(0) = -3$

40. $x'(t) = e^{-7t}$ and $x(0) = 2$

41. $x'(t) = 2/(t + 1)$ and $x(0) = 0$

42. $x'(t) = -3/(t + 2)$ and $x(0) = -2$

43. $x'(t) = 1/\sqrt{1 - t^2}$ and $x(0) = 1$

44. $x'(t) = -2/\sqrt{1 - t^2}$ and $x(0) = 2$

45. $x'(t) = 1/(1 + t^2)$ and $x(0) = 1$

46. $x'(t) = 2/(1 + t^2)$ and $x(0) = -1$

47. $x'(t) = \sin(2t + 3)$ and $x(0) = -1.2$

48. $x'(t) = \sin(3t - 1)$ and $x(0) = 5.9$

49. $x'(t) = \sec^2 t$ and $x(0) = 1$

50. $x'(t) = 3\sec^2 t$ and $x(0) = -2.5$

51. $x'(t) = \sec t \tan t$ and $x(0) = 0$

52. $x'(t) = -3\sec t \tan t$ and $x(0) = -1$

Review/Preview

Exercises 53–60:   Most of these problems concern basic trigonometry. They relate to the following figure, which shows a circle centered at the origin with radius $r > 0$. The angle $\theta \in [0, 2\pi)$ is measured from the positive half of the x-axis, counterclockwise to a ray from the origin to a point P. The coordinates of P are $(r \cos \theta, r \sin \theta)$. Place your calculator in radian mode.

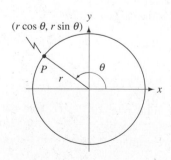

53. Sketch a figure like that shown and calculate the coordinates x and y of the point $P = (r\cos\theta, r\sin\theta)$, where $r = 2$ and $\theta = 2.5$. Use your calculator to show that $x^2 + y^2 \approx r^2$. Give a reason for the close agreement.

54. The point $P = (3/\sqrt{2}, -1/\sqrt{2})$ is on a circle $x^2 + y^2 = r^2$ and determines an angle $\theta \in [0, 2\pi)$. Determine r and θ, sketch the circle, and plot P.

55. Plot the point $(-25, 10)$ in the (x, y)-plane. Find r and θ such that $(r\cos\theta, r\sin\theta) = (-25, 10)$.

56. Plot the point $(15, -5)$ in the (x, y)-plane. Find r and θ such that $(r\cos\theta, r\sin\theta) = (15, -5)$.

57. Describe the result of plotting on the (x, y)-coordinate system the 63 points in the set

$$\{(2\cos\theta, 2\sin\theta) \mid \theta = 0.0, 0.1, 0.2, \ldots, 6.2\}.$$

58. Describe the result of plotting on the (x, y)-coordinate system the 63 points in the set

$$\{(\theta\cos\theta, \theta\sin\theta) \mid \theta = 0.0, 0.1, 0.2, \ldots, 6.2\}.$$

59. With your calculator set in radians, calculate θ if $\theta \in [0, 2\pi)$, $\cos\theta = 0.96$, and $\sin\theta = 0.28$.

60. With your calculator set in radians, calculate θ if $\theta \in [0, 2\pi)$, $\cos\theta = -0.35$, and $\sin\theta = 0.25$.

61. Use the law of cosines in calculating the angle opposite the longest side in a triangle with sides 3.3, 5.4, and 7.9 meters. The law of cosines is summarized in the accompanying figure.

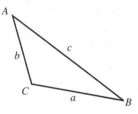

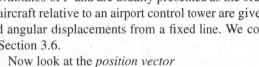

$$c^2 = a^2 + b^2 - 2ab\cos C$$

62. A central angle of 1.2 radians is drawn in a circle of radius 5 cm. What is the length of the arc cut from the circumference of the circle by the sides of this angle?

3.2 VECTORS AND MOTION—POSITION VECTORS

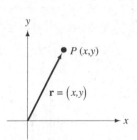

Figure 3.9. A point P in a plane.

In Section 3.1 we discussed the motion of objects in one dimension, that is, on a line. In this section and the next we prepare to study motion in two or more dimensions, so that we can describe the motion of an object moving in a plane or in space. To describe such motion we use vectors. With vectors we can describe and calculate efficiently the object's position, velocity, and acceleration.

Figure 3.9 shows a point P in the (x, y)-plane. The coordinates of P are the numbers x and y, which measure the "signed distances" of P from the y- and x-axes. These *Cartesian coordinates* are usually presented as the ordered pair (x, y). The point P can also be located by giving its distance r from the origin and its angular displacement θ from the positive x-axis. The numbers r and θ are *polar coordinates* of P and are usually presented as the ordered pair (r, θ). The locations of aircraft relative to an airport control tower are given as distances from the tower and angular displacements from a fixed line. We come back to polar coordinates in Section 3.6.

Now look at the *position vector*

$$\mathbf{r} = (x, y)$$

of the point P. The position vector $\mathbf{r}$ is shown in Fig. 3.10 as the arrow directed from the origin to P. We write $\mathbf{r} = (x, y)$ using boldface type to distinguish between the vector $\mathbf{r}$ and the polar distance r between P and the origin. We use larger parentheses to distinguish between the position vector (x, y) and the pair of Cartesian coordinates (x, y) of P. At the same time, we do not intend to emphasize this distinction, as it is often convenient to blur the difference between the position vector and the Cartesian coordinates of P.

Figure 3.10. The position vector $\mathbf{r}$ of P.

The notion of a position vector $\mathbf{r} = (x, y)$ of a point P combines the polar and Cartesian coordinate systems. The visual presentation of $\mathbf{r}$ is based on length and orientation, while symbolically we identify the vector $\mathbf{r}$ with the ordered pair of its Cartesian coordinates.

The boldface notation for vectors works reasonably well on the printed page but is awkward when done by hand. You may wish to use an overhead arrow notation, writing $\vec{r}$ instead of $\mathbf{r}$.

DEFINITION   **position vector and its length and direction**

Let P be a point in the (x, y)-plane with coordinates x and y. The **position vector $\mathbf{r}$** of P is written as $\mathbf{r} = (x, y)$ and represented as an arrow with tail at the origin and head at P.

The **length** of the position vector $\mathbf{r} = (x, y)$ is written as $\|\mathbf{r}\|$ and defined by

$$\|\mathbf{r}\| = \sqrt{x^2 + y^2}.$$

The **direction** of $\mathbf{r}$ is the unique angle $\theta \in [0, 2\pi)$ for which

$$\cos\theta = \frac{x}{\sqrt{x^2 + y^2}} \quad \text{and} \quad \sin\theta = \frac{y}{\sqrt{x^2 + y^2}}.$$

If $\mathbf{r} = (0, 0)$ we assign no direction to $\mathbf{r}$.

■ **EXAMPLE 1**   Sketch the position vectors $\mathbf{a} = (3, 1)$, $\mathbf{b} = (-3, -2)$, and $\mathbf{c} = (-2.5, 1.5)$ and calculate their lengths and directions.

Solution.   To sketch these vectors we draw arrows from the origin to the points $(3, 1)$, $(-3, -2)$, and $(-2.5, 1.5)$, as shown in Fig. 3.11. Their lengths may be calculated from the above definition. We have

$$\|\mathbf{a}\| = \sqrt{3^2 + 1^2} = \sqrt{10}$$
$$\|\mathbf{b}\| = \sqrt{(-3)^2 + (-2)^2} = \sqrt{13}$$
$$\|\mathbf{c}\| = \sqrt{(-2.5)^2 + 1.5^2} = \sqrt{8.5}.$$

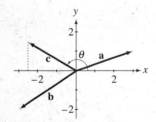

Figure 3.11. Vector lengths and directions.

Calculating the directions of these vectors takes more care. We do one of the calculations and urge you to do the other two. There are several ways of calculating the unique angle $\theta \in [0, 2\pi)$ for which

$$\cos\theta = \frac{x}{\sqrt{x^2 + y^2}} \quad \text{and} \quad \sin\theta = \frac{y}{\sqrt{x^2 + y^2}}.$$

For the vector $\mathbf{c} = (-2.5, 1.5)$ the direction θ satisfies

$$\cos\theta = \frac{-2.5}{\sqrt{8.5}} = -0.85749\ldots \quad \text{and} \quad \sin\theta = \frac{1.5}{\sqrt{8.5}} = 0.51449\ldots.$$

Since the range of the inverse cosine function is $[0, \pi]$, the angle θ can be found with a calculator by first calculating $\cos \theta = -2.5/\sqrt{8.5} = -0.85749\ldots$ and then pressing the ACOS key. In radian mode this gives $\theta \approx 2.6$.

If, instead, you calculate $\sin \theta = 1.5/\sqrt{8.5} = 0.51445\ldots$ and press the ASIN key, you will obtain $0.54041\ldots$. Since θ is a second-quadrant angle, you must subtract this result from π to obtain $\theta = \pi - 0.54041\ldots \approx 2.6$.

Another possibility is to draw a line from $(-2.5, 1.5)$ to the x-axis. This forms a right triangle with sides 2.5 and 1.5. The tangent of the angle opposite the side of length 1.5 is $1.5/2.5 = 0.6$. Pressing the ATAN key gives $0.54041\ldots$. From this we obtain $\theta = \pi - 0.54041\ldots \approx 2.6$.

It is not wrong to calculate these angles in degrees; indeed, this is the "default" in engineering and physics. We shall work in degrees when this seems natural. The reason our default is radians is to avoid complications when we differentiate trigonometric functions. ■

Position vectors are often used to describe motion. We introduce this important topic in an Investigation and then give three examples.

Investigation

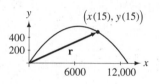

Figure 3.12. Speeding bullet.

Figure 3.12 shows the path of a bullet fired from the origin at an angle of $10°$ to the horizontal and with initial speed of 610 m/s. The angle appears too large because of scaling. Assuming that the bullet is moving in a vacuum and that lengths are measured in meters and time in seconds, the x- and y-coordinates of the bullet at time t are given by

(1)
$$x = x(t) = (610 \cos 10°)t$$
$$y = y(t) = -4.9t^2 + (610 \sin 10°)t.$$

These equations closely approximate the true path of a bullet fired as described. They are yet another variation on Galileo's formula $s = kt^2$. The new feature is that the bullet no longer moves on a line; it moves in a plane.

We said that the equations (1) give the position of the bullet at any time t. To make this more concrete, we calculate the position of the bullet at $t = 0$ and $t = 15$ seconds.

Usually we assume that $t = 0$ marks the beginning of the motion. The coordinates (x, y) of the bullet at $t = 0$ may be calculated by substituting in (1). We obtain

$$x = x(0) = (610 \cos 10°)(0) = 0 \text{ m}$$
$$y = y(0) = -4.9(0)^2 + (610 \sin 10°)(0) = 0 \text{ m}.$$

This shows that the bullet starts from $(0, 0)$.

After 15 seconds the coordinates (x, y) of the bullet are

(2)
$$x = x(15) = (610 \cos 10°)(15) \approx 9011 \text{ m}$$
$$y = y(15) = -4.9(15)^2 + (610 \sin 10°)(15) \approx 486 \text{ m}.$$

The bullet is near the point $(9011, 486)$ after 15 seconds. This position is shown in Fig. 3.12.

By eliminating the dependence on time in the equations (1), we obtain a familiar description of the curve followed by the bullet. We begin by solving the first equation in (1) for t. We find

$$t = \frac{x}{610 \cos 10°}.$$

Substituting this expression for t into the second equation, we obtain

$$y = -4.9\left(\frac{x}{610 \cos 10°}\right)^2 + 610 \sin 10°\left(\frac{x}{610 \cos 10°}\right)$$

$$(3) \qquad y \approx (-1.4 \times 10^{-5})x^2 + (1.8 \times 10^{-1})x.$$

This equation has the form $y = Ax^2 + Bx$, which we recognize as the equation of a parabola. Since $A < 0$, the parabola opens downward.

Equation (3) may be used in describing the curve on which the bullet moves, but we cannot infer from it how x and y depend on time. For this we must use (1).

We show in Fig. 3.12 the position vector $\mathbf{r}$ from the origin to the point $(x(15), y(15))$ through which the bullet passes when $t = 15$. We write

$$\mathbf{r} = \big((610 \cos 10°) \cdot 15,\, -4.9 \cdot 15^2 + (610 \sin 10°) \cdot 15\big).$$

From (2) we have

$$\mathbf{r} \approx (9011, 486).$$

Figure 3.13. Position vector.

This position vector of the bullet is shown in Fig. 3.13, where the scales on the axes are approximately equal.

At $t = 15$ the bullet is at a distance of

$$\|\mathbf{r}\| = \sqrt{9011^2 + 486^2} \approx 9024 \text{ m}$$

from the origin. This is the length of its position vector. Its direction from the origin is the direction of $\mathbf{r}$, which is $\theta \approx \arctan(486/9011) \approx 3.1°$.

The two equations in (1) are *real-valued* functions of t; that is, the ranges of $x = x(t)$ and $y = y(t)$ are subsets of the set R of real numbers. To specify the position of the bullet as a function of t takes two real-valued functions. We put them together in the pair $(x(t), y(t))$, which we regard as a position vector. We write

$$(4) \qquad \mathbf{r} = \mathbf{r}(t) = \big((610 \cos 10°)t,\, -4.9t^2 + (610 \sin 10°)t\big).$$

The value of this *vector-valued* function at $t = 15$ is the vector $\mathbf{r} \approx (9011, 486)$. The domain of the function $\mathbf{r} = \mathbf{r}(t)$ is the interval $[0, b]$ of the t-axis (where b is the time the bullet returns to ground level). Its range is a set of position vectors.

We start with two important and familiar examples of motion: linear and circular motion.

■ **EXAMPLE 2**   An ant is moving in the (x, y)-coordinate system so that at time $t \geq 0$ its position is

$$(5) \qquad \mathbf{r} = (x, y) = (-1 + 5t, 3 - 4t).$$

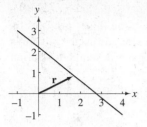

Figure 3.14. Ant on line.

Time is measured in seconds and length in millimeters. Its path for $0 \leq t \leq 1$ is shown in Fig. 3.14. Calculate its position vector at $t = 0$, $t = 1/2$, and $t = 1$. Given that the ant is moving at a constant rate, calculate its speed. Finally, determine an equation of the form $y = mx + b$ for the ant's line of travel.

Solution. Most graphics calculators can plot curves described parametrically. That's what we have here. The x-coordinate is $-1 + 5t$, and the y-coordinate is $3 - 4t$. This might be a good time to learn how to graph curves of this kind. Whether you plot the ant's path with the help of your calculator, plot by hand, or are simply trusting Fig. 3.14, it appears that the ant is moving on a line. We will check on this in another way shortly. Meanwhile we calculate the ant's position vectors at $t = 0$ and $t = 1$. We have

$$\mathbf{r}(0) = \left(-1 + 5 \cdot 0, 3 - 4 \cdot 0\right) = \left(-1, 3\right)$$
$$\mathbf{r}(1) = \left(-1 + 5 \cdot 1, 3 - 4 \cdot 1\right) = \left(4, -1\right).$$

Since the ant is moving at a constant rate, we may use the "distance equals rate times time" formula, $D = RT$. During the one-second interval from $t = 0$ to $t = 1$ the ant has moved a distance of

$$D = \sqrt{(4 - (-1))^2 + (-1 - 3)^2} = \sqrt{41} \approx 6.40 \text{ mm}.$$

Its rate R is

$$R = D/T = \sqrt{41}/1 \approx 6.40 \text{ mm/s}.$$

In the Investigation we obtained an equation $y = Ax^2 + Bx$ for the parabolic trajectory of the bullet by eliminating t. We do the same thing here. In (5) let x and y denote the x- and y-coordinates of the position vector $\mathbf{r}$. We have

$$x = -1 + 5t \quad \text{and} \quad y = 3 - 4t.$$

Solving the first of these equations for t,

$$t = \tfrac{1}{5}x + \tfrac{1}{5}.$$

Substituting in the second equation gives

(6) $$y = 3 - 4\left(\tfrac{1}{5}x + \tfrac{1}{5}\right) = -\tfrac{4}{5}x + \tfrac{11}{5}.$$

The ant's path is on the line with slope $-\tfrac{4}{5}$ and y-intercept $\tfrac{11}{5}$. ∎

■ **EXAMPLE 3** We now consider an ant moving so that for all times $t \geq 0$ its position is

(7) $$\mathbf{r} = \left(x, y\right) = \left(2 \cos t, 2 \sin t\right).$$

Distances are in millimeters and time is in seconds. Show that the ant is moving at a constant speed of 2 mm/s on a circle of radius 2 centered at the origin.

Solution. The last statement above means that we should give mathematical evidence that the ant is moving on a circle. This would be in addition to any graphical

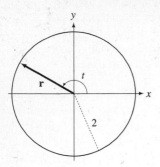

Figure 3.15. Circular motion.

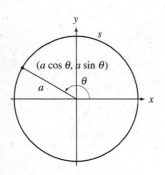

Figure 3.16. Basic trigonometry.

evidence we can produce. There are two ways of showing that the ant is on a circle of radius 2 centered at the origin. First, we may interpret the time t in (7) as the angle t shown in Fig. 3.15 and recall the definition of the sine and cosine functions. For this we show in Fig. 3.16 a circle of radius a, an angle θ in standard position, and a point with coordinates $(a\cos\theta, a\sin\theta)$.

A second way of showing that the ant is moving on a circle is to eliminate the time t. From (7) we have

$$x = 2\cos t \quad \text{and} \quad y = 2\sin t.$$

With the identity $\sin^2 t + \cos^2 t = 1$ in mind, we have

$$x^2 + y^2 = (2\cos t)^2 + (2\sin t)^2 = 4(\sin^2 t + \cos^2 t) = 4 \cdot 1 = 2^2.$$

Dropping the intermediate calculations gives

$$x^2 + y^2 = 2^2.$$

This is the equation of a circle of radius 2 centered at the origin.

To argue that the ant's speed is constant, we recall that the length s of an arc cut from a circle of radius a by a central angle θ is $s = a\theta$ (θ must be measured in radians). See Fig. 3.16. In t seconds the ant moves a distance of $s = 2t$ mm. Since $ds/dt = 2$, the ant is moving with constant speed of 2 mm/s.

Note that although the ant's speed is constant, its direction of travel changes with t, from the positive y direction at $t = 0$ to the negative x direction at $t = \pi/2$. In the next section we discuss the velocity vector, which determines both speed and direction of travel. ∎

Exercises 3.2

Basic

Exercises 1–8: Sketch the vector, and then calculate its length and direction.

1. $\mathbf{r} = (1, 1)$

2. $\mathbf{r} = (2, 1)$

3. $\mathbf{r} = (-3, 4)$

4. $\mathbf{r} = (-1, 7)$

5. $\mathbf{a} = (-4.3, -5.0)$

6. $\mathbf{v} = (-1, -10)$

7. $\mathbf{v} = (2/\sqrt{5}, -1/\sqrt{5})$

8. $\mathbf{r} = (5/\sqrt{29}, -2/\sqrt{29})$

Exercises 9–12: For the given vector equation for the motion and time t_1, eliminate t to find an equation of the form $y = f(x)$ describing the path along which the object is moving, sketch the path, and draw the position vector $\mathbf{r}(t_1)$.

9. $\mathbf{r}(t) = (5t, 2 + 3t)$, $0 \le t \le 1$, $t_1 = 1/2$

10. $\mathbf{r}(t) = (4t - 1, -1 - 3t)$, $0 \le t \le 2$, $t_1 = 1$

11. $\mathbf{r}(t) = (3\cos t, 3\sin t)$, $0 \le t \le 2\pi$, $t_1 = \pi/2$

12. $\mathbf{r}(t) = (2\cos t, 2\sin t)$, $0 \le t \le 2\pi$, $t_1 = 3\pi/2$

Exercises 13–16: Calculate the speed of the object, assuming the speed is constant. Time is measured in seconds and lengths in meters.

13. See Exercise 9.

14. See Exercise 10.

15. See Exercise 11.

16. See Exercise 12.

Growth

Exercises 17–20: Recalling that two points determine a line and that the center and radius determine a circle, sketch the path of the given motion, eliminating the parameter t only if necessary.

17. $\mathbf{r}(t) = (3 + t, -2 + 3t)$, $0 \le t \le 2$

18. $\mathbf{r}(t) = (5 - 4t, -3 + 2t)$, $0 \le t \le 2$

19. $\mathbf{r}(t) = (2 + \cos t, -1 + \sin t)$, $t \ge 0$

20. $\mathbf{r}(t) = (1 + 2\cos t, -1 + 2\sin t)$, $t \ge 0$

Exercises 21 and 22: Assuming constant-speed motion, find the speed of the object with the given position vector. The quantities a, b, c, d, and ω are constants. Units are meters and seconds.

21. $\mathbf{r}(t) = (a + bt, c + dt)$, $t \ge 0$

22. $\mathbf{r}(t) = (a + b\cos\omega t, c + b\sin\omega t)$, $t \ge 0$

Exercises 23–28: Sketch the path of the object with the given position vector.

23. $\mathbf{r}(t) = (|t - 1|, |t - 2|)$, $0 \le t \le 5$

24. $\mathbf{r}(t) = (|2t - 5|, 2|-t + 4|)$, $0 \le t \le 1$

25. $\mathbf{r}(t) = (3\cos t, 2\sin t)$, $0 \le t \le 2\pi$

26. $\mathbf{r}(t) = (\cos t, 2\sin t)$, $0 \le t \le 2\pi$

27. $\mathbf{r}(t) = (t^2, 2t + 1)$, $t \ge 0$

28. $\mathbf{r}(t) = (t, 1/t)$, $1/2 \le t \le 3$

Exercises 29–32: Find a unit vector, that is, a vector of length 1, in the direction of the given vector.

29. $\mathbf{r} = (1, 1)$

30. $\mathbf{r} = (-2, 5)$

31. $\mathbf{r} = (0.29, 1.49)$

32. $\mathbf{r} = (-13/7, 5/7)$

Review/Preview

Exercises 33–45: Differentiate with respect to x.

33. $\sqrt{2x + 1}$

34. $x^2/(x^3 + 1)$

35. $\sin(3x^2)$

36. $\cos(x^3)$

37. e^{2x-3}

38. 2^{5x}

39. $\arcsin \sqrt{x}$

40. $\arctan x^2$

41. $\sqrt{x}/(x^2 + 1)$

42. $(3x + 5)/(7x + 9)$

43. $\ln(x^2 + 9)$

44. $\ln(\sqrt{x} + 1)$

3.3 VECTORS AND MOTION—ADDING VECTORS

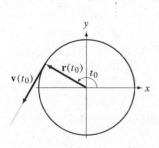

Figure 3.17. Velocity vector.

The position vector $\mathbf{r}(t)$ of an object at time t locates it relative to a given coordinate system. It is natural to base this vector at the origin. The velocity vector $\mathbf{v}(t)$ of an object gives its speed and direction. The velocity vector measures change in position. This vector is most naturally based at the point $\mathbf{r}(t)$ from which the change takes place. In this section we discuss vectors based at points other than the origin so that we can define the velocity vector of an object in motion.

We show in Fig. 3.17 the position and velocity vectors of, say, a proton in a syncrotron. The position vector $\mathbf{r}$ traces the circular path of the proton. The velocity vector $\mathbf{v}$ points in the direction of motion. If the magnetic field is turned off at $t = t_0$, the proton will continue along the line through $\mathbf{r}(t_0)$ and in the direction of the velocity vector $\mathbf{v}(t_0)$.

Equivalent Vectors

We show in Fig. 3.18 a position vector $\mathbf{r} = (x, y)$ and three of the infinitely many vectors having the same length and direction as $\mathbf{r}$. We say that such vectors are equivalent to $\mathbf{r}$.

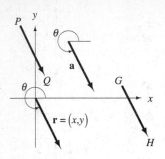

Figure 3.18. Equivalent vectors.

DEFINITION   **equivalent vectors**

Let $\mathbf{r} = (x, y)$ be a position vector. A vector $\overrightarrow{PQ}$, based at the point $P = (p_1, p_2)$ and terminating at point $Q = (q_1, q_2)$, is **equivalent** to $\mathbf{r}$ if

$$q_1 - p_1 = x \quad \text{and} \quad q_2 - p_2 = y.$$

Vectors $\overrightarrow{PQ}$ and $\overrightarrow{GH}$ are **equivalent** if each is equivalent to the same position vector.

The length and direction of a vector based at a point other than the origin are defined in terms of the position vector to which it is equivalent.

DEFINITION   **length and direction of a vector**

Let $\mathbf{r} = (x, y)$ be a position vector and $\mathbf{w}$ a vector equivalent to $\mathbf{r}$. The **length** $\|\mathbf{w}\|$ of $\mathbf{w}$ is defined to be $\|\mathbf{r}\|$, the length of $\mathbf{r}$.

The **direction** of $\mathbf{w}$ is defined to be the direction of $\mathbf{r}$, which is the unique angle $\theta \in [0, 2\pi)$ for which

$$\cos \theta = \frac{x}{\sqrt{x^2 + y^2}} \quad \text{and} \quad \sin \theta = \frac{y}{\sqrt{x^2 + y^2}}.$$

If $\mathbf{r} = (0, 0)$, we assign no direction to $\mathbf{w}$.

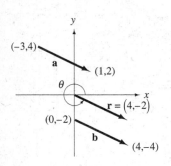

Figure 3.19. Equivalent vectors.

■ **EXAMPLE 1**   In Fig. 3.19 vector $\mathbf{a}$ stretches from $(-3, 4)$ to $(1, 2)$ and $\mathbf{b}$ from $(0, -2)$ to $(4, -4)$. Show that $\mathbf{a}$ and $\mathbf{b}$ are equivalent and that these vectors have the same direction and length.

Solution.   Since $1 - (-3) = 4 - 0 = 4$ and $2 - 4 = -4 - (-2) = -2$, $\mathbf{a}$ and $\mathbf{b}$ are each equivalent to the position vector $\mathbf{r} = (4, -2)$. Hence, $\mathbf{a}$ and $\mathbf{b}$ are equivalent. The vectors $\mathbf{a}$ and $\mathbf{b}$ have the same direction and length since these are calculated from the position vector $\mathbf{r}$ to which both are equivalent. The common length of $\mathbf{a}$, $\mathbf{b}$, and $\mathbf{r}$ is

$$\sqrt{4^2 + (-2)^2} = \sqrt{20} \approx 4.47.$$

The direction of $\mathbf{r}$ is the angle $\theta \in [0, 2\pi)$ for which

$$\cos \theta = \frac{4}{\sqrt{4^2 + (-2)^2}} \quad \text{and} \quad \sin \theta = \frac{-2}{\sqrt{4^2 + (-2)^2}}.$$

Perhaps the easiest way to calculate θ is to focus on the small angle $2\pi - \theta$ between $\mathbf{r}$ and the positive x-axis. The tangent of this angle is $2/4 = 0.5$. Hence

$$\theta = 2\pi - \arctan 0.5 = 2\pi - 0.46364\ldots \approx 5.8. \qquad ■$$

■ **EXAMPLE 2**   Calculate the position of an object if it is moved 7.75 units in the direction 63° south of east from its initial position, $(-9.0, 8.0)$.

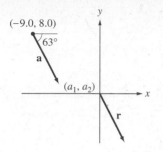

Figure 3.20. Calculating an end point.

Solution.   This problem can be solved using right triangle trigonometry. We give an equivalent solution using vectors. We represent the move by the vector **a**, whose initial point is $(-9.0, 8.0)$ and whose terminal point is (a_1, a_2), which gives the final position of the object. See Fig. 3.20. Letting $\mathbf{r} = (x, y)$ be the position vector equivalent to **a**, we have

$$\mathbf{r} = (7.75 \cos 297°, 7.75 \sin 297°) \approx (3.52, -6.91).$$

We have converted from the direction $63°$ south of east to $360° - 63°$. Since **a** is equivalent to **r**,

$$a_1 - (-9.0) \approx 3.52$$
$$a_2 - (8.0) \approx -6.91.$$

Solving for a_1 and a_2 gives $a_1 \approx -5.48$ and $a_2 \approx 1.09$. The final position of the object is $(-5.48, 1.09)$. ■

Adding Position Vectors and the Parallelogram Law

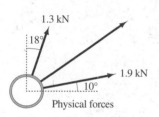

Physical forces

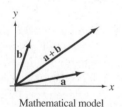

Mathematical model

Figure 3.21. Modeling forces.

The definition of vector addition grew out of geometrical and physical ideas. We show an example of what we mean in Fig. 3.21. We suppose that 1.9 kN and 1.3 kN forces act on the metal ring in the directions shown (kN denotes kilonewtons, a measure of the magnitude or strength of a force; 1 kN ≈ 225 lb). If we model these forces with vectors **a** and **b** having lengths 1.9 and 1.3 and directions as shown in the figure, the vector sum **a** + **b** will model the resultant of the two ring forces. The resultant force is the name given to the single force having the same effect as the two forces acting together.

We define addition of position vectors and then use this to define addition of vectors not based at the origin.

DEFINITION   **adding position vectors**

The sum **u** + **v** of position vectors $\mathbf{u} = (u_1, u_2)$ and $\mathbf{v} = (v_1, v_2)$ is the position vector defined by

(1) $$\mathbf{u} + \mathbf{v} = (u_1 + v_1, u_2 + v_2).$$

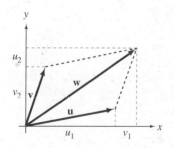

Figure 3.22. Vector addition.

This definition of vector addition corresponds to the geometric interpretation shown in Fig. 3.22. The sum $(u_1 + v_1, u_2 + v_2)$ of **u** and **v** is the diagonal of the parallelogram with sides **u** and **v**. Vector addition, particularly its geometric representation, often is called the *parallelogram law.*

The parallelogram law follows from the observation that the sum of the segments with lengths u_1 and v_1 is equal to the segment of length w_1, where w_1 is the first coordinate of **w**. Similarly, the sum of the segments with lengths v_2 and u_2 is equal to the segment of length w_2.

■ **EXAMPLE 3**   Calculate the sum/resultant of the vectors/forces $\mathbf{F}_1$ and $\mathbf{F}_2$, where $\mathbf{F}_1 = (1, -2)$ and $\mathbf{F}_2 = (2, 3)$. We show these vectors/forces in Fig. 3.23. So well do the idea and arrow graphic of the vector fit our intuitive notion of force that the two ideas tend to merge. If we were dealing with forces $\mathbf{F}_1$ and $\mathbf{F}_2$ of, say,

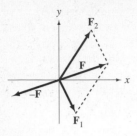

Figure 3.23. Tugs attached to docking ring.

two tugboats pulling on hawsers attached to a docking ring on a pier, we might have described them differently. For example, $\mathbf{F}_2$ could have been given as a force with magnitude $\|F_2\| = \sqrt{2^2 + 3^2} = \sqrt{13}$ and acting in the direction $\arctan(3/2) \approx 56.3°$.

Solution.   Since we may as well put the docking ring at the origin, the two forces are position vectors. Their sum is

$$\mathbf{F} = \mathbf{F}_1 + \mathbf{F}_2 = (1, -2) + (2, 3) = (3, 1).$$

We show the vector $\mathbf{F}$ in the figure. It is the diagonal of the parallelogram with sides $\mathbf{F}_1$ and $\mathbf{F}_2$. In terms of forces, $\mathbf{F}$ is the resultant of $\mathbf{F}_1$ and $\mathbf{F}_2$.

The single force required to just balance $\mathbf{F}_1$ and $\mathbf{F}_2$ is shown in Fig. 3.23 as $-\mathbf{F}$. This force has the same magnitude as $\mathbf{F}$ but points in the opposite direction. Geometrically, it is clear that $-\mathbf{F} = (-3, -1)$.   ■

Adding Vectors and the Parallelogram Law

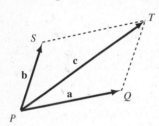

Figure 3.24. Adding vectors based at P.

We have defined addition of position vectors and shown that vector addition corresponds to the parallelogram law. We turn this around in adding vectors based at a point other than the origin. We start with a diagram similar to Fig. 3.22. The diagram is used to illustrate the physical or geometric meaning of the addition. If we need an arithmetic representation, we add the equivalent position vectors. We give two interpretations of adding vectors not necessarily based at the origin.

Referring to Fig. 3.24, let $\overrightarrow{PQ}$ and $\overrightarrow{PS}$ be vectors based at P. We write

$$\overrightarrow{PQ} + \overrightarrow{PS} = \overrightarrow{PT}$$

when we are thinking of these vectors as representing forces pulling on a docking ring at P. This is the parallelogram law based at P. If we require the length and direction of $\overrightarrow{PT}$, we add the position vectors $\mathbf{a}$ and $\mathbf{b}$ equivalent to $\overrightarrow{PQ}$ and $\overrightarrow{PS}$, respectively. The position vector $\mathbf{c} = \mathbf{a} + \mathbf{b}$ is equivalent to the vector $\overrightarrow{PT}$.

The second interpretation may be expressed in terms of displacement. Referring to Fig. 3.25, suppose that we interpret the vector $\overrightarrow{PQ}$ as the displacement of an object at P to Q and $\overrightarrow{QT}$ as the further displacement of that object from Q to T. Taken together, these displacements are the same as the displacement $\overrightarrow{PT}$ of the object from P directly to T. We may write this as

$$\overrightarrow{PQ} + \overrightarrow{QT} = \overrightarrow{PT}.$$

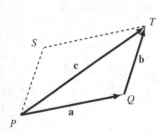

Figure 3.25. Displacements.

If we must calculate the length and direction of the displacement $\overrightarrow{PT}$ or the co-ordinates of T, then, as before, we add the position vectors $\mathbf{a}$ and $\mathbf{b}$ equivalent to $\overrightarrow{PQ}$ and $\overrightarrow{QT}$, respectively. The position vector $\mathbf{c} = \mathbf{a} + \mathbf{b}$ is equivalent to the vector $\overrightarrow{PT}$.

■ **EXAMPLE 4**   Referring to Fig. 3.26, suppose $P = (-1, 2)$, $Q = (1, -1)$, and $S = (1, 6)$. Calculate $\mathbf{a}$, $\mathbf{b}$, and their sum, $\mathbf{c}$. Calculate the coordinates of T.

Solution.   The position vectors $\mathbf{a}$ and $\mathbf{b}$ corresponding to $\overrightarrow{PQ}$ and $\overrightarrow{PS}$ are

$$\mathbf{a} = (1 - (-1), -1 - 2) = (2, -3)$$
$$\mathbf{b} = (1 - (-1), 6 - 2) = (2, 4).$$

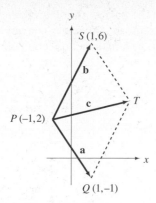

Figure 3.26. Adding vectors based at P.

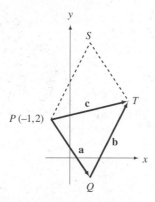

Figure 3.27. Displacements.

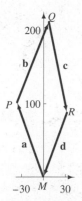

Figure 3.28. Displacements.

Table 1. Quadrilateral survey

	Distance (m)	Bearing
$\overrightarrow{MP}$	107.65	109.7°
$\overrightarrow{PQ}$	118.05	68.5°
$\overrightarrow{QR}$	125.95	281.4°
$\overrightarrow{RM}$	93.10	249.2°

The sum **c** of **a** and **b** is

$$\mathbf{c} = (2, -3) + (2, 4) = (4, 1).$$

To find the coordinates of $T = (t_1, t_2)$, note that **c** is the position vector equivalent to $\overrightarrow{PT}$. Hence

$$\mathbf{c} = (t_1 - p_1, t_2 - p_2) = (4, 1).$$

From this it follows that the coordinates of T are

$$(t_1, t_2) = (4 + p_1, 1 + p_2) = (4 + (-1), 1 + 2) = (3, 3). \quad \blacksquare$$

In the next example we use the same geometry as in Example 4, but we interpret vectors as displacements.

■ **EXAMPLE 5**   An object at P is displaced first to Q and then to T, as in Fig. 3.27. The displacement vectors $\overrightarrow{PQ}$ and $\overrightarrow{QT}$ are equivalent to position vectors $\mathbf{a} = (2, -3)$ and $\mathbf{b} = (2, 4)$. Calculate the single equivalent displacement from P to T and the coordinates of T.

Solution.   We may view the position vectors **a** and **b** as applicable to any point in the plane. They give the direction and length of the displacement. For example, **a** displaces an object at an arbitrary point (x, y) of the plane two units to the right and three units down, that is,

$$(x, y) \xrightarrow{\ \mathbf{a}\ } (x + 2, y + (-3)).$$

Hence $P = (-1, 2)$ is displaced by **a** to $Q = (-1 + 2, 2 + (-3)) = (1, -1)$. Displacing the object at Q by $\mathbf{b} = (2, 4)$ moves it two units to the right and four units up, that is,

$$(x, y) \xrightarrow{\ \mathbf{b}\ } (x + 2, y + 4).$$

Thus **b** displaces $Q = (1, -1)$ to $T = (3, 3)$. The single equivalent displacement is $\mathbf{c} = \mathbf{a} + \mathbf{b} = (2, -3) + (2, 4) = (4, 1)$. We may check by displacing P by **c**. The point P would be moved to $(-1 + 4, 2 + 1) = (3, 3)$. ■

■ **EXAMPLE 6**   A student team surveys a quadrilateral on campus to check on their technique. They start from a bronze marker M designated as origin and measure bearing (in degrees) and distance (in meters) from the marker to point P, from P to Q, from Q to R, and from R back to their starting point, M. See Fig. 3.28. Their data are given in Table 1. By calculating the coordinates of M from their data, show that their error of closure (the distance between the actual point M and the surveyed point M) is approximately 1.4 m.

Solution.   The sum of the displacements $\overrightarrow{MP}$, $\overrightarrow{PQ}$, $\overrightarrow{QR}$, and $\overrightarrow{RM}$ is the zero displacement vector $\overrightarrow{MM}$. If all measurements were accurate, the sum of the position vectors **a**, **b**, **c**, and **d** equivalent to $\overrightarrow{MP}$, $\overrightarrow{PQ}$, $\overrightarrow{QR}$, and $\overrightarrow{RM}$ would be the zero vector, **0**. The distance between the origin and the point with position vector $\mathbf{a} + \mathbf{b} + \mathbf{c} + \mathbf{d}$ is the error of closure.

The vectors in Table 1 are given in polar form. The polar form of a vector $\mathbf{r} = (x, y)$ is (r, θ), where r is the length of $\mathbf{r}$ and θ is its direction. We may convert from polar to rectangular form using the definitions of the length and direction of a vector. Recall that if $\mathbf{r} = (x, y)$ is a vector, its length r and direction θ are

$$\|\mathbf{r}\| = r = \sqrt{x^2 + y^2}$$

and the unique angle $\theta \in [0, 2\pi)$ for which

$$\cos \theta = \frac{x}{\sqrt{x^2 + y^2}} \quad \text{and} \quad \sin \theta = \frac{y}{\sqrt{x^2 + y^2}}.$$

It follows from these definitions that the rectangular form of $\mathbf{r}$ is given by

(2) $$\mathbf{r} = (r \cos \theta, r \sin \theta).$$

Many calculators have keys that convert between the polar and rectangular forms of vectors. Using $\overrightarrow{MP}$ to illustrate, if you enter the values [107.65 ∠109.7] into your calculator (set in degree mode), it should return [−36.29 101.35].

The rectangular forms of the displacements in Table 1 may be calculated by formula or calculator. Check a few of the following calculations using either (1) or the polar/rectangular key on your calculator.

$$\mathbf{a} = 107.65(\cos 109.7, \sin 109.7) \approx (-36.29, 101.35)$$

$$\mathbf{b} = 118.05(\cos 68.5, \sin 68.5) \approx (43.27, 109.84)$$

$$\mathbf{c} = 125.95(\cos 281.4, \sin 281.4) \approx (24.89, -123.47)$$

$$\mathbf{d} = 93.10(\cos 249.2, \sin 249.2) \approx (-33.06, -87.03)$$

From these calculations the coordinates of M are

$$\mathbf{a} + \mathbf{b} + \mathbf{c} + \mathbf{d} \approx (-1.19, 0.69).$$

The error of closure is $\|(-1.19, 0.69)\| \approx 1.37$ m.　　　　　　■

Multiplying a Vector by a Scalar

When it is necessary to distinguish vectors and real numbers, the latter are often called **scalars.** The multiplication of a vector by a scalar occurs naturally. If we form the sum $\mathbf{r} + \mathbf{r}$, we have

$$\mathbf{r} + \mathbf{r} = (x, y) + (x, y) = (2x, 2y).$$

It is reasonable to write the sum $\mathbf{r} + \mathbf{r}$ as $2\mathbf{r}$. The relation $2\mathbf{r} = (2x, 2y)$ suggests a definition of the product of a scalar and a vector, which is often called scalar multiplication.

Doubling a displacement or a force is a commonplace instance of scalar multiplication. We may also increase a force **F** by 150 percent. Presumably, the latter would be written as 1.5**F**. We may multiply a vector by the scalar -2, which doubles the vector and reverses its direction. In terms of displacements:

The vector **a** $= (2, 1)$ displaces a point (x, y) two units to the right and one unit up.

The vector **b** $= (-4, -2) = -2(2, 1)$ displaces a point (x, y) four units to the left and two units down.

The sum of the displacements **a** + **a** + **b** in this example is the zero displacement. And, of course, thinking of the vectors as position vectors,

$$\mathbf{a} + \mathbf{a} + \mathbf{b} = (2, 1) + (2, 1) + (-4, -2) = (0, 0) = \mathbf{0}.$$

From this result we may reasonably write

$$2\mathbf{a} + \mathbf{b} = \mathbf{0} \quad \text{or} \quad \mathbf{b} = -2\mathbf{a}.$$

More generally, if a position vector **a** $= (a_1, a_2)$ and a real number s are given, $s\mathbf{a}$ is defined to be the position vector (sa_1, sa_2). We show in Fig. 3.29 the effect on a typical vector **a** of scalar multiplication by $s = -1$, $s = 2$, and $s = 0.5$. By definition,

$$(-1)\mathbf{a} = (-a_1, -a_2), \quad 2\mathbf{a} = (2a_1, 2a_2), \quad \text{and} \quad 0.5\mathbf{a} = (0.5a_1, 0.5a_2).$$

We form the product of a scalar s and a vector $\overrightarrow{PQ}$ based at a point P by calculating $s\mathbf{r}$, where **r** is a position vector equivalent to $\overrightarrow{PQ}$. The product $s\overrightarrow{PQ}$ is the vector based at P and equivalent to $s\mathbf{r}$. Multiplying a vector **r** by a scalar gives a vector pointing in the same or opposite direction as **r**.

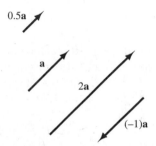

Figure 3.29. Scalar multiplication.

Subtracting Vectors

We have shown that the sum of two vectors based at the same point may be interpreted as one of the two diagonals of the parallelogram they form. The other diagonal is their difference. We show in Fig. 3.30(a) vectors **a** and **b** based at P. Their difference **a** − **b** is defined by

$$\mathbf{a} - \mathbf{b} = \mathbf{a} + (-\mathbf{b})$$

where by −**b** we mean $(-1)\mathbf{b}$, which is a vector equal in length to **b** but in the opposite direction. The vector labeled as **a** + $(-1)\mathbf{b}$ in Fig. 3.30(a) is the sum of **a** and −**b**. It is the diagonal of the parallelogram with sides **a** and −**b**.

The vector labeled **a** − **b** in Fig. 3.30(b) has the same length and direction as **a** + $(-\mathbf{b})$. This vector is one of the two diagonals in the parallelogram formed by **a** and **b**.

It is useful to view subtraction in terms of displacement. In Fig. 3.30(b) the sum of the two displacements **b** and **a** − **b** is equal to the displacement **a**. This corresponds to the algebraic statement **a** = **b** + (**a** − **b**).

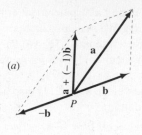

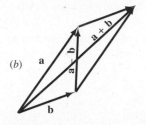

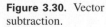

Figure 3.30. Vector subtraction.

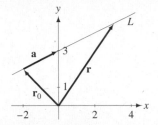

Figure 3.31. Vector equation of a line.

Vector Equation of a Line

We illustrate displacement and scalar multiplication by working out a vector equation of the line L with scalar equation $y = \frac{1}{2}x + 3$. Figure 3.31 includes the graph of this equation, a position vector $\mathbf{r}_0$ of the point $(-2, 2)$ on L, and a vector $\mathbf{a}$ based at this point and pointing in the direction of L. We argue that given any point $\mathbf{r} = (x, y)$ on L, we may choose a scalar t so that

$$\mathbf{r} = (x, y) = \mathbf{r}_0 + t\mathbf{a} = (-2, 2) + t(2, 1) = (-2 + 2t, 2 + t).$$

We take $\mathbf{r}_0 = (-2, 2)$ since $(-2, 2)$ satisfies the equation $y = \frac{1}{2}x + 3$. Any other point on L would do as well. The slope of the given line is $m = \frac{1}{2}$. Since slope is rise over run, if we run 2, the rise would be 1. Thus, $\mathbf{a} = (2, 1)$ points in the direction of the line. In terms of displacements, the displacement of an object from the origin to an arbitrary point $\mathbf{r} = (x, y)$ on L is equivalent to the sum of two displacements. First move from the origin to the point $\mathbf{r}_0$ and then to $\mathbf{r}$ by the displacement $t\mathbf{a}$. The scalar t depends upon $\mathbf{r}$. If, for example, $t = 0$, then $\mathbf{r} = \mathbf{r}_0$. If $t = 1$, then

$$\mathbf{r} = \mathbf{r}_0 + (1)\mathbf{a} = (-2, 2) + (1)(2, 1) = (0, 3).$$

To values of t between 0 and 1 correspond points of L between $\mathbf{r}_0$ and $\mathbf{r}_0 + \mathbf{a}$; to values of t greater than 1 correspond points of L beyond $\mathbf{r}_0 + \mathbf{a}$; and, finally, to negative values of t correspond points of L downward and to the left of $\mathbf{r}_0$.

The vector equation

$$(3) \qquad\qquad\qquad \mathbf{r} = \mathbf{r}_0 + t\mathbf{a}$$

is a standard form of a line, just as the scalar equation $y = mx + b$ is a standard form of a line. For the latter, m is the slope and b the y-intercept of the line under consideration. For the vector equation, $\mathbf{r}_0$ is a position vector of a point on the line and $\mathbf{a}$ is a vector in the direction of the line (in either of the two directions a line determines). Usually, $\mathbf{a}$ is based at $\mathbf{r}_0$.

■ **EXAMPLE 7** Sketch and determine a scalar equation of the line L with vector equation

(4) $$\mathbf{r} = \mathbf{r}_0 + t\mathbf{a} = (4, 3) + t(-3, -1).$$

Solution. Since two points determine both the graph and equation of a line, we may set $t = 0$ and $t = 1$ to obtain two points on L. We have

$$\mathbf{r}(0) = (4, 3) + (0)(-3, -1) = (4, 3)$$
$$\mathbf{r}(1) = (4, 3) + (1)(-3, -1) = (1, 2).$$

We may sketch the line through these two points. In Fig. 3.32 we show the line as well as $\mathbf{r}_0$ and $\mathbf{a}$. From these points we may calculate the slope of L and then substitute into the point-slope form of a line. The slope calculation is

$$m = \frac{3-2}{4-1} = \frac{1}{3}.$$

Substituting into the point-slope form $y - y_1 = m(x - x_1)$, we have

(5) $$y - 2 = \tfrac{1}{3}(x - 1).$$

A second way of determining a scalar equation for L is to eliminate the parameter t. We saw this in Section 3.2. Setting $\mathbf{r} = (x, y)$, we have from (4)

$$x = 4 - 3t$$
$$y = 3 - t.$$

Solving the first of these equations for t gives $t = (4 - x)/3$. Substituting this into the second equation gives

$$y = 3 - (4 - x)/3 = \tfrac{1}{3}x + \tfrac{5}{3} = \tfrac{1}{3}(x - 1) + 2.$$

This equation agrees with (5). ■

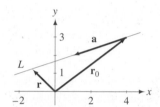

Figure 3.32. Vector equation of a line.

Exercises 3.3

Basic

Exercises 1–23: Include a good, illustrative sketch with your solution.

1. Let $P = (1, 1)$, $Q = (4, -1)$, $S = (7, 5)$, and $T = (10, 3)$. Show that $\overrightarrow{PQ}$ and $\overrightarrow{ST}$ are equivalent.
2. Let $P = (-1, -2)$, $Q = (-4, -1)$, $S = (5, -1)$, and $T = (2, 0)$. Show that $\overrightarrow{PQ}$ and $\overrightarrow{ST}$ are equivalent.
3. Find a vector based at $(4, -4)$ and equivalent to the position vector $(2, 3)$.
4. Find a vector based at $(-2, -5)$ and equivalent to the position vector $(-5, 2)$.

5. Calculate the length and direction of the vector $\overrightarrow{PQ}$ where P and Q are the points $(-3, 1)$ and $(5, 1)$.

6. Calculate the length and direction of the vector $\overrightarrow{PQ}$ where P and Q are the points $(-3, 3)$ and $(1, 1)$.

7. Calculate the length and direction of the vector $\overrightarrow{PQ}$ where P and Q are the points $(0, 1)$ and $(-2, -7)$.

8. Calculate the length and direction of the vector $\overrightarrow{PQ}$ where P and Q are the points $(1, 9)$ and $(-5, -3)$.

9. Calculate the position of an object if it is moved 25.3 meters in the direction of $12°$ west of north from its initial position of $(5.0, -7.0)$.

10. Calculate the position of an object if it is moved 15.9 meters in the direction of $17°$ west of south from its initial position of $(5.7, 7.3)$.

11. Two tugboats exert forces $\mathbf{F}_1 = (4, -7)$ and $\mathbf{F}_2 = (6, 8)$ on a docking ring. Calculate the single force on the ring that would balance $\mathbf{F}_1$ and $\mathbf{F}_2$.

12. Two barges are tied to a docking ring and exert forces $\mathbf{F}_1 = (450, 725)$ and $\mathbf{F}_2 = (700, 390)$. Calculate the single force on the ring that would balance $\mathbf{F}_1$ and $\mathbf{F}_2$.

13. Two tugboats exert forces $\mathbf{F}_1$ and $\mathbf{F}_2$ on a docking ring. Calculate the magnitude and direction of the single balancing force on the ring if the first force has magnitude 45 and acts in the direction $81°$, and the second force has magnitude 65 and acts in the direction $194°$.

14. Two barges are tied to a docking ring and exert forces $\mathbf{F}_1$ and $\mathbf{F}_2$. Calculate the magnitude and direction of the single balancing force on the ring if the first force has magnitude 55 and acts in the direction $340°$, and the second force has magnitude 65 and acts in the direction $194°$.

15. Using Fig. 3.26 for notation, suppose $P = (3, 4)$, $Q = (8, 6)$, and $S = (5, 7)$. Calculate $\mathbf{a}$, $\mathbf{b}$, and their sum, $\mathbf{c}$. Calculate the coordinates of T.

16. Using Fig. 3.26 for notation, suppose $P = (1, -3)$, $Q = (-1, -5)$, and $S = (4, -2)$. Calculate $\mathbf{a}$, $\mathbf{b}$, and their sum, $\mathbf{c}$. Calculate the coordinates of T.

17. An object at $(0, 4)$ is displaced to Q by $(2, 5)$ and then to T by $(-12, 13)$. Calculate the single equivalent displacement and the coordinates of T.

18. An object at $(-4, 9)$ is displaced to Q by $(-4, -5)$ and then to T by $(1, 1)$. Calculate the single equivalent displacement and the coordinates of T.

19. Form the difference $\mathbf{a} - \mathbf{b}$ of the position vectors $\mathbf{a} = (3, 3)$ and $\mathbf{b} = (2, 1)$. First do this geometrically using Fig. 3.30. Check your work by an arithmetic calculation.

20. Form the difference $\mathbf{a} - \mathbf{b}$ of the position vectors $\mathbf{a} = (4, 5)$ and $\mathbf{b} = (1, 2)$. First do this geometrically using Fig. 3.30. Check your work by an arithmetic calculation.

21. Find m so that the position vector $\mathbf{r} = (a, b)$, where $a \neq 0$, is parallel to any line with slope m. Find a position vector $\mathbf{r}$ parallel to a line with slope m.

22. The point $P = (3, 5)$ is displaced by the vector $\mathbf{a}_1 = (4, 7)$, and then displaced further by $\mathbf{a}_2 = (-3, 1)$, $\mathbf{a}_3 = (-1, 0)$, and $\mathbf{a}_4 = (-11, 5)$. What is its final position?

23. Using the displacement vectors given in Exercise 22, what point P should be taken so that its final position is at the origin?

Growth

Exercises 24 and 25:   Find the error of closure of the survey data for the polygon with vertices $P_0, P_2, \ldots, P_n$. Given are the coordinates of P_0 and the length (in meters) and bearing (in degrees) of the "legs" $P_0 P_1, P_1 P_2, \ldots, P_{n-1} P_n, P_n P_0$. Roughly sketch the locations of the points.

24. $P_0 = (2100.00, 3700.00)$; $(180.28, 56.31°)$, $(310.16, 1.85°)$, $(497.69, 292.44°)$, $(669.1, 154.0°)$

25. $P_0 = (0.00, 0.00)$; $(693.11, 223.83°)$, $(110.00, 270.00°)$, $(962.05, 0.60°)$, $(337.26, 56.12°)$, $(715.80, 155.16°)$

Exercises 26–29:   Write a vector equation of the line with equation $y = mx + b$. Use the given $\mathbf{r}_0$.

26. $y = \frac{2}{3}x + 5$, $\mathbf{r}_0 = (3, 7)$

27. $y = -2x + 3$, $\mathbf{r}_0 = (4, -5)$

28. $y = x + 1$, $\mathbf{r}_0 = (4, 5)$

29. $y = \frac{9}{7}x + 3$, $\mathbf{r}_0 = (0, 3)$

Exercises 30–33: Find a vector equation for the line determined by the position vectors **p** and **q**. Sketch the line. (Hint: Their difference determines the direction of the line, and either of them is a point on the line.)

30. $\mathbf{p} = (4, -2)$, $\mathbf{q} = (0, 3)$

31. $\mathbf{p} = (5, 0)$, $\mathbf{q} = (0, -2)$

32. $\mathbf{p} = (-3, 0)$, $\mathbf{q} = (-3, 5)$

33. $\mathbf{p} = (2, 9)$, $\mathbf{q} = (2, -13)$

34. Determine if the point $(-24, -11)$ is on the line with equation $\mathbf{r} = (-3, 4) + t(7, 5)$.

35. Determine if the point $(1, 15)$ is on the line with equation $\mathbf{r} = (-1, 4) + t(1, 6)$.

36. An object is moving downward and to the left at a speed of 3 m/s on the line passing through the points $(2, 1)$ and $(5, 7)$. If its position vector at $t = 0$ is $(-4, -11)$, give the position vector **r** at any time $t \geq 0$.

Review/Preview

Exercises 37–48: Given a point (x, y) in the plane, let r denote its distance from the origin and θ the angle between the positive x-axis and the line from the origin to (x, y). We assume that $(x, y) \neq (0, 0)$ and $0 \leq \theta < 2\pi$. If (x, y) is given, determine r and θ. If r and θ are given, determine (x, y).

37. $(x, y) = (1, 1)$

38. $(x, y) = (-3, 0)$

39. $(x, y) = (4, -1)$

40. $(x, y) = (3, 4)$

41. $(x, y) = (-5, 3)$

42. $(x, y) = (-3, 3)$

43. $r = \sqrt{2}$, $\theta = \pi/4$

44. $r = 1$, $\theta = \pi/3$

45. $r = 3$, $\theta = 2.9$

46. $r = 0.4$, $\theta = 5.3$

47. $r = 3.2$, $\theta = 1.7$

48. $r = \sqrt{1.2}$, $\theta = \pi$

3.4 VECTORS AND MOTION—VELOCITY VECTORS

Now that we have discussed position vectors, vectors based at points other than the origin, the multiplication of a vector by a scalar, and addition and subtraction of vectors, we are ready to define velocity. As in the one-dimensional case, we define velocity as the limit of average velocity.

Average Velocity

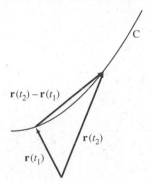

Figure 3.33. Average velocity vector.

Suppose that an object is moving on the path C shown in Fig. 3.33 and has position vector

$$\mathbf{r} = \mathbf{r}(t) = (x(t), y(t)), \quad a \leq t \leq b.$$

Our goal is to define and calculate its velocity at $t = t_1$. As in the one-dimensional case, velocity is the limit of the change in position of the object over time intervals $[t_1, t_2]$, as $t_2 \to t_1$. We show the position of the object at times t_1 and t_2. The vector $\mathbf{r}(t_2) - \mathbf{r}(t_1)$ measures the change in the position of the object during the time interval $[t_1, t_2]$. We divide this difference by $t_2 - t_1$ to get the rate at which **r** is changing with respect to t. In terms of the coordinate functions $x(t)$ and $y(t)$ of $\mathbf{r}(t)$, we have

$$\frac{1}{t_2 - t_1}(\mathbf{r}(t_2) - \mathbf{r}(t_1)) = \left(\frac{x(t_2) - x(t_1)}{t_2 - t_1}, \frac{y(t_2) - y(t_1)}{t_2 - t_1}\right).$$

This is the object's **average velocity** over the time interval $[t_1, t_2]$. The average velocity vector lies along the vector $\mathbf{r}(t_2) - \mathbf{r}(t_1)$ shown in the figure. Note that the units of these two vectors are not the same. If lengths are measured in meters and time in seconds, the units of the vector $\mathbf{r}(t_2) - \mathbf{r}(t_1)$ are meters; those of the average velocity vector are meters per second.

For objects moving with constant speed on a line, velocity and average velocity are equal. Suppose, for example, an object is moving on the line L shown at the top of Fig. 3.34 and that at time t its position is

$$\mathbf{r} = \mathbf{r}(t) = (-2, 2) + t(2, 1), \quad t \geq 0.$$

To show that its speed is constant, we calculate the distance $\|\mathbf{r}(t_2) - \mathbf{r}(t_1)\|$ the object has moved in time $t_2 - t_1$. Referring to the sketch at the bottom of Fig. 3.34, we have

$$
\begin{aligned}
\|\mathbf{r}(t_2) - \mathbf{r}(t_1)\| &= \|(-2, 2) + t_2(2, 1) - (-2, 2) - t_1(2, 1)\| \\
&= \sqrt{(2(t_2 - t_1))^2 + (1(t_2 - t_1))^2} \\
&= |t_2 - t_1| \sqrt{5}.
\end{aligned}
$$

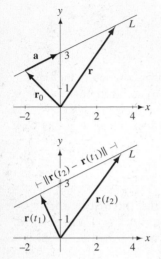

Figure 3.34. Vector equation of a line.

Dividing the distance $\|\mathbf{r}(t_2) - \mathbf{r}(t_1)\|$ by the time $t_2 - t_1$ gives $\sqrt{5}$, which shows that the object's speed is constant.

The calculation of its average velocity is similar. We have

$$
\begin{aligned}
\frac{1}{t_2 - t_1}(\mathbf{r}(t_2) - \mathbf{r}(t_1)) &= \frac{1}{t_2 - t_1}\big((-2, 2) + t_2(2, 1) - (-2, 2) - t_1(2, 1)\big) \\
&= \left(\frac{2(t_2 - t_1)}{t_2 - t_1}, \frac{1(t_2 - t_1)}{t_2 - t_1}\right) \\
&= (2, 1),
\end{aligned}
$$

which shows that the average velocity is constant. Since

$$\lim_{t_2 \to t_1} \frac{1}{t_2 - t_1}(\mathbf{r}(t_2) - \mathbf{r}(t_1)) = (2, 1),$$

the average velocity on any interval and the velocity of the object are equal to $(2, 1)$.

DEFINITION   **average velocity, velocity, and speed**

Suppose an object is moving on the path described by

$$\mathbf{r} = \mathbf{r}(t) = (x(t), y(x)), \quad a \leq t \leq b.$$

The **average velocity** of the object on a subinterval $[t_1, t_2]$ of $[a, b]$ is

$$\mathbf{v}[t_1, t_2] = \frac{1}{t_2 - t_1}(\mathbf{r}(t_2) - \mathbf{r}(t_1)) = \left(\frac{x(t_2) - x(t_1)}{t_2 - t_1}, \frac{y(t_2) - y(t_1)}{t_2 - t_1}\right).$$

If the coordinate functions $x = x(t)$ and $y = y(t)$ of $\mathbf{r} = \mathbf{r}(t)$ are differentiable on $[a, b]$, the **velocity** $\mathbf{v}(t_1)$ of the object at $\mathbf{r}(t_1)$ is the limit of the average velocity vector on $[t_1, t_2]$, as $t_2 \to t_1$.

$$\mathbf{v}(t_1) = \lim_{t_2 \to t_1} \frac{1}{t_2 - t_1}(\mathbf{r}(t_2) - \mathbf{r}(t_1))$$

$$= \lim_{t_2 \to t_1} \left(\frac{x(t_2) - x(t_1)}{t_2 - t_1}, \frac{y(t_2) - y(t_1)}{t_2 - t_1} \right)$$

$$= \left(\lim_{t_2 \to t_1} \frac{x(t_2) - x(t_1)}{t_2 - t_1}, \lim_{t_2 \to t_1} \frac{y(t_2) - y(t_1)}{t_2 - t_1} \right)$$

$$= (x'(t_1), y'(t_1)).$$

We write this result as

$$\mathbf{v}(t_1) = \mathbf{r}'(t_1).$$

The **speed** of the object at time t is defined to be $\|\mathbf{v}(t)\|$, that is, the length of the velocity vector. Speed is often denoted by v, a scalar quantity.

In this definition we have assumed that limits of vectors are taken coordinate-wise; that is, if $\mathbf{w}(t) = (u(t), v(t))$ is a vector-valued function, then

$$\lim_{t_2 \to t_1} \mathbf{w}(t_2) = \lim_{t_2 \to t_1} (u(t_2), v(t_2)) = \left(\lim_{t_2 \to t_1} u(t_2), \lim_{t_2 \to t_1} v(t_2) \right).$$

This reduces the calculation of velocity to two one-dimensional differentiations.

■ **EXAMPLE 1**   The path of an object is described by

$$\mathbf{r} = (t, (t - 1)^2 + 0.5), \quad 0 \le t \le 2.5.$$

A sketch of this path is shown in Fig. 3.35. Lengths are measured in meters and time in seconds.

a. Calculate the velocity vector and speed of the object at $t = 0.5$.

b. Calculate the average velocity vector on the intervals $[0.5, 0.6]$, $[0.5, 0.51]$, $[0.5, 0.501]$, and $[0.5, 0.5001]$. Compare the results with the velocity vector.

c. Eliminate the parameter to obtain an equation of the form $y = f(x)$ for the path of the object.

d. Show that the tangent line to the path at $(0.5, 0.75)$ is parallel to the velocity vector there.

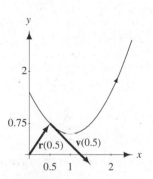

Figure 3.35. Position and velocity vectors.

Solution

a. We differentiate the position vector

$$\mathbf{r}(t) = (t, (t - 1)^2 + 0.5)$$

to obtain the velocity vector. We have

$$\mathbf{v}(t) = \mathbf{r}'(t) = (1, 2(t - 1)).$$

Table 1. Average velocities as $t_2 \to t_1$

t_2	$\mathbf{r}(t_2) - \mathbf{r}(t_1)$	$\mathbf{v}[t_1, t_2]$
0.6	$\left(0.1, -0.09\right)$	$\left(1, -0.9\right)$
0.51	$\left(0.01, -0.0099\right)$	$\left(1, -0.99\right)$
0.501	$\left(0.001, -0.000999\right)$	$\left(1, -0.999\right)$
0.5001	$\left(0.0001, -0.00009999\right)$	$\left(1, -0.9999\right)$

Evaluating $\mathbf{v}(t)$ at $t = 0.5$ gives

$$\mathbf{v}(0.5) = \left(1, -1\right) \text{m/s}.$$

The speed at $t = 0.5$ is the length of $\mathbf{v}(0.5)$. We have

$$\|\mathbf{v}(0.5)\| = \sqrt{1^2 + (-1)^2} \approx 1.4 \text{ m/s}.$$

b. Next we calculate the average velocity vector on the intervals $[t_1, t_2]$, where $t_1 = 0.5$ and $t_2 = 0.6, 0.51, 0.501,$ and 0.5001. We summarize the results in Table 1.

It is apparent from these calculations that at $t = 0.5$ the average velocity is approaching $\mathbf{v}(0.5) = \left(1, -1\right)$ m/s.

c. From the coordinate equations $x = t$ and $y = (t - 1)^2 + 0.5$, we find the equation $y = (x - 1)^2 + 0.5$. The graph of this equation is a parabola, part of which is shown in Fig. 3.35.

d. Since $dy/dx = 2(x - 1)$, the slope at $x = 0.5$ is -1. The velocity vector is $\left(1, -1\right)$ at this point. Rise over run is $-1/1$, so that the tangent line is parallel to the velocity vector at $(0.5, 0.75)$. ∎

■ **EXAMPLE 2** The path of an object is described by

$$\mathbf{r} = \left(x, y\right) = \left(a \cos bt, a \sin bt\right), \quad t \geq 0,$$

where a and b are positive constants. Lengths are in meters and time in seconds. Show that the object is moving counterclockwise on a circle of radius a and with constant speed ab m/s. In Fig. 3.36 we show the path for the special case $a = 2$ and $b = 1$. We show the position and velocity vectors for $t = 150/(2\pi)$, which corresponds to 150°.

Solution. This example repeats in vector form the calculations done in Example 3 of Section 3.2. Since for all t we have

$$x^2 + y^2 = (a \cos bt)^2 + (a \sin bt)^2 = a^2,$$

the object is moving on a circle of radius a. We may calculate its velocity $\mathbf{v}$ and speed $\|\mathbf{v}\|$ by differentiating $\mathbf{r}$. We have

$$\mathbf{v} = \mathbf{r}' = \left(-ab \sin bt, ab \cos bt\right).$$

The length of the velocity vector is the speed of the object. We have

$$\|\mathbf{v}\| = \sqrt{(-ab \sin bt)^2 + (ab \cos bt)^2} = ab \text{ m/s}.$$

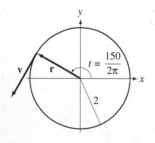

Figure 3.36. Velocity vector.

As the motion begins, $\mathbf{v}$ points upward. The motion is therefore counterclockwise. Since $\|\mathbf{v}\| = ab \neq 0$, the motion remains counterclockwise, for if it were to change direction, the speed would be 0 at some time.

In Fig. 3.36, for which $a = 2$ and $b = 1$, the vectors $\mathbf{r}$ and $\mathbf{v}$ at $t = 150/(2\pi)$ are

$$\mathbf{r}(150/(2\pi)) = (2\cos 150°, 2\sin 150°) \approx (-1.73, 1.00)$$

$$\mathbf{v}(150/(2\pi)) = (-2\sin 150°, 2\cos 150°) \approx (-1.00, -1.730). \qquad \blacksquare$$

Exercises 3.4

Basic

Exercises 1–8: Sketch the path of the motion. Find the velocity and speed of the motion at the given time t_1. Sketch the position and velocity vectors at t_1. Time is measured in seconds and lengths in meters.

1. $\mathbf{r} = (2t, 3t^2)$, $t \geq 0$, $t_1 = 1$

2. $\mathbf{r} = (4t, t^2)$, $t \geq 0$, $t_1 = 1$

3. $\mathbf{r} = (1 - t, 1 + 2t^2)$, $t \geq 0$, $t_1 = 1$

4. $\mathbf{r} = (2 - 3t, 3t^2 - 2)$, $t \geq 0$, $t_1 = 2$

5. $\mathbf{r} = 5(\cos 2t, \sin 2t)$, $t \geq 0$, $t_1 = \pi/6$

6. $\mathbf{r} = 2(\sin 3t, \cos 3t)$, $t \geq 0$, $t_1 = \pi/2$

7. $\mathbf{r} = (1, 2) + t(1, 1/2)$, $t \geq 0$, $t_1 = 2$

8. $\mathbf{r} = (0, 5) + t(-2, 0)$, $t \geq 0$, $t_1 = 1$

Exercises 9–12: Sketch the path of the object; calculate the average velocity vector in the intervals $[t_1, t_1 + 0.1]$, $[t_1, t_1 + 0.01]$, and $[t_1, t_1 + 0.001]$; and compare the results with $\mathbf{v}(t_1)$. Time is measured in seconds and lengths in meters.

9. $\mathbf{r} = (t, \sin t)$, $t \geq 0$; $t_1 = \pi/6$

10. $\mathbf{r} = (t, \cos t)$, $t \geq 0$; $t_1 = \pi/3$

11. $\mathbf{r} = (t, e^t)$, $t \geq 0$; $t_1 = 1.0$

12. $\mathbf{r} = (t, \ln(t + 1))$, $t \geq 0$; $t_1 = 1$

Growth

Exercises 13–18: Give a vector equation describing the given motion.

13. An object is moving to the right with speed 10 m/s on the line passing through $(-5, 1)$ and $(6, 7)$. At $t = 0$ the object is at $(-5, 1)$.

14. An object moving downward with speed 7 m/s on the line passing through $(5, 1)$ and $(6, -10)$. At $t = 0$ the object is at $(5, 1)$.

15. An object is moving counterclockwise with speed 7 m/s on a unit circle centered at $(0, 0)$.

16. An object is moving counterclockwise with speed 10 m/s on a unit circle centered at $(0, 0)$.

17. The velocity of the object for $t \geq 0$ is $\mathbf{v} = (1, 5)$. It was at the point $(0, -2)$ at $t = 0$.

18. The velocity of the object for $t \geq 0$ is $\mathbf{v} = (-3, 3)$. It was at the point $(4, 3)$ at $t = 0$.

Review/Preview

Exercises 19–26: Graph the equation.

19. $x^2 + (y/2)^2 = 1$

20. $x^2 - y^2 = 1$

21. $y = \ln x$

22. $y = e^{2x}$

23. $y = \sin(2x)$

24. $y = \cos(3x)$

25. $y = 1 - \sin x$

26. $y = 3 - 2\cos x$

3.5   PARAMETRIC EQUATIONS

In Sections 3.2–3.4 we used vector-valued functions to describe the paths of objects. We usually wrote these as equations of the form

$$(1) \qquad \mathbf{r} = \mathbf{r}(t) = \big(x(t), y(t)\big), \quad a \leq t \leq b.$$

The independent variable used in describing motion is usually time t. Variables like t are called **parameters** to distinguish them from coordinate variables such as x and y, which are fundamentally related to the chosen coordinate system. An equation like (1) is called a **parametric equation** with parameter t.

We used the strong, intuitive connection between time and motion to introduce parametric equations in a natural way. In this section we discuss parametric equations with parameters other than time. Here we are less interested in the path of an object and more in the geometric properties of a curve described by an equation like (1). The parameter often will have no connection with time, but will correspond to an angle, a length, or the cost of a product.

Although many useful curves can be described as the graph of a real-valued function, there are many curves that cannot be described in this way. One such curve is shown in Fig. 3.37. We give another example after discussing two parabolas, each of which is described by a parametric equation. We urge you to use your graphing calculator to graph most of the curves we discuss in this section. You should use the parametric graphing mode.

Although parametric equations were intrinsic to our first look at motion, we put this earlier discussion aside and start afresh. We begin by investigating two parabolas and a curve that models a cable unwrapping from a spool.

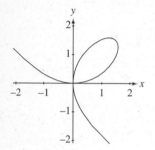

Figure 3.37. Parametric curve.

Investigation 1

A parabola with horizontal axis can be described by an equation of the form $x = ay^2 + by + c$. The curve C shown in Fig. 3.38 is part of such a parabola. Since C does not pass the vertical line test, it cannot be the graph of a function f defined on a subset of the x-axis.

The curve C is part of the graph of the equation

$$(2) \qquad x = \tfrac{1}{4}y^2 + y - \tfrac{13}{4}.$$

By completing the square this equation can be put into the standard form

$$(3) \qquad x - \left(-\tfrac{17}{4}\right) = \tfrac{1}{4}(y - (-2))^2.$$

From this equation, the parabola opens to the right, its horizontal axis is the line with equation $y = -2$ and its vertex is at $(-\tfrac{17}{4}, -2)$.

We outline an argument that the vector function

$$(4) \qquad \mathbf{r} = \mathbf{r}(t) = \big(t^2 - 17/4, 2(t - 1)\big), \quad -3/2 \leq t \leq 9/2$$

describes the curve C.

How did we decide upon the parametrization (4)? From the equation (3) it seemed natural to set $(y + 2)/2$ equal to a new variable t. From this, $y = 2(t - 1)$

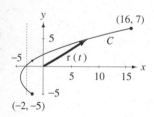

Figure 3.38. Parabola with horizontal axis.

and, from (3), $x = t^2 - \frac{17}{4}$. Using the equation $t = (y + 2)/2$, we determine the domain $[-3/2, 9/2]$ for t from the domain $[-5, 7]$ for y.

As an alternative to the vector notation in (4), we may simply list the coordinates x and y of **r** separately, as in

(5)
$$\begin{aligned} x &= t^2 - \tfrac{17}{4} \\ y &= 2(t - 1) \end{aligned} \qquad -3/2 \le t \le 9/2.$$

As given in (4) or (5), the parameter t is restricted to the interval $[-2, 4]$. We first calculate $\mathbf{r}(t)$ for the end points $t = -2$ and $t = 4$. It seems plausible that these will be the end points of the curve C as well. We have

$$\mathbf{r}(-2) = \big((-2)^2 + (-2) - 4, 2(-2) - 1\big) = \big(-2, -5\big)$$
$$\mathbf{r}(4) = \big((4)^2 + (4) - 4, 2(4) - 1\big) = \big(16, 7\big).$$

It appears that as t increases from -2 to 4, the curve is traced in the direction of the arrow on the curve. This can easily be made clearer by plotting a few more points. Or, if you plotted this parametric curve using your calculator, you will have noticed that the plotting order agrees with the indicated direction. We show the position vector $\mathbf{r}(t)$ in Fig. 3.38. It is useful to imagine this vector tracing the curve as t traces the interval $[-2, 4]$.

Another way of checking the tracing order for this curve is to examine the second or y-coordinate of $\mathbf{r}(t)$. The equation

$$y = 2(t - 1) = 2t - 2$$

shows that y increases as t increases. Indeed, since $dy/dt = 2$, y increases twice as fast as t.

In some parametric equations we may "eliminate the parameter." If we solve the equation $y = 2t - 2$ for t in terms of y, we obtain

$$t = \tfrac{1}{2}(y + 2).$$

Substituting this result into the first equation in (5),

$$x = \tfrac{1}{4}(y + 1)^2 - \tfrac{17}{4} = \tfrac{1}{4}y^2 + y - \tfrac{13}{4}.$$

This is the equation with which we began.

Investigation 2

The second parametric equation is

(6)
$$\mathbf{r} = \mathbf{r}(t) = \big(t - t^2, t + t^2\big), \qquad -2 \le t \le 2.$$

As t varies over $[-2, 2]$, $\mathbf{r}(t)$ traces the curve C shown in Fig. 3.39. We may calculate the end points

$$\mathbf{r}(-2) = \big(-6, 2\big) \quad \text{and} \quad \mathbf{r}(2) = \big(-2, 6\big)$$

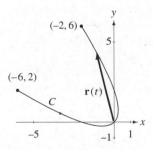

Figure 3.39. Parabola with tilted axis.

and otherwise convince ourselves that the curve C is traced in the direction of the arrow on the curve.

We may eliminate the parameter to obtain an equation in x and y. From (6) we have

$$x = t - t^2$$
$$y = t + t^2.$$

The form of these equations suggests that we add and subtract them. We have

$$x + y = 2t$$
$$x - y = -2t^2.$$

Solving the first of these equations for t and substituting into the second equation gives

(7)
$$x - y = -2\left(\frac{x + y}{2}\right)^2 = -\frac{1}{2}(x + y)^2.$$

Most calculators will graph equation (7) directly, using conic mode. The result will be a curve extending to infinity in the second quadrant (zooming out will give evidence of this). It will include C as a subset. However, this does not show that C is part of a parabola. For this we can do either of two things, both of which we leave as exercises. First, we may show that there is a line (directrix) and a point (focus) for which all points equally distant from the line and point satisfy (7). This uses one of the definitions of the curves called conics. (See Exercise 48.) Second, we may express (7) in terms of a rotated set of axes, reducing it to an equation without an xy term. (See Exercise 49.)

Investigation 3

A cable is unwound from a spool of radius 1 meter, as shown in the top part of Fig. 3.40. If the cable is kept taut as it is unwound, the free end of the cable traces a curve C known as an involute of a circle. This curve is relatively easy to describe with a parametric equation.

Looking at the figure, the free end of the cable is at Q. The cable is kept taut as it is unwound. From the point P at which the cable is just leaving the spool, the unwound cable lies along $\overrightarrow{PQ}$. Let θ be the angle from the positive x-axis counter-clockwise around to P, $\mathbf{u}$ the position vector of P, $\mathbf{r}$ the position vector of Q, and $\mathbf{w}$ the position vector equivalent to $\overrightarrow{PQ}$.

Our goal is to express $\mathbf{r}$ in terms of θ. This will give us a parametric equation for the involute C. Our strategy is to express the vectors $\mathbf{u}$ and $\mathbf{w}$ in terms of θ and then use the equation $\mathbf{r} = \mathbf{u} + \mathbf{w}$ to express $\mathbf{r}$ as a function of θ. This equation is clear from the figure since the displacement $\mathbf{r}$ is equal to the sum of the displacements $\mathbf{u}$ and $\mathbf{w}$.

Since the spool is a unit circle,

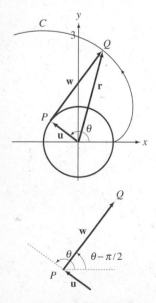

Figure 3.40. Unwrapping a cable.

$$\mathbf{u} = (\cos \theta, \sin \theta).$$

Referring to the sketch at the bottom of Fig. 3.40, since the angle between **w** and the positive *x*-axis is $\theta - \pi/2$, a unit vector in the direction of **w** is

$$\mathbf{v} = \left(\cos(\theta - \pi/2), \sin(\theta - \pi/2)\right) = \left(\sin\theta, -\cos\theta\right).$$

The length of **w**, which is the length of the cable unwound, is θ. This comes from the formula $s = a\theta$ for the length of the arc of the sector with radius a and central angle θ. The radius of the spool is 1 meter. If we multiply the unit vector **v** by θ, we obtain a vector of length θ. Thus

$$\mathbf{w} = \theta\left(\sin\theta, -\cos\theta\right).$$

Putting everything together, we have

$$\mathbf{r} = \mathbf{r}(\theta) = \mathbf{u} + \mathbf{w} = \left(\cos\theta + \theta\sin\theta, \sin\theta - \theta\cos\theta\right), \quad \theta \geq 0.$$

Before giving other examples of curves described parametrically, we define what we mean by **curve** and **tangent vector.**

DEFINITION **curve and tangent vector**

A **curve** C is a set of points with position vector

$$\mathbf{r} = \mathbf{r}(t) = \left(x(t), y(t)\right)$$

as t varies over an interval I. We assume that the vector-valued function **r** is defined on I and that its coordinate functions x and y are continuous on I. We say that the function **r** describes C or is a **parametrization** of the curve C.

If the coordinate functions of **r** are differentiable at a point $t \in I$, we say that **r** is differentiable at t and denote by $\mathbf{r}'(t)$ the vector with coordinate functions $x'(t)$ and $y'(t)$. The vector $\mathbf{r}'(t)$ is a **tangent vector** to the curve C at $\mathbf{r}(t)$.

The definition of tangent vector has the same form as the definition of velocity vector in Section 3.4. We give a geometric interpretation in Fig. 3.41. If we calculate the difference of the two position vectors $\mathbf{r}(t_2)$ and $\mathbf{r}(t_1)$ for t_2 near t_1, we expect that the vector $\mathbf{r}(t_2) - \mathbf{r}(t_1)$ will point very nearly in the direction of a tangent to the curve at $\mathbf{r}(t_1)$. Dividing by $t_2 - t_1$ converts $\mathbf{r}(t_2) - \mathbf{r}(t_1)$ to a rate of change; taking the limit as $t_2 \to t_1$ gives

$$\begin{aligned}
\mathbf{r}'(t_1) &= \lim_{t_2 \to t_1} \frac{\mathbf{r}(t_2) - \mathbf{r}(t_1)}{t_2 - t_1} \\
&= \lim_{t_2 \to t_1} \left(\frac{x(t_2) - x(t_1)}{t_2 - t_1}, \frac{y(t_2) - y(t_1)}{t_2 - t_1}\right) \\
&= \left(x'(t), y'(t)\right).
\end{aligned}$$

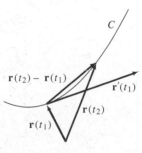

Figure 3.41. Tangent vector to a curve.

■ **EXAMPLE 1** Show that the graph C of the natural logarithm function defined on $(0, \infty)$ is a curve according to the above definition. Show also that the slope of the graph of ln at the point $(0.5, \ln 0.5)$ determines the same direction as a tangent vector to C at this point.

Solution.   A natural parameter for the graph of ln is suggested by the equation $y = \ln x$. From this we see that the position vector

$$\mathbf{r} = \mathbf{r}(x) = (x, \ln x), \quad 0 \le x < \infty$$

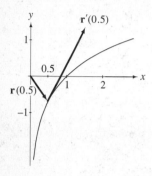

Figure 3.42. x as parameter.

traces the graph of ln as x varies over $(0, \infty)$. Since the coordinate functions x and $\ln x$ are continuous, C is a curve. We show the position vector $\mathbf{r}(0.5)$ in Fig. 3.42. We calculate the tangent vector to C at $\mathbf{r}(0.5)$ by differentiating the coordinate functions of $\mathbf{r}(x)$ with respect to the parameter x. We have

$$\mathbf{r}'(x) = \left(\frac{d}{dx} x, \frac{d}{dx} \ln x \right) = \left(1, \frac{1}{x} \right).$$

Evaluating at $x = 0.5$,

$$\mathbf{r}'(0.5) = \left(1, \frac{1}{0.5} \right) = (1, 2).$$

The relation between the slope of C at $\mathbf{r}(0.5)$ and the tangent vector $\mathbf{r}'(0.5) = (1, 2)$ may be understood by thinking of slope as "rise over run." The coordinates of the tangent vector are run and rise. Thus the slope of C at $\mathbf{r}(0.5)$ is $2/1$.   ■

As preparation for the Tangent Vector and Slope Theorem, we use the above example to relate the ideas of slope and tangent vector at a more basic level. Referring to Fig. 3.43, let P and Q be the points on C with position vectors

$$P = (0.5, \ln 0.5) \quad \text{and} \quad Q = (0.5 + h, \ln(0.5 + h)).$$

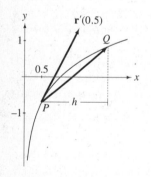

Figure 3.43. Relating slope and a tangent vector.

The slope m_{PQ} of the line segment joining P and Q is

$$(8) \qquad m_{PQ} = \frac{\ln(0.5 + h) - \ln 0.5}{0.5 + h - 0.5} = \frac{\ln\left(\dfrac{0.5 + h}{0.5}\right)}{h} = \frac{\ln(1 + 2h)}{h}.$$

Now using vectors instead of slope, the vector $\overrightarrow{PQ}$ has coordinates

$$\overrightarrow{PQ} = (0.5 + h - 0.5, \ln(0.5 + h) - \ln 0.5)$$
$$(9) \qquad \overrightarrow{PQ} = (h, \ln(1 + 2h)).$$

The number in (8) and the vector in (9) are different objects but carry related information. The slope is a rate of change. The vector $\overrightarrow{PQ}$ is a difference of the position vectors to Q and P, not a rate of change. We may convert $\overrightarrow{PQ}$ to a rate of change by dividing it by h, so that we obtain the change in $\overrightarrow{PQ}$ per unit change in the parameter x. The direction of $(1/h)\overrightarrow{PQ}$ is the same as that of a line with slope m_{PQ}. We show in Table 1 how $(1/h)\overrightarrow{PQ}$ and m_{PQ} vary as $h \to 0$.

Table 1. Tangent vector calculations

h	$\overrightarrow{PQ}$	$\dfrac{1}{h}\overrightarrow{PQ}$	m_{PQ}
1.5000	$(1.5000, 1.3863)$	$(1, 0.9242)$	0.9241
1.0000	$(1.0000, 1.0986)$	$(1, 1.0986)$	1.0986
0.1000	$(0.1000, 0.1823)$	$(1, 1.8232)$	1.8232
0.0100	$(0.0100, 0.0198)$	$(1, 1.9803)$	1.9803
0.0010	$(0.0010, 0.0020)$	$(1, 1.9980)$	1.9980
0.0001	$(0.0001, 0.0002)$	$(1, 1.9998)$	1.9998

The close relation between tangent vector and slope we have observed in these calculations holds more generally. We state the general case as the Tangent Vector and Slope Theorem. Here and elsewhere we often use ′ to denote differentiation with respect to a parameter, relying upon context to make the reference unambiguous.

Tangent Vector and Slope Theorem

Let C be a curve and $\mathbf{r} = \mathbf{r}(t)$, $t \in I$, a parametrization of C. For each $t \in I$ for which $\mathbf{r}$ is differentiable and $x'(t) \neq 0$, the slope of C at $\mathbf{r}(t)$ may be calculated from the tangent vector $\mathbf{r}'(t) = (x'(t), y'(t))$ as

(10)
$$\frac{dy}{dx} = \frac{\dfrac{dy}{dt}}{\dfrac{dx}{dt}} = \frac{y'(t)}{x'(t)}.$$

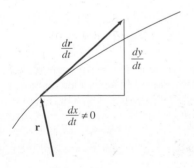

Figure 3.44. Slope of a parametrically defined curve.

To prove this theorem, it would be enough to show that at a point $\mathbf{r}(t)$ of C where $x'(t) \neq 0$, the equation $x = x(t)$ could be solved for t in terms of x and then substituted into $y = y(t)$, thus showing that y is a function of x. This would justify writing dy/dx and would lead to (10). More informally, we use Fig. 3.44 to help make (10) plausible. At a point $\mathbf{r} = \mathbf{r}(t)$ of C where $x'(t) \neq 0$, the x-coordinate is either increasing or decreasing and so the curve is not vertical. The coordinates of the tangent vector $\mathbf{r}'(t)$ are $x'(t)$ and $y'(t)$. The "run" $x' = dx/dt$ is not zero. The slope dy/dx of the line tangent to C at $\mathbf{r}(t)$ can be found by calculating the ratio

$$\frac{dy}{dx} = \frac{\dfrac{dy}{dt}}{\dfrac{dx}{dt}} = \frac{y'(t)}{x'(t)}.$$

We illustrate the Tangent Vector and Slope Theorem using one of the parabola examples given earlier and the cable unwrapping example.

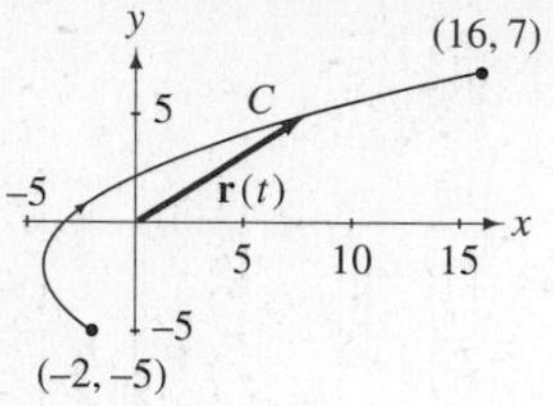

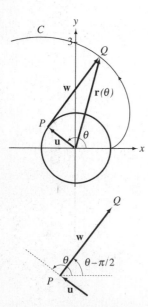

Figure 3.45. Parabola with horizontal axis.

■ **EXAMPLE 2**   We studied the equation $x = \frac{1}{4}y^2 + y - \frac{13}{4}$ earlier. It describes a parabola with horizontal axis. We obtained the parametric form $\mathbf{r} = (t^2 + t - 4, 2t - 1)$ for this curve. Use the equation $x = \frac{1}{4}y^2 + y - \frac{13}{4}$ to calculate the slope of its graph, which is the parabola shown in Fig. 3.45, at the point $(8, 5)$. Compare this result with that obtained using the parametric form $\mathbf{r} = (t^2 + t - 4, 2t - 1)$ and the Tangent Vector and Slope Theorem.

Solution.   The calculation based on the theorem is very easy. We evaluate the tangent vector

$$\mathbf{r}' = (2t + 1, 2) = (x', y')$$

at $t = 3$. This gives $\mathbf{r}' = (x', y') = (7, 2)$. By the Tangent Vector and Slope Theorem, the slope at $\mathbf{r}(3) = (8, 5)$ is $m = y'/x' = 2/7$.

To calculate the slope at $(8, 5)$ using the equation $x = \frac{1}{4}y^2 + y - \frac{13}{4}$, we use implicit differentiation. We have

$$\frac{d}{dx}x = \frac{d}{dx}\left(\frac{1}{4}y^2 + y - \frac{13}{4}\right)$$

$$1 = \frac{1}{2}yy' + y'.$$

Letting $y = 5$ and solving this equation for y',

$$y' = \frac{1}{1 + \frac{1}{2}y} = \frac{1}{1 + \frac{1}{2}5} = \frac{2}{7},$$

which is the slope of the parabola at $(8, 5)$. This agrees with the result based on the Tangent Vector and Slope Theorem.   ■

■ **EXAMPLE 3**   Earlier we showed that the curve C traced by the end of a cable unwrapping from a 1-meter spool is

$$\mathbf{r} = \mathbf{r}(\theta) = \mathbf{u} + \mathbf{w} = (\cos\theta + \theta\sin\theta, \sin\theta - \theta\cos\theta), \quad \theta \geq 0.$$

Figure 3.46 reproduces the earlier figure. Use this equation and the Tangent Vector and Slope Theorem to calculate the slope of C at the point $\mathbf{r}(2.44)$, which is shown as Q in the figure. Also locate the point in the second quadrant at which the tangent line to the curve is horizontal.

Solution.   To answer these questions, we calculate the tangent vector and then use the Tangent Vector and Slope Theorem. We have

$$\mathbf{r}' = (-\sin\theta + \sin\theta + \theta\cos\theta, \cos\theta - \cos\theta + \theta\sin\theta)$$

(11) $$= (\theta\cos\theta, \theta\sin\theta).$$

For the first question we evaluate $\mathbf{r}'(\theta)$ at $\theta = 2.44$. From (11) we have

$$\mathbf{r}'(2.44) \approx (-1.86, 1.57).$$

Figure 3.46. Unwrapping a cable.

From this result the slope is $m = y'/x' \approx -1.86/1.57 \approx -1.18$. To locate the point at which the tangent line is horizontal, we use (11) to write

$$\frac{dy}{dx} = \frac{y'(\theta)}{x'(\theta)} = \frac{\theta \sin \theta}{\theta \cos \theta}$$

$$= \tan \theta.$$

Setting $\theta = 0$ gives $dy/dx = 0$. The tangent line is horizontal at $\mathbf{r}(0) = (1, 0)$, the start of the unwrapping, but this point is not a second-quadrant point. The next value of θ for which $dy/dx = 0$ is $\theta = \pi$. The corresponding point is

$$\mathbf{r}(\pi) = (-1, \pi).$$

This point is just under the label C in Fig. 3.46. ∎

The search in the seventeenth century for accurate clocks, needed for determining longitude at sea, led Christiaan Huygens (1619–1695) to the invention of the pendulum clock. Huygens' clock is not the common pendulum clock, but one based on an isochronal pendulum, which has a period independent of the amplitude of the swing. We show in Fig. 3.47 the curve on which Huygens based his design. Huygens' work was published as *Horologium Oscillatorium* (Paris, 1673).

Also shown in the figure are three points poised to slide on the isochrone. They will be released at the same time. Bets are placed on which point will reach the finish line first. We disregard friction and any possibility that the points will collide. Which point wins? Compare your answer with that of Huygens, who showed that the race ends in a dead heat!

A part of Huygens' isochronal clock is sketched in Fig. 3.48. The flexible cords supporting the spherical mass are constrained to move between two plates, each bent into the shape of an isochrone.

The isochrone is also called a cycloid, based on the fact that it is generated by a point on a rolling circle. We use this property in deriving a parametric equation for the cycloid/isochrone. We note in passing that Huygens also described the involute of a circle, which we discussed in the cable example. You will note similarities in the geometry used in describing these two curves.

■ **EXAMPLE 4**   A circle of radius 1 is rolling on a line. The curve traced by a point on the circle and initially at the origin is called a cycloid and is shown at the top of Fig. 3.49. The bottom of the figure shows a zoom to the first half of the first period of the cycloid. The generating circle is shown in its initial position and after it has turned about its own center through the angle s. Describe C parametrically using s as parameter. Calculate the slope of the cycloid wherever possible.

Solution.   Our goal is to express the position vector $\mathbf{r}$ of the generating point in terms of the angle s. In the cable unwrapping example we expressed $\mathbf{r}$ as a sum of other vectors. Our strategy is the same for the cycloid. Thinking in terms of displacements, we may write

$$\mathbf{r} = \mathbf{u} + \mathbf{v} + \mathbf{w}$$

for any value of $s \geq 0$.

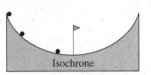

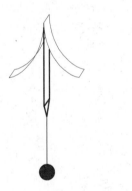

Isochronal pendulum

Figure 3.48. Isochronal clock.

Figure 3.47. Huygens' isochrone.

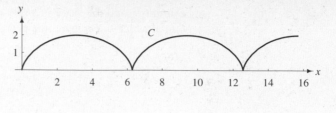

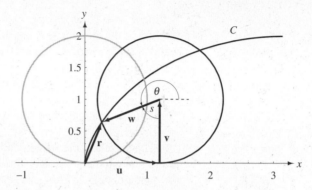

Figure 3.49. Cycloid and generating circle.

The key observation is that the length $\|\mathbf{u}\|$ of $\mathbf{u}$ is the same as the length of the arc of the sector with central angle s. If the circle were rolled back to its initial position, the arc would roll upon the line segment of length $\|\mathbf{u}\|$. Since in a unit circle the measure of an arc is the same as the (radian) measure of the central angle subtending that arc, we have $\mathbf{u} = (s, 0)$.

The vector $\mathbf{v}$ is constant, namely, $\mathbf{v} = (0, 1)$.

If we measure θ counterclockwise from the direction of the positive x-axis, then $\mathbf{w} = (\cos\theta, \sin\theta)$. Since $\theta + s = 3\pi/2$, we have

$$\mathbf{w} = (\cos\theta, \sin\theta) = (\cos(3\pi/2 - s), \sin(3\pi/2 - s)) = (-\sin s, -\cos s).$$

Putting these results together, we have

$$
\begin{aligned}
\mathbf{r} = \mathbf{r}(s) &= \mathbf{u} + \mathbf{v} + \mathbf{w} \\
&= (s, 0) + (0, 1) + (-\sin s, -\cos s) \\
&= (s - \sin s, 1 - \cos s), \quad s \geq 0.
\end{aligned}
$$

(12)

We calculate the tangent vector by differentiating (12). We have

$$\frac{d}{ds}\mathbf{r}(s) = \left(\frac{d}{ds}(s - \sin s), \frac{d}{ds}(1 - \cos s)\right) = (1 - \cos s, \sin s).$$

Using the Tangent Vector and Slope Theorem, we may calculate the slope at any point of the cycloid where $x'(s) \neq 0$. Since $x' = 1 - \cos s$, we must avoid $s = 0, 2\pi, 4\pi, \ldots$. These values of s correspond to the "cusps" of the cycloid, the points of C where the generating point touches the x-axis. At these points the tangent vector is vertical and slope is not defined. The slope is zero at the tops of the arches, where $y' = 0$. Except at $s = 0, 2\pi, 4\pi, \ldots$, the slope dy/dx of the cycloid is

$$\frac{dy}{dx} = \frac{\dfrac{dy}{ds}}{\dfrac{dx}{ds}} = \frac{\sin s}{1 - \cos s}.$$

∎

Exercises 3.5

Basic

Exercises 1–6: Use your calculator to graph the parametric curve. Calculate the end points of the curve. Place an arrowhead on the curve to indicate tracing order. Eliminate the parameter to find an equation in x and y whose graph includes the parametric curve.

1. $\mathbf{r}(t) = (1 + 2t^2, t + 2), \; -2 \le t \le 3$

2. $\mathbf{r}(t) = (t^2 - 4t + 3, t + 1), \; -3 \le t \le 3$

3. $\mathbf{r}(t) = (\frac{1}{2}t^2 + t, t^2 + t), \; -2 \le t \le 2$

4. $\mathbf{r}(t) = (t^2 - t, t^2 + 2t), \; -3 \le t \le 2$

5. $\mathbf{r}(t) = (2t + 1, t - 1), \; -2 \le t \le 3$

6. $\mathbf{r}(t) = (-t + 3, -2t + 5), \; t \ge 0$

Exercises 7–10: Use your calculator to graph the lines with the given parametric equation. Eliminate the parameter to find an equation of the form $y = mx + b$ whose graph includes the parametric curve. Recall from Section 3.3 that $\mathbf{r} = \mathbf{r}_0 + t\mathbf{a}$ describes a line through the point with position vector $\mathbf{r}_0$ and in the direction of the vector $\mathbf{a}$.

7. $\mathbf{r}(t) = (2, 1) + t(1, 3), \; -2 \le t \le 2$

8. $\mathbf{r}(t) = (-3, 1) + t(2, -5), \; 0 \le t \le 2$

9. $\mathbf{r}(t) = t(-2, -2), \; -\infty < t < \infty$

10. $\mathbf{r}(t) = (1, 1) + t(-2, -2), \; t \ge 0$

Exercises 11–20: Write the Cartesian equation $y = f(x)$ as a parametric equation $\mathbf{r} = (x, f(x))$ and calculate the tangent vector and slope at the point with position vector $\mathbf{r}_0$.

11. $y = x^2, \; -2 \le x \le 2, \; \mathbf{r}_0 = (1, 1)$

12. $y = x^3, \; -1 \le x \le 1.5, \; \mathbf{r}_0 = (1, 1)$

13. $y = \tan x, \; 0 \le x \le 1.3, \; \mathbf{r}_0 = (\pi/4, 1)$

14. $y = \sec x, \; -1 \le x \le 1, \; \mathbf{r}_0 = (\pi/6, 2/\sqrt{3})$

15. $y = x^2 - 6x + 11, \; 0 \le x \le 5, \; \mathbf{r}_0 = (4, 3)$

16. $y = x^2 + 4x + 3, \; -4 \le x \le 0, \; \mathbf{r}_0 = (-1, 0)$

17. $y = e^{2x}, \; -1 \le x \le 2, \; \mathbf{r}_0 = (1, e^2)$

18. $y = e^{\sqrt{x}}, \; 0 \le x \le 4, \; \mathbf{r}_0 = (1, e)$

19. $y = (\ln x)^2, \; 0.5 \le x \le 2, \; \mathbf{r}_0 = (1, 0)$

20. $y = \ln \sqrt{x} \; 0 < x \le 4, \; \mathbf{r}_0 = (1, 0)$

Exercises 21–30: Graph either using the parametric form or after eliminating the parameter θ. For curves described by $\mathbf{r} = (x, y) = (a \cos \theta, b \sin \theta)$, form the expression $(x/a)^2 + (y/b)^2$ and use a trigonometric identity. For curves described by $\mathbf{r} = (x, y) = (h + a \cos \theta, k + b \sin \theta)$, form the expression $((x - h)/a)^2 + ((y - k)/b)^2$ and use a trigonometric identity.

21. $\mathbf{r} = (2 \cos \theta, 2 \sin \theta), \; 0 \le \theta \le 2\pi$

22. $\mathbf{r} = (\frac{1}{2} \cos \theta, \frac{1}{2} \sin \theta), \; 0 \le \theta \le 2\pi$

23. $\mathbf{r} = (\cos \theta, \sin \theta), \; 0 \le \theta \le \pi$

24. $\mathbf{r} = (3 \cos \theta, 3 \sin \theta), \; 0 \le \theta \le 3\pi/2$

25. $\mathbf{r} = (1 + 2 \cos \theta, 1 + 2 \sin \theta), \; 0 \le \theta \le 2\pi$

26. $\mathbf{r} = (-1 + 2 \cos \theta, -3 + 2 \sin \theta), \; 0 \le \theta \le 2\pi$

27. $\mathbf{r} = (2 \cos \theta, 3 \sin \theta), \; 0 \le \theta \le 2\pi$

28. $\mathbf{r} = (4 \cos \theta, 1 \sin \theta), \; 0 \le \theta \le 2\pi$

29. $\mathbf{r} = (1 + 2 \cos \theta, 1 + 3 \sin \theta), \; 0 \le \theta \le 2\pi$

30. $\mathbf{r} = (-3 + \cos \theta, -5 + 4 \sin \theta), \; 0 \le \theta \le 2\pi$

Exercises 31–34:   Graph the following *Lissajous figures*. These curves can be produced on a cathode-ray oscilloscope by applying voltages proportional to $x(t)$ and $y(t)$ on the horizontal and vertical plates between which the electron beam passes. The plates deflect the beam as a function of the applied voltage.

31. $\mathbf{r} = (\sin t, \sin 2t), t \geq 0$

32. $\mathbf{r} = (\sin t, \sin 3t), t \geq 0$

33. $\mathbf{r} = (\sin 2t, \sin 3t), t \geq 0$

34. $\mathbf{r} = (\sin 3t, \sin 4t), t \geq 0$

Growth

35. Repeat the tangent vector calculations given in Table 1 of Example 1 for the function $f(x) = e^x$ defined on $(-\infty, \infty)$. Let P be the point $(0, 1)$. Sketch the graph of e^x near $(0, 1)$, including in your sketch $\mathbf{r}'(0)$ and, with $h = 0.25$, the vector $\overrightarrow{PQ}$.

36. Determine a parametric equation for the involute of a circle of radius 2 meters. (See the discussion of unwrapping a cable.) Sketch the involute. Calculate the slope of the curve at the point corresponding to the end of the cable when 0.6 meters have been unwound.

37. Determine a parametric equation for the cycloid curve C generated by a circle of radius 2 rolling on a line. Sketch C. Calculate the slope of C at the point $(\pi/3 - 1, 2 - \sqrt{3})$.

38. Parametrize the graph of $x = y^2$. Can x be used as a parameter? Why or why not?

39. Parametrize the curve with equation $(x + 1)^2 + (y - 2)^2 = 4$.

40. Parametrize the curve with equation $x^2 + y^2 - 2x + 4y = -1$.

41. Graph and parametrize the curve with equation $9x^2 + 4y^2 = 36$.

42. Graph and parametrize the curve with equation $x^2 + 3y^2 - 4x - 30y = -78$.

43. In Example 4 we said "Since $\theta + s = 3\pi/2$, we have...." Correct this statement and comment on whether this affects the equation we got for the cycloid.

44. A point on the cycloid given in Example 4 is on the first arch and is 1.5 meters above the x-axis. How far from the y-axis might it be?

45. Suppose that the wheel generating the cycloid discussed in Example 4 is rolling to the right at 2 meters per second. Describe the motion of a particle that started at $(0, 0)$ and is sticking to the wheel. What is its velocity and speed after $t = 3$ seconds?

46. Graph the *folium of Descartes*, which is described by

$$\mathbf{r}(t) = \frac{1}{t^3 + 1}(3t, 3t^2), \quad t \neq -1.$$

This curve is shown in Fig. 3.37. Find an equation in x and y for the folium by eliminating the parameter. Describe the tracing order and how various subintervals of the t-axis correspond to the three quadrants in which the folium occurs. Locate all points on the folium at which the tangent line is either horizontal or vertical. How many t values correspond to these points?

47. Without calculating the tangent vector, calculate the slope of the tangent line to the cable unwrapping curve at $\mathbf{r}(2.6)$. Locate the point $\mathbf{r}(\theta)$ at which the tangent line is horizontal.

48. Show that the set of all points (x, y) equally distant from the line with equation $y = x - \frac{1}{2}$ and the point $\left(-\frac{1}{4}, \frac{1}{4}\right)$ satisfies the equation

$$x - y = -\tfrac{1}{2}(x + y)^2.$$

Also show that any point (x, y) satisfying this equation is equally distant from the given line and point.

49. In this problem we outline an argument showing that the equation

$$x - y = -\tfrac{1}{2}(x + y)^2$$

describes a parabola. The idea is to rotate the axes to eliminate the xy term. Suppose the x- and y-axes are rotated counterclockwise through an angle θ to give a new set of coordinate axes. Label the rotated x-axis as the x'-axis and the rotated y-axis as the y'-axis. If a point has coordinates (x, y) relative to the "old" system and coordinates (x', y') relative to the "new" system, then

$$x = x' \cos \theta - y' \sin \theta$$
$$y = x' \sin \theta + y' \cos \theta.$$

Substitute these rotation equations into the equation of the conic, expand the result, and collect all $x'y'$ terms together. Choose θ so that the coefficient of the resulting single $x'y'$ term is zero. You should find that $\theta = \pi/4$. Show that with this value of θ the transformed equation reduces to $x'^2 = \sqrt{2}y'$, which has the form of a parabola. Check that the graph of this equation matches that of the given equation.

Review/Preview

Exercises 50–57: The graph of the equation $y = C + A \sin x$ can be obtained from the graph of $y = \sin x$ by adjusting its amplitude and horizontal placement. If $A < 0$, the graph must be reflected about the x-axis in addition to having its amplitude adjusted . To graph $y = 3 - 2 \sin x$, for example, start with the sine curve, double its amplitude to get graph 2, reflect graph 2 about the x-axis to get graph 3, and, finally, move graph 3 upward by three units. In practice, you may be able to combine one or more steps.

Graph by hand one period of each equation. Check by graphing with your calculator.

50. $y = 2 + \sin x$

51. $y = 3 + \sin x$

52. $y = 2 - \sin x$

53. $y = 3 - \sin x$

54. $y = 1 - 2 \sin x$

55. $y = 1 - 3 \sin x$

56. $y = 1 - \sqrt{2} \sin x$

57. $y = 1 - (2/\sqrt{3}) \sin x$

Exercises 58–65: Recall that the polar form of a vector $\mathbf{r} = (x, y)$ is (r, θ), where r is the length of $\mathbf{r}$ and θ its direction. We may convert from polar to rectangular form using the definitions of the length and direction of a vector. Recall that if $\mathbf{r} = (x, y)$ is a vector, its length r and direction θ are $\|\mathbf{r}\| = r = \sqrt{x^2 + y^2}$ and the unique angle $\theta \in [0, 2\pi)$ for which $\cos \theta = x/\sqrt{x^2 + y^2}$ and $\sin \theta = y/\sqrt{x^2 + y^2}$. It follows from these definitions that the rectangular form of $\mathbf{r}$ is $(x, y) = (r \cos \theta, r \sin \theta)$.

If a vector is given in polar form, convert it to rectangular form. If it is in rectangular form, convert it to polar form. Summarize in a sketch, interpreting all four numbers.

58. Polar form is $(2, \pi/6)$.

59. Polar form is $(2, -\pi/6)$.

60. Polar form is $(5, 5\pi/4)$.

61. Polar form is $(1, 1)$.

62. Rectangular form is $(3, 4)$.

63. Rectangular form is $(4, -3)$.

64. Rectangular form is $(-3, 4)$.

65. Rectangular form is $(-3, -4)$.

3.6 POLAR COORDINATES

In earlier sections we used polar coordinates in several discussions. For example, you may recall that in Section 3.3 on vectors we discussed an example of a student survey of a quadrilateral. The field data were given in the form of a list of the lengths and bearings of a series of lines. The length and bearing of a line joining two points are quantities that can be measured directly. To measure the east-west and north-south displacements from one point to another would require more effort. Polar coordinates are the standard choice for survey data.

In tracking a planet or comet about the sun, the preferred description of motion uses polar coordinates, with the pole at the center of the sun.

It remains true, however, that for some curves the most natural description is the graph of a function f or an equation. It's hard to improve on $y = ax^2 + bx + c$, $y = mx + b$, or $y = \sin x$ as descriptions of a parabola with vertical axis, line with slope m and y-intercept b, or standard sine curve, respectively.

Cartesian coordinates for a plane are based on two intersecting lines. A point P is located by the ordered pair (x, y), where x and y are the **Cartesian coordinates** of P measured on these lines. For some purposes it is simpler to locate P with the ordered pair (r, θ) of **polar coordinates,** where r is the distance to P from a fixed point called the **pole** and θ is an angle measured from a fixed direction. We show in Fig. 3.50 both the Cartesian and polar coordinates of a point P.

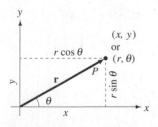

Figure 3.50. Polar and Cartesian coordinates.

A simple relation between the two coordinate systems can be obtained by placing the pole at the origin of a Cartesian coordinate system and measuring the angle θ counterclockwise from the positive x-axis.

Also shown in Fig. 3.50 is a position vector $\mathbf{r}$ of P. The standard polar coordinates of P are the length $r = \|\mathbf{r}\|$ and direction θ of the vector $\mathbf{r}$. We discuss "standard polar coordinates" below. The Cartesian coordinates (x, y) and polar coordinates (r, θ) of P are related by

$$(1) \qquad\qquad x = r\cos\theta \qquad y = r\sin\theta.$$

If we are given polar coordinates r and θ of P, we may use (1) to calculate the Cartesian coordinates x and y of P. Going the other way has a few complications. If we are given x and y, how do we calculate the standard polar coordinates r and θ? For P in the first quadrant, as in Fig. 3.50, we note directly from the figure that

$$r = \sqrt{x^2 + y^2} \quad \text{and} \quad \tan\theta = \frac{y}{x}.$$

If $x = 1$ and $y = 1$, for example, $r = \sqrt{2}$ and $\tan\theta = 1$, so that $\theta = \pi/4$. For P in other quadrants we calculate r in the same way and adjust θ for the quadrant.

Another complication is that it is sometimes useful to relax the condition that $\theta \in [0, 2\pi)$. To illustrate how this can arise, take the case $x = 1$ and $y = -1$. Just as above, $r = \sqrt{2}$ and $\tan\theta = -1$. Suppose we enter -1 into a scientific calculator (in radian mode) and press the ATAN key. We obtain $-0.7853\ldots$. The calculator has used a standard convention and, in effect, decided that $\theta = -\pi/4 = -0.7853\ldots$. We may prefer $\theta = 7\pi/4$ or $315°$. Indeed, we may choose θ from the infinite number of possibilities

$$\theta = -\pi/4 \pm n\pi, \quad n = 0, 1, 2, \ldots.$$

If we substitute into (1) any of these values of θ together with $r = 1$, we find $x = 1$ and $y = -1$.

In Cartesian coordinates the simplest graphs of equations are $x = a$ and $y = b$, where a and b are constants. These are a vertical line through $(a, 0)$ and a horizontal line through $(0, b)$. These "coordinate curves" are shown at the top of Fig. 3.51. At the bottom of Fig. 3.51 we show the coordinate curves in polar coordinates. These have equations $r = a$ and $\theta = b$. The graph of the equation $r = 1$, for example, is a circle of radius 1 centered at the pole. This is the set of all points (r, θ) in the plane for which $r = 1$. The graph of the equation $\theta = \pi/4$, for example, is the ray from the origin making an angle of $\pi/4$ with the positive x-axis. If r can be negative (we said above that r is a distance, so how can $r < 0$?), the graph of $\theta = \pi/4$ is the line through the pole. We discuss this last complication more fully in Example 2. In any case, the graph of $\theta = \pi/4$ is the set of all points (r, θ) in the plane for which $\theta = \pi/4$.

We give several examples of plotting polar equations $r = f(\theta)$. The first is similar to graphing $y = x$ in Cartesian coordinates. The last two examples will relate graphs of polar equations to parametric equations.

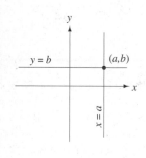

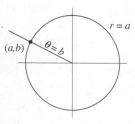

Figure 3.51. Coordinate curves.

■ **EXAMPLE 1**   Graph the polar equation $r = \theta$ for $0 \le \theta \le 2\pi$.

Solution.   We must sketch all points (r, θ) satisfying the equation $r = \theta$. As θ increases from 0 to 2π, we may picture a corresponding point P with polar

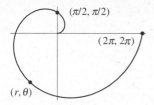

Figure 3.52. The spiral $r = \theta$.

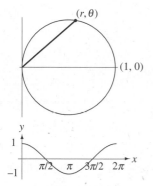

Figure 3.53. A polar circle.

coordinates (r, θ) tracing a curve in the plane. As θ increases, the distance r from the origin increases as well. At $\theta = 0$, $r = 0$ as well. The curve starts at the origin or pole. At $\theta = \pi/2$, P would be on the y-axis and $\pi/2 \approx 1.6$ units from the origin. The entire graph is the spiral shown in Fig. 3.52. ■

■ **EXAMPLE 2**   Graph the polar equation $r = \cos\theta$.

Solution.   First we graph $y = \cos x, 0 \leq x \leq 2\pi$, in Cartesian coordinates to see how $r = \cos\theta$ varies with θ. This is the bottom graph in Fig. 3.53. The decrease in y from 1 to 0 as x increases from 0 to $\pi/2$ on this graph corresponds to the decrease in r from 1 to 0 as θ increases from 0 to $\pi/2$ on the upper graph. If we take r as a distance, so that $r \geq 0$, we would plot no points as θ increases from $\pi/2$ to $3\pi/2$ since the cosine function is negative for these values of θ. As θ increases from $3\pi/2$ to 2π, the cosine function is again nonnegative and increases from 0 to 1. This completes the polar graph of $r = \cos\theta$. Note that the periodicity of the cosine function makes it unnecessary to consider values of θ outside of $[0, 2\pi]$. It appears that $r = \cos\theta$ describes a circle. We verify this below.

If we agree to plot points (r, θ), where $r < 0$, as if they were $(-r, \theta + \pi)$, we may graph the equation $r = \cos\theta$ and obtain an (r, θ)-point for all θ. Said differently, if $r < 0$, we use the ray determined by θ but plot on its backward extension through the origin. This gives a point in the plane for every (r, θ), whether $r \geq 0$ or not. With this agreement we would generate the top half of the upper figure as θ increases from 0 to $\pi/2$ and the lower half as θ increases from $\pi/2$ to π. The figure is retraced as θ increases from π to 2π.

Another way of graphing the polar equation $r = \cos\theta$ is to use (1) to transform $r = \cos\theta$ to Cartesian coordinates. We start with the equation

$$r = \cos\theta$$

and multiply both sides by r, obtaining

$$r^2 = r\cos\theta.$$

From (1) we have

$$x^2 + y^2 = x.$$

Bringing the x term to the left and completing the square on $x^2 - x$ gives

$$x^2 - x + \left(-\tfrac{1}{2}\right)^2 + y^2 = \left(-\tfrac{1}{2}\right)^2$$
$$\left(x - \tfrac{1}{2}\right)^2 + y^2 = \left(\tfrac{1}{2}\right)^2.$$

We recognize the last equation as that of a circle with center $(1/2, 0)$ and radius $1/2$. This is the circle shown in Fig. 3.53. By this argument we have not shown that the polar graph of $r = \cos\theta$ is the same as that of the circle, only that every point (r, θ) satisfying the polar equation is on the circle. We leave as an exercise showing that, conversely, every point on the circle is indeed a point of the polar graph.

Finally, we may graph the polar equation $r = \cos\theta$ by using θ as a parameter. Indeed, any polar equation can be graphed this way. We use the transformation equations (1). We use **r** as a position vector and r as a polar coordinate. This is an ambiguous but common usage. Note that $\|\mathbf{r}\| = r$. We define

$$\mathbf{r} = \mathbf{r}(\theta) = (r\cos\theta, r\sin\theta).$$

Since $r = \cos\theta$, we have

$$\mathbf{r} = (\cos\theta\cos\theta, \cos\theta\sin\theta).$$

Although this equation may be graphed using parametric mode on your calculator, it is useful to rewrite it using two double-angle identities. We have

$$\begin{aligned}\mathbf{r} &= \left(\cos^2\theta, \tfrac{1}{2}\sin 2\theta\right)\\ &= \left(\tfrac{1}{2}(1+\cos 2\theta), \tfrac{1}{2}\sin 2\theta\right)\\ &= \left(\tfrac{1}{2}, 0\right) + \tfrac{1}{2}\left(\cos 2\theta, \sin 2\theta\right)\end{aligned}$$

We recognize that the equation

$$\mathbf{r} = \tfrac{1}{2}\left(\cos 2\theta, \sin 2\theta\right), \quad 0 \le \theta \le \pi$$

describes a circle centered at the origin and having radius $1/2$. Hence, the equation

$$\mathbf{r} = \left(\tfrac{1}{2}, 0\right) + \tfrac{1}{2}\left(\cos 2\theta, \sin 2\theta\right)$$

describes a circle centered at $(1/2, 0)$ and having radius $1/2$. The effect of the vector $(1/2, 0)$ is to translate the circle centered at the origin to a new center at $(1/2, 0)$. ■

In our last example we graph a *limaçon,* one of a class of curves having polar equations of the form $r = a + b\sin\theta$, where a and b are constants. Other cases are given in the exercises.

■ **EXAMPLE 3** Graph the polar curve $r = 1 - \sqrt{2}\sin\theta$.

Solution. To graph polar curves for which r can be negative, solve the equation $r = 0$ first. This gives the values of θ at which r may change sign. Also, you may wish to graph the related (and more familiar) equation

$$y = 1 - \sqrt{2}\sin x$$

in rectangular coordinates. From this the graph of the polar equation $r = 1 - \sqrt{2}\sin\theta$ follows easily.

Starting with the graph of $y = \sin x$, which is shown at the top of Fig. 3.54, the factor $\sqrt{2}$ stretches the sine curve in the y direction; the factor -1 reflects the result about the x-axis; and adding 1 shifts the reflection upward by one unit. The final result is shown in the middle of the figure. For $\pi/4 \le x \le 3\pi/4$, the graph lies below the x-axis. Switching back to r and θ, we note that r is negative for θ between $\pi/4$ and $3\pi/4$. For $\theta = \pi/2$, for example,

$$r = 1 - \sqrt{2} \approx -0.414.$$

Since $r < 0$ we plot the point with polar coordinates $(-1 + \sqrt{2}, \pi/2 + \pi)$.

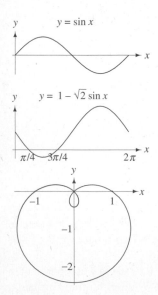

Figure 3.54. A limaçon.

We may draw the limaçon by mentally tracing the middle graph, in which x, now θ, varies over the three intervals $[0, \pi/4]$, $[\pi/4, 3\pi/4]$, and $[3\pi/4, 2\pi]$. This is sufficient since $1 - \sqrt{2}\sin\theta$ is periodic with period 2π.

For $\theta \in [0, \pi/4]$, r decreases to 0 from 1. The curve starts at $(1, 0)$ and, as $\theta \to \pi/4$, becomes tangential to the ray $\theta = \pi/4$. As θ moves through $\pi/4$, r becomes negative. The trace moves into the third quadrant. At $\theta = \pi/2$, $r = 1 - \sqrt{2}$. On the middle graph in Fig. 3.54, this corresponds to the bottom of the dip between $\pi/4$ and $3\pi/4$. At $\theta = 3\pi/4$, r becomes positive. We continue in this way. At $\theta = 3\pi/2$, $r = 1 + \sqrt{2} \approx 2.414$, which is the maximum value of r.

We remarked that this limaçon is tangent to the ray $\theta = \pi/4$ at the origin. To show this by a calculation, we calculate its slope at the point $(0, \pi/4)$. For this we describe the limaçon with the parametric equation

$$\mathbf{r} = \left((1 - \sqrt{2}\sin\theta)\cos\theta, (1 - \sqrt{2}\sin\theta)\sin\theta\right), \quad 0 \le \theta \le 2\pi.$$

Using the Tangent Vector and Slope Theorem, we have

$$\frac{dy}{dx} = \frac{\dfrac{dy}{d\theta}}{\dfrac{dx}{d\theta}}$$

$$= \frac{(1 - \sqrt{2}\sin\theta)\cos\theta + (-\sqrt{2}\cos\theta)\sin\theta}{(1 - \sqrt{2}\sin\theta)(-\sin\theta) + (-\sqrt{2}\cos\theta)\cos\theta}$$

$$= \frac{\cos\theta - 2\sqrt{2}\sin\theta\cos\theta}{-\sin\theta + \sqrt{2}\sin^2\theta - \sqrt{2}\cos^2\theta}.$$

Evaluating this expression at $\theta = \pi/4$, we have

$$\frac{dy}{dx} = 1.$$

Since $r = 0$ and the slope of the limaçon is 1 for $\theta = \pi/4$, it is tangent to the ray $\theta = \pi/4$ at the origin. ■

Exercises 3.6

Basic

Exercises 1–12: If the Cartesian coordinates of P are given, find polar coordinates (r, θ) of P, where $r \ge 0$ and $0 \le \theta \le 2\pi$. If the polar coordinates of P are given, find the Cartesian coordinates (x, y) of P. Use (1), common sense, and a graphing calculator (set in radian mode). Make a reasonably neat sketch.

1. The polar coordinates of P are $(\sqrt{2}, \pi/4)$.
2. The polar coordinates of P are $(\sqrt{2}, 3\pi/4)$.
3. The polar coordinates of P are $(2, 1)$.
4. The polar coordinates of P are $(3, 1.5)$.
5. The polar coordinates of P are $(1, 3.5)$.
6. The polar coordinates of P are $(2, 5.2)$.

7. The Cartesian coordinates of P are $(2, 1)$.
8. The Cartesian coordinates of P are $(1, 5)$.
9. The Cartesian coordinates of P are $(-2, 0)$.
10. The Cartesian coordinates of P are $(0, -5)$.
11. The Cartesian coordinates of P are $(-1.5, 2.5)$.
12. The Cartesian coordinates of P are $(-2.6, -4.3)$.

Exercises 13–24:   Graph the polar equation. Plot points (r, θ) where $r < 0$ using the convention discussed in Example 2. If θ is not restricted, plot all (r, θ) satisfying the equation.

13. $r = \theta^2, 0 \leq \theta \leq \pi$

14. $r = \sqrt{\theta}, 0 \leq \theta \leq 2\pi$

15. $r = \theta(2 - \theta), 0 \leq \theta \leq \pi$

16. $r = \theta(1 - \theta), 0 \leq \theta \leq \pi$

17. $r = 2\cos\theta$

18. $r = -\cos\theta$

19. $r = \sin\theta$. Check your graph by multiplying both sides of the equation by r, using (1) to transform it to Cartesian coordinates, and completing the square.

20. $r = \sin(\pi/2 - \theta)$

21. The limaçon described by $r = 1 - (2/\sqrt{3})\sin\theta$.

22. The limaçon described by $r = \sqrt{2}\sin\theta - 1$.

23. The four-petaled polar curve $r = \cos 2\theta$. It is useful to sketch the Cartesian equation $y = \cos 2x$, as in Example 2 and Fig. 3.53.

24. The four-petaled polar curve $r = \sin 2\theta$. It is useful to sketch the Cartesian equation $y = \sin 2x$, as in Example 2 and Fig. 3.53.

Growth

25. Calculate the slope of the polar curve described by $r = \theta$ at $(r, \theta) = (1, 1)$.

26. Calculate the slope of the limaçon given in Example 3, at the point corresponding to $\theta = 5\pi/4$.

27. Graph together the limaçon described by $r = 1 - 2\sin\theta$ and its tangent line at the point corresponding to $\theta = \pi/4$.

28. The limaçon described by $r = 1 - \sin\theta$ is often called a cardioid. Graph this cardioid.

29. Graph the curve described by $r = \ln\theta, 0 < \theta \leq 2\pi$. Find polar coordinates of the point where the curve crosses itself.

30. We showed in Example 2 that every point (r, θ) satisfying $r = \cos\theta$ lies on the graph of the equation $(x - 1/2)^2 + y^2 = (1/2)^2$. We did not show that every point satisfying $(x - 1/2)^2 + y^2 = (1/2)^2$ lies on the graph of $r = \cos\theta$. Give such an argument.

Review/Preview

31. Line L passes through $(3, 4)$ and has slope 2. Give an equation of a line through $(3, 4)$ and perpendicular to L.

32. Line L has slope $-2/3$. Give an equation of a line perpendicular to L and with y-intercept -3.

33. Give an equation of the line perpendicular to the line segment joining the points $(2, 5)$ and $(-3, 7)$ and passing through its midpoint.

34. In Section 3.3 we discussed the vector form of lines. We wrote $\mathbf{r} = \mathbf{r}_0 + t\mathbf{a}$ to describe a line passing through $\mathbf{r}_0$ and in the direction of $\mathbf{a}$. Suppose we have a line L with slope $m \neq 0$ and passing through (x_0, y_0). Write a vector form of this line. Also write a vector form of the line passing through the same point but perpendicular to L.

3.7   DOT PRODUCT

Referring to Fig. 3.55, the angle θ between vectors $\mathbf{a}$ and $\mathbf{b}$ is easy to calculate if we know their directions α and β. For the vectors shown in the figure,

$$\theta = \beta - \alpha.$$

For other pairs of vectors it may be necessary to adjust $\beta - \alpha$ to give an angle in $[0, \pi]$, which is the conventional range for the angle between two vectors.

To recover α and β from the rectangular forms

$$\mathbf{a} = (a_1, a_2) \quad \text{and} \quad \mathbf{b} = (b_1, b_2)$$

of $\mathbf{a}$ and $\mathbf{b}$, we may recall that α, for example, lies in $[0, 2\pi)$ and satisfies

$$a_1 = \|\mathbf{a}\|\cos\alpha \quad \text{and} \quad a_2 = \|\mathbf{a}\|\sin\alpha.$$

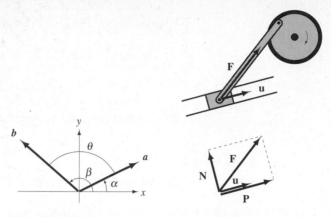

Figure 3.55. Angle between vectors.

Figure 3.56. Resolving a force.

The angles α and β can be calculated in this way and then subtracted to obtain θ. Or, with the help of a calculator, each vector can be converted to its polar form. From the polar forms the angle between **a** and **b** can be easily calculated.

A simpler calculation of the angle between two vectors can be based upon what is called the *dot product* of **a** and **b**.

A second application of the dot product is illustrated in Fig. 3.56. Suppose the vector **F** shown in the figure represents a force acting in a slider-crank mechanism. The direction of the slider relative to **F** is given by the unit vector **u**. In calculating the work done on the slider by the force **F**, we must resolve **F** into a sum of vectors **P** and **N**, where **P** is in the direction of motion of the slider and **N** is perpendicular to that direction. Calculating the dot product of **F** and **u** is a key step in resolving **F** into **P** and **N**.

Finally, to build a roller coaster that is both exciting and safe, engineers must be able to express the acceleration of the car as a sum of its tangential and radial accelerations. The tangential acceleration is in the direction of motion, and the radial acceleration is perpendicular to that direction.

The dot product has a simple algebraic definition.

DEFINITION   **dot product of vectors**

The **dot product** of vectors $\mathbf{a} = (a_1, a_2)$ and $\mathbf{b} = (b_1, b_2)$ is defined by

$$\mathbf{a} \cdot \mathbf{b} = a_1 b_1 + a_2 b_2.$$

The dot product of vectors $\mathbf{a} = (2, 5)$ and $\mathbf{b} = (-3, 7)$ is

$$\mathbf{a} \cdot \mathbf{b} = (2, 5) \cdot (-3, 7) = (2)(-3) + (5)(7) = -6 + 35 = 29.$$

Evidently, the arithmetic of the dot product is easy. Just multiply corresponding coordinates and add. Learning how to make effective use of the dot product takes a little longer. After listing several algebraic properties of the dot product, we give a geometric interpretation and discuss how to resolve a vector **F** into vectors **P** and **N**.

Properties of the Dot Product

1. The dot product is commutative:

$$\mathbf{a} \cdot \mathbf{b} = \mathbf{b} \cdot \mathbf{a}.$$

2. The dot product is distributive over addition and subtraction:

$$\mathbf{a} \cdot (\mathbf{b} \pm \mathbf{c}) = \mathbf{a} \cdot \mathbf{b} \pm \mathbf{a} \cdot \mathbf{c}.$$

3. Within a dot product, a scalar s can be moved to the left:

$$(s\mathbf{a}) \cdot \mathbf{b} = s(\mathbf{a} \cdot \mathbf{b}) \quad \text{and} \quad \mathbf{a} \cdot (s\mathbf{b}) = s(\mathbf{a} \cdot \mathbf{b}).$$

4. If $\mathbf{a} = 0$, then $\mathbf{a} \cdot \mathbf{a} = 0$; otherwise $\mathbf{a} \cdot \mathbf{a} > 0$.
5. The length $\|\mathbf{a}\|$ of a vector $\mathbf{a}$ is given by

$$\|\mathbf{a}\| = \sqrt{\mathbf{a} \cdot \mathbf{a}}.$$

Properties 1–5 are easy to prove. For example, property (1) follows from

$$\mathbf{a} \cdot \mathbf{b} = a_1 b_1 + a_2 b_2 = b_1 a_1 + b_2 a_2 = \mathbf{b} \cdot \mathbf{a}.$$

We leave the proofs of the remaining properties as exercises.

A Test for Perpendicularity

A geometric interpretation of dot product may be based on Fig. 3.57(b). Let $\mathbf{a}$ and $\mathbf{b}$ be given nonzero vectors. We use properties 1 and 2 in calculating $\|\mathbf{a} - \mathbf{b}\|^2$.

$$
\begin{aligned}
\|\mathbf{a} - \mathbf{b}\|^2 &= (\mathbf{a} - \mathbf{b}) \cdot (\mathbf{a} - \mathbf{b}) \\
&= (\mathbf{a} - \mathbf{b}) \cdot \mathbf{a} - (\mathbf{a} - \mathbf{b}) \cdot \mathbf{b} \\
&= \mathbf{a} \cdot (\mathbf{a} - \mathbf{b}) - \mathbf{b} \cdot (\mathbf{a} - \mathbf{b}) \\
&= \mathbf{a} \cdot \mathbf{a} - \mathbf{a} \cdot \mathbf{b} - \mathbf{b} \cdot \mathbf{a} + \mathbf{b} \cdot \mathbf{b} \\
\|\mathbf{a} - \mathbf{b}\|^2 &= \mathbf{a} \cdot \mathbf{a} - 2\mathbf{a} \cdot \mathbf{b} + \mathbf{b} \cdot \mathbf{b}.
\end{aligned}
$$

Using property (5), the last expression may be rewritten as

$$(1) \qquad \|\mathbf{a} - \mathbf{b}\|^2 = \|\mathbf{a}\|^2 - 2\mathbf{a} \cdot \mathbf{b} + \|\mathbf{b}\|^2.$$

If the triangle formed by $\mathbf{a}$ and $\mathbf{b}$ is a right triangle, as shown in Fig. 3.57(b), then by the Pythagorean theorem,

$$(2) \qquad \|\mathbf{a} - \mathbf{b}\|^2 = \|\mathbf{a}\|^2 + \|\mathbf{b}\|^2.$$

Equations (1) and (2) show that if $\mathbf{a}$ and $\mathbf{b}$ are perpendicular, then $\mathbf{a} \cdot \mathbf{b} = 0$. For nonzero vectors the converse holds as well. That is, two nonzero vectors are perpendicular if their dot product is zero.

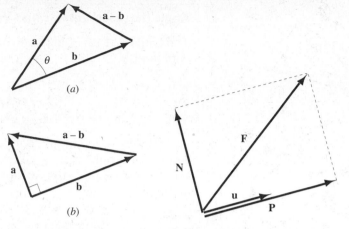

Figure 3.57. Geometric
interpretations of $\mathbf{a} \cdot \mathbf{b}$.

Figure 3.58. Projections of $\mathbf{F}$.

Length of a Projection

We use the law of cosines and (1) to obtain a geometric interpretation of $\mathbf{a} \cdot \mathbf{b}$ when $\mathbf{a}$ and $\mathbf{b}$ are not necessarily perpendicular. From Fig. 3.57(a) we have

$$\|\mathbf{a} - \mathbf{b}\|^2 = \|\mathbf{a}\|^2 + \|\mathbf{b}\|^2 - 2\|\mathbf{a}\|\,\|\mathbf{b}\| \cos \theta, \quad \text{where } 0 \le \theta \le \pi.$$

With (1) this gives

(3) $$\mathbf{a} \cdot \mathbf{b} = \|\mathbf{a}\|\,\|\mathbf{b}\| \cos \theta.$$

This result provides an easy way to calculate the length of the projection of one vector on another.

We return to the slider-crank mechanism shown in Fig. 3.58, which is an enlargement of part of Fig. 3.56. Suppose we wish to write the force $\mathbf{F}$ acting on the slider as the sum

$$\mathbf{F} = \mathbf{P} + \mathbf{N}$$

where $\mathbf{P}$ is in the direction $\mathbf{u}$ of the slider and $\mathbf{N}$ is perpendicular to $\mathbf{u}$. Recall that $\mathbf{u}$ is a unit vector, that is, $\|\mathbf{u}\| = 1$.

Since $\mathbf{P}$ is parallel to $\mathbf{u}$, we may write $\mathbf{P} = t\mathbf{u}$ for some scalar t. To find t, $\mathbf{P}$, and $\mathbf{N}$, we start with

$$\mathbf{F} = t\mathbf{u} + \mathbf{N}.$$

Now dot both sides of the equation with $\mathbf{u}$ to get

$$\mathbf{F} \cdot \mathbf{u} = t(\mathbf{u} \cdot \mathbf{u}) + \mathbf{N} \cdot \mathbf{u}.$$

Since $\mathbf{u} \cdot \mathbf{u} = 1$ and $\mathbf{N}$ is perpendicular to $\mathbf{u}$ (so that $\mathbf{N} \cdot \mathbf{u} = 0$), we have

(4) $$t = \mathbf{F} \cdot \mathbf{u} \quad \text{and} \quad \mathbf{F} = (\mathbf{F} \cdot \mathbf{u})\mathbf{u} + \mathbf{N}.$$

We now summarize these results and establish some conventions.

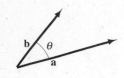

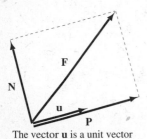

The vector **u** is a unit vector in the direction of **a**.

Figure 3.59. Projections of **F**.

Angle between Vectors and Vector Projections

The angle $\theta \in [0, \pi]$ between nonzero vectors **a** and **b** (see Fig. 3.59) is given by

(5)
$$\cos \theta = \frac{\mathbf{a} \cdot \mathbf{b}}{\|\mathbf{a}\| \|\mathbf{b}\|}.$$

Let **F** be a vector and **a** a nonzero vector. In calculating the projection of **F** onto **a**, it is the direction of **a** that is important, not its length. We express the formulas for projection in terms of the unit vector

$$\mathbf{u} = \frac{1}{\|\mathbf{a}\|}\mathbf{a}.$$

The **projection** of **F** onto **a** is the vector

(6)
$$\mathbf{P} = (\mathbf{F} \cdot \mathbf{u})\mathbf{u}.$$

The length of the projection of **F** onto **a** is the scalar

(7)
$$\|\mathbf{P}\| = \|\mathbf{F} \cdot \mathbf{u}\| = \|\mathbf{F}\| |\cos \theta|.$$

The projection of **F** onto a nonzero vector perpendicular to **a** is the vector

(8)
$$\mathbf{N} = \mathbf{F} - \mathbf{P} = \mathbf{F} - (\mathbf{F} \cdot \mathbf{u})\mathbf{u}.$$

We give several examples illustrating the use of the dot product, including the calculation of the length of a projection, the projection, and the angle between two vectors.

■ **EXAMPLE 1** Referring to Fig. 3.60, find the length d of the projection of $\mathbf{b} = (1, 1)$ onto the unit vector $\mathbf{a} = \left(\sqrt{3}/2, 1/2\right)$.

Solution. Since **a** is a unit vector, $\mathbf{u} = \mathbf{a}$. From (7) we have

$$d = |\mathbf{b} \cdot \mathbf{u}| = 1 \cdot (\sqrt{3}/2) + 1 \cdot (1/2) \approx 1.3660.$$

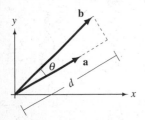

Figure 3.60. Length of a projection.

A second method of calculating d may be used if we know θ. For this we use the other part of (7). Since $\theta = \pi/4 - \pi/6 = \pi/12$, we find

$$d = \|\mathbf{b}\| |\cos \theta| = \sqrt{2} \cos(\pi/12) \approx 1.3660. \qquad ■$$

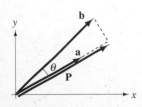

Figure 3.61. A projection vector.

■ **EXAMPLE 2** Find the projection **P** of **b** onto the unit vector **a**, where **b** and **a** are as in Example 1. See Fig. 3.61.

Solution.   To find **P** we use (6).

$$\mathbf{P} = (\mathbf{b} \cdot \mathbf{u})\mathbf{u} \approx 1.3660\mathbf{u} \approx (1.1830, 0.6830).$$

The projection **P** of **b** onto **a** is shown in Fig. 3.61.   ■

■ **EXAMPLE 3**   In Fig. 3.62 are shown vectors $\mathbf{v} = (1, 3)$ and $\mathbf{w} = (-4, -1)$. Use (5) to calculate the angle θ between **v** and **w**.

Solution.   We have

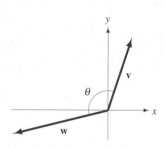

Figure 3.62. Angle between vectors.

$$\cos \theta = \frac{\mathbf{v} \cdot \mathbf{w}}{\|\mathbf{v}\| \|\mathbf{w}\|} = \frac{-4 - 3}{\sqrt{10} \sqrt{17}} = -\frac{7}{\sqrt{170}} \approx -0.5369.$$

In radians, $\theta \approx 2.138$; in degrees, $\theta \approx 122.5$.   ■

■ **EXAMPLE 4**   Under certain conditions, winds are welcomed by owners of recreational vehicles (RVs). A tail wind can decrease fuel costs. However, a strong cross wind can result in pressure sufficient to make driving tricky or even to tip the vehicle onto its side. The velocity vector **v** of an RV traveling on a curve C is shown at the top of Fig. 3.63. At the moment, the RV is moving in the direction 40° north of east at 60 kph. There is a steady 40 kph wind blowing in the direction 30° south of east. Calculate the tail wind **t**, the cross wind **c**, and their speeds.

Solution.   Letting **w** be the velocity vector of the wind, the tail wind **t** is the projection of **w** onto **v** and $\mathbf{c} = \mathbf{w} - \mathbf{t}$. We need **w** and a unit vector **u** in the direction of **v** for the calculation. We have

$$\mathbf{w} = 40(\cos(-30°), \sin(-30°)) \quad \text{and} \quad \mathbf{u} = (\cos 40°, \sin 40°).$$

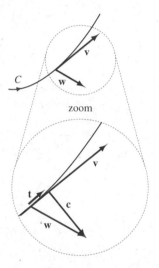

Figure 3.63. Winds on an RV.

The tail wind **t** is given by

$$\mathbf{t} = (\mathbf{w} \cdot \mathbf{u})\mathbf{u} = (40(\cos(-30°), \sin(-30°)) \cdot (\cos 40°, \sin 40°))\mathbf{u}$$
$$\approx 13.7\mathbf{u} = (10.5, 8.8).$$

We may now calculate the cross wind **c**. We have

$$\mathbf{c} = \mathbf{w} - \mathbf{t} = 40(\cos(-30°), \sin(-30°)) - \mathbf{t}$$
$$\approx (24.2, -28.8).$$

The speeds of **t** and **c** are

$$\|\mathbf{t}\| = 13.7 \text{ kph} \quad \text{and} \quad \|\mathbf{c}\| = 37.6 \text{ kph.}$$ ■

■ **EXAMPLE 5**   Referring to Fig. 3.64, decide which of the two forces $\mathbf{F}_1$ and $\mathbf{F}_2$ is more "effective" in moving the sled uphill. The hill has an inclination of 15°, and the forces $\mathbf{F}_1$ and $\mathbf{F}_2$ act at angles of 25° and 50° to the hill, with magnitudes of 150 N and 210 N, respectively. (The abbreviation N denotes the newton, a unit of force. A 1 N force is about $1/4$ pound.)

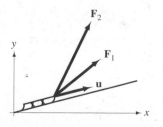

Figure 3.64. Forces on a sled.

Solution.   By assuming that the friction between the sled runners and the snow is negligible, we may compare the two forces by calculating the lengths of their projections onto the unit vector **u**.

The vectors **u**, $\mathbf{F}_1$, and $\mathbf{F}_2$ are

$$\mathbf{u} = \left(\cos 15°, \sin 15°\right)$$

$$\mathbf{F}_1 = 150\left(\cos 40°, \sin 40°\right)$$

$$\mathbf{F}_2 = 210\left(\cos 65°, \sin 65°\right).$$

The lengths d_1 and d_2 of the projections of $\mathbf{F}_1$ and $\mathbf{F}_2$ onto **u** may be calculated using (7). We have

$$d_1 = |\mathbf{F}_1 \cdot \mathbf{u}| = \left|150\left(\cos 40°, \sin 40°\right) \cdot \left(\cos 15°, \sin 15°\right)\right| \approx 136 \text{ N}$$

$$d_2 = |\mathbf{F}_2 \cdot \mathbf{u}| = \left|210\left(\cos 65°, \sin 65°\right) \cdot \left(\cos 15°, \sin 15°\right)\right| \approx 135 \text{ N}.$$

Since d_1 is larger than d_2, the force $\mathbf{F}_1$ is more effective.   ■

Work

A friend might complain, "It's hard work pulling you and that sled up the hill." Such statements can be quantified, through the idea of work used in mechanics. The amount of work done by the friend in moving the sled uphill depends on the magnitude of the force applied, its direction, and the length of the hill. Specifically, the **work** W done by a force **F** acting on an object as the object moves a distance s in the direction of a unit vector **u** is defined as

$$W = (\mathbf{F} \cdot \mathbf{u})s.$$

Applying this to the sled example, if the sled is moved a distance of $s = 6.0$ meters by the force $\mathbf{F}_1$, the work W done by this force is

$$W = (\mathbf{F}_1 \cdot \mathbf{u})(6.0) \approx (136 \text{ N})(6.0 \text{ m}) \approx 816 \text{ N·m} = 816 \text{ J}$$

The amount 1 newton-meter of work is called the **joule** (J). The amount of work done by $\mathbf{F}_1$ in moving the sled 6.0 meters is approximately 816 J. The amount of work done by $\mathbf{F}_2$ in moving the sled 6.0 meters is approximately 809 J. Over the same distance, the force $\mathbf{F}_1$ does more work on the sled.

Exercises 3.7

Basic _____

Exercises 1–6:   Find the dot product of the given vectors.

1. $\mathbf{a} = (2, 9)$, $\mathbf{b} = (3, -2)$

2. $\mathbf{a} = (-13, 4)$, $\mathbf{b} = (2, 5)$

3. $\mathbf{a} = (1, t)$, $\mathbf{b} = (3, -2)$

4. $\mathbf{a} = (2, 9)$, $\mathbf{b} = (3, -2)$

5. $\mathbf{a} = (1, m)$, $\mathbf{b} = (-m, 1)$

6. $\mathbf{a} = (p, q)$, $\mathbf{b} = (-q, p)$

7. Find the length of the projection of $(3, 2)$ onto the unit vector $(1/\sqrt{2}, 1/\sqrt{2})$.

8. Find the length of the projection of $(7, 5)$ onto the unit vector $(\sqrt{3}/2, 1/2)$.

9. Find the length of the projection of $(1, 5)$ onto a unit vector in the direction of $(5, 1)$.

10. Find the length of the projection of $(2, 7)$ onto a unit vector in the direction of $(10, 3)$.

11. Find the length of the projection of $\mathbf{a}$ onto the unit vector $\mathbf{u}$ if the angle θ between $\mathbf{a}$ and $\mathbf{u}$ is $\pi/3$ and $\|\mathbf{a}\| = 3$.

12. Find the length of the projection of $\mathbf{a}$ onto the unit vector $\mathbf{u}$ if the angle θ between $\mathbf{a}$ and $\mathbf{u}$ is $115°$ and $\|\mathbf{a}\| = 1.8$.

13. Find the projection of $(1, 2)$ onto a unit vector in the direction of the vector $(-1, 7)$.

14. Find the projection of $(2, 9)$ in the direction of a unit vector inclined at $45°$ to the x-axis.

15. Find the angle (in radians) between the vectors $(2, 1)$ and $(4, 5)$.

16. Find the angle (in radians) between the vectors $(3, 3)$ and $(2, -1)$.

17. Find the angle (in degrees) between the vectors $(-1, 5)$ and $(3, -2)$.

18. Find the angle (in degrees) between the vectors $(1.4, 1.7)$ and $(-1.2, -7.9)$.

19. Find the angle (in degrees) between the vectors $3(\cos 12°, \sin 12°)$ and $2.5(\cos 87°, \sin 87°)$.

20. Find the angle (in degrees) between the vectors $3(\cos 110°, \sin 110°)$ and $2.5(\cos 333°, \sin 333°)$.

21. Are the vectors $(2.3, 4.7)$ and $(-7.05, 3.45)$ perpendicular?

22. Are the vectors $(-2.88, -1.68)$ and $(-2.1, 3.6)$ perpendicular?

23. Are the vectors $(3.08, 5.17)$ and $(1.88, -1.08)$ perpendicular?

24. Are the vectors $(-0.75, 1.38)$ and $(-2.1, -1.2)$ perpendicular?

25. An RV is headed directly south at 55 mph and there is a wind from the northwest blowing at 28 mph. Calculate the tail and cross wind vectors.

26. An RV is headed $10°$ east of south at 50 mph, and there is a wind from the northeast blowing at 35 mph. Calculate the tail and cross wind vectors.

27. A bicyclist is riding southeast at 18 mph. If the wind is blowing at 20 mph from $5°$ south of east, what are her head and cross wind vectors?

28. A motorcyclist is riding $10°$ south of due west and encounters winds blowing out of the south at 40 mph. What are his head and cross wind vectors?

29. Which of the forces $\mathbf{F}_1$ or $\mathbf{F}_2$ is more effective at pulling a sled up a $15°$ hill, where $\mathbf{F}_1$ and $\mathbf{F}_2$ have respective magnitudes 380 N and 550 N and make angles of $24°$ and $49°$ with the hill?

30. Which of the forces $\mathbf{F}_1$ or $\mathbf{F}_2$ is more effective at pulling a sled up a $10°$ hill, where $\mathbf{F}_1$ and $\mathbf{F}_2$ have respective magnitudes 265 N and 380 N and make angles of $27°$ and $51°$ with the hill?

31. An object is displaced 15 m up a $10°$ ramp by a force of 500 N acting at a $45°$ angle relative to the ramp. Calculate the work done by the force.

32. An object is displaced 7.5 m up a $15°$ ramp by a force of 750 N acting at a $23°$ angle relative to the ramp. Calculate the work done by the force.

Growth

33. Determine a nonzero vector (x, y) perpendicular to $(1, 2)$.

34. Determine a nonzero vector (x, y) perpendicular to $(-4, 3)$.

35. A line has slope m. Determine a vector parallel to this line.

36. A line has slope m. Determine a vector perpendicular to this line.

37. Find the angles (in radians) in the triangle with vertices $A = (2, 1)$, $B = (5, 2)$, and $C = (3, 4)$.

38. Find the angles (in degrees) in the triangle with vertices $A = (-1, 1)$, $B = (5, 3)$, and $C = (-3, 4)$.

39. Find t such that the vectors $(9, -3)$ and $(2t - 3, 4)$ are perpendicular.

40. Find all values of t for which the vectors $(t, -3)$ and $(t + 1, 4)$ are perpendicular.

41. Use the idea of the length of the projection of one vector in the direction of another vector in calculating the distance from the line $\mathbf{r} = (3, 4) + t(1, 1)$ to the point with position vector $(10, 1)$.

42. If for an arbitrary force $\mathbf{F}$ and unit vector $\mathbf{u}$ we define vectors $\mathbf{P}$ and $\mathbf{N}$ by

$$\mathbf{P} = (\mathbf{F} \cdot \mathbf{u})\mathbf{u} \quad \text{and} \quad \mathbf{N} = \mathbf{F} - \mathbf{P},$$

show by calculating the dot product of $\mathbf{P}$ and $\mathbf{N}$ that these vectors are perpendicular.

43. The earth pulls with a vertical force of 140,000 N on a truck parked on a road inclined at $8°$ to the horizontal. What force must be produced by the brakes to resist the pull of the earth?

44. At what angle do the sine and cosine curves cross?

45. At what angle does the cosine curve intersect the graph of the equation $y = x$?

46. Let **a** and **b** be unit vectors with directions α and β. By using the dot product to calculate the angle between **a** and **b**, show that

$$\cos(\alpha - \beta) = \cos \alpha \cos \alpha + \sin \alpha \sin \beta.$$

47. Assume that $\mathbf{e}_1$ and $\mathbf{e}_2$ are perpendicular unit vectors. The coordinates of a vector $\mathbf{w} = (x, y)$ relative to the pair $\mathbf{e}_1$ and $\mathbf{e}_2$ are the numbers x' and y' satisfying the equation

$$\mathbf{w} = x'\mathbf{e}_1 + y'\mathbf{e}_2.$$

Show that $x' = \mathbf{w} \cdot \mathbf{e}_1$ and $y' = \mathbf{w} \cdot \mathbf{e}_2$. (Hint: Calculate $\mathbf{w} \cdot \mathbf{e}_1$.)

48. Referring to the preceding exercise, let

$$\mathbf{e}_1 = (\cos \theta, \sin \theta)$$
$$\mathbf{e}_2 = (\cos(\theta + \pi/2), \sin(\theta + \pi/2)).$$

Show that these vectors are perpendicular unit vectors. Show that the coordinates x' and y' of $\mathbf{w} = (x, y)$ relative to the pair $\mathbf{e}_1$ and $\mathbf{e}_2$ are

$$x' = x \cos \theta + y \sin \theta$$
$$y' = -x \sin \theta + y \cos \theta.$$

These two equations are often used to calculate the coordinates (x', y') of a point (x, y) relative to an (x', y') system obtained from the (x, y) system by rotating it through an angle θ. Explain this viewpoint.

49. Show that the dot product is distributive over addition.

50. Show that within a dot product, scalars can be moved to the left.

51. Show that for all vectors **a**, $\mathbf{a} \cdot \mathbf{a} = \|\mathbf{a}\|^2$.

52. Show that if $\mathbf{a} = \mathbf{0}$ is the zero vector, then $\mathbf{a} \cdot \mathbf{a} = 0$; otherwise $\mathbf{a} \cdot \mathbf{a} > 0$.

53. Derive and give a geometric interpretation to the identity

$$\|\mathbf{a} + \mathbf{b}\|^2 + \|\mathbf{a} - \mathbf{b}\|^2 = 2\|\mathbf{a}\|^2 + 2\|\mathbf{b}\|^2.$$

54. We showed geometrically that

$$\mathbf{a} \cdot \mathbf{b} = \|\mathbf{a}\| \|\mathbf{b}\| \cos \theta$$

for all vectors **a** and **b**. Since $\cos \theta \leq 1$ for all θ, it follows that

$$|\mathbf{a} \cdot \mathbf{b}| \leq \|\mathbf{a}\| \|\mathbf{b}\|, \quad \text{for all vectors } \mathbf{a} \text{ and } \mathbf{b}$$

This result is called the Cauchy-Schwarz inequality. Show that the triangle inequality, which is

$$\|\mathbf{a} + \mathbf{b}\| \leq \|\mathbf{a}\| + \|\mathbf{b}\|, \quad \text{for all vectors } \mathbf{a} \text{ and } \mathbf{b},$$

follows from the Cauchy-Schwarz inequality.

55. (The following question was posted on an Internet news group by a graphics software designer.) One base of a cylinder is centered at $\mathbf{r}_0 = (x_0, y_0, z_0)$, its height is h, the radius of its base is a, and the axis of the cylinder is in the direction of the unit vector **u**. Given any point $\mathbf{r} = (x, y, z)$ in space, how can you determine if **r** is on or within the cylinder?

Review/Preview

56. Write a vector equation for the line through the point with position vector $(3, -1)$ and in the direction of the vector $(2, 1)$.

57. Calculate the velocity and speed at $t = 1$ of the object with position vector $\mathbf{r} = \mathbf{r}(t) = (\sqrt{t^2 + 1}, t^2/5)$ for $t \geq 0$.

58. Express the speed of an object in terms of the dot product.

59. An object at P is displaced first to Q by $\mathbf{a} = (3, -7)$ and then to T by $\mathbf{b} = (9, -1)$. Calculate the single equivalent displacement from P to T. Express the coordinates of T in terms of the coordinates of P.

60. An object is moving on a line and has constant velocity $\mathbf{v} = (2, -1)$ m/s. If at $t = 0$ the object is at the point $(-1, 5)$, find its location after 10 seconds.

Exercises 61–68: For each equation and point, calculate the slope of the tangent line to the graph at the point.

61. $y = \sqrt{x^2 + 5}$; $(2, 3)$

62. $y = \ln(2x + 3)$; $(5, \ln 13)$

63. $x^2/2 + y^2/3 = 1$; $(1, \sqrt{6})$

64. $y = \arctan(x^{1/3})$; $(1, \pi/4)$

65. $y = x \sin x$; $(\pi/2, \pi/2)$

66. $y = \sec(1 - x)$; $(1, 1)$

67. $y = 3^x$; $(-1, 1/3)$

68. $y^2x + \tan(xy) = 1$, where $x \geq 0$; $(2, 3)$

Exercises 69–72: Given the velocity and initial position of an object, calculate its position vector **r**.

69. $\mathbf{v} = (1, t)$; $\mathbf{r}(0) = (0, 1)$

70. $\mathbf{v} = (t^2, 3t^4)$; $\mathbf{r}(0) = (2, 1)$

71. $\mathbf{v} = (\sin 2t, \cos 2t)$; $\mathbf{r}(0) = (-1/2, 1/2)$

72. $\mathbf{v} = (e^t, e^{2t})$; $\mathbf{r}(0) = (1, 1)$

3.8 NEWTON'S LAWS

Mechanics, the branch of physics concerning the motion of bodies, includes both kinematics and dynamics. The first subject is concerned with describing the motion of an object without reference to the causes of that motion. We discussed kinematics in Sections 3.1–3.6, though much of our effort went toward introducing the ideas from mathematics needed to describe motion. We discussed position vectors, displacement vectors, vector arithmetic, velocity vectors, and parametric equations.

Isaac Newton (1642–1727) initiated his great works in mechanics, mathematics, and optics just after he graduated from Trinity College of Cambridge University with a Bachelor of Arts degree in 1665. He worked at his family home in Woolsthorpe, to which he had returned because the plague had caused the university to "disperse." Newton later said of his discoveries during this period, "All this was in the two plague years of 1665 and 1666, for in those days I was in the prime of my age for invention, and minded mathematics and [science] more than at any other time since."

In this section we define the acceleration vector of a moving object and discuss several applications of Newton's second law and his law of universal gravitation.

Newton's Laws of Motion and Universal Gravitation

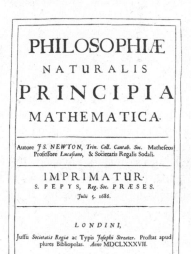

Newton gave his axioms or laws of motion in his *Philosophiae Naturalis Principia Mathematica,* first published in 1686. We state Newton's laws of motion, first in language similar to that used by Newton. After giving some terminology, we restate the second law in modern notation.

Newton's Laws of Motion and Universal Gravitation

First law: Every body continues in its state of rest, or of uniform motion in a line, unless it is compelled to change that state by forces impressed upon it.

Second law: The change in the quantity of motion is proportional to the motive power impressed; and is made in the direction of the line in which that force is impressed.

Third law: To every action there is always opposed an equal reaction.

Universal gravitation: There is a power of gravity pertaining to all bodies proportional to the several quantities of matter which they contain...decreasing always as the inverse square of the distances.

To express Newton's laws quantitatively, we use the concepts of *mass, momentum,* and *force*. We rely upon our intuitive notions of mass and force. The momentum **p** of a body is a measure of its "quantity of motion" and is defined as the product of its mass m and velocity **v**:

$$\mathbf{p} = m\mathbf{v}.$$

The phrase "motive power impressed" is the *net force* acting on a body. Denoting the net force by the vector **F**, Newton's second law becomes

(1) $$\mathbf{F} = \frac{d}{dt}\mathbf{p} = \frac{d}{dt}(m\mathbf{v}).$$

In case the mass m of the body is a constant, Newton's second law can be rewritten as $\mathbf{F} = m(d\mathbf{v}/dt)$. The *acceleration* **a** of the body is defined by

$$\mathbf{a} = \frac{d}{dt}\mathbf{v}.$$

This definition agrees with our perception of acceleration as a change in velocity during a time interval. If you were in a vehicle on a straight stretch of highway, a change in velocity from 100 km/h to 0 km/h in 1 minute would provoke little comment. If this were to happen in 1 second, little comment would be possible. In an automobile, changes in velocity occur when it hits something or is hit by something, or when the brake or accelerator is pressed. Dividing the change $\Delta\mathbf{v}$ in velocity by the time Δt gives the rate of change of velocity, and taking the limit as $\Delta t \to 0$ gives acceleration.

In terms of acceleration, when the mass of a body is constant, we may write Newton's second law (1) as

(2) $$\mathbf{F} = m\frac{d}{dt}\mathbf{v} = m\mathbf{a}.$$

After a brief discussion of units and universal gravitation, we give examples of motion on a line and in a plane. As in Section 3.1, we work on both the forward problem and the inverse problem. The forward problem in Section 3.1 was to infer from the position vector **r** of an object its velocity **v**. We did this by differentiation. We extend the forward problem by differentiating a second time to obtain the acceleration vector **a** of the object. The inverse problem may be extended as well. The extended inverse problem is to infer the position and velocity vectors of an object from its acceleration vector.

Units   We use SI units almost entirely. The abbreviation SI stands for Système International d'Unités (International System of Units). Length is measured in meters (m), mass in kilograms (kg), and time in seconds (s). Units for other quantities are derived from these units. Velocity and speed are measured in meters per second (m/s). Acceleration is measured in meters per second per second (m/s^2). The relation (2) may be used to measure and name a unit of force. The units of force are kg·m/s^2. In honor of Isaac Newton, the force required to accelerate a mass of 1 kg to 1 m/s^2 is called a newton (N).

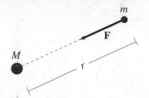

Figure 3.65. Universal gravitation.

Universal Gravitation

Newton proposed the law of universal gravitation to account for the motion of the sun, moon, and planets. To start simply, suppose we have two particles of masses m and M and separated by a distance of r. Newton proposed that the magnitude of the force exerted by either of the particles on the other is proportional to the product of their masses, inversely proportional to the square of the distance between them, and directed along the line joining the particles. In Fig. 3.65 we show the force $\mathbf{F}$ exerted by the particle of mass M on the particle of mass m. Denoting the proportionality constant by G, we have

$$(3) \qquad \mathbf{F} = \frac{GmM}{r^2}\mathbf{u}$$

where $\mathbf{u}$ is a unit vector based at the particle of mass m and directed toward the other particle. The constant G is called the *universal gravitational constant*. Its value is 6.67259×10^{-11}. The units of G are N·m²/kg².

To apply this law to the sun, moon, or earth, Newton had to show that such objects can be regarded as particles. Specifically, Newton had to show that the force exerted on an object by a spherical body S is the same as that exerted on the object by a particle located at the center of S and having the same mass as S. We use this result to simplify the calculation of the position and velocity of objects falling to the surface of the earth.

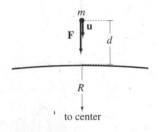

Figure 3.66. Force on a small, nearby object.

Earth's Gravitational Force

Referring to Fig. 3.66, suppose we have an object of mass m at a distance d from the earth's surface, where m and d are small compared, respectively, to the mass M of the earth and its radius R. From the law of universal gravitation, applied to particles of masses m and M and separated by a distance of $R + d$, we have

$$(4) \qquad \mathbf{F} = \frac{GmM}{(R + d)^2}\mathbf{u} \approx \frac{GM}{R^2}m\mathbf{u} = gm\mathbf{u}.$$

In this equation, $\mathbf{u}$ is a unit vector directed toward the center of the earth and g denotes the constant GM/R^2. For small, nearby objects the replacement of $R + d$ by R is inconsequential, and we shall use $\mathbf{F} = gm\mathbf{u}$ without further comment.

The accepted value of the constant g, called the acceleration of gravity, is $g = 9.80665$ m/s². We usually use the value $g = 9.8$ m/s². Generally accepted values of M and R are $M \approx 5.97 \times 10^{24}$ kg and $R \approx 6.378 \times 10^6$ m.

For small, nearby objects we may combine Newton's second law and (4) to obtain

$$m\mathbf{a} = mg\mathbf{u}.$$

Hence

$$(5) \qquad \mathbf{a} = g\mathbf{u}.$$

Thus the object is accelerated toward the center of the earth. The magnitude of the acceleration is $g = 9.8$ m/s² and is independent of the mass of the object.

We use (5) to derive Galileo's formula

$$(6) \qquad x = x(t) = 30 - 4.9t^2, \quad 0 \le t \le \sqrt{30/4.9},$$

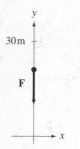

Figure 3.67. The cannonball revisited.

with which we began this chapter. You may recall that this formula modeled a cannonball falling from 30 m above the earth's surface. The ball falls from rest and in a vacuum.

Referring to Fig. 3.67, the force **F** of the earth on the cannonball depends on the mass m of the cannonball and its acceleration **a**. Combining **F** $=$ m**a** with the law of universal gravitation as in (4) and (5) gives

$$(7) \qquad\qquad\qquad\qquad \mathbf{a} = g\mathbf{u}$$

where $g = 9.8$ is the magnitude of the acceleration of gravity and **u** is a unit vector in the direction of **F**. Since the force **F** is directed toward the center of the earth, $\mathbf{u} = (0, -1)$. The coordinate system in Fig. 3.67 models the sketch in Fig. 3.66. An even simpler model is possible since the motion, the force **F**, and the acceleration **a** lie along the y-axis.

Rewriting (7) in terms of velocity, we have

$$(8) \qquad\qquad\qquad \frac{d\mathbf{v}}{dt} = \mathbf{a} = 9.8\mathbf{u} = 9.8(0, -1).$$

This equation shows that when the velocity vector is differentiated, the result is the constant vector $9.8(0, -1)$. It is reasonable (and will be proved in Chapter 4) that the velocity vector must have the form

$$\mathbf{v} = 9.8t(0, -1) + \mathbf{c}_1$$

where $\mathbf{c}_1$ is a constant vector. Note that the derivative of $9.8t(0, -1) + \mathbf{c}_1$ is indeed $9.8(0, -1)$, as required by (8). Since we know that the initial velocity of the cannonball is $\mathbf{v}(0) = (0, 0) = \mathbf{0}$, the constant vector $\mathbf{c}_1$ must be the zero vector. Thus

$$\frac{d\mathbf{r}}{dt} = \mathbf{v} = 9.8t(0, -1).$$

This equation shows that when the position vector **r** of the cannonball is differentiated, the result is t times a constant vector. It is reasonable that

$$\mathbf{r} = \tfrac{1}{2}9.8t^2(0, -1) + \mathbf{c}_2$$

where $\mathbf{c}_2$ is a constant vector. If we differentiate this vector function, we obtain **v**. At $t = 0$ the position of the cannonball is $\mathbf{r}(0) = \mathbf{c}_2 = (0, 30)$. Thus

$$\mathbf{r} = \mathbf{r}(t) = (x(t), y(t)) = (0, -4.9t^2 + 30).$$

The y-coordinate of **r** is Galileo's formula (1), given at the beginning of Section 3.1.

Motion on a Line

Although the motion of real objects takes place in three dimensions, it is often possible to model the motion of particular objects in one or two dimensions. The motion of a planet subject only to the force of the sun can be modeled in two dimensions, in what turns out to be the plane of the orbit. The motion of a subatomic particle in a linear accelerator can be modeled in one dimension. We start with an example of motion on a line.

■ **EXAMPLE 1**   The mass of a fully loaded passenger jet is 3.6×10^5 kg, and its engines provide a force of 7.7×10^5 N. We take this force as the net force and ignore air and runway friction. What is the minimum length of runway needed for this aircraft to reach its takeoff speed of 86 m/s?

Solution.   Figure 3.68 shows the plane, its position $x = x(t)$ on a coordinate system oriented in the direction of motion, and the net force **F** acting on the jet at any time t. We start with Newton's second law, as expressed in (2) for constant mass. (Note that the mass is not in fact constant since jet fuel is being burned. This is another simplifying assumption.)

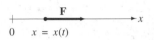

Figure 3.68. Minimum runway length.

$$\mathbf{F} = m\mathbf{a} = m\frac{d\mathbf{v}}{dt}$$

Since the motion is one dimensional, we may use scalar quantities. Dropping the boldface type, we have

$$F = ma = m\frac{dv}{dt}.$$

Denoting the ratio $(7.7 \times 10^5)/(3.6 \times 10^5)$ of the known force and mass by k,

$$\frac{dv}{dt} = k.$$

From our recent discussion about Galileo's formula, as well as from Section 3.1 and recent Review/Preview problems, the function $kt + c_1$, where c_1 is a constant, is the only function whose derivative with respect to t is k. Thus

(9) $$v = kt + c_1.$$

Since at $t = 0$ the velocity v is 0, the constant c_1 must be zero. Thus

(10) $$v = \frac{dx}{dt} = kt.$$

Using similar reasoning we have

$$x = \tfrac{1}{2}kt^2 + c_2, \quad \text{where } c_2 \text{ is a constant.}$$

Since at $t = 0$ the coordinate position x is 0, the constant c_2 must be zero. Thus

(11) $$x = \tfrac{1}{2}kt^2.$$

From (10) we find the time $t = 86/k$ at which $v = 86$ m/s. Substituting this value of t into (11), we have

$$x = \frac{1}{2}kt^2 = \frac{1}{2}k\frac{86^2}{k^2} = \frac{1}{2}\frac{86^2}{k} = \frac{(86^2)(3.6 \times 10^5)}{(2)(7.7 \times 10^5)} \approx 1700 \text{ m.} \quad \blacksquare$$

Projectile Motion

Earlier we used the equation

(12)
$$\mathbf{a} = g\mathbf{u}$$

to derive Galileo's formula

$$x = x(t) = 30 - 4.9t^2, \quad 0 \le t \le \sqrt{30/4.9}.$$

We took $\mathbf{u} = (0, -1)$. We used the initial conditions $\mathbf{v}(0) = (0, 0) = \mathbf{0}$ and $\mathbf{r}(0) = (30, 0)$.

We use (12) in calculating the path of a rifle bullet in a vacuum. Since for a cannonball and a bullet the only force acting is the gravitational force, the calculation will be the same except when we use the initial conditions. We take the initial velocity of the rifle bullet to be the velocity with which the bullet leaves the muzzle of the rifle. The cannonball falls from rest.

■ **EXAMPLE 2** In the year 1900 typical infantry rifles fired bullets with muzzle velocities of 610 m/s. By 1943 muzzle velocities had increased to 700 m/s. We test a rifle from each of these two years. Assume that the rifles are fired at an angle of 10° from the horizontal and calculate the difference in their ranges.

Solution. We put the origin of a Cartesian coordinate system at the muzzle of the rifles. The paths of the two test bullets are shown in Fig. 3.69. The 10° angle at the origin is exaggerated in the figure because of the scaling on the axes.

We calculate the position vectors of the two bullets by solving an inverse problem. Letting $\mathbf{a}$ and $\mathbf{v}$ denote the acceleration and velocity vector of either of the bullets, we have

$$\mathbf{a} = \frac{d\mathbf{v}}{dt} = g(0, -1).$$

The function $\mathbf{v}$ must have the form

$$\mathbf{v} = gt(0, -1) + \mathbf{c}_1$$

where $\mathbf{c}_1$ is a constant vector. Since the initial velocity is $\mathbf{v}(0) = b(\cos 10°, \sin 10°)$, where $b = 610$ or 700, we have

$$\mathbf{v} = gt(0, -1) + b(\cos 10°, \sin 10°).$$

Since $\mathbf{v} = d\mathbf{r}/dt$, we have

$$\frac{d\mathbf{r}}{dt} = gt(0, -1) + b(\cos 10°, \sin 10°).$$

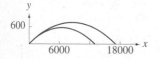

Figure 3.69. Ranges of two bullets.

The function $\mathbf{r}$ must have the form

$$\mathbf{r} = \tfrac{1}{2}gt^2(0, -1) + bt(\cos 10°, \sin 10°) + \mathbf{c}_2$$

where $\mathbf{c}_2$ is a constant vector. Since $\mathbf{r}(0) = 0$ for either bullet, $\mathbf{c}_2 = 0$. Thus the position vector of either bullet is

$$(13) \quad \mathbf{r} = \tfrac{1}{2}gt^2(0, -1) + bt(\cos 10°, \sin 10°) = (bt \cos 10°, -\tfrac{1}{2}gt^2 + bt \sin 10°).$$

To find the range of a bullet with this position vector, we set the y-coordinate equal to zero. We have

$$-\tfrac{1}{2}gt^2 + bt \sin 10° = 0.$$

Solving this equation for the nonzero root, we find

$$t = \frac{2 \cdot b \sin 10°}{g}.$$

Letting $g = 9.8$, $b = 610$, and then $b = 700$, we find the times t_1 and t_2 at which the two bullets hit the x-axis. We have

$$t_1 = \frac{2 \cdot 610 \sin 10°}{9.8} \approx 21.62$$

$$t_2 = \frac{2 \cdot 700 \sin 10°}{9.8} \approx 24.81.$$

We may now calculate the ranges of the two bullets. Rounding the results to two significant figures,

$$\mathbf{r}(t_1) \approx (13{,}000, 0)$$

$$\mathbf{r}(t_2) \approx (17{,}000, 0).$$

The faster bullet took longer to return to earth and has the longer range. ∎

Uniform Circular Motion

An object is in **uniform circular motion** if its position vector has the form

$$(14) \qquad \mathbf{r} = \mathbf{r}(t) = r(\cos \omega t, \sin \omega t), \quad t \geq 0.$$

Referring to Fig. 3.70, the constants r and ω are the radius of the circle on which the motion takes place and the angular speed of the object. To clarify the meaning of ω, note that ωt is the polar angle of the object. The rate of change of the angle ωt is $d(\omega t)/dt = \omega$, so that ω has units of radians per second.

An example of an object in uniform circular motion is a roller coaster car on a level, circular track. The car and passengers may be completing a 90° or a 180° turn as part of their path. To get uniform circular motion we must neglect friction, which would result in a decrease in angular speed.

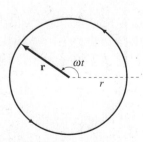

Figure 3.70. Uniform circular motion.

We may calculate the velocity **v** and acceleration **a** of the roller coaster car by differentiating its position vector. From (14) we have

$$(15) \qquad \mathbf{v} = \frac{d\mathbf{r}}{dt} = r\left(-\omega \sin \omega t, \omega \cos \omega t\right) = r\omega\left(-\sin \omega t, \cos \omega t\right)$$

and

$$(16) \qquad \mathbf{a} = \frac{d\mathbf{v}}{dt} = r\omega\left(-\omega \cos \omega t, -\omega \sin \omega t\right) = -r\omega^2\left(\cos \omega t, \sin \omega t\right).$$

The velocity and acceleration of the car are shown in Fig. 3.71. Note that **a** is in the opposite direction of **r**. This follows from equations (14) and (16), which show that

$$\mathbf{a} = -\omega^2 \mathbf{r}.$$

A second property of uniform circular motion is that **v** is perpendicular to **r**. To show this we calculate their dot product. We have

$$\mathbf{r} \cdot \mathbf{v} = r\left(\cos \omega t, \sin \omega t\right) \cdot r\omega\left(-\sin \omega t, \cos \omega t\right)$$
$$= r\omega^2\left(-\cos \omega t \sin \omega t + \sin \omega t \cos \omega t\right) = 0.$$

Since neither **v** nor **a** is the zero vector, these vectors are perpendicular.

The speed v of the car can be calculated from (15). We have

$$(17) \qquad v = \|\mathbf{v}\| = r\omega.$$

The magnitude a of the acceleration vector is $a = \|\mathbf{a}\| = r\omega^2$, from (16). We may write the magnitude of the acceleration in terms of the radius r of the circle and the speed v. From (17) we have

$$\omega = \frac{v}{r}$$

and hence

$$(18) \qquad a = r\omega^2 = \frac{v^2}{r}.$$

According to Newton's first law, there must be a force **F** exerted on a roller coaster car of mass m moving in uniform circular motion, for otherwise it would move in a straight line. From Newton's second law and (16) we have

$$(19) \qquad \mathbf{F} = m\mathbf{a} = m(-r\omega^2)\left(\cos \omega t, \sin \omega t\right).$$

From (18) and (19) we find that the magnitude $F = \|\mathbf{F}\|$ of this force is

$$(20) \qquad F = \|\mathbf{F}\| = \frac{mv^2}{r}.$$

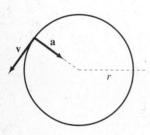

Figure 3.71. Velocity and acceleration vectors.

Everything we have said applies as well to a satellite of mass m in a circular earth orbit of radius r. We use (20) and the universal law of gravity (3) to relate the speed of the satellite to the radius of its orbit. From (3) we have $F = GMm/r^2$, where M is the mass of the earth. Substituting into (20), we have

$$\frac{GMm}{r^2} = \frac{mv^2}{r},$$

from which we obtain

$$(21) \qquad v^2 = \frac{GM}{r}.$$

To end this section, we use Newton's laws, including the universal law of gravity, to calculate the orbital radius of earth's most important satellite, the moon. Close agreement between the calculated radius and the observed radius would be evidence supporting Newton's laws. It may have been this calculation Newton had in mind when commenting 50 years later that

> I thereby compared the force requisite to keep the Moon in her Orb with the force of gravity at the surface of the earth, and found them answer pretty nearly.

■ EXAMPLE 3   Calculate the radius of the moon's orbit, assuming that the moon circles the earth in $27\frac{1}{3}$ days.

Solution.   From (21) we have

$$(22) \qquad v^2 = \frac{GM}{r}$$

where M is the mass of the earth, v the speed of the moon, and r its distance from earth.

Equation (22) is expressed in terms of the moon's speed, whereas we are given its period. Speed v and period T are related by the equation $vT = 2\pi r$. This becomes clear when we recall that the period is the number of time units needed for one revolution. Thus if we multiply the speed of the moon, in meters per second, by its period, measured in seconds per revolution, we get the circumference $2\pi r$.

Rewriting (22) in terms of T, we have

$$\frac{4\pi^2 r^2}{T^2} = \frac{GM}{r}.$$

Recalling from (4) that $g = GM/R^2$, we may write this equation as

$$\frac{4\pi^2 r^2}{T^2} = \frac{gR^2}{r}.$$

Solving for r^3, we have

$$(23) \qquad r^3 = \frac{gR^2 T^2}{4\pi^2}.$$

From (23) we have

$$r^3 \approx \frac{(9.80)(6.378 \times 10^6)^2 \times (27\frac{1}{3} \times 3600 \times 24)^2}{4\pi^2} \approx 5.63 \times 10^{25}.$$

Taking the cube root, we obtain

$$r \approx 3.83 \times 10^8 \text{ m.}$$

This "answer(s) pretty nearly" to $r \approx 3.85 \times 10^8$ m, which is a modern measurement of the mean distance to the moon. ∎

Exercises 3.8

Basic

Exercises 1–10:   Repeat the reasoning given in the relevant example. Avoid simply substituting into formulas as much as possible.

1. In Example 1, suppose that at $t = 0$ the plane has velocity 4 m/s. How does this affect the minimum required length of runway?

2. In Example 1, suppose that the engines are run at 90 percent of the given force. How does this affect the minimum required length of runway?

3. Calculate the range of a rifle made in 1943 and fired as in Example 2, except that the muzzle is 1.5 m above the ground and the angle is 12° above the horizontal.

4. Calculate the range of a rifle made in 1943 and fired as in Example 2, except that the muzzle is 1.5 m above the ground and the angle of fire is 0° above the horizontal.

5. For an object moving with uniform circular motion on a circle with radius 0.5 m, find its speed in meters per second if its angular speed is 2000 rpm (revolutions per minute).

6. For an object moving with uniform circular motion on a circle with radius 0.5 m, find its angular speed in radians per second if its speed is 12 m/s.

7. Assuming a circular orbit, calculate the orbital speed (in km/h) required to maintain a satellite at an altitude of 160 km. What is the period (in minutes) of such a satellite?

8. Assuming a circular orbit, calculate the orbital speed (in km/h) required to maintain a satellite at an altitude of 180 km. What is the period (in minutes) of such a satellite?

9. The speed of the manned USSR spacecraft Soyuz 10 was 7765 m/s in a nearly circular orbit. What is its altitude in kilometers?

10. Calculate the altitude of a satellite moving in a circular orbit and with a period of 89 minutes.

Exercises 11–18:   Solve the inverse problem; that is, find the position $\mathbf{r} = \mathbf{r}(t)$ of the object at any time $t \geq 0$, given its velocity $\mathbf{v} = \mathbf{v}(t)$ and initial position $\mathbf{r}(0)$.

11. $\mathbf{v} = (2, t), \mathbf{r}(0) = (1, 1)$

12. $\mathbf{v} = ((t - 1), t^2), \mathbf{r}(0) = (0, 0)$

13. $\mathbf{v} = ((2t + 5), \sqrt{t}), \mathbf{r}(0) = (2, 2)$

14. $\mathbf{v} = ((-3t + 1)^2, \sqrt{2 + 3t}), \mathbf{r}(0) = (1, 2)$

15. $\mathbf{v} = ((-3t + 1)^2, \sqrt{2 + 3t}), \mathbf{r}(0) = (-2, 3)$

16. $\mathbf{v} = (1/(t^2 + 1), 3/(t + 1)), \mathbf{r}(0) = (0, 0)$

17. $\mathbf{v} = (\sin(2t + 1), \cos(2t + 1)), \mathbf{r}(0) = (2, 2)$

18. $\mathbf{v} = (\sec^2 t, \sec t \tan t), \mathbf{r}(0) = (2, 2)$

Growth

19. Repeat the reasoning used to derive Galileo's formula, but suppose that at $t = 0$ the cannonball is 100 m from the ground and has speed 10 m/s directly upward.

20. Repeat the reasoning used to derive Galileo's formula, but suppose that at $t = 0$ the cannonball is 150 m from the ground and has speed 10 m/s directly downward.

21. Suppose in Example 1 that the runway is not horizontal but slopes upward at $2°$ from the horizontal. How does this affect the required length of the runway?

22. Suppose in Example 1 that the runway is not horizontal but slopes downward at $2°$ from the horizontal. How does this affect the required length of the runway?

23. What muzzle velocity would be required to increase the range to 18,000 m of a rifle fired at an angle of $10°$ to the horizontal?

24. What muzzle velocity would be required to increase the range to 18,000 m of a rifle fired at an angle of $12°$ to the horizontal?

25. Calculate the angle of fire for maximum range and calculate this range for a rifle made in 1943.

26. Calculate the angle of fire for maximum range and calculate this range for a rifle made in 1900.

27. The manned USSR spacecraft Soyuz 9 moved in an orbit that was approximately circular. Find its altitude if it completed 286 orbits in 424 hours 59 minutes.

28. The manned U.S. spacecraft Gemini 3 moved in an orbit that was approximately circular. Find its altitude if it completed three orbits in 4 hours 53 minutes.

29. Sketch the position vector, velocity vector, and acceleration vectors at $t = 3$ s for the object whose position vector for any time t is $\mathbf{r} = \mathbf{r}(t) = 2(\cos(0.8t), \sin(0.8t))$. Find the magnitude of the force required to maintain this circular motion if the object has mass 2 kg.

30. Referring to (4), let

$$F_1 = \|\mathbf{F}_1\| = \frac{GmM}{(R + d)^2}$$

$$F_2 = \|\mathbf{F}_2\| = \frac{GmM}{R^2}.$$

If the forces $\mathbf{F}_1$ and $\mathbf{F}_2$ of (4) are acting on a mass of 80 kg at a distance of d meters above the earth's surface, find the least d such that a 1 percent difference between F_1 and F_2 is observed.

31. A geostationary satellite is one moving so that it stays above a fixed location on the earth. Assuming that such a satellite has a circular orbit, find its altitude.

32. Kepler discovered three laws of planetary motion. The first is that the planets orbit the sun in ellipses, with the sun at one focus. His third law is that the square of a planet's orbital period T is proportional to the cube of the semimajor axis a of its orbit. Excepting only Mercury and Pluto, the orbits of the planets are very nearly circular. First, verify empirically that Kepler's third law holds. Use the periods and semimajor axes of the orbits of Earth, Venus, Jupiter, and Uranus. Data for the last three are given in the table. The periods are measured in tropical years, the time required for the earth to complete one orbit about the sun. The semimajor axes are measured in astronomical units (AU), the length of the semimajor axis of Earth's orbit.

	T	a
Venus	0.61521	0.7233316
Jupiter	11.86224	5.202561
Uranus	84.01247	19.21814

Does Kepler's third law follow from (23)? Why or why not?

Review/Preview

Exercises 33–44: Find the derivative $f'(x)$.

33. $f(x) = x/(4x - 1)$

34. $f(x) = x^2 \tan x$

35. $f(x) = e^{-x^2}$

36. $f(x) = \sqrt{x} \cos x$

37. $f(x) = \ln(3x^2 + 1)$

38. $f(x) = x^3 2^{-5x}$

39. $f(x) = \sin^2(1/x + 1)$

40. $f(x) = (\sqrt{x} + 1/x)^3$

41. $f(x) = 1/(5\sqrt{x} + 2)^2$

42. $f(x) = \sin^3((x/2 + 2/x)^2)$

43. $f(x) = \cos(x)/(x^3 + 1)$

44. $f(x) = \sin^2 x$

Exercises 45–48:   Write the equation of the tangent line to the graph of f at $(x_0, f(x_0))$. Sketch the graph of the function and the tangent line together.

45. $f(x) = \sqrt{x+1}, (3, f(3))$

46. $f(x) = \sin^2 x, (\pi/4, f(\pi/4))$

47. $f(x) = 1/(2x+3), (0, f(0))$

48. $f(x) = \ln(2x+1), (1, f(1))$

REVIEW OF KEY CONCEPTS

This chapter continues the work of Chapter 1 on rates of change. A major goal was to define velocity and acceleration, two important rates of change. To describe the velocity and acceleration of a particle moving in two dimensions, we defined position vectors, vectors based at points other than the origin, vector addition and subtraction, the length of a vector, and scalar multiplication. Parametric equations and polar coordinates were discussed, usually with the help of vector notation. We defined the dot product operation for vectors and discussed how it may be used in calculating angles and projections. We solved inverse problems in several of the sections. We ended the chapter with Newton's three laws and his law of universal gravitation.

We summarize most of the ideas and operations connected with vectors in table form. For a concept such as *position vector* we briefly describe it in words, give a simple geometric interpretation, and describe the associated numerical representation or operation.

Chapter Summary

	Position vector $\mathbf{r}$ of a point (x, y) of the plane; based at the origin and terminating at (x, y). The notation (x, y) is used when the primary interpretation is position vector.	$\mathbf{r} = (x, y)$
	Vector $\overrightarrow{PQ}$ based at P, equivalent to position vector $\mathbf{r}$, where (p_1, p_2) and (q_1, q_2) are the points P and Q of the plane.	$\mathbf{r} = (q_1 - p_1, q_2 - p_2)$
	Sum of position vectors $$\mathbf{a} = (a_1, a_2) \quad \text{and} \quad \mathbf{b} = (b_1, b_2)$$	$\mathbf{a} + \mathbf{b} = (a_1 + b_1, a_2 + b_2)$ Often called the parallelogram law.

	Adding vectors based at the same point. Geometrically, the sum of vectors $\overrightarrow{PQ}$ and $\overrightarrow{PR}$ is the vector $\overrightarrow{PT}$.	Let **a** and **b** be position vectors equivalent to $\overrightarrow{PQ}$ and $\overrightarrow{PR}$, respectively. The vector $\overrightarrow{PT}$ is equivalent to **a** + **b**.
	Adding "tail-to-tip" vectors. Geometrically, the sum of displacements $\overrightarrow{PR}$ and $\overrightarrow{RT}$ is the vector $\overrightarrow{PT}$.	Let **a** and **b** be position vectors equivalent to $\overrightarrow{PR}$ and $\overrightarrow{RT}$, respectively. The vector $\overrightarrow{PT}$ is equivalent to **a** + **b**.
	Multiplication of a vector by a scalar. $$\mathbf{a} = (a_1, a_2)$$ s a real number	$$s\mathbf{a} = s(a_1, a_2) = (sa_1, sa_2)$$
	Subtracting vectors based at the same point. Geometrically, the difference of vectors $\overrightarrow{PR}$ and $\overrightarrow{PQ}$ is the vector $\overrightarrow{QR}$. In terms of displacements, the displacement $\overrightarrow{PQ}$ followed by $\overrightarrow{QR}$ is equal to the single displacement $\overrightarrow{PR}$.	Let **a** and **b** be position vectors equivalent to $\overrightarrow{PR}$ and $\overrightarrow{PQ}$, respectively. The vector $\overrightarrow{RQ}$ is equivalent to **a** − **b**.
	Length	Length may be expressed in terms of the dot product. $$\|\mathbf{a}\| = \sqrt{\mathbf{a} \cdot \mathbf{a}} = \sqrt{a_1^2 + a_2^2}$$
	Angle; perpendicularity	$$\cos\theta = \frac{\mathbf{a} \cdot \mathbf{b}}{\|\mathbf{a}\| \cdot \|\mathbf{b}\|}$$ Nonzero vectors **a** and **b** are perpendicular if and only if $\mathbf{a} \cdot \mathbf{b} = 0$.

	Projection **P** of a vector **F** onto a nonzero vector **a**.	The vector **P** is given by $$\mathbf{P} = (\mathbf{F} \cdot \mathbf{u})\mathbf{u}$$ where **u** is a unit vector in the direction of **a**. The length $\|\mathbf{P}\|$ of the projection of **F** onto **a** is $$|\mathbf{F} \cdot \mathbf{u}| = \|\mathbf{F}\| |\cos \theta|$$ where θ is the angle between **F** and **a**.
	The position, velocity, and acceleration vectors of a moving particle.	$$\mathbf{r} = \mathbf{r}(t), \quad t \geq 0$$ $$\mathbf{v} = \frac{d}{dt}\mathbf{r}$$ $$\mathbf{a} = \frac{d}{dt}\mathbf{v}$$
	A line through a point with position vector $\mathbf{r}_0$ and in the direction of the vector **a**.	$$\mathbf{r} = \mathbf{r}_0 + s\mathbf{a}, \quad -\infty < s < \infty$$
	A curve C and a vector/parametric equation describing it.	$$\mathbf{r} = \mathbf{r}(s), \quad a \leq s \leq b$$
	A polar curve and a polar equation or a vector/parametric equation describing it.	$r = f(\theta), \quad \alpha \leq \theta \leq \beta$ or $$r = (f(\theta)\cos\theta, f(\theta)\sin\theta),$$ $$\alpha \leq \theta \leq \beta$$

CHAPTER REVIEW EXERCISES

1. Give the definition of average velocity, both for objects moving on a line and for objects moving in the plane. An object is moving on the x-axis and its coordinate positions at $t = 36$ s and $t = 39$ s are $x = 35.8$ m and $x = 24.7$ m. What is its average velocity? An object is moving in the (x, y)-plane. At $t = 5.5$ s and $t = 7.2$ s its position vectors are $(-10, 3)$ and $(1, 4)$. What is its average velocity?

2. The position x of a mass attached to a spring is given by

$$x = x(t) = A \sin(Bt + C), \quad t \geq 0.$$

Calculate A, B, and C if the initial position of the mass is $x(0) = -0.5$, its initial velocity is -5 cm/s, and it completes 1.5 oscillations per second.

3. A rifle is fired vertically upward so that at $t = 0$ the bullet is 20 m below ground level and has speed 610 m/s. Calculate the position, velocity, and acceleration of the bullet under the assumption of no air resistance. How long before the bullet returns to ground level? At what speed(s) does it pass ground level on the way up and the way down?

4. Approximately 2.5 hours ago your odometer broke. It read 99,999 miles at the time. From a close analysis of available speedometer data, it appears that the true velocity function is very nearly

$$v(t) = -26t^3 + 117t^2 - 156t + 65, \quad 0 \leq t \leq 2.5.$$

What would the odometer reading be now had it not broke?

5. An object is moving upward and to the right with speed 10 m/s on the line through $(-20.5, -4.8)$ and $(1.5, 12.3)$. If the object was first noticed 15 s ago as it passed through the point $(-20.5, -4.8)$, where is it now?

6. Two barges exert forces

$$\mathbf{F}_1 = (120.5, 100.5)\,\text{kN}$$

$$\text{and} \quad \mathbf{F}_2 = (150.3, -30.1)\,\text{kN}$$

on a docking ring. Calculate the single force on the docking ring that would just balance $\mathbf{F}_1$ and $\mathbf{F}_2$. (A kilonewton (kN) is 1000 newtons (N).)

7. What is the error of closure in the following survey? From the origin walk 1200 m along the line with $0°$ bearing; then walk 1900 m on a bearing of $150°$. From this point walk 1050 m to the origin on a bearing of $295°$.

8. Calculate the coordinates of a unit vector whose direction is 0.5 radians more than that of the vector $(6, 1)$.

9. An object at point P is displaced by the vector $\mathbf{a}_1 = (3, 4)$ and then further displaced by $(-10, 7)$, $(2, -8)$, and $(20, 0)$. What is the single equivalent displacement of P? In what direction and distance from P is the object moved by the four displacements?

10. Sketch the curve described in polar coordinates by $r = 12/(3 - 4 \sin \theta)$.

11. Calculate the velocity and speed of an object whose position vector at any time t is

$$\mathbf{r} = \mathbf{r}(t) = (2 \sin t, 3 \cos t), \quad 0 \leq t \leq 2\pi.$$

Sketch the path followed by the object, indicating the direction of travel on the path.

12. Calculate the angle at which the curves with equations

$$r = \sin 2\theta \quad \text{and} \quad r = \cos \theta$$

meet at the point $(\sqrt{3}/2, \pi/6)$.

13. Both Halley's comet and the earth move on ellipses with the sun at one focus. If we assume these ellipses lie in a common plane, their equations would be

$$r = \frac{1.16}{1 - 0.97 \cos \theta} \quad \text{and} \quad r = \frac{0.997}{1 - 0.017 \cos \theta}.$$

Determine the points where the orbits intersect. Do the comet and earth collide? Could they collide?

14. Let C be the curve described by

$$\mathbf{r} = \mathbf{r}(t) = (\cosh t, \sinh t), \quad t \geq 0.$$

Sketch C. Note that $\cosh^2 t - \sinh^2 t = 1$ for all t.

15. Graph the polar curve C described by $r = 1 + \cos \theta$ and show that the set of points with rectangular coordinates satisfying $x^2 + y^2 = \sqrt{x^2 + y^2} + x$ is the same as C.

16. Locate the point of intersection of the graphs described by $r = \theta$ and $r = \cos \theta$, where $0 \leq \theta \leq \pi/2$.

17. Show that the graph of the polar curve with equation $r = 2/(1 + \cos \theta)$ not only looks like a parabola but is in fact a parabola.

18. Sketch the graph of the curve C described by

$$\mathbf{r} = \mathbf{r}(t) = (1 + \cos t, -1 + \sin t), \quad 0 \leq t \leq 2\pi.$$

Find the coordinates of all points on C at which the slope is $2/3$.

19. Calculate the slope and give the equation of the tangent line to the graph of $r = \sqrt{\theta}, \theta \geq 0$, at the point with polar coordinates $(1, 1)$.

20. Object 1 moves on the path described by $\mathbf{r} = (t, t^2)$, $t \geq 0$, while object 2 moves on the path described by $\mathbf{r} = (2t, 2t + 2)$, $t \geq 0$. Do the paths intersect? Do the objects collide if the variable t in each equation is read from the same clock?

21. Determine two vectors based at $(1, 1)$ and tangent to the graph of $y = x^3$.

22. Show that $\mathbf{a} + \mathbf{b}$ is perpendicular to $\mathbf{a} - \mathbf{b}$ if $\mathbf{a}$ and $\mathbf{b}$ are unit vectors.

23. Calculate the angle between the lines with equations

$$\mathbf{r} = (2, 1) + t(3, 4) \quad \text{and} \quad \mathbf{r} = (-1, 5) + t(-3, 1).$$

24. The position vector of an object is

$$\mathbf{r}(t) = (\cos 2t, \sin 2t), \quad t \geq 0.$$

Calculate its acceleration at $t = 3.1$ s.

25. Solve the inverse problem for an object whose acceleration is $\mathbf{a} = (e^t, e^{-t})$, $t \geq 0$, and for which $\mathbf{r}(0) = (0, 0)$ and $\mathbf{v}(0) = (1, 0)$.

26. A 1.2 kg object initially at rest at the origin is acted on by a force $\mathbf{F} = (2.4, 1.7)$ N. What is the object's acceleration? Where is the object, and how fast is it moving 3.5 s after the force is first applied?

27. The mass of an electron is 9.11×10^{-31} kg. Calculate the gravitational force exerted by one electron on another if they are 1 mm apart.

28. Write a vector equation for the circle with center at $(-2, 3)$, radius 5, and traversed in the counterclockwise direction.

29. Calculate the slope of the curve described by

$$\mathbf{r}(t) = (t^3 + 1, t^2 + t + 1), \quad t \geq 0$$

at the point $\mathbf{r}(2)$. Give a vector equation of the tangent line to C at this point.

30. Let $\mathbf{e}_1 = (3, 1)$, $\mathbf{e}_2 = (1, 2)$, and $\mathbf{v} = (5, 7)$. Express $\mathbf{v}$ as a sum of vectors $\mathbf{v}_1$ and $\mathbf{v}_2$, where $\mathbf{v}_1$ and $\mathbf{v}_2$ are parallel to $\mathbf{e}_1$ and $\mathbf{e}_2$, respectively.

31. For what value of t are the vectors $\mathbf{a} = (3, 1)$ and $\mathbf{b} = (-2, t)$ parallel? Perpendicular?

32. Use vectors to show that any angle inscribed in a semicircle is a right angle.

33. Use the vector identity

$$\|\mathbf{a} + \mathbf{b}\|^2 + \|\mathbf{a} - \mathbf{b}\|^2 = 2\|\mathbf{a}\|^2 + 2\|\mathbf{b}\|^2$$

in calculating $\|\mathbf{a}+\mathbf{b}\|$ and $\|\mathbf{a}-\mathbf{b}\|$, given that $\|\mathbf{a}\| = 5$, $\|\mathbf{b}\| = 8$, and the angle between $\mathbf{a}$ and $\mathbf{b}$ is $2\pi/3$.

34. Sketch the polar curve with equation $r = \cot\theta$. Calculate its slope at the point with polar coordinates $(1, \pi/4)$.

35. Sketch the graph of the curve C described by the parametric equation

$$\mathbf{r} = \mathbf{r}(t) = (t - t^2, t + 2t^2), \quad -3 \leq t \leq 3.$$

Eliminate the parameter to obtain an equation in x and y. Show that the curve described by this equation includes the curve C. Calculate the slope of C at the point $(-2, 10)$.

36. Determine a parametric equation for the cycloid curve C generated by a circle of radius 3 rolling on a line. Sketch C. Calculate the slope of C at the point corresponding to $s = \pi/4$.

37. Let $\mathbf{r} = \mathbf{r}_0 + \mathbf{a}$ describe a line L, and let $\mathbf{q}$ be the position vector of a point Q not on L. Show that the distance d from L to Q can be calculated as follows. Let $\mathbf{p}$ be the vector projection of $\mathbf{q} - \mathbf{r}_0$ onto $\mathbf{a}$ and $\mathbf{n} = \mathbf{q} - \mathbf{r}_0 - \mathbf{p}$. Then $d = \|\mathbf{n}\|$. Use this procedure to calculate the distance from the line through $(1, 5)$ and $(7, 2)$ to the point $(-5, -3)$.

Student Project

Timing a Rifle Bullet

A method for calculating the time taken in seconds for a rifle ball to travel from muzzle to target was published in the July 1893 issue of *Scientific American.* This was reported in the "50 and 100 Years Ago" column of the July 1993 issue.

> It may be of interest to amateur riflemen to know the following simple method for ascertaining the effect of gravity upon a bullet: Sight the rifle upon the target, keeping the sights plumb above the center line of the bore of the rifle. Mark where the ball strikes. Then reverse the rifle, so as to have the sights exactly beneath the line of the bore. In this reversed position sight it on the target as before, and mark where the bullet strikes. Divide the difference in elevation of the two bullet marks by 32 and extract the square root. This will give the time in seconds that it took the ball to travel the distance. The distance divided by the time will give the speed of the bullet per second. —J.A.G., Grand Rapids, Michigan.

Problem 1.   Verify J.A.G.'s assertions and conclusions, and comment on the assumptions you and J.A.G. made. You may wish to use the hints contained in Fig. 3.72.

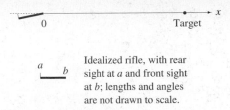

Figure 3.72. Diagram for Problem 1.

Problem 2. The word *extract* used by J.A.G. suggests it was somewhat more difficult to calculate a square root in 1893 than in 1993. Write a short paragraph discussing and contrasting the techniques used by typical students in 1893 and 1993 to calculate, say, $\sqrt{5.73}$.

Student Project

The Quarterback's Problem

Referring to Fig. 3.73, a quarterback and receiver are at points QB and R on a level playing field. The point R is 15 yards downfield from QB and 10 yards to one side. According to plan A, the receiver will run along the line L at 6 yards per second and receive a pass from the quarterback. The quarterback must pass within 5 seconds after the receiver starts running. The dotted line in the figure is the projection onto the field of the trajectory of the ball in a successful pass. For a pass to be successful, we assume that the receiver and ball reach some point P at the same time, that the football leaves the quarterback and arrives at the receiver at the same height, and that the quarterback throws at a speed of 25 yards per second. We use yards and seconds as units.

Problem 1. Assume that the quarterback passes T seconds after the receiver starts running. For each of $T = 1, 3, 5$ s, determine the quarterback's choices for initial velocity vector. Use $\mathbf{g} = (0, 0, -32/3)$ as the acceleration of gravity, and neglect air resistance.

Problem 2. Letting ℓ denote the distance in yards by which the quarterback must "lead" the receiver and θ the angle of elevation (in degrees) of the initial velocity, calculate ℓ and θ for each of the values $T = 1, 3, 5$ s.

Problem 3. On the basis of your calculations, what is your advice to the quarterback? Why?

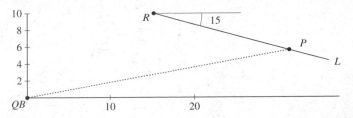

Figure 3.73. Plan A.

A History of Measuring and Calculating Tools
The Sextant

The sextant is used by navigators to measure the angle between the sun or other star and the horizon. The name comes from the Latin *sextans,* meaning "the sixth part." In 1700 Newton designed an instrument based on the same ideas as the sextant, but his invention was not widely known. John Hadley, an Englishman, and Thomas Godrey, an American, invented the sextant independently in 1730.

The sextant shown in Fig. 3.74 has a rigid framework $VWXYZV$ to which are fixed a horizon glass at W and a telescope at T. Pivoting at V is a movable index arm VP with index mirror at V and pointer at P. A scale is scribed on the fixed arc XYZ, running from $0°$ to $120°$. The horizon glass is parallel to VZ. The upper part of the horizon glass is clear glass; the lower part is silvered.

To measure the angle of star S above the horizon H, the navigator sights the horizon through the telescope, so that the horizon line is at the boundary of the clear and silvered halves of the horizon glass. The navigator then moves the index arm so that S is seen through the telescope, at the boundary of the clear and silvered halves of horizon glass. The light from the star reflects at V and then again at W.

We show that $\gamma = \frac{1}{2}\theta$, where γ is the angle between the two mirrors and θ is the angle of the North Star above the horizon. Note that

(1) $\qquad \alpha + \beta + \theta = 90°$

and, since $\angle XVP = \alpha$ (starlight approaches and leaves the index mirror at equal angles),

(2) $\qquad \alpha - \beta = 30°.$

From triangle VUT,

(3) $\quad \gamma + (\beta + 90°) + 60° = 180°.$

By subtracting equations (1) and (2) we obtain $\beta = 30° - \frac{1}{2}\theta$. If we then substitute this result into (3), we obtain $\gamma = \frac{1}{2}\theta$.

The scale on arc XYZ is chosen to obtain a direct reading of θ.

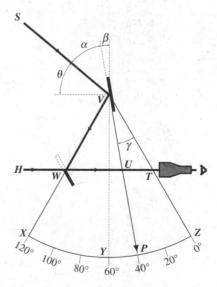

Figure 3.74. Simplified sextant.

Chapter 4

Using the Derivative

Problem Solving with the Derivative

In Chapters 1–3 you became familiar with rates of change of scalar- and vector-valued functions and learned how to differentiate the elementary functions. You used vectors and parametric equations to describe motion.

In this chapter we use the derivatives of scalar- and vector-valued functions to solve several kinds of problems. Some examples are:

- How can we calculate the future positions $\mathbf{r}(t)$ of an object if we know only its present position $\mathbf{r}(t_0)$ and velocity $\mathbf{r}'(t)$ for all $t \geq 0$?

- The value of x for which $x = \cos x$ can be approximated by entering a ball-park guess into your calculator and then pressing the cosine key several times. If you try $x = 0.7$, you should get $0.735\ldots$ after six repetitions (in radian mode). Why does this work, and is there a faster numerical algorithm?

- How can we decide whether a function is increasing or if its graph "holds water"?

- What is meant by $\lim_{x \to a+} f(x) = \infty$ or $\lim_{x \to \infty} f(x) = L$?

- If oil is spilling just offshore at a given rate, how fast is the edge of the slick approaching the beach?

- What is the optimum size of a silicon integrated circuit chip?

4.1 TANGENT LINE APPROXIMATIONS

The angle θ measuring the position of the simple pendulum of mass m shown in Fig. 4.1 is a function of the time t since the pendulum was released. Ignoring air resistance and friction in the axle at the top, it is not difficult to show that θ satisfies

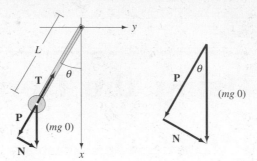

Figure 4.1. Simple pendulum.

the equation

(1)
$$L\frac{d^2\theta}{dt^2} + g\sin\theta = 0.$$

An outline of an argument is given in Exercise 37.

It is not easy to determine a function $\theta = \theta(t)$ satisfying (1) (other than $\theta(t) = 0$ for all t, which describes a pendulum at rest). However, if we assume that the pendulum does not oscillate too far from its downward equilibrium position, so that $|\theta|$ remains small, equation (1) can be (and often is) replaced by the simpler equation

(2)
$$L\frac{d^2\theta}{dt^2} + g\theta = 0.$$

The replacement is based on the tangent line approximation $\sin\theta \approx \theta$.

We show in Fig. 4.2 the graphs of the two functions $\sin\theta$ and θ, $0 \le \theta \le \pi/2 \approx 1.5$. The lower figure shows a zoom of the two graphs to a 0.1×0.1 window centered at $(0.25, 0.25)$, to give an impression of how well θ approximates $\sin\theta$ as $\theta \to 0$. Use your calculator, set in radian mode, to inspect the graphs of these two functions in a 0.1×0.1 zoom window centered at $(0.15, 0.15)$. Are the curves closer? Can you distinguish them? Is the sine curve on the top or bottom?

Evidently, the tangent line approximation θ to $\sin\theta$ is a very good approximation near the origin. It is easy to give numerical evidence as well. If you set your calculator to display 10 decimals, you should be able to verify the data in Table 1.

Why does this approximation work? The tangent line approximation is based on the idea that as we zoom in on a point of the graph of a differentiable function, the graph looks more and more like its tangent line at that point. The approximation $\sin x \approx x$ may be understood by looking at the graph of the sine function at $(0, 0)$. Since the slope of the tangent line through this point is $\cos 0 = 1$, the equation of the tangent line is $y - 0 = 1(x - 0)$, that is, $y = x$. The approximation $\sin x \approx x$ reflects the fact that near $(0, 0)$ the graphs of $\sin x$ and x are very nearly coincident.

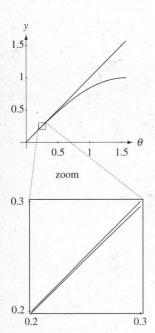

Figure 4.2. Zooming in on graphs of $\sin\theta$ and θ.

Table 1. Tangent line approximation

θ	$\sin\theta$
0.1	0.09983 34166
0.01	0.00999 98333
0.001	0.00099 99998
0.0001	0.00010 00000

Tangent Line Approximation and the Differential

Suppose that we are given a function f and a base point a where f is differentiable. See Fig. 4.3. As x increases from a to $a + h$, f increases from $f(a)$ to $f(a + h)$. This change in f is a function of h and is denoted by $\Delta f(h)$. The Greek letter "delta"

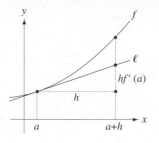

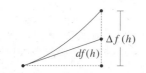

Figure 4.3. Tangent line approximation.

is used since the change in f is a difference, that is, $\Delta f(h) = f(a + h) - f(a)$. We wish to find a simple function df for which

$$\Delta f(h) \approx df(h), \quad \text{for small } h.$$

We show in Fig. 4.3 the tangent line ℓ to the graph of f at $(a, f(a))$. The tangent line is the key to determining the simple function df. If from $(a, f(a))$ we **run** h, the **rise** of the tangent line would be $hf'(a)$ since the slope of the tangent line at $(a, f(a))$ is $f'(a)$. The number $hf'(a)$, which we denote by $df(h)$, is very nearly equal to $\Delta f(h)$ for small values of h. This defines the **differential** df of f at a. Summarizing,

(3) $$\Delta f(h) \approx df(h) = f'(a)h.$$

By replacing $\Delta f(h)$ by $f(a + h) - f(a)$ and transposing the term $f(a)$ to the other side, we may rewrite (3) in the convenient form

(4) $$f(a + h) \approx f(a) + hf'(a).$$

So far, our approach to the tangent line approximation and the differential of f at a has been geometric, depending heavily upon Fig. 4.3. It is useful to approach the differential in a second way, through the definition of the derivative of f at a. Recall that

(5) $$f'(a) = \lim_{h \to 0} \frac{f(a + h) - f(a)}{(a + h) - a} = \lim_{h \to 0} \frac{f(a + h) - f(a)}{h}.$$

Let $E(h)$ denote the difference between the slope $f'(a)$ of the tangent line to the graph of f at $(a, f(a))$ and the slope of the "secant line" through $(a, f(a))$ and $(a + h, f(a + h))$. We use the letter E to suggest the error that would be made if we were to use the slope of the secant line to approximate the slope of the tangent line. The error would be

(6) $$E(h) = \frac{f(a + h) - f(a)}{h} - f'(a).$$

It follows from (5) that $\lim_{h \to 0} E(h) = 0$.

We solve (6) for $f(a + h)$ to obtain

(7) $$f(a + h) = f(a) + hf'(a) + hE(h).$$

This equation has the form of (4) with a small correction term. It shows that $df(h) = hf'(a)$ is a good approximation to $\Delta f(h) = f(a + h) - f(a)$. The differential $df(h)$ is a good approximation to $\Delta f(h)$ since their difference $\Delta f(h) - df(h) = hE(h)$ is small; not only is h small, but it is multiplied by $E(h)$, which goes to zero with h.

DEFINITION   **differential and tangent line approximation**

Let f be a function defined on an interval I and assume that f is differentiable at a point $a \in I$. The **differential** of f at a is the function df

defined for all real numbers h by

(8) $$df(h) = f'(a)h.$$

The **tangent line approximation** to $f(a + h)$ at the base point a is

(9) $$f(a + h) \approx f(a) + df(h) = f(a) + hf'(a),$$

defined for all h for which $a + h \in I$.

For our first example we show that $\sin x \approx x$ for small x. This is what we needed to simplify the equation of motion of the pendulum. We return to the simplified pendulum equation (2) in Example 3.

◼ **EXAMPLE 1**   Use the tangent line approximation to replace the sine function on $[-0.5, 0.5]$ by a simpler function. Discuss the "worst-case scenario" graphically.

Solution.   Given a function f, the tangent line approximation is based at a point a at which we know two things: $f(a)$ and $f'(a)$. Since we wish to approximate the sine function on the interval $[-0.5, 0.5]$ and we know $\sin 0$ and $\sin'(0) = \cos 0$, it makes sense to take $a = 0$. From (9) we have

(10) $$\sin(0 + h) \approx \sin(0) + h\cos(0) = h.$$

Replacing h by x we have $\sin x \approx x$.

The "worst-case scenario" happens at the point $x \in [-0.5, 0.5]$, where the difference is a maximum. We graphed the difference in the window $0 \le x \le 0.5$ and $0 \le y \le 0.03$, obtaining Fig. 4.4. It is clear that the worst error occurs for $x = 0.5$ (and $x = -0.5$; the graph looks the same for $x \le 0$). Specifically, $|\sin 0.5 - 0.5| = 0.02105\ldots$.

To summarize, if for $x \in [-0.5, 0.5]$ we replace $\sin x$ by the simpler function x, we would make an error of less than 0.022.   ◼

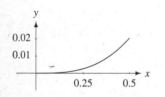

Figure 4.4. Worst case happens at $x = 0.5$.

◼ **EXAMPLE 2**   How long does it take to double your money if you invest it at 6 percent compounded once a year? The "rule of 72" is sometimes used by investors to approximate the doubling time. To apply the rule, divide 72 by the interest rate. For a 6 percent rate, the doubling time would be $72/6 = 12$ years. Why does the rule of 72 work?

Solution.   If you invest P dollars at the beginning of a year at r percent, then at the end of the first year you will have $P + Pr/100 = P(1 + r/100)$ dollars. You have added to your principal P the interest $Pr/100$ earned during the year. At the end of the second year you will have

$$P(1 + r/100) + P(1 + r/100)r/100 = P(1 + r/100)^2 \text{ dollars.}$$

After n years your money will have grown to $P_n = P(1 + r/100)^n$.

We want to choose n so that $P_n = 2P$. This condition gives

$$2P = P(1 + r/100)^n \quad \text{or} \quad 2 = (1 + r/100)^n.$$

We may solve this equation for n. Taking the natural logarithm of both sides and solving for n gives

$$\ln 2 = n \ln(1 + r/100)$$

(11)
$$n = \frac{\ln 2}{\ln(1 + r/100)}.$$

Letting $r = 6$, we find

$$n = \frac{\ln 2}{\ln 1.06} = \frac{0.6931\ldots}{0.0582\ldots} = 11.89.\ldots$$

The exact doubling time is $11.89\ldots$ years. The rule of 72 gave 12, with quite a bit less work.

We use the tangent line approximation to justify the rule of 72. From (11), note that for normal interest rates the number $r/100$ is small. Possibly, then, we may use the tangent line approximation to replace the function $f(x) = \ln(1 + x)$ by a simpler function. We consider taking $a = 0$ since it is near the values of interest and we can calculate $f(a) = \ln(1 + a)$ and $f'(a) = 1/(1 + a)$ easily. From (9) we have

(12)
$$\ln(1 + a + h) = \ln(1 + h) \approx \ln(1 + 0) + h = h.$$

Replacing h by x we have $\ln(1 + x) \approx x$.

Using this result in (11), we have

$$n = \frac{\ln 2}{\ln(1 + r/100)} \approx \frac{0.69}{r/100} \approx \frac{69}{r}.$$

Taking $r = 6$ gives $n \approx 13$. This is not quite as good as the 12 given by the rule of 72. Although the tangent line approximation led us to a rule of 69, it gives some insight as to why the rule of 72 works. The rule of 69 is better than the rule of 72 for $r \le 4$ percent. We discuss some advantages of the rule of 72 in Exercise 29. ■

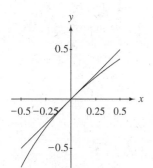

Figure 4.5. Graphs of x and $\ln(1 + x)$, near $x = 0$.

Other Applications of the Tangent Line Approximation

We have used the tangent line approximation to replace relatively complex functions by simpler ones. For a pendulum whose oscillations are not far from the downward equilibrium position, replacing $\sin \theta$ by θ makes possible a simple description of its motion. A second application occurred in discussing the rule of 72. We approximated the function $\ln(1 + x)$ by x. Other approximations are included in the exercises.

We may also use the tangent line approximation to simplify calculations of absolute or relative error. In the next example we calculate the precision required in adjusting the length of a pendulum to achieve, say, an accuracy of 0.001 s in the period of its swing.

In the next section we show that

(13)
$$\theta = \theta_0 \cos(\omega t), \quad \text{where} \quad \omega = \sqrt{g/L},$$

is the solution to the simplified pendulum equation

(14)
$$L\frac{d^2\theta}{dt^2} + g\theta = 0.$$

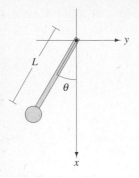

Figure 4.6. Simple pendulum.

The pendulum is displaced to $\theta = \theta_0$ and is released from rest. The period T of the trigonometric function in (13) is 2π divided by the coefficient of t. This gives equation (15) in the following example.

■ **EXAMPLE 3** The period T of the pendulum of a grandfather clock is related to its length L by the equation

$$(15) \qquad T = 2\pi \sqrt{\frac{L}{g}} = k\sqrt{L}, \quad \text{where} \quad k = \frac{2\pi}{\sqrt{g}}.$$

How accurately must we "set" L so that the period of the clock is 2 ± 0.001 s, that is, within 0.001 seconds of 2 seconds?

Solution. We must decide how accurately we must "set" L so that the period of the clock is 2 ± 0.001 s. If we let L_2 denote the length for which the period is exactly 2 s, we must decide how large h can be and yet

$$(16) \qquad |T(L_2 + h) - T(L_2)| < 0.001.$$

The variable h measures the error with which we measure the base point L_2. The quantity $|T(L_2 + h) - T(L_2)|$ is the absolute value of the difference between the true period $T(L_2)$ and the period $T(L_2 + h)$ of a pendulum set with an error of h in its length. The tangent line approximation with base point L_2 is suggested by the difference $T(L_2 + h) - T(L_2)$ appearing in equation (16).
  We have

$$(17) \qquad |T(L_2 + h) - T(L_2)| \approx |T'(L_2)h| = k\frac{1}{2\sqrt{L_2}}|h|.$$

To find the largest allowable error $|h|$ in measuring L, we require that h satisfy

$$(18) \qquad |T(L_2 + h) - T(L_2)| \approx k\frac{1}{2\sqrt{L_2}}|h| < 0.001.$$

Recalling that L_2 is the length that gives $T = 2$, we have

$$(19) \qquad 2 = 2\pi\sqrt{\frac{L_2}{g}}.$$

Hence, $\sqrt{L_2} = \sqrt{g}/\pi \approx 0.996$. (We took $g = 9.8$ m/s^2 and measure L in meters and T in seconds.) Solving the inequality in (18) for $|h|$ and substituting for L_2,

$$(20) \qquad |h| \lesssim \frac{0.001}{k}2\sqrt{L_2} \approx 0.001 \text{ m}.$$

It follows that we must set L to within a millimeter to achieve an accuracy of 0.001 s in the period. Although an accuracy of 0.001 s in the period may appear a bit extreme, note that over a day, which contains $(24 \cdot 3600)/2$ pendulum periods, the clock could be off by approximately 43 seconds. ■

We work through one example featuring the important idea of percentage error. Although the results are shaped by the example we have chosen, the ideas apply more broadly.

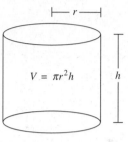

Figure 4.7. Right circular cylinder.

■ **EXAMPLE 4**   To calculate the volumes V of cylinders of fixed height h and variable radius r, we must measure their radii. It is known that in measuring r a percentage error of up to 2 percent can be made. This means that

(21) $$\left| \frac{(r + \Delta r) - r}{r} \right| \times 100 = \left| \frac{\Delta r}{r} \right| \times 100 \le 2,$$

where $r + \Delta r$ and r are measured radius and true radius. The quantity Δr, which may be positive or negative, is the measurement error. Approximate the worst possible percentage error in the volume.

Solution.   We use the tangent line approximation on the function

$$V = V(r) = \pi r^2 h,$$

taking the base point to be r. We regard h as a constant. The percentage error E in the volume is

$$E = \left| \frac{V(r + \Delta r) - V(r)}{V(r)} \times 100 \right|.$$

Using the tangent line approximation $V(r + \Delta r) - V(r) \approx V'(r)\Delta r$, we have

$$E \approx \left| \frac{V'(r)\Delta r}{V(r)} \times 100 \right|.$$

Since $V'(r) = 2\pi r h$, we have

$$E \approx \left| \frac{2\pi r h \Delta r}{\pi r^2 h} \times 100 \right| = 2\left| \frac{\Delta r}{r} \times 100 \right|.$$

From (21) we have

$$E \approx 2\left| \frac{\Delta r}{r} \times 100 \right| \le 4.$$

Although the tangent line approximation is itself subject to error, it appears that the maximum percentage error in the volume is close to 4%.   ■

Tangent Line Approximation for Vector-Valued Functions

The tangent line approximation for vector-valued functions has the same form as for real-valued functions. Letting f be a real-valued function and $\mathbf{r}$ a vector-valued function, we have

$$f(a + h) \approx f(a) + f'(a)h$$
$$\mathbf{r}(a + h) \approx \mathbf{r}(a) + h\mathbf{r}'(a).$$

As is very often the case for vector-valued functions, this result follows by applying the tangent line approximation to the two coordinate functions. Suppose we have a differentiable function $\mathbf{r}(t) = \big(x(t), y(t)\big)$ and wish to approximate $\mathbf{r}(t + h)$. Taking the base point as t and applying (9) to the coordinate functions x and y, we have

$$\mathbf{r}(t + h) = \big(x(t + h), y(t + h)\big)$$
$$\mathbf{r}(t + h) \approx \big(x(t) + hx'(t), y(t) + hy'(t)\big)$$
$$\mathbf{r}(t + h) \approx \big(x(t), y(t)\big) + h\big(x'(t), y'(t)\big)$$
$$(22) \qquad \mathbf{r}(t + h) \approx \mathbf{r}(t) + h\mathbf{r}'(t)$$

The last equation has the same form as (9).

■ **EXAMPLE 5**   An object was seen as it passed through the point $(14.5, 11.0)$ m. At that time its velocity was measured as $(-3.7, -9.0)$ m/s. A possible path of the particle is shown in Fig. 4.8, but all we know is the position and velocity of the object at one time. Use the tangent line approximation in estimating the position of the object 0.5 seconds after it was seen.

Solution.   Let $\mathbf{r} = \mathbf{r}(t) = \big(x(t), y(t)\big)$ be the position vector of the particle at any time t, and suppose it was seen at $t = t_0$. We wish to approximate $\mathbf{r}(t_0 + 0.5)$. From (22) we have

$$\mathbf{r}(t_0 + 0.5) \approx \mathbf{r}(t_0) + h\mathbf{r}'(t_0)$$
$$\mathbf{r}(t_0 + 0.5) \approx \big(14.5, 11.0\big) + 0.5\mathbf{v}(t_0)$$
$$\mathbf{r}(t_0 + 0.5) \approx \big(14.5, 11.0\big) + 0.5\big(-3.7, -9.0\big) = \big(12.65, 6.5\big).$$

Using the tangent line approximation and the observed position $\mathbf{r}(t_0)$ and velocity $\mathbf{v}(t_0)$ at t_0, we predict that its postion 0.5 seconds later will be $\big(12.65, 6.5\big)$. ■

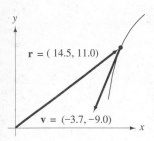

$\mathbf{r} = (14.5, 11.0)$

$\mathbf{v} = (-3.7, -9.0)$

Figure 4.8. Future position?

Exercises 4.1

Basic

1. Use your calculator to make a figure similar to Fig. 4.2, but use the tangent function on $[0, 1]$ instead of sine. Do both parts of the figure.

2. Use your calculator to make a figure similar to Fig. 4.5. Also, zoom in the 0.1×0.1 window centered at $(0.05, 0.05)$.

3. Using base point $x = 0$, is the tangent line approximation to $\sin 0.1$ generally better or worse than the tangent line approximation to $\cos 0.1$? Base your answer on both numerical and graphical data.

4. Using base point $x = \pi/2$, is the tangent line approximation to $\sin 1.4$ generally better or worse than

the tangent line approximation to $\cos 1.4$? Base your answer on both numerical and graphical data.

5. What is the tangent line approximation to e^x near $x = 0$? If we were to replace e^x by this tangent line approximation on the interval $[-0.5, 1.0]$, how big does the difference between e^x and this function become? Your answer may be based on graphical evidence.

6. What is the tangent line approximation to 2^{-x} near $x = 1$? If we were to replace 2^{-x} by this tangent line approximation on the interval $[0, 2]$, how big does the difference between 2^{-x} and this function

become? Your answer may be based on graphical evidence.

7. Determine a simple function g that approximates the square root function in the interval $[3, 6]$. How large does $|g(x) - \sqrt{x}|$ become for $x \in [3, 6]$? Give both numerical and graphical evidence of your error analysis.

8. What is the doubling time if you invest your money at 8 percent compounded once a year? How does this compare with the rule of 72 result?

9. Use the tangent line approximation in deciding how accurately we must set L if we wish the period of a pendulum clock to be within 0.001 seconds of 1 second.

10. Use the tangent line approximation in deciding how accurately we must set L if we wish the period of a pendulum clock to be within 0.001 seconds of 1.5 seconds.

11. An object with velocity $(-9.8, 3.5)$ m/s is observed passing through $(-1.5, 2.4)$. Use the tangent line approximation to calculate its position 2 seconds later.

12. An object with velocity $(5.8, -4.5)$ m/s is observed passing through $(-3.5, -4.4)$. Use the tangent line approximation to calculate its position 2 seconds later.

13. Assuming the object discussed in Example 5 does not increase its speed, how far from the predicted position $(12.65, 6.5)$ could the object be?

14. An object is moving so that at time t its position vector is $\mathbf{r}(t) = (t^2, t^3)$. How far apart are the actual position of this object at $t = 1.1$ and its position $\mathbf{r}(1 + 0.1)$ as predicted by the tangent line approximation?

15. An object is moving on a wire in the shape of the spiral $r = \theta$, so that for $0 \le t \le 1$ its position vector is $\mathbf{r}(t) = (t \cos t, t \sin t)$. Suppose that the wire ends at $\mathbf{r}(1)$. Where will the object be 2 seconds later? Assume that the plane of the spiral is horizontal and the object moves without friction; that is, the net force on the object is $\mathbf{0}$.

Exercises 16–23: Calculate the differential df of the function f at the base point a. Use h as the variable.

16. $f(x) = (2x + 7)^2, a = 3$

17. $f(x) = (x^2 - 1)^{-1}, a = 2$

18. $f(x) = e^x, a = 1$

19. $f(x) = e^{-3x}, a = 1$

20. $f(x) = x/(x^2 + 1), a = -1$

21. $f(x) = \sqrt{x}/(2x - 3), a = 4$

22. $f(x) = \arcsin(\sqrt{x}), a = 1/4$

23. $f(x) = \ln \sqrt{x^2 + 1}, a = 1$

Growth

24. In showing how well x approximates $\sin x$ near 0, it is convenient to graph the difference $x - \sin x$ rather than the graphs of both functions. Graph the difference on the interval $[0, 0.01]$, choosing an appropriate vertical scale.

25. Sketch a figure like that in Fig. 4.3, but draw a graph that lies below all of its tangent lines. (Note, for example, that e^x lies above its tangent lines while its inverse function $\ln x$ lies below its tangent lines.) On your sketch identify $a, a + h, h, (a, f(a))$, $(a + h, f(a + h))$, and $df(h)$.

26. We may use the relation (15) connecting the length L and period T of a pendulum to measure the acceleration of gravity. With a pendulum of a fixed length $L = 1.0$ m, how accurately would the period T need to be measured so that g is within 0.05 of 9.8?

27. Imagine the (x, y)-plane as horizontal. If at $t = 0$ the position and velocity of a grenade are $(21, 13)$ and $(-10.5, -8.1)$ m/s, in which direction should a person at the origin run if the grenade will explode at $t = 1.5$ s?

28. Determine a polynomial $mt + b$ that approximates $f(t) = 4.5e^{-0.32t} \cos(1.7t)$ near $t = 1$. On the interval $[0.8, 1.2]$ where is the approximation best? Worst?

29. For each value of r, from 2 percent to 15 percent, calculate the exact doubling time if the principal is compounded once a year. Also, list with this value of r the approximations $72/r$ and $\ln 2/r$ to the doubling time. Underline the approximation that is closer to the true value. Determine which of the two approximations might be preferred, considering both simplicity and the fact that one of them is larger than the other, and, for relatively long doubling times, tends to overestimate the time required.

30. Let f and g be functions defined for all x and let $w = f \circ g$ be their composition, so that $w(x) = f(g(x))$. Show that $dw(h) = f'(g(x))dg(h)$. If $g(3) = 5$, $g'(3) = 7$, $f'(5) = 9$, and $h = 0.1$, calculate $dw(0.1)$ at the base point $a = 3$.

31. Let $f(x) = x^2$. Let df be the differential of f at $x = 2$. Graph df as a function of h. What relation has this graph to that of the tangent line to f at $(2, 4)$?

32. In approximating $f(a + h)$ by $f(a) + hf'(a)$, the absolute error is the absolute value of the difference of these two quantities. Express the absolute error in terms of $\Delta f(h)$ and $df(h)$. On a figure similar to Fig. 4.3, locate and label the absolute error.

33. A coat of paint 0.04 cm thick is applied to a spherical water tower of diameter 30 m. Approximate the amount of paint required for the job. (There are 264.17 gallons in 1 m^3.)

34. Assume that a new quarter dollar coin has diameter exactly 2.4 cm and that the moon moves in a circular orbit with radius 3.85×10^8 m. At 2.65 m from the eye, the quarter covers the moon exactly. What is the radius of the moon? What percentage error can be made in determining the distance between eye and coin if the moon's radius is to be determined within 5 percent of its true value?

35. The costs of some commercial products drop by a constant percentage each time the cumulative volume of the product doubles. Let V be the cumulative volume and $C = C(V)$ its cost as a function of V. Use the tangent line approximation in arguing that

$$\frac{dC}{dV} \approx -m\frac{C}{V}$$

where m is a constant. (Hint: Consider $C(2V) - C(V)$.)

36. Show that the percentage error in calculating the period of a simple pendulum is half the percentage error made in measuring its length.

37. Referring to Fig. 4.1, let $\mathbf{P}$ be the projection of the force $(mg, 0)$ in the direction of the unit vector $(\cos\theta, \sin\theta)$ and $\mathbf{N} = (mg, 0) - \mathbf{P}$. Show that

$$\mathbf{N} = mg(\sin^2\theta, -\sin\theta\cos\theta).$$

Since $\mathbf{P}$ just balances the tension $\mathbf{T}$ on the bob, the net force $\mathbf{F}$ on the pendulum bob is $\mathbf{N}$. Since the bob swings in a circle of radius L, its position vector at any time t must be

$$\mathbf{r}(t) = L(\cos\theta, \sin\theta)$$

where $\theta = \theta(t)$. We wish to determine this unknown function. To apply Newton's second law $\mathbf{F} = m\mathbf{a}$, calculate the acceleration $\mathbf{a}$ from the above expression for $\mathbf{r}$. Specifically, show that

$$\mathbf{a} = L\theta''(-\sin\theta, \cos\theta) + L\theta'^2(-\cos\theta, -\sin\theta).$$

By equating the coordinates of both sides of $\mathbf{F} = m\mathbf{a}$, show that

$$g\sin^2\theta = -L\theta''\sin\theta - L\theta'^2\cos\theta$$

$$-g\sin\theta\cos\theta = L\theta''\cos\theta - L\theta'^2\sin\theta.$$

Multiply the first of these two equations by $\sin\theta$ and the second by $-\cos\theta$, add the results, and simplify to obtain

$$L\theta'' + g\sin\theta = 0.$$

Review/Preview

38. By reviewing your knowledge of the derivatives of the elementary functions, determine a function, $f(t)$ satisfying the equation $f'(t) = 2f(t)$. Guessing (but verifying your guess) is fine. How would you change your answer if you also know that $f(0) = 3$?

39. Show that if $f(t) = \cos(\omega t)$, then $f''(t) = -\omega^2 f(t)$. Determine a second function with this property.

40. A wind is blowing in the (x, y)-plane. It is a steady wind in the sense that is doesn't change with time. However, it is not from a fixed direction. The direc-

tion of the wind at the point (x, y) is in the direction of the vector $(1, y)$. This is the same as saying that if a molecule of, say, nitrogen, is passing through the point (x, y), its path has slope y there. We are not concerned with the velocity of the wind, just its direction. Choose a sufficient number of points (x, y) and draw short vectors based at (x, y) and pointing in the direction of the wind, so that your drawing conveys a sense of how the wind is blowing in the plane.

41. Repeat the preceding exercise, but suppose that the direction of the wind at (x, y) is $(-y, x)$.

4.2 EULER'S METHOD

In the last section we used the tangent line approximation to approximate the value of a function f at $a+h$ from given values $f(a)$ and $f'(a)$. The base point was $x = a$. In this section we apply the tangent line approximation to estimate $f(a + h)$ as before, with a as base point. Next, using the approximate value of $f(a + h)$ and assuming we can calculate $f'(a + h)$, we approximate $f(a + 2h)$, again using the tangent line approximation, but this time with $a + h$ as base point. This bootstrap procedure, based on the tangent line approximation with a shifting base point, is called Euler's method. It is the simplest method for solving inverse problems requiring a numerical solution.

We discuss Euler's method in several contexts. In an investigation we solve a biological growth problem by exact methods and then by Euler's method. The examples include inverse problems—which are also called initial value problems—relating to the time of death in a homicide; the limiting velocity of a paratrooper, taking air resistance into account; the pendulum problem without the simplification $\sin \theta \approx \theta$; and the model of the spread of a disease discussed in Chapter 2.

Investigation

The bacterium *Escherichia coli,* usually called *E. coli,* is found in the human gut. Under ideal conditions each *E. coli* cell divides into two cells $1/3$ hour after its own "birth." The mass of one *E. coli* cell is approximately $B_0 = 5 \cdot 10^{-13}$ grams (g). If we start with one cell tomorrow morning at 8 A.M. and assume no deaths, adequate food, and ideal habitat, how many years do we have until $t = D$, the time at which the mass B of *E. coli* equals M, the mass of the earth? (See Fig. 4.9.)

The most direct model of the growth of an *E. coli* colony is a *discrete* model, in which time is restricted to the set $t \in \{0, 1/3, 2/3, \ldots\}$. We start by taking a brief look at the discrete model. Next we formulate and solve a *continuous* model, in which times varies over the set $[0, \infty)$. Finally, we obtain an approximate solution to the continuous model, using the exact solution as a check on Euler's method.

The discrete time model is easily solved. After one period ($1/3$ h) the colony has mass $B_1 = 2B_0$ g; after two periods the colony has mass $B_2 = 2^2 B_0$ g; and after n periods the colony has mass

$$B_n = 2^n B_0 \text{ g}, \quad n = 0, 1, 2, \ldots.$$

The average rate of growth of the colony in period $(n + 1)$ is

$$\frac{B_{n+1} - B_n}{1/3} = \frac{2^{n+1}B_0 - 2^n B_0}{1/3} = 3 \cdot 2^n B_0 = 3B_n \text{g}/\text{h}.$$

From this calculation it is clear that the average rate of growth in any period is proportional to the amount present at the beginning of the period. We use this observation in formulating the continuous model.

If we were to place a typical colony on a balance and make a continuous recording of its mass against time, the resulting graph would not be distinguishable from the graph of a continuous function $B = B(t), t \geq 0$. From the discrete model we expect that the rate of change of $B(t)$ would be proportional to $B(t)$, that is,

(1) $$B'(t) = kB(t)$$

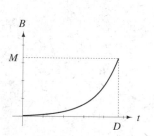

Figure 4.9. Growth of *E. coli.*

for some constant k. Moreover, the unknown function B would satisfy the initial condition

(2) $$B(0) = B_0 = 5 \cdot 10^{-13}.$$

We solve this initial value problem by inferring from (1) the form B must have. You may feel uncomfortable with this, since it depends upon your memory and your ability to recognize a pattern; it is not a cut-and-dried procedure. In the next chapter we give some systematic procedures for determining functions whose derivatives are known; however, these ultimately depend on a table of differentiation formulas, either in your memory, in a book, or built into a calculator or computer.

We seek a function B whose derivative is a constant times B. Since the derivative of e^t is e^t, the exponential function is a possibility. If we suppose that the factor k arose from the chain rule, we soon think of the function e^{kt}. Note that if $B = e^{kt}$, then

$$B' = ke^{kt} = kB.$$

If we multiply e^{kt} by any constant c, thus taking $B = ce^{kt}$, we again have $B' = kB$. Thus we assume that the function B has the form ce^{kt}.

We determine c from the initial condition (2). Replacing t by 0 in the equation

$$B(t) = ce^{kt},$$

we obtain $B(0) = B_0 = 5 \cdot 10^{-13} = c$.

The constant k can be determined from the fact that $B(1/3) = 2B_0$. This gives

$$B(1/3) = 2B_0 = B_0 e^{k/3}.$$

Removing the common factor of B_0 and taking the logarithm of both sides of the equation gives

$$\ln 2 = k/3.$$

We now have the solution to the initial value problem (1) and (2), namely,

(3) $$B(t) = B_0 e^{(3 \ln 2)t}.$$

Next we solve the same initial value problem on the interval from $t = 0$ to $t = 1$ h by Euler's method and check our results against the solution we have just worked out. We set $k = 3 \ln 2$ to keep the notation brief.

We use Euler's method to approximate the solution (3) during the first hour of growth. For convenience we measure the mass in picograms (1×10^{-12} g is 1 picogram). In these units the initial value problem is

$$B'(t) = kB(t)$$
$$B(0) = 0.5.$$

The first step is to use the tangent line approximation to approximate $B(h)$, where $h = 0.1$ is the step size. We have

$$B(h) = B(0 + h) \approx B(0) + hB'(0) = B_0 + h(kB_0) = B_0(1 + kh).$$

In the preceding step $t = 0$ was the base point. We take $t = h$ as the next base point. We know, at least approximately, $B(h)$ and $B'(h) = kB(h)$ at the new base point. Applying the tangent line approximation again, we have

$$B(2h) = B(h + h) \approx B(h) + hB'(h) = B(h) + h(kB(h)).$$

Hence

$$B(2h) \approx B(h)(1 + kh) \approx B_0(1 + kh)^2.$$

In the preceding step $t = h$ was the base point. We take $t = 2h$ as the next base point. We know, at least approximately, $B(2h)$ and $B'(2h) = kB(2h)$ at the new base point. Applying the tangent line approximation again, we have

$$B(3h) = B(2h + h) \approx B(2h) + hB'(2h) = B(2h) + h(kB(2h)).$$

Hence

$$B(3h) \approx B(2h)(1 + kh) \approx B_0(1 + kh)^3.$$

From these calculations it is clear that

(4) $$B(nh) \approx B_0(1 + kh)^n, \quad n = 0, 1, 2, \ldots.$$

Table 1. Approximate and exact solutions for $h = 0.1$

n	$nh = t$	Euler	Exact
0	0.0	0.500	0.500
1	0.1	0.604	0.616
2	0.2	0.730	0.758
3	0.3	0.881	0.933
4	0.4	1.064	1.149
5	0.5	1.286	1.414
6	0.6	1.553	1.741
7	0.7	1.876	2.144
8	0.8	2.266	2.639
9	0.9	2.738	3.249
10	1.0	3.307	4.000

To familiarize yourself with Euler's method and the procedures for entering and evaluating the formulas for the exact and approximate solutions on your calculator or computer, calculate the entries of Table 1. The exact and approximate solutions for $h = 0.1$ are plotted at the top of Fig. 4.10.

If we decrease the step size h from 0.1 to 0.05, so that it takes 20 steps to reach $t' = 1$, the approximate solution becomes closer to the exact solution. The price for increasing accuracy is an increase in the amount of calculation. We show the exact and approximate solutions for $h = 0.05$ at the bottom of Fig. 4.10.

The equation

$$B'(t) = kB(t)$$

satisfied by the *E. coli* mass function has the same general form as the slightly simpler equation

(5) $$y' = y.$$

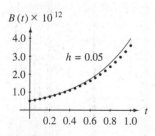

Figure 4.10. Approximate and exact solutions.

Such equations are called **differential equations** since each is an *equation* in an unknown function and its *derivatives*.

Direction Fields

A differential equation such as (5) specifies the slope of the unknown function $y = y(x)$. If the graph of y passes through a point (x, y), then from (5) we know that its slope there is y. For example, if the graph of y passes through $(0, 1)$, then its slope at $(0, 1)$ is 1. If at several points (x, y) we draw a short line with slope y, the resulting figure determines qualitatively the appearance of the graphs of the unknown functions y satisfying (5). We show in Fig. 4.11 such a figure, called a **direction field,** for the differential equation (5). At each of several points (x, y)

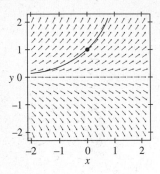

Figure 4.11. Direction field for $y' = y$.

a vector of length 0.2 and in the direction of the vector $(1, y)$ was drawn. This vector indicates the direction of the tangent vector to the graph of $y = y(x)$ with increasing x. Direction fields can be drawn on some calculators and most computer algebra systems. Often they can be drawn rapidly by hand. For (5), points (x, y) at which the slope is 1 are points of the form $(x, 1)$. Hence we draw several short line segments of slope 1 along the line $y = 1$, along $y = 2$ we draw several short line segments of slope 2, and so on.

One solution of $y' = y$ is $y = e^x$. We have included the graph of this solution in the figure. Note how it fits into the direction field. Other solutions of this differential equation have the general form $y = ce^x$, where c is a constant. An initial value problem consists of a differential equation and an *initial condition*. The initial condition specifies that the graph of the solution pass through a given point. An initial value problem ordinarily has one and only one solution. Specifying one point selects exactly one curve from all of the curves that fit the direction field. The solution $y = e^x$ may be selected by the initial condition $(0, 1)$, which is shown as a black dot in the figure.

A direction field helps explain Euler's method. In solving (5) with the initial condition $(0, 1)$, the first step is to use the tangent line approximation at $x = h$. We have

$$y(h) \approx y(0) + hy'(0) = 1 + h \cdot 1 = 1 + h.$$

We started at $(0, 1)$, which is on the graph of the solution $y = e^x$ we are attempting to determine. The second point of the solution is $(h, 1 + h)$. This point is not on the true solution curve since $e^h \neq 1 + h$. The point $(h, 1 + h)$ is, however, on a nearby solution curve. Euler's method uses the tangent line approximation to approximate a nearby point on this new solution curve. As the base points change, the calculated points skip from solution curve to solution curve. By adjusting the step size h, we can control the distance between the calculated solution and the true solution. We outline a discussion of the error in Euler's method in Exercise 16.

Applied to forensic medicine, Newton's law of cooling states that the change in the temperature of a corpse is proportional to the difference between the (ambient) temperature of the room in which the death occurred and the temperature of the corpse. It is assumed that the temperature of the room has been nearly steady since the time of death. Letting $T = T(t)$ denote the Fahrenheit temperature of a corpse t hours after death occurred and A the temperature of the room, Newton's law of cooling is

$$(6) \qquad\qquad T'(t) = k(A - T(t)).$$

The initial condition is usually $T(0) = 98.6$. It is known that $k \approx 0.05$.

■ **EXAMPLE 1**   A body was discovered at 11 A.M., at which time the temperatures of the room and the body were measured and found to be 65°F and 80°F, respectively. Use Euler's method with $h = 1$ hour in calculating the approximate time of death.

Solution.   We use $T(0) = 98.6$ and $T'(0)$ in calculating $T(1)$, using the tangent line approximation. The value of $T'(0)$ comes from the differential equation (6). We have

$$T(1) \approx T(0) + 1 \cdot T'(0) = T(0) + k(65 - T(0)).$$

The next step is

$$T(2) \approx T(1) + 1 \cdot T'(1) = T(1) + k(65 - T(1)).$$

In this way we see that for $n \geq 1$ we have

$$T(n) \approx T(n-1) + 1 \cdot T'(n-1) = T(n-1) + k(65 - T(n-1)).$$

With $k = 0.05$ the equation simplifies to

(7) $T(n) \approx 0.95T(n-1) + 3.25.$

The equation (7), together with $T(0) = 98.6$, can be used to calculate approximate values of $T(1)$, $T(2)$, and so on, continuing until $T(n) \approx 80$. If you know how to handle functions that are defined recursively on your calculator or CAS, you can determine that $T(16) \approx 79.8$ in just a few minutes. If this is something you haven't learned yet, the calculation takes a little longer. Here are a few steps:

$$T(0) = 98.6$$
$$T(1) \approx (0.95)(98.6) + 3.25 = 96.92$$
$$T(2) \approx (0.95)(96.92) + 3.25 = 95.324$$
$$\vdots$$
$$T(16) \approx (0.95)(80.5666) + 3.25 = 79.883$$

It appears that approximately 16 hours after the body began cooling, its temperature had dropped to 80°F. This would put the time of death at approximately 7 P.M. of the prior evening.

We give a "closed form" for calculating $T(n)$ in Exercise 9. In Exercise 5 we pose another forensic law of cooling problem. In Exercise 10 we ask you to determine the exact solution to the differential equation (6), with the initial condition $T(0) = 98.6$.

We show in Fig. 4.12 the exact solution to this problem and the Euler's method solution outlined above. ∎

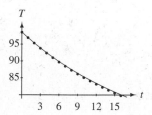

Figure 4.12. Comparison of the exact and Euler's solutions.

∎ **EXAMPLE 2**   A 70 kg paratrooper is falling at 54 m/s at the time his parachute opens. We assume that the magnitude of the retarding force due to the parachute is proportional to $|v(t)|^{1.5}$, where $v(t)$ is his velocity at any time t. It is known from observation that within a few seconds $|v(t)|$ approaches a limiting value of approximately 5.5 m/s. Graph the velocity of the trooper against time.

Solution.   We use Newton's second law of motion,

$$\mathbf{F} = m\mathbf{a} = m \, d\mathbf{v}/dt,$$

where $\mathbf{F}$ is the net force on the trooper and $\mathbf{a} = d\mathbf{v}/dt$ is his acceleration. Since the motion is along a vertical line, we may use scalars instead of vectors for position, velocity, and acceleration. Referring to Fig. 4.13, the net force on the trooper is $mg - kv^{1.5}$, which is the sum of the downward force mg of the earth on the trooper and the upward force $kv^{1.5}$ due to the parachute. We have taken the y-axis with

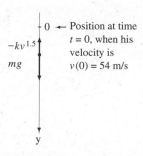

Figure 4.13. Forces on paratrooper.

positive direction downward so that the velocity is positive. (This simplifies the discussion as far as $v^{1.5}$ is concerned. See Exercise 12.) This gives the equation

$$(8) \qquad\qquad 70(9.8) - kv^{1.5} = 70\frac{dv}{dt}.$$

To evaluate k we use the information that the trooper's limiting velocity is 5.5 m/s. This means that as t increases without limit, the velocity approaches 5.5 m/s. We write this as

$$\lim_{t\to\infty} v(t) = 5.5.$$

It also means that as t increases without limit, the change in the velocity approaches 0, that is,

$$\lim_{t\to\infty} dv/dt = 0.$$

If we take the limit of both sides of (8) as $t \to \infty$, we obtain

$$70(9.8) - k5.5^{1.5} = 0.$$

Solving this equation for k gives

$$k = \frac{70(9.8)}{5.5^{1.5}} \approx 53.$$

Dividing both sides of (8) by 70 and replacing k by 53, the problem of graphing the velocity of the trooper against time becomes the initial value problem: Find a function $v = v(t)$ for which

$$(9) \qquad\qquad \frac{dv}{dt} = v'(t) = 9.8 - (53/70)v^{1.5}$$

and

$$(10) \qquad\qquad v(0) = 54.$$

To graph the trooper's velocity, we approximate his velocity at, say, times $0.05, 0.10, \ldots, 2.0$. We write out a few steps of Euler's method to understand the pattern for the calculations. The general step is the tangent line approximation

$$v(t + h) \approx v(t) + hv'(t) = v(t) + h(9.8 - (53/70)v(t)^{1.5}).$$

It is best to put off substituting 0.05 for h, particularly if you wish to change the step size later. Applying Euler's method gives

$$v(h) \approx v(0) + hv'(0)$$
$$v(2h) \approx v(h) + hv'(h)$$
$$v(3h) \approx v(2h) + hv'(2h).$$

The general formula is clear from these results. We have

$$(11) \qquad v(jh) \approx v((j-1)h) + hv'((j-1)h), \quad j = 1, 2, \dots.$$

The value of $v'(jh)$ in (11) is calculated using (9). Specifically,

$$v'((j-1)h) = 9.8 - (53/70)v((j-1)h)^{1.5}.$$

If you and your calculator together can store a formula like

$$v + h(9.8 - (53/70)v^{1.5}),$$

the calculation goes rapidly. Start with the input $v = 54$; the formula returns 39.47. Use this as the next input; the formula returns 30.57. Continue this process, recording as you go, or perhaps you can manage this automatically. In any case you should be able to obtain

$$v(1 \cdot h) = v(0.05) \approx v(0) + hv'(0) = 54 + 0.5(9.8 - (53/70)54^{1.5}) \approx 39.47$$
$$v(2 \cdot h) = v(0.10) \approx v(1 \cdot h) + hv'(1 \cdot h), \quad \text{so that}$$
$$v(2 \cdot h) \approx 39.47 + 0.5(9.8 - (53/70)39.47^{1.5}) \approx 30.57.$$

Continue this for a while and you will eventually obtain

$$v(40 \cdot h) = v(2.00) \approx v(39 \cdot h) + hv'(39 \cdot h), \quad \text{so that}$$
$$v(40 \cdot h) \approx 5.57 + 0.5(9.8 - (53/70)5.57^{1.5}) \approx 5.56.$$

In Fig. 4.14 we show a sketch including both the exact solution and the results of Euler's method. ■

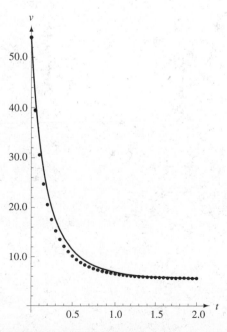

Figure 4.14. Euler's method with $h = 0.05$.

Second-Order Differential Equations

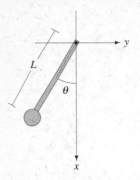

Figure 4.15. Simple pendulum.

Second-order differential equations include the second derivative of the unknown function. Differential equations arising from an application of Newton's second law often are second-order. For example, at the beginning of Section 4.1 we stated that the angle $\theta = \theta(t)$ measuring the position of a pendulum (see Fig. 4.15) satisfies the equation

(12)
$$L\frac{d^2\theta}{dt^2} + g\sin\theta = 0.$$

We used the tangent line approximation $\sin\theta \approx \theta$ to replace this equation with the simpler equation (also second-order)

(13)
$$L\frac{d^2\theta}{dt^2} + g\theta = 0.$$

We assume that $L = 1$ and take $\theta(0) = 0.1$ and $\theta'(0) = 0$ as initial conditions. These conditions mean that we release the bob from rest, having displaced it nearly 6° from the downward equilibrium position. Our goal is to solve these two initial value problems.

The "right" equation, that is, equation (12), must be solved numerically, so that we end up with an approximate solution. Equation (13) can be solved exactly, but, of course, it is not quite the right equation. So again we end up with an approximate solution.

First, we show that

(14)
$$\theta = 0.1\cos(\omega t), \quad \text{where } \omega = \sqrt{g/L},$$

is a solution to equation (13), with $\theta(0) = 0.1$ and $\theta'(0) = 0$.

This function satisfies the differential equation (13). To show this, we calculate the second derivative of θ. We have

$$\frac{d\theta}{dt} = -0.1\omega\sin(\omega t)$$

$$\frac{d^2\theta}{dt^2} = -0.1\omega^2\cos(\omega t).$$

Hence

$$L\frac{d^2\theta}{dt^2} + g\theta = (-0.1\omega^2 L + 0.1g)\cos(\omega t) = 0.$$

This shows that $\theta = 0.1\cos(\omega t)$ satisfies (13).

We return to this solution after we have solved (12) by Euler's method.

■ **EXAMPLE 3**   Use Euler's method to solve the initial value problem

(15)
$$L\theta'' + g\sin\theta = 0$$
$$\theta(0) = 0.1 \qquad \theta'(0) = 0.$$

Note that we have replaced $d^2\theta/dt^2$ by the simpler θ''.

Solution.   Since the pendulum equation is a second-order differential equation, we must reorganize it so that we can apply Euler's method. This problem did not come up in the paratrooper problem since we only wanted to approximate the trooper's velocity, not his position. Here we wish to approximate θ, not θ'. The usual way of proceeding is to give θ' a new name, say, $\theta' = v$. We may then reorganize (15) by defining the vector-valued function

$$\mathbf{r}(t) = \big(\theta(t),\, v(t)\big).$$

If we differentiate $\mathbf{r}$, we obtain

$$\mathbf{r}' = \big(\theta',\, v'\big) = \big(v,\, -(g/L)\sin\theta\big).$$

We used (15) to replace $v' = \theta''$ by $-(g/L)\sin\theta$.

We may solve the initial value problem (15) by solving the initial value problem

(16)
$$\mathbf{r}' = \big(v,\, -9.8\sin\theta\big)$$
$$\mathbf{r}(0) = \big(v(0), \theta(0)\big) = \big(0.1, 0\big).$$

From a notational viewpoint the calculations are the same as in the other Euler's method examples. This is an advantage of vector notation; it reduces two-dimensional problems to what look like one-dimensional problems. Here are several steps in the calculation:

$$\mathbf{r}(1 \cdot h) \approx \mathbf{r}(0) + h\mathbf{r}'(0).$$

In the first step we evaluate $\mathbf{r}(0)$ and $\mathbf{r}'(0)$ from the initial conditions in (16). Next,

$$\mathbf{r}(2 \cdot h) \approx \mathbf{r}(h) + h\mathbf{r}'(h).$$

For this step we obtain $\mathbf{r}(h)$ from the preceding step and calculate $\mathbf{r}'(h)$ from the differential equation in (16). In general, we have

$$\mathbf{r}(n \cdot h) \approx \mathbf{r}((n - 1) \cdot h) + h\mathbf{r}'((n - 1) \cdot h).$$

We give several steps of the calculation with $h = 0.1$. These are plotted together with the approximate solution (14) in Fig. 4.16.

$$\mathbf{r}(0.1) \approx \big(0.1, 0\big) + 0.1\big(0, -0.98\big) = \big(0.10, -0.10\big)$$
$$\mathbf{r}(0.2) \approx \big(0.10, -0.10\big) + 0.1\big(-0.10, -1.96\big) = \big(0.09, -0.20\big)$$
$$\mathbf{r}(0.3) \approx \big(0.09, -0.20\big) + 0.1\big(-0.20, -2.84\big) = \big(0.07, -0.28\big)$$
$$\vdots$$
$$\mathbf{r}(2.0) \approx \big(0.21, 0.38\big) + 0.1\big(0.38, 1.77\big) = \big(0.25, 0.18\big).$$

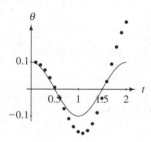

Figure 4.16. Euler's method with $h = 0.1$.

We have calculated only up to $t = 2.0$ since we expect the motion of the pendulum to be periodic with period $2\pi/\sqrt{9.8} \approx 2.007$ s. If we assume that the approximate solution (14) is accurate for pendulums with small swings, it is clear from Fig. 4.16 that the step size $h = 0.1$ is too large.

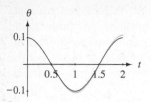

Figure 4.17. Euler's method with $h = 0.01$.

Without some help from a calculator or CAS the calculations for step size $h = 0.01$ become quite tedious. We hope you have gone through the arithmetic for the first three or four steps of the calculation for $h = 0.1$, but we advise against trying Exercise 14 unless you can automate the calculation. We show in Fig. 4.17 the results for step size $h = 0.01$. We have plotted only every other point to improve the sketch.

We leave as Exercise 15 the interesting interpretation of the curve

$$\mathbf{r}(t) = \big(\theta(t), v(t)\big), \quad 0 \leq t \leq 2,$$

in what is called the *phase plane*.   ∎

In our last example of Euler's method we obtain an approximate solution of the three-compartment model for the Hong Kong flu epidemic in New York City, 1968–1969. This model was discussed in Section 2.8 of Chapter 2.

We apply the model to a population of size $P_0 = 7,900,000$. Recall that at any time $t \geq 0$ the population is divided into three parts, with $S(t)$ denoting the number of individuals who are susceptible to the flu, $I(t)$ the number who are infectious, and $R(t)$ the number who have recovered. Since we assume that at any time t the sum of the number of susceptible, infectious, and recovered individuals is P_0, the model may be expressed in terms of two of these variables. The initial value problem given in Chapter 2 was

$$\begin{align} S'(t) &= -aS(t)I(t) \\ I'(t) &= aS(t)I(t) - (1/d)I(t) \\ S(0) &= S_0 \quad \text{and} \quad I(0) = I_0. \end{align} \tag{17}$$

For Hong Kong flu, $a = 0.47/P_0$ and $d = 3$.

■ **EXAMPLE 4**   Apply Euler's method to the initial value problem given in (17). Assume that $S(0) = 0.99P_0$ and $I(0) = 0.01P_0$. Use $h = 1$ day as the step size and track $S(t)$ and $I(t)$ for 80 days. (To justify this set of initial conditions, we would have to assume, for example, that $I(0) = 79,000$ individuals came back from a holiday carrying the flu virus and spread it in an otherwise susceptible population. We have chosen this large initial value for I to avoid the uninteresting slow beginning of the spread of a virus in a population if, at $t = 0$, only a few individuals are infectious.)

Solution.   As in the preceding example, we use the vector form of the tangent line approximation. If we know $\mathbf{r}(t)$ and $\mathbf{r}'(t)$ at the base point t, then

$$\mathbf{r}(t + h) \approx \mathbf{r}(t) + h\mathbf{r}'(t).$$

For our model

$$\mathbf{r}(t) = \big(S(t), I(t)\big).$$

The derivative of $\mathbf{r}$ is given by

$$\mathbf{r}'(t) = \big(S'(t), I'(t)\big) = \big(-aS(t)I(t), aS(t)I(t) - (1/d)I(t)\big).$$

Here are several steps in the calculation, all easily done on a calculator or CAS by defining a few functions for repeated evaluation. See Exercise 17. Recall that $h = 1$. We retained within the calculator all intermediate results to full calculator accuracy for subsequent calculations, but recorded only three significant figures below.

$$\mathbf{r}(0) = \big(S(0), I(0)\big) = \big(7.82 \times 10^6, 7.90 \times 10^4\big)$$

$$\mathbf{r}(h) = \big(S(h), I(h)\big) \approx \mathbf{r}(0) + h\mathbf{r}'(0)$$
$$\approx \mathbf{r}(0) + \big(-aS(0)I(0), aS(0)I(0) - (1/d)I(0)\big)$$
$$\approx \big(7.78 \times 10^6, 8.94 \times 10^4\big)$$

$$\mathbf{r}(2h) = \big(S(2h), I(2h)\big) \approx \mathbf{r}(h) + h\mathbf{r}'(h)$$
$$\approx \mathbf{r}(h) + \big(-aS(h)I(h), aS(h)I(h) - (1/d)I(h)\big)$$
$$\approx \big(7.74 \times 10^6, 1.01 \times 10^5\big)$$

$$\mathbf{r}(3h) = \big(S(3h), I(3h)\big) \approx \mathbf{r}(2h) + h\mathbf{r}'(2h)$$
$$\approx \mathbf{r}(2h) + \big(-aS(2h)I(2h), aS(2h)I(2h) - (1/d)I(2h)\big)$$
$$\approx \big(7.70 \times 10^6, 1.13 \times 10^5\big).$$

After 80 steps (which took less than 10 seconds on our calculator), we obtained

$$\mathbf{r}(80h) = \big(S(80h), I(80h)\big) \approx \mathbf{r}(79h) + h\mathbf{r}'(79h)$$
$$\approx \mathbf{r}(79h) + \big(-aS(79h)I(79h), aS(79h)I(79h) - (1/d)I(79h)\big)$$
$$\approx \big(3.62 \times 10^6, 1.12 \times 10^3\big).$$

We show graphs of these calculations in Figs. 4.18 and 4.19. The separate graphs of S and I came directly from Euler's method, which resulted in the sequence of points

$$(0, S(0), I(0)), (h, S(h), I(h)), \dots, (80h, S(80h), I(80h)).$$

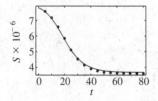

Figure 4.18. $S(t)$, $0 \le t \le 80$.

To graph S against t, we took from this list the points

$$(0, S(0)), (h, S(h)), \dots, (80h, S(80h))$$

and plotted them as dots in Fig. 4.18. We plotted only the points corresponding to $t = 0, 5, 10, \dots, 80$ to avoid clutter. To judge the accuracy of Euler's method we included in this figure a highly accurate graph of S. We graphed I in the same way.

These graphs of S and I fit our expectations of the spread of a disease that, once contracted, provides its own immunity. The number of infectious individuals rises rapidly at first, as the disease spreads among the large number of susceptible individuals. The disease continues to spread, but soon the number of individuals who are infectious or recovered begins to grow rapidly, which causes the number of infectious individuals to peak and then decline rapidly.

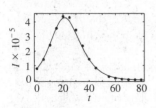

Figure 4.19. $I(t)$, $0 \le t \le 80$.

Exercises 4.2

Basic

1. Referring to the *E. coli* example, use the exact solution to calculate doomsday. That is, starting with one *E. coli* bacterium, how long will it be until the mass of *E. coli* equals the mass of the earth (5.97×10^{24} kg)? How does your answer compare with that obtained using the discrete model $B(n/3) = 2^n B_0$?

2. Do the calculations for $h = 0.05$ for the *E. coli* model. See Fig. 4.10.

3. Sketch the direction field for the differential equation $y' = -x/y$. Roughly sketch in the solution $y = y(x)$ satisfying $y(0) = 1$. (Before reading the next sentence, try to guess the formula for y.) Show that the function $y = \sqrt{a^2 - x^2}$ defined for $-a < x < a$ satisfies the differential equation for any choice of $a > 0$. What value of a is required to satisfy the initial condition $y(0) = 1$? Does $y = -\sqrt{a^2 - x^2}$ satisfy the differential equation?

4. Sketch the direction field for the differential equation $y' = x/y$. Roughly sketch in the solution $y = y(x)$ satisfying $y(0) = 1$. (Before reading the next sentence, try to guess the formula for y.) Show that the function $y = \sqrt{a^2 + x^2}$ defined for $-\infty < x < \infty$ satisfies the differential equation for any choice of $a > 0$. What value of a is required to satisfy the initial condition $y(0) = 1$? Does $y = -\sqrt{a^2 + x^2}$ satisfy the differential equation?

5. The forensic medicine example was adapted from a scene described by P. D. James, an English mystery writer. She gave the "ambient temperature" as 65°. Use Euler's method to solve the same problem with the ambient temperature of 72°.

6. Assuming the paratrooper in Example 2 bailed out at 610 m, estimate how long it will take him to reach earth.

Growth

7. Prepare a brief argument as to why it is reasonable to model the *E. coli* population using continuous variables for both time and mass.

8. Use Euler's method to approximate $y(1)$ for the initial problem

$$y' = y \quad \text{and} \quad y(0) = 1.$$

For $h = 1/n$, show that after n steps

$$y(1) \approx (1 + 1/n)^n.$$

We are concerned here with the sequence of numbers

$$(1 + 1/1)^2, (1 + 1/2)^2, (1 + 1/3)^3,$$
$$(1 + 1/4)^2, \ldots, (1 + 1/n)^n, \ldots.$$

Determine if the successive entries of this sequence tend to increase or decrease. As n gets very large, do the entries of this sequence "settle down" at some number L or do they drift off to infinity? If they settle down, use your calculator or the initial value problem to guess L.

9. Starting with (7) and, for brevity, writing $T(n) = T_n$, show that

$$T_n = 0.95^n T_0$$
$$+ 3.25(1 + 0.95 + 0.95^2 + \cdots + 0.95^{n-1}).$$

Simplify this expression further with the formula for the sum of a finite geometric series.

10. The Newton's law of cooling model (6) with $k = 0.05$ and $A = 65$ becomes

$$T' = 0.05(65 - T).$$

Use your knowledge of derivatives to guess a solution to this equation. If necessary, make the temporary substitution $U = T - 65$, noting that $U' = T'$. Both U and T denote unknown functions of time t. Choose constants so that the initial condition $T(0) = 98.6$ is also satisfied. Use your exact solution to determine the time of death. How does it compare with the value obtained by Euler's method?

11. Assuming that the resistive force of the air is $|v(t)|^{1.6}$, repeat the calculations and graph in the paratrooper example. Assume the limiting velocity is 5.5 m/s when you recalculate k. Compared to the results worked out in the example, will it take more or less time to come within, say 1 m/s of the limiting velocity? Can you predict this from the differential equation?

12. In Example 2 we chose the y-axis downward to avoid a slight complication. If we had taken the positive direction to be upward, show that (9) would have been $-70(9.8) + k(-v)^{1.5} = 70(dv/dt)$. As a separate question, discuss $v^{1.5}$ when $v < 0$. What result does your calculator return when you enter $(-2)^{1.5}$? Why?

13. The initial position of a particle is $(2, 3)$, and its velocity is $(\cos t^2, \sin t^2)$ for all $t \geq 0$. Use Euler's method to approximate the positions of the particle at $t = 0.0, 0.1, \ldots, 1.5$. Graph the results. Also find the time at which the particle is at its rightmost position.

14. In Example 3, show that $\theta(2.0) \approx 0.11$ and $v(2.0) \approx 0.01$ using Euler's method with $h = 0.01$. The calculation is not difficult but is tedious without the help of a calculator or computer. It may be preferable for calculation to think about the problem in the form of the two equations

$$\theta' = v \quad \text{and} \quad v' = (-g/L) \sin \theta.$$

With $\theta(0) = 0.1$ and $v(0) = 0$, calculate $\theta(h)$ and $v(h)$ by the tangent line approximation. Using these results, repeat the calculation. Continue for 199 more steps. If you can program it, then only one step is necessary. Otherwise, 200 steps are required, but with a calculator doing the arithmetic, this isn't too tough.

15. In Example 3 we brought in the vector-valued function

$$\mathbf{r} = \mathbf{r}(t) = (\theta(t), v(t)).$$

We may regard this as describing a curve C in a (θ, v)-plane, called the *phase plane*. Use the exact solution

$$\theta = \theta_0 \cos\left(\sqrt{g/L}\, t\right)$$

to graph C. Interpret the result in terms of the motion of the pendulum. Roughly sketch in the approximate solution obtained with Euler's method with $h = 0.01$. Comment on the result.

16. In first discussing the tangent line approximation in Section 4.1, we wrote an equation of the form

$$f(a + h) = f(a) + hf'(a) + hE(h).$$

The function E goes to 0 as $h \to 0$. The exact form of E depends upon f, h, and the base point a. We use this equation to describe the error in Euler's method. For step 1 of Euler's method, we have

$$f(a + h) = f(a) + hf'(a) + hE_1(h).$$

We use E_1 for the function associated with step 1. Show that for step 2, we have

$$f(a + 2h) = f(a) + h(f'(a) + f'(a + h)) + h(E_1 + E_2).$$

Show that for step n, we have

$$f(a + nh) = f(a) + h(f'(a) + \cdots + f'(a + (n-1)h)) + h(E_1 + \cdots + E_n).$$

The error in Euler's method is the difference

$$|f(a + nh) - f(a) - h(f'(a) + \cdots + f'(a + (n-1)h))|,$$

which is equal to

$$h|h(E_1 + \cdots + E_n)|.$$

Show that if the errors $E_1, \ldots, E_n$ are less than some number E, then for a given step size h, the error in Euler's method is proportional to the number of steps.

17. Do the calculations for Example 4 using the following algorithm. Let f and g be the functions

$$f(x, y) = -axy, \quad g(x, y) = -f(x, y) - (1/d)y.$$

Let $t = 0$, $x = S(0)$, and $y = I(0)$ be starting values. To update the t, x, and y values, use

$$t^* = t + 1$$
$$x^* = f(x, y)$$
$$y^* = g(x, y).$$

The latest output is the new input. As these data are calculated, they can be stored in two vector arrays. On our calculator we used the statistical graph package to plot the matrix data (S and I) for $0 \leq t \leq 80$.

18. Apply the three-compartment model discussed in Example 4 to a German measles epidemic. Take $P_0 = 7{,}900{,}000$, $a = 0.62/P_0$, and $d = 11$. Use $h = 1$. Plot $S(t)$ and $I(t)$ for $t = 0, 1, \ldots, 80$.

Review/Preview

19. A number c is a zero of a function w if $w(c) = 0$. Let f and g be given functions and let $w(x) = f(x) - g(x)$. If the graphs of f and g intersect at a point (p, q), what can be said about w at p? Determine approximately the point where the graphs of $f(x) = x^2$ and $g(x) = x^3 - 1$ cross. What is w in this case?

20. How are the line with equation $y = mx + b$ and the function $f(x) = mx + b$ related? What is the

x-intercept c of the line? What is $f(c)$? If a bead is sliding on the line, where will it cross the x-axis?

21. For at least 4000 years we have had a numerical rule or algorithm for calculating the square root of a positive number. Often the problem has been posed geometrically: Find the side of a square with given area. One such rule is called *divide and average*. To find the side of a square with area 2, start with a rough guess of the side, perhaps 1.5. Divide the area by the guess, and then average the quotient and the guess. This gives a new, much better approximation of the side. Repeat until the side is known with sufficient accuracy. A few steps of this algorithm are: Starting with a square of area 2 (set your calculator to display six decimals), guess 1.5; divide to get 1.333333; average the guess 1.5 with the quotient 1.333333 to get 1.416667 (note that $(1.416667)^2 \approx 2.006944$); divide 2 by 1.416667 to get 1.411765; average the guess with the quotient to get 1.414216 (note that $(1.414216)^2 \approx 2.000006$); and so on. Find $\sqrt{5}$ with this algorithm, perhaps starting with 2 as a guess. Now let a be the number whose square root is wanted. Let x_1 be the first guess. Write a formula for x_2, the second guess. Let $x_3, x_4, \ldots, x_n$ be the third, fourth, $\ldots$, and nth guesses. Write a formula for x_{n+1} in terms of a and x_n.

4.3   NEWTON'S METHOD

A number c is a **zero** of a function f if $f(c) = 0$. Zeros of functions and roots of equations are closely related ideas. To determine the root of the equation $e^x = 2 - x$, for example, we may instead calculate the zero of the function $f(x) = e^x - 2 + x$.

Newton's method is one of several methods or algorithms for approximating a zero of a function. In Exercise 21 of Section 4.2 we discussed the ancient divide-and-average algorithm for calculating the square root of a number. In Exercise 23 we ask you to show that the divide-and-average algorithm for calculating $\sqrt{a}$ is Newton's method applied to the function $f(x) = x^2 - a$. Note that one of the zeros of f is $\sqrt{a}$.

We show in Fig. 4.20 the graph of a function f near c, a zero of f. Newton's method is an application of the tangent line approximation, in which an approximation a to c is replaced by a much better approximation a^*. To calculate a^*, we replace the function f by its tangent line approximation function ℓ, where

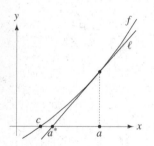

Figure 4.20. One step.

$$(1) \qquad f(x) = f(a + (x - a)) \approx f(a) + f'(a)(x - a) = \ell(x).$$

Now, instead of solving the equation $f(x) = 0$ for x (which we would do if, for example, we were seeking the zeros of $f(x) = ax^2 + bx + c$), we solve the simpler equation $\ell(x) = 0$. Thus we solve the equation

$$(2) \qquad\qquad f(a) + f'(a)(x - a) = 0$$

for x. If we denote the solution by a^*, we have

$$(3) \qquad\qquad a^* = a - \frac{f(a)}{f'(a)}.$$

The number a^* is probably not c, but usually is much closer to c than a. This procedure is repeated until we have determined a sufficiently accurate approximation to c. Figure 4.20 shows the graphs of the functions f and ℓ, the approximation a to c, and the improved approximation a^* to c.

Newton's Method

1. Let a be an approximate value of a zero c of f. To get started, we may choose a by, say, graphing f near a zero c. Otherwise, a is determined in step 3.

2. Calculate

$$a^* = a - \frac{f(a)}{f'(a)}.$$

3. Replace a by a^* and go to step 1.

Newton's method often is written in terms of **iterates** $x_1, x_2, x_3, \ldots$. Referring to Fig. 4.21, x_1 is our first guess a. The geometric counterpart of solving the equation

$$f(x_1) + f'(x_1)(x - x_1) = 0$$

(recall (2) and (3) above) is to determine the x-intercept x_2 of the tangent line to f at $(x_1, f(x_1))$. After this, the procedure repeats, generating the sequence $x_1, x_2, x_3, \ldots$ of approximations to c. Successive iterates are calculated from

$$(4) \qquad x_{n+1} = x_n - \frac{f(x_n)}{f'(x_n)}, \quad n = 1, 2, \ldots.$$

Newton's method usually works very well. Given a good first guess, the sequence $x_1, x_2, x_3, \ldots$ of iterates approaches c (or converges to c) rapidly. In our first examples we concentrate on the positive features of Newton's method, leaving until the end of the section a comment or two on possible difficulties.

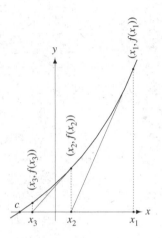

Figure 4.21. Newton's method.

■ **EXAMPLE 1** Use Newton's method to find the positive zero of the function $f(x) = x^2 - 2$.

Solution. We use $x_1 = 2$ as our starting point. The first few steps of Newton's method are shown in Fig. 4.22. The calculations, which you should repeat, are

$$x_2 = x_1 - \frac{x_1^2 - 2}{2x_1} = 2 - \frac{2}{4} \qquad\qquad = 1.5$$

$$x_3 = x_2 - \frac{x_2^2 - 2}{2x_2} = 1.5 - \frac{0.25}{3.0} \qquad\qquad = 1.4166666666$$

$$x_4 = x_3 - \frac{x_3^2 - 2}{2x_3} = 1.41666666667 - \frac{0.00694444445}{2.83333333334} = 1.41421568628$$

The number of correct digits in this calculation doubles with each iteration, which is typical of Newton's method. The positive zero of the function $f(x) = x^2 - 2$ is, of course, $\sqrt{2} = 1.4142135623\ldots$ ■

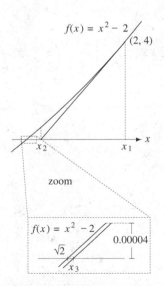

Figure 4.22. Using Newton's method.

In using Newton's method, the decision that an iterate x_n is sufficiently close to a zero c can be made in several different ways. Usually, iterates x_n and x_{n+1} are compared, and when they agree to, say, five decimals, it is assumed that a rounded value of x_{n+1} approximates a zero of f to four or five decimals. In Example 1, if we calculate $x_5 = 1.41421356237$ and compare it to $x_4 = 1.41421568628$, we may feel confident that $|\sqrt{2} - 1.41421| < 0.00001$.

We use this same example to illustrate a second method of checking how close an iterate x_n is to a zero c. This method is certain but requires a calculation, not simply a gut-level feeling that the iterates are close enough.

If we decide to round $x_4 = 1.41421568628$ to four decimals, we can be certain that

$$|1.4142 - \sqrt{2}| < 0.0001$$

provided that

$$f(1.4141)f(1.4143) \approx (-0.00032119)(0.00024449) < 0.$$

Why does this calculation give certainty? Referring to Fig. 4.23, if f changes sign between the end points of the interval [1.4141, 1.4143], then f must have a zero somewhere on this interval. From this it follows that the midpoint $(1.4143 + 1.4141)/2 = 1.4142$ of the interval is within $(1.4143 - 1.4141)/2 = 0.0001$ of a zero of f.

This argument is based on the *Intermediate Value Theorem*.

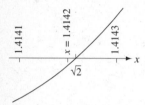

Figure 4.23. Changing sign test.

Intermediate Value Theorem

If f is continuous on an interval $[a, b]$ and $f(a)f(b) < 0$, then f has a zero c in (a, b).

We noted at the beginning of the chapter that the value of x for which $x = \cos x$ can be approximated by entering a guess into your calculator and then pressing the cosine key several times. If you didn't try $x = 0.7$, obtaining $0.735\ldots$ after six repetitions (in radian mode), please try it now. Why does this work, and is there a faster numerical algorithm?

We hope you thought, "Well... Newton's method would probably be a lot faster." You're right; it's better after one iteration! We show this in an example. After the example, we explain why pressing the cosine key also works.

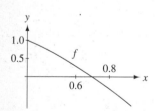

Figure 4.24. Changing signs.

■ **EXAMPLE 2** Find the one zero of the function $f(x) = \cos x - x$.

Solution. First we locate the zero in [0.6, 0.8] with the help of a graph (see Fig. 4.24) or a few calculations. For the latter, showing that $f(0.6) = 0.22\ldots$ and $f(0.8) = -0.10\ldots$ is sufficient, by the Intermediate Value Theorem. We take $x_1 = 0.7$. After calculating $f'(x)$, we have from (4)

$$x_{n+1} = x_n - \frac{f(x_n)}{f'(x_n)} = x_n - \frac{\cos x_n - x_n}{-\sin x_n - 1} = x_n + \frac{\cos x_n - x_n}{\sin x_n + 1}.$$

Table 1. Newton's method calculations

n	x_n
1	0.7
2	0.739436497848
3	0.739085160465
4	0.739085133215
5	0.739085133215

We show in Table 1 the results of continuing the algorithm until no further change in x_n is observed.

Since to calculator accuracy $\cos x_4 = x_4$, x_4 is a good candidate for the zero we seek. ■

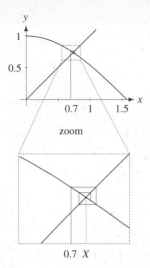

Figure 4.25. A fixed point.

Starting with the input $x = 0.7$, one press of the cosine (COS) key gives the output $0.76484\ldots$. If we leave this number on the calculator screen and press COS again, the calculator takes the former output as new input and gives $0.72149\ldots$. As you continue to press the COS key, the input is changed less and less, eventually resulting in a number X whose cosine is the same as X. Since the cosine function doesn't change X, this number is called a **fixed point** of the cosine function. We show in Fig. 4.25 a graphical interpretation of a fixed point of the cosine function (any point common to the graphs of $y = \cos x$ and $y = x$) and the procedure we used for calculating its value.

Pushing the COS key converts an input x to $\cos x$. We use vertical lines to denote the key press COS, which converts input to output. In the top figure we draw a vertical line from the input $(0.7, 0)$ to the output $(0.7, \cos(0.7))$. To convert output to input, we use a horizontal line. In the bottom figure we draw a line from $(0.7, \cos(0.7))$ to $(\cos(0.7), \cos(0.7))$. Since $\cos(0.7)$ is now an x-coordinate, it is input. Multiple presses of the COS key wraps a web around the fixed point. A few steps are shown in the bottom figure.

As might be expected from the presence of the derivative function f' in the denominator of the Newton algorithm, iterates that come near a zero of f' can lead to unexpected results. See Exercise 19. Even if f' has no zeros, iterates $x_1, x_2 \ldots$ need not converge to a zero of f. We show in Fig. 4.26 a function for which Newton's algorithm does not give an improving sequence of iterates to a zero c of f. A prominent feature of this graph is that it crosses its tangent line at $(0, 0)$.

Another source of unexpected results can occur when a function has at least two zeros. If, say, we are using Newton's method to determine the smaller of the two zeros, it may happen that the iterates converge to the larger zero. See Exercise 19.

Such difficulties can be minimized by graphing the function near the zero of interest and monitoring the progress of the search.

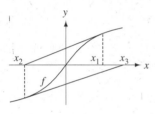

Figure 4.26. Nonconvergence.

Exercises 4.3

Basic

Exercises 1–8: The Intermediate Value Theorem can be stated in the following form: If a continuous function changes sign between a and b, then it has a zero between a and b. Use this result to locate at least one real zero/root of the function/equation between successive "tenths," for example, between 3.7 and 3.8 or between -0.3 and -0.2.

1. $f(x) = x^3 - 2x - 5$

2. $f(x) = 2^x - 4x$

3. $x = 1 - x^3/10$

4. $x + \sin x = 1$

5. $f(x) = x \tan x - 1, \quad 0 < x < 1.5$

6. $f(x) = x^3 - 4x^2 - x + 3$

7. $f(x) = \ln x - x + 2, \quad x > 0$

8. $\tan^3 \theta - 8 \tan^2 \theta + 5 \tan \theta - 4 = 0, \quad 0 \le \theta < 1.5$

Exercises 9–16: Use Newton's method in finding the zero or root to the stated precision. Use your intuition in deciding the required number of iterations. The function cosh in Exercise 15 is the hyperbolic cosine function. Its derivative is the hyperbolic sine function, sinh. These functions were defined as $\cosh x = (e^x + e^{-x})/2$ and $\sinh x = (e^x - e^{-x})/2$ in Chapter 2.

9. Find to four decimal places the real zero of the function $f(x) = x^3 - 5$. Let $x_1 = 2.0$.

10. Find to four decimal places the real zero of the function $f(x) = -x + \cos x$. Let $x_1 = 1.0$.

11. Find to four decimal places the real zero of the function $f(x) = \tan x - x$, where $3.3 \le x \le 4.7$. Let $x_1 = 4.6$.

12. Find to four decimal places the zero of the function $f(x) = 5x^3 - 7x^2 + 9x - 41$ in the interval $[0, 4]$.

13. Find to four decimal places the zero of the function $f(x) = -6x^5 - 11x^4 + 2x + 2$ in the interval $[-1, 0]$.

14. Find to four decimal places the zeros of the function $f(x) = 2^x - 4x$.

15. Find to four decimal places the three smallest positive zeros of the equation

$$\cos(68.617 \sqrt{x}) \cosh(68.617 \sqrt{x}) = -1.$$

16. The equation $x^3 - 2x - 5 = 0$ was used by Wallis in 1685 to illustrate Newton's method. It has been used ever since in works dealing with the numerical solution of equations. Find to four decimal places the real root of this equation.

Growth

17. Locate between successive tenths the zeros of the function

$$f(x) = (x - 2)^{1/3} + 2x^2 - 15.$$

18. Locate between successive integers the first four positive zeros of the function

$$f(x) = \cos x \cosh x + 1.$$

This function becomes very large for relatively small values of x. For example, $f(6) \approx 194.68$. This feature makes it difficult to graphically find approximate locations of zeros. See Exercise 15.

19. The function $f(x) = \cos(\ln x)$, $x > 0$, has a zero between 0 and 1. Explain what happens when the first guess is $x_1 = 0.5$.

20. In Example 2 we discussed the fixed point of the cosine function. Give a similar discussion for the function nex defined by $\text{nex}(x) = e^{-x}$. (The expression "nex" is intended to suggest a negative exponential.) Compare Newton's method and the fixed point iteration. For efficiency you may wish to program your calculator so that it has, in effect, a NEX key.

21. The Intermediate Value Theorem is often stated in this way: If f is continuous on an interval I, u and $v \in I$, and W is between $f(u)$ and $f(v)$, then there is a number w between u and v for which $f(w) = W$. The name of the theorem arises since W can be reasonably described as being a value of f intermediate between $f(u)$ and $f(v)$. A continuous function takes on all of its intermediate values, that is, skips no intermediate value. Show that these statements follow from the statement we gave earlier, namely, if f is continuous on an interval $[u, v]$ and $f(u)f(v) < 0$, then f has a zero c in (u, v). (Hint: Define the function $F(x) = f(x) - W$.)

22. Use either form of the Intermediate Value Theorem (see the preceding exercise) to show that if f and g are continuous functions defined on an interval I and $[f(a) - g(a)][f(b) - g(b)] < 0$ for points a and b of I, then there is a point w between a and b for which $f(w) = g(w)$.

23. By applying Newton's method to the function $f(x) = x^2 - a$, where $a > 0$, justify the ancient divide-and-average rule discussed in Exercise 21 of Section 4.2.

Exercises 24–27: Use the "error bound" E to calculate a number x within E of a zero of f. Recall the discussion in which it was shown that if $f(x - E)f(x + E) < 0$, then x is within E of a zero of f.

24. $f(x) = \ln x - x + 2$, $x \ge 1$; $E = 0.001$

25. $f(x) = \sin^{-1} x - 2x^2$, $0 < x < 1$; $E = 0.001$

26. $f(x) = \sqrt{x} - \tan x$, $0 \le x \le 1$; $E = 0.001$

27. $f(x) = x^5 + x^3 - 1$, $0 \le x \le 1$; $E = 0.001$

Exercises 28–35: Use Newton's method in finding the zero or root to the stated precision. Use your intuition in deciding the required number of iterations.

28. Find to three decimal places the two zeros of $f(x) = (x - 2)^{1/3} + 2x^2 - 15$ in $[-5, 5]$. (Recall Exercise 17.)

29. Find to one decimal place the first four positive zeros of the function

$$f(x) = \cos x \cosh x + 1.$$

(Recall Exercise 18.)

30. Find to within $0.1°$ the real zero of the equation

$$\tan^3 \theta - 4\tan^2 \theta + \tan \theta - 4 = 0.$$

31. Find to within $0.1°$ the three real zeros of the equation

$$\tan^3 \theta - 8\tan^2 \theta + 17\tan \theta - 8 = 0.$$

32. The volume V (in cubic meters) of 1 mole of a gas is related to its temperature T (in kelvins) and pressure P (in atmospheres) by the ideal gas law $PV = RT$. A more accurate equation is van der Waals' equation

$$\left(P + \frac{a}{V^2}\right)(V - b) = RT.$$

The constant R is 0.08207. For carbon dioxide, $a = 3.592$ and $b = 0.04267$. Find to two decimal places the volume V of 1 mole of carbon dioxide if $P = 2.2$ atmospheres and $T = 320$ K.

33. Find to five decimal places all the zeros of the Chebyshev polynomial

$$128x^8 - 256x^6 + 160x^4 - 32x^2 + 1.$$

You may wish to plot this function to choose initial guesses. You can cut your work in half by an observation. You may check your results by using the fact that the zeros x_k are given by

$$x_k = \cos[(2k + 1)\pi/16], \quad k = 0, 1, \ldots, 7.$$

34. Pulleys of radii R and r are connected by a taut belt of total length L, as shown in the following figure. Letting $R = 200$ cm and $r = 100$ cm, and denoting by x the distance between pulley centers, express L

in terms of R, r, and x. Find x to within one decimal place for each of the values $L = 2000$, $2100, \ldots, 2800$ cm.

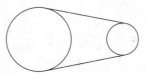

35. Newton's method may be used to find complex zeros. One of the difficulties in finding complex zeros is in locating initial approximations. We illustrate the procedure with the function $f(x) = x^3 - 1$, which we may factor as $f(x) = (x - 1)(x^2 + x + 1)$. The zeros of f are 1 and $-1/2 \pm i\sqrt{3}/2$. We use Newton's method in finding one of the complex zeros. Let $x_1 = -0.4 + i0.8$. Use your calculator to show that $x_2 = -0.516666666667 + i0.866666666667$. Continuing, we find $x_4 = -0.5 + i0.866025403785$. It is clear that the iterates are converging to $-1/2 + i\sqrt{3}/2 = -0.5 + i0.866025403785$. Use Newton's method in finding all of the zeros of the polynomial

$$x^4 - 5x^3 + 21x^2 + 13x + 49.$$

There are zeros near $3 + 4i$ and $-0.5 + 1.3i$.

Review/Preview

36. Divide the interval $0 < x < \infty$ into three parts on each of which $f(x) = 2x - 3\ln x$ is of one sign.

37. Sketch the graph of the sine function for $0 \leq x \leq 2\pi$. Note that the graph goes up or increases on $[0, \pi/2]$ and $[3\pi/2, 2\pi]$ and goes down or decreases on $[\pi/2, 3\pi/3]$. Describe these intervals in terms of the behavior of the cosine function.

38. A particle is moving on the x-axis so that at any time $t \geq 0$ its position is $x(t) = t^2 - 5t + 6$. Describe the times at which the particle is moving to the right.

39. A particle is in simple harmonic motion on the x-axis. At any time $t \geq 0$ its position is $x(t) = 3.75\cos(4.1t - 0.77)$. Find the first time interval in which it is moving to the left.

4.4 MONOTONICITY AND CONCAVITY

The graphs of a function f and its derivative f' are shown in Fig. 4.27. The graph of f is divided into sections by points P, Q, R, S, and T. The graph of f is *decreasing* on intervals (p, q) and (s, t) and *increasing* on (q, s). The point R divides the graph

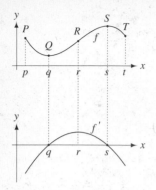

Figure 4.27. Monotonicity and concavity.

Monotone Functions

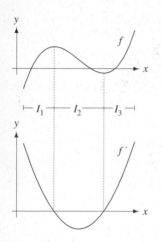

Figure 4.28. f and f'.

into sections PR and RT. The graph is *concave up* on (p, r) and is *concave down* on (r, t). The point R is an *inflection point* of the graph, where the graph changes from concave up to concave down.

Our main goal in this section is to determine intervals on which f is **monotone,** that is, either increasing or decreasing, or is concave up or down. As suggested in the figure, the derivative function is important in locating such intervals. For example, it appears that the points q and s are zeros of f'. What can you say about the point r in relation to the graph of f'? Is the point $(r, f'(r))$ like the point S in any way? If so, what is $f''(r)$?

Monotonicity and concavity are useful in graphing a function or describing its graph. Since we have assumed that you have the means to graph most functions, we place less importance on the mechanics of graphing and more on determining intervals on which f is monotone, concave up, or concave down.

We define what we mean by an increasing or a decreasing function and give an efficient test for these properties. A function f defined on a set including an interval I is *increasing* on I if, for all points u and v of I, $u < v$ implies that $f(u) < f(v)$; f is *decreasing* if, for all points u and v of I, $u < v$ implies that $f(u) > f(v)$.

We use this definition to show that the function $f(x) = x^2$ is increasing on $I = [0, \infty)$. If $0 \le u < v$, then

(1) $$f(v) - f(u) = v^2 - u^2 = (v - u)(v + u) > 0$$

since $v - u > 0$ and $v + u > 0$. It follows that $f(u) < f(v)$. Hence, f is increasing on $[0, \infty)$.

Although this argument is a direct application of the definition, it is often easier to determine intervals on which a function is increasing or decreasing by determining the zeros of f'. Referring to Fig. 4.28, note that f is increasing on intervals I_1 and I_3 and decreasing on I_2. This property of f (or its graph) corresponds to a property of the derivative f' (or its graph). From the figure it appears that the derivative is positive on I_1 and I_3, negative on I_2. Roughly, if the graph of f has positive slope on an interval, then as x increases the graph must rise. Similarly, if f has negative slope on an interval, then as x increases the graph of f must fall. We summarize these observations in the I/D test, so named for quick reference.

I/D Test

Let f be defined on a set including an open interval I and assume that f is differentiable on I.

If $f'(x) > 0$ for $x \in I$, then f is increasing on I.
If $f'(x) < 0$ for $x \in I$, then f is decreasing on I.

Applying this result to $f(x) = x^2$, it follows that f is increasing on $(0, \infty)$ since $f'(x) = 2x > 0$ for $x \in (0, \infty)$. Also, f is decreasing on $(-\infty, 0)$ since $f'(x) = 2x < 0$ for $x \in (-\infty, 0)$. See Fig. 4.29.

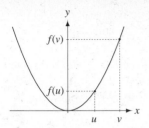

Figure 4.29. f is increasing on $(0, \infty)$.

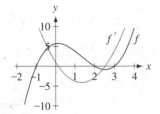

Figure 4.30. Increasing and decreasing.

The function $f(x) = \sqrt{x}$, for $x \geq 0$, is increasing on $(0, \infty)$ since $f'(x) = 1/(2\sqrt{x})$ for $x > 0$. The derivative does not exist at $x = 0$. Since $\sqrt{x} > \sqrt{0}$ for $x > 0$, it follows that f is increasing on $[0, \infty)$.

We give three more examples in which we determine intervals on which a function is increasing or decreasing.

■ **EXAMPLE 1** Use the I/D Test to determine the intervals on which

$$f(x) = x^3 - 4x^2 + x + 6, \quad -\infty < x < \infty$$

is increasing or decreasing.

Solution. Figure 4.30 includes the graphs of f and f'. A glance at the figure shows that, roughly, f is increasing on $(-\infty, 0.1)$, decreasing on $(0.1, 2.5)$, and increasing on $(2.5, \infty)$. We may determine these intervals more accurately by zooming in on the graph or by using the I/D Test. For the latter, we calculate

$$(2) \qquad\qquad f'(x) = 3x^2 - 8x + 1$$

and determine its zeros. We find

$$(3) \qquad\qquad x = \frac{8 \pm \sqrt{64 - 12}}{6} \approx 2.54 \text{ or } 0.13.$$

Since f' is continuous, it can change sign only at its zeros. So we can determine if f' is positive or negative in an interval by testing one point from each of the three intervals determined by the two zeros of f'. Noting that $f'(0) = 1$, for example, shows that f' is positive on $(-\infty, 0.13)$. In this way, we find that f is increasing on $(-\infty, 0.13)$, decreasing on $(0.13, 2.54)$, and increasing on $(2.54, \infty)$. ■

■ **EXAMPLE 2** Determine the intervals on which

$$f(x) = \sin x^2, \quad 0 \leq x \leq 3$$

is increasing or decreasing.

Solution. We may use the definition of increasing/decreasing functions or the I/D Test to determine the intervals on which f is increasing or decreasing. The graph of f is that of the sine function over 9 radians, but compressed into the space above $[0, 3]$. See Fig. 4.31. Using this idea, we may infer from well-known properties of the sine function that:

f is increasing for $x^2 \in [0, \pi/2]$ and $x^2 \in [3\pi/2, 5\pi/2]$.
f is decreasing for $x^2 \in [\pi/2, 3\pi/2]$ and $x^2 \in [5\pi/2, 9]$.

Expressed in terms of x,

f is increasing on $[0, \sqrt{\pi/2}] \approx [0, 1.25]$ and $[\sqrt{3\pi/2}, \sqrt{5\pi/2}] \approx [2.17, 2.80]$.
f is decreasing on $[\sqrt{\pi/2}, \sqrt{3\pi/2}] \approx [1.25, 2.17]$ and $[\sqrt{5\pi/2}, \sqrt{9}] \approx [2.80, 3.00]$.

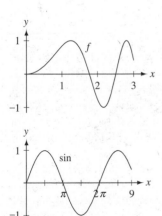

Figure 4.31. f and sin.

The I/D Test gives nearly the same information. The derivative of f is $f'(x) = 2x\cos x^2$. The zeros of f' in $(0, 3)$ are $x = \sqrt{\pi/2}$, $\sqrt{3\pi/2}$, and $\sqrt{5\pi/2}$. These three points determine the open intervals $(0, \sqrt{\pi/2})$, $(\sqrt{\pi/2}, \sqrt{3\pi/2})$, $(\sqrt{3\pi/2}, \sqrt{5\pi/2})$, and $(\sqrt{5\pi/2}, 3)$ on which f' has constant sign. From this we may use the I/D Test to obtain the information listed above, except that we may not infer that f is increasing or decreasing on closed intervals. We may add the end points by testing them separately.   ■

■ **EXAMPLE 3**   Determine the intervals on which

$$f(x) = (x^2 - 1)e^x, \quad -\infty < x < \infty$$

is increasing or decreasing.

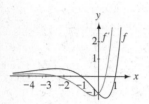

Figure 4.32. f and f'.

Solution.   We show parts of the graphs of f and f' in Fig. 4.32. Depending on how easy it is to produce such a figure and on the accuracy required in describing the intervals on which f is increasing or decreasing, a solution may lean more or less heavily on such a graph. We give an argument independent of the figure but note that the intervals on which f is increasing or decreasing can be read from the graph more or less accurately. If more accuracy is required, we may zoom in on parts of the graph.

For the I/D Test we need the zeros of f'. We have

$$\begin{aligned} f'(x) &= 2xe^x + (x^2 - 1)e^x \\ &= (x^2 + 2x - 1)e^x. \end{aligned}$$

Setting the quadratic factor equal to zero and solving, we find

$$x = -1 \pm \sqrt{2} \approx -2.41 \text{ or } 0.414.$$

Since the leading term of the quadratic factor is positive, $f'(x)$ is positive, negative, and positive for $x \in (-\infty, -2.41)$, $(-2.41, 0.414)$, and $(0.414, \infty)$, respectively. Applying the I/D Test, we conclude that f is increasing on $(-\infty, -2.41)$ and $(0.414, \infty)$ and decreasing on $(-2.41, 0.414)$.   ■

Concavity

A line segment joining two points of the graph of a function is often called a **secant line** of the graph. We show two secant lines tT in Fig. 4.33. A function f is **concave up** on an interval (p, q) if its graph lies below each of its secant lines. A function f is **concave down** on an interval (p, q) if its graph lies above each of its secant lines. Less formally, f is concave up on an interval if its graph "cups up" or "holds water." It is concave down on an interval if its graph "cups down" or "spills water."

These definitions of concavity, like those of increasing and decreasing functions, do not require that f be differentiable. However, although they are geometrically clear, they do not provide an easy test to locate intervals on which a function is concave up or concave down or to locate a point such as $(q, f(q))$, called an **inflection point,** which separates intervals on which f has opposite concavity.

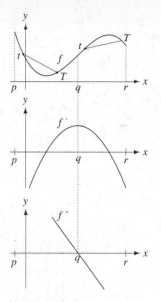

Figure 4.33. Tests for concavity.

The I/D Test is a convenient means to locate intervals on which f is increasing or decreasing. To use it, we must assume that f is differentiable. If we assume this much about f here, we can test for concavity. For, referring to Fig. 4.33, it appears that f is concave up on any interval on which f' is increasing. For if f' is increasing on (p, q), the slopes of the tangent lines to the graph of f are increasing, which means that as we move on the graph from left to right, the tangent lines rotate counterclockwise. Thus the graph of f must "hold water," or lie below its secant lines.

How would this informal argument go if we were on an interval (q, r) on which f' is decreasing?

It appears that if we can determine where f' increases or decreases, then we have a way to test for concavity. For this we may apply the I/D Test to f'. For the I/D Test we must assume that f' is differentiable, that is, that f has a second derivative f''.

The graphs of f, f', and f'' are shown in Fig. 4.33 (the figure does not include the entire graphs of f' and f'' on (p, r)). The function f'' is positive on (p, q) and negative on (q, r). By the I/D Test, this means that the function f' is increasing on (p, q) and decreasing on (q, r). Hence, f is concave up on (p, q) and concave down on (q, r).

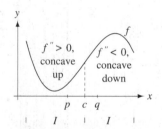

Figure 4.34. Concavity test.

Concavity Test

Let f be defined on a set S including an open interval I and assume that f' is differentiable on I. Referring to Fig. 4.34,

If $f''(x) > 0$ on I, then f is concave up on I.

If $f''(x) < 0$ on I, then f is concave down on I.

A point $c \in S$ is an inflection point of f if f'' changes sign at c, which means that there are intervals $(p, c) \subset S$ and $(c, q) \subset S$ such that f'' is positive on one of these intervals and negative on the other. For most of the functions we consider, if f'' changes sign at c, then $f''(c) = 0$.

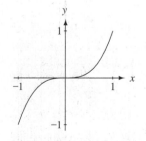

Figure 4.35. Cubic.

■ **EXAMPLE 4**   Describe where $f(x) = x^3$ is concave up or concave down and locate any inflection points.

Solution.   Since $f''(x) = 6x$, we infer from the Concavity Test that f is concave down on $(-\infty, 0)$, concave up on $(0, \infty)$, and has an inflection point at $x = 0$. This agrees with the graph of f shown in Fig. 4.35.   ■

■ **EXAMPLE 5**   In Example 3 we found the intervals on which

$$f(x) = (x^2 - 1)e^x, \quad -2 \le x \le 1.5$$

is increasing or decreasing. The graph of f is shown in Fig. 4.36. Determine the intervals on which f is concave up or concave down and the two inflection points.

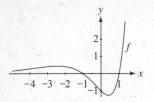

Figure 4.36. Two inflection points?

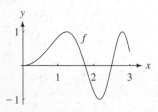

Figure 4.37. Inflection points via Newton's method.

Solution. Attempt to locate the two inflection points from the figure. For their exact values we calculate f''. We have

$$f(x) = (x^2 - 1)e^x$$
$$f'(x) = (x^2 + 2x - 1)e^x$$
$$f''(x) = (x^2 + 4x + 1)e^x$$

The zeros of f'' are easily found to be $-2 \pm \sqrt{3} \approx -3.732$ and -0.268. The quadratic factor of f'' determines the sign of f'' since e^x is always positive. We find that f is concave up on $(-\infty, -3.732)$ and $(-0.268, \infty)$. It is concave down on $(-3.732, -0.268)$. The points -3.732 and -0.268 are inflection points since it is clear that f'' changes sign at each point. ∎

We use the function $f(x) = \sin x^2$, $0 \le x \le 3$, discussed in Example 2, as our third example. The zeros of the second derivative of this function are not so easily determined as in the first two examples. From Fig. 4.37 it appears that there are three inflection points and four intervals on which f is either concave up or concave down.

■ **EXAMPLE 6** Find the leftmost inflection point of the function $f(x) = \sin x^2$, $0 \le x \le 3$.

Solution. We begin by calculating f''. We have

$$f(x) = \sin x^2$$
$$f'(x) = 2x \cos x^2$$
$$f''(x) = 2\cos x^2 - 4x^2 \sin x^2.$$

From Fig. 4.37 it appears that f'' has a zero at $x \approx 0.7$ since the graph appears to change from concave up to concave down there. Setting $f''(x)$ equal to zero and factoring out $4\cos x^2$, we have

$$(4\cos x^2)(0.5 - x^2 \tan x^2) = 0.$$

Since $\cos x^2 \ne 0$ near 0.7, we may simplify this equation to

$$(4) \qquad\qquad x^2 \tan x^2 - 0.5 = 0.$$

Using Newton's method or other means of solving this equation, we find $x = 0.808\ldots$. Since f'' changes sign at this point, we conclude that $x = 0.808\ldots$ is an inflection point of f. ∎

Proof of the I/D Test

The I/D and Concavity Tests have strong connections with the geometry of the graph of a function. We can "see" that f is increasing or is concave up on an interval and can correlate these features of the graph with properties of f' or f''. Nevertheless, mathematics is based on more than visual proofs. We give an argument for the I/D Test, which is the simpler of these two tests. For this we need two basic results, the Max/Min Theorem and the Mean-value Inequality.

Max/Min Theorem

If *f* is continuous on the interval [*a*, *b*], then *f* has a maximum value and a minimum value on [*a*, *b*].

The phrase "has a maximum value and a minimum value on [*a*, *b*]" means that there are numbers $p, q \in [a, b]$ for which

$$f(p) \le f(x) \le f(q), \quad \text{for every } x \in [a, b].$$

See Fig. 4.38. The numbers $f(p)$ and $f(q)$ are minimum and maximum values of f. The points $(p, f(p))$ and $(q, f(q))$ are low and high points on the graph of f. In practice, both the number $f(q)$ and the point $(q, f(q))$ are referred to as the maximum of f.

We do not prove the Max/Min Theorem. Its proof is beyond the level of this text.

The Mean-value Inequality is the second result we need for the proof of the I/D Test. We give two plausibility arguments based on motion and one based on slopes.

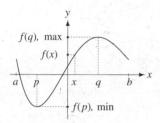

Figure 4.38. Max/min.

Mean-value Inequality

If *f* is continuously differentiable on *I* = [*a*, *b*], then

(5) $(b - a)m \le f(b) - f(a) \le (b - a)M,$

where *m* and *M* are the minimum and maximum values, respectively, of *f*′ on *I*.

For the first two arguments we interpret f as the coordinate position of an automobile. Let $x = f(t)$ be the position of an automobile moving on the x-axis for $t \in [a, a + h]$, where $h = b - a$. We assume the car is moving from left to right, so that $f(a + h) - f(a)$ is the distance traveled in h time units. Let m and M be the minimum and maximum velocities of the automobile. We may rewrite (5) in the form

(6) $$m \le \frac{f(a + h) - f(a)}{h} \le M.$$

This inequality is the statement that the average velocity of the automobile on the time interval [$a, a + h$] is between its minimum and maximum velocities.

The second argument is an interpretation of (5), which we rewrite as

$$h \cdot m \le f(b) - f(a) \le h \cdot M.$$

We may read this as the statement that during a time h, the automobile travels at least as far as it would at its minimum velocity and not further than it would at its maximum velocity.

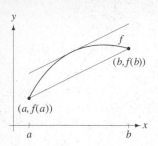

Figure 4.39. Geometry of the Mean-value Inequality.

In Fig. 4.39 we show the graph of a function f and a tangent line to the graph. To draw this particular tangent line, start with the secant line joining the end points $(a, f(a))$ and $(b, f(b))$. Move a copy of this line upward (keeping it parallel to itself) until it is about to lose contact with the graph of f. It is clear that the final position of the moving secant line is tangent to the graph of f. This means that the slope of the secant line, which is

$$\frac{f(b) - f(a)}{b - a},$$

must lie between the smallest and largest slopes of the graph of f, that is,

(7)
$$m \le \frac{f(b) - f(a)}{b - a} \le M.$$

This is equivalent to (5), the Mean-value Inequality.

Proof of a Special Case of the I/D Test

We assume that f is continuously differentiable on an open interval (a, b). We must show that if $f'(x) > 0$ for $x \in (a, b)$, then f is increasing on (a, b). To show that f is increasing, choose any points u and v of (a, b), so that $a < u < v < b$. We must show that $f(u) < f(v)$. We apply the Mean-value Inequality to f on the interval $[u, v]$. We have

$$f(v) - f(u) \ge (v - u)m > 0,$$

where m is the minimum value of f' on $[u, v]$. From this we see that $f(u) < f(v)$. It follows that f is increasing on (a, b). The second part of the I/D Test ($f' < 0$ implies f is decreasing) is proved in the same way.

Exercises 4.4

Basic

Exercises 1–8: Use the definition of increasing/decreasing functions to show that the function is increasing or decreasing on the interval. A graph may be used here and elsewhere in this set of exercises to view the geometric properties of the function, but analytic arguments are expected.

1. x^4, $x > 0$
2. x^3, $-\infty < x < \infty$
3. $1/x$, $x > 0$
4. $1/(x^2 + 1)$, $x > 0$

5. $x^2 + x$, $x < -1/2$
6. $x^2 - x$, $x > 1/2$
7. $(x + 1)/x$, $x > 0$
8. $x/(x - 1)$, $x > 1$

Exercises 9–16: Use the I/D Test to find the intervals on which the function is increasing or decreasing. The functions are the same as in Exercises 1–8 except the domains of the functions may be different.

9. x^4, $-\infty < x < \infty$
10. x^3, $-\infty < x < \infty$
11. $1/x$, $x \ne 0$
12. $1/(x^2 + 1)$, $-\infty < x < \infty$

13. $x^2 + x$, $-\infty < x < \infty$
14. $x^2 - x$, $-\infty < x < \infty$
15. $(x + 1)/x$, $x \ne 0$
16. $x/(x - 1)$, $x \ne 1$

Exercises 17–24:   Use the I/D Test to find the intervals on which the function is increasing or decreasing.

17. $\sin x^3, 0 \le x \le 2$
18. $\cos x^3, 0 \le x \le 2$
19. $\sin^2 x, 0 \le x \le \pi$
20. $x/(x^2 + 1), -\infty < x < \infty$

21. $\ln(x/(x^2 + 1)), x > 0$
22. $\ln(x/(x + 1)), x > 0$
23. $xe^x, -\infty < x < \infty$
24. $xe^{-x}, -\infty < x < \infty$

Exercises 25–30:   Use the Concavity Test to find the intervals on which the function is concave up or concave down. Also, locate any inflection points.

25. $\sin^2 x, 0 \le x \le \pi$
26. $\cos^2 x, 0 \le x \le \pi$
27. $1/(x^2 + 1), -\infty < x < \infty$

28. $1/(x^3 + 1), x \ne -1$
29. $xe^x, -\infty < x < \infty$
30. $xe^{-x}, -\infty < x < \infty$

Growth

31. Use the definition to determine the intervals on which $f(x) = \sin \sqrt{x}, 0 \le x \le 10$, is increasing.
32. Use the definition to determine the intervals on which $f(x) = \cos \sqrt{x}, 0 \le x \le 10$, is increasing.
33. Calculate the rightmost inflection point in Example 6.
34. Calculate the middle inflection point in Example 6.
35. Find the two inflection points of the function $f(x) = (x^2 + 1)e^x, -\infty < x < \infty$. Where is f increasing?
36. Is it contrary to the Concavity Test that for $f(x) = x^4$, $f''(0) = 0$ although $x = 0$ is not an inflection point? Explain.
37. Give specific functions to illustrate the following pairs of properties:
   a. Increasing and concave up on $(0, 1)$.
   b. Increasing and concave down on $(0, 1)$.
   c. Decreasing and concave up on $(0, 1)$.
   d. Decreasing and concave down on $(0, 1)$.
38. Discuss the concepts of increasing, decreasing, concave up, concave down, and inflection point in terms of motion.
39. Let $f(t) = 30/(3 + 7e^{-0.02t}), t \ge 0$. The graph of f is called a *logistic curve* and measures population size $f(t)$ as a function of t. Graph f and find its inflection point. What are the initial and limiting populations? Interpret the inflection point in terms of population growth. If you find a way of using the fact that $f'(t) = 0.002 f(t)(10 - f(t))$, your work will be simplified.
40. Do you thing it possible that on $(0, \infty)$, a function f can be increasing and concave up? Can you give an example? Do you think it possible that on $(0, \infty)$, a function f can be increasing, concave up, and be *bounded above*? "Bounded above" means that a

constant K can be found so that $f(x) \le K$ for all $x \in (0, \infty)$. The Mean-value Inequality can be used to resolve the question.
41. Show that $f(x) = x^2$ is concave up on $(-\infty, \infty)$ by showing that its graph lies below all of its secant lines. Recall that a secant line is a line segment joining two points $(u, f(u))$ and $(v, f(v))$ of the graph of f. The graph of f would lie below one of its secant lines if $f(x) \le g(x)$ for $x \in [u, v]$, where g is the function whose graph is the secant line.
42. Show that if f is increasing (decreasing) on an interval I, then $-f$ is decreasing (increasing) on I.
43. Show that if f is concave up (concave down) on an interval I, then $-f$ is concave down (concave up) on I.
44. Let I be an open interval. Show that if f is differentiable on I and $f'(x) = 0$ on I, then f is constant on I; that is, there is a constant c for which $f(x) = c$ on I. Use this result to show that if f and g are differentiable on I and $f(x) = g(x)$ on I, then f and g "differ by a constant"; that is, there is a constant c for which $f(x) - g(x) = c$ on I.
45. The Mean-value Theorem (not Inequality) is:

   If f is continuous on $[a, b]$ and differentiable on (a, b), then there is a point $c \in (a, b)$ such that

   $$f(b) - f(a) = f'(c)(b - a).$$

   Show that with the help of the Intermediate Value Theorem (see Section 4.3), the Mean-value Theorem follows from the Mean-value Inequality.
46. Use the Mean-value Inequality in showing that $\sin x \le x$ for $x \ge 0$. Show that $|\sin x| \le |x|$ for all x.

Review/Preview

Exercises 47–52:   Each equation defines y as a function of x. Calculate dy/dx at the given point, which satisfies the equation at least approximately. Use implicit differentiation.

47. $x^2y + y^3 = 1$; $(\sqrt{7/4}, 1/2)$

48. $y^3 + 2x^2y + 1 = 0$; $(1, -0.4534)$

49. $x^3 + y^3 - 1 = 0$; $(0.5, 0.9565)$

50. $y^2(x + y) = -x + y$; $(-1, 1.8393)$

51. $y = \sin(x + y)$; $(0.1, 0.7538)$

52. $\sqrt{y - 3x} + y^2 = x + 2$, $-2 \le x \le 0$; $(-0.5, 0.3660)$

4.5   RELATED RATES

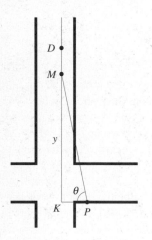

Figure 4.40. dy/dt from $d\theta/dt$?

Referring to Fig. 4.40, a motorist M has passed an intersection, not noticing a police car parked at P. Two officers are testing a device that measures speed without emitting energy detectable by motorists. Their idea is to use existing light to measure the rate at which the angle θ is changing as the motorist passes a given point D. They plan to market the device under the trademark NAB.

NAB requires that two numbers be entered before speeds can be measured. The distance KP from the police car to the center line of the traffic lane, here 50 feet, is one input. The other is the distance between K and D, here 300 feet. The point D for this particular corner is the location of an overhead caution signal. Just after the officers entered the necessary data, motorist M passed the corner. As M passed point D, the officers used NAB to measure $d\theta/dt \approx 0.060$. This number is the rate of change in M's angle from KP, in radians per second. After a brief calculation, NAB printed the time, date, and speed on a blank traffic citation.

Putting aside the difficult optical, mechanical, and electronic design problems the officers overcame in building NAB, what is the mathematical idea on which it is based? Given $d\theta/dt$, how is the motorist's speed dy/dt calculated?

Investigation

As a step toward calculating dy/dt, we try to relate the primary variables y and θ. From triangle MKP in the figure, we have

$$(1) \qquad\qquad y = 50 \tan \theta.$$

It is important to note that this equation relates the motorist variables y and θ at any time t, not just the time at which the motorist passes D. Each of y and θ is an implicit function of t. We say "implicit" since we have not given formulas for y and θ in terms of t. Although it is often possible to give such formulas, this step is usually not necessary.

Since NAB assists in measuring $d\theta/dt$ and we wish to calculate dy/dt, we differentiate both sides of (1) with respect to t. We have

$$\frac{d}{dt}(y) = \frac{d}{dt}(50 \tan \theta).$$

Using the chain rule to differentiate the composite function on the right,

$$(2) \qquad\qquad \frac{dy}{dt} = 50 \sec^2 \theta \, \frac{d\theta}{dt}.$$

At the time t the motorist is passing D, $d\theta/dt = 0.060$. To complete the calculation we need to know θ at that time. From (1) we find

$$\tan \theta = \frac{300}{50} = 6.$$

Substituting into (2) and referring to Fig. 4.41, we obtain

$$\frac{dy}{dt} = 50 \cdot 37 \cdot 0.060 \approx 111 \text{ ft/s} \approx 76 \text{ mph.}$$

Figure 4.41. Side calculation.

Given the inputs KP and KD and, from the optical component of NAB, the value of $d\theta/dt$, the required calculations are not difficult. Indeed, in building NAB the officers were able to program and interface an inexpensive calculator for the necessary calculations and to provide output to the printer.

The NAB problem is a typical related-rates problem. Here is a slightly expanded outline of our approach to the problem.

1. First read and reread the problem until you understand it, including what exactly is given and what is wanted.
2. Next, with the help of a diagram or other graphic overview of the problem, decide on the important variables.
3. Use the diagram or other means to write an equation relating these variables.
4. Differentiate the equation with respect to t to find an equation relating the time derivatives of the variables.
5. Finally, solve the equation relating the time derivatives for the derivative of interest, at the time or place of interest. At this point in the problem, but not before, numerical information specifying the time or place of interest becomes useful.

Try to apply these steps in the following related-rates examples.

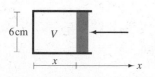

Figure 4.42. Changing volume.

■ **EXAMPLE 1** Referring to Fig. 4.42, suppose a piston is compressing a gas contained in a cylinder. Given that the piston is moving into the cylinder at 10 cm/s, at what rate is the volume of the gas changing?

Solution. The volume V and displacement x of the piston are the variables we wish to relate. The rate of change dx/dt of x with respect to time is given, and we are to calculate dV/dt. Our first goal is to give an equation relating x and V. Since the volume of a cylinder of radius r and height h is $\pi r^2 h$, we have

$$(3) \qquad\qquad V = 9\pi x.$$

Based on the underlying assumption that both V and x are functions of t, and recalling that dx/dt is given and we wish to calculate dV/dt, we differentiate both sides of (3) with respect to t, obtaining

(4)
$$\frac{dV}{dt} = 9\pi \frac{dx}{dt}.$$

Since the piston is moving into the cylinder, x is decreasing with t and, hence, $dx/dt < 0$. Specifically, $dx/dt = -10$. Substituting this value into (4),

(5)
$$\frac{dV}{dt} = 9\pi(-10) \approx -282.7 \text{ cm}^3/\text{s}$$

Since the volume is decreasing, the fact that $dV/dt < 0$ makes sense. ■

■ **EXAMPLE 2**   A particle is moving on the curve shown in Fig. 4.43 and has position vector $\mathbf{r} = \mathbf{r}(t) = (x(t), y(t))$. It is known that its velocity in the x direction is 2 m/s and the coordinates x and y of its position at any time satisfy

(6)
$$x^2 y + y^3 = 1.$$

Calculate its velocity at the time when $x = 3$.

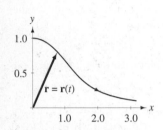

Solution.   The coordinates x and y of the particle are functions of time t. Since the particle is moving on the path defined by (6), the functions $x = x(t)$ and $y = y(t)$ must satisfy that equation. By differentiating (6) with respect to t, we may relate the rates of change of x and y with respect to t.

Before differentiating (6), we review a calculation from Chapter 2. Putting aside for the moment any idea that x and y are functions of t, we say that (6) defines y as a function of x, that is, if x is given, then there is one and only one real number y that satisfies (6) (see Exercise 33). Although we do not have a formula for this function, we may remind ourselves that such a function exists by writing $y = y(x)$. With this in mind, we differentiate both sides of (6) with respect to x. We have

$$\frac{d}{dx}(x^2 y + y^3) = \frac{d}{dx}(1)$$

$$2xy + x^2\frac{dy}{dx} + 3y^2\frac{dy}{dx} = 0$$

The first two terms on the left came from differentiating the product $x^2 y$ of two functions of x, namely x^2 and y. Not having a formula for y, we must write dy/dx when we differentiate it. In our earlier work we solved the last equation for dy/dx, obtaining

$$\frac{dy}{dx} = \frac{-2xy}{x^2 + 3y^2}.$$

Returning to the related-rates problem, in which $x = x(t)$ and $y = y(t)$, we do a similar calculation. What we must bear in mind is that both x and y are symbols

Figure 4.43. Path of particle.

for unknown functions of t. We have

$$\frac{d}{dt}(x^2y + y^3) = \frac{d}{dt} \quad (1)$$

(7)
$$2x\frac{dx}{dt}y + x^2\frac{dy}{dt} + 3y^2\frac{dy}{dt} = 0.$$

The last equation relates dx/dt and dy/dt, which are the velocities of the particle in the x and y directions. If we know one of these velocities, we may solve for the other. For example, since $dx/dt = 2$ m/s we may solve (7) for dy/dt. For brevity, we use the notations x' and y' for dx/dt and dy/dt. From (7) we have

(8)
$$y' = \frac{-2xx'y}{x^2 + 3y^2} = \frac{-4xy}{x^2 + 3y^2}.$$

We may now write out a general expression for the velocity of the particle, though not as an explicit function of t. We have

(9)
$$\mathbf{v} = (x', y') = \left(2, -\frac{4xy}{x^2 + 3y^2}\right).$$

To calculate $\mathbf{v}$ at the time when $x = 3$, we must first calculate y from (6). We ask you to show in Exercise 3 that the equation $y^3 + 9y - 1 = 0$ has exactly one real root, $y \approx 0.11$. Substituting $x = 3$ and $y = 0.11$ into (9), we find $\mathbf{v} = (2, -0.15)$. This result is reasonable since y is decreasing with t, so that y' would be negative. Also, since $x' = 2$ and it appears that the path of the particle is asymptotic to the x-axis, the speed in the y direction must become small. ■

■ **EXAMPLE 3**   Due to a rupture in an undersea pipeline, crude oil is surfacing at a point 1500 m offshore at the rate of 40 m³/min. It has been determined that the average thickness of the slick is 5 mm. As part of a contingency plan, you have been asked to determine roughly the rate at which the edge of the oil slick is approaching the shore when $x = 400$ m, $x = 200$ m, and $x = 100$ m.

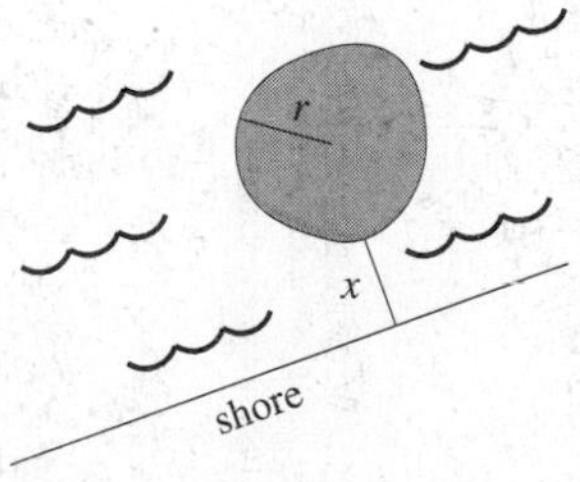

Solution.   For the required rough estimates we assume the sea is calm, with no currents affecting the slick, and that the slick is in the shape of a circle. As shown in Fig. 4.44, let x denote the distance from the edge of the slick to the shore and r the radius of the slick. The volume V of the slick is

$$V = \pi r^2(5/1000) \text{ m}^3.$$

Each of x, r, and V is a function of t. We are given that $V' = 40$ m³/min and are asked to calculate x' when $x = 400$ m, $x = 200$ m, and $x = 100$ m. For this it is sufficient to calculate r' since by differentiating (with respect to t) the equation $x + r = 1500$ we have $x' = -r'$.

   We differentiate the volume equation with respect to t to obtain an equation relating V' and r'. We have

$$V' = 2\pi r r'(5/1000).$$

Figure 4.44. Oil slick.

Replacing V' by 40 and solving for r', we find

$$r' = \frac{40{,}000}{10\pi r} = \frac{4000}{\pi r}.$$

Since we want r' for various values of x, we rewrite this equation in terms of x:

$$r' = \frac{4000}{\pi(1500 - x)}.$$

Setting $x = 400$, $x = 200$, and $x = 100$, we find $r' \approx 1.2$ m/min, $r' \approx 1.0$ m/min, and $r' \approx 0.9$ m/min. Hence the rates at which the edge of the slick is approaching the shore are -1.2 m/min, -1.0 m/min, and -0.9 m/min.

This problem has a more direct solution. We note that after t minutes the volume of the oil slick will be $V = 40t$ m^3. Hence

$$40t = \pi r^2/200.$$

Solving for r^2 and then differentiating with respect to t gives

$$r^2 = \frac{8000t}{\pi}$$

$$2rr' = \frac{8000}{\pi}$$

$$r' = \frac{4000}{\pi r} \text{ m/min.}$$

For $x = 400$, for example, $x' = -r' = -4000/(1100\pi) \approx -1.2$ m/min, as above. ■

■ **EXAMPLE 4**   An underground gasoline storage tank is shown in Fig. 4.45. The tank is a cylinder, with length 6 m and radius of the circular ends 1.5 m. The tank is being filled at the rate of 0.5 m^3/min. Find the rate at which the level of the liquid in the tank is rising when its depth is 2 m.

Solution.   The fill rate of 0.5 m^3/min is the change in the volume of the liquid per minute, that is,

$$V' = 0.5 \text{ m}^3/\text{min.}$$

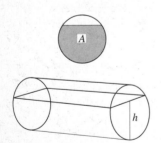

Figure 4.45. Gasoline storage tank.

We are asked to calculate h', where h is the depth of the liquid at time t. We need an equation relating V and h. Letting A denote the area of the submerged segment of the circular end of the tank, we have $V = 6A$. So, to express V in terms of h, we must express A in terms of h. The formula for the area of a segment of a circle of radius a in terms of the height h of the segment is given in most books of formulas:

$$(10) \quad A = \frac{\pi a^2}{2} - a^2 \sin^{-1}(1 - h/a) - (a - h)\sqrt{h(2a - h)}, \quad 0 \le h \le 2a.$$

Despite its complexity, (10) is not difficult to derive. We outline an argument in Exercise 31.

In the following calculation we form $V = 6A$, differentiate V with respect to t, and then simplify the result. We have

$$V = 3\pi a^2 - 6a^2 \sin^{-1}(1 - h/a) - 6(a - h)\sqrt{h(2a - h)}$$

$$V' = -6a^2 \frac{1}{\sqrt{1 - (1 - h/a)^2}}\left(-\frac{h'}{a}\right) + 6h'\sqrt{h(2a - h)}$$

$$- 6(a - h)\frac{1}{2\sqrt{h(2a - h)}}(2ah' - 2hh').$$

This simplifies to

$$V' = 12h'\sqrt{h(2a - h)}.$$

Solving for h' and then replacing a by 1.5, V' by 0.5, and h by 2, we obtain

$$h' = \frac{V'}{12\sqrt{h(2a - h)}}$$

$$h' = \frac{0.5}{12\sqrt{2(2(1.5) - 2)}}$$

$$h' \approx 0.029 \text{ m/min.} \qquad \blacksquare$$

The solutions of these related-rates problems have common features, which we list here as an informal guide to problem solving. These steps reflect the problem solving required whenever physical, biological, or economic phenomena are modeled by mathematical objects. Our comments assume that a problem or subproblem has been identified and stated.

---

1. As you read and **reread** the problem, determine as clearly as possible what is being asked and what a proper solution would involve. Try to capture the essentials of the problem in a sketch or other summarizing diagram, identifying and naming important variables as necessary.

2. Use given information and your own insights to form one or more equations relating the variables you have chosen. For most problem solvers, insights happen as a result of active investigation, not hopeful waiting.

3. Use a mathematical solution procedure to solve the equations. For example, in related-rates problems it is usually appropriate to differentiate the equation or equations with respect to t and solve for the unknown rate.

4. Decide if your solution is reasonable. This important step can help you avoid absurd results and alert you to possible mistakes.

Exercises 4.5

Basic

1. If the speed limit in the NAB problem were 55 mph, what is the largest possible value of $d\theta/dt$ if the motorist is to avoid a ticket? Assume strict enforcement, with no slack given to motorists.

2. If the speed limit is 55 mph, decide if the motorist gets a ticket given that KP is 75 feet, KD is 250 feet, and $d\theta/dt = 0.090$.

3. Use Newton's method in finding the real root of the equation $y^3 + 9y - 1 = 0$, which came up in Example 2. Give evidence that this cubic has only one real root.

4. In Example 3, find the rate at which the oil slick is approaching the shore at the moment it reaches the shore.

5. In Example 3, find the rate at which the oil slick is approaching the shore at the moment it is halfway to the shore.

6. If the thickness of the slick in Example 3 were 4 mm instead of 5 mm, would it approach the shore more slowly or more rapidly? Calculate the rate at which the edge of the slick is approaching the shore when $x = 400$ m.

7. The formula for the area of a circular segment of height h is given in Example 4. Check the accuracy of this formula at $h = 0, a, 2a$. Find the area of a circular segment of height $h = 1/2$ in a circle of radius 1.

8. Give the details in the simplification of V' in Example 4.

9. In Example 4, find the rate at which the level of the gasoline is rising when its depth is 2.9 m.

10. In Example 4, find the rate at which the level of the gasoline is rising when the tank is half full.

11. A particle moves on a circle of radius 5 m. As it passes through (3, 4) the x-coordinate of its velocity is -5 m/s. What is the y-coordinate of its velocity?

12. Variables u and w are functions of time t and are related by the equation $u = \sqrt{w^2 + 1}$. Express dw/dt in terms of du/dt and w.

13. Variables u and w are functions of time t and are related by the equation $uw = \sin w$. Express dw/dt in terms of du/dt and w.

14. A spherical balloon is filled with water. If the water is leaking out at 2 cm³/min, find the rate at which the radius is changing when $V = 1000$ cm³.

15. A blood sample is being drawn from your arm into a cylinder of radius 0.5 cm. If a nurse is withdrawing the piston from the cylinder at the rate of 0.25 cm/s, at what rate (in quarts per hour) are you losing blood? (Fact: 1 cm³ $\approx 1.1 \times 10^{-3}$ qt.)

16. Because of a bumper crop, field corn is being stored outdoors at the local grain elevator. An overhead boom is adding corn to the top of the pile at 0.3 bushels per second. The pile takes the form of a cone whose diameter is four times its height. Find the rate at which the boom must be raised when the pile contains 100,000 bushels. (One bushel is 0.035 m³, approximately.)

17. According to a story for children, a bird visits the top of a mountain once each century and sharpens its beak on the rock. When the mountain has worn away, one day of eternity will have passed. Mt. McKinley rises approximately 5200 m above its base and is roughly conical, with the diameter of its base about four times its height. Suppose that one "sharpening" removes 0.01 mm³ of stone and is done so that the proportions of the cone remain fixed. Calculate the rate of decrease in the height of Mt. McKinley when it has been reduced to half its original volume. What is "one day of eternity," measured in centuries?

18. You are holding a conical cup (height 2 in. and radius 1 in.) full of milk. A "friend" comes along and pushes his cup (same size) into your cup, keeping the axes aligned and pushing at the rate of 3 in./s. At what rate is the milk spilling into your lap when the tip of his cup is 1/2 in. from the bottom of your cup?

Growth

19. Calculate the velocity of the particle in Example 2 at the time the y-coordinate of its position is 3/4.

20. Calculate the velocity of the particle in Example 2 at the time the x-coordinate of its position is 2.0.

21. A particle is moving counterclockwise on the spiral $r = \theta$. If its angular speed is 0.1 radians per second, find its velocity at the time when $r = \pi$.

22. In Example 3, discuss the rate at which the oil slick is approaching the shore "just after" the spill began.

23. In the second part of Example 3 we solved for r^2 and then differentiated. Is this preferable to solving for r and then differentiating? Why or why not?

24. In Example 4, what happens to the rate at which the level of the gasoline is rising as $h \to 3$? Explain why this is reasonable.

25. If the temperature is constant in the piston-cylinder mechanism discussed in Example 1, the pressure and volume of the gas are related through Boyle's law. Assume there is 1 mol of an ideal gas in the cylinder and that the temperature is kept at $0°C$. In this case Boyle's law is $PV = 2271.0$, where V is measured in cubic meters and P in pascals (1 Pa is 1 N/m^2). If the piston is moving into the cylinder at 10 cm/s, find the rate at which the pressure is changing when the pressure is 1000 Pa.

26. A door 0.8 m wide is swinging shut at the rate of 0.25 radians per second. Calculate the rate at which the free edge of the leading edge of the door is approaching the jamb at the time the door is half open.

27. A piston-cylinder arrangement is shown in the following figure. All lengths are measured in centimeters. Assuming that the crankshaft is turning at 250 rpm counterclockwise, find the rate at which the piston is moving when $\theta = 3\pi/4$.

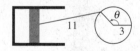

28. A spherical tank of radius 20 m is being filled with liquid propane at the rate of 0.7 m³/min. Calculate the rate at which the propane level is rising when the tank is 75 percent full. The formula for the volume V of a spherical segment of height h, where a is the radius of the sphere and $0 \le h \le 2a$, is given by $V = \pi h^2(a - h/3)$. See the accompanying figure.

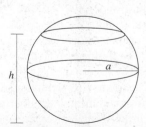

29. A kitchen drinking cup in the shape of a frustum of a cone measures 4.4 cm across its bottom (diameter) and 6.4 cm across its top. Its depth is 7.3 cm. It is being filled with water at the rate of 30 cm³/min (which is, approximately, 1 fluid ounce per minute). At what rate is the water level rising when the depth of the water is 2 cm? The formula for the volume of a frustum of a cone is

$$V = \tfrac{1}{3}\pi H(R^2 + Rr + r^2),$$

where H is the height of the frustum, R the radius of the larger circular base, and r the radius of the smaller circular base.

30. The Great Pyramid at Gizeh has a square base of side 230.4 m and height 146.8 m. The pyramid is being filled with water at a rate of 0.1 m³/min, from a small hole at its top. At what rate is the water rising in the pyramid when it is half full? How long will it take to fill? The formula for the volume of a pyramid is $V = \tfrac{1}{3}a^2h$, where a is the length of a side of its base and h is the height. The formula for the volume of a frustum of a pyramid is $V = \tfrac{1}{3}H(B^2 + Bb + b^2)$, where H is the height of the frustum, B is the side of its lower base, and b is the side of its upper base. See the accompanying figure. A frustum of a pyramid is the shape below a plane through the pyramid and parallel to its base. Auxiliary problem: Test your algebra and geometry skills by deriving the formula for the volume of a frustum from the formula for the volume of a pyramid.

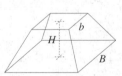

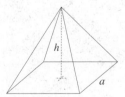

31. Derive (10) in Example 4. Assume as known the formula for the area of a sector of a circle. A sector is a pie-shaped wedge formed by two radii and a part of the circle. The area of a sector with central angle θ in a circle with radius a is $\tfrac{1}{2}a^2\theta$, where θ is measured in radians. (Hint: Divide the segment in half

along the line along which its height is measured.
Divide the half into a sector and a right triangle.)

32. By solving the equation $x^2y + y^3 = 1$ for the values of y corresponding to $x = -3.0, -2.5, \ldots, 2.5,$ 3.0, collect enough data to graph the function $y = y(x)$ implicitly defined by the equation. Cut your work in half by an astute observation.

33. Show that if x is given, there is exactly one real number y satisfying the equation $x^2y + y^3 = 1$. (Hint: With x fixed, consider the function $g(y) = x^2y + y^2 - 1$. Show that $g'(y) > 0$ unless $x = 0$. Hence g is increasing. How many zeros can g have?)

Review/Preview

34. Graph the inverse tangent function. How are the lines $y = \pm\pi/2$ related to the graph? Describe the behavior of the points $(n, \arctan n)$, $n = 1, 2, 3, \ldots$.

35. Graph the hyperbola with equation $x^2/2^2 - y^2 = 1$. Include in your graph the asymptotes of this hyperbola, which are the lines $y = \pm(1/2)x$.

36. Graph (10) in Example 4 as an equation relating A and h, $0 \le h \le 2a$. From the graph and your sense

of how A varies with h in the circle, predict the inflection point. Verify your guess by a calculation.

37. In Exercise 28 a formula was given for the volume of a spherical segment. Taking $a = 1$, graph this equation. From the graph and your sense of how V varies with h in the sphere, predict the inflection point. Verify your guess by a calculation.

4.6   ASYMPTOTIC BEHAVIOR

The American Heritage Dictionary, third edition, defines the word *asymptote* as

> A line considered a limit to a curve in the sense that the perpendicular distance from a moving point on the curve to the line approaches zero as the point moves an infinite distance from the origin. [Ultimately from the Greek *asumptōtos*, not intersecting. . . .]

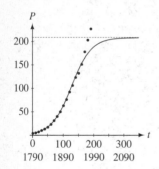

Figure 4.46. U.S. population and logistic equation.

In discussions of world population it is the asymptotic behavior of population that is often the focus of greatest concern and controversy. The **logistic equation** has been used to model population growth. Applied to the population of the United States, it takes the form

$$(1) \qquad P(t) = \frac{210}{1 + 52e^{-0.031t}}, \quad t \ge 0.$$

Time is measured in years, $t = 0$ corresponds to the year 1790, and $P(t)$ is the population in units of millions. This particular equation is based on the census data from the years 1790, 1850, and 1910.

We show in Fig. 4.46 the graph of P, which is the population predicted by (1). The solid dots show the actual population. The graph of the logistic equations fits well up through 1950 or so, after which it starts to level off, approaching the dotted horizontal line as t increases. Try to infer the height of the horizontal asymptote from (1). Calculate $e^{-0.031t}$ for $t = 300, 400,$ and 500.

Asymptotic behavior is common in mathematics and its applications. Hyperbolas are often plotted together with their asymptotes, not only to help in sketching the curve but to provide information about its limiting behavior. We show in Fig. 4.47 the graph of the hyperbola $xy = 1$. The asymptotes are the x- and y-axes. Each axis is an asymptote in the sense of the dictionary definition.

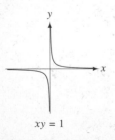

$xy = 1$

Figure 4.47. Vertical and horizontal asymptotes.

Asymptotes provide in visual form information about the limiting behavior of functions. The most common asymptotes are horizontal or vertical lines. Horizontal or vertical asymptotes are specified by numbers h or v; the actual asymptotes are the lines $y = h$ or $x = v$, respectively. Our goal in this section is to help you become familiar with these asymptotes for several kinds of functions.

Vertical Asymptotes

A vertical asymptote is a line marking a point h on the x-axis where the graph of f becomes unbounded as x approaches h. Usually we must specify whether x approaches h from the left or the right since the behavior of $f(x)$ may be quite different as x approaches 0 from the left or the right. We use the notations $x \to 0^-$ and $x \to 0^+$ to specify whether x is approaching 0 from the left (minus) side or the right (plus) side.

What is $\lim_{x \to 0^+} 1/x$? From Fig. 4.48 it appears that $1/x$ becomes and remains larger than any positive number, provided that x is positive and sufficiently close to 0. We write this as

$$\lim_{x \to 0^+} \frac{1}{x} = \infty.$$

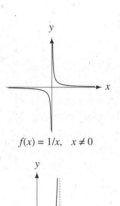

$f(x) = 1/x, \quad x \neq 0$

Numerical evidence is easy to generate. For starters, take $x = 0.1$, $x = 0.01$, and $x = 0.001$. The corresponding values of $f(x) = 1/x$ are 10, 100, and 1000. All of this is consistent with the fact that f is decreasing on $(0, \infty)$, which follows from the I/D Test since $f'(x) = -1/x^2 < 0$.

From Fig. 4.48 we observe that

$$\lim_{x \to 0^-} \frac{1}{x} = -\infty.$$

$f(x) = \tan x,$
$0 \leq x < \pi/2$

This means that $1/x$ becomes and remains smaller than any negative number, provided that x is negative and is sufficiently close to 0. If we take $x = -0.1$, $x = -0.01$, and $x = -0.001$, the corresponding values of $f(x) = 1/x$ are -10, -100, and -1000. This is consistent with the fact that f is decreasing on $(-\infty, 0)$.

We show in Fig. 4.48 the graph of the tangent function on the interval $[0, \pi/2)$. As $x \to \pi/2^-$, the value of $\tan x$ becomes larger than any given positive number. We write this as

$$\lim_{x \to \pi/2^-} \tan x = \infty.$$

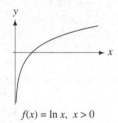

$f(x) = \ln x, \quad x > 0$

Figure 4.48. Vertical asymptotes.

Again, numerical evidence backs up the graphical evidence in the figure.

x	1.50	1.52	1.54	1.56
$\tan x$	14.10	19.67	32.46	92.62

The graph at the bottom of Fig. 4.48 is that of the logarithm function. We write

$$\lim_{x \to 0^+} \ln x = -\infty$$

since $\ln x$ becomes smaller than any given negative number as x approaches 0 from the right. Numerical evidence of this asymptotic behavior of the logarithm function is as close as your calculator.

x	0.1	0.01	0.001	0.0001
$\ln x$	-2.30	-4.61	-6.91	-9.21

It appears that the logarithm function does not approach $-\infty$ rapidly. What value of $x > 0$ gives $\ln x = -100$? (If you must, see Exercise 43.)

Before giving examples of vertical asymptotes, we summarize the four possibilities for a vertical asymptote of the graph of f at $x = h$.

$$\lim_{x \to h^+} f(x) = \infty, \quad \lim_{x \to h^-} f(x) = \infty, \quad \lim_{x \to h^+} f(x) = -\infty, \quad \lim_{x \to h^-} f(x) = -\infty$$

■ **EXAMPLE 1**   Discuss the asymptotic behavior of the function

$$f(x) = \frac{x}{x + 1}, \quad x \neq -1.$$

Solution.   The graph of f is shown in Fig. 4.49. Is this graph essentially correct? Does it show the main features of this function? The function f is an example of a *rational function,* the ratio of two polynomials. The number $x = -1$ was excluded from the domain since the denominator is equal to zero there. The real zeros of the polynomial denominators of rational functions correspond to vertical asymptotes. For the zero $x = -1$,

$$\lim_{x \to -1^-} \frac{x}{x + 1} = \infty \quad \text{and} \quad \lim_{x \to -1^+} \frac{x}{x + 1} = -\infty.$$

These limits are clear from Fig. 4.49. Usually such limits can be determined by a kind of mental limit process. For example, imagine that x is slightly to the left of -1 and is creeping toward -1. The denominator $x + 1$ would be very close to zero and negative, while the numerator would be very close to -1. The ratio of these two negative quantities would be positive and large (ignoring signs, small denominators give large fractions; for example, $1/0.1 = 10$, $1/0.01 = 100$, and so on).

If necessary, we can evaluate f at a sequence of values approaching -1 from the left. For example,

x	-1.1	-1.01	-1.001	-1.0001
$x/(x + 1)$	11	101	1001	10,001

As $x \to -1^-$ the trend of $f(x)$ is clear; the data suggest strongly that

$$\lim_{x \to -1^-} f(x) = \infty.$$

We have now completed our look at the one vertical asymptote (the analysis for $x \to -1^+$ is similar). To save time, we discuss the horizontal asymptotes of this function now instead of returning to this or a similar example later, when we discuss horizontal asymptotes. For rational functions, the asymptotic behavior of the graph for large $|x|$ can be determined by factoring the highest power of x from

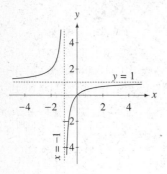

Figure 4.49. Vertical and horizontal asymptotes.

the numerator and denominator and then removing common factors. We have

$$\frac{x}{x+1} = \frac{x(1)}{x(1+1/x)} = \frac{1}{1+1/x}.$$

We observe that $f(x)$ approaches 1 as $|x|$ becomes large. For example,

$$f(-1000) = 1/(0.999) \approx 1.001 \quad \text{and} \quad f(1000) = 1/1.001 \approx 0.999.$$

To complete our analysis of f, we calculate its first two derivatives so that we may apply the I/D and Concavity Tests. We have

$$f'(x) = \frac{(x+1) \cdot 1 - x \cdot 1}{(x+1)^2} = \frac{1}{(x+1)^2}$$

$$f''(x) = -2(x+1)^{-3}.$$

We note that $f'(x) > 0$ for all $x \neq -1$. Hence f is increasing on $(-\infty, -1)$ and $(-1, \infty)$ by the I/D Test. The second derivative is positive to the left of the asymptote and negative to its right. Hence f is concave up on $(-\infty, -1)$ and concave down on $(-1, \infty)$ by the Concavity Test. The graph of f shown in Fig. 4.49 is consistent with all of the above information. For this particular rational function, the viewing window $-5 \leq x \leq 5$ and $-5 \leq y \leq 5$ shows the main features and trends of f. ∎

■ **EXAMPLE 2**   Discuss the asymptotic behavior of the function

$$f(x) = \frac{x^2}{(x+2)(x-1)}, \quad x \neq -2, 1.$$

Solution.   The denominator of this rational function has two real zeros, $x = -2$ and $x = 1$. We expect two vertical asymptotes. Mental calculations give

$$\lim_{x \to -2^-} f(x) = \lim_{x \to -2^-} \frac{x^2}{(x+2)(x-1)} = \infty$$

$$\lim_{x \to -2^+} f(x) = \lim_{x \to -2^+} \frac{x^2}{(x+2)(x-1)} = -\infty$$

$$\lim_{x \to 1^-} f(x) = \lim_{x \to 1^-} \frac{x^2}{(x+2)(x-1)} = -\infty$$

$$\lim_{x \to 1^+} f(x) = \lim_{x \to 1^+} \frac{x^2}{(x+2)(x-1)} = \infty.$$

Here's a sample mental calculation. For the first of these limits we take $x = -2.01$, which is to the left of $x = -2$ and relatively close to it. The numerator of the rational function f is nearly 4, the factor $(x+2)$ of the denominator is -0.01, and the factor $(x-1)$ is nearly -3. This makes it clear that as $x \to -2$ from the left, the numerator approaches 4 and the denominator approaches 0 from the positive side. It follows that $\lim_{x \to -2^-} f(x) = \infty$.

To determine the horizontal asymptote(s), we factor out the highest power of x in the numerator and denominator and simplify. We have

$$f(x) = \frac{x^2}{(x+2)(x-1)} = \frac{x^2 \cdot (1)}{x^2 \cdot (1 + 1/x - 2/x^2)} = \frac{1}{1 + 1/x - 2/x^2}.$$

From this rearrangement it is clear that $f(x)$ approaches 1 as $|x|$ becomes large. If this isn't "clear," calculate $f(x)$ for x equal to the national debt, say 5 trillion dollars or so.

$$f(5.0 \times 10^{12}) \approx 1$$

The authors' calculators gave 1. The true value of $f(5.0 \times 10^{12})$ is a shade under 1. Often, one concrete calculation can improve understanding quite a bit.

Check that the derivatives of f are

$$f'(x) = \frac{x(x-4)}{(x+2)^2(x-1)^2}$$

$$f''(x) = \frac{-2(x^3 - 6x^2 - 4)}{(x+2)^3(x-1)^3}.$$

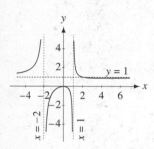

Figure 4.50. Vertical and horizontal asymptotes.

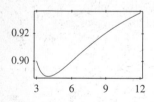

Figure 4.51. A closer view.

For the I/D Test we note that the denominator of f' is always positive where it is defined. From the sign of the numerator we note that f is increasing on $(-\infty, -2)$ and $(-2, 0)$, decreasing on $(0, 1)$ $(1, 4)$, and increasing again on $(4, \infty)$.

To determine where f is concave up or down takes more care for this function. Not only does the denominator of f'' change sign at $x = -2$ and $x = 1$, but we must find the zeros of the cubic in the numerator. We leave as Exercise 11 the problem of showing that $x^3 - 6x^2 - 4$ has a real zero near $x = 6.1$ and that the other zeros are complex. The numerator of f'' is positive to the left of 6.1 and negative to the right (don't forget the factor -2). The function f'' cannot change sign on the intervals $(-\infty, -2)$, $(-2, 1)$, $(1, 6.1)$, and $(6.1, \infty)$, which are determined by the zeros of the numerator and denominator. It follows that f is concave up on $(-\infty, -2)$ and $(1, 6.1)$ and concave down on $(-2, 1)$ and $(6.1, \infty)$. It has one inflection point, at $x \approx 6.1$. The graph is shown in Figs. 4.50 and 4.51. We selected the window for Fig. 4.51 to show more clearly the local minimum at $(4, f(4))$, where there is a horizontal tangent, and the inflection point near $(6.1, f(6.1))$. ∎

These two examples suggest that the asymptotic behavior of rational functions is quite straightforward, except, perhaps, for calculating the zeros of the polynomial numerators and denominators that arise. The same cannot be said for transcendental functions, whose asymptotic behavior is more complex. The difficulty arises not so much in finding zeros as in calculating limits. In the next two examples, and in the exercises, we use graphical or numerical evidence in discussing some of the limits we encounter.

■ **EXAMPLE 3**   Discuss the asymptotic behavior of the function

$$f(x) = 1 - \ln(\cos x), \quad 0 \le x < \pi/2.$$

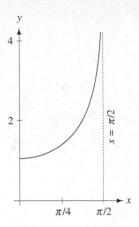

Figure 4.52. Vertical asymptote.

Solution.   At $x = 0$, f is well behaved, but as $x \to (\pi/2)^-$, $\cos x \to 0$; since $\ln x \to -\infty$ as $x \to 0^+$, we conclude that

$$\lim_{x \to (\pi/2)^-} (1 - \ln(\cos x)) = \infty.$$

Thus $x = \pi/2$ is a vertical asymptote.

The derivatives of f are surprisingly simple. We have

$$f'(x) = -\frac{-\sin x}{\cos x} = \tan x$$

$$f''(x) = \sec^2 x.$$

From the I/D and Concavity Tests, f is increasing and convex on $(0, \pi/2)$. The graph of f is shown in Fig. 4.52.   ■

■ **EXAMPLE 4**   Discuss the asymptotic behavior of the function

$$f(x) = \frac{\sqrt{x}}{\ln x}, \quad x > 0 \text{ and } x \neq 1.$$

Solution.   We expect a vertical asymptote at $x = 1$ since $\ln 1 = 0$. We have

$$\lim_{x \to 1^-} \frac{\sqrt{x}}{\ln x} = -\infty \quad \text{and} \quad \lim_{x \to 1^+} \frac{\sqrt{x}}{\ln x} = \infty.$$

For this calculation recall that the logarithm function is increasing on $(0, \infty)$, $\ln 1 = 0$, and hence it goes from negative, to 0, to positive as x varies from, say, 0.5 to 1.5.

The limit of f at the left end of its domain is 0 since the numerator in

$$\lim_{x \to 0^+} \frac{\sqrt{x}}{\ln x}$$

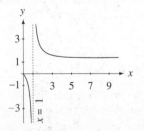

Figure 4.53. Vertical asymptote at $x = 1$.

is approaching 0 and the absolute value of the denominator is becoming large. These "forces" work together to make the fraction approach 0 as $x \to 0^+$. A preliminary graph of f is shown in Fig. 4.53.

The trend of f as x becomes large is not clear from this figure. We give some evidence below that $f(x)$ becomes and remains larger than any specified number for sufficiently large x, that is,

$$\lim_{x \to \infty} \frac{\sqrt{x}}{\ln x} = \infty.$$

We define limits of this form after this example. Meanwhile, we discuss this limit numerically. First, note that as $x \to \infty$, both the numerator and denominator of $\sqrt{x}/\ln x$ become large. The question is which of $\sqrt{x}$ or $\ln x$ is dominant. We may gain a sense that $\sqrt{x}$ is "stronger" by evaluating $f(x)$ at $x_1 = e^2$, $x_2 = e^4$, $x_3 = e^6 \ldots$, which is a sample of x values as $x \to \infty$. We chose this particular sequence to simplify the arithmetic.

x	e^2	e^4	e^6	e^8	e^{10}
$\sqrt{x}/\ln x$	$e/2$	$e^2/4$	$e^4/4$	$e^6/6$	$e^5/5$

The trend of these numbers suggests that although $\sqrt{x}/\ln x$ grows slowly, it eventually becomes larger than any number we might suggest. Note that although $e^{10} \approx 22{,}000$, which is fairly large, the value of $e^5/5$ is only 30 or so. However,

$$\text{for} \quad x = e^{2e^5} \quad \text{we have} \quad \sqrt{x}/\ln x = e^{e^5}/(2e^5) \approx 9.6 \times 10^{61}.$$

The first and second derivatives of f are

(2)
$$f'(x) = \frac{\ln x - 2}{2\sqrt{x}(\ln x)^2}$$

(3)
$$f''(x) = \frac{8 - (\ln x)^2}{4x^{3/2}(\ln x)^3}.$$

The sign of f' is determined by the numerator since the denominator of (2) is always positive. Noting that the numerator changes sign at $x = e^2$ (set $\ln x - 2 = 0$ and solve for x), it follows from the I/D Test that f is decreasing to the left of $e^2 \approx 7.4$ and increasing elsewhere.

The second derivative changes sign three times. The denominator of (3) is negative for $0 < x < 1$ and positive for $x > 1$. So the concavity changes at $x = 1$. The other two points at which f'' changes sign can be determined by setting the numerator of (3) equal to zero and solving the resulting equation. We have

$$\ln x = \pm\sqrt{8}$$
$$x = e^{\pm\sqrt{8}} \approx 0.1, \, 17.$$

It follows from the Concavity Test that f is concave up in the intervals $(0, 0.1)$ and $(1, 17)$ and concave down in $(0.1, 1)$ and $(17, \infty)$. There are two inflection points ($x = 1$ does not qualify since it is not in the domain of f). Using this information we may refine the graph shown in Fig. 4.53. For Fig. 4.54 we chose a viewing window that shows the inflection point near $(17, f(17))$. Try to choose a window to view the change of concavity near $(0.1, f(0.1))$. ∎

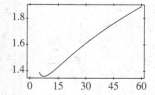

Figure 4.54. Inflection point near $(17, f(17))$.

Horizontal Asymptotes

We discussed horizontal asymptotes in Examples 1 and 3, though our main emphasis was on vertical asymptotes and using the I/D and Concavity Tests as graphing tools. The graph of a function has a horizontal asymptote $y = h$ provided that its domain includes at least one ray, that is, an interval of the form (a, ∞) or $(-\infty, b)$, and

$$\lim_{x \to \infty} f(x) = h \quad \text{or} \quad \lim_{x \to -\infty} f(x) = h.$$

When the limit of a function was first discussed in Chapter 1, we wrote

$$\lim_{x \to a} f(x) = L,$$

where both a and L were real numbers. We said then that L is the limit of $f(x)$ as x approaches a provided that $f(x)$ becomes and remains as close to L as we wish provided that x is sufficiently close to a. Limits in which $a = \infty$ or $L = \infty$ can be described in the same way. The only change is to quantify what it means for x or $f(x)$ to be "close to infinity."

We measure how close x is to infinity by measuring how far x is from the origin. In these terms, the limit

$$\lim_{x \to \infty} \arctan x = \pi/2$$

means that $\arctan x$ becomes and remains as close to $\pi/2$ as we wish provided that x is larger than a number M. For example,

$$|\arctan x - \pi/2| < 0.001 \quad \text{provided that } x > 1000.$$

The limit

$$\lim_{x \to 0^+} \ln x = -\infty$$

means that given any negative number M, $\ln x < M$ provided that x is sufficiently close to 0 and positive.

■ **EXAMPLE 5** Determine the horizontal asymptotes of the graph of the function

$$f(x) = \frac{\sqrt{5x^2 + 3}}{7 - x}, \quad x \neq 7.$$

Solution. The vertical asymptote to this graph is $x = 7$. For the horizontal asymptotes, we start by calculating $\lim_{x \to \infty} f(x)$. We factor the dominant term from numerator and denominator. This gives

$$\lim_{x \to \infty} \frac{\sqrt{5x^2 + 3}}{7 - x} = \lim_{x \to \infty} \frac{x\sqrt{5 + 3/x^2}}{x(7/x - 1)}$$

$$= \lim_{x \to \infty} \frac{\sqrt{5 + 3/x^2}}{(7/x - 1)} = -\sqrt{5}$$

For the limit of f as $x \to -\infty$, we must be careful about factoring x from $\sqrt{5x^2 + 3}$. Take a moment and consider how you would explain the following calculation to a friend. The results are shown in Fig. 4.55.

$$\lim_{x \to -\infty} \frac{\sqrt{5x^2 + 3}}{7 - x} = \lim_{x \to -\infty} \frac{-x\sqrt{5 + 3/x^2}}{x(7/x - 1)}$$

$$= \lim_{x \to -\infty} \frac{-\sqrt{5 + 3/x^2}}{(7/x - 1)} = \sqrt{5} \quad ■$$

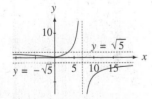

Figure 4.55. Horizontal asymptotes.

Up to this point we have used factoring (dominant term), graphing, and numerical evaluation to determine horizontal and vertical asymptotes. To these tools we add two results comparing the growth of the functions $\ln x$ and e^x to a power of x. We defer a proof of these results to Exercise 48 in Section 5.2.

Asymptotic Behavior of e^x and ln x as $x \to \infty$

For any positive constants a and b,

(4)
$$\lim_{x \to \infty} \frac{(\ln x)^a}{x^b} = 0$$

(5)
$$\lim_{x \to \infty} \frac{x^a}{e^{bx}} = 0$$

■ **EXAMPLE 6** Show that

$$\lim_{x \to \infty} \frac{(\ln x)^{100}}{x^{0.01}} = 0.$$

Solution. A short answer is to take $a = 100$ and $b = 0.01$ in (4). This gives little insight into the growth of the logarithm function relative to a power of x.

For large values of x, the expression $(\ln x)^{100}$ increases much more rapidly than $\ln x$; at the same time, $x^{0.01}$ increases much less rapidly than x. Let's try a few values.

x	10	100	1000
$\dfrac{(\ln x)^{100}}{x^{0.01}}$	1.6×10^{36}	2.0×10^{66}	8.0×10^{83}

This evidence might tempt us to conclude that $f(x) = (\ln x)^{100}/x^{0.01}$ becomes infinite as $x \to \infty$. However, we argued from (4) that $\lim_{x \to \infty} f(x) = 0$. So it can't continue to increase. Can you find two successive powers of 10 between which $f(x)$ starts decreasing toward 0? See Exercises 31 and 32. ■

■ **EXAMPLE 7** Show that $\lim_{x \to 0^+} x \ln x = 0$.

Solution. In (4), $x \to \infty$; here, $x \to 0^+$. To, as it were, convert 0 to ∞, we change variables by letting $y = 1/x$. Note that as $x \to 0^+$, $y \to \infty$. We use (4) to conclude that

$$\lim_{x \to 0^+} x \ln x = \lim_{y \to \infty} (1/y) \ln(1/y) = \lim_{y \to \infty} \frac{-\ln y}{y} = 0.$$ ■

Exercises 4.6

Basic

1. Sketch a graph of one period of the secant function and locate any vertical or horizontal asymptotes.
2. Sketch a graph of the exponential function and locate any vertical or horizontal asymptotes.
3. Sketch a graph of the inverse tangent function and locate any vertical or horizontal asymptotes.
4. Sketch a graph of the equation $x^2 - y^2 = 1$ and locate any lines that are asymptotes.
5. Verify the correctness of f' and f'' in Example 2.
6. Verify the correctness of f' and f'' in Example 4.
7. Graph the function $f(x) = x + 1/x$, $x \neq 0$. Give an informal argument as to why the graphs of $y = x$ and

$y = 1/x$ can be regarded as asymptotes to the graph of f.

8. Graph the function $f(x) = (x^3 + 1)/x$, $x \neq 0$. Give an informal argument as to why the graphs of $y = x^2$ and $y = 1/x$ can be regarded as asymptotes to the graph of f.

9. Find the horizontal asymptotes of the graph of the function

$$f(x) = \frac{\sqrt{2x^2 + 1}}{3x + 1}, \quad x \neq -1/3.$$

10. Find the horizontal asymptotes of the graph of the function

$$f(x) = \frac{x^2 + \sqrt{x + 1}}{\sqrt{x^4 - 1}}, \quad x > 1.$$

11. Use Newton's method or another technique to find the one real zero of the polynomial

$$x^3 - 6x^2 - 4.$$

Give evidence that the other zeros are complex.

Exercises 12–29: As in the examples, graph and discuss the asymptotic behavior of the function.

12. $f(x) = \dfrac{2x - 3}{x + 1}, \quad x \neq -1$

13. $f(x) = \dfrac{3x - 2}{x - 1}, \quad x \neq 1$

14. $f(x) = \dfrac{x^2 + 2x - 4}{x^2}, \quad x \neq 0$

15. $f(x) = \dfrac{x^2 - x - 2}{x^3}, \quad x \neq 0$

16. $f(x) = \dfrac{\sqrt{x} + 1}{\sqrt{x} - 1}, \quad x \geq 0$ and $x \neq 1$

17. $f(x) = \dfrac{2x}{\sqrt{x^2 + 1}}$

18. $f(x) = \dfrac{x^3}{(x + 1)^2(x - 2)}$

19. $f(x) = \dfrac{x^2}{2x^2 + 7x - 4}, \quad x \neq -4, 1/2$

20. $f(x) = 1 - \ln(\sin x), \quad 0 < x \leq \pi/2$

21. $f(x) = 1 + \ln(\tan x), \quad 0 < x < \pi/2$

22. $f(x) = \dfrac{x^2}{\ln x}, \quad x > 0$

23. $f(x) = \dfrac{\sqrt{x}}{\ln x}, \quad x > 0$

24. $f(x) = xe^{-x}$

25. $f(x) = \sqrt{x}e^{-x}, \quad x > 0$

26. $f(x) = e^{-x/4}\sin x, \quad x \geq 0$

27. $f(x) = e^{-x/4}\cos x, \quad x \geq 0$

28. $f(x) = \dfrac{x^{1/3}}{x + 1}, \quad x \neq -1$

29. $f(x) = \dfrac{x^{1/5}}{x + 1}, \quad x \neq -1$

Growth

30. Determine the inflection point of the logistic equation (1) and write a sentence or two about its meaning in terms of population growth.

31. Use your calculator to figure out how big x must be for $e^{0.01x}$ to catch up to x^{100}.

32. Use your calculator to figure out how big x must be for $e^{0.001x}$ to catch up to x^{1000}.

33. How many times do the graphs of $f(x) = 2^x$, $x \geq 0$, and $g(x) = x^{10}$, $x \geq 0$, intersect? Sketch one or more illustrative graphs.

34. What is the asymptotic behavior of the rate at which the oil slick (see Example 3, Section 4.5) is approaching the shore as $t \to 0^+$?

35. What is the asymptotic behavior of the rate at which the liquid is rising (see Example 4, Section 4.5) as the depth approaches 3 m? Why is this reasonable?

36. What happens to the graph of

$$f(x) = \frac{e^{-1/x}}{x^{10}}, \quad x \neq 0$$

as $x \to 0^+$?

37. Evaluate

$$\lim_{x \to \infty} \frac{\ln(\ln x)}{\sqrt{x}}.$$

38. Evaluate

$$\lim_{x \to \infty} \frac{\sqrt{1 + 2^x}}{\sqrt{1 + 3^x}}.$$

39. Graph $f(x) = x^x$, $0 < x < 2$. You may wish to evaluate f at $0.1, 0.01$, and 0.001 to get an idea of its limiting value as $x \to 0^+$. Calculate $\lim_{x\to 0^+} x^x$ by rewriting x^x in terms of the exponential function.

40. Graph $f(x) = x^{1/x}$, $0 < x < \infty$. Calculate $\lim_{x\to 0^+} x^x$ by rewriting x^x in terms of the exponential function. Does the graph have a horizontal asymptote?

41. Show that $y = (b/a)x$ is an asymptote of the graph of the equation $x^2/a^2 - y^2/b^2 = 1$ in the sense of the dictionary definition quoted at the beginning of the section.

42. Use the Mean-value Inequality in showing that if for all $t > 0$ the derivative of a function f exceeds a fixed positive number q, then $\lim_{t\to\infty} f(t) = \infty$.

43. In our discussion of $\lim_{x\to 0^-} \ln x$ we asked for x such that $\ln x = -100$. You should have found that $x = e^{-100} \approx 3.7 \times 10^{-44}$. What happens if this same calculation is done on your calculator for $x = -1000$ and $x = -10,000$? Explain.

Review/Preview

44. Can a rectangle of sides 6 and 8 meters be inscribed in a circle of radius 3 meters? Why or why not?

45. Is it true that the highest point of a smooth and straight highway laid across rolling prairie and joining two town centers is on the crest of a hill? Why or why not? Defend your conjecture.

46. In a few words, what is the difference between the local economy and the global economy?

47. Arrange in order the values of the function $f(x) = \sin x^2$, $0 \le x \le 2$, at 0, 2, and all $x \in (0, 2)$ for which $f'(x) = 0$. Does this have any significance?

4.7   TOOLS FOR OPTIMIZATION

This section and the next concern optimization. To *optimize* means to find the best or most favorable way of designing or affecting a system S. Many optimization problems can be modeled or formulated within mathematics. This means that it is possible to describe a mathematical problem that captures the essentials of the system S. The modeling process would be regarded as successful if we can solve the mathematical problem and then interpret and use the solution to optimize the system S.

Suppose, for example, a sawyer wishes to cut a rectangular beam of maximum strength (least deflection under uniform loading) from a circular log. A beam/log cross-section is shown in Fig. 4.56. If from the mechanics of materials we use the fact that the strength S of such a beam is proportional to the product of its width w and the square of its height h, we may solve the sawyer's problem by choosing w so that S is a maximum, where

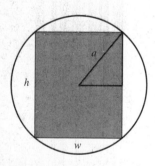

Figure 4.56. Beam from log.

$$(1) \qquad S(w) = kwh^2 = kw(4a^2 - w^2), \quad 0 \le w \le 2a.$$

The mathematical model of the sawyer's problem is: Determine the maximum of the function defined by (1). We return to this problem in Example 1.

We have divided our discussion of optimization into two parts: formulating a mathematical model of a system or process and solving the mathematical model. For now we focus on solving the model; in Section 4.8 we focus on how to formulate or model optimization problems.

A common optimization model of a system S is a function f defined on an interval I, with the idea that if we maximize or minimize f we will optimize S. To maximize or minimize f we must:

1. Find all possible candidates $x \in I$ at which f could possibly take on its maximum or minimum value.

2. Choose among the candidates.

The goal in this section is to help you become familiar with the ideas and technique used to locate and test candidates x at which f could take on its maximum or minimum value.

Local Extrema

We begin by defining what it means for a function f to have a local or global maximum or minimum at a point w. The distinction between *local* and *global* agrees with everyday usage. You may be the tallest person locally (for example, in your home town or neighborhood), but it is not likely that you are the tallest person on the globe. The word *extremum* (plural *extrema*) is used to refer to either a maximum or minimum.

DEFINITION   local or global maximum (minimum)

A function f defined on a set A has a local maximum (minimum) at $w \in A$ if $f(w) \geq f(x)$ $(f(w) \leq f(x))$ for all $x \in A$ sufficiently close to w; f has a global maximum (minimum) at $w \in A$ if $f(w) \geq f(x)$ $(f(w) \leq f(x))$ for all $x \in A$. If f has a local maximum, local minimum, global maximum, or global minimum at $w \in A$, we say that f has an extremum at w and that w is an extreme point. The word "global" is often omitted or replaced by the word "absolute."

The function f whose graph is shown in Fig. 4.57 has local maxima at p and r, a global maximum at p, local minima at q and s, and a global minimum at q.

The most important tool for locating candidates for extreme points is the Candidate Theorem. For most functions it drastically reduces the search for extreme points.

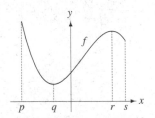

Figure 4.57. Max/min.

Candidate Theorem

If f is continuous on an interval I and has a local maximum or local minimum at $c \in I$, then the only candidates for c are the end points, if any, of I and any interior point x at which $f'(x) = 0$ or at which f is not differentiable.

We prove the Candidate Theorem under the additional assumption that f is differentiable at all interior points of I. This assumption is satisfied for most of the functions we consider.

To prove the Candidate Theorem, we must show that if f has, say, a local maximum at $c \in I$, then c must be an end point or an interior point at which f' is zero. If c is an end point, we have nothing to prove. So suppose that c is not an end point. Since f has a local maximum at c,

$$f(c) \geq f(c + h) \quad \text{for all sufficiently small } |h|.$$

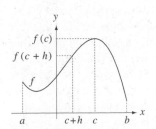

Figure 4.58. Local maximum at c.

See Fig. 4.58. This means that the difference quotient

$$\frac{f(c + h) - f(c)}{h} \quad \text{is} \quad \begin{cases} \geq 0 & \text{for all } h < 0 \quad (|h| \text{ sufficiently small}) \\ \leq 0 & \text{for all } h > 0 \quad (|h| \text{ sufficiently small}). \end{cases}$$

It follows that $f'(c) = 0$, for otherwise the difference quotient would become and remain either positive or negative for all sufficiently small $|h|$.

How do we use the Candidate Theorem? Easy, at least in principle. We round up all candidates, evaluate them one by one, and choose a winner. The candidates are any end points, zeros of f' that lie in the interior of I, and points where f is not differentiable. We then calculate $f(c)$ for all candidates c and choose the local and global extrema by simple comparison. We use the sawyer's problem to illustrate this process.

■ **EXAMPLE 1**   In Fig. 4.56 is shown a beam of height h and width w, cut from a log of radius a. The strength S of such a beam is proportional to the product of its width w and the square of its height h. Find the dimensions of the beam of maximum strength.

Solution.   The strength S of the log is $S = kwh^2$, where k is a constant. (To say that a quantity D is proportional to a quantity W means that $D = kW$ for some constant k. Often it is not necessary to know the numerical value of k; the important thing is that it is a constant.) We must choose w and h so that S is a maximum. In searching for a maximum, we cannot vary h and w independently, for they are related through the right triangle shown in Fig. 4.56, specifically,

$$(2) \qquad (h/2)^2 + (w/2)^2 = a^2.$$

We use this equation to write S as a function of w or as a function of h. Since $S = kwh^2$, it is simplest to solve (2) for h^2 and substitute into $S = kwh^2$. We have

$$(3) \qquad S(w) = kwh^2 = kw(4a^2 - w^2) = k(4a^2w - w^3), \quad 0 \le w \le 2a.$$

We maximize the function in (3). The graph of S is shown in Fig. 4.59.

The function S is everywhere differentiable. Thus we look only at end points 0 and $2a$ and the zeros of S'. Since

$$S' = k(4a^2 - 3w^2),$$

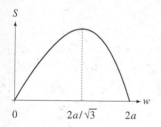

Figure 4.59. Global maximum.

the only zero of S' in $(0, 2a)$ is $w = 2a/\sqrt{3}$. Since

$$S(0) = 0, \quad S(2a) = 0, \quad \text{and} \quad S(2a/\sqrt{3}) > 0,$$

the function S has a global maximum at $a = 2a/\sqrt{3}$. We used the Candidate Theorem for this conclusion. However, we skipped an important point. To make use of the Candidate Theorem, we must know that S actually has a global (and hence local) maximum at some point w in $[0, 2a]$. For this we use the Max/Min Theorem stated in Section 4.4 (we focus on this result in the next example). Since the hypotheses of the Candidate Theorem are satisfied—that f be continuous and have a local maximum or local minimum in $[0, 2a]$—we may use it to conclude that the only candidates for w are 0, $2a$, and $2a/\sqrt{3}$. Since S is zero at the first two and positive at the third, S has a global maximum at $w = 2a/\sqrt{3}$.

In addition to giving the sawyer the rule "measure a and calculate $w = 2a/\sqrt{3}$," we may wish to provide the sawyer with a "rule of thumb." For this we calculate the ratio h/w. We have

$$\frac{h^2}{w^2} = \frac{4a^2 - w^2}{w^2} = \frac{4a^2 - 4a^2/3}{4a^2/3},$$

from which we obtain

$$\frac{h}{w} = \sqrt{2}.$$

A reasonable rule of thumb might be "cut it half again as high as it is wide." ■

In the preceding example we used the Max/Min Theorem to be mathematically certain that S had a global maximum. We restate this result below for convenient reference. That S has a global maximum at a point of its domain follows from the Max/Min Theorem since, as may be seen in (3), S is continuous on the interval $[0, 2a]$.

Max/Min Theorem

If f is continuous on the interval $[a, b]$, then f has a maximum value and a minimum value on $[a, b]$.

■ **EXAMPLE 2**   Find all global extrema of the function

(4) $f(x) = 2x + 1 + 3|2x - 3| - |3x + 1|, \quad -2 \le x \le 3.$

Solution.   The graph of f is shown in Fig. 4.60. We observe a global maximum at $x = -2$ and a global minimum at $x = 1.5$. The Max/Min and Candidate Theorems give the same result. Since the absolute value function value is continuous, it follows that f is continuous. So the function f, defined on $[-2, 3]$, has both a global maximum and minimum. These must occur at the end points -2 or 3, at a zero of f', or at a point where f is not differentiable. Since the absolute value function $|x|$ is differentiable for all x except $x = 0$, we know that f is differentiable except at $x = -1/3$ and $x = 3/2$. It is not difficult to show that f' has no zeros. For example, for $-2 \le x \le -1/3$,

$$f(x) = 2x + 1 + 3(3 - 2x) + 3x + 1 = -x + 11.$$

Thus for $x \in (-2, -1/3)$, $f'(x) = -1$. So the only candidates for a point at which f has a global maximum or minimum are $x = -2$, $x = -1/3$, $x = 3/2$, and $x = 3$. The values of f at these four points are $f(-2) = 13$, $f(-1/3) \approx 11.33$, $f(3/2) = -1.5$, and $f(3) = 6$. It follows that f has a global maximum at $x = -2$ and a global minimum at $x = 1.5$. ■

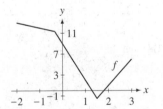

Figure 4.60. f is not everywhere differentiable.

■ **EXAMPLE 3**   Find the global minimum of the function $f(x) = \sqrt{x}\cos x$, where $0 \le x \le 2\pi$.

Solution.   A graph of f is shown in Fig. 4.61. It appears that f has a global minimum near or possibly at $x = \pi$. Since, however, $f(\pi) \approx -1.772$ and

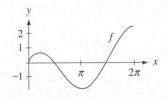

Figure 4.61. Global minimum.

$f(3.2) \approx -1.786$, we can rule out π. Since f is differentiable except at $x = 0$, we calculate the zero of f' near π. By setting

$$f'(x) = \frac{\cos x}{2\sqrt{x}} - \sqrt{x}\sin x$$

equal to zero and rearranging it a little, we obtain the equation

$$2x\sin x - \cos x = 0.$$

We may use Newton's method, for example, to find the root of this equation near π. We find

$$x \approx 3.2923.$$

It is clear that f has a global minimum at $x \approx 3.2923$, where the value of f is -1.79, approximately.   ∎

Global Extrema

It sometimes happens that we wish to optimize functions whose domains are not necessarily *closed and bounded* intervals. An interval is closed if it contains its end points; it is bounded if all of its points are within a fixed distance of the origin. This cannot be said of the interval $(3, \infty)$. It doesn't contain the end point 3, nor are all of its points within a fixed distance of the origin. Nonetheless, functions on such intervals *may* have a global maximum or minimum. We give one result for such functions.

Global Extremum Theorem

If f is defined and concave up (concave down) on an interval I, f is differentiable at $c \in I$, and $f'(c) = 0$, then f has a global minimum (maximum) at c.

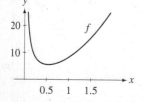

Figure 4.62. Global minimum.

We show in Fig. 4.62 a function that is concave up on $(0, \infty)$. It is clear that such a function must have a global minimum. We prove a special case of the Global Extremum Theorem. We assume that I has the form (a, b), $(-\infty, b)$, (a, ∞), or $(-\infty, \infty)$, where a and b are numbers and $a < b$, and that $f''(x) > 0$ for all $x \in I$. The last assumption implies that f is concave up, by the Concavity Test. Since $f''(x) > 0$ for all $x \in I$, f' is increasing on I by the I/D Test. Since f' is increasing on I and is 0 at $c \in I$, $f'(x) > f'(c) = 0$ for $x > c$. This means that f is increasing on the part of I to the right of c, again using the I/D Test. It follows in the same way that f is decreasing on the part of I to the left of c. Thus, f has a global minimum at c.

■ **EXAMPLE 4**   Find all global extrema of the function

$$f(x) = 2\pi x^2 + \frac{2}{x}, \quad 0 < x < \infty.$$

Solution. This function, whose graph is shown in Fig. 4.62, has no global maximum since $\lim_{x \to 0^+} f(x) = \infty$. We use the Global Extremum Theorem to show that f has a global minimum. For this we calculate f' and f''.

$$f'(x) = 4\pi x - 2x^{-2}$$
$$f''(x) = 4\pi + 4x^{-3}$$

The function f'' is positive on $(0, \infty)$. Hence, by the Concavity Test, it is concave up there. Setting f' equal to zero, we have

$$4\pi x - \frac{2}{x^2} = 0$$

$$4\pi x = \frac{2}{x^2}$$

$$x^3 = 1/(2\pi)$$

$$x \approx 0.5419.$$

It follows from the Global Extremum Theorem that f has a global minimum of approximately 5.536 at $x \approx 0.5419$. ∎

Exercises 4.7

Basic

Exercises 1–6: Sketch the graph of a continuous function defined on $[0, 1]$ and having the given characteristics.

1. Global maximum at 0; global minimum at 1.
2. Local maximum at 0; local minimum at $1/2$; global maximum at 1.
3. Local minimum at 0; global maximum at $1/2$; local minimum at 1.
4. Local minima at $1/4$ and 1; local maxima at 0 and $3/4$.
5. Local maxima at 0, $1/2$, and 1; local minima at $1/4$ and $3/4$.
6. Local maxima at 0 and $1/2$; local minima at $1/4$; global maximum at 1; global minimum at $3/4$.

7. Referring to Example 1, justify the rule of thumb "cut it half again as high as it is wide."
8. In Example 2 we showed that on $[-2, -1/3]$, $f(x) = -x + 11$. Give the details of this calculation.
9. In Example 2, write f in the form $mx + b$ for each of the intervals $[-1/3, 3/2]$ and $[3/2, 3]$.

10. Graph the function $f(x) = |2x - 5| + |x + 2| - 2|x - 4|$ on the interval $[-3, 5]$. On what subintervals of its domain is f concave up? Concave down?
11. Find the local maximum near 0.6 for the function in Example 3.
12. Verify the calculations in Example 4.

Exercises 13–24: Locate all extrema of the function on the given interval. Evaluate the function at any global extrema.

13. $2x^3 - 3x^2 - 12x + 1$, $[-2, 3]$
14. $4x^3 - 3x^2 - 6x - 2$, $[-1, 2]$
15. $x/(x^2 + 1)$, $[-3, 3]$
16. $(x - 1)/(x^2 + 1)$, $[-3, 1]$
17. $x^2/2 + 8/x$, $[1/2, 3]$
18. $x^2 + 5/x + 1$, $[1, 2]$
19. $\arctan(x - \sqrt{x})$, $[0, 5]$
20. $\arctan(x^2 - x)$, $[0, 4]$
21. $2x^3 + 9x^2 + 12x + 1$, $[-1, 1]$
22. $x^4 - 4x + 5$, $[-1, 2]$
23. $\sin^2 x$, $[0, 2\pi]$
24. $x^{5/3} - 5x^{1/3}$, $[0, 2]$

Growth

Exercises 25–36: Locate any global extrema of the function on the given interval. Evaluate the function at any global extrema.

25. $(x^2 + 1)/(x^3 + 1), \quad 0 \le x \le 3$

26. $(x^2 + 1)/(x^4 + 1), \quad 0 \le x \le 3$

27. $|x|/(1 + 2|x|), \quad -1 \le x \le 3$

28. $x^2 \sqrt[3]{x - 1}, \quad -2 \le x \le 2$

29. $-2x^2 + \tan x, \quad 0 \le x \le 1.4$

30. $2xe^{-x}, \quad 0 \le x \le 2$

31. $|x - 1/4| + |x - 1/2| + |x - 3/4|, \quad 0 < x < 1$

32. $\cot x + x^2, \quad (0, 3)$

33. $f(x) = x^x$ for $0 < x \le 2$ and $f(0) = 1$

34. $x^2 + (\sin x - 1)^2, \quad 0 \le x \le \pi/2$

35. $2^x - x - 1, \quad [-1, 2]$

36. $(x - \ln x)/x, \quad [1, 5]$

37. Prove the Candidate Theorem without assuming that f is differentiable on (a, b).

38. As a variation on the sawyer's problem, suppose the logs have roughly elliptical cross-sections. The greatest diameter of the logs is about 10 percent larger than the least diameter. What rule of thumb would you recommend to the sawyer in this forest?

39. Fill in the details in Example 3 of finding a zero of f' by Newton's method.

40. Why can't we conclude from the Global Extremum Theorem that the function e^x defined on $(-\infty, \infty)$ has a global minimum? After all, its second derivative is always positive.

41. Everyone knows that $-\pi/2 < \arctan x < \pi/2$ for all $x \in R$. Write a clear, succinct sentence arguing that the arctan function has no maximum.

42. The Max/Min Theorem has two assumptions on f, namely, that it is continuous and that it is defined on a closed and bounded interval. Give simple examples showing that if either of these assumptions is not satisfied, f may not have a maximum value.

43. The concave up case of the Global Extremum Theorem was argued just after this theorem was stated. Give a similar argument for the concave down case.

44. We proved the Global Extremum Theorem under the assumptions that the domain I of the function was an open interval, that the function f was twice differentiable, and that $f''(x) > 0$ on I. These assumptions made possible a simple argument. The theorem, as stated, assumes much less—only that f is concave up on an interval I and $f'(c) = 0$ for some $c \in I$. Give a proof in this case. (Hint: Suppose f does not have a global minimum at c; then, to take one of two cases, there is a point $z > c$ such that $f(z) < f(c)$. If f is concave up on $[c, z]$, then its graph lies below any of its secant lines. Hence, letting $c + h$ be any point of this interval,

$$f(c + h) \le f(c) + \frac{f(z) - f(c)}{z - c}(c + h - c).$$

Rewrite this to get

$$\frac{f(c + h) - f(c)}{h} \le \frac{f(z) - f(c)}{z - c} < 0.$$

From this show that $f'(c) \ne 0$, a contradiction.)

Review/Preview

Exercises 45–50: At $t = 0$ a particle leaves the point x_0 with velocity v_0. Find its position at t_1 if its acceleration is $a(t)$ for all $t \ge 0$. Units are meters and seconds. These problems are review for the first part of Chapter 5.

45. $x_0 = 0; v_0 = 20; t_1 = 10; a(t) = 2g,$
    where $g = 9.8$

46. $x_0 = 1.5; v_0 = 0; t_1 = 5; a(t) = 3t$

47. $x_0 = -1; v_0 = -3; t_1 = 3; a(t) = t^{3/2}$

48. $x_0 = 0; v_0 = 2; t_1 = 1; a(t) = -4 \sin 2t$

49. $x_0 = 1500; v_0 = -15;$
    $t_1 = 1; a(t) = 0.15e^{-0.01t}$

50. $x_0 = 0; v_0 = 2; t_1 = 2; a(t) = -\sin t$

4.8 MODELING OPTIMIZATION PROBLEMS

Optimization is a major theme in the physical and biological sciences, in economics and business, and in applied mathematics. Here are four examples of optimization.

- What is the optimum size of a silicon chip containing integrated circuits? Here *optimum* means least cost to the manufacturer. As the area of the chip increases, the interconnection and packaging costs decrease. At the same time, costs increase with area since the presence of a few defects on a chip wastes the entire chip.

- It appears that light follows an optimum principle as it reflects from a mirror or moves through a lens system. Among all possible paths, the actual path followed by a ray minimizes transit time.

- Soap bubbles on a wire frame assume shapes that minimize surface tension.

- The fitting of experimental data by least squares or linear regression is based on choosing a line so that the sum of squares of the deviations of the data from the line is a minimum. You may have used least squares in courses in which a straight line is fitted to experimental data. Your calculator probably has a least squares package in its statistics menu.

In this section we discuss a few of the many kinds of optimization problems. How can you optimize your work on these problems? We offer several guidelines we have observed as effective among our students.

- *Energy.* Passive reading of the problem and patient waiting for a flash of inspiration are not recommended. Aggressive reading and active exploration strongly improve your chances for flashes of inspiration or glimmers of understanding. The exploration may be graphical, numerical, verbal, physical, or symbolical. False starts and mistakes are normal. The important thing is to try something and then to modify it as necessary.

- *Objectives and variables.* What, exactly, is the problem? Is it a minimum distance, maximum volume, or minimum cost that is wanted? After assigning a variable to the distance, volume, cost, or whatever is to be optimized, relate that variable to the quantities it depends on. If the problem is to maximize the volume of a cylinder, it is likely that the formula $V = \pi r^2 h$, expressing volume in terms of base radius and height, will be needed. Recall or otherwise obtain all needed formulas, sketch appropriate diagrams, and relate variables as necessary.

- *The mathematical model.* Using these variables, formulas, and relations, formulate a specific extremum problem, including the function to be maximized or minimized and its domain. Usually, common sense or the problem statement provide help in deciding on the domain in which to seek candidates.

- *Solve the mathematical model.* Use the Candidate, Max/Min, and Global Extremum Theorems in solving the mathematical model.

- *Interpretation.* Solve the original problem by interpreting or otherwise adapting the solution of the mathematical model.

We will work through five examples. We calculate the minimum distance from a point to a curve, minimize the cost of a drink container, determine the rightmost position of a particle in motion, derive Fermat's principle, and determine the optimum size of a silicon chip.

■ **EXAMPLE 1** Find the point on the graph of the equation $y = \sqrt{x}, x \geq 0$, closest to the point $(2, 0)$.

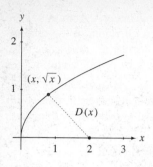

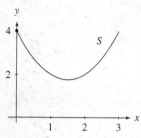

Figure 4.63. Minimum distance problem.

Solution.   The graph of $y = \sqrt{x}$ and the point $(2, 0)$ are shown in Fig. 4.63. Since the distance $D(x)$ between an arbitrary point $(x, \sqrt{x})$ on the curve and the point $(2, 0)$ is

$$\sqrt{(x - 2)^2 + (\sqrt{x} - 0)^2},$$

the minimum distance problem is solved if we can determine $x \geq 0$ that minimizes the function

$$D(x) = \sqrt{(x - 2)^2 + x}, \quad x \geq 0.$$

This is a straightforward problem. The solution is not difficult with the help of the Global Extremum Theorem. However, in problems involving distance, it often simplifies differentiation and evaluation if we minimize or maximize not the distance itself, but its square, that is, determine $x \geq 0$ that minimizes the function

$$S(x) = (x - 2)^2 + x, \quad x \geq 0.$$

The graph of S is shown at the bottom of Fig. 4.63. It is clear that S has a minimum. We may verify this by the Global Extremum Theorem. The derivatives of S are

$$S'(x) = 2(x - 2) + 1$$
$$S''(x) = 2.$$

Since $S''(x) > 0$ on $(0, \infty)$, S is concave up. If S' has a zero $c > 0$, then by the Global Extremum Theorem S has a global minimum at c. From the above calculation, $c = 3/2$ is a zero of D'. It follows that the point $(1.5, \sqrt{1.5}) \approx (1.5, 1.2)$ is the point on the graph of D closest to the point $(2, 0)$.   ■

■ **EXAMPLE 2**   A box in the shape of a pyramid is to contain 12 in.[3] of a fruit drink. The box is to be cut from a piece of foil laminate and folded into a pyramid, as shown in Fig. 4.64. Determine the dimensions x and y that minimize the amount of laminate required.

Solution.   The volume of a pyramid with square base of side a and height h is $V = \frac{1}{3}a^2 h$. Since the volume of the pyramid is to be 12 in.[3], the variables a and h must be related by the equation

$$(1) \qquad\qquad 12 = \tfrac{1}{3}a^2 h.$$

We wish to minimize the amount S of foil laminate, which is the combined area of the square and four triangles. We have

$$(2) \qquad\qquad S = x^2 + 4\left(\tfrac{1}{2}xy\right) = x^2 + 2xy.$$

To minimize S we write it as a function of one variable. For this we rewrite (1) in terms of x and y. Since $a = x$ and $h = \sqrt{y^2 - x^2/4}$, we have

$$(3) \qquad\qquad 12 = \tfrac{1}{3}x^2 \sqrt{y^2 - x^2/4}.$$

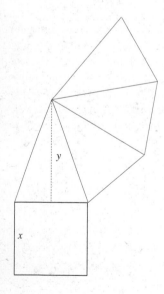

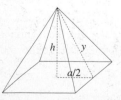

Figure 4.64. Folding a pyramid.

Solving (3) for y and then substituting in (2), we have

$$(4) \qquad S = x^2 + \frac{\sqrt{x^6 + 5184}}{x}, \quad 0 < x < \infty.$$

This function is a valid model of the drink box problem. It has, however, a tough look. To calculate the zeros of S' may require Newton's method or other numerical algorithm. Moreover, since the domain is not a closed and bounded interval, we cannot use the Max/Min Theorem. We may be able to apply the Global Extremum Theorem to verify a "least laminate solution." This would require that we calculate S'' and determine if it is positive on $(0, \infty)$. These calculations also have a tough look.

We discuss several possible solutions. They differ in their dependence on graphical or mathematical evidence and on who or what actually calculates the derivatives. We complete one solution and leave the others as Exercises 17–19.

We start by graphing S, using a calculator or CAS. It appears that $S(x)$ is large for x outside of the interval $[2, 4]$. We may check this by noticing in (4) that both terms become large when $x \ge 4$ and the second term becomes large as x gets close to 0. If we trust Fig. 4.65, it appears that S has a minimum near $x = 3$. Since the machinery that trims foil laminate has limited precision, we need not be overly precise in determining the alleged minimum point. We may choose to approximate the minimum point by zooming on the graph of S. Or we may differentiate (4) to obtain

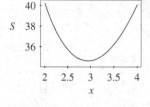

Figure 4.65. Surface area.

$$(5) \qquad S' = \frac{2\left(x^6 + x^3\sqrt{x^6 + 5184} - 2592\right)}{x^2\sqrt{x^6 + 5184}}.$$

Setting the numerator of S' equal to zero, we have

$$(6) \qquad x^6 + x^3\sqrt{x^6 + 5184} - 2592 = 0.$$

We may apply Newton's method or another zero-finding algorithm to show that $x \approx 2.94$ is a zero of S'. Or we may rearrange (6) to obtain

$$x^6 - 2592 = -x^3\sqrt{x^6 + 5184}.$$

Squaring both sides of this equation and simplifying gives

$$(7) \qquad x^6 - 648 = 0.$$

This polynomial has exactly one positive zero, $x \approx 2.94$. Since the zeros of this equation include the zeros of S', we conclude that S' is 0 on $(0, \infty)$ exactly once. Noting from (5) that $S' < 0$ near 0, it follows that $x \approx 2.94$ is a global minimum of S. For, by the I/D Test, S is decreasing to the left of 2.94 and increasing to its right. From (3) we calculate $y \approx 4.42$ in., corresponding to $x \approx 2.94$ in. These dimensions give a container holding 12 in.^{3} of liquid and requiring the least amount of foil laminate. ∎

■ **EXAMPLE 3**   The velocity and initial position of a particle are

$$\mathbf{v}(t) = \left(\cos t^2, \sin t^2\right), \quad t \ge 0, \quad \mathbf{r}(0) = (2, 3).$$

Find the time in $[0, 1.5]$ at which the particle reaches its rightmost position.

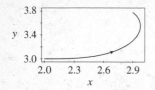

Figure 4.66. Rightmost position.

Solution. We show the path of the particle in Fig. 4.66, which was plotted using Euler's method. Let $x = x(t), t \in [0, 1.5]$, be the x-coordinate of the position of the particle. The rightmost position is the global maximum of this function. Since x has a derivative on $[0, 1.5]$ we know that x is continuous and, hence, has a maximum at $c \in [0, 1.5]$. Since

$$\frac{d^2x}{dt^2} = \frac{d}{dt}\cos t^2 = -2t\sin t^2 < 0, \quad \text{for } x \in (0, 1.5),$$

x is concave down on $(0, 1.5)$. Since $dx/dt = 0$ at $t = \sqrt{\pi/2}$, it follows from the Global Extremum Theorem that x has a global maximum on $(0, 1.5)$. It then follows that x has a global maximum on $[0, 1.5]$. The particle reaches its rightmost position at $t = \sqrt{\pi/2}$. See Exercises 20 and 21. ∎

It appears from observation and experiment that many natural phenomena can be explained in terms of optimality. To explain the refraction of light, Pierre Fermat (1601–1665) assumed that *light travels from a point in one medium to a point in another medium in the least time, and that the velocity is less in the denser medium.* Fermat was able to infer Snell's law from this principle. Willebrord Snellius (1580–1626) had discovered the law of refraction experimentally.

■ **EXAMPLE 4** Referring to Fig. 4.67, a light source at $(0, a)$ in one medium is seen by an observer at (b, c) in another medium. Use Fermat's principle to determine the path taken by the observed ray. Assume that light source, observer, and all light rays are confined to a plane.

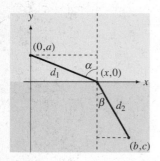

Figure 4.67. Fermat's principle.

Solution. In traveling from $(0, a)$ to (b, c), the observed ray must cross the x-axis. In either of the two media, the light ray travels in a straight line. We use Fermat's principle to determine the crossing point $(x, 0)$. The total time T in transit from $(0, a)$ to (b, c) must be a minimum. Let v_1 be the speed of light in the upper medium, t_1 the time taken by the ray to move from $(0, a)$ to $(x, 0)$, and d_1 the distance from $(0, a)$ to $(x, 0)$. Let v_2, t_2, and d_2 be the corresponding quantities in the lower medium. We wish to minimize the function

$$T(x) = t_1 + t_2.$$

Each of t_1 and t_2 is a function of x. Since in each medium the speed of the ray is constant, we have

$$(8) \qquad \begin{aligned} T(x) &= \frac{d_1}{v_1} + \frac{d_2}{v_2} \\[2mm] T(x) &= \frac{\sqrt{x^2 + a^2}}{v_1} + \frac{\sqrt{(x-b)^2 + c^2}}{v_2}, \quad -\infty < x < \infty. \end{aligned}$$

The function T is always positive and becomes infinite as $|x|$ becomes large. On physical or geometric grounds, it is clear that T has a minimum between 0 and b. Or we may use the Global Extremum Theorem to show that T has a minimum. For this we show that T' has a zero and $T''(x) > 0$ for $x \in (-\infty, \infty)$. We leave the latter calculation as Exercise 24.

To show that T' has a zero, we look at its values near 0 and b. From (8) we have

(9) $$T'(x) = \frac{2x}{2v_1\,\sqrt{x^2 + a^2}} + \frac{2(x - b)}{2v_2\,\sqrt{(x - b)^2 + c^2}}.$$

As $x \to 0$, the first term of $T'(x)$ approaches 0 and the second approaches $-b\big/(v_2\,\sqrt{b^2 + c^2})$, which is negative. So $T'(x)$ is negative for x near 0. As $x \to b$, the second term of $T'(x)$ approaches 0 and the first approaches $b\big/(v_1\,\sqrt{a^2 + b^2})$, which is positive. So $T'(x)$ is positive for x near b. It follows from the Intermediate Value Theorem that T' has a zero between 0 and b.

There are several ways we can determine the path taken by the observed ray. The most direct would be to attempt to determine the crossing point $(x, 0)$ by solving the equation $T'(x) = 0$. As outlined in Exercise 25, this results in a fourth-degree equation. While there is a formula for solving any fourth-degree equation (referred to as Ferrari's method), the result is not useful here. It is better to use the equation $T' = 0$ to infer Snell's law, which gives a relation among the velocities v_1 and v_2 and angles α and β that holds for the observed ray. From the equation

$$T' = 0$$

we have

$$\frac{x}{v_1\,\sqrt{x^2 + a^2}} + \frac{(x - b)}{v_2\,\sqrt{(x - b)^2 + c^2}} = 0.$$

This equation can be rearranged as

$$\frac{x}{v_1 d_1} = \frac{b - x}{v_2 d_2}.$$

From the right triangles in Fig. 4.67 we have

(10) $$\frac{\sin \alpha}{v_1} = \frac{\sin \beta}{v_2}.$$

Equation (10) is Snell's law, which is more often expressed in terms of the indices of refraction, n_1 and n_2, of the two media. These are $n_1 = v/v_1$ and $n_2 = v/v_2$, where v is the speed of light in a vacuum. The angles α and β are called the angles of incidence and refraction, respectively. ■

Our last example concerns the optimum size of the silicon chips containing integrated circuits. It is based on a paper written by Colin A. Warwick and Abbas Ourmazd, "Trends and Limits in Monolithic Integration by Increasing the Die Area" (*IEEE Transactions on Semiconductor Manufacturing,* vol. 6, no. 3, August 1993, pp. 284–289).

■ **EXAMPLE 5**   To predict the main trends in manufacturing silicon chips, AT&T scientists Warwick and Ourmazd modeled the cost in dollars per square centimeter of the area A of a typical chip. Their model closely matched observed costs in 1992. They used the model to anticipate chip size in 2010. They expressed the total manufacturing cost $C = C(A)$ as a sum of processing, assembly, and waste costs. Their equation is

$$(11) \qquad C(A) = k_1 + \frac{k_2}{\sqrt{A}} + \left(k_1\left(1 + \tfrac{1}{2}D_0A\right)^2 - k_1\right), \quad A > 0.$$

In 1992 $k_1 = 10$, $k_2 = 10$, and $D_0 = 0.3$. The first constant, k_1, is the cost in dollars per square centimeter to process a chip, up to the time it is tested for defects; k_2 is the cost of assembly, also measured in dollars per square centimeter; and D_0 is average number of defects per square centimeter. The value 0.3 was the "world class" rate in 1992.

It is easy to see from (11) that $C(A)$ becomes large as A becomes either large or small. We expect, therefore, that C will have a minimum. In explaining their model, Warwick and Ourmazd noted that circuits can be integrated onto a large silicon chip or partitioned onto several smaller chips. The larger chip reduces interconnection and packaging costs but increases the penalty for defects, for if a defect occurs, the entire chip must be scrapped.

Determine the optimum chip size using (11) with the 1992 values of the constants. The model with the 2010 constants appears in Exercise 26.

Solution. Actually, there is not much to do. Warwick and Ourmazd did the hard part, that of formulating a model of the cost of manufacturing silicon chips. In Fig. 4.68 we have reproduced their graph, from which the optimum chip size of approximately 1.3 cm^2 can be read. This size, and its cost of $\$23/$cm^2, are close to actual industrial experience in 1992.

We may check these conclusions by calculating C' and C''. We first simplify the expression for $C(A)$. We have

$$C(A) = k_2A^{-1/2} + k_1(1 + \tfrac{1}{2}D_0A)^2$$
$$C'(A) = -\tfrac{1}{2}k_2A^{-3/2} + k_1D_0(1 + \tfrac{1}{2}D_0A)$$
$$C''(A) = \tfrac{3}{4}k_2A^{-5/2} + \tfrac{1}{2}k_1D_0^2.$$

Setting $C'(A) = 0$ and solving for A, we obtain $A \approx 1.25$. The total cost for this chip size is $C(1.25) \approx 23.0$.

The form of $C''(A)$ shows that $C''(A) > 0$ for all $A > 0$. We conclude from the Global Extremum Theorem that $A \approx 1.25$ is a global minimum. ∎

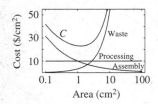

Figure 4.68. Optimum chip size.

Exercises 4.8

Basic

1. Find the point on the graph of the equation $y = \sqrt{2x}$ closest to $(3, 0)$.

2. Find the point on the graph of the equation $y = 3\sqrt{x}$ closest to $(4, 0)$.

3. Find the point on the graph of the equation $y = 2x + 1$ closest to $(1, 1)$.

4. Find the point on the graph of the equation $2x + 3y + 4 = 0$ closest to $(2, 1)$.

5. Find the points on the graph of the equation $y = \sqrt[3]{x}$, $0 \le x \le 2$, furthest and closest to $(2, 0)$.

6. Find the points on the graph of the equation $y = \sqrt[3]{2x}$, $0 \le x \le 2$, furthest and closest to $(1, 0)$.

7. A ray of light is moving from air into a diamond. If its angle of incidence is $10°$, find its angle of refraction. The indices of refraction of air and diamond are 1.000293 and 2.419, respectively.

8. A ray of light is moving from air into a pool of water. If its angle of incidence is $20°$, find its angle of refraction. The indices of refraction of air and water are 1.000293 and 1.333, respectively.

Exercises 9–15:   Identify the key variables, write an expression or function to be optimized, use the problem statement or figure to relate the key variables, specify the domain of the function to be optimized, and use the Candidate Theorem and Global Extremum Theorem to find the maximum or minimum of the function.

9. A box with a square base, rectangular sides, and open top is to contain 1 cubic foot of space. If the material for its base costs $3/\text{ft}^2$ and that for its sides costs $1/\text{ft}^2$, determine its dimensions so that the cost of the materials is a minimum.

10. A rectangle has perimeter 1 meter. What are the height and width of the rectangle with largest possible area?

11. A rectangle is to be cut from an equilateral triangle with side a so that one side is parallel to a side of the triangle. What are the height and width of the rectangle with the largest possible area?

12. A gutter is made by bending a long piece of sheet metal into three equal strips, so that a cross-section is an open trapezoid, as in the following figure. How should the bending angle θ be chosen so that the area of the cross-section is as large as possible?

13. A rectangle is to be cut from a right triangle with sides 3, 5, and $\sqrt{34}$, so that one side is parallel to the side of length 3. What are the dimensions of the rectangle with the largest possible area?

14. Referring to the accompanying figure, an open tray is to be made from a piece of sheet metal 0.6 m by 0.9 m. The four dotted guidelines are drawn parallel to the edges and equally distant from them. A cut of length x is made in a corner square, as shown in the figure. After four such cuts, the light shaded rectangles are bent through $90°$ to form the sides of the tray. The dark shaded rectangles are bent through $90°$ to form the ends of the tray. The corner squares are then bent through $90°$ and soldered into place. Choose x so that the volume of the tray is as large as possible.

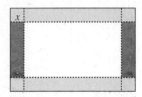

15. Verify all calculations not fully explained in Example 2 up to and including (4).

Growth

16. Show that a nonnegative function f has a minimum/maximum at q if and only if the function g has a minimum/maximum at q, where $g(x) = f(x)^2$.

17. See Example 2. Suppose that upon consulting the marketing department, you find that they have already decided that $2.5 \le a \le 3.75$. Use the Max/Min Theorem and the calculations in the example to solve the drink box problem.

18. See Example 2. Show graphically that $S''(x) > 0$ for $x \in (0, \infty)$.

19. See Example 2. Graph S and zoom as necessary to locate the global minimum to within 0.1 in.

20. In Example 3 we said, "It follows that x has a global maximum on $[0, 1.5]$." We showed that $x'(\sqrt{\pi/2}) = 0$ and x is concave down on $(0, 1.5)$. Use the fact that x is continuous on $[0, 1.5]$ to show that x has a global maximum at $t = \sqrt{\pi/2}$.

21. Use Euler's method to determine approximately the rightmost position of the particle in Example 3.

22. The initial position of a particle is $\mathbf{r}(0) = (2, 3)$ and its velocity is

$$\mathbf{v}(t) = (\cos t^3, \sin t^3).$$

Find the time in $[0, 1.5]$ at which the particle reaches its rightmost position.

23. Continuing the preceding exercise, use Euler's method to determine approximately the rightmost position of the particle.

24. In Example 4 show that

$$T'' = \frac{c^2}{v_2 d_2^3} + \frac{a^2}{v_1 d_1^3}$$

and, hence, $T''(x) > 0$ for all $x \in (0, b)$.

25. In Example 4 show that the equation

$$Ax^4 + Bx^3 + Cx^2 + Dx + E = 0,$$

where $A = v_1^2 - v_2^2$, $B = -2b(v_1^2 - v_2^2)$, $C = (a^2 + b^2)v_1^2 - (b^2 + c^2)v_2^2$, $D = -2a^2bv_1^2$, and $E = a^2b^2v_1^2$, follows from the equation $T'(x) = 0$. (Outline: Combine the two terms into a single fraction, set the numerator equal to 0, and then isolate and square the worst-looking radical expression.)

26. Determine the optimum chip size according to Warwick and Ourmazd's model (see Example 5), for the values $k_1 = 10$, $k_2 = 10$, and $D_0 = 0.01$. Evidently, the number of defects per square centimeter is expected to decrease, thus making larger chips economically feasible.

27. In Example 4, assume that $v_2 = 0.9$, $v_1 = 0.95$, $a = 0.5$, and $(b, c) = (2.0, 1.0)$. Find the point where the observed ray crosses the interface by minimizing T. Find the same point by using Snell's law.

28. A ray of light from a source at $(0, 100)$ is seen at $(200, 50)$ after reflecting from a 200 cm mirror, as in the following figure. Use Fermat's principle to calculate the point on the mirror from which the observed ray reflected. Show that the angles of incidence and reflection are equal.

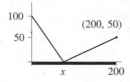

29. The following figure shows a construction for locating the reflecting point of a ray of light moving from $(0, a)$, reflecting from a mirror, and observed at (b, c). The idea is to construct the point $(b, -c)$ and then join $(0, a)$ to $(b, -c)$ by a line segment L. The point where L crosses the mirror is the reflecting point. The dotted ray would be shorter than any other ray and, in particular, the ray shown. The problem of locating the reflecting point is called Heron's problem, after Heron of Alexandria. Show that the construction is a solution to the reflection problem.

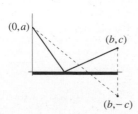

30. A rectangle is inscribed in a semicircle of radius a so that one of its sides lies along the diameter. What are the height and width of the rectangle with largest possible area?

31. A 10 m–wide east-west canal flows into a 15 m–wide north-south canal. Find the length of the longest toothpick that can negotiate the turn.

32. Re-solve Exercise 31, replacing the toothpick with a barge of width 4 m.

33. A soft drink can is to contain 12 fluid ounces and be the standard shape, a cylinder with top and bottom. The bottom and curved sides cost $\$0.02/m^2$, and the top costs $\$0.05/m^2$. What is the ratio of height to base diameter of the can that is cheapest to manufacture? Although you do not need the conversion factor, 1 fluid ounce $\approx 2.957 \times 10^{-5}$ m^3.

34. Light sources are placed at coordinates 0 and 1 on an x-axis. The illumination at any point x between them from either source is proportional to the "strength" of the source and inversely proportional to the square of the distance to that source. Assume that the source at 0 is L times stronger than that at 1. Show that at the "dimmest" point between 0 and 1 the ratio of the illumination received from the source at 0 to the source at 1 is $\sqrt[3]{L}$.

35. Calculate the dimensions of the cylinder of largest volume that can be inscribed in a cone with height 8 m and base radius 5 m. A cross-section is shown in the figure.

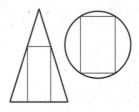

36. Calculate the dimensions of the cylinder of largest volume that can be inscribed in a sphere with radius 5 m. A cross-section is shown in the figure.

37. A conical cup is made by cutting a sector from a paper circle of radius 15 cm and aligning edges AB and BC and fastening with tape. See the following figure. How should the cuts be made if the cup is to have the largest possible volume?

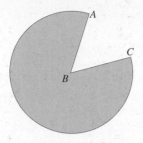

38. The following optimization problem was adapted from *Engineering Economic and Cost Analysis,* second edition, by Courtland A. Collier and William B. Ledbetter (HarperCollins, 1988). One of the key variables in designing electrical transmission lines is the area A of a cross-section of the wire. As the area A decreases, the cost of installation decreases; however, the annual cost of the energy lost due to the electrical resistance of the wire increases. This problem is a simplified design problem for a 200 ft line connecting a transformer to an electrical pump at a municipal water plant. The effect of investing any

savings on installation or subsequent energy costs have, for example, not been considered.

The two costs we consider are that needed to install the line and that due to the electrical resistance of the line. The installation cost is a combination of a fixed cost of $200 for fittings and the cost of the copper. The annual loss of energy due to the electrical resistance of the line is $293.2/A$ kilowatt-hours. It is given that

• The estimated life of the transmission line is 20 years.
• Copper weighs 0.32 lb/in.^{3}.
• Copper costs $0.80 per pound.
• The cost of energy is $0.05 per kilowatt-hour.

Calculate the diameter of the wire that minimizes the annual costs. Report both the diameter and the minimum annual cost.

Review/Preview

Exercises 39–46:   At $t = 0$ a particle leaves the point x_0; for $t \geq 0$ its velocity is $v = v(t)$. Find its position at t_1. Units are meters and seconds.

39. $v = v(t) = t/3, \ x_0 = 0, \ t_1 = 2.0$
40. $v = v(t) = t^2, \ x_0 = -2.5, \ t_1 = 2.5$
41. $v = v(t) = 3t - 5, \ x_0 = 2.0, \ t_1 = 1.0$
42. $v = v(t) = \sqrt{t}, \ x_0 = 1.0, \ t_1 = 31.5$

43. $v = v(t) = \sin \pi t, \ x_0 = -5.6, \ t_1 = 5.0$
44. $v = v(t) = \frac{1}{2}\cos 5t, \ x_0 = 2.3, \ t_1 = 15.0$
45. $v = v(t) = 1/(t + 1), \ x_0 = 0, \ t_1 = 10.9$
46. $v = v(t) = 1/(t^2 + 1), \ x_0 = -12.5, \ t_1 = 1.2$

47. In the first quadrant sketch the triangle whose sides are parts of the lines with equations $y = \frac{3}{4}x, \ y = 0$, and $x = b$ (where $b \geq 0$). Express the area A of the triangle in terms of b. Calculate the rate of change of A with respect to b.

48. Repeat Exercise 47, except that the line is $y = mx$, where $m > 0$. Regard m as fixed.

49. A swimming pool measures 10 m by 50 m. Vertical sections parallel to the 50 m side are trapezoids. The deep end has depth 3 m and the shallow 1 m. Water is flowing into the pool at the deep end at the rate of 0.5 m^3 per minute. Find the rate at which the bottom of the pool is becoming submerged two minutes after the fill began.

REVIEW OF KEY CONCEPTS

This chapter is the fourth and last on "differential calculus" for functions of one variable. It began with the important idea of the tangent line approximation. A major application of the tangent line approximation was to solve initial value problems using Euler's method. A second application was to Newton's method, a rapidly converging algorithm for determining the zeros of functions.

The first and second derivatives were used in the I/D and Concavity Tests for determining whether a function is increasing, decreasing, concave up, or concave down on an interval. The Mean-value Inequality and the Max/Min Theorem were brought in to prove these tests.

Related-rates problems arise in applications in which two or more time-dependent variables are related by one or more equations.

The asymptotic behavior of functions was discussed, including horizontal and vertical asymptotes and the relative growth of the logarithm and exponential functions in comparison with a power function.

The last two sections concerned optimization. The Candidate Theorem, the Max/Min Theorem, and the Global Extremum Theorem are the main tools for solving optimization problems. In the last section the emphasis was on setting up or modeling optimization problems from several areas of applied mathematics.

Chapter Summary

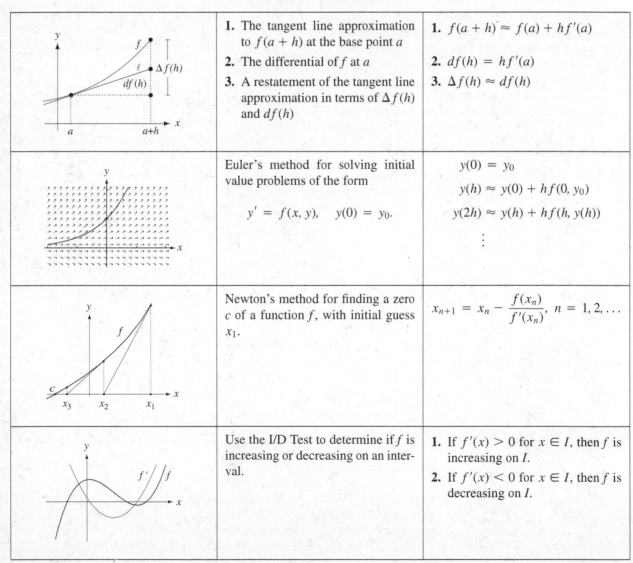

	1. The tangent line approximation to $f(a + h)$ at the base point a	1. $f(a + h) \approx f(a) + hf'(a)$
	2. The differential of f at a	2. $df(h) = hf'(a)$
	3. A restatement of the tangent line approximation in terms of $\Delta f(h)$ and $df(h)$	3. $\Delta f(h) \approx df(h)$
	Euler's method for solving initial value problems of the form $$y' = f(x, y), \quad y(0) = y_0.$$	$y(0) = y_0$ $y(h) \approx y(0) + hf(0, y_0)$ $y(2h) \approx y(h) + hf(h, y(h))$ $\vdots$
	Newton's method for finding a zero c of a function f, with initial guess x_1.	$x_{n+1} = x_n - \dfrac{f(x_n)}{f'(x_n)}, \quad n = 1, 2, \ldots$
	Use the I/D Test to determine if f is increasing or decreasing on an interval.	1. If $f'(x) > 0$ for $x \in I$, then f is increasing on I. 2. If $f'(x) < 0$ for $x \in I$, then f is decreasing on I.

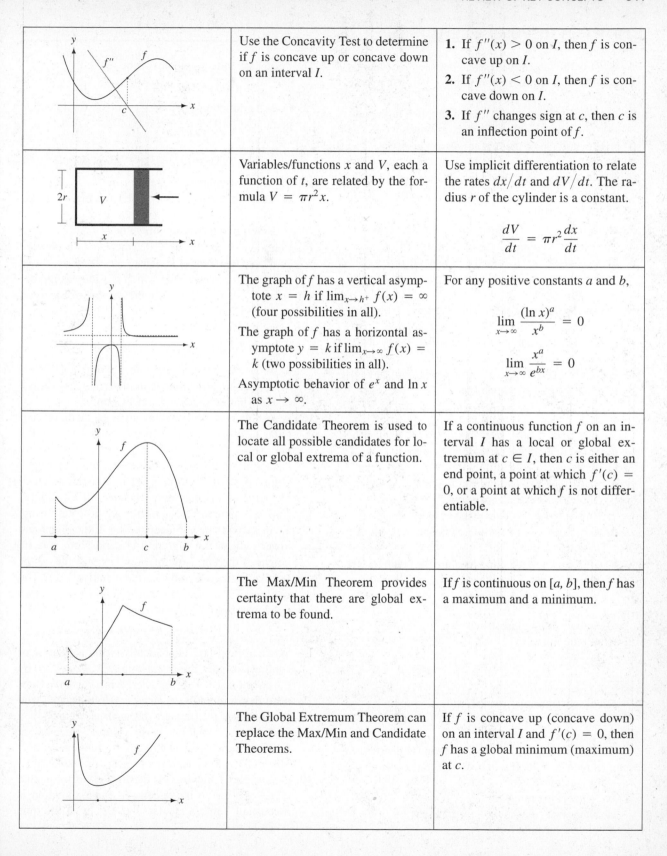

	Use the Concavity Test to determine if f is concave up or concave down on an interval I.	**1.** If $f''(x) > 0$ on I, then f is concave up on I.
		2. If $f''(x) < 0$ on I, then f is concave down on I.
		3. If f'' changes sign at c, then c is an inflection point of f.
	Variables/functions x and V, each a function of t, are related by the formula $V = \pi r^2 x$.	Use implicit differentiation to relate the rates dx/dt and dV/dt. The radius r of the cylinder is a constant. $$\frac{dV}{dt} = \pi r^2 \frac{dx}{dt}$$
	The graph of f has a vertical asymptote $x = h$ if $\lim_{x \to h^+} f(x) = \infty$ (four possibilities in all). The graph of f has a horizontal asymptote $y = k$ if $\lim_{x \to \infty} f(x) = k$ (two possibilities in all). Asymptotic behavior of e^x and $\ln x$ as $x \to \infty$.	For any positive constants a and b, $$\lim_{x \to \infty} \frac{(\ln x)^a}{x^b} = 0$$ $$\lim_{x \to \infty} \frac{x^a}{e^{bx}} = 0$$
	The Candidate Theorem is used to locate all possible candidates for local or global extrema of a function.	If a continuous function f on an interval I has a local or global extremum at $c \in I$, then c is either an end point, a point at which $f'(c) = 0$, or a point at which f is not differentiable.
	The Max/Min Theorem provides certainty that there are global extrema to be found.	If f is continuous on $[a, b]$, then f has a maximum and a minimum.
	The Global Extremum Theorem can replace the Max/Min and Candidate Theorems.	If f is concave up (concave down) on an interval I and $f'(c) = 0$, then f has a global minimum (maximum) at c.

CHAPTER REVIEW EXERCISES

1. What is the largest difference between $\sqrt[3]{x}$ and its tangent line approximation on [6, 10], using base point 8?

2. A coat of paint 0.04 cm thick is applied to the outside of each of 1000 wooden boxes with covers. The bottom and top of the box are 40 cm by 80 cm. The height of the box is 50 cm. Use the tangent line approximation in calculating the amount of paint needed for the job. What percentage error is made in ordering paint?

3. Show that $\sqrt{1+x} \approx 1 + \frac{1}{2}x$ near $x = 0$. How good is this approximation for $-1/2 \le x \le 3/2$?

4. Show that $e^{kx} \approx 1 + kx$ for x near 0. If $k = 0.5$ and $-1 \le x \le 1$, what is the largest error $E(x) = |e^{kx} - (1 + kx)|$?

5. The diameter of a ball bearing is measured as 1.54 cm, with a possible error of 0.01 cm either way. Approximate the maximum error in the mass of the bearing if its density is 7500 kg/m^3.

6. Use Euler's method to obtain an approximate solution to the initial value problem

$$y' = \sqrt{y}, \quad y(0) = 1$$

for $0 \le x \le 1$. Use step size $h = 0.1$. Sketch the direction field and your solution together.

7. Let $p(x) = x^3 + 2x + 1$.
   a. Show that p has an inverse function.
   b. Show that p has a zero between -1 and 0.
   c. How many zeros does p have? Why?
   d. Use Newton's method to calculate the zero of p to within 0.001.

8. Use Newton's method to calculate the positive zero of the function

$$f(x) = x^2 - \cos x$$

to three decimal places accuracy.

9. Graph the function

$$f(x) = \frac{x^2 - 3x + 1}{x^2 + 1}, \quad x \in R.$$

Label all extreme points, inflection points, and asymptotes.

10. Let $f(x) = 2\ln(x^2 + 1)$, $x \ge 0$.
   a. Sketch the graph of f, showing intervals on which it is increasing, decreasing, concave up, or concave down.
   b. What is the range of f?
   c. Show that f has an inverse function.
   d. Give a formula for f^{-1}.

11. Locate the inflection point of the function

$$f(x) = (x - a)(x - b)(x - c).$$

12. Does the function $f(x) = x^4 - x$ have an inflection point at $x = 0$? Why or why not?

13. Let

$$f(x) = (x^2 - 3)e^x, \quad x \in R.$$

   a. On what intervals is f increasing?
   b. Determine the points at which f changes concavity.
   c. Determine the minimum value of f.

14. Let $f(x) = xe^{-ax}$, $x \ge 0$, where a is a positive constant. In terms of a, locate any local or global extrema and intervals on which f is increasing, decreasing, concave up, or concave down. Sketch a generic graph.

15. Sketch the graph of

$$f(x) = x + \cos x, \quad -2\pi \le x \le 2\pi.$$

   a. What are the local and global extrema of f?
   b. If f has an inflection point, determine its value.
   c. On what intervals is f concave up?

16. A particle is moving on the unit circle so that its angular speed is 2π radians per minute. At $t = 0$ the particle is at $(1, 0)$. A second particle, also initially located at $(1, 0)$, moves upward on the line $x = 1$ with a speed of 200 m/min. What is the rate at which the particles are separating when $t = 1/12$ minute?

17. Coffee is poured at a uniform rate of 2 cm^3/s into a cup shaped like a truncated cone. If the upper and lower radii of the cup are 4 cm and 2 cm and the height of the cup is 6 cm, how fast will the coffee level be rising when the coffee is halfway up? The volume of a truncated cone of height h and upper and lower radii R and r is

$$V = \frac{1}{3}\pi h(R^2 + Rr + r^2)$$

18. When air expands adiabatically (without heating or cooling), its pressure P and volume V are related by the equation $PV^{1.4} = k$, where k is a constant. Suppose that a quantity of air is undergoing adiabatic expansion. If the pressure of the air is decreasing at 5 kPa/min, at what rate is the volume increasing at the time the volume is 500 cm^3 and the pressure is 100 kPa? (A kilopascal is a pressure of 1000 N/m^2.)

19. A particle is moving on the curve with equation $x^4y + 2y^3 = 3$. Its velocity in the x direction is a constant 2 m/s. Calculate its velocity at the time when $x = 2$.

20. Gas is pumped into a spherical balloon at 1 ft³/min. How fast is the diameter of the balloon increasing when the balloon contains 36 ft³ of gas?

21. A spherical water tank has many coats of old paint and a radius, with paint, of 20 feet. The paint is sand-blasted off and the tank repainted, with a net decrease of 0.2 in. in the radius. Approximate the net volume of paint removed.

22. A cube of ice is melting so that it stays in the shape of a cube. At the instant the edge of the cube measures 1 cm, how fast is the surface area changing with respect to the volume?

23. Variables u and w are functions of time t and are related by the equation

$$u = \sqrt{w^2 + 1}.$$

Express dw/dt in terms of du/dt and w.

24. Graph the function

$$f(x) = \frac{x}{\sqrt[4]{x^4 + 1}}.$$

Include any horizontal or vertical asymptotes.

25. Calculate $\lim_{x \to 0^+} x \ln(x^2)$.

26. Calculate $\lim_{y \to 0^+} y^{-1} e^{-1/y}$.

27. Does the function $f(x) = (x^2 - 1)/(x^2 + 1)$ have horizontal or vertical asymptotes? Why or why not?

28. Show that

$$\lim_{x \to \infty} (\sqrt{x^2 + 1} - x) = 0.$$

(Hint: Rationalize the numerator.)

29. A wire of length 12 inches can be bent into a circle, bent into a square, or cut into two pieces to make both a circle and a square. How much wire should be used for the circle if the total area enclosed by the figure(s) is to be a minimum? A maximum?

30. Let $m \geq n$. A wire of length L can be bent into a regular m-gon, bent into a regular n-gon, or cut into two pieces to make both a regular m-gon and a reg-

ular n-gon. How much wire should be used for each if the total area enclosed by the figure(s) is to be a minimum? A maximum?

31. Referring to the following figure, choose P so that θ is a maximum.

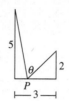

32. Find the maximum and minimum values and the points at which they occur for

$$f(x) = 1 + 3/x - 1/x^2, \quad 1/2 \leq x \leq 10.$$

33. Find the maximum and minimum of the function $f(x) = 3^x - 2^x, -4 \leq x \leq 2$.

34. Determine the dimensions and volume of the right circular cone of the smallest volume that can be circumscribed about a hemisphere of radius a.

35. Locate the maximum and minimum of the function

$$f(x) = \tfrac{1}{2}x^3 - \tfrac{3}{2}x^2 + 5, \quad -1.5 \leq x \leq 2.5.$$

36. What are the local and global extrema of the function

$$f(x) = x^{2/3} - x, \quad -1 \leq x \leq 8.$$

37. A rectangle has one vertex at $(0, 0)$ and the diagonal vertex in the first quadrant and on the line with equation $2x + y = 1$. Determine the rectangle with the largest area.

38. A student hands in a paper with the following statement: "Since $f'(c) = 0$ and c is an interior point of the domain of f, there is a local maximum or minimum at c." If this answer is correct, give a reason why it is correct. If it is false, give a sketch showing why the answer is false.

Student Project

Retrograde Motion of Mars

If at each midnight for the next several years you were to observe the position of Mars against the background of the fixed stars, you would find that its angular position along the ecliptic increases a little each night, but that at certain times it decreases for several weeks. As seen from Earth, the motion of Mars is said to be either direct or retrograde, depending on whether its angular position is increasing or decreasing.

The retrograde motion of Mars was observed by ancient astronomers. The following careful description of the retrograde motion of Mars was taken from a Babylonian clay tablet found in the ruins at Nineveh.

Mars at its greatest power becomes very bright and remains so for several weeks; then its motion becomes retrograde for several weeks, after which it resumes its prescribed course.

In the discussion and questions that follow we have made several simplifying assumptions about the motions of Earth and Mars. These assumptions do not change in any essential way the phenomenon of retrograde motion. We assume

- The plane of Mars' orbit coincides with the plane of Earth's orbit. (The actual plane is tilted 1.8° from that of Earth.) The intersection of the plane of Earth's orbit with the *celestial sphere* is called the *ecliptic*. The celestial sphere may be regarded as a sun-centered sphere. It is the "background of the fixed stars." Its radius is very, very large relative to solar system distances.
- The orbits of Mars and Earth are sun-centered circles. (The actual orbits are ellipses, with eccentricities 0.093 and 0.017, respectively. The eccentricity of a circle is 0.) We use the astronomical unit (AU) as our unit distance. A distance of 1 AU is defined as the mean distance from Earth to the sun ($\approx 1.496 \times 10^8$ km). The radius of Earth's orbit is 1.0 AU and that of Mars 1.52 AU.
- At some time, which we take as $t = 0$, the sun, Earth, and Mars are aligned, in this order.
- The period of Mars is 1.88 Earth years.

From these assumptions it is clear that the orbits of Earth and Mars have the form

(1)
$$\mathbf{r}_E = \mathbf{r}_E(t) = r_E\big(\cos\omega_E t,\ \sin\omega_E t\big), \quad t \geq 0$$
$$\mathbf{r}_M = \mathbf{r}_M(t) = r_M\big(\cos\omega_M t,\ \sin\omega_M t\big), \quad t \geq 0.$$

Problem 1. Assign values to the constants in the vector/parametric equations (1). Graph the two orbits on the same set of axes.

Problem 2. For any fixed $t \geq 0$, let S_t and S_t^* be the intersections of the celestial sphere and the lines with equations

$$\mathbf{r} = \mathbf{r}(u) = \mathbf{r}_E(t) + u\big(\mathbf{r}_M(t) - \mathbf{r}_E(t)\big), \quad u \geq 0, \quad \text{and}$$
$$\mathbf{r} = \mathbf{r}(v) = v\big(\mathbf{r}_M(t) - \mathbf{r}_E(t)\big), \quad v \geq 0.$$

As seen from Earth at time t, the angular displacement $\theta(t)$ of Mars is the angle between the line from $\mathbf{r}_E(t)$ to S_0 and the line from $\mathbf{r}_E(t)$ to S_t. Show that we may instead measure the angular displacement with the angle $\theta^*(t)$ between the line from the origin to S_0 and the line from the origin to S_t^*.

Problem 3. Recall that if $\mathbf{a}$ and $\mathbf{b}$ are vectors, the angle $\theta \in [0, \pi]$ between them may be calculated from

$$\theta = \arccos\left(\frac{\mathbf{a} \cdot \mathbf{b}}{\|\mathbf{a}\|\,\|\mathbf{b}\|}\right).$$

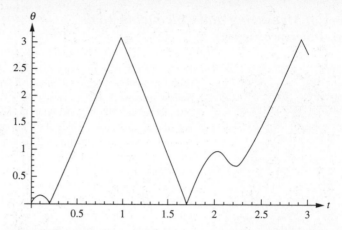

Figure 4.69. Angular position of Mars on ecliptic as seen from Earth.

We have plotted in Fig. 4.69 the angle θ between the vectors $\mathbf{r}_M(t) - \mathbf{r}_E(t)$ and $(1, 0)$, $0 \leq t \leq 3$. Explain and reproduce this graph.

Problem 4.   For what $t \in [0, 3]$ is Mars brightest? Dimmest?

Problem 5.   Explain the comment taken from the Babylonian tablet.

Student Project

Optimal Box Problem

Before cutting, folding, and stapling or gluing, cardboard boxes are flat pieces of cardboard on which are marked cutting and folding lines, as in Fig. 4.70. A company wishes to arrange the cutting and folding lines so that with a fixed amount of cardboard a box of maximum volume results.

Problem.   A piece of cardboard is marked with 10 solid lines and 6 dotted lines along which cuts and folds are to be made. The cuts free two corner pieces, which are not used in the subsequent folding of the cardboard and gluing of the tab to form a box. The area of the piece of cardboard must not exceed 1400 in.^{2}. When glued, the tab overlaps BC by 1.25 in. For strength, the dimensions of the eight pieces forming the top and bottom must satisfy $AB = \frac{1}{2}DE = \frac{1}{3}CD$. Choose dimensions AB and BC so that the volume of the box is a maximum.

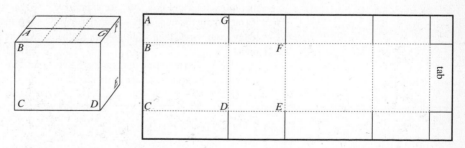

Figure 4.70. After and before.

A History of Measuring and Calculating Tools
Babbage's Difference Engines

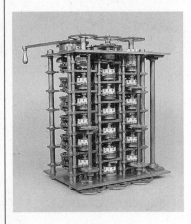

Before World War II mathematicians and scientists used tables of mathematical functions in their work. Before Hewlett-Packard introduced the HP-35 calculator in 1972, calculus students used tables of trigonometric and exponential functions. The importance of accurate tables and, especially, the difficulties in producing them were succinctly expressed by Charles Babbage (1792–1871), designer of Difference Engine No. 1, a small part of which is shown in the photo. Upon being asked to check two independently prepared astronomical tables, Babbage became exasperated by numerous discrepancies and said "I wish to God these calculations had been executed by steam!"

Babbage invented difference engines to calculate tables. We briefly explain *differences* in an example. The polynomial

$$P(x) = x - x^3/3! + x^5/5!$$

can be used to approximate $\sin x$ for small x. We compare $P(x)$ and $\sin x$ in a table.

x	$\sin x$	$P(x)$
0.0	0.000000	0.000000
0.1	0.099833	0.099833
0.2	0.198669	0.198669
0.3	0.295520	0.295520
0.4	0.389418	0.389419
0.5	0.479426	0.479427
0.6	0.564642	0.564648
0.7	0.644218	0.644234

Evidently, we may approximate $\sin x$ by evaluating $P(x)$, although the required multiplications and divisions are expensive if you pay by the hour and, worse, are more likely to go wrong than additions or subtractions.

The key observation to a more efficient calculation—and to the idea at the heart of Babbage's difference engines—is that the fifth differences of $P(x)$ are constant. This is closely related to the fact that

$$\frac{d^5 P(x)}{dx^5} = 1, \text{ a constant.}$$

We explain differences using a simpler polynomial. The table below gives the third differences of x^3. The first column lists 1^3, $2^3, \ldots, 5^3$, a short table of cubes.

1			
8	7		
27	19	12	
64	37	18	6
125	61	24	6

The second column lists the differences of the first column ($7 = 8 - 1$, $19 = 27 - 8, \ldots$), the third column lists the differences of the second, and so on. The third differences are constant. This process starts with cubes and ends with constants. To start with constants and end with cubes, we use the four column heads 6, 12, 7, and 1 and calculate cubes by addition: $12 + 6 = 18$, $7 + 12 = 19$, $1 + 7 = 8$, $18 + 6 = 24$, and so on.

The values of $P(x)$ can be calculated in the same way. Since $P(x)$ is of degree 5, we must calculate the first entries of columns 1 through 5, which are 0.000000, 0.099833, -0.000997, -0.000988, and 0.000010. Upon inserting these starting numbers into a difference engine, steam-powered addition "cranks out" a table of approximations to $\sin x$.

Although Babbage was unable to complete full working models of his difference engines, a full-size difference engine was completed at the Science Museum in London in 1992, the bicentenary of Babbage's birth. Its construction was based on his original designs and flawlessly performed its first major calculation.

Chapter **5**

The Integral

Differential and Integral Calculus

Differential calculus is based on the idea of the derivative of a function at a point, which is a limit of a *difference* quotient. By contrast, integral calculus is based on a limit of a *sum* of function values over an entire interval.

Applications of integral calculus include the calculation of the lengths of curves, areas of regions in the plane or of surfaces in space, and the volumes of solids. Other applications include the calculation of the center of mass of a solid, the probability that a normally distributed random variable is within a given distance of its mean, and the gravitational force exerted on a unit mass by another mass. We give a brief discussion of the last application to show how integrals can arise.

We show in Fig. 5.1 a unit mass of 1 kg and a particle of mass m kg located r meters from the unit mass. Using Newton's law of universal gravitation, the force **F** exerted on the unit mass by the second particle is

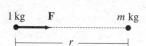

$$\mathbf{F} = \frac{Gm}{r^2}\mathbf{u}$$

where **u** is a unit vector directed from the unit mass to the other mass.

The force on a unit mass arising from three masses m_1, m_2, and m_3 located r_1, r_2, and r_3 meters from the unit mass is given by

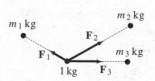

Figure 5.1. Discrete mass models.

$$\mathbf{F} = \frac{Gm_1}{r_1^2}\mathbf{u}_1 + \frac{Gm_2}{r_2^2}\mathbf{u}_2 + \frac{Gm_1}{r_3^2}\mathbf{u}_3$$

where $\mathbf{u}_1$, $\mathbf{u}_2$, and $\mathbf{u}_3$ are unit vectors directed from the unit mass to the three masses.

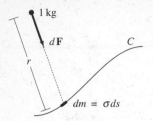

Figure 5.2. Continuous mass model.

So far, this is just simple vector addition. In Fig. 5.2 we show a thin curved rod C with variable density, which is given by a function $\sigma = \sigma(s)$, where s is the distance from one end of the rod. The dimensions of σ are in kilograms/meter. What is the force $\mathbf{F}$ exerted on the unit mass by the rod?

The approach taken in integral calculus is to imagine that the rod has been divided into a large number of very short segments, to calculate the mass of each segment, to add the forces they exert on the unit mass, just as we did above, and then to somehow take the limit of this result as the subdivision of the rod is continued indefinitely. Shown in the figure is a typical segment of mass $dm = \sigma\,ds$, where ds is the length of the segment. To suggest the subdivision process, imagine that the subdivision has progressed to the point at which each segment is a single atom or molecule in length. For most purposes such a segment of a thin rod behaves as a particle.

The mass dm of the segment/particle is its density (kg/m) times its length (m), so that the product dm has units of kilograms. The force $d\mathbf{F}$ on the unit mass due to the particle of mass dm is

$$d\mathbf{F} = \frac{G\,dm}{r^2}\mathbf{u} = \frac{G\sigma\,ds}{r^2}\mathbf{u}.$$

The distance r, the density function σ, and the unit vector $\mathbf{u}$ are functions of s. If we add the forces due to the several segments together and take the limit of the sum as the subdivision is continued, we obtain

$$\int_C \frac{G\sigma}{r^2}\mathbf{u}\,ds.$$

The integral sign $\int$, shaped like an elongated letter S, was chosen by the German mathematician Gottfried Leibniz (1646–1716) to suggest summation.

Newton and Leibniz are usually credited with the discovery of calculus. Although some of the basic ideas of differentiation and integration were known earlier, these two men organized and greatly extended the existing knowledge.

This chapter is divided into three parts. First we describe and define the integral of a function f on an interval $[a, b]$ in terms of its approximating sums. These sums are related to Euler's method. In the second part we use the Fundamental Theorem of Calculus to evaluate the integral of a function f on an interval $[a, b]$ by finding an **antiderivative** F of f, that is, a function F for which $F' = f$. For functions whose antiderivatives can be expressed as a finite combination of elementary functions, we give an introduction to the "integration techniques" used to determine these antiderivatives. For functions whose antiderivatives cannot be so expressed, we discuss numerical integration in the third part.

5.1 THE INTEGRAL

We use the ideas of position and velocity in introducing the integral. The velocity of an object is the rate of change of its position with respect to time, that is, $v = dx/dt$. Calculating an object's velocity v by differentiating its position function x is called the forward problem. The inverse problem is to determine x from $v = dx/dt$.

To introduce and motivate the integral, we solve an inverse problem in two ways:

- Antidifferentiation
- Euler's method

We have discussed both of these before. We review both solution methods below. The connection between these two solutions of the inverse problem is the Fundamental Theorem of Calculus.

Investigation

An object is moving on the x-axis so that for all $t \geq 0$

(1)
$$v = \frac{dx}{dt} = t^2$$

$$x(0) = 0.$$

Time is measured in hours and distance in kilometers. We wish to solve the inverse problem, that is, to determine the position $x(t)$ of the object at any time $t \geq 0$.

Solution by Antidifferentiation
Since $dx/dt = t^2$, the position function x must have the form

$$x = x(t) = \tfrac{1}{3}t^3 + c.$$

An argument for this claim is outlined in Exercises 32–34. We choose c so that the initial condition $x(0) = 0$ is satisfied. Setting $t = 0$, we obtain

$$0 = x(0) = \tfrac{1}{3}0^3 + c, \qquad \text{so that} \quad x = x(t) = \tfrac{1}{3}t^3.$$

The graphs of the velocity and position functions are shown in Fig. 5.3. We discuss the shaded region below.

Solution by Euler's Method
We use Euler's method to argue that for all $t \geq 0$, the position $x(t)$ of the object is numerically equal to the area of the shaded region beneath the velocity graph, as suggested in the figure.

Euler's method is based on the tangent line approximation: if f is a differentiable function at a point t, then for small values of h

$$f(t + h) \approx f(t) + hf'(t).$$

If we replace f by the unknown function $x = x(t)$, then

(2)
$$x(t + h) \approx x(t) + hv(t) = x(t) + ht^2.$$

In words, the position of the object at $t + h$ is approximately the sum of its position at t and its displacement $hv(t)$ during the time interval $[t, t + h]$. If h is sufficiently

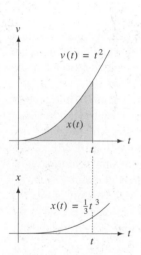

Figure 5.3. Inverse problem.

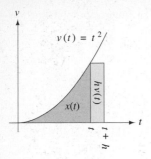

Figure 5.4. Area interpretation of (2).

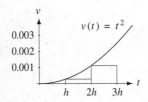

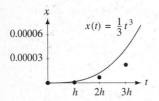

Figure 5.5. Graphical interpretations of Euler's method.

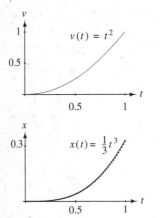

Figure 5.6. Graphical interpretations of Euler's method.

small, the velocity is nearly constant in the interval $[t, t+h]$ and, hence, the distance traveled would be approximately $D = RT = v(t)h$.

We take $h = 1/60$ h. Referring to Fig. 5.4, we show that the approximation (2) can be interpreted in terms of the areas of three regions beneath the velocity curve. Specifically, we show that $x(t)$ is numerically equal to the area of the darker region, $hv(t)$ is numerically equal to the area of the strip, and $x(t + h)$ is numerically equal to the area beneath the velocity curve up to time $t + h$. If these area interpretations are right, the approximation (2) makes good sense, that is, the area beneath the velocity curve up to $t + h$ is approximately equal to the area up to t plus the area of the strip.

All of this emerges from the following calculation, which is just Euler's method for approximating the values of the position function x. Recall that $x(0) = 0$.

$$(3) \qquad x(1 \cdot h) \approx x(0) + hv(0) = h \cdot 0^2$$

$$(4) \qquad x(2 \cdot h) \approx x(h) + hv(h) \approx h \cdot 0^2 + h \cdot h^2 = h^3(1^2)$$

$$(5) \qquad x(3 \cdot h) \approx x(2h) + hv(2h) \approx h^3(1^2 + 2^2)$$

From the observed pattern in these results, we have

$$(6) \qquad x(n \cdot h) \approx h^3(1^2 + 2^2 + \cdots + (n-1)^2), \quad n \geq 2.$$

Referring to the graph of $v = v(t)$ at the top of Fig. 5.5, we may interpret $x(h) \approx h \cdot 0^2$ in (3) as the area of the rectangle with height 0^2 and width h, fitting beneath the velocity curve and above the interval $[0, h]$. The height of this rectangle is the height of the velocity curve at the left end point of $[0, h]$.

We may interpret $x(2h) \approx h \cdot 0^2 + h \cdot h^2$ in (4) as the sum of the area of the first rectangle and the area of the rectangle with height h^2 and width h, which fits beneath the velocity curve and above the interval $[h, 2h]$. The height of this rectangle is the height of the velocity curve at the left end point of $[h, 2h]$.

The first three rectangles beneath the velocity curve are shown in the figure. The value of $x(h)$ is, approximately, the first rectangle; the value of $x(2h)$ is, approximately, the sum of the first two rectangles; and so on. At the bottom of Fig. 5.5 we have plotted these approximations to $x(h)$, $x(2h)$, and $x(3h)$ against the curve $x(t) = \frac{1}{3}t^3$, which we determined earlier by antidifferentiation. Evidently the two solutions of the inverse problem are strongly related.

The general step for the Euler approximation to the position function was given in (6). We show in the upper graph of Fig. 5.6 the 60 rectangles arising in the calculation. In the lower graph are plotted approximations to $x(h), \ldots, x(60h)$ against the curve $x(t) = \frac{1}{3}t^3$. As noted above, $x(kh)$ is the sum of the first k rectangles, approximately.

We give a sample of the calculations in Table 1. We give the Euler approximation to $x(t)$, the exact value of $x(t) = \frac{1}{3}t^3$, and the error E, which is the absolute value of the difference between the Euler approximation and the exact value of $x(t)$. All entries are rounded to three significant figures.

Relating the Two Solutions of the Inverse Problem   From this combination of antidifferentiation, numerical calculation, and geometric interpretation it is reasonable to conjecture that:

Table 1. Numerical analysis of Euler's method

t	$\approx x(t)$	$x(t)$	E
$0 \cdot h$	0	0	0
$1 \cdot h$	0	1.54×10^{-6}	1.54×10^{-6}
$2 \cdot h$	4.63×10^{-6}	1.23×10^{-5}	7.72×10^{-6}
$3 \cdot h$	2.31×10^{-5}	4.17×10^{-5}	1.85×10^{-5}
$4 \cdot h$	6.48×10^{-5}	9.88×10^{-5}	3.40×10^{-5}
$5 \cdot h$	1.39×10^{-4}	1.93×10^{-4}	5.40×10^{-5}
$10 \cdot h$	1.32×10^{-3}	1.54×10^{-3}	2.24×10^{-4}
$20 \cdot h$	1.14×10^{-2}	1.23×10^{-2}	9.10×10^{-4}
$30 \cdot h$	3.96×10^{-2}	4.17×10^{-2}	2.06×10^{-3}
$40 \cdot h$	9.51×10^{-2}	9.88×10^{-2}	3.67×10^{-3}
$50 \cdot h$	1.87×10^{-1}	1.93×10^{-1}	5.75×10^{-3}
$60 \cdot h$	3.25×10^{-1}	3.33×10^{-1}	8.29×10^{-3}

- The position $x(T)$ of an object at time $t = T$ is the area of the region beneath the graph of its velocity function, from $t = 0$ to $t = T$.

- This area can be calculated from an antiderivative of v.

We give an example to strengthen these conjectures.

■ **EXAMPLE 1**   An object is moving on the x-axis so that $x(0) = 0$ and $v(t) = t$ for all $t \geq 0$. Determine an approximate value of $x(1)$ by inscribing 10 rectangles beneath the graph of v, as in the Investigation. Compare your result with that based on finding an antiderivative x of v, using the initial condition $x(0) = 0$, and evaluating $x(1)$.

Solution.   As above, the easier part is finding an antiderivative of v. The position function must have the form $x = x(t) = \frac{1}{2}t^2 + c$. From the initial condition $x(0) = 0$, we see that $c = 0$, giving

$$x = x(t) = \tfrac{1}{2}t^2.$$

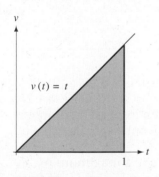

$v(t) = t$

Figure 5.7. $x(1.0) = 0.5.$

Evaluating x at $t = 1$, we have $x = x(1) = 0.5$, that is, the object has moved from $x = 0$ to $x = 0.5$ in 1 second. See Fig. 5.7. From the Investigation, we expect $x(1)$ is equal to the area of the region beneath v and above the t-axis, between $t = 0$ and $t = 1$. We can check our expectation in this case by noting that the region is a triangle and has area $\frac{1}{2} \cdot 1 \cdot 1 = 0.5$.

From Euler's method,

$$x(0) = 0$$

$$x(h) \approx x(0) + hv(0) = 0$$

$$x(2h) \approx x(h) + hv(h) \approx 0 + h \cdot h = h^2(1)$$

$$x(3h) \approx x(2h) + hv(2h) \approx (1)h^2 + h(2h) = h^2(1 + 2)$$

$$x(4h) \approx x(3h) + hv(3h) \approx (1 + 2)h^2 + h(3h) = h^2(1 + 2 + 3).$$

From the pattern observed for $x(0), \ldots, x(4h)$,

$$(7) \qquad x((m + 1)h) \approx h^2(1 + 2 + \cdots + m), \quad m = 1, 2, \ldots.$$

Since we wish to inscribe 10 rectangles, we take $h = 0.1 = (1 - 0)/10$. As in the Investigation, we may interpret expressions such as

$$x(4h) = x(0.4) \approx x(3h) + hv(3h)$$

in terms of area. The term $x(3h)$ is the sum of the areas of the first three inscribed rectangles, approximately, and the term $hv(3h)$ is the area of the fourth inscribed rectangle.

We show in Fig. 5.8 the 10 rectangles. The first rectangle has zero height. From (7) with $m = 9$, we have

$$x(10 \cdot 0.1) = x(1.0) \approx (0.1)^2(1 + 2 + \cdots + 9) = (0.01)(45) = 0.45.$$

The sum of the inscribed rectangles is 0.45, which is 0.05 less than the area of the region beneath the graph of v between $t = 0$ and $t = 1$. We used the formula for the sum $1 + 2 + \cdots + n$ of the first n integers, which is

$$1 + 2 + \cdots + n = \tfrac{1}{2}n(n + 1).$$

If we had inscribed 100 rectangles, so that $h = 0.01 = (1 - 0)/100$,

$$x(100 \cdot 0.01) = x(1.00) \approx (0.01)^2(1 + 2 + \cdots + 99) = (0.01)(4950) = 0.4950.$$

This smaller step size has decreased the error from 0.05 to 0.005. Inscribing 1000 rectangles would give an even closer approximation. It appears likely that $x(1) = 0.5$. ∎

We have interpreted Euler's method as approximating the area of the region beneath a curve by means of inscribed rectangles. From a geometric point of view, the area could as well be approximated using circumscribed rectangles. We use both in the definition of the integral. We give the definition in general terms, not tying it to specific interpretations such as position or area. In this way we are free to interpret the integral as the position of an object, the area of a region beneath a curve, the length of a curve, or the volume of a solid.

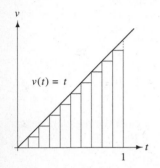

$v(t) = t$

Figure 5.8. $x(1.0) \approx 0.45$.

DEFINITION   **the (definite) integral**

Let f be a continuous function defined on $[a, b]$. Let n be a positive integer and $h = (b - a)/n$. The points $x_0 = a$, $x_1 = a + h$, $x_2 = a + 2h$, ..., $x_i = a + ih$, ..., $x_n = a + nh = b$ divide the interval $[a, b]$ into n equal subintervals of length h. The set of these subintervals is called a **regular subdivision** of $[a, b]$.

In the ith subinterval, f has a minimum m_i and a maximum M_i, by the Max/Min Theorem. The lower and upper sums L_n and U_n of f are

$$(8) \qquad L_n = h(m_1 + m_2 + \cdots + m_n) = \sum_{i=1}^{n} m_i h$$

$$(9) \qquad U_n = h(M_1 + M_2 + \cdots + M_n) = \sum_{i=1}^{n} M_i h.$$

If there is exactly one number I such that $L_n \leq I \leq U_n$ for $n = 1, 2, \ldots$, we say that f is integrable on $[a, b]$ and that I is the value of the integral of f on $[a, b]$, and we write

(10)
$$\int_a^b f(x) \, dx = I.$$

The end points a and b of the **interval of integration** $[a, b]$ are the **lower and upper limits of integration**. The function f is the **integrand**.

It can be shown that every function f defined and continuous on an interval $[a, b]$ is integrable on $[a, b]$. We assume this result without further comment or proof.

In the above definition of the integral we used summation notation, which provides a compact, useful way of describing a sum of terms. The notation gives the formula for the general term and specifies the starting and ending terms. We give a few examples of summation notation here and include several problems involving summation in the exercises.

If we wish to communicate to another person the idea of summing the squares of the first 10 integers, we have several options. Most directly, we could write

$$1^2 + 2^2 + 3^2 + 4^2 + 5^2 + 6^2 + 7^2 + 8^2 + 9^2 + 10^2.$$

If this seems a bit tedious, we may use the *ellipsis* symbol $\cdots$ to denote terms not explicitly written but understood to follow the pattern given in the first several terms:

$$1^2 + 2^2 + \cdots + 10^2.$$

Or we may use the notation

$$\sum_{i=1}^{10} i^2.$$

The formula for the general term, here i^2, is usually expressed in terms of a variable called the **index of summation**. The letters i, j, or k are often used for this purpose. Below and above the summation sign $\sum$ are given the starting and ending values of the index of summation. It is assumed that the index of summation advances in steps of 1 from the starting value to the ending value.

We form the upper sum U_{25} for the function $f(x) = \sqrt{x}, 0 \leq x \leq 1$, to illustrate summation notation in context. A graphical interpretation of U_{25} is shown in Fig. 5.9. Suppose that the interval $[0, 1]$ is divided into 25 subintervals with the points $x_0, x_1, \ldots, x_{25}$, each of length $1/25$. The end of the first subinterval $[x_0, x_1]$ is $x_1 = 1/25$, the end of the second subinterval $[x_1, x_2]$ is $x_2 = 2/25$, and, in general, the end of the ith subinterval $[x_{i-1}, x_i]$ is $x_i = i/25$. Let $h = 1/25$, the common length of the subintervals. The upper sum U_{25} of $\sqrt{x}$, as defined in the definition of integral, is the sum of the values

(11) $U_{25} = \sqrt{x_1}(x_1 - x_0) + \sqrt{x_2}(x_2 - x_1) + \cdots + \sqrt{x_{25}}(x_{25} - x_{24}).$

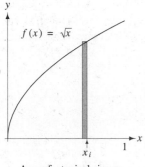

$f(x) = \sqrt{x}$

x_i

Area of a typical circumscribed rectangle is
$\sqrt{x_i}\,(x_i - x_{i-1})$

Figure 5.9. Circumscribed rectangles.

Replacing $x_1, x_2, \ldots, x_{25}$ by their values $1/25, 2/25, \ldots, 25/25$ and each of $(x_1 - x_0), (x_2 - x_1), \ldots, (x_{25} - x_{24})$ by h we have

(12) $$U_{25} = \sqrt{1/25}h + \sqrt{2/25}h + \cdots + \sqrt{25/25}h$$

Factoring the common factor h from these terms gives

(13) $$U_{25} = h\left(\sqrt{1/25} + \sqrt{2/25} + \cdots + \sqrt{25/25}\right).$$

The expressions (11)–(13) can be written more briefly as (14)–(16) using summation notation.

(14) $$U_{25} = \sum_{i=1}^{25} \sqrt{x_i}(x_i - x_{i-1})$$

Replacing x_i by its value $i/25$ and $(x_i - x_{i-1})$ by h, we have

(15) $$U_{25} = \sum_{i=1}^{25} \sqrt{i/25}h.$$

Factoring h from these terms, we have

(16) $$U_{25} = h\sum_{i=1}^{25} \sqrt{i/25}.$$

■ **EXAMPLE 2**   In the Investigation we worked with the function $v(t) = t^2$, using antidifferentiation and Euler's method to determine $x(t) = \frac{1}{3}t^3$. In this example we continue to work with the squaring function, although we take the more general point of view of the definition of the integral. For this, let $f(x) = x^2$. We calculate an upper sum and a lower sum for f on the interval $[0, 1.5]$.

Solution.   We give a graphical interpretation of the lower and upper sums L_{15} and U_{15} in Fig. 5.10. The lower sum may be seen as the sum of 15 inscribed rectangles and the upper sum as the sum of 15 circumscribed rectangles. The interval $[a, b] = [0, 1.5]$ is divided into $n = 15$ equal subintervals, each of length $h = (b - a)/n = 1.5/15 = 0.1$. A zoom to a typical subinterval is shown at the bottom of the figure. According to the definition, we must calculate the minimum m_i of f and the maximum M_i of f on the subinterval $[x_{i-1}, x_i]$. Since $f(x) = x^2$ is increasing on $[0, 1.5]$,

$$m_i = f(x_{i-1}) = x_{i-1}^2 \quad \text{and} \quad M_i = f(x_i) = x_i^2.$$

From (8) and (9), the lower and upper sums are

(17) $$L_{15} = \sum_{i=1}^{15} m_i h = h\sum_{i=1}^{15} x_{i-1}^2$$

(18) $$U_{15} = \sum_{i=1}^{15} M_i h = h\sum_{i=1}^{15} x_i^2.$$

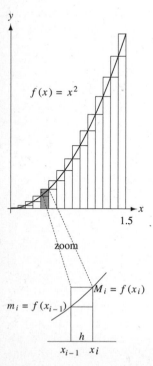

$f(x) = x^2$

$M_i = f(x_i)$

$m_i = f(x_{i-1})$

zoom

Figure 5.10. Typical circumscribed and inscribed rectangles.

The numbers $m_i h$ and $M_i h$ may be interpreted as the areas of the inscribed and circumscribed rectangles shown in the figure. To calculate L_{15} and U_{15}, note that $x_0 = 0$, $x_1 = 1 \cdot h$, $x_2 = 2 \cdot h$, and so on. We may rewrite (17) and (18) as

$$(19) \qquad L_{15} = h \sum_{i=1}^{15} (i-1)^2 \cdot h^2 = h^3(0^2 + 1^2 + \cdots + 14^2)$$

$$(20) \qquad U_{15} = h \sum_{i=1}^{15} i^2 \cdot h^2 = h^3(1^2 + 2^2 + \cdots + 15^2).$$

We may evaluate these sums with a calculator (many have a SUM command) or by using the known formula (see Exercise 36)

$$1^2 + 2^2 + \cdots + m^2 = \tfrac{1}{6}m(m+1)(2m+1),$$

which holds for $m = 1, 2, \ldots$. For $m = 14$ and $m = 15$, we have

$$1^2 + 2^2 + \cdots + 14^2 = \tfrac{1}{6}14(15)(29) = 1015$$
$$1^2 + 2^2 + \cdots + 15^2 = \tfrac{1}{6}15(16)(31) = 1240.$$

Substituting these results into (19) and (20), we have

$$L_{15} = (0.1)^3(1015) = 1.015$$
$$U_{15} = (0.1)^3(1240) = 1.240.$$

If we were to subdivide the interval $[0, 1.5]$ into smaller pieces by taking n larger, we would get lower and upper sums that squeeze down around the integral I of f on $[0, 1.5]$ even more closely. If $n = 1500$, for example, $h = 1.5/1500 = 0.001$ and

$$(21) \qquad L_{1500} = (0.001)^3(1^2 + 2^2 + \cdots + 1499^2) = 1.12387525$$

$$(22) \qquad U_{1500} = (0.001)^3(1^2 + 2^2 + \cdots + 1500^2) = 1.12612525.$$

It appears that $I \approx 1.125$. Recalling the ideas of the Investigation, we expect $I = 1.125$ since the antiderivative of x^2 is $\tfrac{1}{3}x^3$ and, thinking about this in terms of position and velocity, the position after 1.5 seconds would be $\tfrac{1}{3}(1.5)^3$, which is 1.125 exactly. ∎

Exercises 5.1

Basic

1. Do the calculations for the lines of Table 1 corresponding to $t = 1/6$ and $t = 1/3$. State as clearly as you can the point of the Investigation and Example 1.

2. If an object has velocity $v = t^3$ and initial position $x(0) = 0$, what is its position function $x(t)$?

3. In Example 2 calculate L_{15000} and U_{15000}. Mark the positions of $L_{15}, L_{1500}, L_{15000}, U_{15}, U_{1500}, U_{15000}$, and I on an axis. What happens to lower sums as the number of subdivisions increases? What happens to upper sums?

4. As in Example 2, calculate L_{1500} and U_{1500} for the function $f(x) = x^3$. You will need the formula

$$\sum_{i=1}^{m} i^3 = \tfrac{1}{4}(m(m+1))^2.$$

Exercises 5–16:   Expand each sum and evaluate. For example, $\sum_{i=1}^{4} i^3 = 1^3 + 2^3 + 3^3 + 4^3 = 100$.

5. $\sum_{i=1}^{5} i^2$

6. $\sum_{i=1}^{3} i^4$

7. $\sum_{i=1}^{6} (i^2 + 1)$

8. $\sum_{i=1}^{3} (2i^2 - 5i + 1)$

9. $\sum_{j=1}^{2} \sqrt{j+1}$

10. $\sum_{j=1}^{4} \sqrt{3j - 2}$

11. $\sum_{k=0}^{5} \sin(k/10)$

12. $\sum_{k=0}^{7} \cos(k/10)$

13. $\sum_{i=0}^{5} 1/(2i + 1)$

14. $\sum_{i=0}^{5} (3i - 5)/3$

15. $\sum_{j=1}^{3} j^0$

16. $\sum_{j=1}^{3} 1$

Exercises 17–26:   Rewrite using summation notation. Factor out any common factors.

17. $(1^2 + 2^2 + \cdots + 10^2)h^3$

18. $(1^2 + 2^2 + \cdots + 7^2)h^2$

19. $2/3 + 2/4 + 2/5 + \cdots + 2/35$

20. $7/13 + 8/13 + 9/13 + \cdots + 27/13$

21. $\sqrt{h^2(1/19)} + \sqrt{h^2(2/19)} + \cdots + \sqrt{h^2(21/19)}$

22. $\sqrt{3/h^2} + \sqrt{4/h^2} + \cdots + \sqrt{37/h^2}$

23. $e^{1+1} + e^{1+2} + e^{1+3} + \cdots + e^{1+21}$

24. $0.01 \cdot \ln(1) + 0.01 \cdot \ln(2) + \cdots + 0.01 \cdot \ln(50)$

25. $\frac{1}{2} + \frac{2}{3} + \frac{3}{4} + \cdots + \frac{15}{16}$

26. $x^1 + x^2 + \cdots + x^n$

Growth

27. Suppose that in the Investigation we had specified $x(0) = 1$ instead of $x(0) = 0$. Redraw Fig. 5.5 to illustrate this initial value problem. What effect would this change have on the calculations in Table 1?

28. In Example 1, for the 10-rectangle case, show by actual calculation that the difference between 0.5 and 0.45, which is the difference between the area of the triangle and the sum of the areas of the 10 inscribed rectangles, is exactly equal to the sum of the areas of the small triangles beneath the graph of $v(t) = t$ and atop the inscribed rectangles.

29. In the Investigation, show that the sum of the first three rectangles is $5h^3$ and the true area is $9h^3$. Hence, the error is $4h^3$. To improve the approximation, we may add to the rectangles, not the small triangle-like shapes atop the three rectangles and beneath the curve with equation $v = t^2$, but the triangles formed by replacing the parabolic segments

with straight lines. Show that the sum of the three rectangles with adjoined triangles is $\frac{19}{2}h^3$. Finally, calculate the error if we approximate the position $x(3h)$ (or area) with the average of $5h^3$ (too small) and $\frac{19}{2}h^3$ (too large).

30. Calculate U_{25} in (16) and the corresponding lower sum L_{25}, and estimate the area of the region beneath the graph of the square root function and above the segment [0, 1] of the x-axis. Use antidifferentiation to calculate the area exactly.

31. Approximate $\int_0^1 \sin x^2 \, dx$ by averaging L_{10} and U_{10}. How far off could the average be? (Hint:

$$\left| \int_0^1 \sin x^2 \, dx - 0.5(L_{10} + U_{10}) \right|$$

$$\leq 0.5(U_{10} - L_{10}).$$

32. Use the Mean-value Inequality to show that if $f'(x) = 0$ for all x, then $f(x) = c$, a constant. More exactly, show that if f is continuously differentiable on $I = [a, b]$ and $f'(x) = 0$ for all $x \in I$, then $f(x) = f(a)$ for all $x \in I$.

33. Use Exercise 32 in showing that if functions f and g have the same derivative, then they differ by a constant. (Hint: Consider the function $f - g$ and apply Exercise 32.)

34. Use Exercise 33 to show that the only continuously differentiable function whose derivative is t^2 has the form $\frac{1}{3}t^3 + c$, where c is a constant.

35. Use mathematical induction to prove that

$$1 + 2 + \cdots + n = n(1 + n)/2.$$

36. Use mathematical induction to prove that

$$1^2 + 2^2 + \cdots + n^2 = n(n + 1)(2n + 1)/6.$$

37. Use mathematical induction to prove that

$$1^3 + 2^3 + \cdots + n^3 = (n(n + 1))^2/4.$$

38. Show that for any $b > 0$, $\int_0^b x^2\, dx = \frac{1}{3}b^3$. Use the following adaptation of the argument used in Ex-

ample 2. (1) Divide the interval $[0, b]$ into n equal pieces and let $h = b/n$. (2) Show that

$$L_n = \sum_{i=1}^{n} ((i - 1)h)^2 h$$

$$U_n = \sum_{i=1}^{n} (ih)^2 n.$$

(3) Using the formula for the sum of squares given in Example 2, show that

$$L_n = \frac{b^3}{6} \frac{2n^3 - 3n^2 + n}{n^3}$$

$$U_n = \frac{b^3}{6} \frac{2n^3 + 3n^2 + n}{n^3}.$$

(4) Rewrite

$$(2n^3 \pm 3n^2 + n)/n^3 \quad \text{as} \quad 2 \pm 3/n + 1/n^2.$$

(5) Show that both L_n and U_n approach $b^3/3$ as n becomes large. (6) Conclude that $\int_0^b x^2\, dx = \frac{1}{3}b^3$.

Review/Preview

Exercises 39–58: Find an antiderivative of each function.

39. x^5

40. $\frac{2}{7}x^3$

41. $(2x + 1)^2$

42. $(3x - 5)^3$

43. $\sqrt{x}$

44. $x^{3/2}$

45. $\sin(x - 1)$

46. $\cos(2x + 3)$

47. $\sec^2 x$

48. $\sec x \tan x$

49. 2^x

50. 3.5^x

51. e^{2x}

52. $e^{x/2}$

53. $1/(x^2 + 1)$

54. $1/\sqrt{1 - x^2}$

55. $x^2 + \sqrt{x}$

56. $x - \cos x$

57. $1/x$

58. $(x + 1)/x$

5.2   THE FUNDAMENTAL THEOREM

The Fundamental Theorem of Calculus is a context-free statement of the result that the position of an object can be inferred from its velocity by antidifferentiation or by calculating the area of the region beneath the velocity curve. Generalizations of the Fundamental Theorem of Calculus will be discussed in later chapters, when we define line integrals, surface integrals, and multiple integrals. Understanding and proving this theorem are the main goals of this section.

We begin by discussing two properties of the integral that we assume without proof. These properties can be easily understood by interpreting them in terms of area or motion. We end the section by listing two additional properties of the integral. These follow easily from the Fundamental Theorem of Calculus and are left as exercises. All functions considered in this chapter are assumed to be continuous.

Area

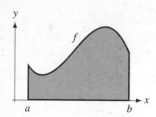

Figure 5.11. Generic area.

The shaded portion of Fig. 5.11 is often described as the "area of the region beneath the curve f." This is a convenient but imprecise way of referring to the number that measures the area of a certain subset S of the plane. We may describe S in words by saying something like "S is the set of points lying beneath the graph of f, above the x-axis, and between the lines $x = a$ and $x = b$." More precisely,

$$S = \{(x, y) : a \le x \le b, 0 \le y \le f(x)\}.$$

In Example 2 of Section 5.1 we linked the integral $\int_a^b f(x)\,dx$ with S. We used the terms *inscribed* and *circumscribed rectangles* to connect the idea of area with lower and upper sums. We state this as a definition.

Definition 1: Area   Let f be a continuous, nonnegative function on $[a, b]$. The area of the region S beneath the graph of f is the number $\int_a^b f(x)\,dx$.

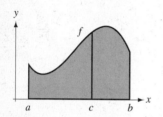

Figure 5.12. Additivity of the integral.

Property 1: Additivity   In terms of area and referring to Fig. 5.12, **additivity** is the property that the sum of the areas of the regions beneath the graph of f, from a to c and from c to b, is equal to the area of the region from a to b. In terms of the integral, for all c such that $a < c < b$,

$$(1) \qquad \int_a^c f(x)\,dx + \int_c^b f(x)\,dx = \int_a^b f(x)\,dx.$$

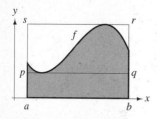

Figure 5.13. Mean-value Inequality for Integrals.

Property 2: Mean-value Inequality for Integrals   The second property is also easy to understand if we interpret the integral in terms of area. From Fig. 5.13 we see that the area of the shaded set lies between the areas of the rectangles with vertices a, b, q, p and a, b, r, s. This observation is equivalent to the inequalities

$$(2) \qquad m(b - a) \le \int_a^b f(x)\,dx \le M(b - a),$$

where the values m and M are the minimum and maximum of f on the interval $[a, b]$. Note, for example, that the height of the rectangle $abqp$ is m and its width is $(b - a)$.

We use additivity and the Mean-value Inequality for Integrals in discussing the Fundamental Theorem of Calculus. For this we need to have a clear understanding of integrals of the form

$$(3) \qquad \int_a^x f(w)\,dw, \quad \text{where } a < x < b.$$

Note that in (3) we have used w, not x, as the *integration variable* to avoid confusion with the upper limit of integration. The integration variable plays much the same role as an index of summation. Just as we recognize that both of the sums

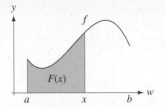

Figure 5.14. Area as function.

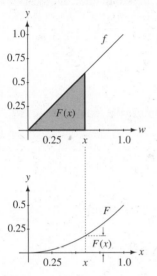

Figure 5.15. Area as function.

$$\sum_{i=1}^{3} i^2 \quad \text{and} \quad \sum_{j=1}^{3} j^2$$

are equal to $1^2 + 2^2 + 3^2$, the values of the integrals

$$\int_a^b f(x)\,dx \quad \text{and} \quad \int_a^b f(w)\,dw$$

are also equal.

The value of the integral in (3) is a function of x. Denoting this function by F and referring to Fig. 5.14, $F(x) = \int_a^x f(w)\,dw$ is the area of the region beneath the graph of f, between $w = a$ and $w = x$.

As a specific illustration, suppose $f(w) = w, 0 \le w \le 1$. The graph of f is at the top in Fig. 5.15. The area $F(x)$ of the triangular region beneath f from $w = 0$ to $w = x$ is $\frac{1}{2}x \cdot x = \frac{1}{2}x^2$. The area varies with x; that is, the area is a function of x. We may write the area function as the integral

$$F(x) = \int_0^x f(w)\,dw = \tfrac{1}{2}x^2, \quad 0 \le x \le 1.$$

The value $F(0.25) = \frac{1}{2}(0.25)^2 = 0.03125$ is the area beneath f up to $w = 0.25$.

Did you notice that $F'(x) = f(x)$, that is, $\left(\frac{1}{2}x^2\right)' = x$? One way of determining the area beneath f is to find an antiderivative of f.

For the statement of the Fundamental Theorem of Calculus, it is convenient to make the convention that $\int_a^a f(x)\,dx = 0$. In terms of area, this amounts to agreeing that between $x = a$ and $x = a$ the area beneath the graph of f is 0.

Definition 2 For any continuous function f defined on an interval $[a, b]$,

$$(4) \qquad \int_a^a f(w)\,dw = 0.$$

The Fundamental Theorem of Calculus

Part I: If f is continuous on $[a, b]$ and the function F is defined by

$$(5) \qquad F(x) = \int_a^x f(w)\,dw, \quad x \in [a, b],$$

then F is an antiderivative of f; that is,

$$(6) \qquad F'(x) = \frac{d}{dx} \int_a^x f(w)\,dw = f(x), \quad x \in [a, b].$$

Part II: If f is defined on $[a, b]$ and G is any antiderivative of f on $[a, b]$, then

$$(7) \qquad \int_a^b f(x)\,dx = G(b) - G(a).$$

Proof of the Fundamental Theorem of Calculus

Part I We prove (6) for $x \in (a, b)$. Recall that the definition of the derivative of F at x is

$$(8) \qquad F'(x) = \lim_{h \to 0} \frac{F(x + h) - F(x)}{h}.$$

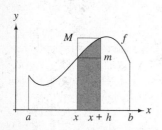

The area of the shaded subset shown in Fig. 5.16 is the numerator of (8). This comes from the additivity property by letting $c = x$ and $b = x + h$ in (1). We have

$$(9) \qquad \int_a^x f(w)\,dw + \int_x^{x+h} f(w)\,dw = \int_a^{x+h} f(w)\,dw$$

Figure 5.16. $F(x + h) - F(x)$ is the area of the shaded region.

If this isn't really clear, just think about it in terms of area. We may rearrange (9) and rewrite the result in terms of F to obtain

$$(10) \qquad F(x + h) - F(x) = \int_x^{x+h} f(w)\,dw.$$

We use the Mean-value Inequality for Integrals on the last integral in (10). Letting $a = x$ and $b = x + h$ in (2), we have

$$(11) \qquad mh \le \int_x^{x+h} f(w)\,dw \le Mh,$$

where m and M are the minimum and maximum values of f on $[x, x + h]$. Equation (10) and inequality (11) show that the difference quotient in (8) is squeezed between m and M as $h \to 0$, that is,

$$(12) \qquad m = \frac{mh}{h} \le \frac{F(x + h) - F(x)}{h} \le \frac{Mh}{h} = M.$$

Since m and M approach $f(x)$ as $h \to 0$, (6) now follows.

An argument for the cases $x = a$ and $x = b$ is outlined in Exercise 42.

Part II The second part of the Fundamental Theorem of Calculus depends on two results. The first result is that the function F defined in (5) is an antiderivative of f. We just proved this. The second result, that if two differentiable functions have the same derivative, then they differ by a constant, was stated in Exercise 33 in Section 5.1. Since $F' = G' = f$, the functions F and G differ by a constant, that is,

$$(13) \qquad F(x) - G(x) = c, \quad x \in [a, b].$$

Equation (7) follows from (13) after we evaluate c. For this, let $x = a$ in (13). We have

$$F(a) - G(a) = c.$$

Since $F(a) = 0$ by (4),

(14) $$c = -G(a).$$

The second part of the Fundamental Theorem of Calculus now follows. We have

(15) $$\int_a^b f(x)\,dx = F(b) = G(b) + c = G(b) - G(a).$$

We apply the Fundamental Theorem to several examples to reinforce the main ideas.

■ **EXAMPLE 1**   Find the area of the region beneath the first arch of the sine curve.

Solution.   The set whose area we wish to calculate is shaded in Fig. 5.17. The area of this set is the value of the integral $\int_0^\pi \sin x\,dx$. From (7), the value of the integral is $G(\pi) - G(0)$, where G is an antiderivative of $\sin x$. It is easy to find G in this case, namely, $G(x) = -\cos x$. Note that $G'(x) = \sin x$. From (7) we have

$$\int_0^\pi \sin x\,dx = G(\pi) - G(0) = -\cos\pi - (-\cos 0) = -(-1) - (-1) = 2. \quad ■$$

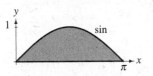

Figure 5.17. First arch of sine.

The preceding calculation depended on both the idea of the integral and the Fundamental Theorem of Calculus. The first step was our recognition that the integral can be used to calculate an area. The second step was the use of the Fundamental Theorem to evaluate the integral by determining an antiderivative. The integral and the Fundamental Theorem working in tandem provide us with a powerful tool. They make possible calculations of areas, volumes, lengths of curves, areas of surfaces, centers of mass, moments of inertia, and probabilities.

■ **EXAMPLE 2**   Find the area of the region beneath the graph of $f(x) = (x - 1)^2, 0 \le x \le 3$.

Solution.   A good first step is to sketch or otherwise consider the geometric object of interest. Time and effort invested here often clarify the problem, expose subproblems whose solutions are necessary, and help avoid major blunders. We show in Fig. 5.18 a sketch that includes the essentials. The size is sufficient for easy interpretation. We added the line with equation $x = 3$ to complete the boundary of the subset whose area is wanted. The area is the value of the integral

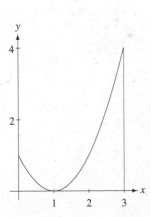

Figure 5.18. The area under a parabola.

$$\int_0^3 (x - 1)^2\,dx.$$

An antiderivative of the integrand is $\frac{1}{3}(x-1)^3$. Thus

$$\int_0^3 (x-1)^2\, dx = \frac{1}{3}(x-1)^3 \Big|_0^3 = \frac{1}{3}(3-1)^3 - \frac{1}{3}(0-1)^3 = \frac{8}{3} + \frac{1}{3} = 3.$$

The vertical bar evaluation symbol used here is defined below. ■

If a function is followed by a vertical bar with attached "limits," it is intended that the function be evaluated at each of the two limits and the resulting values subtracted. More formally, if G is a function, then

$$G(x)\Big|_a^b = G(b) - G(a).$$

Part II of the Fundamental Theorem of Calculus was used in these two examples. In the next two examples we explore applications of part I.

■ **EXAMPLE 3** Let A be the area function

(16)
$$A(x) = \int_0^x \sin w\, dw, \quad 0 \le x \le \pi/2.$$

Determine $A'(\pi/4)$.

Solution. We calculated the area beneath the first arch of the sine function in Example 1. Here we are asked to consider the area function A defined in (16) and illustrated in Fig. 5.19. The value of $A(x)$ for a typical x is the area of the region beneath the sine curve between the lines $w = 0$ and $w = x$. In Example 1 we calculated $A(\pi) = 2$ by noting that $-\cos w$ is an antiderivative of $\sin w$. From part II of the Fundamental Theorem of Calculus,

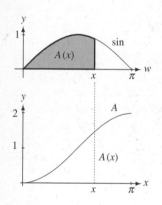

Figure 5.19. Area beneath sine.

(17)
$$A(x) = \int_0^x \sin w\, dw = -\cos w \Big|_0^x = 1 - \cos x.$$

We may calculate $A'(\pi/4)$ from this result. We have

(18)
$$A'(x) = 0 - (-\sin x) = \sin x$$
$$A'(\pi/4) = \sin(\pi/4) = \sqrt{2}/2.$$

This calculation can be done more easily by using part I of the Fundamental Theorem of Calculus. We have

(19)
$$A'(x) = \frac{dA}{dx} = \frac{d}{dx}\left(\int_0^x \sin w\, dw \right) = \sin x.$$

This agrees with (18) and avoids the antidifferentiation step in (17).

While (19) is still fresh in your mind, we note that it is an example of the fact that integration and differentiation are inverse operations. Starting with the sine function, we integrated to form the area function A and then differentiated the area function to end with the sine function. Symbolically,

$$\sin \quad \xrightarrow{\int_0^x} \quad A \quad \xrightarrow{\frac{d}{dx}} \quad \sin$$

More generally, from (6) in part I of the Fundamental Theorem of Calculus,

$$f \quad \xrightarrow{\int_a^x} \quad F \quad \xrightarrow{\frac{d}{dx}} \quad f$$

Hence integration and differentiation are indeed inverse operations. ■

Our last example is another problem that can be solved using part II of the Fundamental Theorem of Calculus, but, as we'll see, applying part I makes it a lot easier.

■ **EXAMPLE 4**   The position of a valve in a circular pipe of radius 1 meter is a function $x = x(t)$ of time t. As shown in Fig. 5.20, the valve opens to the right. The flow L through the valve, measured in cubic meters per minute, is directly proportional to the area of the shaded region. Given that $x = x(t) = 2t^2$, $0 \le t \le 1$, determine the rate of change dL/dt in the flow when the valve is half open.

Solution.   We are given that the flow L through the valve is

(20) $$L = kA(x),$$

where k is a constant and $A(x)$ is the area of the shaded region of the circle with equation

(21) $$(w - 1)^2 + y^2 = 1.$$

Note that we have used variables w and y to avoid possible confusion with the variable x measuring the valve opening. The area $A(x)$ is twice the area of the region beneath the top half of this circle and between the lines $w = 0$ and $w = x$. An equation for the top half of the circle may be found by solving (21) for y in terms of w. We have

$$y = \sqrt{1 - (w - 1)^2} = \sqrt{2w - w^2}.$$

Thus we may rewrite (20) as

(22) $$L = kA(x) = k \cdot 2 \int_0^x \sqrt{2w - w^2}\, dw.$$

A straightforward step toward our goal of calculating dL/dt would be to come up with an antiderivative of $\sqrt{2w - w^2}$ and then apply part II of the Fundamental Theorem. Leaving the details to Exercise 40, we obtain

(23) $$L = \frac{k}{2}\left(\pi + 2(x - 1)\sqrt{2x - x^2} + 2\arcsin(x - 1)\right).$$

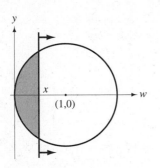

Figure 5.20. Closing valve.

To calculate dL/dt, we recall that $x = x(t) = 2t^2$ and use the chain rule.

$$(24) \qquad \frac{dL}{dt} = \frac{dL}{dx}\frac{dx}{dt} = \left(2k\sqrt{2x - x^2}\right)\cdot 4t.$$

Since the valve is half open when $t = 1/\sqrt{2}$ and $x = x(1/\sqrt{2}) = 1$, we obtain $dL/dt = 4k\sqrt{2}$ for the flow per minute when the valve is half open.

The hardest step in the above calculation was calculating the antiderivative. Part I of the Fundamental Theorem makes this entirely unnecessary! In the equation $L = kA(x)$, x is a function of t. From the chain rule, we have

$$(25) \qquad \frac{dL}{dt} = \frac{dL}{dx}\frac{dx}{dt}.$$

Since $x = x(t) = 2t^2$, we know that $dx/dt = 4t$. This leaves dL/dx, which we calculate using (6) of part I of the Fundamental Theorem. We have

$$\frac{dL}{dx} = \frac{d}{dx}\left(k\cdot 2\int_0^x \sqrt{2w - w^2}\,dw\right) = 2k\sqrt{2x - x^2}.$$

Putting this result together with (25), we have

$$\frac{dL}{dt} = \left(2k\sqrt{2x - x^2}\right)(4t).$$

Since the valve is half open when $t = 1/\sqrt{2}$ and $x = x(1/\sqrt{2}) = 1$, we have

$$\frac{dL}{dt} = \left(2k\sqrt{1}\right)\left(4\cdot 1/\sqrt{2}\right) = 4k\sqrt{2}. \qquad \blacksquare$$

We will need several other properties and definitions relating to the integral. We give these in the summary table on page 365. For completeness, we have included earlier properties and definitions as well.

Definition 3   We put off until Section 5.3 a discussion of the convenient convention given in Definition 3 that, for example, $\int_1^0 x^2dx = \int_0^1 x^2dx$. However, we discuss the definition of the average value of a function (Definition 4) from several viewpoints and argue that Properties 3 and 4 are plausible.

Definition 4: Average Value   Although the definition of the average value of a function f on $[a, b]$ is straightforward, understanding its connection with the average

$$\overline{y} = \frac{1}{n}(y_1 + y_2 + \cdots + y_n) = \frac{1}{n}\sum_{j=1}^{n} y_j$$

of a list of numbers $y_1, y_2, \ldots, y_n$ requires some explanation.

Imagine a machine cutting nails from a coil of wire. To ensure that the lengths of the nails stay within a certain tolerance, the output of the machine is sampled every 30 minutes. Suppose, in particular, that a nail is selected and its length measured at times $m_1, m_2, \ldots, m_n$.

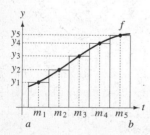

Figure 5.21. Average value of a function.

Definitions and Properties of the Integral

Assume that f and g are continuous on $[a, b]$, $c \in R$, and $r, s, t \in [a, b]$.

Definition 1: Area. Assuming that f is nonnegative on $[a, b]$, the area of the region S beneath the graph of f is the number $\int_a^b f(x) \, dx$.

Definition 2:

$$\int_a^a f(w) \, dw = 0$$

Definition 3:

$$\int_b^a f(x) \, dx = -\int_a^b f(x) \, dx$$

Definition 4: Average value. The average value of f on $[a, b]$ is

$$\frac{1}{b-a} \int_a^b f(x) \, dx.$$

Property 1: Additivity

$$\int_r^s f(x) \, dx + \int_s^t f(x) \, dx = \int_r^t f(x) \, dx$$

Property 2: Mean-value Inequality for Integrals

$$m(b-a) \leq \int_a^b f(x) \, dx \leq M(b-a),$$

where m and M are the minimum and maximum of f on $[a, b]$.

Property 3: Linearity

$$\int_a^b (f(x) + g(x)) \, dx = \int_a^b f(x) \, dx + \int_a^b g(x) \, dx$$

and

$$\int_a^b cf(x) \, dx = c \int_a^b f(x) \, dx$$

Property 4: If $f(x) \geq 0$ on $[a, b]$, then $\int_a^b f(x) \, dx \geq 0$.

We show in Fig. 5.21 the results from $n = 5$ observations. The actual behavior of the machine over a time interval $[a, b]$ is represented by the graph of the function f. The data $y_1, y_2, \ldots, y_n$ represent a sample of this function during $[a, b]$.

The sum

$$M_n = \frac{b-a}{n}(y_1 + y_2 + \cdots + y_n)$$

of the rectangles shown in the figure lies between the lower sum L_n and upper sum U_n for the subdivision of $[a, b]$ shown. Hence we expect that M_n is approximately equal to $\int_a^b f(t)\, dt$, that is,

$$\int_a^b f(t)\, dt \approx \frac{b - a}{n}(y_1 + y_2 + \cdots + y_n) = (b - a)\bar{y}$$

Dividing by $(b - a)$ gives

(26)
$$\frac{1}{b - a}\int_a^b f(t)\, dt \approx \bar{y}$$

In words, the average value of f on $[a, b]$ is approximately the average of a sample of values of f on $[a, b]$.

We show in Fig. 5.22 a shaded rectangle with height H equal to the average value of the sine function on $[0, \pi]$. From Example 1,

$$H = \frac{1}{\pi}\int_0^\pi \sin x\, dx = 2/\pi.$$

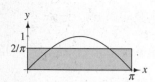

Figure 5.22. Average value of sine.

The area of the rectangle is equal to the area beneath the sine function on $[0, 2\pi]$.

One final comment on average value. If you were to program your calculator or CAS to select randomly 100 numbers in the interval $[0, \pi]$, to calculate the sines of these numbers, and then to average the results, what could you say about the result? What if 20 students did this and the results were averaged? See Exercise 50.

Property 3: Linearity   The two parts of Property 3 may be stated in words: the integral of a sum is the sum of the integrals, and the integral of a scalar times a function is the scalar times the integral of the function. These express the linearity of the integral. The derivative satisfies a similar property: the derivative of a sum of two functions is the sum of their derivatives, and the derivative of a scalar times a function is the product of the scalar and the derivative of that function. In symbols, if c is any scalar and f and g are differentiable functions, then

(27)
$$\frac{d}{dx}(f + g) = \frac{d}{dx}f + \frac{d}{dx}g$$

and

(28)
$$\frac{d}{dx}(cf) = c\frac{d}{dx}f.$$

We show in Exercises 43 and 44 that the integral satisfies Property 3.

Property 4   This property states a fact that is geometrically obvious. In words, if the graph of f lies on or above the x-axis, then the area beneath the graph of f is positive or zero.

Exercises 5.2

Basic

1. Check that the integral is additive for a particular case. Specifically, use part II of the Fundamental Theorem to show that the sum of the values of the two integrals

$$\int_0^1 x^2 \, dx \quad \text{and} \quad \int_1^2 x^2 \, dx \quad \text{add to} \quad \int_0^2 x^2 \, dx.$$

Sketch a figure analogous to Fig. 5.12.

2. Verify that the integral is additive. Specifically, use part II of the Fundamental Theorem to show that the sum of the values of the two integrals

$$\int_0^1 x^3 \, dx \quad \text{and} \quad \int_1^2 x^3 \, dx \quad \text{add to} \quad \int_0^2 x^3 \, dx.$$

Sketch a figure analogous to Fig. 5.12.

3. Check the Mean-value Inequality for Integrals for a special case. Calculate

$$m(b - a), \quad \int_a^b f(x) \, dx, \quad \text{and} \quad M(b - a).$$

Let $f(x) = (x - 3)^2 + 3, 2 \le x \le 6$. Draw a graph similar to Fig. 5.13.

4. Check the Mean-value Inequality for Integrals for a special case. Calculate

$$m(b - a), \quad \int_a^b f(x) \, dx, \quad \text{and} \quad M(b - a).$$

Let $f(x) = -2 \sin x + 3, \pi/6 \le x \le \pi$. Draw a graph similar to Fig. 5.13.

Exercises 5–16: Sketch each curve and use part II of the Fundamental Theorem of Calculus in calculating the area of the region beneath the graph of the function, between the given values of $x = a$ and $x = b$.

5. $f(x) = \cos x, a = 0, b = \pi/2$
6. $f(x) = \sin x, a = \pi/6, b = \pi/2$
7. $f(x) = (x - 1)^3, a = 1, b = 3$
8. $f(x) = \frac{1}{3}(x - 3)^4, a = 2, b = 5$
9. $f(x) = e^x, a = 0, b = 1$
10. $f(x) = 2^x, a = 0, b = 1$
11. $f(x) = \sec^2 x, a = 0, b = \pi/4$
12. $f(x) = \sec x \tan x, a = 0, b = \pi/3$
13. $f(x) = 1/(x^2 + 1), a = 0, b = \pi/4$
14. $f(x) = 1/\sqrt{1 - x^2}, a = 0, b = 1/2$
15. $f(x) = 1/x, a = 1, b = 2$
16. $f(x) = 1/\sqrt{x}, a = 1, b = 4$

Exercises 17–26: Use part I of the Fundamental Theorem of Calculus in differentiating the integrals. Use the chaining variable idea in Example 4 (see (25)) if this clarifies the use of the chain rule when the upper limit is not simply x.

17. $\int_1^x \ln w \, dw$
18. $\int_0^x \sin w^2 \, dx$
19. $\int_1^x \sqrt{1 + t^2} \, dt$
20. $\int_0^x \sqrt{1 - t^2} \, dt$
21. $\int_1^{x^2} \ln w \, dw$
22. $\int_0^{x^3} \cos \sqrt{w} \, dw$
23. $\int_1^{\sqrt{x}} \sqrt{1 + t^2} \, dt$
24. $\int_0^{1/x} \tan t^2 \, dt$
25. $\int_0^{e^x} (\arctan t - t^2) \, dt$
26. $\int_0^{\sin x^2} \sin t^2 \, dt$

Growth

Exercises 27–38: Find an antiderivative of the function f defined for $a \le x \le b$ and evaluate $\int_a^b f(x) \, dx$.

27. $f(x) = \sqrt{x}, 0 \le x \le 4$
28. $f(x) = x^{4/3}, 0 \le x \le 8$
29. $f(x) = x^3 - 7x, 0 \le x \le 1$
30. $f(x) = x - 2x^4, 0 \le x \le 1$
31. $f(x) = x(3x - 1), 0 \le x \le 4$
32. $f(x) = (x + 1)(x^2 + 1)$, $0 \le x \le 2$
33. $f(x) = (x^2 + 1)/x, 1 \le x \le 3$
34. $f(x) = (x^3 + 2x + 1)/x^2$, $1 \le x \le 4$
35. $f(x) = x^{1/3} - 2x^{-2/3}$, $1 \le x \le 8$
36. $f(x) = x^{-1/4} - 7\sqrt{x}$, $1 \le x \le 16$
37. $f(x) = 2x \sin x^2, 0 \le x \le 1$
38. $f(x) = 2xe^{x^2}, 0 \le x \le 1$

39. What is the average value of the cosine function on $[0, \pi/2]$? Determine the height of a rectangle with base resting on $[0, \pi/2]$ and having area equal to the average value of the sine function on $[0, \pi/2]$. Sketch the graph of the cosine function and the rectangle. Write a sentence or two describing the sketch in layperson's terms.

40. This problem relates to Example 4. First show that

$$\tfrac{1}{2}(w - 1)\sqrt{2w - w^2} + \tfrac{1}{2}\arcsin(w - 1)$$

is an antiderivative of $\sqrt{2w - w^2}$. Use this result to check the correctness of (23). Finally, fill in the details leading up to (24).

41. In Example 4 show that the maximum value of dL/dt for $0 \le t \le 1$ is $32k/(3\sqrt{3})$.

42. Complete the proof of part I of the Fundamental Theorem of Calculus. The given argument may be used with only minor changes. For the $x = a$ case, the limit as $h \to 0$ is replaced by the limit as $h \to 0^+$ and we replace $\int_a^a f(w)\,dw$ by 0. The $x = b$ case is similarly handled.

43. Use the Fundamental Theorem of Calculus in verifying the first part of Property 3. (Hint: Let F and G be antiderivatives of f and g, and write an antiderivative of $f + g$ in terms of F and G.)

44. Use the Fundamental Theorem of Calculus in verifying the second part of Property 3. (Hint: Let F be an antiderivative of f, express $c \int_a^b f(x)\,dx$ in terms of F, and, finally, express an antiderivative of cf in terms of F.)

45. Let f be continuous on an interval I. Show that Property 1 of integrals is true for all $a, b, c \in I$. Note that as stated in (1), $a < c < b$. Given Definitions 2 and 3, this restriction is no longer needed.

46. Use the definition of the integral in proving Property 4.

47. Use Property 4 (and other properties as needed) in showing that if p and q are continuous on $[a, b]$ and $p(x) \ge q(x)$ for all $x \in [a, b]$, then $\int_a^b p(x)\,dx \ge \int_a^b q(x)\,dx$.

48. Fill in the details of the following proof that for all $a, b > 0$

(29) $$\lim_{x \to \infty} = \frac{(\ln x)^a}{x^b} = 0.$$

Let $p > 0$; then for $w \ge 1$, $w^{-1} \le w^{-1+p}$. By Property 4,

$$\int_1^x \frac{1}{w}\,dw \le \int_1^x w^{-1+p}\,dw.$$

Hence

$$\ln x \le \frac{x^p}{p} - \frac{1}{p} < \frac{x^p}{p}.$$

Using this result, we have

$$\frac{\ln^a x}{x^b} \le \frac{x^{pa}}{p^a x^b}.$$

After showing that p can be chosen so that $pa - b < 0$, let $x \to \infty$ and infer (29).

49. Let f be defined on $[0, 2]$ by

$$f(x) = \int_0^{x^2} \frac{\sin t}{t + 1}\,dt.$$

Where does the maximum value of f occur? It is not necessary to evaluate f at any point other than $x = 0$.

50. Use your calculator or CAS to generate 100 numbers in the interval $[0, \pi]$. Calculate the sines of these numbers and average the 100 function values thus generated. This number approximates the average value of the sine function on $[0, \pi]$. Most calculators have a built-in random number generator that returns numbers randomly chosen from $[0, 1]$. A labor-intensive procedure is to press the RAND button and then the SIN button, and add the result to a variable whose initial value is 0. Repeat this 100 times. Divide the sum by 100. If you can program your calculator, this procedure can be automated. If you repeat the program, say, 20 times and average the outcomes, you should obtain $2/\pi = 0.6366\ldots$, more or less. It may be necessary to arrange for a different "seed" for the random number generator as you repeat the program.

Review/Preview

Exercises 51–59: Calculate $f'(x)$ or dy/dx.

51. $f(x) = \cos\sqrt{x}$

52. $f(x) = \sin x^2$

53. $y = x3^{2x}$

54. $y = \ln(x^2 + 1)$

55. $f(x) = \int_1^x \tan w^2\,dw$

56. $f(x) = \int_0^x \sqrt{1 + w^2}\,dw$

57. $y = \arctan(1/x)$

58. $y = \arcsin(2x + 1)$

59. $y = e^{2x}\sin(1 - 3x)$

Exercises 60–62: Determine the roots of the quadratic equation by completing the square. Recall that after factoring out the coefficient of the x^2 term, the square of one-half the coefficient of the x term is added and subtracted:

$$ax^2 + bx + c = a\left(x^2 + \frac{b}{a}x\right) + c = a\left(x^2 + \frac{b}{a}x + \frac{b^2}{4a^2}\right) - \frac{b^2}{4a} + c$$

$$= a\left(x + \frac{b}{2a}\right)^2 - \frac{b^2 - 4ac}{4a}.$$

60. $x^2 - 6x + 3 = 0$ **61.** $2x^2 - 5x + 1 = 0$ **62.** $5x^2 - x - 2 = 0$

5.3 INTEGRATION BY SUBSTITUTION

We may evaluate the integral $\int_a^b f(x)\,dx$ if we know an antiderivative F of f. For, according to part II of the Fundamental Theorem of Calculus,

$$\int_a^b f(x)\,dx = F(b) - F(a).$$

Among the methods commonly used to determine F from f are substitution, integration by parts, and partial fractions. In this section we discuss substitution.

Substitution

Substitution is used to change an integral from an unfamiliar form to one that we recognize or can look up in a table. Substitution is based on a change-of-variable equation, the Fundamental Theorem of Calculus, and the chain rule.

If we know that a given function f has an antiderivative F, then by the Fundamental Theorem we can evaluate the integral $\int_c^d f(u)\,du$. We have

(1) $$\int_c^d f(u)\,du = F(d) - F(c).$$

Now if we make a substitution $u = g(x)$, then by the chain rule we have

(2) $$\frac{d}{dx}F(g(x)) = F'(g(x))g'(x) = f(g(x))g'(x).$$

From this we see that the composite function $F \circ g$ is an antiderivative of the function $(f \circ g) \cdot g'$. Hence, by the Fundamental Theorem,

(3) $$\int_a^b f(g(x))g'(x)\,dx = F(g(b)) - F(g(a)).$$

Comparing equations (1) and (3) leads to the following conclusion.

Integration by Substitution

If the function f is continuous on an interval $[c, d]$, the substitution function $u = g(x)$ is differentiable on $[a, b]$, $g(x) \in [c, d]$ for $x \in [a, b]$, and $g(a) = c$ and $g(b) = d$, then

(4)
$$\int_a^b f(g(x))g'(x)\,dx = \int_{g(a)}^{g(b)} f(u)\,du.$$

How can we use equation (4) to evaluate integrals? Here's one way: suppose we wish to evaluate the integral $\int_a^b f(g(x))g'(x)\,dx$ and happen to know an antiderivative of f. From (4) we have

$$\int_a^b f(g(x))g'(x)\,dx = F(u)\Big|_{g(a)}^{g(b)}.$$

■ **EXAMPLE 1** Evaluate the integral

$$\int_0^{\pi/2} \sin^2 x \cos x\,dx.$$

Solution. The form of the integrand suggests the substitution $u = \sin x$. Why? The answer is—the chain rule. To integrate by substitution, you should look for products that remind you of the chain rule. For the integrand $\sin^2 x \cos x$, you should note that $\cos x$ is the derivative of $\sin x$. Referring to (4), g and g' have now been identified. It is useful to express this in the form of the substitution $u = g(x) = \sin x$. Next you should note that the outer function f is the squaring function and, writing $f(u) = u^2$, recall that an antiderivative of f is $F(u) = \frac{1}{3}u^3$. Hence

$$\int_0^{\pi/2} \sin^2 x \cos x\,dx = \int_0^1 u^2\,du = \frac{1}{3}u^3\Big|_0^1 = \frac{1}{3}.$$

The limits on the second integral were found by noticing that $a = 0$ and $b = \pi/2$ and then calculating $g(0) = \sin 0 = 0$ and $g(\pi/2) = 1$.

This calculation fits (4) well but is more formal than necessary. We rearrange the calculation so that the notation helps us along. Start with the given integral, write down a possible substitution equation $u = g(x)$, and calculate the differential du of u. This gives

$$u = \sin x$$

$$du = \cos x\,dx.$$

From these equations we may rewrite the integrand as

$$\sin^2 x \cos x\,dx = (\sin x)^2 (\cos x\,dx) = u^2\,du.$$

The new limits may be calculated by looking at the substitution equation $u = \sin x$ and saying to yourself something like, "When x is 0, u is $\sin 0 = 0$, when x is $\pi/2$, u is $\sin(\pi/2) = 1$."

Putting everything together, we have

$$\int_0^{\pi/2} \sin^2 x \cos x \, dx = \int_0^1 u^2 \, du = \left. \tfrac{1}{3} u^3 \right|_0^1 = \frac{1}{3}.$$

Once you settle on a promising substitution, the substitution technique is self-guiding. It is not necessary to substitute mechanically into (4).   ■

■ **EXAMPLE 2**   Evaluate the integral

(5)
$$\int_0^2 x \sqrt{x^2 + 1} \, dx.$$

Solution.   The key to this integral is that the derivative $2x$ of the inner function $x^2 + 1$ is, apart from a factor of 2, a factor of the integrand. This suggests the substitution $u = x^2 + 1$. A second important feature is that the outer function $\sqrt{u}$ has the known antiderivative $\tfrac{2}{3} u^{3/2}$. To "fix up" the factor x, we may multiply it by 2 provided we multiply the entire integral by $1/2$. By Property 3 of Section 5.2, the combination of these two multiplications leaves the integral unchanged in value. From (4) we have

$$\int_0^2 x \sqrt{x^2 + 1} \, dx = \tfrac{1}{2} \int_0^2 \sqrt{x^2 + 1}(2x) \, dx = \tfrac{1}{2} \int_1^5 \sqrt{u} \, du.$$

In the first step we replaced x by $2x$, adjusting for this by multiplying the integral by $\tfrac{1}{2}$. In the last step the limits changed according to the substitution equation $u = x^2 + 1 = g(x)$. We calculated $g(a) = g(0) = 0^2 + 1$ and $g(b) = g(2) = 2^2 + 1 = 5$. The last integral is easily evaluated. We have

$$\tfrac{1}{2} \int_1^5 \sqrt{u} \, du = \tfrac{1}{2} \cdot \left. \left(\tfrac{2}{3} u^{3/2} \right) \right|_1^5 = \tfrac{1}{3} \left(5^{3/2} - 1^{3/2} \right) \approx 3.39.$$

Here is the self-guided calculation. Starting with the original integral

$$\int_0^2 x \sqrt{x^2 + 1} \, dx$$

and the substitution equation $u = x^2 + 1$, we calculate $du = 2x \, dx$, which we may rewrite as $x \, dx = \tfrac{1}{2} \, du$. Then

$$\int_0^2 x \sqrt{x^2 + 1} \, dx = \int_0^2 \sqrt{x^2 + 1} \, (x \, dx) = \int_1^5 \sqrt{u} \left(\tfrac{1}{2} \, du \right).$$

The limits on the last integral came from the substitution equation $u = x^2 + 1$. When $x = 0$, $u = 1$; when $x = 2$, $u = 2^2 + 1 = 5$. Finally,

$$\int_0^2 x \sqrt{x^2 + 1} \, dx = \tfrac{1}{2} \int_1^5 \sqrt{u} \, du = \tfrac{1}{2} \cdot \left. \left(\tfrac{2}{3} u^{3/2} \right) \right|_1^5 = \tfrac{1}{3} \left(5^{3/2} - 1^{3/2} \right) \approx 3.39.$$ ■

"Reverse" Substitution

Up to this point we have identified a given integral with the left side of (4). For some integrals it is better to use (4) in reverse. For this we rewrite (4) as

$$(6) \qquad \int_{g(a)}^{g(b)} f(x)\,dx = \int_a^b f(g(u))g'(u)\,du.$$

Note that we have interchanged x and u. We interchange x and u in the substitution equation as well, giving $x = g(u)$. After the substitution, we may be able to recognize an antiderivative of $f(g(u))g'(u)$ or locate this form in a table of integrals.

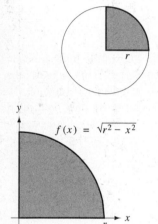

$f(x) = \sqrt{r^2 - x^2}$

Figure 5.23. Quarter-circle.

■ **EXAMPLE 3**   Calculate the area Q of the quarter-circle shown in Fig. 5.23. This calculation gives a proof of the formula $A = \pi r^2$ for the area A of a circle of radius r. The search for this result and a way of proving it has been part of mathematics for more than 3500 years. Some mathematicians continue to work on ideas related to the circle. For example, toward the end of 1991 Gregory and David Chudnovsky calculated π to over 2 billion decimal places, looking for patterns. For their calculations they built a supercomputer in their New York apartment, largely from mail order parts. In approximately 1650 B.C. an Egyptian scribe named Ahmes wrote in the Rhind papyrus that the formula $A = \left(\frac{8}{9}d\right)^2$ is used to calculate the area of a circle of diameter d. The Rhind papyrus begins with the words "Directions for Obtaining Knowledge of All Dark Things."

Solution.   The shaded region at the bottom of the figure lies beneath the graph of $f(x) = \sqrt{r^2 - x^2}$. The area of this region is

$$(7) \qquad Q = \int_0^r \sqrt{r^2 - x^2}\,dx.$$

We use the substitution $x = g(u) = r\sin u$, one of the *trigonometric substitutions*. To substitute into (7), we note that

$$dx = r\cos u\,du, \qquad g(0) = 0, \quad \text{and} \quad g(\pi/2) = r.$$

From (6) we have

$$\int_0^r \sqrt{r^2 - x^2}\,dx = \int_0^{\pi/2} \sqrt{r^2 - r^2\sin^2 u}\,(r\cos u\,du).$$

To simplify the last integral, we factor r from the square root, replace $1 - \sin^2 u$ by $\cos^2 u$, and, noting that $\cos u \geq 0$ for $u \in [0, \pi/2]$, replace $\sqrt{\cos^2 u}$ by $\cos u$. This gives

$$\int_0^r \sqrt{r^2 - x^2}\,dx = r^2 \int_0^{\pi/2} \cos^2 u\,du.$$

The last integral can be evaluated by recalling the trigonometric identity $\cos^2 u = (1 + \cos 2u)/2$.

$$\int_0^r \sqrt{r^2 - x^2}\, dx = \tfrac{1}{2}r^2 \int_0^{\pi/2} (1 + \cos 2u)\, du.$$

An antiderivative of $1 + \cos 2u$ is $u + \tfrac{1}{2}\sin 2u$. We have, then,

$$\int_0^r \sqrt{r^2 - x^2}\, dx = \tfrac{1}{2}r^2 \left(u + \tfrac{1}{2}\sin 2u\right)\Big|_0^{\pi/2}$$

$$= \tfrac{1}{2}r^2\left((\pi/2 + 0) - (0 + 0)\right) = \tfrac{1}{4}\pi r^2.$$

This calculation gives us the area Q of the quarter-circle and the area A of the entire circle of radius r.

$$A = 4Q = 4\int_0^r \sqrt{r^2 - u^2}\, du = 4\left(\tfrac{1}{4}\pi r^2\right) = \pi r^2. \qquad \blacksquare$$

We referred to the substitution $x = r\sin u$ as a trigonometric substitution. This substitution is known to be useful in simplifying expressions having a factor of $\sqrt{r^2 - x^2}$. We explore this as well as other trigonometric substitutions in Exercises 48–60.

■ **EXAMPLE 4**   Evaluate the integral

$$(8) \qquad\qquad \int_0^1 x\sqrt{1 - x}\, dx.$$

Solution.   How do we go about finding a useful substitution for this integral? An answer is to try a substitution that takes advantage of the form of the integrand. We use $1 - x = u^2$, which simplifies the radical expression. We must be careful, though, since it is easy to make errors in simplifying radical expressions. The calculation

$$\sqrt{1 - x} = \sqrt{u^2} = |u| = \begin{cases} u & \text{if } u \geq 0 \\ -u & \text{if } u \leq 0 \end{cases}$$

shows that we must decide if we want

$$u = \sqrt{1 - x} \geq 0 \quad \text{or} \quad u = -\sqrt{1 - x} \leq 0.$$

Our decision will also affect the limits. We choose $u = \sqrt{1 - x}$. Taking the differential of both sides of the equation $1 - x = u^2$ gives $-dx = 2u\, du$. The subsequent calculation is

$$\int_0^1 x\sqrt{1 - x}\, dx = \int_1^0 (1 - u^2)u(-2u\, du)$$

$$= 2\int_1^0 (u^4 - u^2)\, du.$$

We use Definition 3 of Section 5.2 to interchange the limits on this integral. This gives

$$\int_0^1 x\sqrt{1-x}\,dx = -2\int_0^1 (u^4 - u^2)\,du = -2\left(\tfrac{1}{5}u^5 - \tfrac{1}{3}u^3\right)\Big|_0^1$$

$$= -2\left[\left(\tfrac{1}{5}1^5 - \tfrac{1}{3}1^3\right) - \left(\tfrac{1}{5}0^5 - \tfrac{1}{3}0^3\right)\right]$$

$$= -2\left(\frac{1}{5} - \frac{1}{3}\right) = \frac{4}{15}.$$

We leave as Exercise 36 a similar calculation, but one based on the substitution $u = -\sqrt{1-x}$. Since $u^2 = 1 - x$ as before, much of the calculation will be the same. However, this substitution does not invert the limits. The fact that it leads to the same result as above gives a reason for Definition 3. ∎

Short Integral Table and Indefinite Integrals

We end the section by giving a short table of integrals and some examples of its use. The eight integrals listed in the table are among the most basic and most frequently occurring integrals. We shall use them without comment in the remainder of the text. Many integrals are just one substitution away from these integrals. The same arguments can be given for memorizing these eight integrals as for memorizing such facts as $\cos(\pi/3) = \sqrt{3}/2$, even though a calculator will return $0.866025\ldots$ with a few key strokes.

The simplest integrals are those based directly on a differentiation formula. For example, from the formula

$$\frac{d}{dx}x^{n+1} = (n+1)x^n \quad \text{we have} \quad \int_a^b x^n dx = \frac{x^{n+1}}{n+1}\Big|_a^b.$$

Such antidifferentiation formulas are often given in the form of an **indefinite integral,** in which limits of the integration are not displayed. The variable of integration can be x, u, w, or any other variable.

(9) $$\int x^n dx = \frac{x^{n+1}}{n+1} + C \quad \text{or} \quad \int u^n du = \frac{u^{n+1}}{n+1} + C$$

Indefinite integrals are usually given, as here, with an added constant C to indicate that there is an entire family of antiderivatives for a given function, differing among themselves at most by an additive constant. However, if we evaluate an integral with limits—such integrals are often called **definite integrals**—the constant is irrelevant. Note what happens to the constant in the following application of (9).

$$\int_1^4 \sqrt{x}\,dx = \int_1^4 x^{1/2}dx = \frac{x^{3/2}}{3/2} + C\Big|_1^4$$

$$= \left(\frac{4^{3/2}}{3/2} + C\right) - \left(\frac{1^{3/2}}{3/2} + C\right) = \frac{14}{3}$$

The integrals (A)–(H) found below are included in the larger table of integrals in the inside back cover of this book. Integral tables are available in most libraries. We note, however, that the fact that many CASs include powerful integration algorithms makes the question of such tables moot.

Table 1. Short table of indefinite integrals

(A) $\displaystyle\int x^n dx = \frac{x^{n+1}}{n+1} + C$	(B) $\displaystyle\int \frac{1}{x} dx = \ln|x| + C$
(C) $\displaystyle\int \sin x\, dx = -\cos x + C$	(D) $\displaystyle\int \cos x\, dx = \sin x + C$
(E) $\displaystyle\int e^x dx = e^x + C$	(F) $\displaystyle\int a^x dx = \frac{a^x}{\ln a} + C$
(G) $\displaystyle\int \frac{dx}{x^2 + a^2} = \frac{1}{a}\arctan\frac{x}{a} + C$	(H) $\displaystyle\int \frac{dx}{\sqrt{a^2 - x^2}} = \arcsin\frac{x}{a} + C$

Combined with a substitution, such tables can be used to evaluate many integrals. Some examples follow.

■ **EXAMPLE 5** Evaluate the integrals

$$\int_0^2 \sqrt{2x + 3}\, dx \quad \text{and} \quad \int \sqrt{2x + 3}\, dx.$$

Solution. The substitution $u = 2x + 3$ changes the given definite integral into one matching formula (A) from the table, with $n = 1/2$. Since $du = 2\, dx$, $dx = \frac{1}{2}\, du$ and we have

$$\int_0^2 \sqrt{2x + 3}\, dx = \frac{1}{2}\int_3^7 u^{1/2}\, du$$

$$= \frac{1}{2} \cdot \frac{u^{3/2}}{3/2}\bigg|_3^7 = \frac{1}{3}\left(7\sqrt{7} - 3\sqrt{3}\right).$$

We use the same substitution for the indefinite integral but retain the constant and express the result in terms of the original variable. We have

$$(10) \qquad \int \sqrt{2x + 3}\, dx = \frac{1}{2}\int u^{1/2}\, du\bigg|_{u=2x+3}$$

$$(11) \qquad = \frac{1}{2} \cdot \frac{u^{3/2}}{3/2} + C\bigg|_{u=2x+3} = \frac{1}{3}(2x + 3)^{3/2} + C.$$

The last expression is the most general function whose derivative is $\sqrt{2x + 3}$. In working with an indefinite integral with integrand $f(x)$, the idea is to determine the most general function $F(x)$ for which $F' = f$. The evaluation notation in (10) and (11) was used as a reminder that we must return to the original variable, that is, replace the substitution variable u by $2x + 3$ in the final step.

We mentioned that many CASs have integration algorithms. To illustrate, we give the input and output from the CAS known as *Mathematica,* applied to the integrals in this example.

For the definite integral $\int_0^2 \sqrt{2x + 3}\, dx$,

Input: Integrate[Sqrt[2x + 3], {x, 0, 2}]

Output: $-\text{Sqrt}[3] + \dfrac{7\text{Sqrt}[7]}{3}$

For the indefinite integral $\int \sqrt{2x + 3}\, dx$,

Input: Integrate[Sqrt[2x + 3], x]

Output: $(1 + \dfrac{2x}{3})\text{Sqrt}[3 + 2x]$

These results are equivalent to those obtained earlier.   ■

Before giving other examples we add two comments on the evaluation of an indefinite integral by substitution. If we start with an indefinite integral $\int f(x)\, dx$ and make a substitution $x = g(u)$, then we write

$$\int f(x)\, dx = \int f(g(u))g'(u)\, du \bigg|_{u = g^{-1}(x)} = H(u)\bigg|_{u = g-1(x)} + C$$

where $H'(u) = f(g(u))g'(u)$. This calculation requires that the function g in the substitution equation be invertible. In the above example, $x = g(u) = \frac{1}{2}(u - 3)$. This function is certainly invertible.

Although we have made a point of using the evaluation notation in the example and just above, we drop it in what follows and instead keep in mind that in the last step we must return to the original variable. In any case, we have an absolute check on an evaluation of an indefinite integral: just differentiate the final expression. We demonstrate this in the next example.

■ **EXAMPLE 6**   Evaluate the integrals

$$\int_2^4 \frac{dx}{2x - 3} \quad \text{and} \quad \int \frac{dx}{2x - 3}.$$

Solution.   On the range of integration $[2, 4]$ the function $f(x) = 1/(2x - 3)$ is positive and continuous. We mention this for two reasons. First, at $x = 3/2$ the function f has an infinite discontinuity, as shown in Fig. 5.24. The line $x = 3/2$ is a vertical asymptote to the graph of f. If the interval of integration were to include $x = 3/2$, the Fundamental Theorem may not be applicable since it assumes that the integrand is continuous on the interval of integration. The second reason is that the substitution $u = 2x - 3$ leads to $\ln u$, for which we must have $u > 0$.

We use the substitution $u = 2x - 3$, which changes the given definite integral into one matching formula (B) from the table. Since $du = 2\, dx$,

$$\int_2^4 \frac{dx}{2x - 3} = \int_1^5 \frac{\frac{1}{2}\, du}{u} = \frac{1}{2}\int_1^5 \frac{du}{u}$$

$$= \frac{1}{2}\ln|u|\Big|_1^5 = \frac{1}{2}(\ln 5 - \ln 1) = \frac{1}{2}\ln 5 \approx 0.805.$$

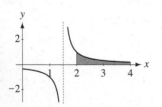

Figure 5.24. Vertical asymptote at $x = 3/2$.

The calculation for the indefinite integral is similar.

(12)
$$\int \frac{dx}{2x-3} = \int \frac{\frac{1}{2}\,du}{u} = \frac{1}{2}\int \frac{du}{u}$$

$$= \frac{1}{2}\ln|u| + C = \frac{1}{2}\ln|2x-3| + C.$$

We check this result by differentiation after discussing formula (B). This formula is actually two formulas written together. If $x > 0$, an antiderivative of $1/x$ is $\ln x$, for

$$\frac{d}{dx}\ln x = \frac{1}{x}.$$

If $x < 0$, an antiderivative of $1/x$ is $\ln(-x)$, for, by the chain rule,

$$\frac{d}{dx}\ln(-x) = \frac{1}{-x}\cdot(-1) = \frac{1}{x}.$$

Formula (B) combines these results, showing that for all $x \neq 0$

$$\frac{d}{dx}\ln|x| = \frac{1}{x}.$$

Using this result together with the chain rule,

$$\frac{d}{dx}\left(\frac{1}{2}\ln|2x-3| + C\right) = \frac{1}{2}\cdot\frac{1}{2x-3}\cdot 2 = \frac{1}{2x-3}.$$

Hence the result in (12) is correct.   ■

■ **EXAMPLE 7**   Evaluate the integral

$$\int_0^1 \frac{dx}{2x^2+3}.$$

Solution.   The integrand most resembles formula (G). The main difference is that we have $2x^2$ and not x^2. The constant term can be written as a^2 by letting $a = \sqrt{3}$. There are two ways of changing the form of the integrand so that formula (G) applies, either by factoring or by a change of variable. First we do it by factoring.

In the first step, with the form $x^2 + a^2$ in mind, we change the form of the denominator:

$$2x^2 + 3 = 2(x^2 + 3/2) = 2\left(x^2 + \left(\sqrt{3/2}\right)^2\right).$$

Using this result, we have

$$\int_0^1 \frac{dx}{2x^2+3} = \frac{1}{2}\int_0^1 \frac{dx}{\left(x^2 + \left(\sqrt{3/2}\right)^2\right)}.$$

From Formula (G) with $a = \sqrt{3/2}$,

$$\int_0^1 \frac{dx}{2x^2 + 3} = \frac{1}{2} \cdot \frac{1}{\sqrt{3/2}} \arctan \frac{x}{\sqrt{3/2}} \Big|_0^1 = \frac{1}{\sqrt{6}} \arctan \sqrt{2/3} = 0.2795\ldots.$$

For the solution based on a change of variable, the first step is

$$2x^2 + 3 = \left(\sqrt{2}x\right)^2 + \left(\sqrt{3}\right)^2.$$

Letting $u = \sqrt{2}x$, we have $du = \sqrt{2}\,dx$ and

$$\int_0^1 \frac{dx}{2x^2 + 3} = \frac{1}{\sqrt{2}} \int_0^{\sqrt{2}} \frac{du}{u^2 + (\sqrt{3})^2}.$$

From Formula (G) with $a = \sqrt{3}$, we have

$$\int_0^1 \frac{dx}{2x^2 + 3} = \frac{1}{\sqrt{2}} \cdot \frac{1}{\sqrt{3}} \arctan \frac{u}{\sqrt{3}} \Big|_0^{\sqrt{2}} = \frac{1}{\sqrt{6}} \arctan \sqrt{2/3} = 0.2795\ldots. \quad \blacksquare$$

In the next example a useful change of variable becomes clear after "completing the square" on an expression of the form $ax^2 + bx + c$.

■ **EXAMPLE 8**   Evaluate

$$\int \frac{dx}{\sqrt{3 + 2x - x^2}}.$$

Solution.   Scanning formulas (A)–(H), the only possibility is formula (H), which shows a difference of two squares beneath a radical. By completing the square, the expression $3 + 2x - x^2$ can be written as the difference of two squares. We have

$$\begin{aligned} 3 + 2x - x^2 &= 3 - (x^2 - 2x) \\ &= 3 - (x^2 - 2x + 1) + 1 \\ &= 4 - (x - 1)^2. \end{aligned}$$

By comparing this result to formula (H), we see that a useful change of variable is $u = x - 1$. Noting that $du = dx$ and applying formula (H),

$$\int \frac{dx}{\sqrt{3 + 2x - x^2}} = \int \frac{dx}{\sqrt{4 - (x - 1)^2}} = \int \frac{du}{\sqrt{2^2 - u^2}}$$

(13)
$$= \arcsin \frac{u}{2} + C = \arcsin \left(\frac{x - 1}{2}\right) + C. \quad \blacksquare$$

Several exercises on completing the square are given as problems 61–70. Also see the Review/Preview exercises at the end of Section 5.2.

Exercises 5.3

Basic

Exercises 1–10: Integrate using the suggested substitution.

1. $\displaystyle\int_0^1 x\sqrt{x^2 + 1}\, dx; \quad u = x^2 + 1$

2. $\displaystyle\int_1^2 x^2(x^3 - 1)^2\, dx; \quad u = x^3 - 1$

3. $\displaystyle\int_0^{\pi/2} \sin^3 x \cos x\, dx; \quad u = \sin x$

4. $\displaystyle\int_0^{\pi/2} \cos^3 x \sin x\, dx; \quad u = \cos x$

5. $\displaystyle\int_1^4 \left(e^{\sqrt{x}}/\sqrt{x}\right) dx; \quad u = \sqrt{x}$

6. $\displaystyle\int_1^4 2^{\sqrt{x}}/\sqrt{x}\, dx; \quad u = \sqrt{x}$

7. $\displaystyle\int_0^1 x\sqrt{3 + 2x}\, dx; \quad u = 3 + 2x$

8. $\displaystyle\int_1^3 (x - 1)\sqrt{x + 1}\, dx; \quad u = x + 1$

9. $\displaystyle\int_1^3 (1/(x + 1))\, dx; \quad u = x + 1$

10. $\displaystyle\int_3^7 (1/(2x - 5))\, dx; \quad u = 2x - 5$

Exercises 11–34: Reduce each integral to one of the integrals in Table 1 by a substitution.

11. $\displaystyle\int \sqrt{2x - 1}\, dx$

12. $\displaystyle\int \sqrt{5 + 2x}\, dx$

13. $\displaystyle\int \frac{1}{(5 - x)^{1/3}}\, dx$

14. $\displaystyle\int \frac{1}{(3x - 8)^{7/5}}\, dx$

15. $\displaystyle\int \frac{1}{2x + 1}\, dx$

16. $\displaystyle\int \frac{9}{3x + 2}\, dx$

17. $\displaystyle\int \sin(1 - 2x)\, dx$

18. $\displaystyle\int \sin\left(\tfrac{1}{2}x + 3\right) dx$

19. $\displaystyle\int \cos\left(\tfrac{1}{3} + \tfrac{2}{7}x\right) dx$

20. $\displaystyle\int \cos(-x + 1)\, dx$

21. $\displaystyle\int e^{2x+3}\, dx$

22. $\displaystyle\int e^{-0.1x+0.6}\, dx$

23. $\displaystyle\int 2^{-x+1}\, dx$

24. $\displaystyle\int 5^{7x+3}\, dx$

25. $\displaystyle\int \frac{dx}{2x^2 + 1}$

26. $\displaystyle\int \frac{dx}{3x^2 + 2}$

27. $\displaystyle\int \frac{dx}{(2x + 1)^2 + 9}$

28. $\displaystyle\int \frac{dx}{(5x - 2)^2 + 9}$

29. $\displaystyle\int \frac{dx}{\sqrt{1 - 2x^2}}$

30. $\displaystyle\int \frac{dx}{\sqrt{1 - 3x^2}}$

31. $\displaystyle\int \frac{dx}{\sqrt{7 - 2x^2}}$

32. $\displaystyle\int \frac{dx}{\sqrt{5 - 3x^2}}$

33. $\displaystyle\int \frac{dx}{\sqrt{1 - (2x + 1)^2}}$

34. $\displaystyle\int \frac{dx}{\sqrt{9 - (7 + 2x)^2}}$

Growth

35. How good is the Egyptian formula for the area of a circle given in Example 3? What value of π are the Egyptians in effect using?

36. Use the substitution $u = -\sqrt{1-x}$ in evaluating the integral given in Example 4. Give a reason why this substitution does not invert the limits and the substitution $u = \sqrt{1-x}$ does.

37. Use the substitution $u = 1 - x$ to work out the integral in Example 4.

38. Differentiate the expressions on the right side of formulas (F)–(H) and show they simplify to the integrands of the corresponding integrals.

39. Calculate $\int_{-3}^{-2}(1/x)\,dx$.

40. Extend the short integral table at the end of the section. For example, formulas for

$$\int \sec^2 x \, dx, \qquad \int \sec x \tan x \, dx, \qquad \int \sinh x \, dx$$

could be added.

41. Evaluate the integral

$$\int \frac{x \, dx}{\sqrt{1-x^2}}$$

by using a substitution to change it to formula (A) in Table 1.

42. Use formula (B) from Table 1 and a substitution to evaluate $\int_0^1 \tan x \, dx$.

43. Unthinking application of an integral formula similar to those listed in Table 1 or in the table of integrals (see inside back cover) can lead to incorrect results. Criticize the following calculation.

$$\int_{-1}^{1} \frac{1}{x^2} \, dx = \left. \frac{x^{-1}}{-1} \right|_{-1}^{1} = -1 - 1 = -2$$

44. See Exercise 43. Comment on $\int_{-3}^{0}(1/(x+1))\,dx$.

45. Keeping Definition 3 of Section 5.2 in mind, show that part II of the Fundamental Theorem is true if $a \geq b$.

46. Assume that f is continuous on an interval $[-a, a]$. Make the substitution $x = -u$ in showing that

$$\int_{-a}^{0} f(x)\,dx = F(a) - F(0).$$

Use this result to show that if f is an even function, that is, $f(-x) = f(x)$ for all $x \in [-a, a]$, then

$$\int_{-a}^{a} f(x)\,dx = 2\int_{0}^{a} f(x)\,dx.$$

Apply this to calculate $\int_{-\pi/2}^{\pi/2} \cos x \, dx$.

47. See Exercise 46. Assume that f is continuous on an interval $[-a, a]$. Show that if f is an odd function on $[-a, a]$, that is, $f(-x) = -f(x)$ for all $x \in [-a, a]$, then

$$\int_{-a}^{a} f(x)\,dx = 0.$$

Apply this to calculate $\int_{-\pi/2}^{\pi/2} \sin x \, dx$.

Exercises 48–60: Evaluate the integral using the suggested substitution. These problems are similar to the trigonometric substitution discussed in Example 3, which was based on the trigonometric identity $1 - \sin^2 u = \cos^2 u$. The substitutions suggested in the exercises are based on this identity or on $\tan^2 u + 1 = \sec^2 u$ or $\sec^2 u - 1 = \tan^2 u$.

48. $\int_0^{1/2} \sqrt{1 - 4x^2}\,dx, \ x = (\sin u)/2$

49. $\int_0^{1/\sqrt{2}} \sqrt{1 - 2x^2}\,dx, \ x = (\sin u)/\sqrt{2}$

50. $\int_0^{\sqrt{3}/2} \sqrt{3 - 2x^2}\,dx, \ x = \sqrt{3/2}\sin u$

51. $\int_0^{\sqrt{2/3}} \sqrt{2 - 3x^2}\,dx, \ x = \sqrt{2/3}\sin u$

52. $\int_0^1 \frac{dx}{(x^2 + 1)^{3/2}}, \ x = \tan u$

53. $\int_0^1 \frac{dx}{(x^2 + 4)^{3/2}}, \ x = 2\tan u$

One of the new limits will be $\arctan(1/2)$. Note that $\sin(\arctan(1/2)) = 1/\sqrt{5}$.

54. $\int_0^1 \sqrt{1 + x^2}\,dx, \ x = \tan u$

You will find the integral $\int \sec^3 x \, dx$ in the table of integrals (see inside back cover).

55. $\int_1^2 \frac{dx}{x^2 \sqrt{9 - x^2}}, \ x = 3\sin u$

What is the derivative of $-\cot u$?

56. $\int_{1/\sqrt{3}}^{1} \dfrac{dx}{x(x^2+1)}$, $x = \tan u$

You will find the integral $\int \cot u\, du$ in the table of integrals (see inside back cover).

57. $\int_{\sqrt{5}}^{\sqrt{10}} \dfrac{x^2\,dx}{\sqrt{x^2-1}}$, $x = \sec u$

You will find the integral $\int \sec^3 u\, du$ in the table of integrals (see inside back cover).

58. Find the area of one-quarter of the ellipse with equation $x^2/4 + y^2/9 = 1$.

59. Find the area of one-quarter of the ellipse with equation $x^2 + y^2/2 = 1$.

60. Show that the area of the ellipse with equation $x^2/a^2 + y^2/b^2 = 1$ is πab. Discuss the case in which $a = b$.

Exercises 61–71: Evaluate the integral by completing the square and making a change of variable. The resulting integral will have the form of one of the formulas (A)–(H) or, for Exercises 67–70, can be done by a trigonometric substitution (see Exercises 48–60).

61. $\int \dfrac{dx}{2x^2 - 6x + 5}$

62. $\int \dfrac{dx}{9x^2 - 12x + 5}$

63. $\int \dfrac{dx}{\sqrt{24 - 10x - 25x^2}}$

64. $\int \dfrac{dx}{\sqrt{-40 + 28x - 4x^2}}$

65. $\int \dfrac{dx}{x^2 + x + 1}$

66. $\int \dfrac{dx}{x^2 - x + 1}$

67. $\int_{0}^{1} \dfrac{dx}{(2x^2 + 6x + 5)^{3/2}}$

68. $\int_{-1/2}^{1/2} \dfrac{dx}{\sqrt{x^2 + x + 1}}$

You will find the integral $\int \sec u\, du$ in the table of integrals (see inside back cover). Your final answer should be $\ln(\sqrt{7} + 2) - \ln\sqrt{3}$.

69. $\int_{-1}^{1} \dfrac{x+1}{\sqrt{x^2 + 2x + 5}}\,dx$

70. $\int_{2}^{3} \dfrac{\sqrt{3 + 2x - x^2}}{x - 1}\,dx$

Review/Preview _____

Exercises 71–74: Determine the intersection point(s) of the graphs corresponding to the given equations.

71. $y - 3 = 2(x - 5)^2$, $y - 5 = 4(x - 4)$
72. $x + 2 = \frac{1}{2}(y - 3)^2$, $y - 7 = \frac{1}{3}(x - 6)$

73. $x^2 + y^2 = 1$, $x^2 + (y - 1.2)^2 = 0.3^2$
74. $x^2/2^2 + y^2/3^2 = 1$, $x^2 - y^2 = 1$

5.4 AREAS BETWEEN CURVES

In Definition 1 of Section 5.2, we defined the area of the region S beneath the graph of a nonnegative function f to be the value of the integral $\int_a^b f(x)\,dx$. Although we calculated the areas of several regions in the preceding sections, our focus has been on how to evaluate integrals, not on regions and their areas. In this section we discuss how to calculate the area of a region R like that in Fig. 5.25, whose upper and lower boundaries are graphs of functions f and g.

We may calculate the area A of the shaded region R in three closely related ways. First, we may calculate A by subtraction, using what we already know. The area A is the difference between A_f and A_g, where A_f is the area of the region beneath the graph of f, and A_g is the area of the region beneath the graph of g. Thus

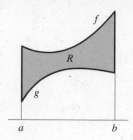

Figure 5.25. Region bounded by graphs of f and g.

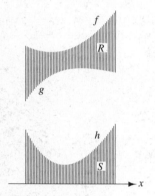

Figure 5.26. Cavalieri's principle.

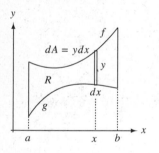

Figure 5.27. An element of area.

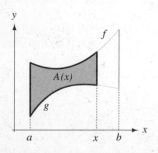

Figure 5.28. Area function $A(x)$.

$$A = A_f - A_g = \int_a^b f(x)\,dx - \int_a^b g(x)\,dx.$$

We may also calculate the area of R by imagining that it is filled with inscribed rectangles, which in the copy of R shown at the top of Fig. 5.26 appear as line segments. We slide the rectangles directly downward and align them on a horizontal line, which we regard as a new x-axis. In their new positions, the rectangles define a region S. Our intuitive understanding of area supports the assertion that the areas of regions R and S are equal. This idea is often called **Cavalieri's principle**, after (Francesco) Bonaventura Cavalieri (1598–1647), an Italian mathematician and contemporary of Galileo.

The tops of the bottom-aligned rectangles lie along the graph of the function h, where $h(x) = f(x) - g(x)$, $x \in [a, b]$. The value $h(x)$ is the length of the inscribed rectangle at position x.

These two ways of calculating A are related by the linearity of the integral, that is, Property 3 of Section 5.2. We have

$$A = \int_a^b f(x)\,dx - \int_a^b g(x)\,dx = 1 \cdot \int_a^b f(x)\,dx + (-1) \cdot \int_a^b g(x)\,dx$$

$$= \int_a^b (1 \cdot f(x) + (-1) \cdot g(x))\,dx = \int_a^b (f(x) - g(x))\,dx$$

$$= \int_a^b h(x)\,dx.$$

The form of the last integral fits the calculation of the area of the region occupied by the bottom-aligned rectangles.

The third way of calculating A depends directly on the geometry of the region R. We sketch a figure like that in Fig. 5.27 and draw an *element of area* dA. The element of area is a thin rectangle of length $y = h(x) = f(x) - g(x)$ and width dx. The coordinate variable x locates the element. Its length y varies with x. We have

$$(1) \qquad dA = y\,dx = (f(x) - g(x))\,dx.$$

If we think about dA as the area of a typical inscribed or circumscribed rectangle and the integral as the limit of approximating sums, then the area A of the region R is

$$(2) \qquad A = \int_a^b dA = \int_a^b y\,dx = \int_a^b (f(x) - g(x))\,dx.$$

The step from (1) to (2) covers a lot of ground. Keeping Fig. 5.27 in mind, it suggests the division of $[a, b]$ into subintervals, the formation of lower and upper sums, and the integral to which these sums converge. We used the geometry of the region R to write (1). We used the integral to sum the elements dA. This process can be justified in another way.

We show in Fig. 5.28 the area function $A(x)$, which is the area of R up to the line located by x. The value of $A(x)$ is given by

$$(3) \qquad A(x) = \int_a^x h(t)\,dt.$$

We note that the area A of R is $A(b)$. By part I of the Fundamental Theorem,

$$(4) \qquad dA = \frac{d}{dx}A(x)\,dx = \left(\frac{d}{dx}\int_a^x h(t)\,dt\right)dx = h(x)\,dx = y\,dx.$$

At the point x where we drew the element of area $dA = y\,dx$, the differential dA of the area function A is $dA = y\,dx$. This supports the step from dA to A, that is, from (1) to (2).

■ **EXAMPLE 1** Find the area between the graphs of $f(x) = x^2 + 1$ and $g(x) = 2x$, $0 \le x \le 1$.

Solution. The graphs of these functions are shown in Fig. 5.29. Note that $f(x) \ge g(x)$ throughout $[0, 1]$. Draw a typical element dA, as shown in the figure. The coordinate x locates the element, y is its length, and dx is its width. The area of the element is $dA = y\,dx$, where $y = f(x) - g(x)$. The integral sums the elements dA as x varies from 0 to 1, giving

$$A = \int_0^1 (f(x) - g(x))\,dx = \int_0^1 (x^2 + 1 - 2x)\,dx$$

$$= \tfrac{1}{3}x^3 + x - x^2 \Big|_0^1$$

$$= \frac{1}{3} + 1 - 1 = \frac{1}{3}. \qquad ■$$

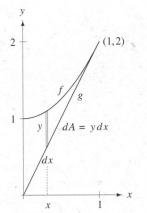

Figure 5.29. Area between curves.

■ **EXAMPLE 2** During a one-minute race the velocities of two cars were

$$v_1(t) = \frac{80}{3t^2 + 1} - 20 \text{ mph}, \quad 0 \le t \le 1$$

$$v_2(t) = 200(t - t^2) \text{ mph}, \quad 0 \le t \le 1.$$

Graphs of v_1 and v_2 are shown in Fig. 5.30. If at the beginning of the race both cars were at the mile 0 mark, who won?

Solution. Before you look at the solution but after you have inspected Fig. 5.30, who, in your judgment, won? If you looked at the regions between the two graphs and judged that the area of the region in which car 2 is going faster is larger than that for which car 1 is going faster, you have a good eye.

Given that the initial position of each car is $x = 0$, their positions $x_1(1)$ and $x_2(1)$ at $t = 1$ are proportional to the areas beneath their respective velocity graphs between $t = 0$ and $t = 1$. We say *proportional* since the velocities were given in miles per hour and the time in minutes. We include a factor of $k = 1/60$ so that the positions are measured in miles. From the difference $x_1(1) - x_2(1)$ we can determine which car won. We have

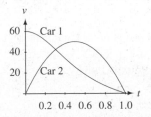

Figure 5.30. Velocity profiles.

$$\frac{x_1(1) - x_2(1)}{k} = \int_0^1 \left(\frac{80}{3t^2 + 1} - 20\right)dt - \int_0^1 200(t - t^2)\,dt.$$

Using formula (A) and anticipating the use of formula (G), we have

$$\frac{x_1(1) - x_2(1)}{k} = \frac{80}{3} \int_0^1 \frac{dt}{t^2 + (1/\sqrt{3})^2}\, dt - 20t \Big|_0^1 - 200 \left(\tfrac{1}{2} t^2 - \tfrac{1}{3} t^3 \right) \Big|_0^1.$$

Evaluating and using formula (G), we have

$$\frac{x_1(1) - x_2(1)}{k} = \frac{80}{3} \cdot \sqrt{3} \arctan\left(\sqrt{3} x \right) \Big|_0^1 - 20 - \frac{200}{6}$$

$$= \frac{80}{3} \cdot \sqrt{3} \cdot \frac{\pi}{3} - \frac{160}{3} \approx -5.0.$$

Hence $x_1(1) - x_2(1) < 0$ or $x_1(1) < x_2(1)$. Car 2 went further in the same time and hence won. ∎

Curves That Cross

The preceding example brings up the question of how the area between two curves can be calculated in case they cross. As we saw,

$$x_1(1) - x_2(1) = \int_0^1 v_1(t)\, dt - \int_0^1 v_2(t)\, dt = \int_0^1 (v_1(t) - v_2(t))\, dt$$

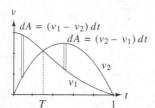

measures the difference in position of the two cars at $t = 1$. To determine the area *A between the two curves,* we must rearrange the calculation. Referring to Fig. 5.31 and letting $(T, v_1(T))$ be the crossing point, the length of the area element in the region to the left of the line with equation $t = T$ is $v_1 - v_2$; to the right of this line the length is $v_2 - v_1$. The area elements are $dA = (v_1 - v_2)\, dt$ and $dA = (v_2 - v_1)\, dt$. The area between the curves is

Figure 5.31. Area between curves.

(5) $$A = \int_0^T (v_1(t) - v_2(t))\, dt + \int_T^1 (v_2(t) - v_1(t))\, dt.$$

Apart from signs, determining the antiderivatives is the same as in Example 2. However, the evaluation step is more complicated. We leave to Exercise 23 the calculation of $T \approx 0.305898$ and $A \approx 22.8393$.

■ **EXAMPLE 3**   You and four friends are sharing a last cookie. Unable to wait for further discussion, one friend took a bite, as shown in Fig. 5.32. Assuming that the radius of the cookie is 4 cm and her bite is closely approximated by the parabola $y = -0.4x^2 + 3.2$, was she fair or not?

Solution.   We might agree that a fair share is between 18 percent and 22 percent of the cookie. The percentage eaten in one bite is $(2A/(\pi 4^2)) \times 100$, where A is the area in the first quadrant bounded above by the bite curve and below by the perimeter of the cookie. We must find A. The element of area $dA = y\, dx$ is shown in Fig. 5.32. Its length y is the difference of the "y of the bite" and the "y of the cookie." Noting that the equation of the circle is $x^2 + (y - 4)^2 = 4^2$, we have

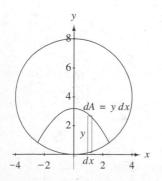

Figure 5.32. Chocolate chip cookie.

$$y_{bite} = -0.4x^2 + 3.2$$

$$y_{cookie} = 4 - \sqrt{16 - x^2}.$$

From these equations we have

(6) $dA = (y_{\text{bite}} - y_{\text{cookie}})\,dx = \left((-0.4x^2 + 3.2) - \left(4 - \sqrt{16 - x^2}\right)\right)dx.$

To find the upper limit of integration, we eliminate y from the equations

(7) $x^2 + (y - 4)^2 = 4^2$ and $y = -0.4x^2 + 3.2$

by substituting from the second equation into the first. After simplification we obtain

$$4x^4 + 41x^2 - 384 = 0.$$

Although this equation is of fourth degree, it is quadratic in x^2. From the quadratic formula we obtain

$$x^2 = \frac{-41 \pm \sqrt{41^2 + 4 \cdot 4 \cdot 384}}{8}.$$

Taking the square root of the positive root, we obtain

(8) $x = \sqrt{\dfrac{-41 + \sqrt{41^2 + 4 \cdot 4 \cdot 384}}{8}} \approx 2.436.$

All that remains is to evaluate A. From (6) and (8) we have

$$A \approx \int_0^{2.436} \left((-0.4x^2 + 3.2) - \left(4 - \sqrt{16 - x^2}\right)\right)dx$$

By Property 3 we may write this in the form

(9) $A \approx \displaystyle\int_0^{2.436} \sqrt{16 - x^2}\,dx - \int_0^{2.436} (0.4x^2 + 0.8)\,dx$

The first of these integrals can be solved by the trigonometric substitution $x = 4\sin\theta$, as in Example 3 in Section 5.3, or found as formula (12) in the table of integrals (see inside back cover). The second integral is straightforward. We obtain $A \approx 5.23$. From this, the percentage bitten off is approximately 21 percent, which, by the criterion that a fair bite lies between 18 percent and 22 percent, is fair. ■

Horizontal Area Elements

We show in Fig. 5.33 a region for which a horizontal area element may be preferred to the vertical elements we have been using.

■ **EXAMPLE 4**   Calculate the area of the region bounded by the graphs of the equations $y^2 = x$ and $y = x - 2$, as shown in Fig. 5.33.

Solution.   If the region is rotated $90°$ counterclockwise, it is bounded on the top by a line and on the bottom by a parabola. Calculating its area can be done as

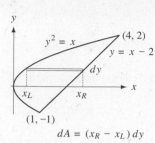
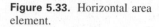
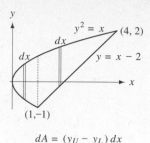

$$dA = (x_R - x_L)\,dy \qquad\qquad dA = (y_U - y_L)\,dx$$

Figure 5.33. Horizontal area element. **Figure 5.34.** Vertical area elements.

in Example 1. Alternatively, we may reflect the region about the 45° line, which amounts to interchanging x and y in the given equations. We leave these two approaches to Exercises 21 and 27 and instead discuss two ways to calculate the area while keeping the given coordinate system and equations.

We show a horizontal area element in Fig. 5.33. The width of the element is dy and its length is $x_R - x_L$, where x_R is "the x of the line" and x_L is "the x of the parabola." Its area is

$$dA = (x_R - x_L)\,dy = \big((y + 2) - y^2\big)\,dy.$$

We use the integral to sum these elements. It is useful to visualize the element dA as sweeping out the area from $y = -1$ to $y = 2$. We have

$$A = \int_{-1}^{2} dA = \int_{-1}^{2} \big((y+2) - y^2\big)\,dy = \tfrac{1}{2}y^2 + 2y - \tfrac{1}{3}y^3 \Big|_{-1}^{2} = \frac{9}{2}.$$

We show vertical area elements in Fig. 5.34. We show two elements since to the left of $x = 1$ the element goes from parabola to parabola while to the right of $x = 1$ the element goes from line to parabola. The formula for dA will be different for $0 \le x \le 1$ and $1 \le x \le 4$. We write

$$dA = (y_U - y_L)\,dx.$$

for both cases, where y_U denotes the upper curve and y_L the lower curve. To the left of the line $x = 1$ the upper curve is the parabola and the lower curve is the parabola. To the right of the line $x = 1$ the upper curve is the parabola and the lower curve is the line $y = x - 2$. This gives

$$\text{(10)} \qquad
\begin{aligned}
dA &= (y_U - y_L)\,dx = \Big(\sqrt{x} - \big(-\sqrt{x}\big)\Big)\,dx, \quad \text{for } 0 \le x \le 1 \\
dA &= (y_U - y_L)\,dx = \Big(\sqrt{x} - (x - 2)\Big)\,dx, \quad \text{for } 1 \le x \le 4.
\end{aligned}$$

The area of the entire parabolic segment is the sum of the areas of the two regions separated by $x = 1$. From (10) we have

$$A = \int_0^1 \Big(\sqrt{x} - \big(-\sqrt{x}\big)\Big)\,dx + \int_1^4 \Big(\sqrt{x} - (x - 2)\Big)\,dx = \frac{9}{2}. \qquad \blacksquare$$

We summarize in Fig. 5.35 the two kinds of area elements we have discussed.

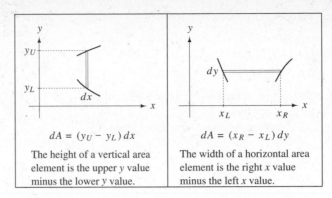

Figure 5.35. Summary of vertical and horizontal area elements.

Exercises 5.4

Basic

Exercises 1–16:   Sketch the bounded region between the curves and calculate its area. Use substitution or integral tables as appropriate.

1. $y = (x - 2)^2$, $y = 2x + 4$
2. $y = x^2$, $y = 2x$
3. $y = \sqrt{x}$, $y = \sqrt[3]{x}$, where $x \geq 0$
4. $y = \sqrt[3]{x}$, $y = \sqrt[4]{x}$, where $x \geq 0$
5. $y = 2x + 1$, $y = e^x$
6. $y = 2x$, $y = \tan x$, where $x \in [0, \pi/2)$
7. $x = y^2$, $x = \frac{1}{2}y^2 + 2$
8. $y = x^2$, $(y - 2)^2 = x - 1$

9. $x^2 + y^2 = 4$, $xy = 1$, where $x > 0$
10. $x^2 + y^2 = 4$, $y = 1/x^2$, where $x > 0$
11. $y = x^5 - x^3$, $y = x - x^3$, where $0 \leq x \leq 1$
12. $y = x - x^5$, $y = x^2 - x^4$
13. $y = 1/(x^2 + 1)$, $y = x^2$
14. $y = 1/(x^2 + 1)$, $y = x^4$
15. $x^2 + y^2 = 1$, $(x - 1.1)^2 + y^2 = 0.5^2$
16. $x^2 - y^2 = 1$, $x^2 + y^2 = 2^2$, where $x \geq 1$

17. The area of a region is $\int_0^{2\pi} |\sin x - \cos x|\, dx$. Sketch such a region and calculate its area.
18. The area of a region is $\int_0^{2\pi} |e^{-x} - \sin x|\, dx$. Sketch such a region and calculate its area.
19. Find the upper limit of integration in Example 3 by eliminating x from the equations (7). This avoids a fourth-degree equation, but x must be calculated after solving a quadratic.
20. Determine the value of the first integral in (9). Given that the value of the second integral is approximately 3.876, check your answer by recalling that $A \approx 5.23$.

21. In Example 4 we mentioned that the problem could be reduced to an earlier kind by reflection in the 45° line and that this could be done by interchanging x and y in the defining equations. Explain, using the equations of the example. Calculate the area in this way.
22. Referring to Example 2, does car 2 catch up to car 1 at the time their velocities are equal? If not, determine this time.
23. Referring to Fig. 5.31, determine T and the area A between these curves.

Growth

24. Referring to Example 2, argue that
$$A = \int_0^1 |v_1(t) - v_2(t)|\, dt.$$
Note that in this formulation the value of T is apparently not needed. However, argue further that unless this integral is somehow evaluated numerically, T is in fact needed.

25. Give a physical interpretation to $v_1 - v_2$ in Example 2.
26. At what time have the two cars in Example 2 covered an equal distance?
27. Work out the details of finding the area described in Example 4 using a 90° rotation. (Hint: It is not necessary to recall the rotation equations
$$x = x' \cos\theta - y' \sin\theta, \quad y = x' \sin\theta + y' \cos\theta.)$$

28. During a one-minute race the velocities of two cars are

$$v_1(t) = 60(t^2 - t^3) + 25, \quad 0 \le t \le 1$$
$$v_2(t) = 200(t - t^2), \quad 0 \le t \le 1.$$

If at the beginning of the race both cars were at the mile 0 mark, who won?

29. During a one-minute race the velocities of two cars are

$$v_1(t) = 75(1 - \cos(\pi t)), \quad 0 \le t \le 1$$
$$v_2(t) = 75t, \quad 0 \le t \le 1.$$

If at the beginning of the race both cars were at the mile 0 mark, who won?

30. Referring to Example 3, if the friend had shoved the cookie into her mouth another 0.8 cm, what percentage would she have gotten?

31. In Example 4, Section 5.2, we asserted that

$$2k \int_0^x \sqrt{2w - w^2}\, dw$$
$$= \frac{k}{2} \left(\pi + 2(x - 1)\sqrt{2x - x^2} + 2\arcsin(x - 1) \right)$$

Write $2w - w^2 = 1 - (w - 1)^2$ and use the substitution $w - 1 = \sin u$ in verifying this result.

Review/Preview

Exercises 32–36:   Show that the function F is an antiderivative of f.

32. $F(x) = (x - 1)e^x;\ f(x) = xe^x$
33. $F(x) = \sin x - x\cos x;\ f(x) = x\sin x$
34. $F(x) = \cos x + x\sin x;\ f(x) = x\cos x$
35. $F(x) = -x + x\ln x;\ f(x) = \ln x$
36. $F(x) = \sqrt{1 - x^2} + x\arcsin x;\ f(x) = \arcsin x$

5.5   INTEGRATION BY PARTS

The second integration technique we discuss is integration by parts, which is based on the product formula

$$(uv)' = uv' + vu',$$

where $u = u(x)$ and $v = v(x)$ are differentiable functions on an interval $[a, b]$. From the viewpoint of the Fundamental Theorem of Calculus, the product formula states that uv is an antiderivative of $uv' + vu'$. Hence

$$\int_a^b (uv' + vu')\, dx = uv \Big|_a^b.$$

Using the linearity of the integral, this result may be rearranged as

$$\int_a^b uv'\, dx = uv \Big|_a^b - \int_a^b vu'\, dx,$$

or, in terms of the differentials $du = u'\, dx$ and $dv = v'\, dx$,

$$\int_a^b u\, dv = uv \Big|_a^b - \int_a^b v\, du.$$

We give Newton's geometric demonstration of this formula at the end of the section. For indefinite integrals the parts formula is

$$\int u\, dv = uv - \int v\, du.$$

To "integrate by parts" means to use one of these formulas—highlighted below for easy reference—to evaluate an integral or to change its form.

Integration by Parts

For differentiable functions $u = u(x)$ and $v = v(x)$ on an interval $[a, b]$,

(1) $$\int_a^b u\, dv = uv \Big|_a^b - \int_a^b v\, du \quad \text{(for definite integrals)}$$

(2) $$\int u\, dv = uv - \int v\, du \quad \text{(for indefinite integrals).}$$

We give a typical application of (1) in Example 1. Suppose that we wish to evaluate an integral $\int_a^b f(x)\, dx$. If we are able to choose "parts" u and dv so that $f(x)\, dx = u\, dv$, then we may use (1) to write

$$\int_a^b f(x)\, dx = uv \Big|_a^b - \int_a^b v\, du.$$

This improves our chances for solving the problem since it provides an alternative. We may hope, for example, that the integral $\int_a^b v\, du$ is easier than the one we began with, $\int_a^b u\, dv = \int_a^b f(x)\, dx$.

■ **EXAMPLE 1** Evaluate the integral $\int_0^1 xe^x\, dx$.

Solution. The key step in using integration by parts is to set the given integrand—here $xe^x\, dx$—equal to $u\, dv$ in (1) and then decide which part should be u and which part should be dv. It is sometimes useful to set u equal to any polynomial factor of the integrand. If, accordingly, we set $u = x$, then dv must be everything else. So, $dv = e^x\, dx$. To use (1) we must work out the two expressions on the right. We need the term uv, for which we know that $u = x$. To obtain v, we integrate dv, that is,

$$v = \int dv = \int e^x\, dx = e^x.$$

(If we are working with an indefinite integral we add a constant C later. For now it is enough to determine any v that works. For definite integrals the constant may be ignored both now and later.) From (1) we have

$$\int_0^1 xe^x\, dx = uv \Big|_0^1 - \int_0^1 v\, du$$

$$= xe^x \Big|_0^1 - \int_0^1 e^x\, dx = e - e^x \Big|_0^1$$

$$= e - (e - 1) = 1.$$

The calculation is similar for the indefinite integral $\int xe^x dx$. We use the same choice of parts and find

$$\int xe^x \, dx = uv - \int v \, du$$

$$= xe^x - \int e^x \, dx = xe^x - e^x + C$$

$$= (x - 1)e^x + C.$$

The last expression is the most general antiderivative of xe^x. ∎

Setting u equal to a polynomial factor is useful in that when we calculate $\int v \, du$, the factor $du = u' \, dx$ is a polynomial of degree 1 less than the degree of u. Often this simplifies the remaining calculations. In any case, in choosing dv we must be able to evaluate $\int dv$. Otherwise we may be stuck and it's not clear how we may use equation (1).

In the next example we show that it is not always useful to set u equal to a polynomial factor.

■ **EXAMPLE 2**   Evaluate the integral $\int x \ln x \, dx$.

Solution.   If we were to set $u = x$ and $dv = \ln x \, dx$, then $du = dx$ and $v = \int \ln x \, dx$. Although we could find an antiderivative of $\ln x$, this problem is no easier than the one we began with. So we try again. This time we set $u = \ln x$ and $dv = x \, dx$. We start by differentiating to get du and integrating to get v. We have

$$du = u' \, dx = (\ln x)' \, dx = \frac{1}{x} \, dx \quad \text{and} \quad v = \int x \, dx = \tfrac{1}{2}x^2.$$

Substituting these results into (2),

$$\int x \ln x \, dx = uv - \int v \, du$$

$$= (\ln x)\left(\tfrac{1}{2}x^2\right) - \int \tfrac{1}{2}x^2 \cdot \frac{1}{x} \, dx$$

$$= \tfrac{1}{2}x^2 \ln x - \tfrac{1}{4}x^2 + C$$

$$= \frac{x^2}{4}(2 \ln x - 1) + C.$$ ∎

The essential feature of this parts assignment is that du, the derivative of the natural logarithm function, is an algebraic function, namely, $1/x$. The product of this algebraic function and dv gives an expression whose antiderivative is known. Since the derivatives of the functions arcsin and arctan are also algebraic, it would be worth attempting integrals of the form

$$\int x^n \arcsin x \, dx \quad \text{or} \quad \int x^n \arctan x \, dx$$

with the same choice of parts, that is, let $dv = \arcsin x$ or $dv = \arctan x$ and set u equal to everything else.

Reduction Formulas

Integration by parts is used to find reduction formulas for integrals. An example of a reduction formula is

(3)
$$\int_0^\pi \sin^n x \, dx = \frac{n-1}{n} \int_0^\pi \sin^{n-2} x \, dx, \quad n \geq 2.$$

Reduction formulas are used to evaluate integrals by applying them as many times as necessary to "reduce" the integral to one with known antiderivative. Suppose, for example, we wish to evaluate the integral

$$\int_0^\pi \sin^5 x \, dx.$$

Setting $n = 5$ in (3) we have

$$\int_0^\pi \sin^5 x \, dx = \frac{4}{5} \cdot \int_0^\pi \sin^3 x \, dx.$$

To continue the reduction to lower powers of the sine function we apply (3) again, this time with $n = 3$. This gives

$$\int_0^\pi \sin^5 x \, dx = \frac{4}{5} \cdot \frac{2}{3} \cdot \int_0^\pi \sin x \, dx$$

$$\int_0^\pi \sin^5 x \, dx = \frac{8}{15} \cdot (-\cos x) \Big|_0^\pi = \frac{8}{15} \cdot 2 = \frac{16}{15}.$$

Formulas (25)–(39) in the table of integrals are reduction formulas. Each may be derived by integration by parts.

■ **EXAMPLE 3** Use integration by parts to derive (3) (which comes from formula (28) in the table of integrals).

Solution. We choose the parts $u = \sin^{n-1} x$ and $dv = \sin x \, dx$. This choice, whatever the value of $n \geq 2$, has the advantage that $\int dv$ can be evaluated. We have

$$du = (n-1)\sin^{n-2} x \cos x \quad \text{and} \quad v = -\cos x.$$

Substituting into (1), we have

$$\int_0^\pi \sin^n x \, dx = -\sin^{n-1} x \cos x \Big|_0^\pi + (n-1) \int_0^\pi \sin^{n-2} \cos^2 x \, dx.$$

Evaluating the first term on the right and replacing $\cos^2 x$ by $1 - \sin^2 x$ in the second term, we have

$$\int_0^{\pi} \sin^n x \, dx = 0 + (n-1) \int_0^{\pi} (\sin^{n-2} x - \sin^n x) \, dx$$

$$\int_0^{\pi} \sin^n x \, dx = (n-1) \int_0^{\pi} \sin^{n-2} x \, dx - (n-1) \int_0^{\pi} \sin^n x \, dx$$

Solving this equation for $\int_0^{\pi} \sin^n x \, dx$ gives (3). ∎

Recursion Formulas

We can simplify our earlier calculation of $\int_0^{\pi} \sin^5 x \, dx$ by changing notation. If we write

$$F_n = \int_0^{\pi} \sin^n x \, dx, \quad n \geq 2,$$

then the reduction formula (3) can be written as

(4) $$F_n = \frac{n-1}{n} F_{n-2}, \quad n \geq 2.$$

If we set

$$F_1 = \int_0^{\pi} \sin x \, dx = 2,$$

then

$$F_5 = \frac{4}{5} \cdot F_3 = \frac{4}{5} \cdot \frac{2}{3} \cdot F_1 = \frac{4}{5} \cdot \frac{2}{3} \cdot 2 = \frac{16}{15}.$$

If we want to calculate F_4, for example, we need to define F_0. From two applications of (4), we have

$$F_4 = \frac{3}{4} \cdot F_2 = \frac{3}{4} \cdot \frac{1}{2} \cdot F_0.$$

We set $F_0 = \pi$ since

$$\int_0^{\pi} \sin^0 x \, dx = \int_0^{\pi} 1 \, dx = x \Big|_0^{\pi} = \pi.$$

With this definition, we have

$$F_4 = \frac{3}{4} \cdot \frac{1}{2} \cdot \pi = \frac{3\pi}{8}.$$

The *recursion formula*

(5)
$$F_0 = \pi$$
$$F_1 = 2$$
$$F_n = \frac{n-1}{n} F_{n-2}, \quad n \geq 2,$$

is a convenient form of the reduction formula (3). Indeed, the three equations listed in (5) can be entered more or less as given into a calculator or CAS. This makes it possible to evaluate integrals like $\int_0^\pi \sin^n x \, dx$ easily. On our calculator it took less than 1 second to calculate

$$\int_0^\pi \sin^{20} x \, dx = F_{20} = \frac{19}{20} \cdot \frac{17}{18} \cdot \cdots \cdot \frac{5}{6} \cdot \frac{3}{4} \cdot \frac{1}{2} \cdot \pi \approx 0.55354.$$

The recursion formula (5) resembles in a general way the formula for Fibonacci numbers, which you may have studied earlier. Are you thinking about the rabbit connection? The first several Fibonacci numbers are

$$1, 1, 2, 3, 5, 8, 13, 21, 34, \ldots.$$

Starting with the third Fibonacci number, each number is the sum of the two numbers preceding it. The recursive definition of the Fibonacci numbers is

(6)
$$F_0 = 1$$
$$F_1 = 1$$
$$F_n = F_{n-1} + F_{n-2}, \quad n \geq 2.$$

More Integration by Parts

In the derivation of the reduction formula (3), integration by parts led to an equation in which it was possible to solve for the unknown integral. This idea is used in the next example.

■ **EXAMPLE 4**   Derive the integration formula

(7)
$$\int e^{ax} \sin bx \, dx = \frac{e^{ax}}{a^2 + b^2}(a \sin bx - b \cos bx),$$

where a and b are nonzero constants.

Solution.   We start by choosing $u = e^{ax}$ and $dv = \sin bx \, dx$ for the parts formula $\int u \, dv = uv - \int v \, du$. Calculating du and v, we obtain

$$du = ae^{ax} \, dx \quad \text{and} \quad v = -\frac{1}{b} \cos bx.$$

Substituting into the parts formula, we have

$$\int e^{ax} \sin bx \, dx = -\frac{1}{b} e^{ax} \cos bx + \frac{a}{b} \int e^{ax} \cos bx \, dx.$$

The integral $\int e^{ax} \cos bx \, dx$ has the same form as the integral $\int e^{ax} \sin bx \, dx$ in (7). If we apply the parts formula again, perhaps we will obtain an equation that can be solved for the original integral. Applying the parts formula to $\int e^{ax} \cos bx \, dx$, we choose $u = e^{ax}$ and $dv = \cos bx$, obtaining

$$du = ae^{ax}dx \quad \text{and} \quad v = \frac{1}{b} \sin bx.$$

This gives (we repeat one line of the calculation started above)

$$\int e^{ax} \sin bx \, dx = -\frac{1}{b}e^{ax} \cos bx + \frac{a}{b} \int e^{ax} \cos bx \, dx$$

$$\int e^{ax} \sin bx \, dx = -\frac{1}{b}e^{ax} \cos bx + \frac{a}{b} \left(\frac{1}{b}e^{ax} \sin bx - \frac{a}{b} \int e^{ax} \sin bx \, dx \right)$$

$$\int e^{ax} \sin bx \, dx = -\frac{1}{b}e^{ax} \cos bx + \frac{a}{b^2}e^{ax} \sin bx - \frac{a^2}{b^2} \int e^{ax} \sin bx \, dx.$$

We solve the last equation for the "unknown" $\int e^{ax} \sin bx \, dx$. We have

$$\left(1 + \frac{a^2}{b^2} \right) \int e^{ax} \sin bx \, dx = \frac{e^{ax}}{b^2}(a \sin bx - b \cos bx).$$

If we divide both sides of this equation by $(1 + a^2/b^2)$ and simplify, we obtain

$$\int e^{ax} \sin bx \, dx = \frac{e^{ax}}{a^2 + b^2}(a \sin bx - b \cos bx) + C. \qquad \blacksquare$$

Newton's Geometric Argument

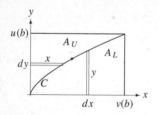

Figure 5.36. Newton's argument.

Newton gave a simple geometric argument for the integration-by-parts formula. His diagram resembles that in Fig. 5.36. We suppose that a curve C lies in a rectangular region, starting and ending at opposite corners of the rectangle, and is given parametrically by $\mathbf{r} = (v(t), u(t))$, $a \le t \le b$. For simplicity we consider the special case in which the curve starts at $(v(a), u(a)) = (0, 0)$. The area of the rectangle is $u(b)v(b)$. The area of the rectangle is also equal to the sum of the areas A_L and A_U of the regions beneath and above the curve C. Since

(8) $$dA_L = y \, dx = u(t)v'(t) \, dt$$

and

(9) $$dA_U = x \, dy = v(t)u'(t) \, dt,$$

we have

(10) $$u(b)v(b) = A_L + A_U = \int_a^b u(t)v'(t) \, dt + \int_a^b v(t)u'(t) \, dt.$$

This result may be rearranged to give a special case of the integration-by-parts formula (1):

$$(11) \qquad \int_a^b u \, dv = u(b)v(b) - \int_a^b v \, du.$$

Although this argument is based on a straightforward geometric idea, some calculations were made with little in the way of justification. We outline a more careful argument in Exercise 34. Removing the restriction that $(v(a), u(a)) = (0, 0)$ is left as Exercise 35.

Exercises 5.5

Basic

Exercises 1–14:   Evaluate by either of the integration-by-parts formulas, (1) or (2).

1. $\int_0^1 xe^{2x} \, dx$

2. $\int_1^2 xe^{-x} \, dx$

3. $\int \ln x \, dx$

4. $\int x^2 \ln 2x \, dx$

5. $\int_0^1 x2^x \, dx$

6. $\int_{-1}^2 x10^x \, dx$

7. $\int \sqrt{x} \ln x \, dx$

8. $\int x^{\frac{1}{3}} \ln x \, dx$

9. $\int x \sin x \, dx$

10. $\int x \cos x \, dx$

11. $\int \arcsin x \, dx$

After parts, try the substitution $x = u^2$ in the second integral.

12. $\int \arctan x \, dx$

After parts, try the substitution $x = u^2$ in the second integral.

13. $\int e^{ax} \cos bx \, dx$

14. $\int e^{\sqrt{x}} \, dx$

After parts, try the substitution $u = \sqrt{x}$ and then use formula (27) in the table of integrals.

Growth

15. Derive the reduction formula

$$\int x^n e^x \, dx = x^n e^x - n \int x^{n-1} e^x \, dx$$

This is formula (27) in the table of integrals, with $a = 1$.

16. Use the reduction formula in Exercise 15 to evaluate $\int x^3 e^x \, dx$.

17. Evaluate $\int x \arctan x \, dx$. In the second integral note that $x^2/(1 + x^2) = 1 - 1/(1 + x^2)$.

18. Evaluate $\int x \arcsin x \, dx$. Make the substitution $x = \sin u$ in the second integral or use (15) in the table of integrals.

19. Evaluate $\int 2^x \sin x \, dx$.

20. Derive the reduction formula

$$\int (\ln x)^n \, dx = x(\ln x)^n - n \int (\ln x)^{n-1} \, dx.$$

21. Use the reduction formula in Exercise 20 to evaluate $\int_1^e (\ln x)^4 \, dx$.

22. Find an antiderivative of $\sin(\ln x)$.

23. Show that in Example 4 we could have taken the parts as $u = \cos bx$ and $dv = e^{ax}\,dx$ provided that in each of the two integrations by parts we continue this pattern, with the exponential as the dv part, but if we reverse the pattern in the two integrations by parts we get an identity.

24. Show that the recursion formula given in (5) may be given in the nonrecursive form

$$F_n = \frac{n-1}{n}\frac{n-3}{n-2}\cdots\frac{1}{2}\pi, \quad \text{if } n \text{ is odd}$$

$$F_n = \frac{n-1}{n}\frac{n-3}{n-2}\cdots\frac{1}{2}2, \quad \text{if } n \text{ is even.}$$

Use these formulas for calculating $\int_0^\pi \sin^8 x\,dx$ and $\int_0^\pi \sin^9 x\,dx$.

25. Show that for $n \geq 2$

$$\int_0^{\pi/2}\cos^n x\,dx = \frac{n-1}{n}\int_0^{\pi/2}\cos^{n-2}x\,dx.$$

Show that this formula may be written as the recursion formula

$$F_0 = \pi/2,\ F_1 = 1,\ F_n = ((n-1)/n)F_{n-2}.$$

Use the recursion formula to show that $F_{15} = 2048/6435$.

26. Use integration by parts in deriving the reduction formula

$$\int (k^2 - x^2)^n\,dx = \frac{x(k^2 - x^2)^n}{2n+1}$$

$$+ \frac{2k^2 n}{2n+1}\int (k^2 - x^2)^{n-1}\,dx.$$

27. Does the reduction formula in Exercise 26 hold for $n = 1$? Why?

28. Use the reduction formula in Exercise 26 to evaluate $\int_0^1(1 - x^2)^5\,dx$.

29. Derive the reduction formula

$$\int \sec^n x\,dx = \frac{\sec^{n-2}x\tan x}{n-1}$$

$$+ \frac{n-2}{n-1}\int \sec^{n-2}x\,dx, \quad n \geq 2.$$

To use this formula for odd n, see formula (4) in the table of integrals.

30. Use the reduction formula in Exercise 29 to evaluate the integral $\int \sec^3 x\,dx$.

31. Use the reduction formula in Exercise 29 to evaluate the integral $\int_0^{\pi/4}\sec^5 x\,dx$.

32. Derive the reduction formula

$$\int \frac{dx}{(x^2 + k^2)^n} = \frac{x}{2k^2(n-1)(x^2+k^2)^{n-1}}$$

$$+ \frac{2n-3}{2k^2(n-1)}\int \frac{dx}{(x^2+k^2)^{n-1}}.$$

(Hint: Let $dv = dx$.)

33. Use the reduction formula in Exercise 32 to evaluate

$$\int_0^1 \frac{dx}{(x^2 + 2)^5}.$$

34. Fill in the following outline of an argument that the area A_L of the region beneath the curve C described parametrically by

$$\mathbf{r} = \big(v(t), u(t)\big),\, t \in [a, b]$$

is given by the integral

$$(12) \qquad A_L = \int_a^b u(t)v'(t)\,dt.$$

Assume that u is nonnegative and v' is continuous and positive on $[a, b]$. This guarantees that the equation $x = v(t)$ can be solved for t in terms of x to give $t = v^{-1}(x)$, where the inverse function v^{-1} is differentiable and continuous on $[a, b]$. Defining $y = y(x) = u(v^{-1}(x))$ for $x \in [0, v(b)]$, the area A_L is

$$A_L = \int_0^{v(b)} y(x)\,dx.$$

Now use the substitution $x = v(t)$. The critical step is

$$y(x) = y(v(t)) = u(v^{-1}(v(t))) = u(t).$$

This proves (12). An argument that $A_U = \int_a^b v(t)u'(t)\,dt$ would be similar.

35. Remove the restriction that $(v(a), u(a)) = (0, 0)$ in Newton's geometric argument for the parts formula. Sketch a figure and reduce to the special case.

Review/Preview

36. Use any method to solve the system of equations

$$2A + B + C = 3$$
$$2A + B - C = -1$$
$$A + B + 2C = 2$$

37. Provided that $a \neq 0$, graphs of equations of the form $y = ax^2 + bx + c$ are parabolas with vertical axes. Find the parabola containing the points $(-1, -2)$, $(2, 7)$, and $(3, 18)$.

38. The fraction $f_1 = 1/((x - 3)(x - 5))$ may be formed by adding the fractions $f_2 = A/(x - 3)$ and $f_3 = B/(x - 5)$, for certain values of A and B. To calculate A and B, form the equation $f_1 = f_2 + f_3$ and "clear fractions" by multiplying both sides of the equation by $(x - 3)(x - 5)$. Since we want this equation to hold for all values of x, we may equate the constant terms from each side of the equation to get one equation in the unknowns A and B and equate the coefficients of x to get a sec-

ond equation. Solve this system of two equations to find the required values of A and B. Equating coefficients in this way is based on the theorem that if two polynomials in x are equal for all values of x, then they must be identical polynomials.

39. Use the procedure discussed in the preceding problem to determine A and B so that

$$\frac{5x - 1}{(x - 2)(x + 1)} = \frac{A}{x + 1} + \frac{B}{x - 2}.$$

5.6 INTEGRATION BY PARTIAL FRACTIONS

Unlike integration by substitution or by parts, each of which offers rich possibilities but no certainty, integration by partial fractions is guaranteed. In fact, all integrals of the form

$$\int \frac{p(x)}{q(x)} \, dx,$$

where p and q are polynomials, can be evaluated by partial fractions.

Ratios of the form $R(x) = p(x)/q(x)$ are called **rational functions.** The technique of integrating a rational function R by partial fractions is based on an algebraic procedure that decomposes R into a sum of integrals having known antiderivatives.

Investigation 1

Suppose we wish to evaluate the integral

(1)
$$\int \frac{x^2 + x - 1}{x^2 - x} \, dx.$$

If, somehow, we knew that

(2)
$$\frac{x^2 + x - 1}{x^2 - x} = 1 + \frac{1}{x} + \frac{1}{x - 1},$$

then (1) may be replaced by the sum of three simpler integrals.

$$\int \frac{x^2 + x - 1}{x^2 - x} \, dx = \int 1 \, dx + \int \frac{1}{x} \, dx + \int \frac{1}{x - 1} \, dx$$

Substituting $u = x - 1$ in the third integral,

$$\int \frac{x^2 + x - 1}{x^2 - x} \, dx = x \ln |x| + \int \frac{du}{u}$$

$$= x + \ln |x| + \ln |u| + C$$

$$= x + \ln |x| + \ln |x - 1| + C = x + \ln |x(x - 1)| + C.$$

What calculations lie behind the decomposition (2)? There are three main steps, all algebraic:

---

1. *Divide* the numerator of the fraction by its denominator to obtain quotient and remainder terms.
2. *Factor* the denominator of the remainder term.
3. *Decompose* the remainder term.

From the long *division* shown in Fig. 5.37, which gives a quotient of 1 and a remainder of $2x - 1$, we have

$$\frac{x^2 + x - 1}{x^2 - x} = 1 + \frac{2x - 1}{x^2 - x}.$$

Figure 5.37 Long division.

Next we *factor* the denominator of the remainder term. Factoring the denominator and determining its zeros are equivalent. The factors of $x^2 - x$ are x and $x - 1$; the zeros 0 and 1 correspond to these factors. We have

(3)
$$\frac{x^2 + x - 1}{x^2 - x} = 1 + \frac{2x - 1}{x(x - 1)}.$$

We put the quotient term aside while we *decompose* the remainder term. The decomposition must take the form

(4)
$$\frac{2x - 1}{x(x - 1)} = \frac{A}{x} + \frac{B}{x - 1}$$

where A and B are unknown constants. This is mostly a matter of common sense; we give a detailed description of the "standard decomposition" later. Next, "clear fractions" by multiplying both sides of (4) by the factored denominator. This gives

(5)
$$2x - 1 = A(x - 1) + Bx.$$

We determine the unknowns A and B from (5) by setting x equal to as many different values as the number of unknowns. With each such value we obtain one equation. The best values to try are the zeros of the factored denominator. The zeros of $x(x - 1)$ are 0 and 1. Substituting these into (5) gives

$$x = 0: \quad 2 \cdot 0 - 1 = A(0 - 1) + B \cdot 0$$

$$x = 1: \quad 2 \cdot 1 - 1 = A(1 - 1) + B \cdot 1.$$

This gives $A = 1$ and $B = 1$ without further calculation. Substituting these values into (4) gives

$$\frac{2x - 1}{x(x - 1)} = \frac{1}{x} + \frac{1}{x - 1}.$$

From this result and (3) we have

$$\frac{x^2 + x - 1}{x^2 - x} = 1 + \frac{1}{x} + \frac{1}{x - 1}.$$

This is (2). We are now ready for the integration following (2).

We work through a second example before giving a more general description of a partial fractions procedure.

Investigation 2

Calculate the value of the integral

(6)
$$\int_1^3 \frac{x + 2}{2x^3 - x^2 + 2x - 1} \, dx.$$

Since the degree of the numerator of the rational function

(7)
$$\frac{x + 2}{2x^3 - x^2 + 2x - 1}$$

is smaller than that of the denominator, we may skip the division step. The factoring and decomposition steps require more effort than in the first example, although this depends on your algebraic skills or your access to a calculator or CAS. The denominator polynomial factors as

(8)
$$2x^3 - x^2 + 2x - 1 = (2x - 1)(x^2 + 1).$$

The factor $2x - 1$ corresponds to the zero $1/2$, and the factor $x^2 + 1$ corresponds to the complex zeros i and $-i$. All polynomials with real coefficients can be factored in this way, as a product of *real* factors. The complex zeros are wrapped up in *irreducible quadratic* factors like $x^2 + 1$. A quadratic factor is said to be irreducible if it has no real zeros.

The decomposition for irreducible quadratic factors such as $x^2 + 1$ is

(9)
$$\frac{x + 2}{(2x - 1)(x^2 + 1)} = \frac{A}{2x - 1} + \frac{Bx + C}{x^2 + 1}.$$

Clearing fractions by multiplying both sides of (9) by $(2x - 1)(x^2 + 1)$ gives

(10)
$$x + 2 = A(x^2 + 1) + (Bx + C)(2x - 1).$$

We generate three equations in the unknowns A, B, and C by replacing x in (10) by three different numbers. We use the real zero $1/2$ for one equation and $x = 0$ and $x = 1$ for the other two equations (each of the complex zeros i and $-i$ leads to two equations—see Exercise 39). These substitutions give

$$x = 1/2 : \quad 5/2 = A(5/4)$$
$$x = 0 : \quad\quad 2 = A(1) + (C)(-1)$$
$$x = 1 : \quad\quad 3 = A(2) + (B + C)(1)$$

From the first equation, $A = 2$; substituting this into the second equation gives $C = 0$; substituting $A = 2$ and $C = 0$ into the third equation gives $B = -1$. Hence

$$\frac{x + 2}{(2x - 1)(x^2 + 1)} = \frac{2}{2x - 1} - \frac{x}{x^2 + 1}.$$

For the integration we make the substitution $u = 2x - 1$ in the first integral and $u = x^2 + 1$ in the second integral. This gives

$$\int_1^3 \frac{x + 2}{2x^3 - x^2 + 2x - 1} \, dx = \int_1^3 \frac{2 \, dx}{2x - 1} - \int_1^3 \frac{x \, dx}{x^2 + 1}$$

$$= \int_1^5 \frac{du}{u} - \frac{1}{2} \int_2^{10} \frac{du}{u}$$

$$= \ln 5 - \frac{1}{2}(\ln 10 - \ln 2)$$

$$= \frac{1}{2} \ln 5.$$

In Investigations 1 and 2 we gave the main steps—divide, factor, and decompose—in integrating a rational function. We give a more general description of the procedure and then give examples to illustrate each step.

The rational function $R(x)$ $p(x)/q(x)$ can be written as

$$R(x) = \frac{p(x)}{q(x)} \quad \text{or} \quad R(x) = Q(x) + \frac{r(x)}{q(x)}.$$

The first form is used when the degree of $p(x)$ is less than the degree of $q(x)$. The second form arises when the degree of $p(x)$ is greater than or equal to the degree of $q(x)$ and we have divided $p(x)$ by $q(x)$, giving a quotient polynomial $Q(x)$ and a remainder term in which the degree of $r(x)$ is less than that of $q(x)$. Since the quotient polynomial is easily integrated, we concentrate on $p(x)/q(x)$ or $r(x)/q(x)$. The following discussion applies to either of these fractions.

Factor $q(x)$ into a product of linear or quadratic factors, so that

(11)
$$\frac{r(x)}{q(x)} = \frac{r(x)}{f_1(x)f_2(x) \cdots f_n(x)}.$$

Each of the factors $f_1(x), \ldots, f_n(x)$ must have the form $g(x)^m$, where $g(x)$ is either a linear expression $rx + s$ or a quadratic expression $ax^2 + bx + c$ and m is a positive integer. The quadratics must be irreducible, that is, $b^2 - 4ac < 0$, so that their zeros are complex numbers.

From algebra it is known that fractions of the form (11) can be decomposed into a sum of terms. The terms corresponding to the factor $f_1(x) = g(x)^m$ are

(12)
$$\frac{w_1(x)}{g(x)^1} + \frac{w_2(x)}{g(x)^2} + \cdots + \frac{w_m(x)}{g(x)^m}.$$

If g is a linear factor, the polynomials $w_1, \ldots, w_m$ are constants $a_1, \ldots, a_m$; if g is a quadratic factor, $w_1, \ldots, w_m$ are polynomials of degree 1, that is, $w_1(x) = b_1 x + c_1, \ldots, w_m = b_m x + c_m$. The constants $a_1, \ldots, a_m$ or $b_1, \ldots, b_m, c_1, \ldots, c_m$ are "unknowns" to be determined.

The remaining factors $f_2(x), \ldots, f_n(x)$ of the denominator of (11) lead to similar groups of terms. We refer to (12) or the sum of such expressions as the **standard decomposition.** The terms in (12) are called **partial fractions.**

A Partial Fractions Procedure

To evaluate $\int R(x)\, dx$, where $R(x) = p(x)/q(x)$ is a rational function and $p(x)$ and $q(x)$ have no common factors, use the following steps:

1. Write $R(x)$ as

$$R(x) = Q(x) + \frac{r(x)}{q(x)},$$

   where $Q(x)$ and $r(x)$ are the quotient and remainder of the division of $p(x)$ by $q(x)$. If the degree of $p(x)$ is less than that of $q(x)$, then $Q(x)$ is the zero polynomial and $p(x) = r(x)$.

2. Factor $q(x)$ into linear and irreducible quadratic factors

$$q(x) = f_1(x) \cdots f_n(x).$$

3. Guided by the discussion accompanying (12), write out the standard decomposition for each of the factors $f_1(x), \ldots, f_n(x)$. Denote each unknown coefficient by a distinct letter. Add all of the decompositions together and equate the result to $r(x)/q(x)$.

4. Clear fractions in the equation formed in step 3 and evaluate the resulting equation at conveniently chosen values of x. If the decomposition has k unknown coefficients, then k values of x will be required.

5. Solve the resulting system of k equations in k unknowns.

6. Integrate any quotient term and all partial fractions.

If you have the impression that this procedure is labor-intensive, you're right. We note, however, that the real work in integrating a rational function $R(x)$ is in factoring and in solving systems of equations. The actual integration generally takes less time. In the remainder of the section we give examples of the standard decompositions of several integrands and the subsequent integration. We say little about factoring or the solution of systems of equations, assuming that you have sufficient algebraic skills, have access to a calculator or CAS, or have put together an optimal combination of these two resources.

From the form of the standard decomposition, it can be seen that only six formulas are needed for the integration. We collect these in Table 1 for easy reference. We label the six formulas to match the corresponding formulas in the table of integrals, adding a prime (') or two ('') if the formula is rearranged or slightly generalized.

Table 1. Six integrals for partial fractions

(**B'**) $\int \dfrac{dx}{x+a} = \ln|x+a| + C$ (**A'**) $\int \dfrac{dx}{(x+a)^n} dx = \dfrac{(x+a)^{-n+1}}{-n+1} + C$

(**G**) $\int \dfrac{dx}{x^2+a^2} = \dfrac{1}{a}\tan^{-1}\dfrac{x}{a} + C$ (**B''**) $\int \dfrac{x\,dx}{x^2+a^2} = \tfrac{1}{2}\ln(x^2+a^2) + C$

(**37**) $\int \dfrac{dx}{(x^2+a^2)^n} = \dfrac{x}{2a^2(n-1)(x^2+a^2)^{n-1}} + \dfrac{2n-3}{2a^2(n-1)}\int \dfrac{dx}{(x^2+a^2)^{n-1}}$

(**A''**) $\int \dfrac{x\,dx}{(x^2+a^2)^n} = \dfrac{(x^2+a^2)^{-n+1}}{2(-n+1)} + C$

■ **EXAMPLE 1**   Evaluate

$$\int_5^{10} \frac{dx}{(x-1)(x-2)}.$$

Solution.   The denominator has two factors, f_1 and f_2, each linear and to the first power. The standard decomposition of the integrand is

(13) $$\frac{1}{(x-1)(x-2)} = \frac{A}{x-1} + \frac{B}{x-2}$$

where A and B are unknowns. Clearing fractions by multiplying both sides of this decomposition equation by $(x-1)(x-2)$ gives

$$1 = A(x-2) + B(x-1).$$

Setting $x = 2$ gives $B = 1$ since

$$1 = A(2-2) + B(2-1) = B.$$

Setting $x = 1$ gives $A = -1$. From (13) we have

$$\int_5^{10} \frac{dx}{(x-1)(x-2)} = \int_5^{10} \left(\frac{-1}{x-1} + \frac{1}{x-2} \right) dx$$

$$= -\int_5^{10} \frac{dx}{x-1} + \int_5^{10} \frac{dx}{x-2}.$$

Each of these integrals has the form of formula (**B'**) in Table 1. We have

$$= -\ln|x-1|\Big|_5^{10} + \ln|x-2|\Big|_5^{10}$$

$$= -\ln 9 + \ln 4 + \ln 8 - \ln 3 = \ln(32/27) \approx 0.170.$$ ■

■ **EXAMPLE 2**   Evaluate

$$\int_1^2 \frac{2x+1}{x(x^2+1)}dx.$$

Solution. The denominator has a linear factor to the first power and an irreducible quadratic factor, also to the first power. The standard decomposition of the integrand is

(14)
$$\frac{2x + 1}{x(x^2 + 1)} = \frac{A}{x} + \frac{Bx + C}{x^2 + 1}$$

where A, B, and C are unknowns. Clearing fractions gives

$$2x + 1 = A(x^2 + 1) + (Bx + C)x.$$

Setting $x = 0$ in this equation gives $A = 1$. Setting $x = 1$ and $x = -1$ gives

(15)
$$2A + B + C = 3$$
$$2A + B - C = -1.$$

Since $A = 1$, this system reduces to two equations in two unknowns. Adding the two equations gives $4 + 2B = 2$, or $B = -1$. Substitution of $A = 1$ and $B = -1$ into either equation of (15) gives $C = 2$. From (14) we have

$$\int_1^2 \frac{2x + 1}{x(x^2 + 1)}\, dx = \int_1^2 \left(\frac{1}{x} + \frac{-x + 2}{x^2 + 1}\right) dx$$

$$= \int_1^2 \frac{dx}{x} - \int_1^2 \frac{x\, dx}{x^2 + 1} + 2\int_1^2 \frac{dx}{x^2 + 1}.$$

Applying formulas (B′), (B″), and (G) from Table 1, we have

$$\int_1^2 \frac{2x + 1}{x(x^2 + 1)}\, dx = \ln x \Big|_1^2 - \tfrac{1}{2}\ln(x^2 + 1)\Big|_1^2 + 2\tan^{-1} x \Big|_1^2$$

$$= \ln 2 - \left(\tfrac{1}{2}\ln 5 - \tfrac{1}{2}\ln 2\right) + 2(\tan^{-1} 2 - \tan^{-1} 1)$$

$$= \tfrac{3}{2}\ln 2 - \tfrac{1}{2}\ln 5 + 2\tan^{-1} 2 - \frac{\pi}{2} \approx 0.879. \qquad \blacksquare$$

■ **EXAMPLE 3** Evaluate

$$\int_{-5}^{-2} \frac{7x^2 + 12x + 4}{(2x + 3)^2(x^2 + x + 1)}\, dx.$$

Solution. This rational function has a repeated linear factor (the linear factor $2x + 3$ has a power of 2) and an irreducible quadratic factor (the zeros of $x^2 + x + 1$ are complex). Recall that a quadratic $ax^2 + bx + c$ has complex zeros if $b^2 - 4ac < 0$. The standard decomposition is

(16)
$$\frac{7x^2 + 12x + 4}{(2x + 3)^2(x^2 + x + 1)} = \frac{A}{2x + 3} + \frac{B}{(2x + 3)^2} + \frac{Cx + D}{x^2 + x + 1}.$$

Clearing fractions gives

$$7x^2 + 12x + 4 = A(2x + 3)(x^2 + x + 1) + B(x^2 + x + 1) + (Cx + D)(2x + 3)^2.$$

To determine the unknowns $A, B, C,$ and D we choose four values of x, perhaps starting with the one real zero ($x = -3/2$) of the denominator. We have

$$7\left(-\tfrac{3}{2}\right)^2 + 12\left(-\tfrac{3}{2}\right) + 4 = B\left(\left(-\tfrac{3}{2}\right)^2 + \left(-\tfrac{3}{2}\right) + 1\right)$$

This gives $B = 1$. To determine $A, C,$ and D we choose $x = 0, 1, -1$. This gives the system of equations

$$
\begin{aligned}
3A \qquad\quad + 9D &= 3 \\
15A + 25C + 25D &= 20 \\
A - \quad C + \quad D &= -2.
\end{aligned}
$$

(17)

The solution of this system is $A = -2$, $C = 1$, and $D = 1$. From (16) we have

$$\int_{-5}^{-2} \frac{7x^2 + 12x + 4}{(2x + 3)^2(x^2 + x + 1)}\, dx = \int_{-5}^{-2}\left(\frac{-2}{2x + 3} + \frac{1}{(2x + 3)^2} + \frac{x + 1}{x^2 + x + 1}\right) dx.$$

The last integral may be written as the sum

$$\int_{-5}^{-2} \frac{-2\,dx}{2x + 3} + \int_{-5}^{-2} \frac{dx}{(2x + 3)^2} + \int_{-5}^{-2} \frac{x + 1}{x^2 + x + 1}\, dx.$$

(18)

These integrals may be evaluated using formulas (B′), (A′), (B″), and (G) in Table 1. We calculate these integrals one at a time. Using formula (B′) to evaluate the first integral gives

$$\int_{-5}^{-2} \frac{-2}{2x + 3}\, dx = -\int_{-5}^{-2} \frac{dx}{x + 3/2}\, dx = \left. -\ln|x + 3/2| \right|_{2}^{5} = \ln 7.$$

(19)

For the second integral from (18), note that $2x + 3 = 2(x + 3/2)$.

$$
\begin{aligned}
\int_{-5}^{-2} \frac{1}{(2x + 3)^2}\, dx &= \frac{1}{4} \cdot \int_{-5}^{-2} \frac{1}{(x + (3/2))^2}\, dx \\
&= \frac{1}{4} \cdot \left. \frac{(x + 3/2)^{-2+1}}{-2 + 1} \right|_{-5}^{-2} = \frac{3}{7}.
\end{aligned}
$$

(20)

To evaluate the third integral from (18) we apply formulas (B″) and (G). For this we must complete the square. Recall that to complete the square of an irreducible quadratic $Q(x) = x^2 + Bx + C$, we add and subtract the square of one-half the coefficient of x. This gives

$$Q(x) = x^2 + Bx + \frac{B^2}{4} - \frac{B^2}{4} + C.$$

We may now write

$$Q(x) = \left(x + \frac{B}{2}\right)^2 + C - \frac{B^2}{4} = \left(x + \frac{B}{2}\right)^2 + \frac{4C - B^2}{4}.$$

Applying this to the third integral from (18), we have

$$\int_{-5}^{-2} \frac{x+1}{x^2+x+1}\,dx = \int_{-5}^{-2} \frac{x+1}{\left(x+\frac{1}{2}\right)^2 + \frac{3}{4}}\,dx.$$

We prepare to apply formulas (B″) and (G) by making the substitution $u = x + \frac{1}{2}$ and setting $a = \sqrt{3}/2$. This gives

$$\int_{-5}^{-2} \frac{x+1}{x^2+x+1}\,dx = \int_{-9/2}^{-3/2} \frac{u+\frac{1}{2}}{u^2+a^2}\,du$$

(21)

$$= \int_{-9/2}^{-3/2} \frac{u}{u^2+a^2}\,du + \frac{1}{2}\int_{-9/2}^{-3/2} \frac{du}{u^2+a^2}$$

$$= \frac{1}{2}\ln(u^2+a^2)\Big|_{-9/2}^{-3/2} + \frac{1}{2a}\tan^{-1}\frac{u}{a}\Big|_{-9/2}^{-3/2}$$

$$\int_{-5}^{-2} \frac{x+1}{x^2+x+1}\,dx = -\frac{1}{2}\ln 7 - \frac{\pi}{3\sqrt{3}} + \frac{1}{\sqrt{3}}\tan^{-1}3\sqrt{3}.$$

Putting (20), (19), and (21) together, we have

$$\int_{-5}^{-2} \frac{7x^2+12x+4}{(2x+3)^2(x^2+x+1)}\,dx$$

$$= \ln 7 + \frac{3}{7} - \frac{1}{2}\ln 7 - \frac{\pi}{3\sqrt{3}} + \frac{1}{\sqrt{3}}\tan^{-1}3\sqrt{3}$$

$$= \frac{1}{2}\ln 7 + \frac{3}{7} - \frac{\pi}{3\sqrt{3}} + \frac{1}{\sqrt{3}}\tan^{-1}3\sqrt{3} \approx 1.594. \qquad \blacksquare$$

Exercises 5.6

Basic

Exercises 1–6: Apply step 1 of the partial fractions procedure to the rational function.

1. $\dfrac{3x^3+5x^2+x-1}{x^2+x+1}$

2. $\dfrac{x^3+2x-1}{(x-1)(2x+1)}$

3. $\dfrac{x^3+3}{(x+1)(x-2)(x+5)}$

4. $\dfrac{x^3+1}{x^3-1}$

5. $\dfrac{2x-1}{(x^2-1)^2}$

6. $\dfrac{x^3}{(x^2+1)^2}$

Exercises 7–12: Factor, as in step 2 of the partial fractions procedure. All zeros are either rational or complex.

7. $x^4+x^3-x^2+x-2$

8. $2x^5-7x^4-3x^3-3x^2-5x+4$

9. x^4-1

10. x^6-1

11. x^4+x^3-x-1

12. $x^4-3x^3-x^2-3x+18$

Exercises 13–18:   Factor, as in step 2 of the partial fractions procedure. Some zeros may require a numerical algorithm.

13. $x^3 + x^2 - 5x + 1$

14. $x^4 - 3x^3 - x^2 - 3x + 17$

15. $x^4 + 3x^3 + x^2 + 3x + 2$

16. $x^4 - 16x^3 + 72x^2 - 96x + 24$

17. $16x^4 - 48x^2 + 12$

18. $192x^5 - 160x^3 + 24x$

Exercises 19–30:   Apply steps 3–5 of the partial fractions procedure.

19. $\dfrac{-2x + 7}{(x + 4)(x + 1)}$

20. $\dfrac{x + 13}{(x + 5)(x - 3)}$

21. $\dfrac{x^3 - x^2 - 7x - 7}{(x + 2)(x - 3)}$

22. $\dfrac{2x^3 - x^2 - x + 6}{(x^2 - 1)}$

23. $\dfrac{x + 4}{(x + 3)^2}$

24. $\dfrac{3x + 8}{(x + 1)^2}$

25. $\dfrac{3x^2 - 7x + 7}{(x^2 + 1)(x - 2)}$

26. $\dfrac{x^2 + 7x + 4}{(x^2 + x + 1)(2x + 1)}$

27. $\dfrac{x^2 + 18x + 30}{(2x + 3)(x + 3)(x - 2)}$

28. $\dfrac{-5x^2 + 9x + 20}{6(x + 2)(x + 1)(x - 1)}$

29. $\dfrac{1}{(x^2 + 3x + 1)^2(x + 1)}$

30. $\dfrac{1}{(x^2 - 3x - 3)^2(x + 1)}$

Exercises 31–38:   Integrate using Table 1.

31. $\displaystyle \int_0^1 \left(\frac{1}{x + 1} + \frac{1}{x + 4} \right) dx$

32. $\displaystyle \int_2^5 \left(\frac{2}{x} + \frac{1}{2x - 3} \right) dx$

33. $\displaystyle \int \left(\frac{1}{(x + 1)^2} - \frac{3}{x^2 + 4} \right) dx$

34. $\displaystyle \int \left(\frac{1}{x^2 + 25} + \frac{5}{(x + 2)^3} \right) dx$

35. $\displaystyle \int \left(\frac{1}{x^2 + 2x + 3} + \frac{1}{x + 1} \right) dx$

36. $\displaystyle \int \left(\frac{1}{x^2 + 6x + 13} - \frac{2}{x + 9} \right) dx$

37. $\displaystyle \int \left(\frac{x}{(x^2 + 4)^2} + \frac{1}{x^2 + 4} \right) dx$

38. $\displaystyle \int \frac{x + 1}{(x^2 + 2)^2} \, dx$

Growth

39. In Example 2 the unknowns A, B, and C were found by setting $x = 0$, 1, -1. Try setting $x = 0$ and $x = i$. For the latter you should obtain the equation $2i + 1 = -B + Ci$. By equating the real and imaginary parts of these two complex numbers, determine B and C.

40. Show that each of the formulas (B′), (A′), (B″), and (A″) can be reduced to the corresponding formula in the table of integrals by a substitution.

41. The integrand of the integral

$$\int_0^4 \frac{dx}{(x - 1)(x - 3)}$$

is the same as that of the integral in Example 1. Comment on the calculation

$$\int_0^4 \frac{dx}{(x - 1)(x - 2)} = \ln |(x - 2)/(x - 1)| \Big|_0^4$$

$$= \ln |2/3| - \ln |2|$$

$$= -\ln 3.$$

Exercises 42–66: Evaluate the integral by partial fractions or a formula from Table 1.

42. $\int \dfrac{x}{x^2 - 1}\, dx$

43. $\int \dfrac{x}{x^2 - 5x + 6}\, dx$

44. $\int \dfrac{dx}{x^4 - 1}$

45. $\int \dfrac{dx}{x^4 - 16}$

46. $\int \dfrac{dx}{(x + 1)^2(x - 1)}$

47. $\int \dfrac{x^2 - 1}{(x + 2)^2(x + 3)}\, dx$

48. $\int \dfrac{x^2 - 1}{x^2 + 5x + 6}\, dx$

49. $\int \dfrac{2x^2 + 1}{2x^2 - x - 1}\, dx$

50. $\int \dfrac{dx}{x^2 + 2x + 2}$

51. $\int \dfrac{dx}{x^2 - 4x + 13}$

52. $\int \dfrac{x^3}{x - 1}\, dx$

53. $\int \dfrac{x^4}{(x - 1)^2}\, dx$

54. $\int \dfrac{dx}{(x + 1)^4}$

55. $\int \dfrac{dx}{(2x + 3)^2}$

56. $\int \dfrac{x}{x^2 + x + 1}\, dx$

57. $\int \dfrac{3x - 2}{2x^2 - x + 2}\, dx$

58. $\int \dfrac{dx}{x^4 - 2x^3 + x^2 - 4x + 4}$

59. $\int \dfrac{x}{x^3 + 1}\, dx$

60. $\int \dfrac{x - 1}{x^2 - 4x + 13}\, dx$

61. $\int \left(\dfrac{x + 2}{x - 1}\right)^2 dx$

62. $\int_0^1 \dfrac{x + 2}{(x^2 + 1)^4}\, dx$

63. $\int \dfrac{x^3 - 1}{x^3 + 1}\, dx$

64. $\int \dfrac{x^2 - 3x + 2}{x^3 + 2x^2 + x}\, dx$

65. $\int \dfrac{x^4}{x^6 - 1}\, dx$

66. $\int \dfrac{x^2(2x^2 - x + 3)}{(x^2 + 1)^2(x - 1)}\, dx$

Review/Preview

67. Show that the area of a trapezoid with height h and bases of length b and B is $h(B + b)/2$.

68. A parabola with equation $y = ax^2 + bx + c$ passes through the points $(-1, p)$, $(0, q)$, and $(1, r)$. Show that

$$\int_{-1}^{1} (ax^2 + bx + c)\, dx = \tfrac{1}{3}(p + 4q + r).$$

5.7 NUMERICAL INTEGRATION

To calculate the value of an integral $\int_a^b f(x)dx$ by the Fundamental Theorem of Calculus we must know an antiderivative F of f. If we are able to determine F (by substitution, integration by parts, partial fractions, or other integration technique), then we can use the formula

$$\int_a^b f(x)\, dx = F(b) - F(a).$$

If neither human nor CAS can determine F, this method won't work. The length of the orbit of Mars, for example, can be written as a definite integral, but the integral yields to no known integration technique. Calculating the position of a simple pendulum depends on a similar integral. Some calculations in statistical quality control depend on the integral of the normal distribution function, which must be determined numerically.

Many functions describing physical phenomena are known only at discrete points of their domains. For example, the velocity of an object may be known only at times 0.0, 0.1, 0.2,..., 1.0 s. Although we have no formula for v, we may approximate its position at these times. The profile of a hillside is usually not specified by an algebraic formula, but rather as data in a surveyor's notebook. If a hillside is to be leveled for a housing development, the contractor (in preparing a bid on the job) may need to estimate the volume of earth to be moved.

Numerical integration is the name associated with methods for approximating the integral $\int_a^b f(x)\,dx$ of a continuous function f. We discuss three of these: the trapezoid rule, the midpoint rule, and Simpson's rule. We use each of these rules to approximate the integral

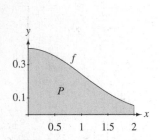

Figure 5.38. Part of a normal distribution function.

$$(1) \qquad P = \int_0^2 f(x)\,dx, \quad \text{where} \quad f(x) = \frac{1}{\sqrt{2\pi}}e^{-x^2/2}.$$

The number P is the probability that a normally distributed variable (with mean 0 and variance 1) lies between 0 and 2. Referring to Fig. 5.38, this number can be interpreted as the area of the region beneath the graph of f on the interval $[0, 2]$.

Trapezoid Rule

As in defining the integral $\int_a^b f(x)\,dx$, we divide the interval $[a, b]$ into n equal subintervals with the points $a = x_0, x_1, \ldots, x_n = b$. Each subinterval has length $h = (b - a)/n$. We zoom in on a typical subinterval $[x_i, x_{i+1}]$ in Fig. 5.39. We approximate the area beneath f by a trapezoid instead of an inscribed or circumscribed rectangle. We use the trapezoid with height h and bases $y_i = f(x_i)$ and $y_{i+1} = f(x_{i+1})$. The area of this trapezoid is $h(y_i + y_{i+1})/2$.

Adding the trapezoid approximations on the n subintervals gives

$$\int_a^b f(x)\,dx \approx T_n = \sum_{i=0}^{n-1} h\frac{f(x_i) + f(x_{i+1})}{2} = \frac{h}{2}\sum_{i=0}^{n-1}(y_i + y_{i+1}).$$

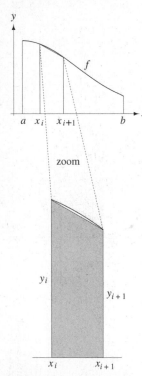

Figure 5.39. The trapezoid approximation.

We rewrite T_n in a more efficient form for calculation:

$$(2) \qquad \int_a^b f(x)\,dx \approx T_n = \frac{h}{2}(y_0 + 2(y_1 + y_2 + \cdots + y_{n-1}) + y_n).$$

■ **EXAMPLE 1**   Use the trapezoid rule (2) to approximate the number P in (1) to within 0.01.

Solution.   We calculate T_5, T_{10}, and T_{20}. Data for T_5 are given in Table 1. We used the full internal accuracy of our calculator for all calculations. From (2) we have

$$(3) \qquad T_5 = \frac{h}{2}(y_0 + 2(y_1 + y_2 + y_3 + y_4) + y_5)$$

Table 1. Data for T_5

i	x_i	y_i
0	0.0	0.3989...
1	0.4	0.3682...
2	0.8	0.2896...
3	1.2	0.1941...
4	1.6	0.1109...
5	2.0	0.0539...

where $h = (b - a)/n = (2 - 0)/5 = 0.4$.

We obtain $T_5 \approx 0.4758$ from (3). To judge the accuracy of this value, we calculate $T_{10} \approx 0.4769$ and $T_{20} \approx 0.4772$. Although proof is lacking, it appears that P is within 0.01 of the number 0.48. The change from T_5 to T_{10} is 0.0011, and the change from T_{10} to T_{20} is 0.0003. It appears that further increases in n will not affect the second decimal place.   ■

The Midpoint Rule

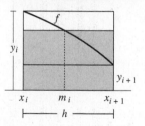

Figure 5.40. The midpoint approximation.

The following discussion uses Fig. 5.40 and assumes that f is a decreasing function. If on a typical subinterval $[x_i, x_{i+1}]$ of a subdivision of $[a, b]$ we approximate the area beneath the graph of f with the area hy_{i+1} of the inscribed rectangle, the approximation is too small. The area hy_i of the circumscribed rectangle is too big. The trapezoid rule attempts to improve on these two approximations by averaging their values. The trapezoid contribution on $[x_i, x_{i+1}]$ is

$$\frac{hf(x_{i+1}) + hf(x_i)}{2} = h\frac{f(x_i) + f(x_{i+1})}{2} = h\frac{y_i + y_{i+1}}{2}.$$

The midpoint rule is based on the idea that it may be better to average x_i and x_{i+1} than y_i and y_{i+1}, as in the trapezoid rule. We evaluate f at the midpoint $m_i = (x_i + x_{i+1})/2$ of the interval $[x_i, x_{i+1}]$ and approximate the area beneath f with the number $hf(m_i)$. For many functions the midpoint rule is indeed more accurate than the trapezoid rule. We give some numerical evidence for this in Examples 2 and 3. For functions that are concave up or down on $[a, b]$, we outline a proof in Exercise 16.

If we let the midpoints of the n subintervals be

$$m_1 = (x_0 + x_1)/2, \ldots, m_n = (x_{n-1} + x_n)/2,$$

the formula for the midpoint rule is

$$(4) \qquad \int_a^b f(x)\,dx \approx M_n = h(f(m_1) + f(m_2) + \cdots + f(m_n)).$$

■ **EXAMPLE 2** Compare the trapezoid and midpoint rules by approximating $\int_1^2 \sqrt{x}\,dx$ using one subinterval.

Solution. Letting $n = 1$ for one subinterval, we have $h = 1$, $x_0 = 1$, and $x_1 = 2$. We calculate the exact value of this integral as well as T_1 and M_1:

$$\int_1^2 \sqrt{x}\,dx = \frac{2}{3}x^{3/2}\Big|_1^2 = 1.2189\ldots$$

$$\int_1^2 \sqrt{x}\,dx \approx T_1 = \frac{h}{2}(y_0 + y_1) = 0.5\left(\sqrt{1} + \sqrt{2}\right) = 1.2071\ldots$$

$$\int_1^2 \sqrt{x}\,dx \approx M_1 = h\sqrt{(x_0 + x_1)/2} = \sqrt{(1 + 2)/2} = 1.2247\ldots.$$

The trapezoid approximation is too small by approximately 0.01 and the midpoint approximation too big by approximately 0.006. For this integral, the error in the midpoint rule is about half that in the trapezoid rule. ■

■ **EXAMPLE 3** From tables of the normal distribution the value of

$$P = \int_0^2 f(x)\,dx, \quad \text{where} \quad f(x) = \frac{1}{\sqrt{2\pi}}e^{-x^2/2}$$

Table 2. Comparison of the midpoint and trapezoid rules

n	T_n	M_n	$T_n - P$	$M_n - P$	$\dfrac{T_n - P}{M_n - P}$
5	0.475814	0.477966	−0.001436	0.000716	−2.004
6	0.476252	0.477748	−0.000998	0.000498	−2.003
7	0.476516	0.477616	−0.000734	0.000366	−2.002
8	0.476688	0.477531	−0.000562	0.000281	−2.002
9	0.476806	0.477472	−0.000444	0.000222	−2.001
10	0.476890	0.477430	−0.000360	0.000180	−2.001

is $P \approx 0.47724987$. Using this value, compare the trapezoid and midpoint approximations to P for $n = 5, 6, \ldots, 10$.

Solution. Calculate several of the lines in Table 2. The effort will pay off both in increased understanding of this example and in the forthcoming discussion of Simpson's rule. For your calculations, note that since the interval has length 2, the midpoints are easily calculated. This is illustrated in Fig. 5.41 for $n = 5$.

The midpoint approximations $M_5, \ldots, M_{10}$ to P are shown in column 3 of the table. They approach the value $P \approx 0.47724987$ of the integral more quickly than $T_5, \ldots, T_{10}$. This is shown in columns 4 and 5. We discuss the sixth column below. ■

Figure 5.41. The midpoints for $n = 5$.

Simpson's Rule

The data in Table 2 leads to Simpson's rule. Included in the table are the differences $T_n - P$ and $M_n - P$ between P and several trapezoid and midpoint approximations. In this instance the T_n values are a little too small and the M_n values are too big. The midpoint rule appears to give more accuracy for roughly the same amount of calculation. The most notable thing in the table is the sixth column, which is nearly constant. This regularity of the error suggests Simpson's rule, which improves on both the trapezoid and midpoint rules. From the observation that

$$\frac{T_n - P}{M_n - P} \approx -2,$$

we have

$$T_n - P \approx -2M_n + 2P.$$

Solving for P, we have

(5) $$P \approx \frac{T_n + 2M_n}{3} = S_n.$$

This result is the heart of Simpson's rule. Since we have taken account of the error, we expect that S_n will better approximate P than M_n or T_n. Data to test this expectation are given in Table 3. We include parts of Table 2 so that the errors $|T_n - P|$, $|M_n - P|$, and $|S_n - P|$ can be compared.

Although this approach to Simpson's rule is based on a special case, the result holds more generally. A geometric approach to Simpson's rule is outlined in

Table 3. Comparison of Simpson's rule with the midpoint and trapezoid rules

n	S_n	$S_n - P$	$T_n - P$	$M_n - P$
5	0.47724887	−0.00000100	−0.001436	0.000716
6	0.47724939	−0.00000048	−0.000998	0.000498
7	0.47724961	−0.00000026	−0.000734	0.000366
8	0.47724972	−0.00000015	−0.000562	0.000281
9	0.47724978	−0.00000009	−0.000444	0.000222
10	0.47724981	−0.00000006	−0.000360	0.000180

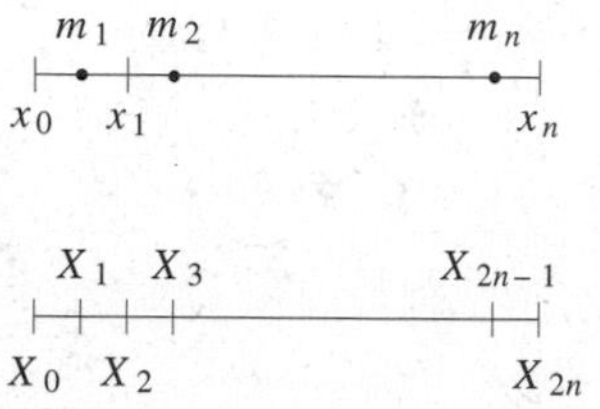

Figure 5.42. New subdivision.

Exercise 17. Finally, Simpson's rule may be given without reference to the midpoint and trapezoid rules. Suppose we have a subdivision $x_0, x_1, \ldots, x_n$ or $[a, b]$, with $h = (b - a)/n$ the common length of the subintervals. We form a new subdivision $X_0, X_1, \ldots, X_{2n-1}, X_{2n}$ of $[a, b]$ by adjoining to the points $x_0, x_1, \ldots, x_n$ the midpoints $m_1, \ldots, m_n$. We show this in Fig. 5.42. Letting $H = (b - a)/(2n)$ be the common length of the new subintervals, we have from (5), (2), and (4) that

$$S_n = \frac{T_n + 2M_n}{3} = \frac{1}{3}\frac{h}{2}\big(y_0 + 2(y_1 + y_2 + \cdots + y_{n-1}) + y_n\big)$$

$$+ \frac{2}{3}h\big(f(m_1) + f(m_2) + \cdots + f(m_n)\big)$$

$$= \frac{h}{6}\Big(f(X_0) + 2(f(X_2) + \cdots + f(X_{2n-2})) + f(X_{2n})$$

$$+ 4\,(f(X_1) + \cdots + f(X_{2n-1}))\Big)$$

$$= \frac{H}{3}\Big(Y_0 + 4(Y_1 + \cdots + Y_{2n-1}) + 2(Y_2 + \cdots + Y_{2n-2}) + Y_{2n}\Big).$$

To write this result in standard form, suppose we have a subdivision $x_0, x_1, x_2, \ldots, x_{2n}$ of $[a, b]$, with $h = (b - a)/(2n)$. We have

$$(6)\quad \int_a^b f(x)\,dx \approx S_n^* = \frac{h}{3}\Big(y_0 + 4(y_1 + \cdots + y_{2n-1}) + 2(y_2 + \cdots + y_{2n-2}) + y_{2n}\Big).$$

Note that while the approximations T_n and M_n are based on subdivisions with n subintervals, S_n^* is based on $2n$ subintervals.

Error Analysis for the Trapezoid Rule

We used the trapezoid rule to approximate

$$(7)\qquad P = \int_0^2 \frac{1}{\sqrt{2\pi}}e^{-x^2/2}\,dx$$

in Examples 1 and 3, but we said little about choosing the number n of subdivisions so that T_n is within some error tolerance E of P. We end the section by showing how to choose n so that

$$|P - T_n| < 0.001 = E.$$

Our discussion is based on the inequality

$$(8) \qquad \left| \int_a^b f(x)\,dx - T_n \right| \le \frac{b-a}{12} h^2 M$$

where $h = (b-a)/n$. The number M is chosen so that the graph of $|f''|$ lies entirely below the line with equation $y = M$. See, for example, Fig. 5.43, where one choice for M is shown. We use the inequality (8) without proof. Full discussions of this inequality—and similar inequalities for the midpoint rule and Simpson's rule—may be found in books on numerical analysis.

To determine a value of M, we calculate f'' and then graph $|f''|$. Differentiating

$$f(x) = \frac{1}{\sqrt{2\pi}} e^{-x^2/2}$$

twice we obtain

$$f''(x) = \frac{x^2 - 1}{\sqrt{2\pi}} e^{-x^2/2}.$$

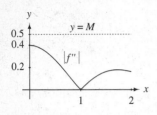

Figure 5.43. Determining a suitable M.

A graph of $|f''|$ is shown in Fig. 5.43. On the interval $[0, 2]$, it is clear from the figure that the line with equation $y = 0.5$ lies entirely above the graph of $|f''|$. Perhaps $y = 0.4$ would work, but to be on the safe side (and not spend too much time or energy on choosing M), we take $M = 0.5$.

Next, we choose n so that

$$(9) \qquad |P - T_n| \le \frac{2-0}{12} h^2 M < 0.001.$$

The first inequality in (9) depends on (8). We force the second inequality to hold by choosing n sufficiently large so that

$$(10) \qquad |P - T_n| \le \frac{2-0}{12} h^2 \cdot 0.5 = \frac{1}{12} \left(\frac{2}{n} \right)^2 = \frac{1}{3n^2} < 0.001.$$

How do we choose n so that the last inequality holds? To discuss the question we take $n = 5$. Substituting into (10), we have

$$(11) \qquad |P - T_5| \le \frac{2-0}{12} h^2 \cdot 0.5 = \frac{1}{12} \left(\frac{2}{5} \right)^2 = \frac{1}{3 \cdot 5^2} = 0.0133\ldots.$$

From this we are unable to say for sure that T_5 is within $E = 0.001$ of P. We do know T_5 is within $0.0133\ldots$ of P. This trial calculation shows that we should try to determine a value of n so that $1/(3n^2) < E = 0.001$. This inequality can be solved for n, or we can experiment with several values of n. The least value of n for which $1/(3n^2) < E = 0.001$ is $n = 19$. This follows from several trials on your calculator or the inequality

$$n > \sqrt{1/(3E)} \approx 18.3.$$

Exercises 5.7

Basic

1. Use the natural logarithm key on your calculator to obtain a good approximation to $\ln 3$. Since $\ln 3 = \int_1^3 dx/x$, we may approximate $\ln 3$ by numerical integration. Calculate T_5 and T_{10} and the differences between them and the calculator value of $\ln 3$. Retain six decimals.

2. Repeat Exercise 1 but use the midpoint rule.

3. Repeat Exercise 1 but use Simpson's rule.

4. Approximate the value of

$$P = \frac{1}{\sqrt{2\pi}} \int_0^1 e^{-x^2/2} dx$$

with T_5, T_{10}, and T_{15}. It is given that $P \approx 0.3413. \ldots$ Would you have been confident that $P \approx 0.34$ on the basis of your calculations?

5. Repeat Exercise 4 but use the midpoint rule.

6. Repeat Exercise 4 but use Simpson's rule.

7. Use Simpson's rule to calculate $\sin^{-1} 0.5$. Determine n empirically so that S_n is within 0.0001 of $\sin^{-1} 0.5$. Recall that

$$\sin^{-1} 0.5 = \int_0^{0.5} \frac{dx}{\sqrt{1-x^2}}.$$

8. The length L of the orbit of a planet moving on an ellipse $x^2/a^2 + y^2/b^2 = 1$ is given by the integral

$$L = 4a \int_0^{\pi/2} \sqrt{1 - e^2 \sin^2 x}\, dx.$$

where e is the eccentricity of the ellipse. The eccentricity is related to the lengths a and b of the semimajor and semiminor axes by $e^2 = 1 - b^2/a^2$. Calculate L to within 0.0001 by using the midpoint rule. Use Pluto's orbit data: $a = 39.78$ AU and $e = 0.2539$ (1 AU $= 149.6 \times 10^6$ km).

Growth

9. Acceleration data (ft/s^2) for a small rocket with burnout time of 40 s are given below. Calculate the maximum height reached and the total flight time. (This problem comes from *Applied Methods for Digital Computation*, third edition, 1985, by M. L. James et al., HarperCollins Publishers.)

(0.0, 38.4), (2.0, 39.3), (4.0, 39.3), (6.0, 38.1),

(8.0, 35.7), (10.0, 35.2), (12.0, 28.7), (14.0, 24.7),

(16.0, 20.7), (18.0, 17.1), (20.0, 14.0),

(22.0, 11.3), (24.0, 9.2), (26.0, 7.5),

(28.0, 6.2), (30.0, 5.2), (32.0, 4.5),

(34.0, 4.0), (36.0, 3.7), (38.0, 3.4), (40.0, 3.2)

10. The range of hills shown in the following figure, which is not drawn to scale, has uniform cross-sections perpendicular to the horizontal line joining B and A. A cross-section was surveyed, with elevations made every 100 feet along CB. Starting at C, with all measurements given in feet, the survey data were recorded as (x_i, y_i) pairs.

(0, 0.0), (100, 20.5), (200, 29.2), (300, 30.8),

(400, 28.9), (500, 25.7), (600, 22.8), (700, 20.9),

(800, 20.1), (900, 20.1), (1000, 20.3), (1100, 20.0),

(1200, 18.8), (1300, 16.1), (1400, 11.9),

(1500, 6.9), (1600, 2.2), (1700, 0.0)

A cut for a road is to made through the hills, perpendicular to line BA. The base of the cut is to lie in the plane determined by A, B, and C. The width of the cut is to be 75 feet. Use the trapezoid rule in determining the volume of the cut.

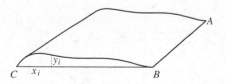

11. The function

$$\text{Si}(x) = \int_0^x \frac{\sin t}{t}\, dt, \quad -\infty < x < \infty$$

is called the **sine integral.** Based on the fact that $\lim_{\theta \to 0}(\sin \theta / \theta) = 1$, the integrand is defined to be 1 at $t = 0$. With this understanding, the integrand is continuous everywhere. Duplicate Table 2 for Si(1). Does Simpson's rule follow from these data as in Example 3? Using your table, calculate Si(1) using Simpson's rule. An accurate value of Si(1) is 0.94608307....

12. Put together a few sentences to convince a friend that if a dog is lost somewhere on a straight stretch of interstate highway, between mile markers 225 and 237, she cannot be further away than $(237 - 225)/2 = 6$ miles from mile marker $(225 + 237)/2 = 231$. Relate this to the following result. Although the value of the integral $I = \int_a^b f(x)\,dx$ is unknown, it is certainly true that $L_n \le I \le U_n$, where L_n and U_n are lower and upper sums of f. Moreover,

$$\left| I - \frac{L_n + U_n}{2} \right| \le \frac{U_n - L_n}{2}.$$

13. Continuing Exercise 12, for a decreasing function f, the trapezoid rule is related to the lower and upper sums of f by

$$T_n = \frac{L_n + U_n}{2}.$$

Moreover,

(12) $$|I - T_n| \le \frac{U_n - L_n}{2} = \frac{b - a}{n} \frac{f(a) - f(b)}{2}.$$

What would change if f were increasing?

14. The result in the preceding exercise may be used to calculate how big n must be to achieve a given accuracy. In Example 1, for example, we used the trapezoid rule to approximate an integral whose value was denoted by P. Suppose we want to find T_n so that $|P - T_n| < 0.01$. From (12) we have

$$|P - T_n| \le \frac{2 - 0}{n} \frac{f(a) - f(b)}{2}$$

$$\le \frac{2}{n}\left(\frac{1}{\sqrt{2\pi}} \cdot 1 - \frac{1}{\sqrt{2\pi}e^{-2}} \right) < \frac{0.7}{n}.$$

Thus, if we require that $0.7/n < 0.01$, then T_n will be within 0.01 of P. How big does n need to be to achieve this? Judging from the results in Example 1, is this value of n too conservative?

15. Apply the results of the preceding exercises in finding a value of n so that $|\text{Si}(1) - T_n| < 0.01$. See Exercise 11.

16. In the text we gave some numerical evidence that the midpoint rule is better than the trapezoid rule for some functions. Fill in the details of the following argument that for a continuous function f that is concave down on $[a, b]$,

(13) $$0 \le M_n - \int_a^b f(x)\,dx \le \int_a^b f(x)\,dx - T_n.$$

In the following figure, let Q_1 denote the area of triangle APB, Q_2 the area of triangle PCD, T the area of trapezoid $GFCA$, Q_3 the area of triangle PCE, and Q_4 the area of triangle ACP. Let $I = \int_a^b f(x)\,dx$. Step 1: Show that $Q_1 + Q_2 = Q_3 = Q_4$. This is easy geometry. Step 2: Show that the midpoint approximation M on $[x_i, x_{i+1}]$ is equal to the area beneath any line passing through P. (This is also geometry.) Step 3: Observe that $I \ge T + Q_4$ and, hence,

$$0 \le M - I \le M - T - Q_4 = Q_1 + Q_2.$$

Step 4: Verify the inequality

$$0 \le M - I \le Q_1 + Q_2 = Q_3 = Q_4 \le I - T$$

Since this inequality holds on each subinterval, (13) follows. A similar argument holds for functions that are concave up.

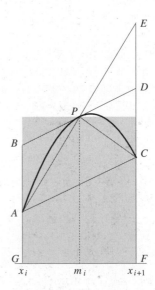

17. Let the points $\{a = x_0, x_1, \ldots, x_{2n-1}, x_{2n} = b\}$ subdivide $[a, b]$ into $2n$ subintervals with $h = (b - a)/(2n)$ their common length. On a typical subinterval $[x_i, x_{i+1}]$, the trapezoid rule in effect replaces f by a linear function L_i and approximates $\int_{x_i}^{x_{i+1}} f(x)\,dx$ by $\int_{x_i}^{x_{i+1}} L_i(x)\,dx$. The graph of L_i is

the line joining (x_i, y_i) to (x_{i+1}, y_{i+1}). Simpson's rule uses quadratic functions instead of linear functions, giving a closer fit in most cases. To determine a parabola, we need three points, not two as for a line. Simpson's rule follows from fitting a parabola on each of the n three-point subintervals, integrating the resulting quadratic functions, and adding the results. Complete the following argument.

We show in the following figure three successive points x_{i-1}, x_i, and x_{i+1} of the subdivision. We wish to approximate $\int_{x_{i-1}}^{x_{i+1}} f(x)\,dx$ by $\int_{x_{i-1}}^{x_{i+1}} g(x)\,dx$, where g is the quadratic function whose graph passes through the points (x_{i-1}, y_{i-1}), (x_i, y_i), and (x_{i+1}, y_{i+1}). We take advantage of the fact that we may translate f to the left or right without changing the value of the integral. We show the interval $[x_{i-1}, x_{i+1}]$ translated to $[-h, h]$ in the left sketch in the figure. We have relabeled the ordinates as y_{-1}, y_0, and y_1. Let the parabola through the points $(-h, y_{-1})$, $(0, y_0)$, and (h, y_1) have the equation $y = ax^2 + bx + c$. Simpson's rule is based on the approximation

$$(14) \quad \int_{x_{i-1}}^{x_{i+1}} f(x)\,dx \approx \int_{-h}^{h} (ax^2 + bx + c)\,dx.$$

The value of the last integral is $\frac{2}{3}ah^3 + 2ch$, which does not depend on b. We find that $c = y_0$ from the condition that the point $(0, y_0)$ must lie on the parabola. We obtain the equations (which we can solve for a and b)

$$y_{-1} = ah^2 - bh + c$$
$$y_1 = ah^2 + bh + c$$

from the conditions that the points $(-h, y_{-1})$ and (h, y_1) must also lie on the parabola. Adding these equations, we find $y_{-1} + y_1 = 2ah^2 + 2c$. From this result and (14) we have

$$\int_{x_{i-1}}^{x_{i+1}} f(x)\,dx \approx \frac{h}{2}(2ah^2 + 2c + 4c)$$
$$\approx \frac{h}{3}(y_{-1} + y_1 + 4y_0)$$
$$\approx \frac{h}{3}(y_{i-1} + y_{i+1} + 4y_i).$$

Adding the approximations for all n pairs of subintervals gives Simpson's rule (6).

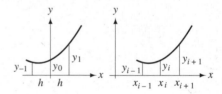

18. Show by examples that Simpson's rule is exact for polynomials of degree 3 or less. Why is this true?

Review/Preview

19. A metal washer is stamped from a steel plate 0.125 cm thick. The inner radius of the washer is 0.5 cm and the outer radius is 1.5 cm. Calculate its volume.
20. Find an antiderivative of $\int x/(1 + x)\,dx$.
21. Find an antiderivative of $\int xe^x\,dx$.
22. Find an antiderivative of $\int (1 - (\pi/2)x)^3\,dx$.
23. Find the area bounded by the curves with equations $y = 2 + 3(x - 1)^2$ and $y = 3x + 5$.

REVIEW OF KEY CONCEPTS

This chapter introduced integration and presented some applications. We defined the definite integral $\int_a^b f(x)\,dx$ as a limit of lower and upper sums. In subsequent sections we discussed the area of the region beneath the graph of a function or between the graphs of two functions. We also discussed how the integral can be used to calculate the position of an object from its velocity. We listed several properties of the integral and discussed the Fundamental Theorem of Calculus.

Much of the remainder of the chapter was spent on techniques for evaluating $\int_a^b f(x)\,dx$. We first discussed techniques that depend on the Fundamental Theorem. For functions f for which an antiderivative F is more or less easily found, we discussed antidifferentiation by searching in a table of integrals, by substitution, by parts, and by partial fractions. If we must evaluate $\int_a^b f(x)\,dx$ and it is not possible or practical to determine F, we must use a numerical integration rule. We discussed the trapezoid rule, midpoint rule, and Simpson's rule.

Chapter Summary

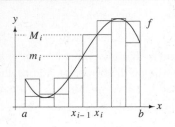

Let f be a continuous function on $[a, b]$. The integral $\int_a^b f(x)\, dx$ is the unique number which, for any subdivision $x_0, x_1, \ldots, x_n$ of $[a, b]$, lies between the sum L_n of the inscribed rectangles and the sum U_n of the circumscribed rectangles.

For any subdivision $x_0, x_1, \ldots, x_n$ of $[a, b]$

$$L_n \le \int_a^b f(x)\, dx \le U_n$$

where m_i and M_i are the minimum and maximum of f on $[x_{i-1}, x_i]$ and

$$L_n = h(m_1 + m_2 + \cdots + m_n)$$
$$U_n = h(M_1 + M_2 + \cdots + M_n).$$

Definitions and Properties of the Integral

Assume that f and g are continuous on $[a, b]$, $c \in R$, and $r, s, t \in [a, b]$.

Definition 1: Area. Assuming that f is nonnegative on $[a, b]$, the area of the region S beneath the graph of f is the number $\int_a^b f(x)\, dx$.

Definition 2
$$\int_a^a f(w)\, dw = 0$$

Definition 3
$$\int_b^a f(x)\, dx = -\int_a^b f(x)\, dx$$

Definition 4: Average value. The average value of f on $[a, b]$ is

$$\frac{1}{b - a} \int_a^b f(x)\, dx$$

Property 1: Additivity

$$\int_r^s f(x)\, dx + \int_s^t f(x)\, dx = \int_r^t f(x)\, dx$$

Property 2: Mean-value Inequality for Integrals

$$m(b - a) \le \int_a^b f(x)\, dx \le M(b - a),$$

where m and M are the minimum and maximum of f on $[a, b]$.

Property 3: Linearity

$$\int_a^b (f(x) + g(x))\, dx = \int_a^b f(x)\, dx + \int_a^b g(x)\, dx$$

and

$$\int_a^b c f(x)\, dx = c \int_a^b f(x)\, dx$$

Property 4. If $f(x) \ge 0$ on $[a, b]$, then $\int_a^b f(x)\, dx \ge 0$.

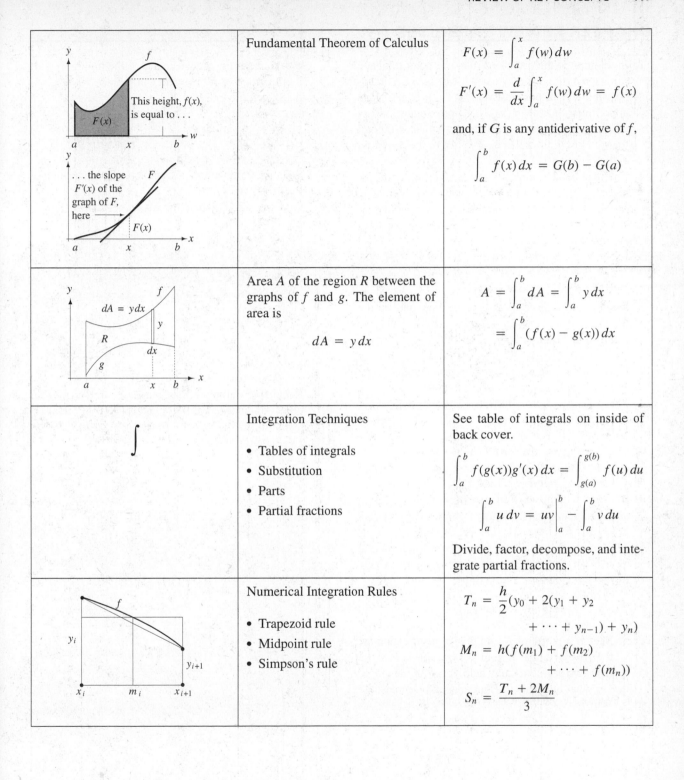

	Fundamental Theorem of Calculus	$F(x) = \displaystyle\int_a^x f(w)\,dw$
		$F'(x) = \dfrac{d}{dx}\displaystyle\int_a^x f(w)\,dw = f(x)$
		and, if G is any antiderivative of f,
		$\displaystyle\int_a^b f(x)\,dx = G(b) - G(a)$
	Area A of the region R between the graphs of f and g. The element of area is $$dA = y\,dx$$	$A = \displaystyle\int_a^b dA = \int_a^b y\,dx$ $= \displaystyle\int_a^b (f(x) - g(x))\,dx$
	Integration Techniques • Tables of integrals • Substitution • Parts • Partial fractions	See table of integrals on inside of back cover. $\displaystyle\int_a^b f(g(x))g'(x)\,dx = \int_{g(a)}^{g(b)} f(u)\,du$ $\displaystyle\int_a^b u\,dv = uv\Big|_a^b - \int_a^b v\,du$ Divide, factor, decompose, and integrate partial fractions.
	Numerical Integration Rules • Trapezoid rule • Midpoint rule • Simpson's rule	$T_n = \dfrac{h}{2}(y_0 + 2(y_1 + y_2 + \cdots + y_{n-1}) + y_n)$ $M_n = h(f(m_1) + f(m_2) + \cdots + f(m_n))$ $S_n = \dfrac{T_n + 2M_n}{3}$

CHAPTER REVIEW EXERCISES

1. In addition to the trapezoid rule, midpoint rule, and Simpson's rule, we may use a lower sum L_n, an upper sum U_n, or, perhaps, $A_n = (L_n + U_n)/2$ as an approximation to $\int_a^b f(x)\,dx$. Approximate $\int_0^1 x^4\,dx$ with A_5. Compare A_5 to the exact value of this integral. How is A_5 related, if at all, to the trapezoid approximation T_5?

2. Calculate L_6, U_6, and $A_6 = 0.5(L_6 + U_6)$ for the integral $\int_1^3 \sqrt[3]{x}\,dx$. Compare A_6 with the actual value of this integral. Sketch the graph of $\sqrt[3]{x}$ and the inscribed and circumscribed rectangles on the third subinterval.

3. Express in summation notation: $3/2 + 5/4 + 7/8 + \cdots + 21/1024$.

4. Calculate $\sum_{j=1}^{6}(j-2)^2$.

5. Determine the area of the region bounded by the graphs of $y = 3x$ and $y = 3(x-2)^2$.

6. Determine the area of the region between the curves with equations $y = 4 - (x-2)^2$ and $y = x$.

7. Determine the area of the region between the x-axis and the graph of $f(x) = x - x^2$.

8. The velocity of an object is $v(t) = t/(t+1)$ for $t \geq 0$ and its initial position is $x(0) = 1$. Graph its position for $0 \leq t \leq 5$.

9. The velocity of the end E of a spring is $v(t) = e^{-t}\sin t$ for $t \geq 0$. The initial position of E is $x(0) = 0$. Plot the position of E for $0 \leq t \leq 5$.

10. If $F(x) = \int_0^x \sin(w^2)\,dw$, $x \geq 0$, write the equation of the tangent line to the graph of F at $(1, F(1))$. It is given that $F(1) \approx 0.31$.

11. The graph of a function f defined on the interval $[0, 6]$ is shown in the accompanying figure. A new function F is defined on $[0, 6]$ by

$$F(x) = \int_0^x f(w)\,dw.$$

   a. Sketch a graph of F and give a justification for the general shape of your sketch. Answers to the next two problems may help you with the graph.

   b. On what intervals is F increasing? Decreasing? Why?

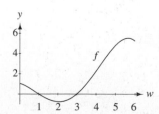

   c. What are the x-coordinates of the high points of the graph of F? How do you know?

   d. Sketch the graph of F'.

12. Let

$$G(x) = \int_0^{1/x} e^{w^2}\,dw, \quad 1 \leq x \leq 10.$$

   Sketch a rough graph of G and calculate the slope of the graph at $x = 2$.

13. Give the standard decomposition of the function.

   a. $\dfrac{2x - 1}{(x^2 - 5x + 6)}$

   b. $\dfrac{x(x^3 + 1)}{x^4 - 1}$

   c. $\dfrac{1}{(5x - 3)^2(x^2 + 1)^3}$

   d. $\dfrac{x^2 + x + 1}{(x^2 + 1)^2(x + 1)(x^2 - 2)^2}$

   e. $\dfrac{5x^2 + 4x + 2}{(2x^2 + 2x + 1)(x + 1)}$

14. Recall the definition of the average velocity of an object on an interval $[a, b]$ and relate it to the average value of the velocity function on $[a, b]$.

15. Evaluate the integral.

   a. $\displaystyle\int_0^1 (2x + 1)^5\,dx$

   b. $\displaystyle\int \frac{dx}{2x + 3}$

   c. $\displaystyle\int \frac{dx}{2x^2 + 1}$

   d. $\displaystyle\int \frac{dx}{\sqrt{7 - x^2}}$

   e. $\displaystyle\int_1^{\pi^2/4} \frac{\sin \sqrt{x}}{\sqrt{x}}\,dx$

   f. $\displaystyle\int_1^e x^{-2} \ln x\,dx$

   g. $\displaystyle\int_3^5 \frac{dx}{x^2 - 2x}$

   h. $\displaystyle\int_2^3 (x^2 - 2x + 1)^{-1}\,dx$

   i. $\displaystyle\int_0^{\pi/4} \frac{dx}{\cos^2 x}$

   j. $\displaystyle\int \frac{dx}{\sqrt{x}(1 + x)}$

k. $\displaystyle\int_0^{\pi/2} x \sin x \, dx$

l. $\displaystyle\int \frac{1 + e^x}{1 - e^x} \, dx$

m. Use the trigonometric substitution $x = \sin u$ for the integral

$$\int_0^{1/2} \frac{x^2}{\sqrt{1 - x^2}} \, dx.$$

n. For positive constants a and b, show that

$$\int_{-b}^{b} \sqrt{1 + \sinh(x/a)} \, dx = 2a \sinh(b/a).$$

Recall the identity $\cosh^2 x - \sinh^2 x = 1$.

16. The integral

$$\int_{-1}^{(\sqrt{5}-2)/2} \frac{dx}{\sqrt{5 - (x + 1)^2}}$$

is transformed to $\int_c^d F(u) \, du$ by the substitution $x + 1 = \sqrt{5} \sin u$. Express $F(u)$, c, and d in simplest form.

17. Using either the tangent line approximation or numerical integration, approximate $s(3.1)$ if $s(3) \approx 9.74709$ and

$$s(x) = \int_0^x \sqrt{1 + 4w^2} \, dx.$$

18. Find a recursion formula for the integral

$$I_n = \int_1^e x(\ln x)^n \, dx.$$

and then use it to evaluate I_3.

19. If a stream of water were running down the graph of

$$f(x) = 2x^3 - 15x^2 + 35x - 21,$$

a pool would fill and overflow. Calculate the area of the filled pool.

20. Establish the identity

$$\arctan x + \arctan \frac{1}{x} = \frac{\pi}{2}, \quad x > 0$$

by first showing that

$$\int_x^1 \frac{dw}{1 + w^2} = \int_1^{1/x} \frac{dw}{1 + w^2}.$$

(Hint: Let $w = 1/u$ in one of these integrals and then evaluate each integral.)

21. A reduction formula for evaluating integrals containing powers of the logarithm function is

$$\int x^b (\ln ax)^n \, dx = \frac{x^{b+1}}{b + 1} (\ln ax)^n$$
$$- \frac{n}{b + 1} \int x^b (\ln ax)^{n-1} \, dx.$$

where n is a positive integer, $a > 0$, and $b \neq -1$.
a. Use integration by parts to derive this formula.
b. Use the reduction formula to evaluate $\int \sqrt{x} (\ln x)^2 \, dx$. Explain how you use the reduction formula at each step.

22. Let

$$I_n = \int_0^x w^n (w^2 + a^2)^{-1/2} \, dw.$$

Use integration by parts to show that

$$nI_n = x^{n-1} \sqrt{x^2 + a^2}$$
$$- (n - 1)a^2 I_{n-2}(x), \quad n \geq 2.$$

23. Use the trapezoid rule with $n = 10$ to approximate $\pi/4$. Note that

$$\pi/4 = \int_0^1 \frac{dx}{1 + x^2}.$$

How accurate is the approximation?

24. Use the midpoint rule with $n = 10$ to approximate

$$\int_0^{0.5} x^2 e^{-x^2} \, dx.$$

25. Use Simpson's rule with $n = 10$ to approximate $\ln 3$. Note that

$$\ln 3 = \int_1^3 (1/x) \, dx.$$

How accurate is the approximation?

26. Use the inequality

$$\left| \int_a^b f(x) \, dx - T_n \right| \leq \frac{b - a}{12} h^2 M$$

in approximating $\int_0^\pi \sin(x^2) \, dx$ to within 0.001. Show all calculations.

Student Project

Improved Calculation of "Hyperbola-Areas"

In the Autumn of 1665 Isaac Newton, newly graduated from Cambridge, discovered how to calculate the area beneath a hyperbola. He recorded his discovery in a notebook. We reproduce part of this notebook below, with only minor changes in his notation. We note that "valor" means "value" and "&c." means "et cetera." We have made use of *The Mathematical Papers of Isaac Newton*, D. T. Whiteside (ed.), vol. I, Cambridge University Press, 1967, pp. 134–141.

Improved Calculation of Hyperbola-Areas

If $ea \parallel vb \parallel dc \perp ac \parallel ev = vb = a.\ \&\ bc = x.\ \&\ dc = y = \dfrac{aa}{a+x}$. Then is vd a Hyperbola &c:

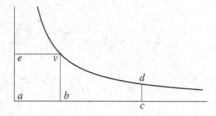

And if $\dfrac{aa}{a+x}$ bee divided as in decimall fractions ye product is

$$\frac{aa}{a+x} = y = a - x + \frac{xx}{a} - \frac{x^3}{aa} + \frac{x^4}{a^3} - \frac{x^5}{a^4} + \frac{x^6}{a^5} - \frac{x^7}{a^6} + \frac{x^8}{a^7}$$

$$- \frac{x^9}{a^8} + \frac{x^{10}}{a^9} - \frac{x^{11}}{a^{10}} + \frac{x^{12}}{a^{11}} - \frac{x^{13}}{a^{12} + a^{11}x}\ \&c.$$

Which valor of y being each terme thereof multiplyed by x & divided by ye number of its dimensions: The product will bee ye area vbcd. viz:

$$vbcd = ax - \frac{xx}{2} + \frac{x^3}{3a} - \frac{x^4}{4a^2} + \frac{x^5}{5a^3} - \frac{x^6}{6a^4} + \frac{x^7}{7a^5} - \frac{x^8}{8a^6} + \frac{x^9}{9a^7}$$

$$- \frac{x^{10}}{10a^8} + \frac{x^{11}}{11a^9} - \frac{x^{12}}{12a^{10}}\ \&c.$$

As for example. If $a = 1.\ \&\ x = 0.1$. The calculation is as Followeth. [We show on the next page a few of the calculations Newton did by hand, copied out by hand, and added by hand.]

$$0{:}10000; 00000; 00000; 00000; 00000; 00000; 00000; 00000; 00000; 00000; 00000 = ax$$

$$0{:}00033; 33333; 33333; 33333; 33333; 33333; 33333; 33333; 33333; 33333 = \frac{x^3}{3a}$$

$$0{:}00000; 20000; 00000; 00000; 00000; 00000; 00000; 00000; 00000; 00000; 00000 = \frac{x^5}{5a^3}$$

$$0{:}00000; 00142; 85714; 28571; 42857; 14285; 71428; 57142; 85714; 28571; 42857 = \frac{x^7}{7a^5}$$

$$\vdots$$

$$0{:}00000; 00000; 00000; 00000; 00000; 00000; 00000; 00000; 00000; 00000; 00196 = \frac{x^{51}}{51a^{49}}$$

$$0{:}00000; 00000; 00000; 00000; 00000; 00000; 00000; 00000; 00000; 00000; 00001 = \frac{x^{53}}{53a^{51}}$$

$$\vdots$$

$$0{:}00500; 00000; 00000; 00000; 00000; 00000; 00000; 00000; 00000; 00000; 00000 = \frac{1}{2}xx$$

$$0{:}00002; 50000; 00000; 00000; 00000; 00000; 00000; 00000; 00000; 00000; 00000 = \frac{x^4}{4a^2}$$

$$0{:}00000; 01666; 66666; 66666; 66666; 66666; 66666; 66666; 66666; 66666 = \frac{x^6}{6a^4}$$

$$0{:}00000; 00012; 50000; 00000; 00000; 00000; 00000; 00000; 00000; 00000; 00000 = \frac{x^8}{8a^6}$$

$$\vdots$$

$$0{:}00000; 00000; 00000; 00000; 00000; 00000; 00000; 00000; 00000; 00000; 02000 = \frac{x^{50}}{50a^{48}}$$

$$0{:}00000; 00000; 00000; 00000; 00000; 00000; 00000; 00000; 00000; 00000; 00019 = \frac{x^{52}}{52a^{50}}$$

The [Difference of these two summes] is equall to ye area $bcdv$, viz:

$$bcdv = 0.09531, 01798, 04324, 86004, 39521,$$
$$23280, 76509, 22206, 05365, 30864, 41992$$

This ends Newton's calculation. Note that Newton knew, in 1665, that the area beneath the graph of $y = x^n$, $0 \le x \le b$, is $b^{n+1}/(n+1)$.

Problem 1.   Explain and verify the details of Newton's work.

Problem 2.   Calculate Newton's result using a "modern" function, one Newton made no mention of in his notebook. Explain your work.

Problem 3.   Did Newton include sufficiently many terms to obtain 52 decimal places of accuracy, as he claimed? Explain your answer.

Student Project

How to Discover an Integration Algorithm

The trapezoid rule for approximating $\int_a^b f(x)dx$ is given by

$$(1) \qquad \int_a^b f(x)\,dx \approx T_n = \frac{1}{2}h\left(f(a) + f(b) + 2\sum_{i=1}^{n-1} f(a + ih)\right),$$

where $h = (b - a)/n$. In applying the trapezoid rule, we attempt to control the error $E_n = T_n - \int_a^b f(x)\,dx$ by choosing n. By studying how E_n depends on n,

Werner Romberg discovered in 1955 a way to improve the efficiency of the trapezoid rule. Romberg showed that several applications of the trapezoid rule can be combined to obtain a strong increase in accuracy at the cost of relatively little additional calculation.

For a function f and interval $[a, b]$, Romberg looked at the approximation sequence $T_2, T_4, T_8, \ldots$. This sequence is efficient since in calculating T_4 roughly half of necessary function evaluations are included in T_2. In the interval $[0, 1]$, for example, T_4 requires that we evaluate $f(0)$, $f(1/4)$, $f(2/4)$, $f(3/4)$, and $f(1)$. Of these, three are included in T_2. Similarly, for T_8 we may use T_4 and the additional values $f(1/8)$, $f(3/8)$, $f(5/8)$, and $f(7/8)$.

In general, the calculation of T_4 from the function data needed for T_2, the calculation of T_8 from T_4, and so on, is given by the formula

$$(2) \qquad R(m + 1, 1) = \frac{1}{2} R(m, 1) + \frac{b - a}{2^{m+1}} \sum_{i=1}^{2^m} f\left(a + (2i - 1)\frac{b - a}{2^{m+1}}\right),$$

where $R(m, 1)$ denotes T_{2^m}, $m = 1, 2, 3, \ldots$. We ask you to verify (2) in Problem 1.

For Problems 2–5 we shall need the value I of the integral

$$(3) \qquad I = \int_0^\pi \frac{\sin x}{x} dx = 1.85193705198.\ldots$$

Problem 1. Verify (2) for the special cases $m = 1, 2, 3$. Do this for arbitrary f, a, and b. This problem is to be done "by hand."

Problem 2. Use (2) in calculating the data for Table 1.

Problem 3. Show that the results in Table 1 lead to the conjecture

$$I \approx \frac{4R(m + 1, 1) - R(m, 1)}{4 - 1}.$$

Problem 4. Calculate the data for Table 2, where

$$R(m + 1, 2) = \frac{4R(m + 1, 1) - R(m, 1)}{4 - 1}.$$

Table 1

m	2^m	$R(m, 1)$	$E_m = R(m, 1) - I$	E_m / E_{m+1}
1	2	1.78539816340	−0.0665389	4.05011
2	4	1.83550812328	−0.0164289	4.01220
3	8	1.84784230644	−0.00409475	4.00303
4	16	1.85091414037	−0.00102291	4.00076
5	32	1.85168137241	−0.00025568	4.00019
6	64	1.85187313511	−0.0000639169	4.00005
7	128	1.85192107295	−0.000015979	4.00001
8	256	1.85193305724	−0.00000399475	4.00000
9	512	1.85193605330	−0.000000998686	4.00000
10	1024	1.85193680231	−0.000000249671	

Table 2

m	2^m	$R(m, 1)$	$R(m, 2)$	$R(m, 2) - I$
1	2	1.78539816340		
2	4	1.83550812328	1.85221144324	2.7×10^{-4}
3	8	1.84784230644	1.85195370083	1.7×10^{-5}
4	16	1.85091414037	1.85193808501	1.0×10^{-6}
5	32	1.85168137241	1.85193711643	6.4×10^{-8}
6	64	1.85187313511	1.85193705601	4.0×10^{-9}
7	128	1.85192107295	1.85193705223	2.5×10^{-10}
8	256	1.85193305724	1.85193705200	1.6×10^{-11}
9	512	1.85193605330	1.85193705198	9.8×10^{-13}
10	1024	1.85193680231	1.85193705198	6.1×10^{-14}

Problem 5.   These results suggest forcibly that the arithmethic combination

$$R(m + 1, 2) = \frac{4R(m + 1, 1) - R(m, 1)}{4 - 1}$$

of $R(m + 1, 1)$ and $R(m, 1)$ better approximates the integral (3) than either $R(m, 1)$ or $R(m+1, 1)$ alone. Table 2 gives evidence that the sequence $R(2, 2)$, $R(3, 2)$, . . . of numbers converges to I faster than $R(1, 1)$, $R(2, 1)$, ... Provide evidence that the *convergence factor* of the former sequence is $4^2 = 16$ while, as shown above, the trapezoid sequence has convergence factor 4. The above reasoning can be repeated to obtain sequences having even larger convergence factors. The general case is

$$R(m + 1, k + 1) = \frac{4^k R(m + 1, k) - R(m, k)}{4^k - 1}, \quad m = k, k + 1, \ldots .$$

The calculations can be displayed in tabular form as follows:

$R(1, 1)$

$R(2, 1) \quad R(2, 2)$

$R(3, 1) \quad R(3, 2) \quad R(3, 3)$

$R(4, 1) \quad R(4, 2) \quad R(4, 3) \quad R(4, 4)$

$\vdots \qquad \vdots \qquad \vdots \qquad \vdots \qquad \ddots$

Such a table is called a Romberg table. The numbers in the first column are the trapezoid rule estimates corresponding to 2^1, 2^2, . . . subintervals. For the integral (4) a partial Romberg table is

1.78539816340

1.83550812328   1.85221144324

1.84784230644   1.85195370083   1.85193651801

1.85091414037   1.85193808501   1.85193704395   1.85193705230

1.85168137241   1.85193711643   1.85193705186   1.85193705198

The accuracy in the fourth column is striking. These estimates arise from the trapezoid estimates T_2, T_4, . . . , T_{32}. Most calculations using the Romberg algorithm do not go beyond three or four columns. This provides sufficient accuracy for most purposes and avoids errors due to rounding that may occur in larger Romberg tables.

Problem 6. Use the Romberg algorithm to calculate to 9 decimal places the value of the elliptic integral

$$\int_0^1 \frac{1}{\sqrt{1 - 0.5 \sin^2 x}}\, dx.$$

Use your judgment as to when you have attained nine places of accuracy.

Problem 7. Give an empirical argument that the Romberg algorithm is more efficient than the trapezoid rule. Measure efficiency by the number of function evaluations required to achieve a given accuracy.

Problem 8. Show algebraically that $R(m + 1, 2)$ is Simpson's rule with 2^{m+1} intervals.

A History of Measuring and Calculating Tools
Bush's Differential Analyzer

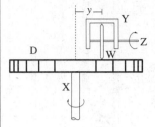

Figure 5.44. Disk-wheel integrator.

The need to calculate integrals $\int_a^b f(x)\,dx$ arises in calculations of the center of mass of an object or the area bounded by a given curve, in the preparation of ballistic tables, and in solving the differential equations describing complex power systems. Although numerical methods such as Simpson's rule could be used, the size or complexity of many problems made this approach impractical before digital computers came into common use after World War II. The idea of a mechanical integrator was introduced as early as 1825, but it took more than a century to refine the basic idea and to construct a device capable of solving large-scale problems. Vannevar Bush led the team at MIT that built the differential analyzer. (Bush is shown here with the machine.) The machine was used for solving differential equations.

We describe the idea of a mechanical integrator, which with other mechanical analogues of mathematical operations make up the differential analyzer.

Referring to Fig. 5.44, disk D, seen from the side, is driven by input shaft X, which has turned through x radians from a fixed position. Rolling on D is a wheel W of radius a, held in a yoke Y, which moves along a diameter of D. The position y of the yoke is a given function f of the input x to X.

The wheel W drives output shaft Z. Denoting the rotation of Z from a fixed position by z, suppose that $z = 0$ when $x = 0$.

For input x, suppose x increases by a small amount dx. Wheel W will turn on Z through dz and roll a distance $a\,dz$ upon D, which is approximately equal to $y\,dx$ on D. From this we have

$$a\,dz \approx y\,dx.$$

The solution of the initial value problem

$$\frac{dz}{dx} = y \quad \text{and} \quad z(0) = 0$$

is

$$z = z(x) = \int_0^x f(x)\,dx.$$

The output $z(x)$ of shaft Z for input x can be used to drive a plotter to graph the antiderivative z of f. The disk-wheel mechanism is an integrator.

Small industrial integrator/calculators were built containing six integrators, two plotters, six constant multipliers, and four adders. These machines were the size of an office desk, weighed about 500 pounds, and were capable of 0.1 percent accuracy. Currently, calculators occupying less than one-thousandth of the volume of an office desk, weighing about one-thousandth of 500 pounds, and having at least one thousand times more computational power are commonly available for less than the cost of a pair of running shoes.

Chapter 6

Applications
of the Integral

Volumes, Lengths,
Forces, Work, . . .

Many new ideas in the physical, mathematical, and engineering sciences succeed or fail depending on whether they reduce classical problems to routine calculations, how many hitherto unsolved problems they settle, and the kinds of new ideas and problems they open up.

By these criteria, the idea of the integral introduced by Newton and Leibniz shortly after 1650 has been a spectacular success. Their work greatly extended that of Greek geometers, who by 300 B.C. knew how to calculate the areas of rectangles, circles, parabolic segments, and a spiral; the volumes of parallelepipeds, spheres, and pyramids; and the lengths of lines and circles.

Newton and Leibniz's work made it possible to calculate the areas, volumes, and lengths of a huge variety of geometric objects and to formulate and solve new problems. A modern area in which integration plays a key role is CAT scan technology, in which an object is irradiated from many directions and exposure data are integrated to construct a view of the interior of the object.

We take up seven applications of the integral that are of continuing interest:

- The volumes of solids with variable cross-sections
- The arc length of a curve
- The areas of regions bounded by vector/parametric or polar curves
- Line integrals and the work done by a force field on an object
- The mass and the center of mass of an object
- Curvature and components of acceleration
- Improper integrals

6.1 VOLUMES BY CROSS-SECTION

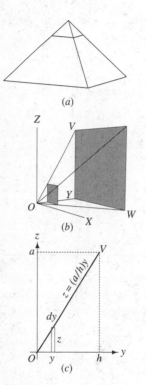

(a)

(b)

(c)

Figure 6.1. The Great Pyramid, stacked.

Ultramicrotomes are used by scientists in preparing very thin cross-sections of plants or animals for study. For examination in a transmission electron microscope, a cross-section must be sufficiently thin that the electron beam can penetrate. The small advances of the diamond knife used to cut successive sections are achieved by thermal expansion of the support of the knife, which typically moves the knife 20–100 nanometers.

We imagine a student setting an ultramicrotome to work on the Great Pyramid at Gizeh (see Fig. 6.1(a)), cutting 30-nanometer sections parallel to its base. Since 30 nanometers is 0.00000003 m and the Great Pyramid's height is approximately 146.8 m, she gets slightly more than $m = 4,893,333,333$ sections. As the sections are cut, she stacks them in a convenient corner O, as shown in Fig. 6.1(b). The lines OX, OY, and OZ forming the corner are mutually perpendicular. Shown in the figure are two shaded sections, a small one and the base. Each section rests on one of its edges, with a corner touching OY.

With the idea of approximating the volume of the Great Pyramid by adding the volumes of the nearly 5 billion sections, she measures $a = YW = 230.4$ m and $h = OY = 146.8$ m. The volume of the sections ought to be close to the volume of the Great Pyramid since, apart from a little dust, the cutting and stacking hasn't changed the volume.

Since each section is so thin, she uses $z^2\,dy$ as a close approximation to its volume, where z is the side of a square section, dy is its thickness, and the section is located y meters along OY. Such a section is seen edgewise in Fig. 6.1(c). The student knows that the equation of line OV in Fig. 6.1(c) is $z = (a/h)y$ since its slope is a/h. This determines the side z of the section cut y meters down from the top. Now it's just a matter of adding the volumes of the 5 billion sections. The volume of the first would be $((a/h) \cdot dy)^2\,dy$, the second $((a/h) \cdot 2 \cdot dy)^2\,dy$, and the last $((a/h) \cdot m \cdot dy)^2\,dy$. She now writes

$$V \approx \frac{a^2}{h^2} \cdot dy^2 \cdot (1^2 + 2^2 + \cdots + m^2)\,dy.$$

With the help of the formula $1^2 + 2^2 + \cdots + m^2 = \frac{1}{6}m(m+1)(2m+1)$ and recalling that $dy = 3 \times 10^{-8}$ m, she obtains

$$V \approx \frac{a^2}{h^2} \cdot 9 \times 10^{-16} \cdot (3.90565 \times 10^{28}) \cdot 3 \times 10^{-8} \approx 2.59759 \times 10^6 \text{ m}^3.$$

Just before she collapses from exhaustion she suddenly realizes that in the limit, as the thickness of the sections approaches zero, she may sum the volume elements $dV = z^2\,dy$ with the integral. She writes

$$V = \int_0^h z^2\,dy = \int_0^h (a/h)^2 y^2\,dy = \frac{a^2}{h^2}\frac{1}{3}y^3\Big|_0^h = \frac{1}{3}a^2h.$$

She checks her result with this formula (also known to the Egyptians), obtaining $V \approx 2.59758 \times 10^6$ m^3. Astonishing!

This short story illustrates the main idea of this section. If we can write a formula for the area of parallel cross-sections of a solid as a function of their position along a line, then we can determine the volume of that solid by integration. The slicing and stacking of the pyramid into a corner without changing its volume illustrates Cavalieri's principle (a two-dimensional version of which we used in Chapter 5 to calculate the area of the region between two curves): "If two solids have equal altitudes and if sections made by planes parallel to the bases and at equal distances from them are always in the same ratio, then the volumes are also in this ratio." The two solids are the Great Pyramid, in place, and the sectioned pyramid, stacked into a corner.

As we show in the first example, often we may slice the solid in place and calculate its volume without rearranging the sections.

■ **EXAMPLE 1** Calculate the volume of a pyramid with height h and square base with side a.

Solution. We take cross-sections of the pyramid, parallel to its base. Let s be the length of the side of the section that is z units from the top of the pyramid. From the thickness dz and side s of the section we may write its volume dV as $dV \approx s^2 dz$. We calculate the side s of the section in terms of z. For this draw triangle OZW. Figure 6.2 shows this triangle below the pyramid as well, to minimize distractions. Using similar triangles, we have

$$\frac{z}{y} = \frac{h}{a/2}.$$

Solving this equation for y gives $s = 2y = az/h$. Hence

$$dV \approx s^2 dz = \frac{a^2 z^2}{h^2} dz,$$

and the volume of the pyramid is

$$V = \int_0^h \frac{a^2 z^2}{h^2} dz = \frac{a^2}{h^2} \tfrac{1}{3} z^3 \Big|_0^h = \tfrac{1}{3} a^2 h.$$

■

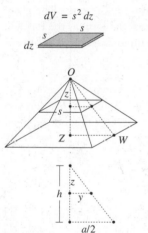

Figure 6.2. Sectioning a pyramid.

Disks and Washers

Many solids have circular cross-sections or may be generated by revolving a two-dimensional region about an axis in the plane of the region. For example, a (solid) ball can be generated by revolving a disk (circle plus interior) about one of its diameters. A (solid) right circular cone can be generated by revolving a right triangle (plus its interior) about one of its sides. We calculate the volumes of two solids of revolution after discussing the general case.

A set of three mutually perpendicular coordinate axes is shown in Fig. 6.3. The generating curve C for the solid of revolution is the graph of an equation $y = f(x)$, where $a \le x \le b$. The graph lies in the (x, y)-plane.

As C is revolved about the x-axis, it generates (the surface of) a solid of revolution. The solid is suggested by the simple line drawing shown in Fig. 6.3. The figure consists of three axes, the curve C, three quarter circles, and a foreshortened view of C in the (x, z)-plane. Only one quarter of the solid is sketched so that the essential distances and variables can be seen and identified. Practice on this kind of figure will pay off not only in the exercises for this section, but later as well,

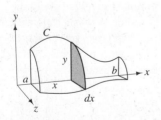

Figure 6.3. Solid of revolution.

when you study multidimensional calculus. A subdivision of the interval $[a, b]$ on the x-axis divides the solid into thin slices. One quarter of a typical slice is shown in Fig. 6.3. It has thickness dx and is located by the coordinate x. The slice is a circular cross-section or disk with radius $y = f(x)$. The volume dV of the disk is $dV \approx \pi y^2 \, dx$, which comes from the formula $V = \pi r^2 h$ for the volume of a right circular cylinder of height h and base radius r.

The volume of the solid of revolution may be calculated by summing the slices, letting the number of subdivisions increase without bound, and evaluating the limiting integral. This procedure can be summarized by

$$(1) \qquad V \approx \sum dV = \sum \pi y^2 \, dx \to \int_a^b \pi y^2 \, dx = \int_a^b \pi (f(x))^2 \, dx = V.$$

■ EXAMPLE 2   Show that the volume V of a right circular cone of height h and base radius r is given by $V = \frac{1}{3}\pi r^2 h$. (This has the same form as the formula $V = \frac{1}{3}a^2 h$ for the volume of a pyramid in that each formula is $\frac{1}{3}$ times the area of the base times the height.)

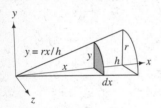

$y = rx/h$

Figure 6.4. Cone.

Solution.   A line drawing of one quarter of the cone is shown in Fig. 6.4. The cone is generated by revolving the line $y = rx/h$ around the x-axis. We show in the figure two quarter-circle cross-sections, one at $x = h$ and the other at a typical point x. The radius of the circular cross-section or disk at x is $y = f(x) = rx/h$. For the volume dV of the disk at x we have

$$dV \approx \pi y^2 \, dy = \pi(r^2 x^2 / h^2) \, dx.$$

By integrating between $x = 0$ and $x = h$ we take in the entire cone.

$$V = \int_0^h \pi(r^2 x^2 / h^2) \, dx = \frac{\pi r^2}{h^2} \int_0^h x^2 \, dx = \frac{\pi r^2}{h^2} \frac{1}{3} x^3 \Big|_0^h = \frac{1}{3}\pi r^2 h \qquad ■$$

■ EXAMPLE 3   A shaded cross-section of a "bead" of height h is shown in Fig. 6.5. Such a bead may be made by drilling a hole of radius r through a ball of radius a, along a diameter. Show that the volume V of the bead is $V = \frac{1}{6}\pi h^3$.

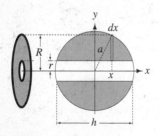

Figure 6.5. Bead.

Solution.   The volume of the bead may be calculated in several ways. We could calculate the volume of the drilled hole and subtract this from the volume of the sphere. Or we may view the bead as generated by revolving about the x-axis the region bounded above by the circle $x^2 + y^2 = a^2$ and below by the line $y = r$. We show in this region an element at position x, of width dx, and lying between the line $y = r$ and the circle $x^2 + y^2 = a^2$. If this area element is revolved about the x-axis, it forms a volume element resembling a metal washer. A washer element is shown on the left side of the figure. Its volume is the difference in volume of two disks (actually cylinders of height dx)—a larger one with radius R, which varies with x, and a smaller one with radius r, which is a constant. Noting that $x^2 + R^2 = a^2$, we have

$$dV \approx \pi R^2 \, dx - \pi r^2 \, dx = \pi(R^2 - r^2) \, dx$$
$$dV \approx \pi((a^2 - x^2) - r^2) \, dx.$$

Summing the volume elements between $x = -h/2$ and $x = h/2$, we have

$$V = \pi \int_{-h/2}^{h/2} (a^2 - r^2 - x^2)\, dx = \pi \left((a^2 - r^2)x - \tfrac{1}{3}x^3\right)\Big|_{-h/2}^{h/2}$$

$$V = \pi \left((a^2 - r^2)h - \tfrac{1}{3}\tfrac{1}{4}h^3\right).$$

We may relate the height h of the bead and the radius r of the drilled hole by noting that the point $(\tfrac{1}{2}h, r)$ lies on the circle. Hence, $\tfrac{1}{4}h^2 + r^2 = a^2$. Using this result, the volume of the bead of height h is

$$V = \pi \left((\tfrac{1}{4}h^2)h - \tfrac{1}{3}\tfrac{1}{4}h^3\right) = \tfrac{1}{6}\pi h^3. \qquad \blacksquare$$

You may have noticed that the volume of the bead depends only on its height and not on the radius of the sphere from which it is cut. So if we wish to manufacture a bead of height, say, $h = 0.75$ cm, we may drill either the earth (radius 6378 km) or a cranberry (average radius 0.5 cm). The volume will be $\tfrac{1}{6}\pi(0.75)^3$ in either case. Hmmm.

We gave in (1) a "formula" for the disk method. We think it best not to emphasize such formulas since they tend to work against the advantages in adapting a calculation to a specific problem. We think it is more important to gain a working understanding of a reasonably general procedure for calculating volumes by cross-section. We outline such a procedure before working through one more example.

For this, let S be a solid sectioned by planes perpendicular to a line L and suppose that the areas of the cross-sections depend on the position of the plane with respect to L.

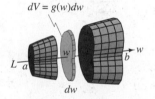

$dV = g(w)\,dw$

Figure 6.6. Volumes by cross-section.

To calculate the volume V of S by cross-section:

1. Regard L as a coordinate axis, and choose a coordinate variable—say, w—and an interval $[a, b]$ on L so that as w varies over $[a, b]$, the solid S can be completely sectioned by planes perpendicular to L.

2. Use supplementary sketches and geometry as needed, and express the area of the cross-section at w in terms of a function $g(w)$.

3. Express the volume dV of the cross-section of thickness dw at w as $dV \approx g(w)\, dw$.

4. Evaluate $\int_a^b g(w)\, dw$ to sum the volumes of the cross-sections.

This procedure is based on recognizing a continuous function g defined on an interval $[a, b]$ such that:

- If we subdivide $[a, b]$ into n equal subintervals, the volume V lies between the lower and upper sums for this subdivision.
- These sums approach $\int_a^b g(w)\, dw$ as $n \to \infty$.

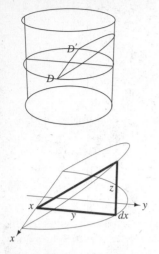

Figure 6.7. Tree wedge.

■ EXAMPLE 4   A 30° wedge is cut from a cylindrical tree, as in Fig. 6.7. One plane of the wedge is perpendicular to the axis of the tree/cylinder. The planes intersect along a diameter DD' of the circular cross-section of the cylinder. Calculate the volume of the wedge if DD' has length $2a$.

Solution.   Cross-sections of the wedge by planes perpendicular to DD' are triangles, one of which is shown in the enlarged view of the wedge at the bottom of the figure. The diameter DD' has become the x-axis. (There are other possibilities. For example, cross-sections by planes perpendicular to the diameter that became the y-axis are rectangles.)

The triangular cross-sections have area $\frac{1}{2}yz$. Since the wedge angle is 30°, $z = (\tan 30°)y = y/\sqrt{3}$. The volume dV of the cross-section at x, with thickness dx and area $y^2/(2\sqrt{3})$, is

$$dV \approx \frac{y^2\,dx}{2\sqrt{3}}.$$

Since $y^2 = a^2 - x^2$, we have

$$dV \approx \frac{(a^2 - x^2)\,dx}{2\sqrt{3}}.$$

Making use of symmetry, the volume of the wedge is

$$V = 2\int_0^a \frac{(a^2 - x^2)\,dx}{2\sqrt{3}} = \frac{1}{\sqrt{3}}\int_0^a (a^2 - x^2)\,dx$$

$$= \frac{1}{\sqrt{3}}\left(a^2 x - \tfrac{1}{3}x^3\right)\Big|_0^a = \frac{2a^3}{3\sqrt{3}}.$$ ■

Water Quality

An estuary is the part of the wide lower course of a river where its current is influenced by the tides. Because estuaries are important in the ecology of the contiguous lands, bays, and oceans, their waters are closely monitored for contaminants. Among the data used to determine water quality are the average net movement of water in and out of the estuary and the total volume of water passing a given point. The calculations of such quantities are closely related to those used to determine the volume of a solid by cross-section.

We use data taken from the paper *Observations of Tidal Flow in the Delaware River,* by E. G. Miller (U.S. Geological Survey Water-Supply Paper 1586–C). The Delaware River flows past Philadelphia and Wilmington and empties into Delaware Bay in the Atlantic Ocean. In Table 1 we give data collected at the Delaware Memorial Bridge, just below Wilmington, on August 21, 1957. The third column gives the area of the cross-section of the water in the river channel at the point at which the mean velocity and flow were determined. The positive direction is downstream, so that at 7:00 A.M., for example, the movement of the tide was upstream (this is flood tide; during ebb tide the water flows downstream) at a speed of 2.48 ft/s. At the same time, some 424,000 ft^3 of water was passing the metering point each second.

From the data given in Table 1 we calculate the average net movement of water and the total volume of water passing the metering point during one tide cycle. We

Table 1. Tidal flow data

Time	Gauge height (ft)	Area (ft²)	Mean velocity (ft/s)	Flow (ft³/s)
7:00 A.M.	4.8	171,000	−2.48	−424,000
8:00	4.8	171,000	−1.63	−280,000
8:56	4.2	168,000	0.00	0
9:00	4.1	167,000	0.17	28,000
10:00	3.0	161,000	1.52	245,000
11:00	2.1	156,000	2.24	348,000
12:00	1.3	151,000	2.61	394,000
1:00 P.M.	0.7	148,000	2.51	370,000
2:00	0.3	145,000	2.03	296,000
3:00	0.7	148,000	0.49	721,000
3:13	1.1	150,000	0.00	0
4:00	2.6	158,000	−2.03	−322,000
5:00	3.8	166,000	−3.16	−523,000
6:00	5.0	173,000	−3.44	−594,000
7:00	5.8	177,000	−3.26	−578,000
8:00	6.2	180,000	−2.44	−418,000
9:00	6.0	178,000	−1.23	−220,000
9:37	5.5	175,000	0.00	0

leave some details to Exercise 38 and, for a second data set, pose a similar question. (See Exercise 39.)

A tide cycle is the period of time between two ebb tides, measured from the "slack water" times. We note in the table that slack water occurred at 8:56 A.M., just prior to an ebb tide. The slack water preceding the next ebb tide occurred at 9:37 P.M.

To calculate the average net movement of water on the basis of observations at one point, we imagine a "filament of water" flowing past the metering point, upstream or downstream depending upon the velocity data. The filament is assumed to have a fixed length, so that its progress is governed by its velocity at the metering point. We imagine the filament moving on an x-axis with origin at the metering point and positive direction downstream. Letting $x = x(t)$ and $v = v(t)$ be the position and velocity, respectively, of the filament at any time t, we know that

$$(2) \qquad x(t_{15}) = \int_{t_0}^{t_{15}} v(t)\, dt,$$

where t_0 and t_{15} are measured in seconds and correspond to 8:56 and 9:37. The position of the filament at t_{15} is the "net area" beneath the velocity graph. We show the velocity graph in Fig. 6.8. We use the trapezoid rule to approximate $x(t_{15})$. Because the time intervals are not equal, we calculate the trapezoid approximation on each subinterval. We have

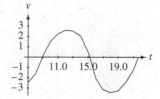

Figure 6.8. Mean velocity.

$$(3) \qquad \begin{aligned} x(t_{15}) &= 3600 \int_{t_0}^{t_{15}} v(t)\, dt \\ &\approx 3600 \cdot \tfrac{1}{2} \cdot \sum_{i=0}^{14} (v(t_i) + v(t_{i+1}))(t_{i+1} - t_i) \\ &\approx -13{,}700 \text{ ft.} \end{aligned}$$

Depending on how closely you looked at the data in Table 1, this result may be surprising. For it appears that the filament went some 13,700 ft ($\approx$ 2.6 miles) upstream during the tide cycle. We would expect that the net movement of the water would be downstream. We put this aside for a moment and calculate the net flow past the metering point.

The net flow F, which is the volume of water passing the metering point during one tide cycle, depends on both the velocity and the area of the cross-section at the metering point. Otherwise, the calculation is similar to that in (2). Letting $A(t)$ denote the cross-section of the river at time t at the metering point, we have

$$F = \int_{t_0}^{t_{15}} A(t)v(t)\,dt$$

(4)
$$\approx 3600 \cdot \tfrac{1}{2} \cdot \sum_{i=0}^{14} \big(A(t_i)v(t_i) + A(t_{i+1})v(t_{i+1})\big)(t_{i+1} - t_i)$$

$$\approx -3{,}190{,}000{,}000 \text{ ft}^3.$$

Again, this result may be surprising. During this tide cycle there is a net upstream flow of water past the metering point.

What does this mean? Although the net flow is just one of the many measurements made to monitor the water quality in an estuary, this finding suggests that during this particular tide cycle contaminants are accumulating in the estuary. The calculation may be used to order an increase in direct sampling of the water.

Exercises 6.1

Basic

Exercises 1–10:   Use disks in calculating the volume of the solid of revolution generated by revolving a region about the x-axis. The region is in the (x, y)-plane and is bounded above by a curve and below by the x-axis. Make a simple sketch of the region and volume, including a typical cross-section.

1. $y = \sqrt{a^2 - x^2}, -a \le x \le a$
2. $y = b\sqrt{1 - x^2/a^2}, -a \le x \le a$
3. $y = x^2, 0 \le x \le 2$
4. $y = x^3, 0 \le x \le 3/2$
5. $y = \sqrt{x}, 0 \le x \le 2$

6. $y = \sqrt[3]{x}, 0 \le x \le 2$
7. $y = \sin x, 0 \le x \le \pi$
8. $y = \cos x, 0 \le x \le \pi/2$
9. $y = \sec x, 0 \le x \le \pi/4$
10. $y = \tan x, 0 \le x \le \pi/4$

Exercises 11–18:   Use washers in calculating the volume of the solid of revolution generated by revolving a region about the x-axis. The region is in the (x, y)-plane and is bounded by two curves. Make a simple sketch of the region and solid, including a typical cross-section and washer.

11. $y = x^2, y = x^3, 0 \le x \le 1$
12. $y = \sqrt{x}, y = \sqrt[3]{x}, 0 \le x \le 1$
13. $y = -x(x - 5), y = x$
14. $y = x + 1, y = (x - 1)^2$
15. $y = \sin x, y = (2/\pi)x, 0 \le x \le \pi/2$ (useful identity: $\sin^2 x = (1 + \cos 2x)/2$)

16. $y = \sec x, \pi\sqrt{3}(y - 2) = 4(\sqrt{3} - 1)(x - \pi/3), |x| \le \pi/2$
17. $y = \ln x, y = -(x - 1)(x - 3), x \ge 1$
18. $y = e^x, y = -5x(x - 3)$

19. Work through the Egyptian calculus student's calculations, given at the beginning of the section, but take the sections 0.001 m (1 millimeter) thick.

20. We stated that the builders of the Great Pyramid at Gizeh used the formula $V = \frac{1}{3}a^2h$. Although this is quite probable, all that is known for sure is that they used the formula $V = \frac{1}{3}h(a^2 + ab + b^2)$ for the volume of a frustum of a pyramid, where h is the height of the frustum and a and b its bases. Derive the formula for the volume of a frustum from that for the volume of a pyramid.

21. Show that the volume of the oblique cone shown is $V = \frac{1}{3}\pi r^2 h$. Note that this result is evidence that Cavalieri's principle is true.

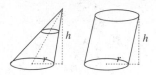

22. Show that the volume of the oblique cylinder shown above is $V = \pi r^2 h$. Note that this result is evidence that Cavalieri's principle is true.

23. The volume of the washer element in Example 3 is $dV = \pi(R^2 - x^2)\,dx$. A common mistake in using washers is to write this as $dV = \pi(R - r)^2\,dx$. Show that these are different.

24. Generalize Example 4 by considering a wedge with angle $\theta \in (0, \pi/2)$.

25. Calculate the volume of the solid formed by rotation of the line joining $(0, b)$ and (H, B), where $0 < b < B$ and $H > 0$, about the x-axis.

Growth

26. Calculate directly the volume of the drilled hole in Example 3. Note that part of the drilled hole is a right circular cylinder.

27. Example 3 gives a calculation of the volume of a bead of "height" h. At the end of the example we mentioned drilling a bead of height 0.75 cm from either the earth or a cranberry. Determine the radius of the drills needed for these two beads. Describe all of this in words that a layperson could understand.

28. Show that the volume of an elliptical bead of height $h < 2a$ is $\frac{1}{6}\pi h^3(b^2/a^2)$, where the ellipse has equation $x^2/a^2 + y^2/b^2 = 1$. The bead is drilled along the x-axis.

29. An alternate approach to Example 4 was mentioned in the solution. Calculate the volume of the wedge using rectangular cross-sections.

30. Express the volume $V(h)$ of punch in a 16 in.–diameter hemispherical bowl as a function of its depth.

31. (Continuation of Exercise 30) Given that 1 quart is 57.75 in.^{3}, at what depth does the bowl contain 8 quarts of punch?

32. (Continuation of Exercise 30) "Calibrate" the punch bowl by engraving marks on a great (semi)circle. If the marks are to be located by measuring from the bottom of the bowl along this circle, at what distances should the 4, 8, 12, and 16 quart marks be placed?

33. Find the volume of the torus formed by revolving the circle

$$x^2 + \left(y - \frac{b + a}{2}\right)^2 = \left(\frac{b - a}{2}\right)^2$$

about the x-axis, where $b > a \geq 0$.

34. A solid has a circular base of radius r. Sections of the solid perpendicular to a designated diameter of the base are squares. Calculate the volume of the solid. Two sections of this solid are shown below. Also shown are the curves along which the top corners of the square sections fall.

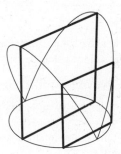

35. The region bounded on the left by the graph of the equation $(y - 3)^2 = x - 4$ and on the right by the line joining the points $(5, 2)$ and $(8, 5)$ is revolved about the x-axis. Calculate the volume of the resulting solid.

36. The larger of the two regions bounded by the circle with equation $(x - 2)^2 + (y - 2)^2 = 1$ and the line through $(1, 2)$ and $(2, 3)$ is revolved about the x-axis. Calculate its volume.

37. A solid has a right triangle as base, with sides a and b. Sections perpendicular to the side OB of length b are right triangles with the vertex of the right angle on OB. For a typical cross-section, the ratio of the length of the vertical side to the horizontal side is c/a. Sketch the solid and calculate its volume.

38. Check the problem formulation and arithmetic for the tidal flow data in Table 1.

39. Table 2 lists data collected at the Burlington-Bristol Bridge, just above Philadelphia, on August 24, 1956. Use these data to calculate the average net movement of water and the total volume of water passing the metering point during one tide cycle. Compare these results with those calculated for the Delaware Memorial Bridge.

Table 2. Tidal flow data

Time	Gauge height (ft)	Area (ft²)	Mean velocity (ft/s)	Flow (ft³/s)
4:46 A.M.	7.2	34,500	0.00	0
5:00	7.0	34,200	0.68	23,200
6:00	6.0	33,000	2.02	66,500
7:00	4.9	31,600	2.09	65,800
9:00	2.8	29,600	1.61	46,600
11:00	1.0	26,900	1.36	36,600
12:00	0.7	26,600	0.15	3,980
12:03 P.M.	0.9	26,800	0.00	0
1:00	2.6	28,700	-2.18	-62,700
2:00	4.0	30,500	-2.31	-70,200
3:00	5.5	32,300	-1.67	-54,200
4:00	6.2	33,200	-1.00	-33,200
4:42	6.1	33,000	0.00	0
5:00	5.8	32,700	0.79	25,900

40. Assume that at a distance r from the center of a pipe of radius R the velocity and density of a fluid are

$$v(r) = v_c(1 - r^2/R^2) \text{ m/s}$$

$$\rho(r) = \rho_c(1 + \alpha r^2/R^2) \text{ kg/m}^3,$$

where v_c, ρ_c, and α are positive constants. Explain why the integral

$$\frac{dm}{dt} = \int_0^R v\rho 2\pi r \, dr$$

gives the mass flow rate through the pipe.

41. To calculate the volume of the solid generated by revolving the region bounded by the parabola with equation $(y - 3)^2 = x - 1$ and line with equation $x + y = 6$ about the x-axis, we may use washer elements. As the accompanying sketch of the region shows, however, the description of the washers changes at $x = 2$. From $x = 1$ to $x = 2$, the outer and inner radii of the washer come from the parabola. From $x = 2$ to $x = 5$, the outer radius of the washer comes from the parabola and the inner radius comes from the line. In practice this means that the volume will be expressed as the sum of two integrals. As a warmup exercise, calculate the volume of the solid in this way. The volume is $45\pi/2 \approx 70.69$.

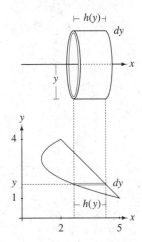

42. (Continuation of Exercise 41) As shown in the preceding problem, the volume of the solid generated by revolving the region between the parabola and the line about the x-axis can be done by the washer method. However, the calculation requires two integrals. The *shell method* provides an alternative approach to this calculation, and for this region and axis of revolution is easier.

The element of area is drawn parallel to the axis of revolution. As shown in the figure, the thin rectangular element is located by y, has thickness dy, and has length $h(y)$. Show that

$$h(y) = (6 - y) - (1 + (y - 3)^2).$$

Imagine that the element has been revolved about the x-axis. The element generates a cylindrical shell, as shown at the top of the figure. The volume dV of the shell is the difference between the volumes of two cylinders, one with radius y and the other with radius $y + dy$. Recalling that the volume of a cylinder with radius r and height h is $\pi r^2 h$, we have

$$dV = \pi(y + dy)^2 h(y) - \pi y^2 h(y)$$
$$= \pi\big(2y\,dy + (dy)^2\big)h(y).$$

Since $(dy)^2$ is small relative to dy, we may write

$$dV \approx 2\pi y h(y)\,dy.$$

This result suggests that the volume may be calculated with the integral

$$V = 2\pi \int_1^4 y h(y)\,dy$$
$$= 2\pi \int_1^4 (-y^3 + 5y^2 - 4y)\,dy.$$

Evaluate this integral, showing that its value is $45\pi/2$. This agrees with the washer method calculation given above.

43. (Continuation of Exercise 42) The volume of the cylindrical shell was calculated by subtracting the volumes of two cylinders. An alternative calculation is based on cutting the cylindrical shell along a line parallel to its axis and then unrolling it to form a rectangular parallelepiped, approximately. The volume of the latter is length times width times height. Hence the volume of the cylindrical shell is $dV \approx 2\pi y h(y)\,dy$. Fill in the details of this discussion.

44. Calculate the volume of the solid generated by revolving the region between the curves described by $y = \sqrt{x}$ and $y = x$ about the x-axis. First use washers and then try shells.

Exercises 45–48: These problems concern the prismoidal formula and its use in such applications as using survey data to approximate the amount of earth to be moved in constructing a highway.

45. The prismoidal formula is: If $p(x) = Ax^3 + Bx^2 + Cx + D$ is a polynomial of degree 3 or less, then

$$\int_a^b p(x)\,dx = \frac{b - a}{6}\big(p(a) + 4p(m) + p(b)\big),$$

where $m = (a + b)/2$.
   a. Verify the prismoidal formula for the special cases $p(x) = x^n$, $n = 0, 1, 2, 3$.
   b. Show that the formula fails for $p(x) = x^4$.
   c. Show that the formula holds for $p(x) = Ax^3 + Bx^2 + Cx + D$.

46. A **prismoid** is a solid S lying between two parallel planes P and P'. Assume that P and P' lie along a line L perpendicular to both planes, with P at $x = a$ and P' at $x = b$. The cross-sections of S at $x = a$ and $x = b$ are called bases of S. Assume also that p is a polynomial of degree 3 or less such that the area $A(x)$ of the cross-section of S at coordinate position x is numerically equal to $p(x)$, where $a \le x \le b$. Show that the volume V of such a prismoid is given by

(5) $V = \frac{1}{6}h(B + 4M + B')$,

where B and B' are the areas of the bases of S, and M is the area of the cross-section at $x = (a + b)/2$.

47. Show that the following solids are prismoids and calculate their volumes by the prismoidal formula. Note that a and c are special cases of b and d.
   a. S is bounded by a cone.
   b. S is bounded by a cone and two parallel planes.
   c. S is bounded by a sphere.
   d. S is bounded by a sphere and two parallel planes.
   e. S is a pyramid.

48. We show later that the prismoid formula holds for prismoids S resembling that shown in the following figure, where the bases B and B' are bounded by convex polygons and the lateral surfaces of S are bounded by convex quadrilaterals or triangles. Apply (5) to calculate the amount of earth to be moved for the highway cut shown in the figure. The site survey contains the necessary data. The polygons B, M, and B' lie in parallel planes, which are 550 feet

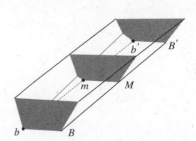

apart (all lengths are given in feet). The vertices of polygons B, M, and B' labeled as b, m, and b' have coordinates $(0, 0)$. These points lie along the dotted line. Counterclockwise from this line, the remaining vertices of B are $(90, 0)$, $(130, 70)$, and $(-20, 70)$; those of M are $(90, 0)$, $(135, 70)$, $(-35, 65)$, and $(-15, 10)$; and those of B' are $(90, 0)$, $(140, 70)$, $(-50, 60)$, and $(-30, 20)$.

Review/Preview

49. During the time interval $[0, 3]$ the position vector of an object is given by $\mathbf{r}(t) = (2 + 3t, 1 + t)$. How far did the object travel?

50. During the time interval $[0, 1]$ the position vector of an object is $\mathbf{r}(t) = (t, t^2)$. By dividing $[0, 1]$ in half and assuming the object traveled from $(0, 0)$ to $(0.5, 0.25)$ and from $(0.5, 0.25)$ to $(1, 1)$ on straight line segments, approximate how far the object traveled. Is your answer too big or too small? How could you obtain a better approximation?

51. Look up an antiderivative of $\sec^3 x$. Use this together with the substitution $2x = \tan \theta$ to evaluate the integral $\int_0^1 \sqrt{1 + 4x^2}\, dx$.

6.2 ARC LENGTH AND UNIT TANGENT

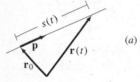

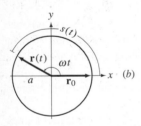

Figure 6.9. Length of path.

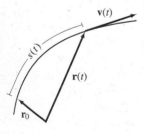

Figure 6.10. Length of path.

Calculating the distance traveled by an object moving with uniform motion on a line or circle is straightforward. Suppose, for example, that the position of an object is

$$\mathbf{r}(t) = \mathbf{r}_0 + t\mathbf{p}, \quad t \geq 0.$$

The line on which it moves is shown in Fig. 6.9(a). At $t = 0$ the object is at $\mathbf{r}_0$ and moves in the direction of $\mathbf{p}$. Uniform motion on a line means that the object has constant velocity $\mathbf{v} = d\mathbf{r}/dt = \mathbf{p}$. Its speed is $v = \|\mathbf{v}\| = \|\mathbf{p}\| = p$. Hence up to time t the distance traveled by the object is pt, which is equal to the length $s(t)$ of the curve on which it is moving.

For the uniform circular motion described by

$$\mathbf{r} = \mathbf{r}(t) = a(\cos \omega t, \sin \omega t), \quad t \geq 0,$$

the object starts at $\mathbf{r}_0 = a(1, 0)$. After t seconds its position angle is ωt and the object has moved a distance $a\omega t$ on the circle. See Fig. 6.9(b). The distance traveled by the object is equal to the length $s(t)$ of the curve on which it is moving.

Referring to Fig. 6.10, suppose the position of an object at any time $t \geq 0$ is

$$\mathbf{r} = \mathbf{r}(t) = (x(t), y(t)), \quad t \geq 0.$$

Assuming that the object does not retrace its path, the distance it traveled up to time t is equal to the length $s(t)$ of the curve on which it is moving. The rate of change ds/dt is the speed $\|v(t)\|$ of the object. This means that s is an antiderivative of $\|v(t)\|$ and hence

(1)
$$s(t) = \int_0^t \|v(w)\|\, dw.$$

In this section we discuss the length of curves C not necessarily associated with the motion of an object. We use the term *arc length* instead of distance traveled or length of curve. We shall always assume that the position vector $\mathbf{r}$ given by the vector/parametric equations used to describe a curve does not retrace the curve.

Arc Length of a Curve

The formula for the length of a path given in (1) was based on our intuition and understanding of motion. To understand length more deeply, we shall think of it as a number associated with a curve C, just as we regard area as a number associated with a region R. *Curve* and *region* are geometric objects.

We suppose C is described by a vector/parametric equation

$$\mathbf{r} = \mathbf{r}(t) = \big(x(t), y(t)\big), \quad a \le t \le b.$$

In earlier chapters we assumed that the functions x and y were continuous on $[a, b]$. For calculating arc length we shall need to assume that these functions have continuous derivatives on $[a, b]$. Curves described by such functions are said to be **smooth curves.** A representative curve C is shown near the top of Fig. 6.11.

Subdivide $[a, b]$ with equally spaced points $a = t_0, t_1, \ldots, t_n = b$ and let $h = (b - a)/n$ denote the length of each subinterval. The points with position vectors $\mathbf{r}(t_0), \mathbf{r}(t_1), \ldots, \mathbf{r}(t_n)$ subdivide C as shown in Fig. 6.11. The bottom part of the figure shows a zoom to the segment of C corresponding to a typical subinterval $[t_{i-1}, t_i]$. Evidently, the arc length of the segment of C between $\mathbf{r}(t_{i-1})$ and $\mathbf{r}(t_i)$ is closely approximated by

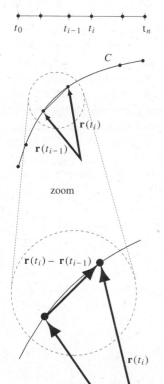

$$(2) \qquad \begin{aligned} ds_i &= \|\mathbf{r}(t_i) - \mathbf{r}(t_{i-1})\| \\ ds_i &= \sqrt{\big(x(t_i) - x(t_{i-1})\big)^2 + \big(y(t_i) - y(t_{i-1})\big)^2}, \end{aligned}$$

which is the length of the vector $\mathbf{r}(t_i) - \mathbf{r}(t_{i-1})$ or the length of the line segment joining the points $\mathbf{r}(t_{i-1})$ and $\mathbf{r}(t_i)$.

We use the tangent line approximation to simplify ds_i before adding these approximations together to approximate the arc length s of C.

Recall that if f is a differentiable function on an interval I, then

$$(3) \qquad f(x + k) - f(x) \approx kf'(x)$$

for $x, x + k \in I$. This is the tangent line approximation.

Since $x = x(t)$ and $y = y(t)$ are differentiable, we may use (3) to rewrite the differences $x(t_i) - x(t_{i-1})$ and $y(t_i) - y(t_{i-1})$ in (2). We have

$$\begin{aligned} ds_i &= \|\mathbf{r}(t_i) - \mathbf{r}(t_{i-1})\| = \|\mathbf{r}(t_{i-1} + h) - \mathbf{r}(t_{i-1})\| \\ &= \sqrt{\big(x(t_{i-1} + h) - x(t_{i-1})\big)^2 + \big(y(t_{i-1} + h) - y(t_{i-1})\big)^2} \\ &\approx \sqrt{\big(hx'(t_{i-1})\big)^2 + \big(hy'(t_{i-1})\big)^2} \\ &\approx \sqrt{x'(t_{i-1})^2 + y'(t_{i-1})^2} \, h. \end{aligned}$$

Figure 6.11. Approximating the arc length of C.

To approximate the arc length s of C we add the lengths ds_i to obtain

$$(4) \qquad s \approx \sum_{i=1}^{n} ds_i \approx \sum_{i=1}^{n} \sqrt{\big(x'(t_{i-1})\big)^2 + \big(y'(t_{i-1})\big)^2} \, h.$$

You may have noticed the two "approximately equal to" symbols in equation (4). The first was needed since the length ds_i of the vector $\mathbf{r}(t_i) - \mathbf{r}(t_{i-1})$ is not in general equal to the arc length of C between the points $\mathbf{r}(t_{i-1})$ and $\mathbf{r}(t_i)$. The second was needed since the tangent line approximations used to simplify ds_i are in general not exact. However, we expect the approximations in (4) to improve as we subdivide $[a, b]$ more and more finely. Indeed, based on the preceding discussion, the form of (4), and the definition of the integral, it appears that the arc length s of C is given by

$$s = \int_a^b \|\mathbf{r}'(t)\| \, dt = \int_a^b \sqrt{x'(t)^2 + y'(t)^2} \, dt.$$

This result agrees with (1), which was the result of our earlier discussion on motion.

We have not considered the question of alternative descriptions of the same curve. This could arise, for example, if two objects traverse the same curve but have different velocities. We might have hare and tortoise objects. One of them starts fast but slacks off later, while the other plods steadily over the same path. The position vectors of these two objects would trace the same curve, but their coordinate functions or domains would necessarily be different. We would expect, however, that the distances traveled would be the same unless one of the objects reversed its direction one or more times, retracing its path. If they were on a circular curve, one might go around twice. In this case the distance traveled would be twice the length of the curve. As mentioned earlier, we shall always assume that the position vector $\mathbf{r}$ given by the vector/parametric equations used to describe a curve does not retrace the curve.

DEFINITION   arc length of a smooth curve

Let C be a smooth curve described by

$$\mathbf{r} = \mathbf{r}(t) = \big(x(t), y(t)\big), \quad a \le t \le b.$$

The **arc length** s of C is defined as

(5)
$$s = \int_a^b \|r'(t)\| \, dt = \int_a^b \sqrt{x'(t)^2 + y'(t)^2} \, dt.$$

The arc length function $s = s(t)$ is defined to be the arc length of C from $\mathbf{r}(a)$ to $\mathbf{r}(t)$, so that

(6)
$$s(t) = \int_a^t \|r'(u)\| \, du.$$

■ **EXAMPLE 1**   Calculate the arc length of the cycloid described by

(7)
$$\mathbf{r}(t) = \big(t - \sin t, 1 - \cos t\big), \quad 0 \le t \le 2\pi.$$

Solution.   The parametric equation (7) for the cycloid was derived in Section 3.5. One arch of a cycloid is shown in Fig. 6.12. Before substituting into (5), we calculate and simplify $x'(t)^2 + y'(t)^2$. We use the trigonometric identity

$$\sin^2 \theta = \frac{1 - \cos 2\theta}{2}$$

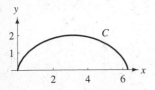

Figure 6.12. Cycloid arch.

in the simplification. Calculating $x'(t)$ and $y'(t)$, we have

$$x'(t)^2 + y'(t)^2 = (1 - \cos t)^2 + (\sin t)^2 = 1 - 2\cos t + \cos^2 t + \sin^2 t$$
$$= 2(1 - \cos t).$$

Letting $t = 2\theta$ in the identity, we have

$$x'(t)^2 + y'(t)^2 = 2 \cdot 2\sin^2 \tfrac{1}{2}t.$$

From this we may calculate the arc length of the cycloid. We have

$$s = \int_0^{2\pi} \sqrt{x'(t)^2 + y'(t)^2}\, dt$$
$$= 2 \int_0^{2\pi} \sin \tfrac{1}{2}t\, dt = 8.$$

A glance at Fig. 6.12 shows that this result is in the right ballpark. ■

We may use the arc length formula (5) to calculate the arc length of the graph of an equation $y = f(x)$ on an interval $[a, b]$ or a polar equation $r = f(\theta)$ on an interval $[\alpha, \beta]$. For this we describe the graphs parametrically. Since we have required for arc length calculations that the parametric curve be smooth, we assume that the defining function f and its derivative are continuous.

To calculate the arc length s of the graph C of the function f on $[a, b]$, recall that C may be described parametrically by regarding x as a parameter. We have

$$\mathbf{r}(x) = \big(x, f(x)\big), \quad a \le x \le b.$$

Before substituting into (5), we calculate $\mathbf{r}'(x)$ and $\|\mathbf{r}'(x)\|$. We have

$$\mathbf{r}'(x) = \big(1, f'(x)\big)$$
$$\|\mathbf{r}'(x)\| = \sqrt{1 + (f'(x))^2}.$$

From (5), the arc length of a curve C described by $y = f(x)$, $a \le x \le b$, is:

Arc Length Formula for Cartesian Curves

(8)
$$s = \int_a^b \sqrt{1 + (f'(x))^2}\, dx = \int_a^b \sqrt{1 + y'^2}\, dx$$

■ **EXAMPLE 2**   Calculate the arc length of the parabolic arc described by $y = x^2$, for $0 \le x \le 1$.

Solution.   The parabolic arc is the curve C shown in Fig. 6.13. From (8) we have

(9)
$$s = \int_0^1 \sqrt{1 + y'^2}\, dx = \int_0^1 \sqrt{1 + 4x^2}\, dx.$$

Suggestions for evaluating this integral were given in Exercise 51 at the end of Section 6.1. Alternatively, we may look for the form $\sqrt{a^2 + x^2}$ in the table of

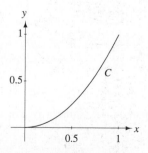

Figure 6.13. Parabolic arc.

integrals. In either case we find

$$s = \tfrac{1}{2}\sqrt{5} + \tfrac{1}{4}\ln(\sqrt{5} + 2) \approx 1.48.$$ ■

The arc length of a curve C described by a polar equation $r = f(\theta), \alpha \le \theta \le \beta$, may be calculated by describing C parametrically. We show in Fig. 6.14 a polar curve C and a point P on this curve. The point P has polar coordinates (r, θ), and rectangular coordinates $(r\cos\theta, r\sin\theta)$. The curve C is traced by the position vector

$$\mathbf{r}(\theta) = \big(f(\theta)\cos\theta, f(\theta)\sin\theta\big), \quad \alpha \le \theta \le \beta,$$

which may be written more briefly as

$$\mathbf{r}(\theta) = \big(r\cos\theta, r\sin\theta\big), \quad \alpha \le \theta \le \beta.$$

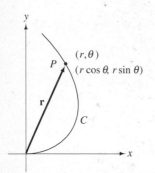

Figure 6.14. Polar curve.

First we calculate $\mathbf{r}'(\theta)$ and its length $\|\mathbf{r}'(\theta)\|$. Keeping in mind that $r = f(\theta)$,

$$\mathbf{r}'(\theta) = \big(r'\cos\theta + r(-\sin\theta), r'\sin\theta + r\cos\theta\big).$$

The square of the length of this vector is

$$\begin{aligned}
\|\mathbf{r}'(\theta)\|^2 &= (r'\cos\theta + r(-\sin\theta))^2 + (r'\sin\theta + r\cos\theta)^2 \\
&= r'^2\cos^2\theta - 2r'r\cos\theta\sin\theta + r^2\sin^2\theta \\
&\quad + r'^2\sin^2\theta + 2r'r\sin\theta\cos\theta + r^2\cos^2\theta \\
&= r'^2(\cos^2\theta + \sin^2\theta) + r^2(\sin^2\theta + \cos^2\theta) \\
&= r'^2 + r^2.
\end{aligned}$$

From (5), the arc length of a curve C described by the polar equation $r = f(\theta), \alpha \le \theta \le \beta$, is

---

Arc Length Formula for Polar Curves

(10) $$s = \int_a^b \sqrt{r^2 + r'^2}\, d\theta$$

■ **EXAMPLE 3** Calculate the arc length of the Archimedean spiral described by the polar equation $r = \theta$, for $0 \le \theta \le 2\pi$.

Solution. Applying (10), we have

$$s = \int_0^{2\pi} \sqrt{r^2 + r'^2}\, d\theta = \int_0^{2\pi} \sqrt{\theta^2 + 1^2}\, d\theta.$$

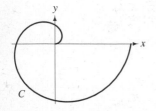

Figure 6.15. The spiral $r = \theta$.

This integral has the same form as (9) and may be evaluated in the same way. We have

$$s = \int_0^{2\pi} \sqrt{\theta^2 + 1^2}\, d\theta = \pi\sqrt{4\pi^2 + 1} + \tfrac{1}{2}\ln\!\left(2\pi + \sqrt{4\pi^2 + 1}\right) \approx 21.26.$$ ■

Unit Tangent

In the definition of arc length we defined the arc length function $s = s(t)$ to be the arc length of C from $\mathbf{r}(a)$ to $\mathbf{r}(t)$, so that

$$(11) \qquad s(t) = \int_a^t \|\mathbf{r}'(u)\|\, du.$$

From the Fundamental Theorem of Calculus, we see that

$$(12) \qquad \frac{ds}{dt} = \|\mathbf{r}'(t)\|.$$

If t is time and $\mathbf{r}(t)$ describes the motion of an object on C, then

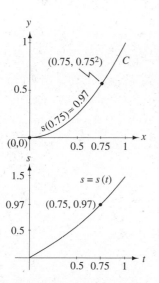

$$(13) \qquad \frac{ds}{dt} = \|\mathbf{r}'(t)\| = v,$$

where v is the speed of the object. This makes sense, since speed is the change in position per unit time.

It makes sense also that the position vector $\mathbf{r}$ is a function of s. If we know the distance s we have walked on C, we should be able to calculate our position $\mathbf{r}$. Consider the curve C described by $\mathbf{r} = (t, t^2)$, $0 \le t \le 1$, shown at the top of Fig. 6.16. The arc length from $(0, 0)$ to $(0.75, 0.75^2)$ is approximately 0.97. From the number 0.97 we should be able to determine $t = 0.75$. We show the invertible arc length function $s = s(t)$ for C at the bottom of Fig. 6.16.

More technically, since the arc length function $s = s(t)$ is increasing, it is invertible. Thus we can solve the equation $s = s(t)$ for t in terms of s, obtaining $t = g(s)$. We may then write

$$\mathbf{r} = \mathbf{r}(t) = \mathbf{r}(g(s)),$$

which shows that $\mathbf{r}$ is a function of s.

A simple curve C that can be described explicitly with the arc length parameter is the line shown in Fig. 6.17 and described by

Figure 6.16. s as parameter.

$$\mathbf{r} = \mathbf{r}(t) = \mathbf{a} + t\mathbf{b}, \qquad t \ge 0.$$

The arc length function is

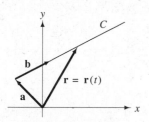

$$s = s(t) = \int_0^t \|\mathbf{r}'(u)\|\, du = \int_0^t \|\mathbf{b}\|\, du = tb,$$

where $b = \|\mathbf{b}\|$. Solving the equation $s = tb$ for t in terms of s gives

Figure 6.17. Arc length as parameter.

$$t = g(s) = \frac{s}{b}.$$

Using this result, we may describe C by the equation

$$(14) \qquad \mathbf{r} = \mathbf{r}(s/b) = \mathbf{a} + (s/b)\mathbf{b} \qquad s \ge 0.$$

If we differentiate $\mathbf{r}$ with respect to the parameter s, we obtain a tangent vector to C. We have

$$\frac{d\mathbf{r}}{ds} = \mathbf{0} + \frac{1}{b}\mathbf{b} = \frac{1}{b}\mathbf{b}.$$

It follows that $d\mathbf{r}/ds$ is a **unit tangent vector.**

We show that this result holds more generally. For this let

$$\mathbf{r} = \mathbf{r}(t), \quad a \le t \le b,$$

describe a curve C and regard t as a function of arc length s. Differentiating $\mathbf{r}$ with respect to s using the chain rule,

(15)
$$\frac{d\mathbf{r}}{ds} = \frac{d\mathbf{r}}{dt}\frac{dt}{ds} = \frac{d\mathbf{r}}{dt}\frac{1}{ds/dt} = \frac{1}{ds/dt}\frac{d\mathbf{r}}{dt}.$$

This result depends on the following result from Chapter 2: If $y = f(x)$ and $x = f^{-1}(y)$, then

$$\frac{dx}{dy} = f^{-1\prime}(y) = \frac{1}{f'(f^{-1}(y))} = \frac{1}{f'(x)} = \frac{1}{dy/dx}.$$

From (15) and (12) we have

(16)
$$\frac{d\mathbf{r}}{ds} = \frac{1}{\|\mathbf{r}'(t)\|}\frac{d\mathbf{r}}{dt}.$$

It now follows that $d\mathbf{r}/ds$ is a unit vector.

(17)
$$\left\|\frac{d\mathbf{r}}{ds}\right\| = \frac{1}{\|\mathbf{r}'(t)\|}\left\|\frac{d\mathbf{r}}{dt}\right\| = 1.$$

We summarize these results in a definition.

DEFINITION   **unit tangent vector**

Let C be a smooth curve described by

(18)
$$\mathbf{r} = \mathbf{r}(t) = \big(x(t), y(t)\big), \quad a \le t \le b.$$

At all points $\mathbf{r}(t)$ of C at which $\|\mathbf{r}'(t)\| \ne 0$, the unit tangent vector $\mathbf{T}$ is defined by

(19)
$$\mathbf{T} = \frac{d\mathbf{r}}{ds}.$$

The unit tangent usually is calculated in terms of the given parameter t.

(20)
$$\mathbf{T} = \mathbf{T}(t) = \frac{1}{\|\mathbf{r}'(t)\|}\mathbf{r}'(t).$$

■ **EXAMPLE 4** Find the unit tangent **T** for the cycloid given in Example 1.

Solution. From Example 1 the equation describing the cycloid is

$$(21) \qquad \mathbf{r}(t) = (t - \sin t, 1 - \cos t), \quad 0 \le t \le 2\pi.$$

To find the unit tangent **T** we calculate **r**′ and its length $\|\mathbf{r}'\|$. We have

$$\mathbf{r}' = (1 - \cos t, \sin t)$$

$$\|\mathbf{r}'\| = \sqrt{(1 - \cos t)^2 + (\sin t)^2}$$

$$= \sqrt{2(1 - \cos t)}.$$

Substituting **r** and **r**′ into (20), we have

$$\mathbf{T} = \frac{1}{\|\mathbf{r}'\|}\mathbf{r}' = \left(\frac{1 - \cos t}{\sqrt{2(1 - \cos t)}}, \frac{\sin t}{\sqrt{2(1 - \cos t)}} \right).$$

As might be expected, the unit tangent is not defined at the "cusp" points $\mathbf{r}(0)$ and $\mathbf{r}(2\pi)$, where $\mathbf{r}' = 0$. The expression for **T** can be simplified by noting that $1 - \cos t = 2\sin^2(t/2)$ and $\sin t = 2\sin\frac{1}{2}t\cos\frac{1}{2}t$. Leaving the details to Exercise 30, we obtain

$$\mathbf{T} = \left(\sin\tfrac{1}{2}t, \cos\tfrac{1}{2}t \right).$$

We show in Fig. 6.18 the cycloid, the position vector $\mathbf{r}(4\pi/3)$, and the unit tangent vector $\mathbf{T}(4\pi/3)$ at this point. To improve the sketch, the unit tangent was not drawn to scale. ■

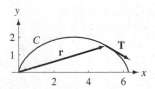

Figure 6.18. Unit tangent.

Exercises 6.2

Basic

Exercises 1–6: An object is traveling on a described curve C. Sketch C and calculate the distance traveled by the object. Assume units are meters and seconds.

1. $\mathbf{r} = (3, -4) + t(5, 6), \quad 0 \le t \le 25$

2. $\mathbf{r} = (1, 1) - 3t(5, 6), \quad 0 \le t \le 25$

3. $\mathbf{r} = 3(\cos 100\pi t, \sin 100\pi t), \quad 0 \le t \le 60$

4. $\mathbf{r} = 5(\sin 90\pi t, \cos 90\pi t), \quad 0 \le t \le 60$

5. $\mathbf{r} = 3(2t - \sin 2t, 1 - \cos 2t), \quad 0 \le t \le \pi$

6. $\mathbf{r} = -2(t - \sin t, 1 - \cos t), \quad 0 \le t \le 4\pi$

Exercises 7–18: Sketch the described curve and calculate its arc length. Assume units are meters.

7. $y = x^2, \quad 0 \le x \le 2$

8. $y = \frac{1}{2}x^2, \quad 0 \le x \le 5$

9. $y = x^{3/2}, \quad 0 \le x \le 5$

10. $y = 1 + \frac{1}{2}x^{3/2}, \quad 0 \le x \le 1$

11. $r = 1 + \cos\theta, \quad 0 \le \theta \le \pi$

12. $r = 1 - \sin\theta, \quad -\pi/2 \le \theta \le \pi/2$

13. $r = e^{\theta}, \quad 0 \le \theta \le \pi$

14. $r = 2^{\theta}, \quad 0 \le \theta \le \pi$

15. $\mathbf{r} = (\cos\theta + \theta\sin\theta, \sin\theta - \theta\cos\theta), \quad 0 \le \theta \le \pi$

16. $\mathbf{r} = (\sin\theta + \theta\cos\theta, \cos\theta - \theta\sin\theta), \quad 0 \le \theta \le \pi$

17. $\mathbf{r} = (t\cos t, t\sin t), \quad 0 \le t \le 1$

18. $\mathbf{r} = (t\sin t, t\cos t), \quad 0 \le t \le 1$

Exercises 19–22: The arc length integrals of many curves must be evaluated numerically, using either the midpoint rule discussed in Section 5.7 or the numerical integration algorithm on your calculator or CAS. The given value of n is sufficient for M_n to be within 0.01 of the arc length. Sketch the curve and approximate its arc length to within 0.01. Assume units are meters.

19. $\mathbf{r} = (3\cos t, 2\sin t)$, $\quad 0 \le t \le \pi/2; n = 5$

20. $\mathbf{r} = (\cos t, 2\sin t)$, $\quad 0 \le t \le \pi/2; n = 7$

21. $y = x^3$, $\quad 0 \le x \le 1; n = 7$

22. $y = 1/x$, $\quad 1 \le x \le 2; n = 5$

Exercises 23–26: Calculate the unit tangent $\mathbf{T}$ at the given point of the curve. Sketch the curve and unit tangent at the given point.

23. $\mathbf{r} = (3\cos t, 2\sin t)$, $\quad 0 \le t \le 2\pi; \quad \mathbf{r}(\pi/4)$

24. $\mathbf{r} = (\cos t, 2\sin t)$, $\quad 0 \le t \le \pi/2; \quad \mathbf{r}(\pi/6)$

25. $y = x^3$, $\quad 0 \le x \le 2; \quad (1, 1)$

26. $y = 1/x$, $\quad 1/2 \le x \le 2; \quad (1, 1)$

27. Approximate the arc length of the curve C described by $\mathbf{r}' = (t, \ln t)$, $1/2 \le t \le 2$. Use (4). Divide the interval $[1/2, 2]$ into five parts. Is the approximation too big or too small? Would the approximation improve if the interval were replaced by $[3/2, 3]$? Why?

28. The equations

$$\mathbf{r} = (\cos(2\pi t), \sin(2\pi t)), \quad 0 \le t \le 1$$
$$\mathbf{r} = (\cos(2\pi t^2), \sin(2\pi t^2)), \quad 0 \le t \le 1$$

describe the same curve C. Show that the two arc length calculations result in the same number.

29. Fill in the integration details for Example 2.

30. Fill in the details of the simplification of $\mathbf{T}$ in Example 4.

Growth

31. In Exercise 28 a curve was described by two different vector/parametric equations. Assume that these equations describe the motion of hare and tortoise objects. Which equation describes the motion of the hare? Why? Are the velocities and speeds of the two objects the same or different at $t = 0.5$? At $t = 0.75$?

32. In Example 4 we found $\mathbf{T} = (\sin \frac{1}{2}t, \cos \frac{1}{2}t)$, which exists for all $t \in [0, 2\pi]$. We ignored the fact that $\|\mathbf{r}'\| = 0$ at $t = 0, 2\pi$. Give a reason why the calculation appears to work for all $t \in [0, 2\pi]$. Do you think that $\mathbf{T}$ varies continuously for $t \in [0, 3\pi]$? What about $\mathbf{r}'$ for $t \in [0, 3\pi]$? Explain.

33. Find the length of the polar circle described by $r = \cos\theta, 0 \le \theta \le \pi$. Does it matter that r is nonpositive in the interval $[\pi/2, \pi]$?

34. Referring to (4), calculate approximations to the length of the cycloid in Example 1 corresponding to the subdivisions of $[0, 2\pi]$ with $n = 2, 4, 8$.

Review/Preview

35. Shade the region that is inside both polar curves $r = \sin\theta$ and $r = \cos\theta$. Find polar and rectangular coordinates of the points in which these curves intersect.

36. Shade the region that is inside both polar curves $r = 3\sin\theta$ and $r = 1 + \sin\theta$. Find polar and rectangular coordinates of the points in which these curves intersect.

37. Find the rectangular and polar coordinates of the points where the curves $r = \theta$ and $r = \cos\theta$ intersect, where $0 \le \theta \le \pi/2$.

38. Graph the polar curve $r = 1 - 2\sin\theta$.

6.3 AREAS OF REGIONS BOUNDED BY CLOSED CURVES

Areas of Regions Whose Boundaries Are Described by Parametric Equations

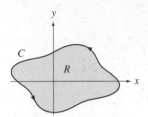

Figure 6.19. Region bounded by a closed curve.

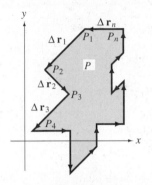

Figure 6.20. Polygonal regions.

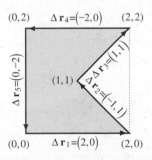

Figure 6.21. Five-sided polygon.

In Chapter 5 we discussed methods for calculating the area of a region beneath the graph of a function or between the graphs of two functions. Referring to Fig. 6.19, we now give a general method for calculating the area of a region R bounded by a closed curve C. The method is based on a formula used by surveyors (see *Elementary Surveying,* Paul R. Wolf and Russell C. Brinker, eighth edition, 1989, HarperCollins Publishers).

Investigation

Referring to Fig. 6.20, suppose we have surveyed a polygonal region P, determining the coordinates $(x_1, y_1), (x_2, y_2), \ldots, (x_n, y_n)$ of its vertices $P_1, P_2, \ldots, P_n$ as we walk in a counterclockwise direction on the boundary. The area A of the region P is

$$(1) \qquad A = \tfrac{1}{2} \sum_{j=1}^{n} (x_j \Delta y_j - y_j \Delta x_j),$$

where

$$\Delta \mathbf{r}_1 = \left(x_2 - x_1, y_2 - y_1 \right) = \left(\Delta x_1, \Delta y_1 \right)$$

$$\Delta \mathbf{r}_2 = \left(x_3 - x_2, y_3 - y_2 \right) = \left(\Delta x_2, \Delta y_2 \right)$$

$$\vdots$$

$$\Delta \mathbf{r}_n = \left(x_1 - x_n, y_1 - y_n \right) = \left(\Delta x_n, \Delta y_n \right).$$

We show in Fig. 6.21 an example of a five-sided polygon. The area of the enclosed region can be calculated from (1). With $P_1 = (0, 0)$,

$$(2) \qquad A = \tfrac{1}{2} \big[\left(0 \cdot 0 - 0 \cdot 2 \right) + \left(2 \cdot 1 - 0 \cdot (-1) \right) + \left(1 \cdot 1 - 1 \cdot 1 \right)$$
$$+ \left(2 \cdot 0 - 2 \cdot (-2) \right) + \left(0 \cdot (-2) - 2 \cdot 0 \right) \big] = 3.$$

This result agrees with the calculation $2 \cdot 2 - \tfrac{1}{2} \cdot 2 \cdot 1 = 3$, which is the difference between the areas of the square and triangle in the figure.

The surveyor's formula (1), a partial proof of which is outlined in Exercises 33 and 34, leads in a natural way to a general formula for calculating the areas of regions bounded by closed curves. Referring to Fig. 6.22, let C be the smooth curve described by

$$\mathbf{r} = \mathbf{r}(t) = \left(x(t), y(t) \right), \quad a \le t \le b.$$

We assume that as t varies from a to b, the position vector $\mathbf{r}$ traces C in a counterclockwise direction. We also assume that C does not self-intersect, so that the only

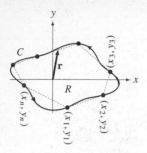

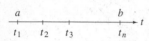

Figure 6.22. Region bounded by a simple closed curve.

point traced twice is $\mathbf{r}(a) = \mathbf{r}(b)$. Such curves are called **simple closed curves.** Let R denote the region bounded by C. We wish to calculate the area A of R. At the bottom of the figure is a typical subdivision of the interval $[a, b]$. For convenience, we have numbered the subdivision from 1 to n instead of (the customary) 0 to n.

The points $\mathbf{r}(t_1)$, $\mathbf{r}(t_2)$, ..., $\mathbf{r}(t_n)$ of C corresponding to $t_1, t_2, \ldots, t_n$ may be regarded as the vertices of a polygon, where

$$\mathbf{r}(t_j) = \big(x(t_j), y(t_j)\big) = \big(x_j, y_j\big), \quad j = 1, 2, \ldots, n.$$

As the interval $[a, b]$ is divided more finely, the area of the corresponding polygon approaches the area of the region R.

Let A_n denote the area of the polygon with vertices $(x_1, y_1), \ldots, (x_n, y_n)$. From (1) we have

$$(3) \qquad A_n = \tfrac{1}{2} \sum_{j=1}^{n} \big(x_j \Delta y_j - y_j \Delta x_j\big).$$

From the tangent line approximation,

$$\Delta y_j \approx y'(t_j) \Delta t \quad \text{and} \quad \Delta x_j \approx x'(t_j) \Delta t,$$

where $\Delta t = (b - a)/(n - 1)$. With these replacements (3) becomes

$$A_n \approx \tfrac{1}{2} \sum_{j=1}^{n} \big(x(t_j) y'(t_j) - y(t_j) x'(t_j)\big) \Delta t.$$

As the number n of subdivisions increases without bound, we obtain

$$(4) \qquad A_n = \tfrac{1}{2} \int_a^b \big(x(t) y'(t) - y(t) x'(t)\big) dt.$$

We refer to (4) as the **general area formula.**

■ **EXAMPLE 1** Use (4) in showing that the area of an ellipse with semimajor axis a and semiminor axis b is πab. See Fig. 6.23.

Solution. The ellipse C can be described by the equation

$$\mathbf{r} = \mathbf{r}(t) = \big(a \cos t, b \sin t\big), \quad 0 \le t \le 2\pi.$$

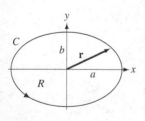

Figure 6.23. Elliptical region.

With this parametrization, C is traced in counterclockwise direction, as required in the argument leading to (4). Applying the general area formula, the area of the ellipse is

$$A = \tfrac{1}{2} \int_0^{2\pi} \big((a \cos t)(b \cos t) - (b \sin t)(-a \sin t)\big) dt$$

$$= \tfrac{1}{2} ab \int_0^{2\pi} \big(\cos^2 t + \sin^2 t\big) dt = \tfrac{1}{2} ab \int_0^{2\pi} 1 \, dt = \pi ab. \qquad ■$$

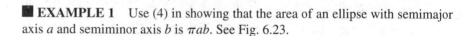

The general area formula (4) can be applied to many area calculations. For example, (4) includes the standard formula

$$A = \int_a^b f(x)\,dx$$

for the area of the region beneath the graph of f. See Exercise 35. To derive this familiar result it is necessary to broaden our discussion to include **piecewise smooth curves.** A curve C is piecewise smooth if it can be divided into a finite number of smooth curves $C_1, \ldots, C_n$. At the points of division, C need not have a tangent vector. The boundary of the region R in Fig. 6.24 is a piecewise smooth curve. It is useful to write the curve C as

$$C = C_1 + C_2 + \cdots + C_n.$$

We give an example of an application of the general area formula to a region bounded by a piecewise smooth curve.

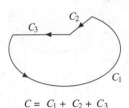

$$C = C_1 + C_2 + C_3$$

Figure 6.24. Piecewise smooth curve.

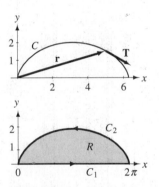

Figure 6.25. Area of one arch.

■ **EXAMPLE 2**   Calculate the area beneath one arch of the cycloid C shown in Fig. 6.25 and described by

$$(5) \qquad \mathbf{r} = \mathbf{r}(t) = \big(x(t), y(t)\big) = (t - \sin t, 1 - \cos t), \quad 0 \le t \le 2\pi.$$

Solution.   The unit tangent vector $\mathbf{T}$ gives the order in which C is traced. To apply the general area formula (4) it is necessary to describe the boundary of R as a piecewise smooth closed curve with a counterclockwise orientation. We show in the lower figure the piecewise smooth curve $C_1 + C_2$, where C_2 is the cycloid C reoriented (which is often indicated by writing $C_2 = -C$) and C_1 is the line segment from 0 to 2π. Applying the general area formula, we have

$$A = \tfrac{1}{2}\int_{C_1+C_2} \big(xy' - yx'\big)\,dt.$$

This integral can be written as the sum of two integrals, corresponding to the curves C_1 and C_2.

$$A = \tfrac{1}{2}\int_{C_1} \big(xy' - yx'\big)\,dt + \tfrac{1}{2}\int_{C_2} \big(xy' - yx'\big)\,dt.$$

This result depends upon the additivity of the integral: if $a \le c \le b$, then

$$\int_a^b f(x)\,dx = \int_a^c f(x)\,dx + \int_c^b f(x)\,dx.$$

The curve C_1 is described by

$$\mathbf{r} = \mathbf{r}(t) = \big(x(t), y(t)\big) = (t, 0), \quad 0 \le t \le 2\pi.$$

Since $y(t) = 0$ for all $t \in [0, 2\pi]$, $xy' - yx' = x \cdot 0 - 0 \cdot x' = 0$. Hence the first integral is 0.

As for the second integral, we were given (5), which describes C, not C_2. As outlined in Exercise 36, it is not difficult to modify the description of C to reverse its tracing order. Instead, and more simply, we use the result that

$$\int_{C_2} (xy' - yx')\,dt = -\int_C (xy' - yx')\,dt.$$

This is related to the fact that if the vertices of a polygon are traversed in the opposite order, the vectors $\Delta \mathbf{r}_j$ are reversed. It is also related to the definition

$$\int_b^a f(x)\,dx = -\int_a^b f(x)\,dx.$$

Putting everything together,

$$A = \tfrac{1}{2}\int_{C_2} (xy' - yx')\,dt$$

$$= -\tfrac{1}{2}\int_0^{2\pi} \big((t - \sin t)(\sin t) - (1 - \cos t)(1 - \cos t)\big)\,dt$$

$$= -\tfrac{1}{2}\int_0^{2\pi} (t\sin t - 2 + 2\cos t)\,dt$$

$$= -\tfrac{1}{2}(-6\pi) = 3\pi. \qquad \blacksquare$$

Areas of Regions Whose Boundaries Are Described by Polar Equations

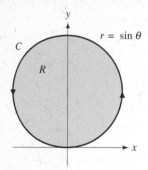

Figure 6.26. Polar circle.

We use (4) in calculating the area of a region whose boundary is a closed polar curve C, for example, the polar circle shown in Fig. 6.26. For this, instead of describing C by the polar form

$$r = f(\theta), \quad \alpha \le \theta \le \beta,$$

we use the parametric/vector form

$$\mathbf{r} = (r\cos\theta,\, r\sin\theta), \quad \alpha \le \theta \le \beta,$$

where $r = f(\theta)$. We assume that C is traced counterclockwise. (Care must be taken with the letter "r" in this context. The boldface $\mathbf{r}$ is used for position vectors and nonboldface r for polar coordinates. Both of these notations are common.)

We first calculate the integrand $xy' - yx'$ in (4). Since the parameter is θ, the ' indicates differentiation with respect to θ. We have

$$xy' - yx' = (r\cos\theta)(r'\sin\theta + r\cos\theta) - (r\sin\theta)(r'\cos\theta - r\sin\theta)$$
$$= r^2(\cos^2\theta + \sin^2\theta) = r^2.$$

We substitute this result into (4), obtaining

(6)
$$A = \tfrac{1}{2}\int_\alpha^\beta r^2\,d\theta.$$

We use (6) to calculate the area of the region R shown in Fig. 6.26. The polar circle C is described by

$$r = \sin\theta, \quad 0 \le \theta \le \pi.$$

From (6) we have

$$A = \tfrac{1}{2}\int_0^\pi \sin^2\theta\, d\theta = \frac{\pi}{4}.$$

This is the area of a circle of radius $1/2$.

Referring to Fig. 6.27, a **polar sector** is a region R whose boundary consists of a polar curve C described by

$$r = f(\theta), \quad \alpha \le \theta \le \beta,$$

the segment of the ray $\theta = \alpha$ between $r = 0$ and $r = f(\alpha)$, and the segment of the ray $\theta = \beta$ between $r = 0$ and $r = f(\beta)$. For a "standard" polar sector, C is traversed in a counterclockwise direction and $r = f(\theta) \ge 0$ for $\alpha \le \theta \le \beta$. If C is a closed curve, the ray segments, which we denote by A and B, are not needed.

We show that the formula (6) may be used for calculating the area of standard polar sectors. The boundary of R is $C + A + B$. Note that the orientation of the boundary is counterclockwise. From (4) we have

$$(7) \qquad\qquad A = \tfrac{1}{2}\int_{C+A+B}(xy' - yx')\,dt.$$

This integral can be written as the sum of three integrals. We show that the integrals on A and B are 0. Since A is a line through the origin and in the direction $(\cos\alpha, \sin\alpha)$, it is described by

$$\mathbf{r} = (x, y) = (0,0) + t(\cos\alpha, t\sin\alpha) = (t\cos\alpha, t\sin\alpha), \quad 0 \le t \le f(\alpha).$$

Hence

$$\int_A (xy' - yx')\,dt = \int_0^{f(\alpha)}(t\cos\alpha\sin\alpha - t\sin\alpha\cos\alpha)\,dt = 0.$$

Similarly, the integral on B is zero. To evaluate the remaining integral we describe C with the parametric/vector form

$$\mathbf{r} = (r\cos\theta, r\sin\theta), \quad \alpha \le \theta \le \beta,$$

where $r = f(\theta)$. Simplifying the integrand $xy' - yx'$ first, we have

$$xy' - yx' = (r\cos\theta)(r'\sin\theta + r\cos\theta) - (r\sin\theta)(r'\cos\theta - r\sin\theta)$$
$$= r^2(\cos^2\theta + \sin^2\theta) = r^2.$$

Substituting this result into (7) gives

$$(8) \qquad\qquad A = \tfrac{1}{2}\int_\alpha^\beta r^2\, d\theta.$$

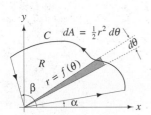

Figure 6.27. Polar sector.

We have shown that the formula (8) for calculating the area of a polar sector comes from the general area formula (4). Part of the point of this calculation was to help you become familiar with parametrizing curve segments like the rays A and B and with evaluating integrals on them. These skills will be needed in the next section.

Formula (8) for the area of a polar sector has a direct geometric interpretation, which is suggested by the shaded polar element of area shown in Fig. 6.27. Recall that the area A of a sector of a circle with central angle ω (measured in radians) and radius r is $A = \frac{1}{2}r^2\omega$. See Fig. 6.28. Hence the area dA of the shaded polar element of area in Fig. 6.27 is

$$dA = \tfrac{1}{2}r^2\, d\theta.$$

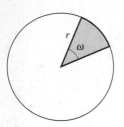

Figure 6.28. Sector of circle.

We may sum these elements with the integral, obtaining

$$A = \int_{\alpha}^{\beta} \tfrac{1}{2}r^2\, d\theta.$$

This agrees with (8).

■ **EXAMPLE 3**   Calculate the area of the region bounded by the curve described by $r = \cos\theta$, $0 \le \theta \le \pi/2$, and the $\theta = 0$ ray.

Solution.   The polar curve described by $r = \cos\theta$, $0 \le \theta \le \pi/2$, is a semicircle. As θ varies from 0 to $\pi/2$, the value of $\cos\theta$ varies from 1 to 0; thus the curve starts at the point with polar coordinates $(1, 0)$ and ends at $(0, \pi/2)$. Shown in Fig. 6.29 is the semicircle and a typical area element $dA = \tfrac{1}{2}r^2\, d\theta$, where $r = \cos\theta$. As θ varies from $\alpha = 0$ to $\beta = \pi/2$, the area element sweeps through the entire semicircle. The integral sums the area swept by the element.

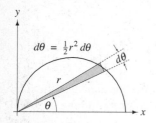

Figure 6.29. Polar circle.

$$
\begin{aligned}
A &= \int_0^{\pi/2} \tfrac{1}{2}r^2\, d\theta = \tfrac{1}{2}\int_0^{\pi/2} \cos^2\theta\, d\theta \\
&= \tfrac{1}{2}\int_0^{\pi/2} \frac{1 + \cos 2\theta}{2}\, d\theta \\
&= \tfrac{1}{4}\left(\theta + \tfrac{1}{2}\sin 2\theta\right)\Big|_0^{\pi/2} \\
&= \pi/8
\end{aligned}
$$

From the formula $A = \pi a^2$ for a circle, the area of the semicircle of radius $1/2$ is $\pi/8$.   ■

Finally, we calculate the area of a region lying between two polar curves. We use an area element analogous to the element used for the region between two Cartesian curves.

■ **EXAMPLE 4**   Calculate the area of the region in the first quadrant and between the polar curves described by $r = \theta$ and $r = \sin\theta$.

Solution.   The two curves are shown in Fig. 6.30. The region between them is bounded by the graphs of the circle, the spiral, and the one ray segment $\theta = \beta = \pi/2$, from $r = 1$ to $r = \pi/2$. We take the element of area dA to be the difference

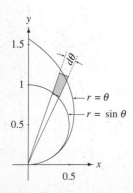

Figure 6.30. Spiral and circle.

of two sector elements. We have

$$dA = \tfrac{1}{2}\theta^2 \, d\theta - \tfrac{1}{2}\sin^2\theta \, d\theta = \tfrac{1}{2}(\theta^2 - \sin^2\theta) \, d\theta.$$

The region with area dA is shaded in the figure. As θ varies from 0 to $\pi/2$, this area element sweeps the region between the two curves. We have

$$A = \int_\alpha^\beta \tfrac{1}{2}(\theta^2 - \sin^2\theta) \, d\theta$$

$$= \tfrac{1}{2}\int_0^{\pi/2} (\theta^2 - \sin^2\theta) \, d\theta$$

$$= \tfrac{1}{2}\left(\tfrac{1}{3}\theta^3 - \tfrac{1}{2}\theta + \tfrac{1}{4}\sin 2\theta\right)\Big|_0^{\pi/2}$$

$$A = \frac{\pi}{48}(\pi^2 - 6) \approx 0.253.$$

Exercises 6.3

Basic

1. Use the surveyors' formula for calculating the area of the triangle with vertices $(2, 1)$, $(5, 1)$, and $(5, 5)$. Check your work by calculating the area in another way.

2. Use the surveyors' formula for calculating the area of the triangle with vertices $(1, 1)$, $(22, 1)$, and $(1, 21)$. Check your work by calculating the area in another way.

3. Use the surveyors' formula for calculating the area of the parallelogram with vertices $(-1, 2)$, $(5, 3)$, $(7, 10)$, and $(1, 5)$. Check your work by calculating the area in another way.

4. Use the surveyors' formula for calculating the area of the parallelogram with vertices $(-2, 2)$, $(-6, -2)$, $(-8, -5)$, and $(-4, -1)$. Check your work by calculating the area in another way.

Exercises 5–20: Sketch the described region R and calculate its area.

5. The boundary of R is

$$\mathbf{r} = \left(-\sin t - 2\cos t, -\sin t + 2\cos t\right),$$
$$0 \le t \le 2\pi.$$

6. The boundary of R is

$$\mathbf{r} = \left(-\sin t - 3\cos t, -\sin t + 3\cos t\right),$$
$$0 \le t \le 2\pi.$$

7. The boundary of R is

$$\mathbf{r} = \left(t - t^2, t^2 - t^3\right), \quad 0 \le t \le 1.$$

8. The boundary of R is

$$\mathbf{r} = \left(2t^2 - 4t^3, t - 2t^2\right), \quad 0 \le t \le 1/2.$$

9. R is bounded above by the curve

$$\mathbf{r} = \left(e^t, \sin t\right) \quad 0 \le t \le \pi,$$

and below by the x-axis. Formulas (23) and (24) in the table of integrals are needed.

10. R is bounded above by the curve

$$\mathbf{r} = \left(t^2, \sin t\right), \quad 0 \le t \le \pi$$

and below by the x-axis. Formulas (25) and (26) in the table of integrals are needed.

11. R is bounded by the spiral $r = \theta/2, 0 \le \theta \le 2\pi$, and the ray segment $\theta = 2\pi$ between $r = 0$ and $r = \pi$.

12. R is bounded by the spiral $r = 2^\theta, 0 \le \theta \le 2\pi$, and the ray segment $\theta = 2\pi$ between $r = 0$ and $r = 2^{2\pi}$.

13. R is bounded by the cardioid $r = 1 + \cos\theta$.

14. R is inside the cardioid $r = 1 + \cos\theta$ and outside the circle $r = 3\cos\theta$.

15. R is inside both circles $r = \cos\theta$ and $r = \sin\theta$.

16. R is inside both circles $r = a$ and $r = 2a\cos\theta$.

17. R is inside the smaller loop of the curve $r = 1 + 2\cos\theta$.

18. R is inside the smaller loop of the curve $r = 1 + 2\sin\theta$.

19. R is inside $r = 4$ and above $r = 3\csc\theta, 0 < \theta < \pi$.

20. R is inside $r = 2$ and to the right of $r = \sec\theta$, $-\pi/2 < \theta < \pi/2$.

Growth

21. The area of the region R in Exercise 9 may be calculated by describing the curved boundary with an equation $y = f(x)$. To determine f, solve $x = e^t$ for t in terms of x and substitute into $y = \sin t$. In the integral for area make the substitution $u = \ln x$.

22. The area of the region R in Exercise 10 may be calculated by describing the curved boundary with an equation $y = f(x)$. To determine f, solve $x = t^2$ for t in terms of x and substitute into $y = \sin t$. In the integral for area make the substitution $x = u^2$.

23. Use the general area formula in determining the area between the curves with equations $y = 2 - x^2$ and $y = x$. Check your work by calculating the area in another way.

24. With $0 < \theta < \pi/2$, determine the area of the region bounded by the graphs of $r = \cos\theta$ and $r = \theta$.

25. Show that the area of the region inside the cardioid $r = 1 + \sin\theta$ and outside the circle $r = 3\sin\theta$ is $\pi/4$. (Hint: The region that is inside the cardioid and outside the circle is the union of two crescent moon shapes. Calculate the area of one crescent and double the result.)

26. With $0 < \theta < \pi/2$, calculate the area of the region "outside" $r = 2\theta$ and to the right of $r = \sec\theta$.

27. The polar curve with equation $r = \csc\theta - \sqrt{2}$ has a loop. Determine the area of the region inside the loop.

28. Let R be the region

$$\{(r, \theta) : 0 \le \theta \le 1 \text{ and } \theta/2 \le r \le \arcsin\theta\}.$$

Sketch R and calculate its area. The integral for the area can be done by parts. Alternatively, evaluate it numerically, either using a built-in numerical integrator or, say, the midpoint rule.

29. Sketch and calculate the area of the region inside the loop of the curve $r = \ln\theta, 0 < \theta < 3\pi/2$.

30. Why isn't the value of the integral

$$\int_0^{2\pi} \tfrac{1}{2} \cos^2\theta \, d\theta$$

equal to the area enclosed by the curve with equation $r = \cos\theta$?

31. The calculation in Example 2 of the area beneath the cycloid can be simplified. The idea is suggested in the following figure, which shows the area element $dA = y \, dx$. First, from the equation

$$\mathbf{r} = (x, y) = (t - \sin t, 1 - \cos t), \quad 0 \le t \le 2\pi$$

describing the cycloid, calculate $dA = y \, dx$ and then integrate. Show that

$$dA = y \, dx = (1 - \cos t)(1 - \cos t) \, dt$$

and

$$A = \int_0^{2\pi} dA = \int_0^{2\pi} (1 - \cos t)(1 - \cos t) \, dt = 3\pi.$$

A justification for this formal calculation is given in the following exercise.

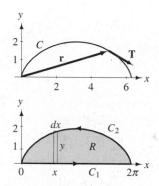

32. (Continuation of Exercise 31) Let R be a region bounded on the top by a curve C described by the vector/parametric equation

$$\mathbf{r} = \mathbf{r}(t) = (x(t), y(t)), \quad c \le t \le d.$$

Assume that $a = x(c)$, $b = x(d)$, $x'(t)$ is continuous on $[c, d]$, and positive on (c, d). We may solve $x = x(t)$ for t in terms of x, giving $t = g(x)$. Define f by substituting this into $y = y(t)$, giving $y = f(x) = y(g(x))$. Use the reverse substitution formula in Section 5.3 to show that the substitution $x = x(t)$ (which may also be written as $t = g(x)$) gives

$$\int_c^d y(t)x'(t) \, dt = \int_a^b f(x) \, dx = A.$$

33. Referring to the following figure, show that the area A of the region T is given by

(9) $A = \tfrac{1}{2}(x_1 y_2 + x_2 y_3 + x_3 y_1$

$\qquad - x_2 y_1 - x_2 y_1 - x_3 y_2 - x_1 y_3).$

A good start on the calculation is to use the triangle shown in the figure. Draw a line through P_3 and parallel to the x-axis. Drop perpendiculars to this line from P_1 and P_2. The area of the given triangle can be written as a combination of the areas of a big right triangle, a small right triangle, and a trapezoid.

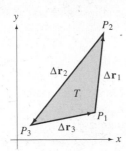

34. Given the result of Exercise 33, the surveyors' formula (1) for the case $n = 4$ follows by noticing that any quadrilateral can be divided into two triangles by joining two vertices. Referring to the figure below, write the area of the quadrilateral by using (9) twice, once for the triangle $P_1P_2P_3$ and once for $P_3P_4P_2$. The result can be written in the form of (1) for $n = 4$.

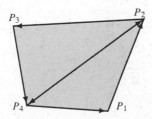

35. Referring to the following figure, show that the formula $\int_a^b f(x)\,dx$ for the area of R may be inferred from the general area formula (4). Here is an outline of an argument. The boundary of R is $C_1 + C_2 - C + C_3$. From (4) we have

$$\int_{C_1+C_2-C+C_3} (xy' - yx')\,dt$$
$$= \int_{C_1} (xy' - yx')\,dt + \cdots.$$

Describe each of the four curves and evaluate each integral separately. Describe C_1 by $x = t$, $y = 0$, for $a \le t \le b$, and show that the integral on C_1 is 0. Show that the integral on C_2 is $bf(b)$. Describe $-C_3$ by $x = a$, $y = y$, for $0 \le y \le f(a)$. Show that the integral on C_3 is $-af(a)$. Describe C by $x = x$, $y = f(x)$, for $a \le x \le b$. Evaluating the integral for $-C$ requires an integration by parts. Add the four integrals together.

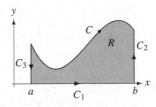

36. Referring to equation (5) in Example 2, show that if in the coordinate functions

$$x(t) = t - \sin t \text{ and } y(t) = 1 - \cos t$$

the variable t is replaced by $2\pi - \tau$, the resulting vector equation describes C_2. The parameter τ (tau) varies over $[0, 2\pi]$.

Exercises 37–44: This group of problems concerns the polar equation $r = p/(1 + e\cos\theta)$, which with suitable choices of p and e can describe any of the conic sections. Most of the problems concern ellipses.

37. The equation

$$\frac{x^2}{a^2} + \frac{y^2}{b^2} = 1, \quad a > b > 0,$$

describes an ellipse with its center at the origin. See the following figure. The major axis of this ellipse runs between the x-intercepts $(-a, 0)$ and $(a, 0)$; its minor axis runs between the y-intercepts $(0, -b)$ and $(0, b)$. The lengths of the semimajor and semiminor axes are a and b, respectively. The foci of this ellipse are at $(-c, 0)$ and $(c, 0)$, where $c^2 = a^2 - b^2$. The eccentricity of the ellipse is $e = c/a$.

Graph the ellipse with center at the origin, $c = 1$, and $e = 3/5$. Would the ellipse with $e = 4/5$ be more or less circular?

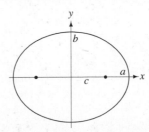

38. (Continuation of Exercise 36) The equation

$$\frac{(x-h)^2}{a^2} + \frac{(y-k)^2}{b^2} = 1, \quad a > b > 0,$$

describes an ellipse with center at (h, k). The numbers a, b, c, and e have the same meaning as before. By completing the square, write the equation

$$x^2 - 2x + 4y^2 + 16y + 13 = 0$$

in standard form, graph the equation, locate the foci of the ellipse, and calculate its eccentricity.

39. The polar equation

$$r = \frac{p}{1 + \varepsilon \cos \theta}, \quad \text{where } |\varepsilon| < 1 \text{ and } p > 0,$$

describes an ellipse with one focus at the origin, eccentricity equal to $|\varepsilon|$, semimajor axis of length $p/(1 - \varepsilon^2)$, semiminor axis of length $p/\sqrt{1 - \varepsilon^2}$, and center at $(-p\varepsilon/(1 - \varepsilon^2), 0)$. Verify these claims by transforming the polar equation to a rectangular equation. Recall that $x = r\cos\theta$, $y = r\sin\theta$, and $r^2 = x^2 + y^2$. Start by rewriting the polar equation as $r = p - r\varepsilon\cos\theta$ and then squaring both sides.

40. For the ellipse with equation

$$r = 1/(1 - 0.6\cos\theta), \quad 0 \le \theta \le 2\pi,$$

locate the other focus and the center of the ellipse. Also calculate the length of the semimajor and semiminor axes.

41. Transform the ellipse with equation

$$r = 1/(1 - 0.6\cos\theta), \quad 0 \le \theta \le 2\pi$$

by replacing θ by $\pi - \theta$. Show that the equation becomes $r = 1/(1 + 0.6\cos\theta)$. What happens to the graph? Check your answer by graphing the transformed equation.

42. Transform the ellipse with equation

$$r = 1/(1 - 0.6\cos\theta), \quad 0 \le \theta \le 2\pi$$

by replacing θ by $\pi/2 - \theta$. Show that the equation becomes $r = 1/(1 - 0.6\sin\theta)$. What happens to the graph? Check your answer by graphing the transformed equation.

43. The polar equation

$$r = \frac{p}{1 + \varepsilon \cos \theta}, \quad \text{where } |\varepsilon| = 1 \text{ and } p > 0$$

describes a parabola with its focus at the origin. If $\varepsilon = -1$, the parabola opens to the right. Graph the equation $r = 1/(1 - \cos\theta)$.

44. The polar equation

$$r = \frac{p}{1 + \varepsilon \cos \theta}, \quad \text{where } |\varepsilon| > 1 \text{ and } p > 0$$

describes a hyperbola with one focus at the origin. Graph the equation $r = 1/(1 + 2\cos\theta)$. Locate the center of the hyperbola.

Review/Preview

45. A force **F** of magnitude 10 N in the 190° direction is given. Resolve it into forces **P** and **N**, where **P** is in the 210° direction and **N** is perpendicular to **P**.

46. A force of $(4.5, 0)$ N acts on an object, moving it from $(1, 0)$ m to $(3, 0)$ m. How much work has been done?

6.4   WORK

In Section 3.7 we briefly discussed the concept of work and calculated the work W done by a constant force **F** acting on an object moving a distance s on a line. We used the formula

$$(1) \qquad\qquad W = (\mathbf{F} \cdot \mathbf{u})s$$

where **u** is a unit vector in the direction of the line. Work is measured in joules (J), where 1 J is a force of 1 newton acting over a distance of 1 meter. We show in Fig. 6.31 a constant force **F** acting on an object over a distance s. The magnitude and direction of **F** are constant although the point at which it acts moves with the object. The work done by this force is the component of **F** in the direction of motion times the distance s. In this section we calculate the work done by a variable force acting on an object moving on a smooth or piecewise smooth curve.

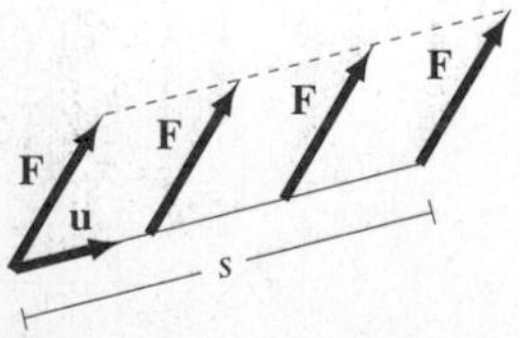

Figure 6.31. Constant force.

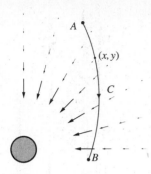

Figure 6.32. Earth's field.

We show in Fig. 6.32 the curved path C of an object moving in the earth's gravitational field $\mathbf{F}$. If at any point (x, y) of this field there were an object of unit mass, the gravitational force $\mathbf{F}(x, y)$ on the object would be directed toward the center of the earth and would have magnitude inversely proportional to the square of the distance from that point to the center of the earth.

To calculate the work done by the gravitational field acting on an object of unit mass as it moves on C from A to B, we describe C by an equation

$$(2) \qquad \mathbf{r} = \mathbf{r}(t) = (x, y) = (x(t), y(t)), \quad a \le t \le b,$$

where $\mathbf{r}(a)$ and $\mathbf{r}(b)$ are position vectors of A and B. At any point $\mathbf{r}(t) = (x(t), y(t))$ of C the force acting on the object would be $\mathbf{F}(x(t), y(t))$. We often shorten this notation to $\mathbf{F}(\mathbf{r}(t))$.

On a short segment of the curve, say of length 1 angstrom (10^{-10} m), the force field is nearly constant and the curve is nearly coincident with its tangent line. The work done by the field as the object moves on such a segment can be closely approximated with the help of formula (1). We then sum these work elements and use a limiting argument.

In more detail, we subdivide the interval $[a, b]$, choose a typical point t and subinterval of length dt, and consider the corresponding point $\mathbf{r}(t)$ and segment of C from $\mathbf{r}(t)$ to $\mathbf{r}(t + dt)$. See Fig. 6.33. The object has moved a distance of

$$\Delta s = \int_a^{t+dt} \|\mathbf{r}'(t)\| \, dt - \int_a^t \|\mathbf{r}'(t)\| \, dt$$

along the curve. Applying the tangent line approximation

$$\Delta s \approx ds = \|\mathbf{r}'(t)\| \, dt,$$

we observe that the object has moved a distance of ds, approximately, along the line segment from the tail of the vector $\mathbf{T} \, ds$ to its tip, while being acted on by the (nearly) constant force $\mathbf{F}(\mathbf{r}(t))$. From (1), the work dW done by this force is

$$(3) \qquad dW = (\mathbf{F}(\mathbf{r}(t)) \cdot \mathbf{T}) \, ds = \mathbf{F} \cdot \mathbf{T} \, ds.$$

The total work W done by the force field $\mathbf{F}$ as the object moves on C from $A = \mathbf{r}(a)$ to $B = \mathbf{r}(b)$ would be approximated by a sum of such work elements. As the number of subdivisions of $[a, b]$ grows without bound, we are led to an integral for W. We give this integral after agreeing on notation and rewriting dW in a convenient form.

To rewrite dW in (3), we recall from Section 6.2 that at $\mathbf{r}(t)$ on C the unit tangent vector is given by

$$\mathbf{T} = \mathbf{T}(t) = \frac{1}{\|\mathbf{r}'(t)\|}\mathbf{r}'(t).$$

Since $ds/dt = \|\mathbf{r}'(t)\|$, we may write $\mathbf{T}$ in the form

$$\mathbf{T} = \frac{1}{ds/dt}\mathbf{r}'(t).$$

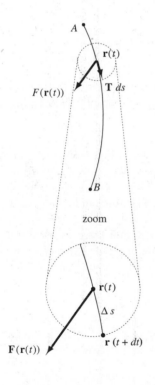

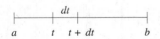

Figure 6.33. Variable force.

Thus from (3) the element dW of work at $\mathbf{r}(t)$ may be rewritten as

$$dW = \mathbf{F}(\mathbf{r}(t)) \cdot \mathbf{T}\,ds = \mathbf{F}(\mathbf{r}(t)) \cdot \frac{1}{ds/dt}\mathbf{r}'(t)\,ds.$$

And, finally, since $ds = (ds/dt)\,dt$,

(4) $$dW = \mathbf{F}(\mathbf{r}(t)) \cdot \mathbf{r}'(t)\,dt.$$

DEFINITION   work

The **work** W done by a force $\mathbf{F}$ acting on an object moving on a smooth curve C is defined as

(5) $$W = \int_C \mathbf{F} \cdot d\mathbf{r} = \int_a^b \mathbf{F}(\mathbf{r}(t)) \cdot \mathbf{r}'(t)\,dt,$$

where C is described by the equation $\mathbf{r} = \mathbf{r}(t), a \leq t \leq b$.

The notation $\int_C \mathbf{F} \cdot d\mathbf{r}$ included in this definition is commonly used in discussions of work in physics and engineering texts. This integral often is called a **line integral.** We comment on it after a few examples.

■ **EXAMPLE 1**   The gravitational force field $\mathbf{F}$ shown in Fig. 6.32 is not constant. Indeed, both the direction and magnitude of $\mathbf{F}(x, y)$ vary with (x, y). However, near the earth's surface the gravitational force on an object of mass m is nearly constant, with $\mathbf{F}(x, y) = (0, -mg)$ for all (x, y). This force field is shown in Fig. 6.34.

Calculate the work done by the (constant) gravitational force as an object moves from $(0, 3)$ to $(2, 0)$ on the line C connecting these points. Also calculate the work done as the object moves from $(0, 3)$ to $(2, 0)$ on the graph of any function f for which f' is continuous. See Fig. 6.35.

Solution.   We use (5) in calculating the work done by the gravitational force. For this we describe C by $\mathbf{r} = \mathbf{r}(t) = (0, 3) + t(2, -3)$, where $0 \leq t \leq 1$. The vector $(2, -3)$ points in the direction of motion. It is the difference between the position vectors $(2, 0)$ and $(0, 3)$. We have

$$W = \int_a^b \mathbf{F}(\mathbf{r}(t)) \cdot \mathbf{r}'(t)\,dt = \int_0^1 (0, -mg) \cdot (2, -3)\,dt$$

$$= \int_0^1 3mg\,dt = 3mgt \Big|_0^1 = 3mg.$$

If the object moves on the graph C^* of a continuously differentiable function f, we may describe C^* by $\mathbf{r} = \mathbf{r}(x) = (x, f(x))$, where $0 \leq x \leq 2$. The calcula-

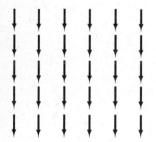

Figure 6.34. Constant field.

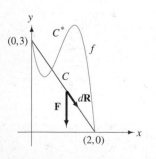

Figure 6.35. Work.

tion of work in this case is

$$W = \int_a^b \mathbf{F}(\mathbf{r}(t)) \cdot \mathbf{r}'(t)\,dt = \int_0^2 (0, -mg) \cdot (1, f'(x))\,dx$$

$$= -mg \int_0^2 f'(x)\,dx = -mg f(x)\Big|_0^2 = -mg(f(2) - f(0))$$

$$= 3mg.$$

The constant force field $\mathbf{F} = (0, -mg)$ is an example of a **conservative force field.** You may have noticed that the work done by the gravitational force on the object did not vary with the path between $(0, 3)$ and $(2, 0)$. Indeed, if C^* is any smooth curve connecting $(0, 3)$ and $(2, 0)$, then

$$\int_{C^*} \mathbf{F} \cdot d\mathbf{r} = 3mg.$$

See Exercise 10. A conservative field is one for which the work done by the field is *independent of path*. We discuss the nonconstant case of the earth's field in Example 5, showing that it also is conservative. ■

The notation $\int_C \mathbf{F} \cdot d\mathbf{r}$ for a line integral is used when a brief notation is wanted and attention is on the curve C as a geometric object, not on its description by a particular vector function $\mathbf{r} = \mathbf{r}(t)$, $a \le t \le b$. The notation includes $d\mathbf{r}$ but does not name a parameter; it stays neutral and uncommitted.

A variation on this notation can make some calculations easier. For this, let C be described by

$$\mathbf{r} = \mathbf{r}(t) = (x(t), y(t))$$

and define the differential $d\mathbf{r}$ by

$$d\mathbf{r} = \mathbf{r}'(t)\,dt.$$

Using the differentials $dx = x'(t)\,dt$ and $dy = y'(t)\,dt$, we may rewrite $d\mathbf{r}$ as

$$d\mathbf{r} = \mathbf{r}'(t)\,dt = (x'(t), y'(t))\,dt = (dx, dy).$$

Writing $\mathbf{F} = (F_1, F_2)$ in terms of its coordinate functions,

$$\int_C \mathbf{F} \cdot d\mathbf{r} = \int_C (F_1, F_2) \cdot (dx, dy) = \int_C F_1\,dx + F_2\,dy.$$

We use this notation in calculating the work done by the constant gravitational field $\mathbf{F} = (0, -mg)$ as an object of mass m moves from $(0, 3)$ to $(2, 0)$ on the piecewise smooth curve $C_1 + C_2$ shown in Fig. 6.36. We write the work integral in the form

$$W = \int_{C_1+C_2} \mathbf{F} \cdot d\mathbf{r} = \int_{C_1+C_2} 0\,dx + (-mg)\,dy$$

$$= \int_{C_1} -mg\,dy + \int_{C_2} -mg\,dy.$$

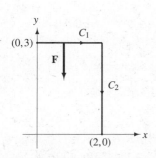

Figure 6.36. Piecewise smooth curve.

On C_1, y is identically 3, no matter how we describe C_1. Hence $dy = 0$ and the first integral is 0. For the second integral, the parametric equations $x = 2$ and $y = y$, for $0 \le y \le 3$, describe $-C_2$. Hence

$$W = \int_{C_2} -mg\,dy = -\int_0^3 -mg\,dy = 3mg.$$

The sketch in Fig. 6.36 is typical in its representation of a force field, showing only a representative vector, acting at a typical point of the curve on which the object moves.

In the next example we calculate the work done in stretching a spring. This is a variable-force problem since a spring resists stretching or compression with a force proportional to the distance it is stretched or compressed from its relaxed state. This is Hooke's law.

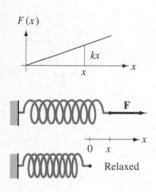

Figure 6.37. Work and Hooke's law.

■ **EXAMPLE 2**   Calculate the work done in stretching a spring 0.25 m from its relaxed state. See Fig. 6.37. It is given that a force of magnitude 20 N is required to stretch the spring 0.1 m. Also calculate the work done in compressing the spring 0.25 m from its relaxed state.

Solution.   The relaxed spring is shown at the bottom of the figure. Denote the force required to stretch the spring x meters by $F(x)$. (We may drop the boldface notation here since the force is one-dimensional.) By Hooke's law, $F(x) = kx$, where k is the spring constant. We know that $F(0.1) = 20$ N. Hence $20 = k(0.1)$, $k = 200$ N/m, and the force required to stretch the spring x meters is $F(x) = 200x$. A plot of $F(x)$ against x is shown at the top of Fig. 6.37.

The idea of an object moving on a curve is less explicit in this example. We may, however, imagine an object at the point of application of the stretching force. This object moves along a line from $x = 0$ to $x = 0.1$.

The amount of work done by the stretching force as the object moves from x to $x + dx$ is $dW = F(x)\,dx = 200x\,dx$. Hence the work done in stretching the spring 0.25 m is

$$W = \int_0^{0.25} 200x\,dx = (200)\tfrac{1}{2}x^2 \Big|_0^{0.25} = 6.25 \text{ J}.$$

We may choose to use the formula (5) for the calculation. For this we take $F(x) = 200x$ and describe the curve by $r(x) = x, 0 \le x \le 0.25$. The element of work is $dW = \mathbf{F} \cdot \mathbf{r}'\,dx = 200x\,dx$, as above.

For compressing the spring 0.25 m, the force is negative and the work will be negative. For $-0.25 \le x \le 0$, we have $dW = F(x)\,dx = 200x\,dx$. The work done in compressing the spring is

$$\int_{-0.25}^0 200x\,dx = (200)\tfrac{1}{2}x^2 \Big|_{-0.25}^0 = -6.25 \text{ J}. \qquad ■$$

■ **EXAMPLE 3**   An object moves counterclockwise on a semicircle C of radius a, from $(a, 0)$ to $(-a, 0)$, acted on by a force that attracts the particle toward the point $(-a, 0)$ and has magnitude proportional to the distance between $(-a, 0)$ and the object. Calculate the work done by this force.

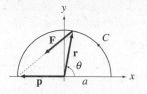

Figure 6.38. Attractive force.

Solution. A typical position of the object is shown in Fig. 6.38. Let $\mathbf{p} = (-a, 0)$ and describe C by

$$\mathbf{r} = a(\cos\theta, \sin\theta), \quad 0 \le \theta \le \pi.$$

At $\mathbf{r}$ the force $\mathbf{F}$ is in the direction of $\mathbf{p} - \mathbf{r}$. Its magnitude is $k\|\mathbf{p} - \mathbf{r}\|$, where k is a proportionality constant. To find $\mathbf{F}$ we divide the vector $\mathbf{p} - \mathbf{r}$ by its own length and then multiply by the magnitude of $\mathbf{F}$. This gives

$$\begin{aligned}
\mathbf{F} &= k\|\mathbf{p} - \mathbf{r}\|\frac{1}{\|\mathbf{p} - \mathbf{r}\|}(\mathbf{p} - \mathbf{r}) \\
&= k(\mathbf{p} - \mathbf{r}) = k((-a, 0) - a(\cos\theta, \sin\theta)) \\
&= -ka(1 + \cos\theta, \sin\theta).
\end{aligned}$$

We calculate $dW = \mathbf{F} \cdot \mathbf{r}'\, d\theta$ and then W.

$$\begin{aligned}
\mathbf{F} \cdot \mathbf{r}'\, d\theta &= -ka(1 + \cos\theta, \sin\theta) \cdot a(-\sin\theta, \cos\theta) \\
&= -ka^2(-\sin\theta - \sin\theta\cos\theta + \sin\theta\cos\theta) \\
&= ka^2 \sin\theta.
\end{aligned}$$

Hence

$$\begin{aligned}
W &= \int_0^\pi \mathbf{F} \cdot \mathbf{r}'\, d\theta = ka^2 \int_0^\pi \sin\theta\, d\theta \\
&= -ka^2 \cos\theta \Big|_0^\pi = 2ka^2.
\end{aligned}$$

■

Pumping Fluid from a Tank

To calculate the amount of energy needed to pump liquid from a tank or reservoir, we calculate the work required to lift the water to the top of the tank. To lift a given mass of water, we must balance the gravitational force and displace the mass to the top of the tank. We give an example of pumping a reservoir with trapezoidal cross-section.

■ **EXAMPLE 4** A reservoir with trapezoidal cross-section is filled with water. The cross-section has height 10 m, and the lengths of its two bases are 50 m and 35 m. The length of the reservoir is 200 m. Calculate the work required to pump the reservoir dry. The density of water is $\sigma = 1000$ kg/m³.

Solution. First, we calculate the work required to move a thin, horizontal layer of water to the top of the reservoir. A cross-section of a layer y meters from the bottom of the reservoir and dy thick is shown in Fig. 6.39. The reason for considering a horizontal layer of water is that every molecule in the layer must be displaced by the same amount, $10 - y$, against the gravitational force to reach the top of the reservoir. This is a constant-force work problem. After solving it, we use the integral to sum the work for the layers.

The mass dm of this section, which is a rectangular parallelepiped, is $dm = \sigma(2x)(200)\, dy$, where $2x$ is the width of the section. To lift this mass of water requires a force $dF = g\, dm$, which just balances the gravitational force. Since dF

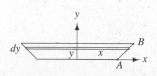

Figure 6.39. Reservoir.

acts over $(10 - y)$ m, the work required to pump out this cross-section is equal to $dw = (10 - y)\,dF$ joules. Hence the total amount of work required to empty the reservoir is given by

$$W = \int_0^{10} dw = \int_0^{10} (10 - y)\,dF = \int_0^{10} (10 - y)g\,dm.$$

Substituting $dm = \sigma(2x)(200)\,dy$ gives

$$W = g\int_0^{10} (10 - y)\sigma(2x)(200)\,dy.$$

To express x in terms of y, we note that the coordinates of points A and B are $(35/2, 0)$ and $(25, 10)$. The equation of the line through these two points is $x = (3y + 70)/4$. Hence

$$W = 400g\sigma \int_0^{10} (10 - y)x\,dy = 100g\sigma \int_0^{10} (10 - y)(3y + 70)\,dy$$

$$W \approx 3.92 \times 10^9 \text{ J.} \qquad\blacksquare$$

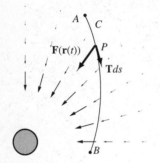

Figure 6.40. Conservative field.

In Example 1 we showed that the constant force field $\mathbf{F} = (0, -mg)$, often used for objects near earth, is conservative. In our final example we show that the force field shown in Fig. 6.40 and detailed below is also conservative. Aside from the influence of the sun, moon, and other planets, the field suggested in the figure is reasonably realistic.

What do we have to do to show that this field is conservative? As discussed in Example 1, we must show that the work done by the field on an object of mass m is "independent of path." This means that if we choose any two points A and B and a smooth curve C connecting them, the work done depends only on the end points of C, not its shape.

■ **EXAMPLE 5**   Show that the earth's gravitational field is conservative.

Solution.   With the origin at the center of the earth, let A and B be arbitrary points and C the smooth curve described by

$$\mathbf{r} = \mathbf{r}(t) = (x(t), y(t)), \quad a \le t \le b,$$

where $\mathbf{r}(a)$ and $\mathbf{r}(b)$ are position vectors of A and B. The object of mass m is at P on the curve, where P has position vector $\mathbf{r}(t)$. The field $\mathbf{F}(\mathbf{r}(t))$ at P points in the opposite direction of $\mathbf{r}(t)$. According to Newton's universal law of gravitation, the magnitude of $\mathbf{F}$ is GMm/r^2, where $r = \|\mathbf{r}(t)\|$, M is the mass of the earth, and G is a constant. Hence

$$\mathbf{F}(\mathbf{r}(t)) = -\frac{GMm}{r^2}\frac{1}{r}\mathbf{r}(t) = -\frac{GMm}{r^3}\mathbf{r}(t).$$

From (5), the work done by the gravitational field on the object is

$$W = \int_C \mathbf{F} \cdot d\mathbf{r} = -GMm \int_a^b \frac{1}{r^3}\mathbf{r}(t) \cdot \mathbf{r}'(t)\,dt.$$

Replacing $\mathbf{r}(t)$ by (x, y), $\mathbf{r}'(t)$ by (x', y'), and r by $(x^2 + y^2)^{1/2}$, W becomes

(6)
$$W = -GMm \int_a^b \frac{1}{(x^2 + y^2)^{3/2}} (x, y) \cdot (x', y') \, dt$$
$$= -GMm \int_a^b \left((x^2 + y^2)^{-3/2} (xx' + yy') \right) dt.$$

This looks difficult, but a remarkable thing happens, which is characteristic of conservative fields. We note that

$$\frac{d}{dt}(x^2 + y^2)^{-1/2} = -\frac{1}{2}(x^2 + y^2)^{-3/2}(2xx' + 2yy')$$

(7)
$$\frac{d}{dt}(x^2 + y^2)^{-1/2} = -(x^2 + y^2)^{-3/2}(xx' + yy').$$

Substituting (7) into (6), we obtain

$$W = GMm \int_a^b \frac{d}{dt}(x^2 + y^2)^{-1/2} \, dt.$$

By the Fundamental Theorem of Calculus,

$$W = GMm(x^2 + y^2)^{-1/2} \Big|_a^b = GMm \left(\frac{1}{\|\mathbf{r}(a)\|} - \frac{1}{\|\mathbf{r}(b)\|} \right).$$

This shows that the work W depends only on the end points of C. Hence the earth's gravitational field is conservative.   ■

Kinetic Energy

The definition of kinetic energy and a relation between work and kinetic energy follow from an application of Newton's second law to the integral for the work done on an object by a force field. Suppose that we have an object of mass m moving in a force field $\mathbf{F}$. If the object moves from A to B on a curve C, the work done by the field is

$$W = \int_C \mathbf{F} \cdot d\mathbf{r}.$$

If at any point on C the net force on the object is $\mathbf{F}$, then $\mathbf{F} = m\mathbf{a}$ or, writing the acceleration $\mathbf{a}$ in terms of velocity, $\mathbf{F} = m\mathbf{v}'$. With this replacement we may rewrite the work integral as

$$W = \int_a^b m\mathbf{v}' \cdot \frac{d\mathbf{r}}{dt} \, dt = \int_a^b m\mathbf{v}' \cdot \mathbf{v} \, dt,$$

where $\mathbf{r} = \mathbf{r}(t)$, $a \le t \le b$ describes C, which connects A and B. Writing $\mathbf{v} = (x', y')$ and $\mathbf{v}' = (x'', y'')$,

$$W = m \int_a^b (x'x'' + y'y'') \, dt.$$

The form of the integrand $x'x'' + y'y''$ suggests the chain rule. Specifically,

$$\frac{1}{2}\frac{d}{dt}\left(x'^2 + y'^2\right) = \frac{1}{2}(2x'x'' + 2y'y'') = x'x'' + y'y''.$$

It now follows that

$$W = \frac{1}{2}m\int_a^b \frac{d}{dt}\left(x'^2 + y'^2\right)dt = \frac{1}{2}m\left(x'^2 + y'^2\right)\Big|_a^b$$

$$= \frac{1}{2}mv_b^2 - \frac{1}{2}mv_a^2.$$

If we define the kinetic energy of an object with mass m and speed $v = \|\mathbf{v}\|$ as $T = \frac{1}{2}mv^2$, then W may be written as

(8) $$W = T_b - T_a.$$

In words, equation (8) states that the change in the object's kinetic energy as it moves from A to B is equal to the work done by the force field.

To see how this result can simplify some kinds of calculations, we use the physics lesson Jason gave to Paige in a 1994 *Fox Trot* comic strip. Jason, with calculator in hand, accompanies Paige on the Death Demon 2 coaster. Referring to Fig. 6.41, suppose a roller coaster car moves from A to B, a vertical drop of h meters. Jason teaches Paige how to calculate their speed $\mathbf{v}_b$ at point B, assuming that their speed at A was "essentially nothing." From (8),

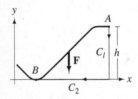

Figure 6.41. Death Demon 2.

(9) $$W = T_b - T_a = \frac{1}{2}mv_b^2.$$

If we knew W, we could solve this equation for v_b. Jason assumes that the gravitational force is $(0, -mg)$. If this were the net force, we could calculate W by using the fact that the gravitational force is conservative. On real coasters, however, the net force includes friction, and the resulting force field is not conservative.

Jason goes for a "quick-and-dirty" solution by neglecting friction. He calculates the work W by choosing the simple path $C_1 + C_2$. The work is

$$W = \int_{C_1+C_2} (0, -mg)\cdot(dx, dy) = -mg\int_{C_1} dy = mgh.$$

(Jason notes that mgh is their potential energy at A.) From (9),

$$mgh = \frac{1}{2}mv_b^2.$$

Solving for v_b, we obtain

$$v_b = \sqrt{2gh}.$$

For $h = 25$ m (estimated from the comic strip), $v_b \approx 22$ m/s, which is about 50 mph.

Exercises 6.4

Basic

1. Referring to Example 1, use $W = \int_C \mathbf{F} \cdot d\mathbf{r}$ in calculating the work done by the constant gravitational force on an object of mass m as it moves from $(0, 3)$ to $(2, 0)$ on the line described by the equations $x = x$ and $y = -3x/2 + 3$, where $0 \le x \le 2$.

2. Calculate the line integral $\int_C \mathbf{F} \cdot d\mathbf{r}$ where $\mathbf{F}$ is the field given in Example 1 and C is the piecewise smooth curve from $(0, 3)$ to $(0, 0)$ and then from $(0, 0)$ to $(2, 0)$. See Fig. 6.36 and the associated calculations.

3. Calculate the work done by the constant gravitational force on an object of mass 2 kg as it moves from $(1, 10)$ to $(8, 1)$. Choose any convenient smooth or piecewise smooth curve joining these two points.

4. Calculate the work done by the constant gravitational force on an object of mass 2 kg as it moves from $(0, 3)$ to $(2, 0)$ on the graph of the function $f(x) = -3x^2/4 + 3, 0 \le x \le 2$.

5. A 25 N force is required to compress a spring 0.05 m. Calculate the work done in compressing this spring 0.30 m from its relaxed state.

6. A 30 N force is required to compress a spring 0.05 m. Calculate the work done in compressing this spring 0.20 m from its relaxed state.

7. An object moves from $(-1, 0)$ to $(1, 0)$ on the graph of $y = 1 - x^2$, acted on by a force that attracts it toward the point $(1, 0)$ and has magnitude proportional to the distance between the object and $(1, 0)$. Calculate the work done by the force.

8. An object moves from $(-1, 0)$ to $(1, 0)$ on the graph of $y = x^3 - x$, acted on by a force that attracts it toward the point $(1, 0)$ and has magnitude proportional to the distance between the object and $(1, 0)$. Calculate the work done by the force.

9. In the reservoir example, show that $W \approx 3.92 \times 10^9$ J, as claimed.

Growth

10. Assuming that the smooth curve C^* in Example 1 is described by $\mathbf{r}(t) = \big(x(t), y(t)\big), a \le t \le b$, show that the work done by the gravitational force is $3mg$.

11. Calculate the work done by the earth's gravitational force $\mathbf{F} = -GMm/x^2$ on an object of mass $m = 50{,}000$ kg as it is boosted radially from earth's surface to 160 km. The coordinate x is measured on a line with origin at the center of the earth and positive direction outward.

12. A 500 m steel cable with density 15 kg/m is coiled at the bottom of a vertical shaft. Calculate the work required to lift the cable so that it hangs straight.

13. An elevator of mass 1500 kg hangs at the end of a 500 m steel cable with density 15 kg/m. Calculate the work required to raise the elevator 490 m.

14. Referring to Example 5 for general context, show that the field

$$\mathbf{F}(\mathbf{r}(t)) = -\frac{GMm}{r^2}\mathbf{r}(t)$$

is conservative.

15. A tank in the shape of a right circular cylinder with radius 3 m and height 10 m is placed so that its axis is horizontal. If the tank is full of gasoline, with density 680 kg/m^3, calculate the work required to empty the tank through a hole at the top of one of the circular ends.

16. A tank in the shape of a sphere with radius 10 m is full of gasoline, with density 680 kg/m^3. Calculate the work required to empty the tank through a hole at the top.

17. Assume that the Great Pyramid at Gizeh is solid granite, with density $\sigma = 25{,}000$ kg/m^3. Calculate the work required to put this granite in place. Start with volume element dV given in Example 1 of Section 5.1 and form a mass element dM by multiplying dV by σ. To get the mass dM in place, a constant force $-9.8\,dM$ must act on the mass (the positive direction is downward in Example 1; hence the force will be directed upward). Since the mass must move from ground level to height $h - z$, $dW = -9.8(h - z)\,dM$.

18. Referring to the following figure and Jason Fox's calculations reported in the text, what is the speed of the car when it reaches B? Why?

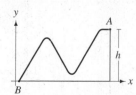

19. This problem was adapted from *Engineering Mechanics*, by Andrew Pytel and Jaan Kiusalaas (HarperCollins Publishers, 1994). Referring to the following figure, a force **F** acts on the end of a spring as the end moves on a curve C. At point A the spring is relaxed. Show that the work done by the force on the spring is independent of the curve C joining A and B. Let the spring constant be k and assume that C can be described by

$$\mathbf{r} = \mathbf{r}(h), \quad c \le h \le d,$$

where h is the distance from O to the point of application of **F**, $\|\mathbf{r}(c)\| = a = \|\overrightarrow{OA}\|$, and $\|\mathbf{r}(d)\| = b = \|\overrightarrow{OB}\|$. Fill in the following outline. (i) Show that $\|\mathbf{r}(h)\| = h$. (ii) Show that the force at $\mathbf{r}(h)$ is

$$\mathbf{F}(\mathbf{r}(h)) = (k(h - a)/h)\mathbf{r}.$$

(iii) Show that

$$dW = (k(a - h)/h)\mathbf{r} \cdot \mathbf{r}' \, dh.$$

(iv) Show that $\mathbf{r} \cdot \mathbf{r}' = h$. From parts (iii) and (iv), show that $dW = k(a - h) \, dh$. Now complete the problem.

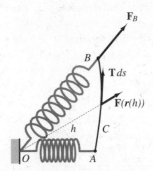

Review/Preview

20. A triangle has vertices $(0, 0)$, $(b, 0)$, and (a, c), where $b, c > 0$. Show that the medians of this triangle meet at the point $((a + b)/2, c/2)$. Recall that a median is a line joining a vertex and the midpoint of the opposite side.

21. (Continuation of Exercise 20) Divide the triangle into two pieces of equal area by drawing a line parallel to its base. If the triangle were held horizontally, would it balance on this line?

22. Divide a semicircle of radius 1 into two pieces of equal area by drawing a line parallel to its diameter. If the semicircle were held horizontally, would it balance on this line?

6.5 CENTER OF MASS

Unless a hammer is thrown without spin, most of its constituent atoms will follow complex paths. A relatively simple throw is shown in Fig. 6.42, where the spin is confined to a vertical plane. Both experimental evidence and theoretical analysis show that for any throw of the hammer there is one point/atom of the hammer that follows a simple path. That point is called the **center of mass** of the hammer.

A system of particles in motion can be modeled by a one-particle system located at the center of mass of the particles and having their total mass. In studying the motion of the earth in the solar system, for example, the earth may be modeled by a particle with the earth's mass and located at the earth's center of mass. For tracking a thrown hammer it may be sufficient to apply Newton's second law to the

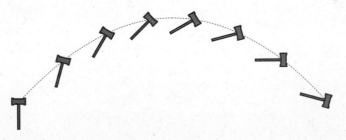

Figure 6.42. Tracking the center of mass.

one-particle system consisting of a particle with mass equal to that of the hammer and located at its center of mass. This would be useful even if, for example, we were asked which part of the spinning hammer hit the ground first.

In this section we study both finite and continuous systems. Although a hammer and the earth are finite systems of atoms, it is often easier to model such objects with regions of Euclidean 3-space consisting of an infinite number of triples (x, y, z) of real numbers. Line segments, segments of curves, and regions of the plane or of space are often called *continuous* since they have no gaps.

Finite Systems

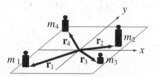

Figure 6.43. Point masses.

By a system S of **point masses** we mean a set of physical objects located at points on a line, in a plane, or in space. The objects have masses $m_1, m_2, \ldots, m_n$ and position vectors $\mathbf{r}_1, \mathbf{r}_2, \ldots, \mathbf{r}_n$. Such a system can model the planets in the solar system, the pellets of shot emerging from a shotgun, the fragments of a subatomic collision, or several heavy pieces of machinery arranged on the floor of a factory. We show in Fig. 6.43 a system S of four point masses in a plane.

The system S of point masses is completely characterized or described by the pairs

$$\langle m_1, \mathbf{r}_1 \rangle, \langle m_2, \mathbf{r}_2 \rangle, \ldots, \langle m_n, \mathbf{r}_n \rangle,$$

where the ith mass m_i has position vector $\mathbf{r}_i$ relative to some coordinate system. We use the angle brackets $\langle \cdots \rangle$ to distinguish the mass/position pairs from vectors or points. If the system is on a line, the coordinate system may be chosen so that the position vectors are simply coordinate numbers $x_1, x_2, \ldots, x_n$. If the system is not collinear, the position vectors have at least two coordinates. In a plane

$$\mathbf{r}_1 = (x_1, y_1), \mathbf{r}_2 = (x_2, y_2), \ldots, \mathbf{r}_n = (x_n, y_n)$$

DEFINITION   **center of mass of a system of point masses**

The center of mass of the system S

(1) $$\langle m_1, \mathbf{r}_1 \rangle, \langle m_2, \mathbf{r}_2 \rangle, \ldots, \langle m_n, \mathbf{r}_n \rangle$$

is the point $\mathbf{R}$ with position vector

(2) $$\mathbf{R} = \frac{1}{m} \sum_{i=1}^{n} m_i \mathbf{r}_i,$$

where $m = \sum_{i=1}^{n} m_i$ is the total mass of the system.

On a line we set $\mathbf{R} = X$ and write (2) in the simpler form

(3) $$X = \frac{1}{m} \sum_{i=1}^{n} m_i x_i.$$

In a plane we set $\mathbf{R} = (X, Y)$. We may write (2) in the form

(4)
$$X = \frac{1}{m} \sum_{i=1}^{n} m_i x_i, \quad Y = \frac{1}{m} \sum_{i=1}^{n} m_i y_i.$$

For some calculations these equations are preferable to (2).

■ **EXAMPLE 1**   Find the center of mass of the system

$$\langle 6, -5 \rangle, \langle 2, -3 \rangle, \langle 5, 2 \rangle, \langle 7, 4 \rangle,$$

which models four masses on a (massless) seesaw, as shown in Fig. 6.44.

Solution.   The center of mass may be calculated using (3), which is the one-dimensional case of (2). The mass of the system is $m = 6 + 2 + 5 + 7 = 20$ and

$$X = \frac{1}{m} \sum_{i=1}^{n} m_i x_i = \frac{1}{20} \big((6)(-5) + (2)(-3) + (5)(2) + (7)(4) \big) = 0.1.$$

The center of mass of this system is the point with coordinate $X = 0.1$. The seesaw would be in equilibrium if the fulcrum were placed at this point.   ■

■ **EXAMPLE 2**   The masses shown on a plane in Fig. 6.43 are modeled by the system

$$\langle 2, (-4.0, -3.2) \rangle, \langle 3, (4.0, 1.0) \rangle, \langle 1, (4.4, -2.6) \rangle, \langle 2, (-4.6, 3.4) \rangle.$$

Calculate the center of mass of the system. The units are kilograms and meters.

Solution.   We use (4) to calculate $\mathbf{R} = (X, Y)$. Noting that $m = 8$ kg,

$$X = \frac{1}{8} \big((2)(-4.0) + (3)(4.0) + (1)(4.4) + (2)(-4.6) \big) = -0.1 \text{ m}$$

$$Y = \frac{1}{8} \big((2)(-3.2) + (3)(1.0) + (1)(-2.6) + (2)(3.4) \big) = 0.1 \text{ m}.$$

The center of mass of this system is the point with position vector

$$\mathbf{R} = (X, Y) = (-0.1, 0.1) \text{ m}.$$

If the plane containing this system were horizontal, as suggested in the figure, the point $(-0.1, 0.1)$ would be the "balance point."   ■

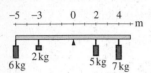

-5 -3   0   2   4

Figure 6.44. Center of mass.

Continuous Mass

Although it is possible to imagine calculating the center of mass of a solid object by regarding it as a finite system of its constituent atoms and then using (2), it is usually simpler to regard a solid object as a limiting case of a finite system of point masses. We consider three "continuous masses": rods, curved rods, and thin plates or laminas. These are one- and two-dimensional cases of continuous mass.

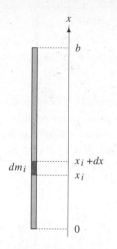

Figure 6.45. Metal rod.

One-Dimensional Continuous Mass We wish to calculate the center of mass of the metal rod of length b shown in Fig. 6.45. For a continuous rod, the mass of the rod per unit length is used instead of the mass at a point. Mass per unit length is called **lineal density** and is measured in kilograms per meter. We assume at first that the density is constant along the rod and equal to σ kg/m. We divide the rod into n mass elements corresponding to a subdivision of $[0, b]$, look upon the mass elements as a system of objects, and calculate the center of mass X_n of the system using (3). The element of mass between x_i and $x_i + dx$ has mass $dm_i = \sigma\, dx$. From (3) we have

$$X_n = \frac{\sum_{i=1}^{n} dm_i x_i}{\sum_{i=1}^{n} dm_i} = \frac{\sum_{i=1}^{n} x_i(\sigma\, dx)}{\sum_{i=1}^{n} \sigma\, dx} = \frac{\sum_{i=1}^{n} x_i\, dx}{\sum_{i=1}^{n} dx}.$$

In the second and third steps we replaced dm_i by $\sigma\, dx$ and removed the common factor of σ. The number X_n is the center of mass of the system of n mass elements. We expect that as the number of subdivisions increases, X_n will approach the center of mass X of the rod. Thus

$$(5) \qquad X = \frac{\int_0^b x\, dx}{\int_0^b dx} = \frac{\frac{1}{2}b^2}{b} = \frac{1}{2}b.$$

This is hardly surprising, for we expect the center of mass to be at the midpoint of a rod of constant density.

This calculation may be used with very little change to find the center of mass of a rod with variable density. Assuming the density of the rod at x is $\sigma(x)$, the mass dm_i of the element of mass between x_i and $x_i + dx$ is $dm_i = \sigma(x_i)\, dx$. From (3) we have

$$X_n = \frac{\sum_{i=1}^{n} dm_i x_i}{\sum_{i=1}^{n} dm_i} = \frac{\sum_{i=1}^{n} x_i \sigma(x_i)\, dx}{\sum_{i=1}^{n} \sigma(x_i)\, dx}.$$

The number X_n is the center of mass of the system of n mass elements. We expect that as the number of subdivisions increases, X_n will approach the center of mass X of the rod. Thus

$$(6) \qquad X = \frac{\int_0^b x\sigma(x)\, dx}{\int_0^b \sigma(x)\, dx}.$$

■ **EXAMPLE 3** A 1.5 m rod of bronze is cast, with the density varying linearly from 2.5 kg/m to 3.5 kg/m from end to end. Find the mass of the rod and its center of mass. Bronze is an alloy of tin and copper, with the proportions of the two metals varying fairly widely, depending on the intended use.

Solution. To say that the density varies linearly means that if we plot density σ against position x on the rod, the graph will be a line; that is, the density function will have the form $\sigma(x) = Ax + B$. Referring to Fig. 6.46, if we place the origin of a coordinate system at the less dense end, then since $\sigma(0) = 2.5$ and $\sigma(1.5) = 3.5$, we have

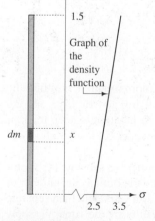

Figure 6.46. Bronze rod.

$$\sigma(x) = 2.5 + (1/1.5)x, \quad 0 \le x \le 1.5.$$

From (6) we have

$$X = \frac{\int_0^{1.5}(2.5 + (1/1.5)x)x\,dx}{\int_0^{1.5}(2.5 + (1/1.5)x)\,dx} = \frac{3.5625}{4.5} \approx 0.79.$$

The mass of the rod is 4.5 kg, and the center of mass is approximately 0.79 m from the origin. Since the denser end is farther from the origin, this result is reasonable.

Rather than solve a problem by substituting into a formula, as we have just done, it often increases understanding to go through a stripped-down version of the argument leading to the formula. We divided the rod into n point masses by subdividing the interval $[0, 1.5]$. The point mass at x_i has mass $dm_i \approx \sigma(x_i)\,dx_i$. From (3) we have

$$X \approx \frac{\sum_{i=1}^n dm_i x_i}{\sum_{i=1}^n dm_i} = \frac{\sum_{i=1}^n x_i\sigma(x_i)\,dx_i}{\sum_{i=1}^n \sigma(x_i)\,dx_i}.$$

Taking the limit as $n \to \infty$, we have

$$X = \frac{\int_0^{1.5} x\sigma(x)\,dx}{\int_0^{1.5} \sigma(x)\,dx}. \qquad \blacksquare$$

Curved Rods   Next we calculate the center of mass of a curved rod. Although such a rod is a set of points in Euclidean 2-space, the rod can be described by a function $\mathbf{r} = \mathbf{r}(t)$ of one variable. This makes it possible to use one-dimensional integrals in the calculation. Referring to the curved rod C in Fig. 6.47, it is natural to express the density of C at a point P in terms of the length of arc from one end of the rod, say A, to P. However, to do this for even simple curves C often requires rather heavy numerical integration and other numerical procedures. To eliminate some of these problems we assume that the density is constant. This assumption makes it possible to regard center of mass as a geometric concept, so that instead of locating the center of mass, we locate a "center of length" or "center of area." In a constant-density context the center of mass is often called the **centroid** of the line segment, curve, or region.

We use (2) in determining the center of mass of the curved rod C shown in Fig. 6.47 and described by $\mathbf{r} = \mathbf{r}(t)$, $a \le t \le b$. We shall need the mass element $dm = \sigma\,ds$, where σ is the density of the rod at P and ds the arc length element. We use (2) to approximate the center of mass $\mathbf{R}$ of the rod. We obtain

(7)
$$\mathbf{R} \approx \frac{1}{\sum dm}\sum dm\,\mathbf{r}(t) = \frac{1}{\sigma\sum ds}\sigma\sum \mathbf{r}(t)\,ds.$$

We used a bare minimum of notation in writing this approximation, omitting the subscripts associated with a subdivision of $[a, b]$ and using generic mass and arc length elements. It is often an advantage to use enough notation to keep track of what's going on, but not so much that the physical concepts get lost in a flurry of subscripts.

At this point we have a choice. To determine $\mathbf{R}$, either we separate the vector sum on the right of (7) into its coordinates and then take the limit as the number of subdivisions of $[a, b]$ becomes infinite, or we take the limit first and then separate the resulting vector integral into its coordinates. The results are the same. We show both calculations.

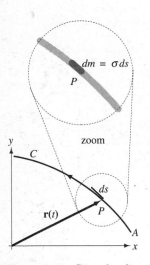

Figure 6.47. Curved rod.

Recalling that $ds = (ds/dt)\,dt = \|\mathbf{r}'(t)\|\,dt$ and removing the common factor of σ, we take the limit in (7) first and obtain

$$(8) \qquad \mathbf{R} = \frac{1}{\int_a^b \|\mathbf{r}'(t)\|\,dt} \int_a^b \|\mathbf{r}'(t)\|\mathbf{r}(t)\,dt.$$

Before separating the integral into its coordinates, we note that the denominator of this expression for the centroid of C is the arc length L of C. With this notation we can write $\mathbf{R}$ as

$$\mathbf{R} = \frac{1}{L}\int_a^b \|\mathbf{r}'(t)\|\big(x(t),\, y(t)\big)\,dt$$

$$(9) \qquad \mathbf{R} = (X, Y) = \left(\frac{1}{L}\int_a^b \|\mathbf{r}'(t)\|x(t)\,dt,\ \frac{1}{L}\int_a^b \|\mathbf{r}'(t)\|y(t)\,dt\right).$$

Separating the right side of (7) into its coordinates first, we obtain

$$\mathbf{R} = (X, Y) \approx \frac{1}{\sigma\sum ds}\sigma\sum \mathbf{r}(t)\,ds = \frac{1}{\sum ds}\left(\sum x(t)\,ds,\ \sum y(t)\,ds\right).$$

As $n \to \infty$, we obtain

$$(10) \qquad X = \frac{1}{L}\int_a^b \|\mathbf{r}'(t)\|x(t)\,dt, \qquad Y = \frac{1}{L}\int_a^b \|\mathbf{r}'(t)\|y(t)\,dt.$$

This agrees with (9).

■ **EXAMPLE 4** Determine the centroid of the semicircular rod of radius a shown in Fig. 6.48.

Solution. The equation

$$\mathbf{r} = a(\cos t, \sin t), \quad 0 \le t \le \pi$$

describes the rod. For a semicircle, the arc length part of the calculation is easy. Noting that $\|\mathbf{r}'(t)\| = a$ and $L = \pi a$, from (10) we have

$$X = \frac{1}{\pi a}\int_0^\pi ax(t)\,dt = \frac{1}{\pi a}\int_0^\pi a^2\cos t\,dt = 0$$

$$Y = \frac{1}{\pi a}\int_0^\pi ay(t)\,dt = \frac{1}{\pi a}\int_0^\pi a^2\sin t\,dt = \frac{2a}{\pi}.$$

For a semicircle of radius 1, the centroid would be $\mathbf{R} = (0, 2/\pi)$, a point not on the rod. If the rod were embedded in a rigid piece of plastic with negligible mass and held in a horizontal plane, the balance point would be at the centroid. In future calculations we take advantage of such symmetries as that of the semicircle about the y-axis. We may infer directly from this symmetry that $X = 0$. ■

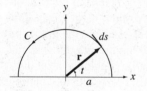

Figure 6.48. Semicircular rod.

In the last section, most of the calculations of the work done by a force field on an object were done under the assumption that the object was a point mass. For calculating the work done by a field on a real, irregularly shaped object, it is often possible to model such an object by a point mass located at its center of mass. We illustrate this idea in the next example.

■ **EXAMPLE 5** A piece of string falls to the x-axis. The string initially is in the shape of the graph of a function f on an interval $[a, b]$ and has constant lineal density σ. Show that the work done by the gravitational field on the string is equal to $-gmY$, where m is the mass of the string and Y is the y-coordinate of the center of mass of the string in its initial position.

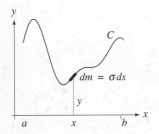

Figure 6.49. Falling string.

Solution. Referring to Fig. 6.49, the work done by the gravitational field $\mathbf{F} = \langle 0, -g \rangle$ on the string element of mass dm is $dW = (-g\,dm)y$. Hence the work done by the field on the entire string is

$$W = -\int_a^b gy\,dm = -g\sigma \int_a^b y\,ds.$$

From (10) we have

$$Y = \frac{1}{L}\int_a^b \|\mathbf{r}'(x)\|y(x)\,dx = \frac{1}{L}\int_a^b y\,ds.$$

Since the mass of the string is $m = \sigma L$, it follows that $W = -gmY$. ■

Two-Dimensional Continuous Mass

We consider thin metal plates, which are called laminas, whose shapes can be described by two functions on an interval. The lamina in Fig. 6.50 is bounded on the top by the graph of f, on the bottom by the graph of g, and on the sides by the lines $x = a$ and $x = b$. We assume that its areal density σ, with units kg/m^2, is constant. We calculate the mass and center of mass (or centroid) of the lamina much as we did in the one-dimensional case, by dividing the lamina into n mass elements dm_i corresponding to a subdivision of the interval $[a, b]$ into n equal pieces, applying (2) or (4), and then letting $n \to \infty$. The length and width of the mass element are $f(x_i) - g(x_i)$ and dx_i. Hence $dm_i = \sigma(f(x_i) - g(x_i))\,dx$.

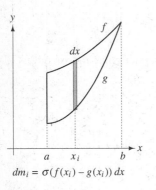

$dm_i = \sigma(f(x_i) - g(x_i))\,dx$

Figure 6.50. Metal lamina and element of mass.

The calculation is different in one important respect. For rods and curved rods, the mass element dm_i could be imagined as a point mass since both of its dimensions were small. For the lamina, however, the mass element dm_i has only one small dimension. To accommodate this difference, we regard the mass element as a rod and use our earlier result that the center of mass of a rod with constant density is at its center. This makes it possible to replace the n mass elements dm_i by the system

$$\langle dm_1, \mathbf{r}_1 \rangle, \langle dm_2, \mathbf{r}_2 \rangle, \ldots, \langle dm_n, \mathbf{r}_n \rangle$$

of point masses, where $\mathbf{r}_i$ is a position vector of the center of the strip with mass dm_i. We relate the center of mass of this system to that of the lamina by the Mass Subdivision Theorem. To minimize notation, we state the theorem for two subsystems but note that the statement is true for any (finite) number of subsystems.

Mass Subdivision Theorem

If a system S of point masses is divided into two subsystems A and B, then the center of mass $\mathbf{R}$ of S is equal to the center of mass of the system of two point masses described by $\langle m_A, \mathbf{R}_A \rangle$ and $\langle m_B, \mathbf{R}_B \rangle$, where m_A and m_B are the total masses of the two subsystems and $\mathbf{R}_A$ and $\mathbf{R}_B$ are the centers of mass of the two subsystems.

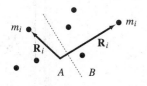

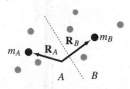

Figure 6.51. Mass Subdivision Theorem.

We show in Fig. 6.51 a system of seven point masses, divided into subsystems A and B. If we calculate the centers of mass $\mathbf{R}_A$ and $\mathbf{R}_B$ of the two subsystems and put at these points the total masses m_A and m_B of the subsystems, we have a new system of two point masses. This is shown in the lower sketch in the figure. The Mass Subdivision Theorem states that the center of mass of the two-mass system is the same as the center of mass of the original seven-mass system.

The theorem is proved by rearranging $\sum_S m_i \mathbf{R}_i$ and $m = \sum_S m_i$ into two sums, an A part and a B part. Letting $m = m_A + m_B$, it follows from

$$\mathbf{R}_A = \frac{1}{m_A} \sum_A m_i \mathbf{R}_i \quad \text{and} \quad \mathbf{R}_B = \frac{1}{m_B} \sum_B m_i \mathbf{R}_i$$

that

$$\mathbf{R} = \frac{1}{m} \sum_S m_i \mathbf{R}_i = \frac{1}{m_A + m_B} \left(\sum_A m_i \mathbf{R}_i + \sum_B m_i \mathbf{R}_i \right)$$

$$= \frac{1}{m_A + m_B} (m_A \mathbf{R}_A + m_B \mathbf{R}_B).$$

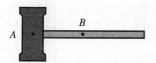

Figure 6.52. Subsystems of a hammer.

This gives a simple way of calculating the center of mass of a complex system from simpler subsystems. For example, the center of mass of a hammer similar to that shown in Fig. 6.52 can be calculated by replacing the hammer by the two particle system

$$\langle m_1, \mathbf{r}_1 \rangle, \langle m_2, \mathbf{r}_2 \rangle$$

where m_1 and $\mathbf{r}_1$ are the mass and center of mass of the head, and m_2 and $\mathbf{r}_2$ are the mass and center of mass of the handle. See Exercise 14.

We use the Mass Subdivision Theorem in calculating the center of mass of a lamina. For the lamina shown in Fig. 6.53, we use the position vector

$$\mathbf{r}_i = (x_i, y_i) = \left(x_i, \tfrac{1}{2}(f(x_i) + g(x_i)) \right)$$

to mark the center of the mass element dm_i. Recall that its length is $f(x_i) - g(x_i)$. The Mass Subdivision Theorem shows how we can approximate the center of mass $\mathbf{R}$ of the lamina by replacing it by the system

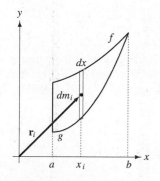

Figure 6.53. Metal lamina.

$$\langle dm_1, \mathbf{r}(x_1) \rangle, \langle dm_2, \mathbf{r}(x_2) \rangle, \ldots, \langle dm_n, \mathbf{r}(x_n) \rangle.$$

We use the coordinate form (4) for the calculation. We have

$$X \approx \frac{\sum_{i=1}^{n} dm_i x_i}{\sum_{i=1}^{n} dm_i} = \frac{\sigma \sum_{i=1}^{n} x_i (f(x_i) - g(x_i)) \, dx}{\sigma \sum_{i=1}^{n} (f(x_i) - g(x_i)) \, dx}$$

$$Y \approx \frac{\sum_{i=1}^{n} dm_i y_i}{\sum_{i=1}^{n} dm_i} = \frac{\sigma \sum_{i=1}^{n} \frac{1}{2}(f(x_i) + g(x_i))(f(x_i) - g(x_i)) \, dx}{\sigma \sum_{i=1}^{n} (f(x_i) - g(x_i)) \, dx}.$$

Taking the limit as $n \to \infty$ and removing the constant density factor,

(11)
$$X = \frac{\int_a^b x(f(x) - g(x)) \, dx}{\int_a^b (f(x) - g(x)) \, dx}$$

(12)
$$Y = \frac{\frac{1}{2} \int_a^b x(f(x)^2 - g(x)^2) \, dx}{\int_a^b (f(x) - g(x)) \, dx}.$$

The mass of the lamina is

(13)
$$M = \sigma \int_a^b (f(x) - g(x)) \, dx.$$

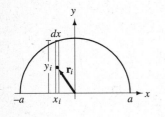

Figure 6.54. Semicircular lamina.

■ **EXAMPLE 6** Find the center of mass of a semicircular lamina with radius a and constant density σ. See Fig. 6.54.

Solution. As with the variable-density rod, we go through a stripped-down version of the above discussion rather than simply substitute into the formulas (11) and (12) for X and Y. We apply the Mass Subdivision Theorem to the semicircular lamina after dividing it into strips corresponding to a subdivision of the interval $[-a, a]$. The point x_i of the subdivision locates a typical element. Its width and height are dx and y_i. The centers of mass of these elements are at their centers, located by the position vector $\mathbf{r} = (x_i, \frac{1}{2}y_i)$. The mass dm_i of a typical element is $\sigma y_i \, dx$. This is the area $y_i \, dx$ of the element times its density σ. From the Mass Subdivision Theorem we have

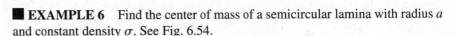

$$X \approx \frac{\sum dm_i x_i}{\sum dm_i} = \frac{\sigma \sum x_i y_i \, dx}{\sigma \sum y_i \, dx} \to \frac{\sigma \int_{-a}^{a} xy \, dx}{\sigma \int_{-a}^{a} y \, dx} = X$$

$$Y \approx \frac{\sum dm_i \frac{1}{2} y_i}{\sum dm_i} = \frac{\sigma \sum (\frac{1}{2} y_i) y_i \, dx}{\sigma \sum y_i \, dx} \to \frac{\sigma \int_{-a}^{a} \frac{1}{2} y^2 \, dx}{\sigma \int_{-a}^{a} y \, dx} = Y.$$

We used the symbol $\to$ to indicate the limiting values of these sums as the number of subdivisions becomes infinite. Substituting $y = \sqrt{a^2 - x^2}$ and noting that the integral in the denominators of X and Y is equal to the area $\pi a^2 / 2$ of a semicircle, we have

$$X = \frac{\int_{-a}^{a} x \sqrt{a^2 - x^2} \, dx}{\pi a^2 / 2} = 0$$

$$Y = \frac{\int_{-a}^{a} \frac{1}{2}(a^2 - x^2) \, dx}{\pi a^2 / 2} = \frac{4a}{3\pi}.$$

We have shown that the center of mass of a semicircle is on its line of symmetry, 42 percent of the distance from the center. It would have saved us time in this calculation simply to have inferred that $X = 0$ from the symmetry of the semicircle about the line $x = 0$.   ■

Exercises 6.5

Basic

1. Find the center of mass of the system of point masses $\langle 5, -3 \rangle$, $\langle 4, -1 \rangle$, $\langle 6, 2 \rangle$, $\langle 3, 7 \rangle$.

2. Five objects are arranged on a 10 m board with negligible mass. Starting from one end, a 5 kg mass is 1 m from the end, a 15 kg mass is 3 m, a 5 kg mass is 4 m, a 20 kg mass is 8 m, and a 10 kg mass is 9 m from the end. Find the center of mass of this system.

3. Find the center of mass **R** of the system
$$\langle 4, (-3, 5) \rangle, \langle 6, (-5, -6) \rangle, \langle 7, (5, -1) \rangle,$$
$$\langle 8, (1, 1) \rangle, \langle 1, (6, 6) \rangle.$$

4. Find the center of mass **R** of the system
$$\langle 4.5, (-2.4, 5.1) \rangle, \langle 3.7, (-4.9, -5.8) \rangle,$$
$$\langle 5.5, (4.2, -1.0) \rangle, \langle 7.7, (1.3, 1.4) \rangle,$$
$$\langle 1.5, (7.2, 6.9) \rangle.$$

5. Find the mass and center of mass of a 0.5 m rod whose density varies linearly from 3.0 kg/m to 3.7 kg/m.

6. Find the mass and center of mass of a 0.7 m rod whose density varies linearly from 2.6 kg/m to 2.9 kg/m.

7. Assuming that the center of mass of a square lamina with uniform density is at its center, use the Mass Subdivision Theorem to locate the center of mass of the lamina shown below. The edge of the larger square is twice that of the smaller square. Place the origin at the lower left corner of the larger square and assume it has side $2a$.

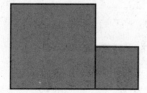

8. Assuming that the center of mass of a triangle with constant density is at the intersection of its medians (lines drawn from a vertex to the midpoint of the opposite side) and that of a square with constant density is at its center, use the Mass Subdivision Theorem to locate the center of mass of the lamina shown below. Place the origin at the right angle of the triangle and assume the square has side a.

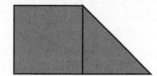

9. Use the Mass Subdivision Theorem together with the result from Example 6 to calculate the center of mass of a quarter circle of radius a. Check your work by calculating the center of mass directly.

10. Calculate the mass and center of mass of the lamina of constant density σ and bounded by $x = 0$, $x = 1$, $y = 0$, and $y = e^x$.

11. Calculate the mass and center of mass of the lamina of constant density σ and bounded by $x = 0$, $x = \pi$, $y = 0$, and $y = \sin x$.

12. Show that the mass and center of mass of the "second-degree parabolic spandrel" (a shape used to reinforce a joint between beams; it is welded into place) bounded by $x = 0$, $x = b$, $y = h - hx^2/b^2$, and $y = h$ are $M = bh/3$, $X = 3b/4$, and $Y = 7h/10$. Assume that the density is 1 kg/m^2.

13. Show that the mass and center of mass of the "third-degree parabolic spandrel" (a shape used to reinforce a joint between beams; it is welded into place) bounded by $x = 0$, $x = b$, $y = h - hx^3/b^3$, and $y = h$ are $M = bh/4$, $X = 4b/5$, and $Y = 5h/7$. Assume that the density is 1 kg/m^2.

Growth

14. Determine the center of mass of a hammer. The steel head is a cylinder of diameter 7 cm and height 14 cm; its density is $\sigma = 7800$ kg/m^3. The wooden handle is a cylinder of diameter 2.5 cm and height 35 cm; its density is $\sigma = 600$ kg/m^3.

15. Show that for a finite system of masses the calculation of the center of mass does not depend on the choice of coordinate system. Specifically, let S be the system $\langle m_1, \mathbf{r}_1 \rangle, \dots, \langle m_n, \mathbf{r}_n \rangle$ and suppose that S becomes $\langle m_1, \mathbf{r}_1' \rangle, \dots, \langle m_n, \mathbf{r}_n' \rangle$, relative to the translation of axes $\mathbf{r}_i' = \mathbf{r}_i + (h, k)$, where h and k are constants. Show that the centers of mass $\mathbf{R}$ and $\mathbf{R}'$ calculated in the two coordinate systems are related by $\mathbf{R}' = \mathbf{R} + (h, k)$.

16. The idea of a bar of variable density can be used to calculate the center of mass of a tapered flagpole made from a material of constant density. Find the mass and center of mass of a 10 m aluminum flagpole, which tapers uniformly from a diameter of 15 cm to 10 cm. Assume the pole is a tube whose walls are 3 cm thick. The density of aluminum is 2702 km/m^3.

17. Determine the centroid of a rod in the shape of the parabola described by $y = x^2$, $-1 \le x \le 1$. If you use the parametric description $\mathbf{r} = (x, x^2)$, you will encounter an integral of the form $\int x^2 \sqrt{1 + 4x^2} \, dx$. Use the substitution $2x = \tan u$ and then formula (31) from the table of integrals. Numerical integration is also a possibility.

18. Determine the centroid of the cycloid curve described by

$$\mathbf{r} = (t - \sin t, 1 - \cos t), \quad 0 \le t \le 2\pi.$$

The identities $1 - \cos t = 2\sin^2 \frac{1}{2}t$, $\sin t = 2\sin \frac{1}{2}t \cos \frac{1}{2}t$, and $\cos t = 1 - 2\sin^2 \frac{1}{2}t$ will be useful, together with formula (32) from the table of integrals. Numerical integration is also a possibility.

19. Show that the centroid of a triangle with vertices $(0, 0)$, $(0, b)$ and (a, c), where $b, a > 0$, is at the point where the medians of the triangle meet. Recall that a median is a line joining a vertex and the midpoint of the opposite side.

20. How would you go about locating the center of mass of a rod 1 meter in length, the first third of which has density 1 kg/m and the remainder density 2 kg/m?

21. A circular lamina with diameter 3 cm has a hole of diameter 1 cm stamped from its center. It is then cut along a diameter to form a half-washer. Use the result of Example 6 in calculating the centroid of the half-washer.

22. Find the center of mass of the lamina bounded by the graphs of $y = x$ and $y = x^2 - 3x$.

23. Find the mass and centroid of the infinite lamina bounded by $x = 0$, $y = 0$, and $y = e^{-x}$.

24. Find the mass and centroid of the lamina bounded by $x = 0$, $x = \pi/2$, $y = 1$, and $y = \sin x$.

25. Find the mass and centroid of the lamina bounded by $x = 0$, $y = 0$, and the first-quadrant part of the ellipse with equation $x^2/a^2 + y^2/b^2$.

26. Describe in detail how the Mass Subdivision Theorem may be used in determining the center of mass of a lamina formed by punching a circular hole through a circular lamina.

27. Referring to Fig. 6.42, the center of mass of the hammer follows the curve

$$\mathbf{r}(t) = \tfrac{1}{2}(0, -g)t^2 + 10(\cos 50°, \sin 50°)t, \quad t \ge 0.$$

If the end of the handle is 0.3 m below the center of mass $t = 0$ and the hammer has rotated 105° by the time its center of mass returns to the level from which it started, determine the path followed by the end of the handle.

Review/Preview

28. Determine the unit tangent $\mathbf{T}$ at the point $(1, 3\sqrt{3}/2)$ of the ellipse described by

$$\mathbf{r} = (2\cos t, 3\sin t), \quad 0 \le t \le 2\pi.$$

29. What is the direction of the acceleration vector at $t = 2$ for the object whose position at t is

$$\mathbf{r} = (t^2, t^3), \quad t \ge 0?$$

6.6 CURVATURE, ACCELERATION, AND KEPLER'S SECOND LAW

The unit tangent vector $\mathbf{T}$ determines the direction of a curve. Curvature measures the rate at which the curve bends away from $\mathbf{T}$. Curvature is used in expressing the acceleration of a moving object in terms of tangential and normal acceleration vectors. For planetary motion, which is often described in polar coordinates, the acceleration vector can be written as the sum of radial and transverse acceleration vectors. This facilitates a derivation of Kepler's second law, which states that the line from the sun to a planet sweeps out equal areas in equal times.

Curvature

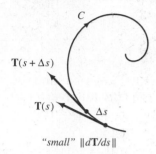

"small" $\|d\mathbf{T}/ds\|$

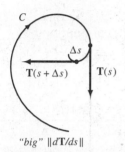

"big" $\|d\mathbf{T}/ds\|$

Figure 6.55. Big and small curvature.

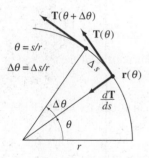

Figure 6.56. Circle case.

At what rate does a curve bend away from its unit tangent and how can this be measured? Let C be a smooth curve described by $\mathbf{r} = \mathbf{r}(t)$, $a \le t \le b$, and assume that $\mathbf{r}'(t) \ne 0$ for all $t \in [a, b]$. (This assumption ensures that the unit tangent $\mathbf{T}$ is defined on $[a, b]$ and is a function of arc length s.) Since the length of $\mathbf{T}$ is fixed, only the direction of $\mathbf{T}$ can change with s. The vector $d\mathbf{T}/ds$ is called the **curvature vector** of C. Based at $\mathbf{r}(t)$, it points in the direction in which C "bends away from" its unit tangent vector $\mathbf{T}$. The **curvature** of C is the length of the curvature vector. We show in Fig. 6.55 a curve whose curvature is (relatively) small on one part of the curve and (relatively) big on another.

Investigation

We start by exploring the ideas of curvature vector and curvature in a simple setting. We calculate the change in $\mathbf{T}$ relative to s for a circle C of radius r, part of which is shown in Fig. 6.56. The circle is described by

$$\mathbf{r} = \mathbf{r}(\theta) = r\big(\cos\theta, \sin\theta\big), \quad 0 \le \theta \le 2\pi.$$

Denoting differentiation with respect to the parameter θ by $'$, the unit tangent vector $\mathbf{T}$ at $\mathbf{r}(\theta)$ is

$$\mathbf{T} = \mathbf{T}(\theta) = \frac{\mathbf{r}'}{\|\mathbf{r}'\|} = \big(-\sin\theta, \cos\theta\big).$$

The arc length s of this circle, measured from $(r, 0)$, is related to the parameter θ by $s = r\theta$. We use this relation and the chain rule to calculate the curvature vector. We have

$$\frac{d\mathbf{T}(\theta)}{ds} = \frac{d}{ds}\mathbf{T}(s/r) = \frac{d}{ds}\big(-\sin(s/r), \cos(s/r)\big)$$

$$= \frac{1}{r}\big(-\cos(s/r), -\sin(s/r)\big).$$

Next we show that $\mathbf{T}$ and $d\mathbf{T}/ds$ are perpendicular:

$$\mathbf{T} \cdot \frac{d\mathbf{T}}{ds} = \big(-\sin(s/r), \cos(s/r)\big) \cdot \frac{1}{r}\big(-\cos(s/r), -\sin(s/r)\big) = 0.$$

This shows that the vector $d\mathbf{T}/ds$ points to the center of the circle, which is the direction in which C bends away from $\mathbf{T}$. The length $1/r$ of $d\mathbf{T}/ds$ is the curvature of C at $\mathbf{r}(\theta)$.

We summarize these results. At each point $\mathbf{r}(\theta)$ of a circle:

- The curvature vector $d\mathbf{T}/ds$ points in the direction in which the circle is bending away from $\mathbf{T}$.
- The curvature vector is perpendicular to the unit tangent vector.

- The curvature, which is the length of the curvature vector, is $1/r$.
- The reciprocal of the curvature is the radius of the circle.

DEFINITION   curvature vector, curvature, and unit normal

Let C be a smooth curve described by

$$(1) \qquad \mathbf{r} = \mathbf{r}(t) = \big(x(t), y(t)\big), \quad a \le t \le b.$$

Assume that the functions $x = x(t)$ and $y = y(t)$ have continuous second derivatives on $[a, b]$. If $\mathbf{r}'(t) \ne \mathbf{0}$ at $\mathbf{r}(t)$, the curvature vector to C at $\mathbf{r}(t)$ is defined to be

$$(2) \qquad \frac{d\mathbf{T}}{ds}.$$

The curvature $\kappa(t)$ and unit normal $\mathbf{N}(t)$ at $\mathbf{r}(t)$ are defined by

$$(3) \qquad \kappa(t) = \left\| \frac{d\mathbf{T}}{ds} \right\| \quad \text{and} \quad \mathbf{N}(t) = \frac{1}{\kappa(t)} \frac{d\mathbf{T}}{ds}.$$

The reciprocal of $\kappa(t)$ is the **radius of curvature** of C at $\mathbf{r}(t)$.

In the first example we use these definitions. It will become apparent almost immediately that curvature calculations are somewhat lengthy. Some relief is provided after the example in the form of formulas.

■ EXAMPLE 1   The cycloid described by

$$(4) \qquad \mathbf{r}(t) = \big(t - \sin t, 1 - \cos t\big), \quad 0 \le t \le 2\pi,$$

is shown in Fig. 6.57. Calculate the curvature and unit normal of the cycloid at $\mathbf{r}(4\pi/3)$.

Solution.   We calculated the arc length and unit tangent vector for the cycloid in Examples 2 and 4 in Section 6.2. We found

$$(5) \qquad \|\mathbf{r}'\| = \frac{ds}{dt} = \sqrt{x'(t)^2 + y'(t)^2} = 2 \sin \tfrac{1}{2}t$$

$$(6) \qquad \mathbf{T} = \big(\sin \tfrac{1}{2}t, \cos \tfrac{1}{2}t\big), \quad t \ne 0, 2\pi.$$

The curvature vector and curvature are $d\mathbf{T}/ds$ and $\kappa = \|d\mathbf{T}/ds\|$, as defined in (2) and (3). We calculate $d\mathbf{T}/ds$ first, using (5) and (6). The chain rule is needed since the curvature vector is $d\mathbf{T}/ds$ and $\mathbf{T}$ is given in terms of t.

$$\frac{d\mathbf{T}}{ds} = \frac{d\mathbf{T}}{dt}\frac{dt}{ds} = \frac{1}{ds/dt}\frac{d}{dt}\big(\sin \tfrac{1}{2}t, \cos \tfrac{1}{2}t\big)$$

$$(7) \qquad \frac{d\mathbf{T}}{ds} = \frac{1}{2 \sin \tfrac{1}{2}t}\big(\tfrac{1}{2} \cos \tfrac{1}{2}t, -\tfrac{1}{2} \sin \tfrac{1}{2}t\big).$$

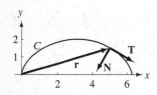

Figure 6.57. Normal vector.

The length of $d\mathbf{T}/ds$ is the curvature κ. We have

$$(8) \qquad \kappa = \frac{1}{4\sin\frac{1}{2}t}\sqrt{\cos^2\tfrac{1}{2}t + \sin^2\tfrac{1}{2}t} = \frac{1}{4\sin\frac{1}{2}t}.$$

Dividing $d\mathbf{T}/ds$ by its length κ gives the unit normal $\mathbf{N}(t)$. From (7) and (8) we have

$$\mathbf{N}(t) = \left(\cos\tfrac{1}{2}t, -\sin\tfrac{1}{2}t\right).$$

At the point $\mathbf{r}(4\pi/3)$ the curvature and unit normal are

$$\kappa = \frac{1}{4\sqrt{3/2}} \approx 0.29$$

$$\mathbf{N}(4\pi/3) = \left(-\frac{1}{2}, -\frac{\sqrt{3}}{2}\right) \approx (-0.5, -0.87).$$

We show in Fig. 6.57 the unit tangent vector $\mathbf{T}(4\pi/3)$ and the unit normal vector $\mathbf{N}$ to the cycloid at the point $\mathbf{r}(4\pi/3)$. It is easy to check that at $\mathbf{r}(4\pi/3)$ the vectors $\mathbf{T}$ and $\mathbf{N}$ are perpendicular. Indeed, these vectors are perpendicular everywhere they are defined. ∎

Before giving a second example, we give two formulas for curvature.

Formulas for Curvature

Let C be a smooth curve described by

$$\mathbf{r} = \mathbf{r}(t) = \left(x(t), y(t)\right), \quad a \le t \le b.$$

If $\mathbf{r}'(t) \ne \mathbf{0}$ at $\mathbf{r}(t)$, then the curvature at $\mathbf{r}(t)$ is

$$(9) \qquad \kappa(t) = \frac{|x''(t)y'(t) - x'(t)y''(t)|}{\left(x'(t)^2 + y'(t)^2\right)^{3/2}}.$$

If C is the graph of a function f defined on $[a, b]$, so that C can be described by

$$\mathbf{r}(x) = \left(x, f(x)\right), \quad a \le x \le b,$$

then the curvature at $\mathbf{r}(x)$ is (a short form is also given)

$$(10) \qquad \kappa(x) = \frac{|f''(x)|}{\left(1 + f'(x)^2\right)^{3/2}} = \frac{|y''|}{\left(1 + y'^2\right)^{3/2}}.$$

The second of the formulas is a special case of the first. If C can be described as the graph of a function f defined on an interval $[a, b]$, then a vector/parametric description of C is

$$\mathbf{r} = \mathbf{r}(t) = (t, f(t)), \quad a \leq t \leq b.$$

Applying (9), we have

$$\kappa(t) = \frac{|x''(t)y'(t) - x'(t)y''(t)|}{(x'(t)^2 + y'(t)^2)^{3/2}} = \frac{|0 \cdot f'(t) - 1 \cdot f''(x)|}{(1^2 + f'(t)^2)^{3/2}}.$$

If we replace t by x and simplify, equation (10) follows.

We outline two proofs of (9) in Exercises 20 and 21.

■ **EXAMPLE 2** At $(1/2, 1/4)$ on the graph C of the equation $y = x^2$, calculate the curvature and the unit tangent and normal vectors. Sketch the results.

Solution. We describe C by the vector equation $\mathbf{r} = (x, x^2)$. Starting with

$$\mathbf{r} = (x, x^2),$$

$$\frac{d\mathbf{r}}{dx} = (1, 2x) \quad \text{and} \quad \left\| \frac{d\mathbf{r}}{dx} \right\| = \frac{ds}{dx} = \sqrt{1 + 4x^2}.$$

The unit tangent vector $\mathbf{T}$ is

$$\mathbf{T} = \frac{1}{\sqrt{1 + 4x^2}}(1, 2x).$$

The curvature vector is $d\mathbf{T}/ds$, which we calculate using the chain rule.

$$\frac{d\mathbf{T}}{ds} = \frac{d\mathbf{T}}{dx}\frac{dx}{ds} = \frac{1}{\sqrt{1 + 4x^2}}\frac{d}{dx}\left(\frac{1}{\sqrt{1 + 4x^2}}(1, 2x)\right)$$

After differentiating and simplifying this expression—see Exercise 13—we obtain

$$\frac{d\mathbf{T}}{ds} = \frac{2}{(1 + 4x^2)^2}(-2x, 1).$$

The curvature κ is the length of the curvature vector.

$$\kappa = \frac{2}{(1 + 4x^2)^{3/2}}$$

At the point $(1/2, 1/4)$ we find

$$\mathbf{T} = \left(\frac{1}{\sqrt{2}}, \frac{1}{\sqrt{2}}\right)$$

$$\frac{d\mathbf{T}}{ds} = \left(-\frac{1}{2}, \frac{1}{2}\right)$$

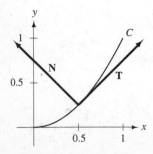

Figure 6.58. Unit tangent and normal vectors.

$$\kappa = \frac{1}{\sqrt{2}}$$

$$\mathbf{N} = \frac{1}{\kappa}\frac{d\mathbf{T}}{ds} = \left(-\frac{1}{\sqrt{2}}, \frac{1}{\sqrt{2}}\right).$$

We show in Fig. 6.58 the curve C and the vectors $\mathbf{T}(1/2)$ and $\mathbf{N}(1/2)$ at $(1/2, 1/4)$. ∎

Acceleration Vectors

The excitement of a roller coaster depends on changes in either the tangential or normal acceleration vectors. The acceleration vector on a straight incline points in the tangential direction; on circular turns it points in the normal direction, toward the center of the circle. On most wooden coasters riders feel these alternately, on different sections of the track. Steel coasters include sections on which riders feel both accelerations simultaneously. Coaster engineers make the turns tighter toward the end of the ride. This increases the magnitude of the normal acceleration, which helps maintain excitement as the coaster loses speed.

In analyzing motion it is often useful to separate the net force acting on an object (or, through Newton's second law $\mathbf{F} = m\mathbf{a}$, the acceleration) into forces in the tangential and normal directions or, for motion best described in polar coordinates, in the radial (toward the polar origin) and transverse directions. For circular motion about the polar origin, the radial and normal directions are the same and the transverse and tangential directions are the same.

The calculation of these acceleration vectors becomes easier if we give two differentiation formulas for vector functions. These formulas are easy to remember since they are very much like the formula for differentiating a product of scalar functions, that is, $(fg)' = f'g + fg'$.

Differentiation Formulas

Suppose that $\mathbf{f}$ and $\mathbf{g}$ are vector functions and h a scalar function, all defined and differentiable on $[a, b]$. The following formula shows how to differentiate the product of a scalar and a vector function.

(11)
$$(h\mathbf{f})' = h'\mathbf{f} + h\mathbf{f}'$$

The following formula shows how to differentiate the dot product of two vector functions.

(12)
$$(\mathbf{f} \cdot \mathbf{g})' = \mathbf{f}' \cdot \mathbf{g} + \mathbf{f} \cdot \mathbf{g}'$$

We noted in several examples that the unit tangent and normal vectors are perpendicular. This follows easily from (12). Since $\mathbf{T}$ is a unit vector, $1 = \mathbf{T} \cdot \mathbf{T}$ wherever it is defined. Differentiating both sides of this identity, we have

$$0 = 1' = (\mathbf{T} \cdot \mathbf{T})' = \mathbf{T}' \cdot \mathbf{T} + \mathbf{T} \cdot \mathbf{T}' = 2\mathbf{T} \cdot \mathbf{T}'.$$

Assuming neither $\mathbf{T}$ nor $\mathbf{T}'$ is the zero vector, it follows from this equation that $\mathbf{T}$ and $\mathbf{T}'$ are perpendicular since their dot product is zero. Since the unit normal is in the same direction as $\mathbf{T}'$, it follows that $\mathbf{T}$ and $\mathbf{N}$ are perpendicular.

Tangential and Normal Acceleration   The motion of an object on a curve C is often described by a vector/parametric equation

$$\mathbf{r} = \mathbf{r}(t) = \big(x(t), y(t)\big), \quad a \leq t \leq b.$$

It is understood that $\mathbf{r}(t)$ is the position vector of the object at time t. The velocity, speed, and acceleration of the object are given by

$$\mathbf{v} = \mathbf{v}(t) = \mathbf{r}'(t), \quad v = v(t) = \|\mathbf{r}'(t)\|, \quad \mathbf{a} = \mathbf{a}(t) = \mathbf{v}'(t).$$

The tangential and normal acceleration vectors are the projections of the acceleration vector in the $\mathbf{T}$ and $\mathbf{N}$ directions. To determine these vectors, we start with

(13) $$\mathbf{a} = \mathbf{v}' = (v\mathbf{T})'.$$

Applying (11) to (13), we have

$$\mathbf{a} = (v\mathbf{T})'$$
$$= v'\mathbf{T} + v\mathbf{T}' = v'\mathbf{T} + v\frac{d\mathbf{T}}{ds}\frac{ds}{dt}$$
$$= v'\mathbf{T} + v^2\frac{d\mathbf{T}}{ds}$$

since $v = \|\mathbf{r}'(t)\| = ds/dt$. Since $\mathbf{N} = (1/\kappa)d\mathbf{T}/ds$, we have

(14) $$\mathbf{a} = v'\mathbf{T} + v^2\kappa\mathbf{N}.$$

The tangential and normal acceleration vectors of an object moving on a curve C are shown in Fig. 6.59. Since $v = ds/dt$, the magnitude of the tangential acceleration is

$$\|v'\mathbf{T}\| = |v'| = \left|\frac{d}{dt}\left(\frac{ds}{dt}\right)\right| = \left|\frac{d^2s}{dt^2}\right|,$$

which is the magnitude of the acceleration along the curve. The magnitude of the normal acceleration is

(15) $$\|v^2\kappa\mathbf{N}\| = v^2\kappa = \frac{v^2}{\rho}$$

where $\rho = 1/\kappa$ is the radius of curvature. Recall that for the uniform circular motion

$$\mathbf{r} = r\big(\cos\omega t, \sin\omega t\big)$$

the acceleration vector is $\mathbf{r}'' = -(v^2/r)\mathbf{r}$, which is consistent with (15).

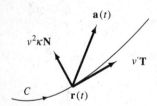

Figure 6.59. Tangential and normal acceleration vectors.

Radial and Transverse Acceleration

For motion best described in polar coordinates, for example, the motion of a planet about the sun, the natural way of writing the acceleration vector is as a sum of the radial and transverse acceleration vectors.

We assume in this discussion that the motion of an object on a curve C is described by an equation in the polar form

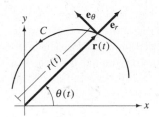

Figure 6.60. The vectors $\mathbf{e}_r$ and $\mathbf{e}_\theta$.

$$(16) \qquad \mathbf{r} = \mathbf{r}(t) = r(t)\big(\cos\theta(t), \sin\theta(t)\big), \quad \alpha \leq t \leq \beta,$$

where the magnitude $r(t)$ and direction $\theta(t)$ of $\mathbf{r}(t)$ are functions of t. See Fig. 6.60.

The radial and transverse directions are defined to be the unit vectors

$$(17) \qquad \mathbf{e}_r = \big(\cos\theta(t), \sin\theta(t)\big), \qquad \mathbf{e}_\theta = \big(-\sin\theta(t), \cos\theta(t)\big).$$

These vectors are based at $\mathbf{r}(t)$. The vector $\mathbf{e}_r$ points in the direction of $\mathbf{r}(t)$, which is the **radial direction**; $\mathbf{e}_\theta$ is perpendicular to $\mathbf{e}_r$ and points in the direction of increasing θ, which is the **transverse direction.** The vectors $\mathbf{e}_r$ and $\mathbf{e}_\theta$ are perpendicular, which follows from

$$\begin{aligned}\mathbf{e}_r \cdot \mathbf{e}_\theta &= \big(\cos\theta, \sin\theta\big) \cdot \big(-\sin\theta, \cos\theta\big) \\ &= (\cos\theta)(-\sin\theta) + (\sin\theta)(\cos\theta) = 0.\end{aligned}$$

For use below we calculate the derivatives of the vectors $\mathbf{e}_r$ and $\mathbf{e}_\theta$. As above, we shorten $\theta(t)$ to θ but keep in mind that θ is a function of t. We have

$$(18) \quad \begin{aligned}\mathbf{e}_r' &= \big(\cos\theta, \sin\theta\big)' = \big((-\sin\theta)\theta', (\cos\theta)\theta'\big) = \theta'\mathbf{e}_\theta \\ \mathbf{e}_\theta' &= \big(-\sin\theta, \cos\theta\big)' = \big((-\cos\theta)\theta', (-\sin\theta)\theta'\big) = -\theta'\mathbf{e}_r.\end{aligned}$$

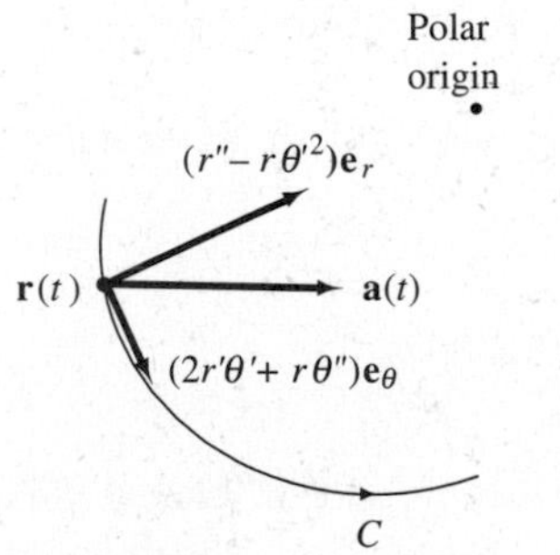

With this calculation done, expressing $\mathbf{a}$ in terms of the radial and transverse accelerations takes just a few steps. We differentiate $\mathbf{r} = r\mathbf{e}_r$ twice to get acceleration, using (17), (18), and the differentiation formula (11) as necessary. We have

$$\begin{aligned}\mathbf{r} &= r\mathbf{e}_r \\ \mathbf{v} &= \mathbf{r}' = r'\mathbf{e}_r + r\mathbf{e}_r' = r'\mathbf{e}_r + r\theta'\mathbf{e}_\theta \\ \mathbf{a} &= \mathbf{v}' = r''\mathbf{e}_r + r'\theta'\mathbf{e}_\theta + (r\theta'' + r'\theta')\mathbf{e}_\theta + r\theta'(-\theta'\mathbf{e}_r)\end{aligned}$$

$$(19) \qquad \mathbf{a} = \big(r'' - r\theta'^2\big)\mathbf{e}_r + (2r'\theta' + r\theta'')\mathbf{e}_\theta.$$

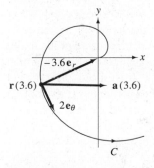

Figure 6.61. The radial and transverse acceleration vectors.

Figure 6.61 shows the radial and transverse acceleration vectors at a point $\mathbf{r}(t)$. We calculate the radial and transverse acceleration vectors for the object moving on a spiral.

■ **EXAMPLE 3** Calculate and sketch the acceleration vector $\mathbf{a}(3.6)$ and the radial and transverse acceleration vectors for the object whose position vector at any time $t \geq 0$ is

$$(20) \qquad \mathbf{r} = \mathbf{r}(t) = \big(t\cos t, t\sin t\big).$$

Solution. Comparing (16) and (20), we see that $r(t) = t$ and $\theta(t) = t$. Since $r' = 1$, $r'' = 0$, $\theta' = 1$, and $\theta'' = 0$, it follows from (19) that

$$\begin{aligned}\mathbf{a} &= \big(r'' - r\theta'^2\big)\mathbf{e}_r + (2r'\theta' + r\theta'')\mathbf{e}_\theta \\ &= -t\mathbf{e}_r + 2\mathbf{e}_\theta.\end{aligned}$$

Figure 6.62. Acceleration vectors at $\mathbf{r}(3.6)$.

Evaluating at $t = 3.6$, we have

$$\mathbf{a} = -3.6\mathbf{e}_r + 2\mathbf{e}_\theta.$$

The radial and transverse acceleration vectors of the object are shown in Fig. 6.62. We note that in circular motion the acceleration is entirely in the normal/radial direction; this spiral motion has a transverse acceleration as well. The combination of these accelerations results in the object moving around the pole and simultaneously outward. ◼

Kepler's Second Law

Johannes Kepler (1571–1630) inferred his three laws of planetary motion from astronomical data. These laws are:

1. Each planet moves on an ellipse with the sun at one focus.
2. The line from the sun to a planet sweeps out equal areas in equal times.
3. The square of the period of a planet is proportional to the cube of its greatest distance from the sun.

Kepler's laws suggest that polar coordinates may be a natural choice for describing planetary motion. Indeed, if we place the sun at the pole or origin of a coordinate system, the elliptical orbits of the planets can be described by (relatively) simple equations. We use our earlier work with the polar area element and the recent discussion of radial and transverse acceleration in inferring Kepler's second law.

We show in Fig. 6.63 a planet P moving on an elliptical orbit, with the sun at one focus. The polar equation for the ellipse shown is

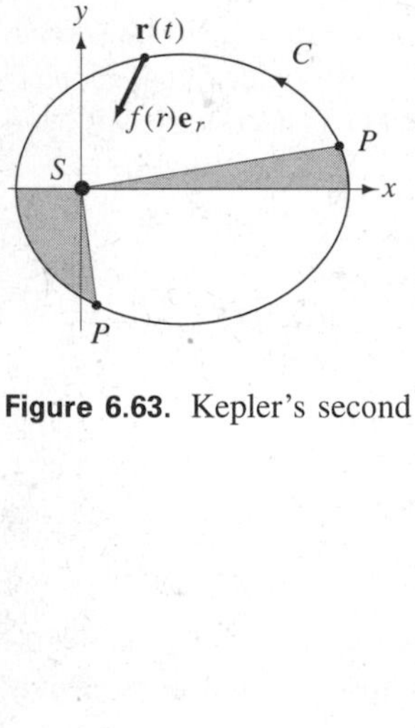

Figure 6.63. Kepler's second law.

$$(21) \qquad r = \frac{1}{1 - 0.6\cos\theta}, \qquad 0 \le \theta \le 2\pi.$$

Actual planetary orbits are much more circular than the one shown. Indeed, this ellipse has eccentricity 0.6, which is larger even than that of Pluto, which is the most eccentric of the known planets. The eccentricity of earth's orbit is 0.017, which means that the orbit is very nearly circular. Try to reproduce the graph of the ellipse with equation (21) with the help of your calculator.

If a planetary year is divided into, say, 12 equal months, then, according to Kepler's second law, a line drawn from sun to planet would sweep out $1/12$ of the area of the entire ellipse each month. We show in the figure the regions corresponding to two equal months. The shaded region on the right side of the figure is swept out during the month just after aphelion (the point on the orbit at which the planet is furthest from the sun and is moving with the least speed). The shaded area on the left is swept out during the month just after perihelion (the point on the orbit at which the planet is closest to the sun and is moving with the greatest speed). The areas of the two regions are equal.

Suppose that the position vector of a planet P moving on a curve C is described by the polar form

$$\mathbf{r} = \mathbf{r}(t) = r(t)\big(\cos\theta(t), \sin\theta(t)\big).$$

We show that Kepler's second law follows from the assumption that the force exerted by the sun at $\mathbf{r}(t)$ is entirely in the radial direction, that is,

$$\mathbf{F}(\mathbf{r}(t)) = f(r)\mathbf{e}_r.$$

The force $\mathbf{F}$ is related to the acceleration $\mathbf{a}$ by Newton's second law, $\mathbf{F} = m\mathbf{a}$. Replacing $\mathbf{F}$ by $f(r)\mathbf{e}_r$ and writing the acceleration as the sum of the radial and transverse accelerations, we have

(22) $$f(r)\mathbf{e}_r = m(r'' - r\theta'^2)\mathbf{e}_r + m(2r'\theta' + r\theta'')\mathbf{e}_\theta.$$

Since the left side of this equation has no $\mathbf{e}_\theta$ term, it follows that

(23) $$m(2r'\theta' + r\theta'') = 0$$

for all t. Also see Exercise 15. After removing the factor of m, this expression may be rewritten as a derivative by dividing by 2 and multiplying by r. We have

(24) $$0 = rr'\theta' + \tfrac{1}{2}r^2\theta'' = \frac{d}{dt}(\tfrac{1}{2}r^2\theta').$$

Recalling that the area element in polar coordinates is $dA = \tfrac{1}{2}r^2 d\theta$, which may be written as $dA/dt = \tfrac{1}{2}r^2\theta'$, it now follows from (24) that dA/dt is a constant (its derivative is 0); that is, equal areas are swept out in equal times, which is Kepler's second law.

Exercises 6.6

Basic

Exercises 1–8: Calculate the unit tangent **T**, unit normal **N**, and curvature κ at the given point of the curve. Sketch the curve, drawing in the unit tangent at the given point. Where on the curve is the curvature the greatest? The least?

1. $\mathbf{r} = (2t, t^2)$, $-1 \le t \le 2$; $\mathbf{r}(0)$

2. $\mathbf{r} = (\tfrac{1}{2}t^2, t)$, $-1 \le t \le 2$; $\mathbf{r}(0)$

3. $\mathbf{r} = (3\cos t, 2\sin t)$, $0 \le t \le 2\pi$; $\mathbf{r}(\pi/4)$

4. $\mathbf{r} = (\cos t, 2\sin t)$, $0 \le t \le \pi/2$; $\mathbf{r}(\pi/6)$

5. $y = x^3$, $0 \le x \le 2$; $(1, 1)$

6. $y = 1/x$, $1/2 \le x \le 2$; $(1, 1)$

7. $r = 4\cos\theta$, $0 \le \theta \le \pi/2$; $(2\sqrt{2}, \pi/4)$

8. $r = e^\theta$, $0 \le \theta \le \pi/2$; $(3, \ln 3)$

9. The motion of an object is given by

$$\mathbf{r} = \mathbf{r}(t) = (t - 1, 2t - t^2), \quad t \ge 0.$$

Show that

$$\mathbf{a} = \frac{4(t - 1)}{\sqrt{4(1 - t)^2 + 1}}\mathbf{T}$$
$$+ \frac{2}{\sqrt{4(1 - t)^2 + 1}}\mathbf{N}.$$

10. The motion of an object is given by

$$\mathbf{r} = \mathbf{r}(t) = (\cos e^t, \sin e^t), \quad t \ge 0.$$

Show that

$$\mathbf{a} = e^t\mathbf{T} + e^{2t}\mathbf{N}.$$

11. Show that the radial and transverse acceleration vectors at $\mathbf{r}(\pi/6)$ for the motion described by

$$r(t) = \sin(3t)(\cos(3t), \sin(3t)), \quad 0 \le t \le \pi/3,$$

are $(0, -18)$ and $(0, 0)$. Sketch the curve and these two vectors. What is the acceleration at this point?

12. Show that the radial and transverse acceleration vectors at $\mathbf{r}(\pi/2)$ for the motion described by

$$\mathbf{r}(t) = e^t(\cos(2t), \sin(2t)), \quad 0 \le t \le \pi,$$

are $\left(3e^{\pi/2}, 0\right)$ and $\left(0, -4e^{\pi/2}\right)$. Sketch the curve and these two vectors. What is the acceleration at this point?

13. In Example 2 we skipped the details in the calculation

Growth

14. On the curve described by $f(x) = e^x$ where is the curvature a maximum? What is the maximum curvature?

15. Show that (23) follows from (22) by dotting both sides of the latter equation by $\mathbf{e}_\theta$.

16. A polar curve C is described by $r = r(\theta)$. Show that the curvature $\kappa = \kappa(\theta)$ of C at (r, θ) is given by

$$\kappa(\theta) = \frac{\left|r^2 + 2r'^2 - rr''\right|}{\left(r^2 + r'^2\right)^{3/2}}.$$

17. (Continuation of Exercise 16) What is the curvature of the polar curve described by $r = 2^\theta$ at $(1, 2)$?

18. Use formula (12) in showing that for a differentiable vector function $\mathbf{r} = \mathbf{r}(t)$ defined on an interval $[a, b]$,

$$rr' = \mathbf{r}' \cdot \mathbf{r},$$

where $r^2 = \mathbf{r} \cdot \mathbf{r} = \|\mathbf{r}\|^2$.

19. The unit tangent for the cycloid described by

$$\mathbf{r} = \mathbf{r}(t) = (t - \sin t, 1 - \cos t), \quad 0 \le t \le 2\pi,$$

is given by $\mathbf{T} = \left(\sin \frac{1}{2}t, \cos \frac{1}{2}t\right)$, which exists for all $t \in [0, 2\pi]$. This ignores the restriction in the definition $\mathbf{T} = \mathbf{r}'/\|\mathbf{r}'\|$ that $\|\mathbf{r}'\| \ne 0$. Give a reason why the calculation appears to work for all $t \in [0, 2\pi]$. Does $\mathbf{T}$ vary continuously for $t \in [0, 3\pi]$? Does $\mathbf{r}'$ vary continuously for $t \in [0, 3\pi]$? Explain.

20. Shown in the accompanying figure are a curve C with arc length $s = s(t)$ from $\mathbf{r}(a)$ to $\mathbf{r}(t)$, the unit tangent vector $\mathbf{T} = \mathbf{T}(t)$, and the angle $\alpha = \alpha(t)$ of inclination of $\mathbf{T}$ at $\mathbf{r}(t)$. Since $\mathbf{T}$ is a unit vector we have

$$(25) \qquad \mathbf{T} = (\cos \alpha, \sin \alpha).$$

Assuming that the arc length function $s = s(t)$ can be solved for t in terms of s, say $t = g(s)$, we may differentiate (25) with respect to s to obtain

$$\frac{d\mathbf{T}}{ds} = (-\sin \alpha, -\cos \alpha)\frac{d\alpha}{ds}.$$

$$\frac{d}{dx}\left(\frac{1}{\sqrt{1 + 4x^2}}(1, 2x)\right) = \frac{1}{(1 + 4x^2)^{3/2}}(-4x, 2).$$

Check the calculation in two ways, directly and with the help of formula (11).

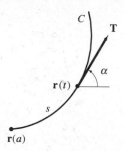

The curvature κ at $\mathbf{r}$ is the length of the curvature vector $d\mathbf{T}/ds$ and hence

$$(26) \qquad \kappa = \left\|\frac{d\mathbf{T}}{ds}\right\| = \left|\frac{d\alpha}{ds}\right|.$$

Use (26) and the fact that $\alpha = \tan^{-1}(y'/x')$ to show that

$$\kappa = \left|\frac{d\alpha}{ds}\right| = \frac{|x'y'' - x''y'|}{(x'^2 + y'^2)^{3/2}},$$

where $dx/dt = x'$ and $dy/dt = y'$.

21. Starting with

$$\mathbf{r} = \mathbf{r}(t) = (x(t), y(t)), \quad a \le t \le b,$$

derive the curvature formula (9). The calculation is straightforward but on the long side. You will save time if you simplify and check your calculations as you go.

22. In machining curves with a milling machine, machinists mill a section of the curve by selecting a point on the curve and milling along a circle instead of the curve. Near the point $(1, 1)$ of the parabola described by $y = x^2$, what are the center and radius of the circle along which the machinist should mill? Now suppose this parabolic shape is to be cut from a sheet of metal by setting a milling machine to follow the parabola (or, possibly, a number of approximating circles). A single bit must be chosen for the job. A bit is a cylinder with cutting edges on its lateral surface and axis perpendicular to the sheet.

If the parabola extends from $(-2, 4)$ to $(1, 1)$, what is the maximum possible radius of the bit?

23. Assume that the "excitement" of a roller coaster ride is directly proportional to the magnitude of the acceleration (this ignores other sensory inputs—how high the car is, the pressure of the air on one's face, the screams of other passengers, and so on). Suppose the acceleration was $g/\sqrt{2}$ on a 45° straight incline at the beginning of the ride. What radius would be required on a level circular curve to get 80 percent of the excitement if the speed at that time is 15 m/s? What about 10 m/s? Explain the comment made earlier about roller coasters, that turns are tighter toward the end of the ride to maintain excitement.

24. An object has constant speed v_0 on the curve described by $r = b \cos 3\theta$, $-\pi/2 < \theta < \pi/2$. Show that its acceleration at $(b, 0)$ is $\mathbf{a} = (-10 v_0^2/b)\mathbf{e}_r$.

25. Referring to Fig. 6.63, determine the coordinates of both points labeled P.

Review/Preview

26. Let C_a be the region beneath the curve with equation $y = 1/x$ and above the interval $[a, 1]$. The area of C_a is

$$A(a) = \int_a^1 x^{-1}\, dx.$$

Calculate $A(0.1)$, $A(0.01)$, and $A(0.001)$. Determine a so that $A(a) > 1000$ square units. What can be said about the area of the region C_0?

27. Let C_b be the region beneath the curve with equation $y = 1/x^2$ and above the interval $[1, b]$. The area of C_b is

$$A(b) = \int_1^b x^{-2}\, dx.$$

Calculate $A(10)$, $A(100)$, and $A(1000)$. What can be said about the area of the region C_∞?

6.7 IMPROPER INTEGRALS

Improper Integrals of the First Kind

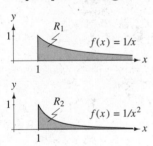

A "proper" integral $\int_a^b f(x)\, dx$ is one for which f is continuous and the range of integration $[a, b]$ is finite. We discussed proper integrals in Chapter 5. If the interval of integration has the form $[a, \infty)$, the integral is an improper integral of the first kind. To motivate the definition of this kind of integral, we investigate three instances of the question as to whether a region of infinite extent can have finite area.

Figure 6.64. Two regions of infinite extent.

Investigation 1

A region of the plane has "infinite extent" if it cannot be contained in any circle centered at the origin. The question is, can a region of infinite extent have finite area and, if so, how can we calculate its area? Referring to Fig. 6.64, which shows regions R_1 and R_2, which we suppose to have infinite extent, does either region have finite area?

In Fig. 6.65 we show regions $R_1(b)$ and $R_2(b)$ of finite extent. In some sense, these regions approximate the regions R_1 and R_2, particularly as b becomes large. The advantage of the regions $R_1(b)$ and $R_2(b)$ is that their areas $A_1(b)$ and $A_2(b)$ can be calculated with a proper integral.

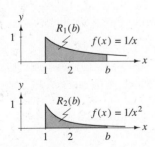

Figure 6.65. Two regions of finite extent.

(1) $$A_1(b) = \int_1^b \frac{dx}{x^1} = \ln x \Big|_1^b = \ln b - \ln 1 = \ln b$$

(2) $$A_2(b) = \int_1^b \frac{dx}{x^2} = \frac{x^{-1}}{-1}\Big|_1^b = 1 - \frac{1}{b}$$

Since

$$\lim_{b \to \infty} A_1(b) = \infty \quad \text{and} \quad \lim_{b \to \infty} A_2(b) = 1,$$

we have reason to say that region R_1 has infinite area and R_2 has finite area. Evidently, the graph of $f(x) = 1/x^2$ approaches its asymptote $y = 0$ quickly enough that the area beneath its graph on the interval $[1, \infty)$ is finite.

In Fig. 6.66 we show a third region, R, having infinite extent. This region is bounded on top by the function

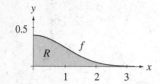

Figure 6.66. A third region of infinite extent.

(3) $$f(x) = \frac{1}{\sqrt{2\pi}} e^{-x^2/2}, \quad -\infty < x < \infty.$$

As above, we may think of calculating the area of a region like $R(b)$ shown in Fig. 6.67. This region includes all of R except for the "tail" extending to infinity. There is a difficulty, however, since the function defined in (3) has no antiderivative expressible in terms of elementary functions.

You may remember from the numerical integration section that we approximated the area $A(2)$ of the region $R(2)$, determining that $A(2) \approx 0.477$. If, as we remarked in Section 5.7, this integral is the probability that the normally distributed variable x lies between 0 and 2, then our question about the third region is: what is the probability that x lies between 0 and ∞? Since the probability of any event lies between 0 and 1, the area of R—or, better, the limit $\lim_{b \to \infty} R(b)$—ought to be finite.

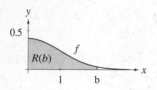

Figure 6.67. A region of finite extent.

To tighten up this part of the investigation, let

(4) $$A(b) = \int_0^b f(x)\,dx.$$

The number $A(b)$ measures the area of the region $R(b)$ beneath the graph of f for $0 \le x \le b$. We want to know if $\lim_{b \to \infty} A(b)$ is finite.

To answer this question we bring in the test function $g(x) = e^{-x}$, whose graph is shown in Fig. 6.68. Since

$$f(x) \le g(x), \quad \text{for all } x \ge 0,$$

it follows that

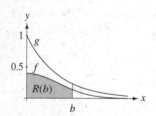

Figure 6.68. A test function.

(5) $$A(b) = \int_0^b f(x)\,dx \le \int_0^b e^{-x}\,dx = -e^{-x}\Big|_0^b = -e^{-b} + 1 = G(b).$$

The graphs of the area functions A and G are shown in Fig. 6.69. Since

$$\lim_{b \to \infty} G(b) = \lim_{b \to \infty} \left(1 - e^{-b}\right) = 1,$$

the graph of G has $y = 1$ as a horizontal asymptote. Since $A(b) \le G(b)$ for all $b \ge 0$ and A is increasing on $[0, \infty)$, the limit of $A(b)$ as $b \to \infty$ exists. This is made plausible in Fig. 6.69, where, moreover, it appears that $\lim_{b \to \infty} A(b) = 0.5$.

Putting everything together, it is reasonable to write

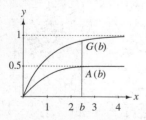

Figure 6.69. The function A is bounded and increasing.

$$\lim_{b \to \infty} A(b) = \lim_{b \to \infty} \int_0^b f(x)\,dx = \int_0^\infty f(x)\,dx = 0.5.$$

DEFINITION **convergent and divergent improper integrals of the first kind**

If f is continuous on the infinite interval $[a, \infty)$, the improper integral of f on $[a, \infty)$ is defined to be

$$\int_a^\infty f(x)\, dx = \lim_{b \to \infty} \int_a^b f(x)\, dx$$

when the limit is a number L. In this case we write $\int_0^\infty f(x)\, dx = L$ and say that the improper integral is convergent (or converges to L). If the limit is ∞ (or $-\infty$), we write $\int_0^\infty f(x)\, dx = \infty$ (or $-\infty$) and say that the improper integral is divergent (or diverges to infinity or minus infinity). Otherwise we say that the improper integral is divergent (diverges).

■ **EXAMPLE 1** Does the improper integral

$$\int_0^\infty \frac{x}{1 + x^2}\, dx$$

converge or diverge? If it converges, calculate its value.

Solution. This improper integral diverges since for large x

$$\frac{x}{1 + x^2} = \frac{1}{\dfrac{1}{x} + x} \approx \frac{1}{x},$$

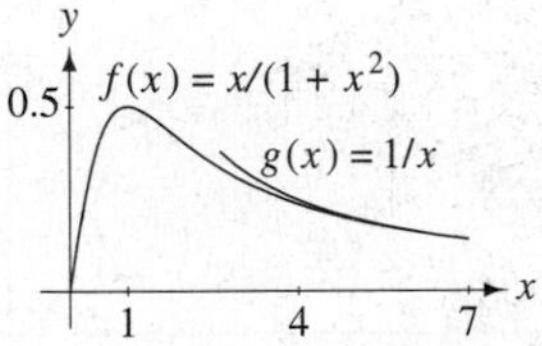

and, from Investigation 1, $\int_1^\infty x^{-1}\, dx = \infty$. Figure 6.70 gives an indication of the asymptotic relation between the integrand f and $1/x$ as x becomes large. Just after the next example, we give a result justifying this kind of argument. Meanwhile, we use the substitution $u = 1 + x^2$ and obtain

$$\lim_{b \to \infty} \int_0^b \frac{x}{1 + x^2}\, dx = \lim_{b \to \infty} \int_1^{1+b^2} \frac{du}{2u} = \lim_{b \to \infty} \tfrac{1}{2} \ln(1 + b^2) = \infty.$$

The given improper integral diverges to ∞. ■

Figure 6.70. The integrand is asymptotic to $1/x$ for large x.

■ **EXAMPLE 2** Does the improper integral

$$\int_0^\infty \frac{dx}{1 + x^2}$$

converge or diverge? If it converges, calculate its value.

Solution. We expect this improper integral to converge since for large x,

$$\frac{1}{1 + x^2} \approx \frac{1}{x^2},$$

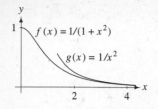

Figure 6.71. The integrand is asymptotic to $1/x^2$ for large x.

and, from Investigation 1, $\int_1^\infty x^{-2}\,dx = 1$. Figure 6.71 shows something of the relation between the integrand f and $1/x^2$ for large x. Using formula (G) from the table of integrals,

$$\lim_{b\to\infty}\int_0^b \frac{dx}{1+x^2} = \lim_{b\to\infty}\arctan b = \pi/2.$$

The given improper integral converges to $\pi/2$. ∎

Although we gave informal arguments bearing on the convergence or divergence of the integrals in these two examples, we ultimately depended on the limit of an evaluated proper integral. When it is either not possible or not convenient to do this, we must depend on a comparison with an improper integral that is known to converge or diverge. For this we give a theorem and a stock of test integrals.

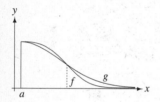

Figure 6.72. Eventually, $0 \le f \le g$.

> **Comparison Theorem I**
>
> Assume that f and g are continuous on $[a, \infty)$ and, for all sufficiently large x, $0 \le f(x) \le g(x)$, as shown in Fig 6.72. For improper integrals
>
> $$\int_a^\infty f(x)\,dx \quad \text{and} \quad \int_a^\infty g(x)\,dx$$
>
> of the first kind:
>
> **1.** If $\int_a^\infty g(x)\,dx$ converges, then so does $\int_a^\infty f(x)\,dx$.
>
> **2.** If $\int_a^\infty f(x)\,dx$ diverges, then so does $\int_a^\infty g(x)\,dx$.

We used the idea of this theorem earlier when we argued that the improper integral

$$(6) \qquad \frac{1}{\sqrt{2\pi}}\int_0^\infty e^{-x^2/2}\,dx$$

converges. The argument depended on the inequality

$$\frac{1}{\sqrt{2\pi}}e^{-x^2/2} \le e^{-x}$$

holding for at least all large x, for which we gave graphical evidence. Using this inequality and Comparison Theorem I, it follows that the improper integral (6) is convergent provided that $\int_0^\infty e^{-x}\,dx$ is convergent. This integral converges since

$$\lim_{b\to\infty}\int_0^b e^{-x}\,dx = \lim_{b\to\infty}(1 - e^{-b}) = 1.$$

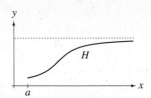

Figure 6.73. Bounded and nondecreasing functions have horizontal asymptotes.

Comparison Theorem I is an application of the following fact. Referring to Fig. 6.73, if H is a function that is defined, bounded above, and nondecreasing on an interval $[a, \infty)$, then the graph of H has a horizontal asymptote, that is, $\lim_{x\to\infty} H(x)$ exists and is finite.

Test Integrals   We used $\int_0^\infty e^{-x}\,dx$ as a test integral in Investigation 1 and hinted at $\int_1^\infty x^{-1}\,dx$ and $\int_1^\infty x^{-2}\,dx$ in some examples. We give a class of test integrals based on simple power functions. The improper integral

(7)
$$\int_1^\infty \frac{dx}{x^p} \quad \text{is} \quad \begin{cases} \text{convergent if } p > 1 \\ \text{divergent if } p \le 1. \end{cases}$$

In the next example we show how Comparison Theorem I and a test integral can be teamed up to subdue a relatively difficult integral.

■ **EXAMPLE 3**   Classify as convergent or divergent the improper integral

(8)
$$\int_1^\infty \frac{dx}{\sqrt{x^4 + x^2 + 1}}.$$

Solution.   We expect this integral to converge since as x becomes very large, the terms x^2 and 1 are small compared to x^4. Hence the integrand is like $(x^4)^{-1/2} = x^{-2}$. Since the test integral $\int_1^\infty x^{-2}\,dx$ converges, we expect the given integral to converge as well. With this in mind, we look at the integrand more closely. We have

$$\sqrt{x^4 + x^2 + 1} = x^2\sqrt{1 + x^{-2} + x^{-4}} \ge x^2$$

since x^{-2} and x^{-4} are positive. Taking reciprocals, we have

$$f(x) = \frac{1}{\sqrt{x^4 + x^2 + 1}} \le \frac{1}{x^2} = g(x).$$

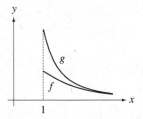

Figure 6.74. $f(x) \le 1/x^2$.

The graphs of f and g are shown in Fig. 6.74. It follows from Comparison Theorem I that the improper integral (8) is convergent.   ■

Integrating on $(-\infty, b]$ and $(-\infty, \infty)$ If f is continuous on $(-\infty, b]$, we define

$$\int_{-\infty}^b f(x)\,dx = \lim_{a\to-\infty}\int_a^b f(x)\,dx$$

and say that this improper integral is convergent or divergent depending on whether the limit is finite or not. A corresponding version of Comparison Theorem I holds for these integrals.

If f is continuous on $(-\infty, \infty)$, the improper integral $\int_{-\infty}^{\infty} f(x)\,dx$ is convergent provided that both improper integrals $\int_{-\infty}^{0} f(x)\,dx$ and $\int_{0}^{\infty} f(x)\,dx$ are convergent; otherwise it is divergent. If $\int_{-\infty}^{\infty} f(x)\,dx$ is convergent, we write

$$\int_{-\infty}^{\infty} f(x)\,dx = \int_{-\infty}^{0} f(x)\,dx + \int_{0}^{\infty} f(x)\,dx.$$

Improper Integrals of the Second Kind

For improper integrals of the first kind, the interval of integration is not bounded. For improper integrals of the second kind, the interval of integration is bounded but the integrand is not bounded on this interval.

Investigation 2

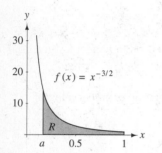

Figure 6.75. The second kind.

We show in Fig. 6.75 the graph of $f(x) = x^{-3/2}$ for $0 < x \le 1$. The line $x = 0$ is a vertical asymptote to this graph. From what we have seen with improper integrals of the first kind, it makes geometric sense to write

$$(9) \qquad \int_{0}^{1} \frac{dx}{x^{3/2}} = \lim_{a \to 0^+} \int_{a}^{1} \frac{dx}{x^{3/2}}.$$

This expresses the idea that the area of the region R beneath f and above the interval $(0, 1]$ is finite or infinite depending on the limit in (9). Since $-2x^{-1/2}$ is an antiderivative of $x^{-3/2}$,

$$\lim_{a \to 0^+} \int_{a}^{1} \frac{dx}{x^{3/2}} = \lim_{a \to 0^+} -2x^{-1/2}\bigg|_{a}^{1} = \lim_{a \to 0^+} \left(-2 + \frac{2}{\sqrt{a}}\right) = \infty.$$

We say, then, that the region R has infinite area and the improper integral in (9) is divergent.

Following the pattern for integrals of the first kind, we define improper integrals of the second kind, state a comparison theorem, and give some test integrals.

DEFINITION   convergent and divergent improper integrals of the second kind

If f is continuous on the interval $(a, b]$ and $\lim_{x \to a^+} f(x) = \pm\infty$ or the limit does not exist, the improper integral of f on $(a, b]$ is defined to be

$$\int_{a}^{b} f(x)\,dx = \lim_{w \to a^+} \int_{w}^{b} f(x)\,dx$$

when the limit is a number L. In this case we write $\int_{a}^{b} f(x)\,dx = L$ and say that the improper integral is convergent (or converges to L). If the limit is $\pm\infty$, we write $\int_{a}^{b} f(x)\,dx = \pm\infty$ and say that the improper integral is divergent (or diverges to $\pm\infty$). Otherwise we say that the improper integral is divergent (diverges).

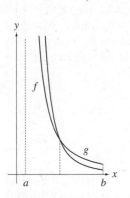

Figure 6.76. Near a, $0 \le f(x) \le g(x)$.

> ## Comparison Theorem II
>
> Assume that f and g are continuous on $(a, b]$ and, for all x near a, $0 \le f(x) \le g(x)$, as in Fig 6.76. For improper integrals
>
> $$\int_a^b f(x)\, dx \quad \text{and} \quad \int_a^b g(x)\, dx$$
>
> of the second kind:
>
> **1.** If $\int_a^b g(x)\, dx$ converges, then so does $\int_a^b f(x)\, dx$.
>
> **2.** If $\int_a^b f(x)\, dx$ diverges, then so does $\int_a^b g(x)\, dx$.

Test Integrals The improper integral

$$(10) \qquad \int_0^1 \frac{dx}{x^p} = \lim_{a \to 0^+} \int_a^1 \frac{dx}{x^p} \quad \text{is} \quad \begin{cases} \text{convergent if } p < 1 \\ \text{divergent if } p \ge 1. \end{cases}$$

In comparing this result with the test integrals in (7), we note that each of $\int_1^\infty x^{-p}\, dx$ and $\int_0^1 x^{-p}\, dx$ is divergent for $p = 1$, but for all other values of p each is convergent/divergent when the other is divergent/convergent.

■ **EXAMPLE 4** Classify the improper integral

$$(11) \qquad \int_0^1 x^{-1/2} e^{-x}\, dx.$$

A graph of the integrand is shown in Fig. 6.77.

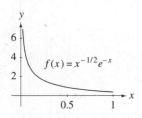

Figure 6.77. $\int_0^1 f(x)\, dx < \infty$.

Solution. For $x \in (0, 1]$, e^{-x} is between $e^0 = 1$ and $e^{-1} \approx 0.37$. Hence we expect the integral to converge since the integrand is like $x^{-1/2}$. Recall that the test integral $\int_0^1 x^{-p}\, dx$ is convergent for $p < 1$.

More formally, since the function e^{-x} is decreasing, we have

$$x^{-1/2} e^{-x} \le x^{-1/2} e^0, \quad 0 < x \le 1.$$

Since the Test Integral corresponding to $p = 1/2$ is convergent, it follows from Comparison Theorem II that the improper integral (11) is convergent. ■

Other Improper Integrals of the Second Kind If f is continuous on $[a, b)$ but its limit as $x \to b^-$ is infinite or does not exist, the definition of the corresponding improper integral and the modification of Comparison Theorem II are similar to the definition and theorem just given. If f is continuous on a finite interval (a, b) but fails to have a finite limit at both a and b, the improper integral $\int_a^b f(x)\,dx$ is defined as the sum of two improper integrals, one on $(a, c]$ and the other on $[c, b)$, where c is any point between a and b. The improper integral $\int_a^b f(x)\,dx$ is convergent only when both improper integrals in the sum are convergent. In this case, its value is the sum of the values of the convergent integrals.

Other Improper Integrals   The improper integral

$$(12) \qquad \int_0^\infty x^{-1/2} e^{-x}\,dx$$

has both an infinite interval of integration and an integrand having a vertical asymptote at $x = 0$. A sketch of the integrand is shown in Fig. 6.78. This improper integral is defined to be the sum of two improper integrals:

$$(13) \qquad \int_0^\infty x^{-1/2} e^{-x}\,dx = \int_0^c x^{-1/2} e^{-x}\,dx + \int_c^\infty x^{-1/2} e^{-x}\,dx$$

where c is any positive number. We say that (12) is convergent provided that both improper integrals in the sum in (13) are convergent. In Example 4 we showed that the first of these is convergent. We leave as Exercise 38 the classification of the second improper integral.

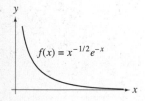

$f(x) = x^{-1/2} e^{-x}$

Figure 6.78. Mixed kind.

Exercises 6.7

Basic

Exercises 1–20: Classify each improper integral as convergent or divergent. If it is convergent, calculate its value.

1. $\displaystyle\int_1^\infty \frac{dx}{x+1}$

2. $\displaystyle\int_1^\infty \frac{x}{x^2+1}\,dx$

3. $\displaystyle\int_1^\infty x^{-3/2}\,dx$

4. $\displaystyle\int_1^\infty x^{-5/2}\,dx$

5. $\displaystyle\int_1^\infty \frac{dx}{\sqrt[3]{x^2}}$

6. $\displaystyle\int_1^\infty \frac{dx}{\sqrt[5]{x^6}}$

7. $\displaystyle\int_0^1 \frac{dx}{x^{5/3}}$

8. $\displaystyle\int_1^2 \frac{dx}{(x-1)^2}$

9. $\displaystyle\int_0^1 \frac{dx}{\sqrt{1-x^2}}$

10. $\displaystyle\int_1^2 \frac{dx}{\sqrt{x^2-1}}$

11. $\displaystyle\int_1^2 \frac{dx}{x\sqrt{x^2-1}}$

12. $\displaystyle\int_0^1 \frac{dx}{x\sqrt{1-x^2}}$

13. $\displaystyle\int_0^{\pi/2} \tan x\,dx$

14. $\displaystyle\int_0^\infty \sin x\,dx$

15. $\displaystyle\int_1^\infty \frac{dx}{x(x+1)}$

16. $\displaystyle\int_1^\infty \frac{dx}{x(2x-1)}$

17. $\displaystyle\int_0^1 \ln x \, dx$

18. $\displaystyle\int_1^\infty \frac{\ln x}{x} \, dx$

19. $\displaystyle\int_{-\infty}^\infty \frac{dx}{x^2 + 1}$

20. $\displaystyle\int_{-\infty}^0 xe^{-x} \, dx$

21. Show that $\int_1^\infty x^{-p} \, dx$ is convergent for all $p > 1$ and divergent otherwise. This verifies a statement in (7).

22. Show that $\int_0^1 x^{-p} \, dx$ is convergent for all $p < 1$ and divergent otherwise. This verifies a statement in (10).

23. Calculate the area of the region between the graphs of $\cosh x = (e^x + e^{-x})/2$ and $\sinh x = (e^x - e^{-x})/2$ on the interval $[0, \infty)$.

24. Calculate the area of the region between the graphs of $y = x/(x^2 + 1)$ and $y = 1/x$ on the interval $[1, \infty)$.

25. Classify as convergent or divergent the improper integral

$$\int_0^\infty \frac{dx}{\sqrt{2x^3 + x + 1}}.$$

26. Classify as convergent or divergent the improper integral

$$\int_0^\infty \frac{dx}{\sqrt{x^4 + 1}}.$$

27. Classify as convergent or divergent the improper integral

$$\int_0^1 x^{-2/3} e^{-x} \, dx.$$

28. Classify as convergent or divergent the improper integral

$$\int_0^1 x^{-4/3} e^{-x} \, dx.$$

29. Show that the integral $\int_0^\infty e^{-2x} \sin 3x \, dx$ is convergent and calculate its value.

Growth

Exercises 30–35: Classify each improper integral as convergent or divergent.

30. $\displaystyle\int_0^\infty 3^{-x^2} \, dx$

31. $\displaystyle\int_1^\infty \frac{\ln x}{x^3} \, dx$

32. $\displaystyle\int_0^{\pi/2} \frac{dx}{\sqrt{x + \sin x}}$

33. $\displaystyle\int_0^1 \frac{dx}{\ln x \sqrt{1 - x}}$

34. $\displaystyle\int_{-1}^\infty \frac{dx}{\sqrt{1 + x^3}}$

35. $\displaystyle\int_0^1 \frac{\cos x}{x} \, dx$

36. How much work does the gravitational field of earth do on a mass of 1 kg that is moved from the surface of the earth to the vicinity of Alpha Centauri, a triple star about 4.3 light-years from earth? (Hint: Use an improper integral for the calculation.)

37. Determine the area in the loop of the "folium of Descartes" described by

$$\mathbf{r} = \left(3t/(t^3 + 1), 3t^2/(t^3 + 1)\right), \quad -\infty < t < \infty.$$

38. Show that the improper integral given in (13) is convergent.

39. Referring to the improper integral (12), show that its convergence does not depend on the particular choice of c.

40. Show first that for $0 \le x < 3$,

$$\sqrt{\frac{81 - 5x^2}{9 - x^2}} = \sqrt{\frac{81 - 5x^2}{3 + x}} \cdot \frac{1}{\sqrt{3 - x}}$$

$$\le 3 \cdot \frac{1}{\sqrt{3 - x}}.$$

Then show that

$$\int_0^3 \sqrt{\frac{81 - 5x^2}{9 - x^2}} \, dx$$

is convergent. Show that the substitution $x = 3 \sin \theta$ transforms the improper integral into a proper integral.

41. Give a divergent improper integral that diverges to neither ∞ nor $-\infty$.

42. Show that if $x \geq 2$, then $x^2 \geq 2x$. Use this in showing that $e^{-x} \geq e^{-x^2/2}$ and $e^{-x} \geq (1/\sqrt{2\pi})e^{-x^2/2}$. This verifies, at least for $x \geq 2$, that $g(x) \geq f(x)$ as shown in Fig. 6.68.

43. Figure 6.68 in Investigation 1 shows a shaded area $R(b)$. Since the graph of f is very close to its horizontal asymptote for $x \geq 3$, we may expect that $A(3)$ is a good approximation to $\int_0^\infty f(x)\,dx$. See Fig. 6.69. Show that $A(3) \approx 0.499$ by calculating T_{31}. This value of n was chosen so that T_n is within 0.001 of $A(3)$. In the next problem we ask you to verify this choice of n.

44. (Continuation of Exercise 43) For the function f defined in (3), use the trapezoid rule error formula in showing that $|T_{31} - A(3)| < 0.001$. The formula is given in Exercise 48.

45. Referring to Fig. 6.69, show that both functions G and A are increasing.

46. Show that if $b > a$, then $\int_a^\infty f(x)\,dx$ is convergent if and only if $\int_b^\infty f(x)\,dx$ is convergent. Also show that if $\int_a^\infty f(x)\,dx$ is convergent, then for any $b > a$ we have

$$\int_a^\infty f(x)\,dx = \int_a^b f(x)\,dx + \int_b^\infty f(x)\,dx.$$

47. Using appropriate definitions, show that

$$\int_{-\infty}^\infty xe^{-x^2}\,dx = 0.$$

From the symmetry of the graph of f about the origin we may be tempted to argue that, obviously, the value of this integral is 0. If, however, we argued similarly that the improper integral

$$\int_{-\infty}^\infty \frac{x}{1+x^2}\,dx$$

is 0, we would be mistaken. Why?

48. Verify and fill in the details of the following outline for approximating the improper integral

$$\int_0^\infty \sqrt{x}e^{-x^2}\,dx$$

to within 0.01 of its value. Start by making a change of variable (let $x = u^2$) to simplify the integrand. Split the resulting integral into two parts by finding a

number b such that the last integral (which we refer to as the "tail") in

$$\int_0^\infty \sqrt{x}e^{-x^2}\,dx = \int_0^\infty 2u^2 e^{-u^4}\,du$$

$$= \int_0^b 2u^2 e^{-u^4}\,du + \int_b^\infty 2u^2 e^{-u^4}\,du$$

is small. We then use the trapezoid rule to approximate $\int_0^b 2u^2 e^{-u^4}\,du$. We must choose n for the trapezoid rule and b for the tail so that

$$\left| \int_0^\infty 2u^2 e^{-u^4}\,du - T_n \right| = \left| \int_0^b 2u^2 e^{-u^4}\,du - T_n \right.$$

$$\left. + \int_b^\infty 2u^2 e^{-u^4}\,du \right| < 0.01.$$

We choose b so that

$$\int_b^\infty 2u^2 e^{-u^4}\,du < 0.005.$$

With b chosen, choose n so that

$$\left| \int_0^b 2u^2 e^{-u^4}\,du - T_n \right| < 0.005.$$

To bound the tail we use the fact that $u^2 e^{-u^4} < u^2 e^{-u^3}$ for $u > 1$. We have

$$\int_b^\infty 2u^2 e^{-u^4}\,du < \int_b^\infty 2u^2 e^{-u^3}\,du = \frac{2}{3}e^{-b^3}.$$

The inequality $\frac{2}{3}e^{-b^3} < 0.005$ is satisfied for $b = 1.7$. To find n so that

$$\left| \int_0^b 2u^2 e^{-u^4}\,du - T_n \right| < 0.005,$$

we recall the trapezoid error formula

$$\left| \int_a^b f(x)\,dx - T_n \right| \leq \frac{(b-a)^3}{12n^2}M,$$

where M is at least as large as the maximum value of $|f''|$ on $[a, b]$. Show that we may take $M = 7$, perhaps by graphing the second derivative of $f(u) = 2u^2 e^{-u^4}$. It follows that $n = 24$ and $T_{24} = 0.6126\ldots$.

49. See Exercise 48. The exponential integral function is defined as

$$E_1(x) = \int_x^\infty \frac{e^{-t}}{t}\,dt, \quad x > 0.$$

Sketch a graph of E_1 by calculating and plotting $E_1(x)$, for $x = 0.2, 0.4, \ldots, 1.2$. For graphing, finding $E_1(x)$ within 0.1 is sufficient. Begin by finding b such that the tail $\int_b^{\infty} e^{-t}/t \, dt < 0.05$. Next, find a single value of n for the trapezoid rule that is suitable for all of the x values.

Review/Preview

50. Polynomials of degree 0, 1, and 2 have the forms

$$a, \quad ax + b, \quad ax^2 + bx + c.$$

    Describe in a general but useful way the graphs of these polynomials. Choose nondegenerate cases of the three forms and graph the results.

51. If the graph of a function at a specific point, say, $(a, f(a))$, is to be described by one or more numbers, then the height $f(a)$, slope $f'(a)$, and curvature $\kappa(a)$ of the graph would be among those considered. Recall that for a function f, the curvature is given by

$$\kappa(a) = \frac{|f''(a)|}{(1 + f'(a)^2)^{3/2}}.$$

Suppose we are attempting to describe a function f for which we know the height, slope, and curvature of its graph at a certain point. Specifically, for $a = 1$, suppose the height, slope, and curvature are 2, 1, and 3. Find polynomials of degree 0, 1, and 2 that describe f as well as possible with the information given. Graph your results on the same set of axes.

REVIEW OF KEY CONCEPTS

The main goal of this chapter was to show how the integral is used to measure such aspects of geometric or physical objects as

- The volume of certain three-dimensional solids
- The arc length of smooth curves
- The areas of regions whose boundaries can be described by parametric or polar curves
- The work done by a variable force on an object moving on a curve
- The center of mass of rods with variable density or of a lamina
- The curvature of a curve

The section on improper integrals was less an application of the integral than an extension of it. The discussions of normal, tangential, radial, and transverse acceleration vectors were applications of the integral only in that the idea of arc length was a prerequisite. Kepler's second law followed from the ideas of polar area element and radial and transverse acceleration.

A major theme was the use of elements of volume, arc length, work, mass, or polar area. Using an element helps break complex problems into simpler geometric or physical problems. Through the summing and limiting operations implicit in the transition from element to integral, we can calculate the volume of a solid of revolution, the length of a curve, the work done by a variable force, or the center of mass of a rod or lamina.

Chapter Summary

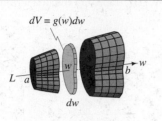

Let S be a solid sectioned by planes perpendicular to a line L and suppose that the area of the cross-section at w on L is $g(w)$; then the volume element is

$$dV = g(w)\,dw.$$

In the general case, the volume is

$$V = \int_a^b g(w)\,dw.$$

For volumes of revolution the volume is

$$V = \int_a^b \pi f(w)^2\,dw$$

where $f(w)$ is the radius of the circular cross-section at w. Here, $g(w) = \pi f(w)^2$.

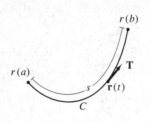

Let C be a smooth curve described by $\mathbf{r} = \mathbf{r}(t), a \le t \le b$. The unit tangent at $\mathbf{r}(t)$ is

$$\mathbf{T} = \mathbf{T}(t) = \frac{\mathbf{r}'(t)}{\|\mathbf{r}'(t)\|}$$

and the element of arc length is

$$\begin{aligned}
ds &= \|\mathbf{r}'(t)\|\,dt \\
&= \sqrt{x'(t)^2 + y'(t)^2}\,dt.
\end{aligned}$$

If C is described by:

- $\mathbf{r} = \mathbf{r}(t), a \le t \le b$, then

$$s = \int_a^b \|\mathbf{r}'\|\,dt.$$

- $y = f(x), a \le x \le b$, then

$$s = \int_a^b \sqrt{1 + y'^2}\,dx.$$

- $r = f(\theta), \alpha \le \theta \le \beta$, then

$$s = \int_\alpha^\beta \sqrt{r^2 + r'^2}\,d\theta.$$

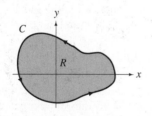

Let R be a region bounded by a smooth curve C described by

$$\mathbf{r} = \mathbf{r}(t) = \big(x(t), y(t)\big),$$
$$a \le t \le b.$$

Assume that C is traced in a counterclockwise direction.

The area A of R is

$$A = \tfrac{1}{2} \int_a^b \big(x(t)y'(t) - y(t)x'(t)\big)\,dt.$$

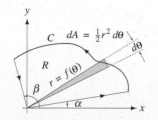

The polar area element for a region R whose boundary C is described by $r = f(\theta), \alpha \le \theta \le \beta$, and the rays $\theta = \alpha$ and $\theta = \beta$ is

$$dr = \tfrac{1}{2} r^2\,d\theta.$$

The area A of R is

$$V = \int_\alpha^\beta \tfrac{1}{2} r^2\,d\theta = \int_\alpha^\beta \tfrac{1}{2} f(\theta)^2\,d\theta.$$

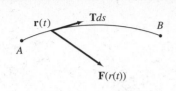	The element of work done by a force field $\mathbf{F}$ in moving an object a distance ds on a smooth curve C described by $\mathbf{r} = \mathbf{r}(t), a \le t \le b$, is $$dW = \mathbf{F} \cdot \mathbf{T}\,ds.$$	The work W done by the force $\mathbf{F}$ is $$W = \int_a^b \mathbf{F}(\mathbf{r}(t)) \cdot \mathbf{r}'(t)\,dt.$$
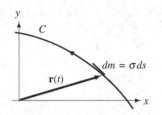	For the mass M and center of mass $\mathbf{R}$ of a curved (or straight) rod C of constant density σ and described by $\mathbf{r} = \mathbf{r}(t), a \le t \le b$, the element of mass is $$dm = \sigma\,ds.$$	The mass of the rod and its center of mass are $$M = \sigma \int_a^b \|\mathbf{r}'(t)\|\,dt$$ $$\mathbf{R} = \frac{\int_a^b \|\mathbf{r}'(t)\| \mathbf{r}(t)\,dt}{\int_a^b \|\mathbf{r}'(t)\|\,dt}.$$
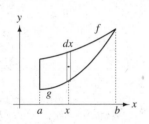	For the mass M and center of mass $\mathbf{R} = (X, Y)$ of a plane lamina with constant density σ (set $\sigma = 1$ to locate the centroid), the element of mass is $$dm = \sigma\big(f(x)^2 - g(x)^2\big)dx.$$	The mass of the lamina and the coordinates of the center of mass are $$M = \sigma \int_a^b (f(x) - g(x))\,dx$$ $$X = \frac{1}{M/\sigma} \int_a^b x(f(x) - g(x))\,dx$$ $$Y = \frac{1}{M/\sigma} \int_a^b \tfrac{1}{2}\big(f(x)^2 - g(x)^2\big)dx.$$
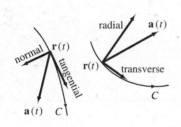	Let C be a smooth curve described by $\mathbf{r} = \mathbf{r}(t), a \le t \le b$. Curve geometry: unit tangent $\mathbf{T}$, curvature vector $d\mathbf{T}/ds$, curvature κ, and unit normal $\mathbf{N}$. If $\mathbf{r}(t)$ is the position vector of an object in motion, its acceleration can be written as a sum of tangential and normal accelerations or, in polar coordinates, radial and transverse accelerations.	$$\mathbf{T} = \frac{d\mathbf{r}/dt}{\|d\mathbf{r}/dt\|}$$ $$\frac{d\mathbf{T}}{ds} = \frac{d\mathbf{T}/dt}{ds/dt} \quad \text{(curvature vector)}$$ $$\kappa = \|d\mathbf{T}/ds\|$$ $$\mathbf{N} = \frac{d\mathbf{T}/ds}{\kappa}$$ $$\mathbf{a} = (dv/dt)\mathbf{T} + v^2\kappa\mathbf{N}$$ $$\mathbf{a} = (r'' - r\theta'^2)\mathbf{e}_r \\ + (2r'\theta' + r\theta'')\mathbf{e}_\theta$$
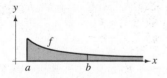	Improper integral of the first kind: $$\int_a^\infty f(x)\,dx = \lim_{b \to \infty} \int_a^b f(x)\,dx$$ where f is continuous on $[a, \infty)$. If the limit exists, the improper integral is convergent; otherwise it is divergent.	Test integrals for improper integrals of the first kind: $$\int_1^\infty \frac{dx}{x^p} \quad \text{is} \quad \begin{cases} \text{convergent if } p > 1 \\ \text{divergent if } p \le 1. \end{cases}$$

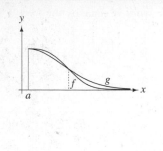	**Comparison Theorem I** If $0 \le f(x) \le g(x)$ for all sufficiently large x, then $$\int_a^\infty g < \infty \Longrightarrow \int_a^\infty f < \infty$$ $$\int_a^\infty f = \infty \Longrightarrow \int_a^\infty g = \infty.$$	Since $0 \le e^{-x^2} \le e^{-x}$ for $x \ge 1$ and $\int_0^\infty e^{-x}\, dx < \infty$, then $\int_0^\infty e^{-x^2}\, dx < \infty$. Since $0 \le 1/x < 1/\ln x$ for $x \ge 2$ and $\int_2^\infty (1/x)\, dx = \infty$, then $\int_2^\infty (1/\ln x)\, dx = \infty$.
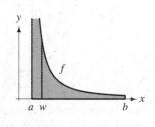	**Improper integral of the second kind:** $$\int_a^b f(x)\, dx = \lim_{w \to a^+} \int_w^b f(x)\, dx$$ where f is continuous on $(a, b]$. If the limit exists, the improper integral is convergent; otherwise it is divergent.	**Test integrals for improper integrals of the second kind:** $$\int_0^1 \frac{dx}{x^p} \text{ is } \begin{cases} \text{convergent if } p < 1 \\ \text{divergent if } p \ge 1. \end{cases}$$
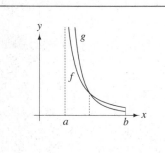	**Comparison Therorem II** If $0 \le f(x) \le g(x)$ for all x near a, then $$\int_a^b g < \infty \Longrightarrow \int_a^b f < \infty$$ $$\int_a^b f = \infty \Longrightarrow \int_a^b g = \infty.$$	Since $0 \le e^{-x}/\sqrt{x} \le 1/\sqrt{x}$ for $x > 0$ and $\int_0^1 (1/\sqrt{x})\, dx < \infty$, then $\int_0^1 (e^{-x}/\sqrt{x})\, dx < \infty$. Since $0 \le 1/x \le 1/(x \sin x)$ for $x \in (0, \pi/2)$ and $\int_0^{\pi/2}(1/x)\, dx = \infty$, then $\int_0^{\pi/2}(1/(x \sin x))\, dx = \infty$.

CHAPTER REVIEW EXERCISES

1. A solid has a circular base of radius 10 m. Each cross-section of the solid by a plane perpendicular to a fixed diameter of the base is an equilateral triangle. Calculate the volume of the solid.

2. A solid has a rectangular base with sides 2 m and 5 m. Cross-sections of the solid by planes perpendicular to one of the long sides of the rectangle are trapezoids, with their bases perpendicular to the rectangle. At a distance x from one of the short sides of the rectangle, the bases have lengths $\frac{1}{2}x$ and $2x$. Sketch the solid and calculate its volume.

3. Calculate the volume of the solid resulting from revolving a sector of a circle about one of its radii. Assume the radius of the sector is a and its central angle is θ.

4. Calculate the volume of the solid generated by revolving about the x-axis the region bounded by the curves described by $y = \sqrt{4 - x}$ and $x/4 + y/2 = 1$.

5. Calculate the volume of a tetrahedron all of whose sides have length a. A tetrahedron has four triangular faces.

6. The region beneath the graph of $y = \sec x$, $x \in [0, \pi/4]$, is revolved about the x-axis. Calculate its volume.

7. The polar curve $r = \theta$ has a vertical tangent at $\theta_1 \in (0, \pi/2)$. The region bounded by the curve $r = \theta$, $0 \le \theta \le \theta_1$, the vertical tangent, and the x-axis

is revolved about the *x*-axis. Calculate its volume. Use a vector/parametric form of the polar curve, and start with the volume element $dV = \pi y^2\, dx$.

8. Evaluate numerically an integral for the volume of a doughnut with circular sections and inner radius 1 cm and outer radius 5 cm.

9. Give a simplified but not evaluated integral whose value is the arc length of the curve described by $x^2 - y^2 = 1$, from $(2, \sqrt{3})$ to $(3, 2\sqrt{2})$. Repeat this, but use the points $(1, 0)$ and $(2, \sqrt{3})$. Discuss how the "difficulty" may be resolved by a substitution.

10. Show that the integral for finding the arc length of the curve described by $y = 3x^{2/3}, 0 \le x \le 8$, is improper. Rewrite the integrand to remove negative exponents and use the substitution $u = x^{2/3} + 4$ to show that the arc length integral converges to the value $16\sqrt{2} - 8$.

11. What is the arc length of the catenary curve with equation $y = a\cosh(x/a), -b \le x \le b$?

12. Determine the area of the region bounded by the loop of the curve described by

$$\mathbf{r}(t) = \left(3(t^2 - 3), t(t^2 - 3)\right), \quad t \in R.$$

13. Calculate the area of the region bounded by the rays $\theta = -\pi/4$ and $\theta = \pi/4$ and the graph of $r = 1 + \sin\theta$.

14. Calculate the area of the region bounded by the lemniscate described by $r^2 = \cos 2\theta$.

15. Calculate the area of the region lying inside the curve described by $r = 3\cos\theta$ and outside that described by $r = 1 + \cos\theta$.

16. Determine the area of the region in the first quadrant and bounded by the curves with equations $r = \theta$ and $r = \sec\theta$.

17. The curve with equation $r = \theta$ divides the first-quadrant portion of the circle with equation $r = \cos\theta$ into two regions. Calculate the area of the smaller region.

18. Graph the polar curve with equation $r = 1/(1 - \sin\theta)$ and calculate the area bounded by this curve and the rays $\theta = 0$ and $\theta = \pi/4$.

19. A force of 890 N is required to stretch a spring 1 m from its natural length. Calculate the work required to stretch the spring 2 m from its natural length.

20. Before liftoff a launch vehicle includes a 4.3×10^3 kg permanent structure, a 2.6×10^3 kg payload, and 43.1×10^3 kg of fuel. Assuming no air resistance and complete burning of fuel, how much work is required for the vehicle to attain an orbit with height 160 km?

21. A small cup in the shape of a frustum of a cone is filled with a soft drink whose density is 1010 kg/m^3. The top and bottom diameters of the cup are 6 cm and 4.6 cm; its height is 7 cm. Calculate the work required to move this liquid to a height 13 cm above the top of the cup.

22. Measuring from one end of a 12-meter beam, there are masses at the 0 m, 4 m, 9 m, and 12 m marks. The masses are 50 kg, 40 kg, 55 kg, and 45 kg, respectively. Assuming the beam has negligible mass, find the balance point.

23. Six masses rest on a plane. Their masses are 18, 10, 15, 17, 19, and 19 kg, and the corresponding position vectors locating them are $(-2.2, -4.7)$, $(-2.1, 3.1)$, $(-4.0, 0)$, $(4.8, -5.4)$, $(6.3, 10.5)$, and $(0.0, 0.0)$. All lengths are in meters. Calculate the center of mass of this system. Recalculate the center of mass by dividing the system into two systems. Suppose system *A* includes the first three masses and system *B* includes the last three masses. Calculate the center of mass of these two systems, and then calculate the center of mass of the two-mass system.

24. Calculate the mass and the center of mass of a 2.0 m rod of bronze, with the density varying linearly from 2.8 kg/m to 3.2 kg/m from end to end.

25. A rod of lineal density 2 kg/m has the shape of the graph of $y = \sin x, 0 \le x \le \pi$. Determine its mass and center of mass.

26. Calculate the center of mass of a lamina in the shape of a rectangle 3 m by 4 m, with a semicircle of diameter 4 m attached along one of the long sides of the rectangle. Assume the density if $\sigma \text{ kg/m}^2$. Avoid integration; use the Mass Subdivision Theorem together with a result worked out in an example.

27. Calculate the mass and center of mass of a lamina with constant density $\sigma \text{ kg/m}^2$ and in the shape of a sector of a circle. Assume the radius of the circle is a and the central angle of the sector is θ, where $0 \le \theta \le \pi/2$. If this problem is solved, how could you find the center of mass of a sector with central angle $2\pi/3$?

28. Calculate the unit tangent, curvature, and unit normal to the curve at the given point. Also calculate $\mathbf{T} \cdot \mathbf{N}$ as a partial check on your work.
   a. The curve described by $\mathbf{r} = (2t, t^2); (-2, 1)$.
   b. The curve described by the polar equation $r = \theta^2; (1, 1)$.
   c. The graph of the function $f(x) = e^x; (2, e^2)$.
   d. The curve described by $\mathbf{r} = (2\cos t, \sin t), 0 \le t \le 2\pi; (2, 0)$.

29. The position vector of a particle is $\mathbf{r} = (t\cos t, t\sin t)$ for all $t \geq 0$. Show that the magnitudes of its tangential and normal acceleration vectors are $t/\sqrt{1 + t^2}$ and $(2 + t^2)/(1 + t^2)^{3/2}$. Sketch the path.

30. How would you convince a skeptical, intelligent friend who has not had calculus that a region with infinite extent may have finite area?

31. Give examples of improper integrals of the first and second kinds. Give an example of an improper integral which is not of the first or second kind and explain how questions of convergence or divergence for such integrals are handled.

32. Determine if each of the integrals

$$\int_3^{12} (x - 3)^{-3/2}\, dx \quad \text{and} \quad \int_{12}^{\infty} (x - 3)^{-3/2}\, dx$$

is convergent or divergent. If it is convergent, determine its value.

33. For each of the integrals $\int_0^{\infty} 1/(1 + x^2)\, dx$, $\int_0^{\infty} 1/(1 + x^3)\, dx$, and $\int_0^2 1/(4x - x^2 - 3)\, dx$, determine if it is improper. If it is, decide if it is convergent or divergent.

34. Show that the integral

$$\int_{-\infty}^{\infty} \frac{dx}{(x^2 + a^2)^{3/2}}$$

is convergent and evaluate it.

35. For each integral, determine if it is convergent or divergent. If it is convergent, determine its value.

a. $\displaystyle\int_0^{\pi/2} \tan x\, dx$

b. $\displaystyle\int_0^1 \ln x\, dx$

c. $\displaystyle\int_1^3 (x/\sqrt{x - 1})\, dx$

d. $\displaystyle\int_0^{\infty} (x/(x^2 + 1))\, dx$

e. $\displaystyle\int_0^{\infty} xe^{-x}\, dx$

Student Project

Tsar-Kolokol (Tsar Bell)

Among the Old Believers in Russia are those who believe that on the "Day of the Last Judgment Tsar-Kolokol will rise slowly from its granite pedestal and begin to toll." Tsar-Kolokol is the largest bell ever cast. This massive bell is presently on display inside the Kremlin. It has never tolled. Shortly after it was cast in 1735, the Tsar Bell was badly damaged in a major Moscow fire. Firemen, afraid that the bell might melt in the heat of the fire, sprayed it with cold water. This resulted in a 13-ton section of the bell breaking from its base.

Bell making is a highly complex art. The size, shape, and tone of the bell must be chosen and designed; the molds in which the bell is cast must be designed and built; the alloy must be chosen and metal ordered; furnaces for melting the metal must be planned and built; the bell must be cast without major error; the bell must be separated from its mold, decorated, and tuned; and, especially difficult for very large bells, the bell must be lifted to its final resting place. (Immediately before large bells were cast in Russian bell foundries, icons were placed, candles lighted, prayers offered, and all workmen removed their hats and crossed themselves. This information, the quote at the beginning, and the data for this problem were adapted from the book *The Bells of Russia,* by Edward V. Williams (Princeton University Press, 1985).) In this problem we are concerned with one phase of the making of the Tsar Bell.

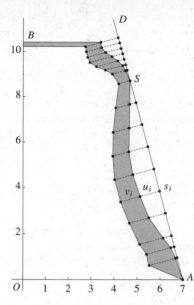

Figure 6.79. Profile and layout of the Tsar Bell.

The profile of the Tsar Bell is shown in Fig. 6.79 The bottom diameter was divided into 14 equal parts, whose common length was the unit used in the foundry. A line AD was drawn from point A on the bell through S on its shoulder. The bell's profile was given in terms of the lengths of line segments u_i and v_i perpendicular to AD at points s_i on AD, $i = 1, \ldots, 18$, all measured in "foundry units."

These data are given in Table 1. The point s_i shown in the figure corresponds to $i = 6$. The point s_6 is four foundry units from A, and the lengths of segments u_6 and v_6 are $7/8$ and $7/8$ foundry units, respectively. It is not clear from Williams' description which other data were specified. In any case, we take OA and angle OAD to be 2.9 meters and $75°$, respectively.

Problem 1.   Use the trapezoid rule (with unequal subdivisions) in calculating the volume (in cubic meters) of the Tsar Bell.

Problem 2.   Calculate the mass of the Tsar Bell, given that it is 85 percent copper and 15 percent tin by mass. The densities of copper and tin are 8.96 g/cm^3 and 7.30 g/cm^3, respectively.

Table 1. Tsar Bell profile data

i	s_i	u_i	v_i	i	s_i	u_i	v_i
1	0	0	0	10	8	$9/32$	$5/8$
2	1	$1/16$	$39/32$	11	9	0	$1/2$
3	1.5	$5/32$	$35/32$	12	9.5	$1/16$	$19/32$
4	2	$3/8$	$17/16$	13	9.75	$1/8$	$25/32$
5	3	$3/4$	$7/8$	14	10	$9/32$	$31/32$
6	4	$7/8$	$7/8$	15	10.25	$15/32$	1
7	5	$15/16$	$13/16$	16	10.5	$11/16$	$13/16$
8	6	$27/32$	$23/32$	17	10.75	$13/16$	$11/16$
9	7	$9/16$	$23/32$	18	11.0	$25/32$	$11/16$

Problem 3.   Allowing for a 5 percent loss from melting and casting, calculate the amounts of copper and tin that must be ordered.

Problem 4.   Determine the height of the Tsar Bell.

Student Project

Optimal Sprayer

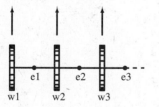

Figure 6.80. Sprayer schematic.

A mobile field irrigation system is shown in Fig. 6.80. Pipe is supported on a framework attached to water-driven wheels w1, w2, w3, ... Water emitters e1, e2, e3, ... are spaced evenly along the pipe. The entire apparatus moves through the field at a uniform speed. We assume that each emitter distributes water uniformly in a circular pattern. Ignoring the special cases of the first and last emitters, determine the spacing of the emitters to optimize the uniformity of coverage. No point in the field is to receive water from more than two emitters.

We may assume that the emitter spray radius is 1 unit. Since no point of the field receives water from more than two emitters, the problem reduces to determining the spacing parameter c. See Fig. 6.81. Note that $1 \le c \le 2$.

Problem 1.   Explain why it is useful to find an expression $w(x)$ whose value is proportional to the water received by points along the vertical line through $(x, 0)$,

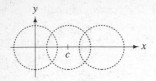

Figure 6.81. Emitter coverage.

where $0 \leq x \leq c/2$. Show that $w(x)$ is given by

$$w(x) = \begin{cases} \sqrt{1 - x^2}, & 0 \leq x \leq c - 1 \\ \sqrt{1 - x^2} + \sqrt{1 - (x - c)^2}, & c - 1 \leq x \leq c/2. \end{cases}$$

Why may the domain of w be restricted to $[0, c/2]$? Evidently, we have set the proportionality constant equal to 1. Give a justification for this simplification.

Problem 2. The total amount of water received on $[0, c/2]$ is equal to $\int_0^{c/2} w(x)\,dx$. Show by integration that the total amount of water received on $[0, c/2]$ is $\pi/4$. Show also that this result follows directly from Fig. 6.81 by "common sense." The average amount of water received on $[0, c/2]$ is the average value of the function $w(x)$ on this interval.

Problem 3. From the above calculation of total water, the average value of w on $[0, c/2]$ is $\pi/(2c)$. Give an argument as to why we may optimize the uniformity of coverage by minimizing the "average variation" function

$$g(c) = (2/c) \int_0^{c/2} (w(x) - \pi/(2c))^2\,dx, \quad 1 \leq c \leq 2.$$

Problem 4. Show that $g(c)$ may be written in the form

$$g(c) = \frac{k_1}{c} + \frac{k_2}{c^2} + \frac{4}{c} \int_{c-1}^{c/2} \sqrt{1 - x^2}\sqrt{1 - (x - c)^2}\,dx, \quad 1 \leq c \leq 2,$$

where k_1 and k_2 are constants.

Problem 5. Use a numerical integration algorithm to calculate g at a sufficient number of points so that the point at which g takes its minimum may be found, accurate to one decimal place.

In 1929 K. E. Gould, a graduate student working with MIT professor Norbert Wiener, designed a machine "which performs certain integrations which are important in engineering...." Gould's machine could graph the convolution

$$f * g(t) = \int_0^t f(\tau)g(t - \tau)d\tau$$

of given functions f and g in a fraction of the time required by other methods, which depended on repeated numerical integration.

A schematic of the integrator is shown in Fig. 6.82. Parallel frames at O and P and collector QR are equally spaced and centered on an axis perpendicular to the planes and QR. A $2m \times h$ infrared source is placed behind the frame at O. The radiation can pass through the frames only below the graphs of f and G, where $G(x) = g(2x)$. The collector QR measures the to-tal radiation received through the two apertures. We sketch an argument showing that the received radiation is proportional to $\int_0^{2m} f(x)g(x)\,dx$.

Divide the received radiation into wedges with vertical planes through QR. The wedge W shown in the figure meets frames O and P in the darkened strips. Within W the radiation received at frame P is proportional to $f(x)\,dx$. Of the amount $kf(x)\,dx$ received at P, only the fraction $(g(x)/h)kf(x)\,dx$ gets through. The radiation qr received at the collector QR is proportional to $(g(x)/h)kf(x)\,dx$. Summing over all wedges, it follows that the radiation received by the collector is proportional to

$$\int_0^{2m} f(x)g(x)\,dx.$$

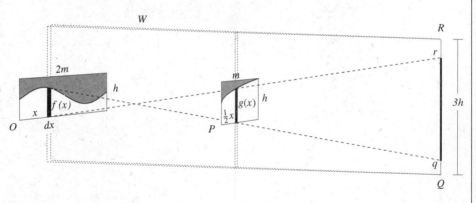

Figure 6.82. Infrared integrator.

Chapter 7

Infinite Series, Sequences, and Approximations

Approximation Using Series

Use your calculator to find the sine of 1.18. Two things are apparent when you do this: the answer 0.924606012408 appears very quickly, and the 12-decimal-place accuracy is impressive. How does the calculator determine the value to 12 decimal places, and how does it do it so quickly? The procedure that your calculator uses is based on a formula or algorithm that is known to provide a good approximation to the sine function. This formula or algorithm is programmed into the calculator so that it runs very quickly and gives 12-decimal-place accuracy.

We have already studied many approximation procedures: the tangent line approximation to a function, the Euler method for approximating the solution to a differential equation, Newton's method for approximating the solution to an equation, the trapezoid rule for approximating the value of a definite integral. Many of these processes (or refinements thereof) are used by calculators and computer algebra systems. In this chapter we will study power series. We will see that power series provide a way to obtain good approximations to functions, definite integrals, and solutions to equations. We will also use power series to solve some differential equations.

7.1 TAYLOR POLYNOMIALS

As early as Chapter 1 we saw that the equation of a line tangent to the graph of a function can be used to provide a good approximation to the function. As an example, let $f(x) = e^x$. The line tangent to the graph of $y = e^x$ at the point $(0, 1)$ has equation

$$L(x) = f(0) + f'(0)(x - 0) = e^0 + e^0(x - 0) = 1 + x.$$

If x is close to 0, then $1 + x$ is a good approximation to e^x:

$$e^x \approx 1 + x.$$

This can be seen by looking at the graphs of $y = e^x$ and $y = 1 + x$ on a small interval centered at 0, and noting that the two graphs are very close together. This suggests that e^x and $L(x)$ are close when x is near 0. (See Fig. 7.1.)

When we sketch the graphs of $y = e^x$ and $y = L(x)$ on a larger interval centered at 0, we see that when x is not close to 0, $L(x)$ is not very close to e^x (see Fig. 7.2). For example, when $x = 2.3$, the value of e^x and $L(x) = 1 + x$ differ by

$$e^{2.3} - (1 + 2.3) \approx 6.674182$$

which is about 67% of the value of $e^{2.3}$. But suppose we want to approximate $e^{2.3}$. What can we do? We could try to find the equation for a line tangent to the graph at a point close to $x = 2.3$ and then use this new tangent line. But this approach is effective only if we can find a point (a, e^a) close to $(2.3, e^{2.3})$ for which the tangent line equation

$$y = e^a + e^a(x - a)$$

is easy to work with. This is unlikely since e^a may be as hard to compute as $e^{2.3}$ and, if it is not, then a will not be easy to work with or compute. The point $(0, 1)$ is the only point on the graph of $y = e^x$ with a tangent line equation that is easy to work with.

Our other option is to continue basing our work at the point $(0, 1)$ and to try to find a simple function that provides a better approximation than the tangent line.

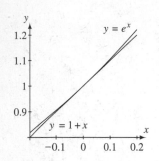

Figure 7.1. For $-0.2 \le x \le 0.2$, $1 + x$ is a good approximation to e^x.

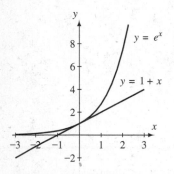

Figure 7.2. The line tangent to the graph of $y = e^x$ at $(0, 1)$ is not close to the graph when x is not close to 0.

Polynomial Approximations

Investigation 1

The tangent line approximation discussed above works well because near $(0, 1)$ the graph of $y = e^x$ looks very much like the graph of the tangent line equation $y = L(x)$. This is because

a. $L(x)$ and e^x both take the value 1 at $x = 0$.

b. $L(x)$ and e^x have the same derivative at $x = 0$.

We know, however, that the graph of $y = e^x$ is not a straight line, so it is not surprising that the graphs of $y = e^x$ and $y = L(x)$ are not close when x is not close to 0. We could get a better approximation if we could replace the tangent line equation by a simple equation whose graph "bends" like the graph of $y = e^x$. Because the curvature of the graph is related to the second derivative, we design our approximation to have the same second derivative as e^x at $x = 0$. At the same time, we want to retain the features listed above that made the tangent line a good approximation for x near 0. Thus for our next try at approximating e^x we seek a quadratic expression

$$(1) \qquad Q(x) = a_0 + a_1 x + a_2 x^2$$

satisfying the following requirements:

a. $Q(x)$ and e^x take the same value at $x = 0$.

b. $Q(x)$ and e^x have the same first derivative at $x = 0$.

c. $Q(x)$ and e^x have the same second derivative at $x = 0$.

We can use these requirements to find the values of a_0, a_1, and a_2. Requirement **a** implies

$$a_0 = Q(0) = e^0 = 1,$$

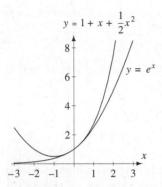

$y = 1 + x + \frac{1}{2}x^2$

$y = e^x$

Figure 7.3. The quadratic gives a good approximation to e^x for x near 0.

and **b** says

$$a_1 = Q'(0) = (e^x)'\big|_{x=0} = e^0 = 1.$$

From **c** we obtain

$$2a_2 = Q''(0) = (e^x)''\big|_{x=0} = e^0 = 1,$$

so that

$$a_2 = \frac{1}{2}.$$

Substituting the values for a_0, a_1, and a_2 into (1), we have

$$Q(x) = 1 + x + \frac{1}{2}x^2.$$

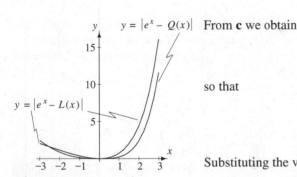

$y = |e^x - Q(x)|$

$y = |e^x - L(x)|$

Figure 7.4. On most of the interval $-3 \le x \le 3$, $Q(x)$ approximates e^x better than does $L(x)$.

Figure 7.3 shows the graphs of $y = e^x$ and $y = Q(x)$ for $-3 \le x \le 3$. Comparing this picture with Fig. 7.2, we see that the quadratic approximation does a better job than the tangent line approximation. Neither one is particularly good when x is far from 0, but the quadratic is close to e^x on a larger interval than was the case for the tangent line approximation.

For another perspective, look at the graphs of

$$y = |e^x - L(x)| \quad \text{and} \quad y = |e^x - Q(x)|.$$

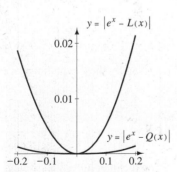

$y = |e^x - L(x)|$

$y = |e^x - Q(x)|$

Figure 7.5. The quadratic approximation is much better than the tangent line approximation on $-0.2 \le x \le 0.2$.

These graphs illustrate how much $L(x)$ and $Q(x)$ differ from e^x, giving us a picture of the error for these approximations. Figure 7.4 shows the graphs of these errors on the interval $-3 \le x \le 3$. For most x values in this interval the error in using $Q(x)$ is less than the error in using $L(x)$. However, we also see that for $x < -2.5$, $L(x)$ does a better job than $Q(x)$. Figure 7.5 shows the graphs of the errors for $-0.2 \le x \le 0.2$. For x in this interval, the error in using $Q(x)$ to approximate e^x is much less than the error in using $L(x)$.

For a more practical test, check the values of e^x, $L(x)$, and $Q(x)$, for $x = 0.15$:

$$e^{0.15} \approx 1.16183424273, \quad L(0.15) = 1.15, \quad Q(0.15) = 1.16125.$$

The error in the approximation $e^{0.15} \approx Q(0.15)$ is about one-twentieth of the error in the approximation $e^{0.15} \approx L(0.15)$.

In the investigation just completed, we saw that the quadratic approximation $Q(x)$ is a better approximation to e^x than is $L(x)$ in that

a. $Q(x)$ is much closer to e^x than is $L(x)$ for x close to 0.

b. $Q(x)$ is close to e^x on a larger interval than is the case for $L(x)$.

It is true that $L(x)$ does better than $Q(x)$ for $x < -2.5$, but neither approximation is good for such x values, so it is unlikely that either would be used in practice.

Because of the success with $Q(x)$, it seems natural to see what happens when we construct polynomials of higher degree that agree with e^x and several of its derivatives at $x = 0$.

■ **EXAMPLE 1** Find a polynomial $C(x)$ of degree 3 that agrees with e^x and its first three derivatives at $x = 0$. Graphically investigate the error when $C(x)$ is used to approximate e^x on $-1 \le x \le 1$ and compare this with the error when $Q(x)$ is used.

Solution. Let

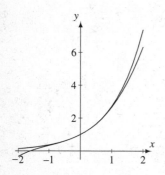

Figure 7.6. The graph of $y = e^x$ and $y = C(x) = 1 + x + x^2/2 + x^3/6$ are close for $-2 \le x \le 2$.

(2) $$C(x) = a_0 + a_1 x + a_2 x^2 + a_3 x^3.$$

First note that e^x and its first three derivatives all have value 1 at $x = 0$. The conditions on $C(x)$ and its derivatives are

$$a_0 = C(0) = e^0 = 1$$
$$a_1 = C'(0) = e^0 = 1$$
$$2a_2 = C''(0) = e^0 = 1$$
$$6a_3 = C'''(0) = e^0 = 1.$$

Substituting into (2) the values for a_0, a_1, a_2, and a_3 given by these equations, we have

$$C(x) = 1 + x + \frac{1}{2}x^2 + \frac{1}{6}x^3.$$

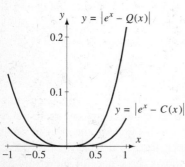

Figure 7.7. For $-1 \le x \le 1$, we see $|e^x - C(x)|$ is much less than $|e^x - Q(x)|$.

Figure 7.6 shows the graphs of $y = e^x$ and $y = C(x)$ on the interval $-2 \le x \le 2$. This figure illustrates how well $C(x)$ approximates e^x. In Fig. 7.7 we see the graphs of

$$y = |e^x - C(x)| \quad \text{and} \quad y = |e^x - Q(x)|$$

for $-1 \leq x \leq 1$. The figure shows that for these x values $C(x)$ approximates e^x better than $Q(x)$.

When we compare the values of $Q(x)$ and $C(x)$ for $x = -0.2$, we see evidence that for small x, the approximation $e^x \approx C(x)$ is better than the approximation $e^x \approx Q(x)$:

$$e^{-0.2} \approx 0.818730753078, \quad Q(-0.2) = 0.82, \quad C(-0.2) \approx 0.818667. \quad \blacksquare$$

Taylor Polynomials

Investigation 2

We now generalize the idea hinted at in the previous example and investigation. Suppose that we wish to approximate a function f near a point c in its domain. If f has n derivatives at $x = c$, then we seek a polynomial $P(x)$ of degree n that has the same value as f at $x = c$ and such that P and f have the same first derivative, the same second derivative, $\ldots$, and the same nth derivative at $x = c$.

Since we will be evaluating P and several of its derivatives at $x = c$, it is convenient to express $P(x)$ in powers of $x - c$. Hence assume

$$(3) \qquad P(x) = a_0 + a_1(x - c) + a_2(x - c)^2 + \cdots + a_n(x - c)^n.$$

Substituting $x = c$ into (3), we see immediately see that

$$(4) \qquad\qquad P(c) = a_0.$$

It also follows from (3) that

$$P'(x) = a_1 + 2a_2(x - c) + 3a_3(x - c)^2 + \cdots + na_n(x - c)^{n-1}$$

so that

$$(5) \qquad\qquad P'(c) = a_1.$$

The second derivative of P is

$$P''(x) = 2 \cdot 1 a_2 + 3 \cdot 2 a_3(x - c) + 4 \cdot 3 a_4(x - c)^2 + \cdots + n(n - 1)a_n(x - c)^{n-2},$$

which leads to

$$(6) \qquad\qquad P''(c) = (2 \cdot 1)a_2.$$

Continuing in this fashion, we see that for integer k between 1 and n the kth derivative of P is

$$P^{(k)}(x) = [k(k - 1)(k - 2) \cdots 1]a_k + [(k + 1)k \cdots 2]a_{k+1}(x - c) + \cdots$$
$$+ [n(n - 2)(n - 2) \cdots (n - k + 1)]a_n(x - c)^{n-k}.$$

Putting $x = c$ in this expression, we have

$$(7) \qquad\qquad P^{(k)}(c) = [k(k - 1)(k - 2) \cdots 1]a_k = k!a_k.$$

Equating the derivatives of P at $x = c$ to the derivatives of f at $x = c$ and taking (4), (5), (6), and (7) into account, we obtain the equations

$$f(c) = P(c) \quad = a_0$$
$$f'(c) = P'(c) \quad = a_1$$
$$f''(c) = P''(c) \quad = 2!a_2$$
$$\vdots$$
$$f^{(k)}(c) = P^{(k)}(c) = k!a_k$$
$$\vdots$$
$$f^{(n)}(c) = P^{(n)}(c) = n!a_n.$$

Solving these equations for the coefficients $a_0, a_1, \ldots, a_n$, we find

$$a_0 = f(c), \qquad a_1 = \frac{f'(c)}{1!}, \qquad a_2 = \frac{f''(c)}{2!}, \qquad \cdots,$$

$$a_k = \frac{f^{(k)}(c)}{k!}, \qquad \cdots, \qquad a_n = \frac{f^{(n)}(c)}{n!}.$$

Substituting the values for these coefficients into (3), we have

$$P(x) = f(c) + \frac{f'(c)}{1!}(x - c) + \frac{f''(c)}{2!}(x - c)^2 + \cdots$$

$$+ \frac{f^{(k)}(c)}{k!}(x - c)^k + \cdots + \frac{f^{(n)}(c)}{n!}(x - c)^n$$

$$= \sum_{k=0}^{n} \frac{f^{(k)}(c)}{k!}(x - c)^k$$

This polynomial is called a **Taylor polynomial.**

DEFINITION **Taylor polynomial**

Let f be a function, and suppose that the nth derivative of f exists at $x = c$. The nth Taylor polynomial for f at $x = c$ is the polynomial

$$T_n(x; c) = f(c) + \frac{f'(c)}{1!}(x - c) + \frac{f''(c)}{2!}(x - c)^2 + \cdots$$

(8)
$$+ \frac{f^{(k)}(c)}{k!}(x - c)^k + \cdots + \frac{f^{(n)}(c)}{n!}(x - c)^n$$

$$= \sum_{k=0}^{n} \frac{f^{(k)}(c)}{k!}(x - c)^k.$$

In (8) we take $f^{(0)}(c) = f(c)$ and $0! = 1$.

The Taylor polynomial of degree n for f at $x = c$ is designed to closely fit f at $x = c$. Thus Taylor polynomials often provide good approximations to functions.

■ **EXAMPLE 2**   Let $f(x) = \tan x$. Find $T_3(x; 0)$ and estimate the maximum possible error if we use $T_3(x; 0)$ to approximate $\tan x$ on $-0.5 \le x \le 0.5$.

Solution.   We begin by finding the first three derivatives of $\tan x$.

$$f^{(0)}(x) = f(x) = \tan x$$

$$f^{(1)}(x) = f'(x) = (\tan x)' = \sec^2 x$$

$$f^{(2)}(x) = f''(x) = (\sec^2 x)' = 2 \sec x (\sec x)' = 2 \tan x \sec^2 x$$

$$f^{(3)}(x) = f'''(x) = (2 \tan x \sec^2 x)' = 2(\tan x)'(\sec^2 x) + 2(\tan x)(\sec^2 x)'$$
$$= 2 \sec^4 x + 4 \tan^2 x \sec^2 x$$

Substitute $x = 0$ in these derivatives to obtain the coefficients of the Taylor polynomial:

$$\frac{f^{(0)}(0)}{0!} = \tan 0 = 0$$

$$\frac{f^{(1)}(0)}{1!} = \sec^2 0 = 1$$

$$\frac{f^{(2)}(0)}{2!} = \frac{1}{2}(2 \tan 0 \sec^2 0) = 0$$

$$\frac{f^{(3)}(0)}{3!} = \frac{1}{6}(2 \sec^4 0 + 4 \tan^2 0 \sec^2 0) = \frac{1}{3}$$

Hence

$$T_3(x; 0) = \frac{f(0)}{0!} + \frac{f'(0)}{1!}(x - 0) + \frac{f''(0)}{2!}(x - 0)^2 + \frac{f^{(3)}(0)}{3!}(x - 0)^3$$

$$= x + \frac{1}{3}x^3.$$

When we use $T_3(x; 0)$ to approximate $\tan x$, the absolute value of the error is

$$|\tan x - T_3(x; 0)|.$$

In Fig. 7.8 we see the graph of

$$y = |\tan x - T_3(x; 0)|.$$

This graph shows that on $-0.5 \le x \le 0.5$, the error is less than 0.005 in absolute value. In Fig. 7.9 we have graphed $y = \tan x$ and $y = T_3(x; 0)$ for $-1 \le x \le 1$. The two graphs are very close; in fact they are almost indistinguishable on the interval $-0.5 \le x \le 0.5$.   ◼

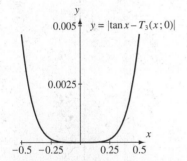

Figure 7.8. When $-0.5 \le x \le 0.5$, the difference between $\tan x$ and $T_3(x; 0)$ is no more than 0.005 in absolute value.

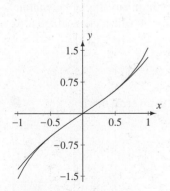

Figure 7.9. The graphs of $y = \tan x$ and $y = T_3(x; 0)$. Which graph goes with which function?

■ **EXAMPLE 3**   Let $p(x) = -2x^3 - 3x^2 + 4x - 5$. Find all of the Taylor polynomials $T_n(x; -2)$ for p.

Solution.   We start by finding the values of the derivatives of p at $x = -2$.

$$p(x) = -2x^3 - 3x^2 + 4x - 5 \qquad p(-2) = -9$$
$$p'(x) = -6x^2 - 6x + 4 \qquad p'(-2) = -8$$
$$p''(x) = -12x - 6 \qquad p''(-2) = 18$$
$$p'''(x) = -12 \qquad p'''(-2) = -12$$
$$p^{(4)}(x) = p^{(5)}(x) = \cdots = 0 \qquad p^{(4)}(-2) = p^{(5)}(-2) = \cdots = 0$$

The Taylor polynomials are

$$T_0(x; -2) = \frac{p^{(0)}(-2)}{0!}$$
$$= -9$$

$$T_1(x; -2) = \frac{p^{(0)}(-2)}{0!} + \frac{p^{(1)}(-2)}{1!}(x - (-2))$$
$$= -9 - 8(x + 2)$$

$$T_2(x; -2) = \frac{p^{(0)}(-2)}{0!} + \frac{p^{(1)}(-2)}{1!}(x - (-2)) + \frac{p^{(2)}(-2)}{2!}(x - (-2))^2$$
$$= -9 - 8(x + 2) + 9(x + 2)^2$$

$$T_3(x; -2) = \frac{p^{(0)}(-2)}{0!} + \frac{p^{(1)}(-2)}{1!}(x - (-2))$$
$$\qquad + \frac{p^{(2)}(-2)}{2!}(x - (-2))^2 + \frac{p^{(3)}(-2)}{3!}(x - (-2))^3$$
$$= -9 - 8(x + 2) + 9(x + 2)^2 - 2(x + 2)^3$$

Because $p^{(k)}(-2) = 0$ for $k = 4, 5, 6, \ldots$, we see that in any Taylor polynomial $T_n(x; -2)$ with $n \geq 4$, all terms $(x + 2)^k$ with $k \geq 4$ will have coefficient 0. Hence for $n = 4, 5, 6, \ldots$ we have

$$T_n(x; -2) = T_3(x; -2) = -9 - 8(x + 2) + 9(x + 2)^2 - 2(x + 2)^3$$

In fact, for $n \geq 4$ we have $T_n(x; -2) = p(x)$. Can you verify this?   ■

■ **EXAMPLE 4**   Let $\ell(x) = \ln x$.

a. Let n be a nonnegative integer. Find $T_n(x; 1)$ for ℓ.

b. Analyze the error in the approximation

$$\ln x \approx T_5(x; 1)$$

for $0.8 \leq x \leq 1.2$.

Solution

a. First note that $\ell(1) = \ln 1 = 0$, so the constant term in the Taylor series is 0. Hence when $n = 0$, we have

$$T_0(x; 1) = 0.$$

Next compute the coefficients for the other terms of the Taylor polynomial. We have

$$\ell^{(1)}(x) = (\log x)' = x^{-1} \qquad\qquad \frac{\ell^{(1)}(1)}{1!} = 1$$

$$\ell^{(2)}(x) = (x^{-1})' = (-1)x^{-2} \qquad\qquad \frac{\ell^{(2)}(1)}{2!} = -\frac{1}{2}$$

$$\ell^{(3)}(x) = ((-1)x^{-2})' = 2x^{-3} \qquad\qquad \frac{\ell^{(2)}(1)}{3!} = \frac{1}{3}$$

$$\ell^{(4)}(x) = (2x^{-3})' = (-3!)x^{-4} \qquad\qquad \frac{\ell^{(2)}(1)}{4!} = -\frac{1}{4}$$

$$\vdots \qquad\qquad\qquad\qquad \vdots$$

$$\ell^{(k)}(x) = ((-1)^k(k-2)!x^{-(k-1)})' \qquad \frac{\ell^{(k)}(1)}{k!} = (-1)^{k+1}\frac{1}{k}$$

$$= (-1)^{k+1}(k-1)!x^{-k}$$

$$\vdots \qquad\qquad\qquad\qquad \vdots$$

The coefficients for the Taylor polynomial follow a predictable pattern, so we can write down the polynomial of degree n for $n \geq 1$:

$$T_n(x; 1) = \sum_{k=0}^{n} \frac{\ell^{(k)}(1)}{k!}(x-1)^k$$

(9)

$$= \sum_{k=1}^{n} \frac{(-1)^{k+1}}{k}(x-1)^k$$

$$= (x-1) - \frac{1}{2}(x-1)^2 + \frac{1}{3}(x-1)^3 - \cdots + \frac{(-1)^{n+1}}{n}(x-1)^n.$$

b. Setting $n = 5$ in (9), we find

$$T_5(x; 1) = (x-1) - \frac{(x-1)^2}{2} + \frac{(x-1)^3}{3} - \frac{(x-1)^4}{4} + \frac{(x-1)^5}{5}.$$

The graph of

$$y = |\ln x - T_5(x; 1)|$$

is shown in Fig. 7.10. From this graph we see that when $0.8 \leq x \leq 1.2$, the error in the approximation

$$\ln(x) \approx T_5(x; 1)$$

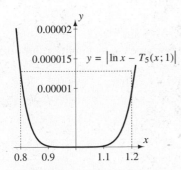

Figure 7.10. For $0.8 \leq x \leq 1.2$, the difference between $\ln x$ and $T_5(x; 1)$ is less than 0.000015 in absolute value.

is less than 0.000015 in absolute value. For another way of looking at the error, the table below shows the values of $\ln x$, $T_5(x; 1)$, and $y = |\ln x - T_5(x; 1)|$ for several values of x between 0.8 and 1.2. ■

x	$\ln x$	$T_5(x; 1)$	$\lvert \ln x - T_5(x; 1)\rvert$
0.80	-0.2231435513	-0.2231306667	1.288×10^{-5}
0.85	-0.1625189295	-0.1625167500	2.179×10^{-6}
0.90	-0.1053605157	-0.1053603333	1.823×10^{-7}
0.95	-0.0512932944	-0.0512932917	2.721×10^{-9}
1.00	0.0	0.0	0.0
1.05	0.0487901642	0.0487901667	2.497×10^{-9}
1.10	0.0953101798	0.0953103333	1.535×10^{-7}
1.15	0.1397619424	0.1397636250	1.683×10^{-6}
1.20	0.1823215568	0.1823306667	9.110×10^{-6}

Exercises 7.1

Basic

Exercises 1–6: Find $T_0(x; 0)$, $T_1(x; 0)$, $T_2(x; 0)$, and $T_3(x; 0)$ for the indicated function, and then calculate $f(0.1)$ and $T_3(0.1; 0)$.

1. $\dfrac{1}{1+x}$

2. $(3 + x)^{10}$

3. $\sin x$

4. $\cos x$

5. $\arctan x$

6. e^{2x}

Exercises 7–12: Find $T_0(x; c)$, $T_1(x; c)$, $T_2(x; c)$, and $T_3(x; c)$ for the indicated function and the given value of c. Then calculate $f(z)$ and $T_3(z; c)$ for the given value of z.

7. $\dfrac{1}{1+x}$, $\quad c = -2$, $\quad z = -1.95$

8. $\tan x$, $\quad c = \dfrac{\pi}{4}$, $\quad z = \dfrac{\pi}{4} - 0.03$

9. $x^2 \ln x$, $\quad c = 3$, $\quad z = 2.95$

10. $\sin x$, $\quad c = \dfrac{3\pi}{4}$, $\quad z = \dfrac{3\pi}{4} + 0.1$

11. $\arctan x$, $\quad c = 1$, $\quad z = 1.05$

12. e^x, $\quad c = \ln 3$, $\quad z = 1$

13. In Fig. 7.6, which graph is the graph of $y = e^x$?

14. In Fig. 7.9, which graph is the graph of $y = \tan x$?

15. In Example 3, verify that $T_n(x; -2) = p(x)$ for $n \geq 4$.

16. Verify at least four of the entries in the table shown at the end of Example 4.

17. At the beginning of this section we remarked that the point $(0, 1)$ was the only point on the graph of $y = e^x$ for which the tangent line equation is easy to work with. Explain the sense in which this remark is true.

18. Let $I = [-\pi, \pi]$.
   a. Find $T_6(x; 0)$ for $\sin x$.
   b. Graph $y = \sin x$ and $y = T_6(x; 0)$ on I.
   c. Graph $y = |T_6(x; 0) - \sin x|$ on I.
   d. Use the graph to estimate the maximum value of $|T_6(x; 0) - \sin x|$ on I.
   e. Calculate $\sin x$ and $T_6(x; 0)$ for $x = -0.2, 0, 0.5,$ and 1.

19. Let $I = (-\infty, \infty)$.
   a. Find $T_{17}(x; 2)$ for $x^3 - 3x^2 + 4x - 7$.
   b. Graph $y = x^3 - 3x^2 + 4x - 7$ and $y = T_{17}(x; 0)$ on I.
   c. Graph $y = |T_{17}(x; 0) - (x^3 - 3x^2 + 4x - 7)|$ on I.

20. Let $I = \left[-\dfrac{3}{4}, \dfrac{3}{4}\right]$.

  a. Find $T_6(x; 0)$ for $\dfrac{1}{1-x}$.

  b. Graph $y = \dfrac{1}{1-x}$ and $y = T_6(x; 0)$ on I.

  c. Graph $y = \left| T_6(x; 0) - \dfrac{1}{1-x} \right|$ on I.

  d. Use the graph to estimate the maximum value of
  $$\left| T_6(x; 0) - \dfrac{1}{1-x} \right| \text{ on } I.$$

  e. Calculate $\dfrac{1}{1-x}$ and $T_6(x; 0)$ for $x = -0.9$, $-0.05, 0.5,$ and 0.7.

21. Let $I = [0, 2]$.

  a. Find $T_4(x; 1)$ for $\sqrt{x}$.

  b. Graph $y = \sqrt{x}$ and $y = T_4(x; 1)$ on I.

  c. Graph $y = |T_4(x; 1) - \sqrt{x}|$ on I.

  d. Use the graph to estimate the maximum value of $|T_4(x; 1) - \sqrt{x}|$ on I.

  e. Calculate $\sqrt{x}$ and $T_4(x; 0)$ for $x = 0.4, 0.8, 1,$ $1.4,$ and 1.8.

Growth

Exercises 22–26: Let n be a nonnegative integer. Write out the general Taylor polynomial $T_n(x; c)$.

22. $\dfrac{1}{1-x}$, $c = 0$

23. $\sin x$, $c = 0$

24. $\dfrac{2}{3+x}$, $c = -1$

25. $\sqrt{x}$, $c = 1$

26. $f(x) = \dfrac{1}{x}$, $c \neq 0$

27. Let $g(x) = \sin x$, let c be a real number, and let n be a nonnegative real number. Write out the general Taylor polynomial $T_n(x; c)$ for g.

28. Let $h(x) = \sqrt{x}$, let $c > 0$, and let n be a nonnegative real number. Write out the general Taylor polynomial $T_n(x; c)$ for h.

29. Let $f(x) = \sin x$, $g(x) = \cos x$, and $h(x) = \sin x + \cos x$. Let $T_4(x; 0)$ be the fourth-degree Taylor polynomial at 0 for f, $P_4(x; 0)$ be the fourth-degree Taylor polynomial at 0 for g, and $Q_4(x; 0)$ be the fourth-degree Taylor polynomial at 0 for h. Is it true that
$$Q_4(x; 0) = T_4(x; 0) + P_4(x; 0)?$$

30. Let f, g, and h be functions with $h(x) = f(x) + g(x)$ for all real x. Let $T_n(x; c)$ be the nth-degree Taylor polynomial at c for f, $P_n(x; c)$ the nth-degree Taylor polynomial at c for g, and $Q_n(x; c)$ the nth-degree Taylor polynomial at c for h. Show that
$$Q_n(x; c) = T_n(x; c) + P_n(x; c).$$

31. Let $f(x) = \dfrac{1}{1-x}$ and $g(x) = f(5x)$. Let $T_3(x; 0)$ be the third-degree Taylor polynomial at 0 for f and $P_3(x; 0)$ the third-degree Taylor polynomial at 0 for g. Is it true that
$$P_3(x; 0) = T_3(5x; 0)?$$

32. Let f and g be functions and a a real number. Suppose that $g(x) = f(ax)$ for all real x. Let $T_n(x; 0)$ be the nth-degree Taylor polynomial at 0 for f and $P_n(x; 0)$ the nth-degree Taylor polynomial at 0 for g. Show that
$$P_n(x; 0) = T_n(ax; 0).$$

33. Let $f(x) = \sin x$ and $g(x) = f(\pi x)$. Let T_3 be the third-degree Taylor polynomial for f and $P_3(x; 1)$ the third-degree Taylor polynomial for g. Is it true that
$$P_3(x; 1) = T_3(\pi x; 1)?$$

Is it true that
$$P_3(x; 1) = T_3(\pi x; \pi)?$$

34. Let f and g be functions and a a real number. Suppose that $g(x) = f(ax)$ for all real x. Let $T_n(x; ac)$ be the nth-degree Taylor polynomial at ac for f and $P_n(x; c)$ the nth-degree Taylor polynomial at c for g. Show that
$$P_n(x; c) = T_n(ax; ac).$$

35. It can be shown that for all real x,
$$\lim_{n \to \infty} \left(1 + \dfrac{x}{n}\right)^n = e^x.$$

Hence when n is large, $P_n(x) = \left(1 + \dfrac{x}{n}\right)^n$ can be used as an approximation for e^x. Investigate how $P_5(x)$ and $T_5(x; 0)$ each approximate e^x on the interval $[-2, 2]$. Which approximation do you think is better? Why?

36. *A quadratic Newton's method.* Consider the following procedure for approximating a solution to an equation $f(x) = 0$.

   1. Make an initial guess r_0 for the root of the equation.

   2. Assume that r_k is our kth guess for the root. Find the second-degree Taylor polynomial $T_2(x; r_k)$

for f at r_k. Find a nearby root of $T_2(x; r_k) = 0$ and let this root be r_{k+1}.

   3. Repeat step 2 until you have approximated the desired root of $f(x) = 0$ to the needed accuracy.

Comment on this procedure. What are its advantages over Newton's method? What are its disadvantages? Use the technique described here to approximate the root of $\cos x - 0.6 = 0$ with initial guess $r_0 = 1$. How does the result after three iterations compare with the result for Newton's method after three iterations?

Exercises 37–40: For the indicated function f and value of c find $T_n(x; c)$ for $n = 0, 1, 2, 3, 4$. For each of these values of n graph $y = |f(x) - T_n(x; c)|$ on the given interval I. Use the graphs to find E_n, the maximum value of $|f(x) - T_n(x; c)|$ on I. Record E_0, E_1, E_2, E_3, E_4, and then write a short paragraph describing how E_n changes as n gets larger.

37. $f(x) = e^x$, $c = 0$, $I = [-0.5, 0.5]$

38. $f(x) = \dfrac{1}{2 - x}$, $c = 1$, $I = [0.5, 1.5]$

39. $f(x) = \sin x$, $c = \dfrac{\pi}{4}$, $I = \left[0, \dfrac{\pi}{2}\right]$

40. $f(x) = \arcsin x$, $c = \dfrac{1}{2}$, $I = [0, 1]$

Review/Preview

41. Let $s(n) = n^2$. Find $s(1)$, $s(2)$, $s(3)$, and $s(4)$.

42. Let $r(n) = \cos(\pi n)$. Find $r(1)$, $r(2)$, $r(3)$, and $r(4)$. For positive integer n, what is $r(n)$?

43. Let $x_0 = 1$ and for $n \geq 0$ define

$$x_{n+1} = \frac{x_n^2 + 2}{2x_n}.$$

Find x_1, x_2, x_3 and x_4.

44. Let $s_1 = 2$ and for $n \geq 1$ define

$$s_{n+1} = \sqrt{2 + \sqrt{s_n}}.$$

Find s_1, s_2, s_3, and s_4.

45. Let $f(x)$ be a function defined on an interval $[a, b]$. Suppose that $|f'(x)| \leq 2$ on $[a, b]$. Tell how the mean value inequality (Section 4.3) can be used to conclude that $|f(b) - f(a)| \leq 2|a - b|$.

46. Apply the Mean-value Inequality (Section 4.4) to show that for any real numbers x and t,

$$|\sin x - \sin t| \leq |x - t|.$$

47. Let $r(\theta) = \sin 2\theta$. What is $\dfrac{d^{21} r}{d\theta^{21}}$?

48. Let $h(t) = t^{25} - 2t^{24} + t^{14} - 3t^{10} - 8t + 7$. Find

$$\frac{d^{23}h}{dt^{23}}, \quad \frac{d^{25}h}{dt^{25}}, \quad \frac{d^{26}h}{dt^{26}}.$$

49. Let $f(x) = e^{ax}$ and let n be a positive integer. Find $f^{(n)}(1)$.

50. Let $F(x) = 2(x - 3)^4 + (x - 3)^3 - a(x - 3)^2 + 4(x - 3) - 8$. Find a if $F''(3) = 7$.

51. Let $G(t) = 8(t + 2)^7 + b(t + 2)^6 + c(t + 2)^4 - 4(t + 2)^3$. Find b and c if $G^{(4)}(-2) = 0$ and $G^{(6)}(-2) = -\pi$.

7.2   APPROXIMATIONS AND ERROR

Use your calculator to find the approximate values of

$$2^{-3.1415}, \quad \tan 16.8, \quad \text{and} \quad \ln 3456.78.$$

In each case your calculator gives an answer with at least six significant digits. You are probably confident that the answers given by your calculator are correct to the number of digits displayed. It's very easy in this day and age to take the capabilities of calculators and computers for granted. But have you ever wondered how calculators can take such a wide range of inputs for so many different functions and quickly come up with answers that are correct to six or more digits?

When mathematicians and engineers design a program that a calculator can use to estimate, say, log x for $x > 0$, they not only need a fast method for approximating log x but must also know in advance what the error in the approximation will be. In particular, they have to know that for all allowable inputs, the calculator will give an answer correct to at least six significant digits. This means that the difference in the calculator value for log x and the true value for log x must be less than 0.00005 percent of the actual value of log x. If the people designing the calculator did not know about the error inherent in the procedure used to approximate log x, then the process could not be used in a calculator. The calculator market is highly competitive. Any company selling a machine that was accurate only 99 percent of the time would soon go out of business. The risks inherent in implementing an approximation procedure without knowledge about the error are not worth it.

The Remainder

As mentioned above, it can be unwise to use an approximation without knowledge of the error. Since we plan to use Taylor polynomials to approximate functions, we now study the error for such approximations.

DEFINITION **Taylor remainder**

Let $f(x)$ be a function and let $T_n(x; c)$ be the nth-degree Taylor polynomial for f at $x = c$. The remainder associated with $T_n(x; c)$ is

$$f(x) - T_n(x; c).$$

We denote this remainder by $R_n(x; c)$. Hence

(1) $$R_n(x; c) = f(x) - T_n(x; c).$$

The remainder $R_n(x; c)$ is also called the error in the approximation by the Taylor polynomial of degree n.

We will usually be interested in knowing how well $T_n(x; c)$ approximates $f(x)$ on an interval

$$c - r \leq x \leq c + r,$$

that is, on an interval centered at c. See Fig. 7.11. For the most part we will be interested in the magnitude of the error, not the sign of the error. Thus we will usually want to say something about

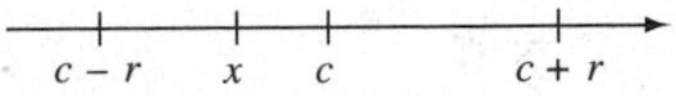

$$|R_n(x; c)| = |f(x) - T_n(x; c)|$$

Figure 7.11. The x-axis interval $c - r \leq x \leq c + r$.

In many of our applications we will be interested in finding an answer to one of two problems:

a. Given a Taylor polynomial $T_n(x; c)$, find the maximum value of $|R_n(x; c)|$ on $c - r \leq x \leq c + r$.

b. Given a number $\epsilon > 0$ and an interval $c - r \leq x \leq c + r$, find a Taylor polynomial $T_n(x; c)$ with

$$|R_n(x; c)| = |f(x) - T_n(x; c)| < \epsilon$$

for $c - r \leq x \leq c + r$.

From *a* we see how accurate the approximation $f(x) \approx T_n(x; c)$ is for $c - r \leq x \leq c + r$. From *b* we find out which Taylor polynomial can be used to approximate $f(x)$ with error less than ϵ.

■ **EXAMPLE 1**   Let $f(x) = \ln x$. Find n so that on $0.5 \leq x \leq 1.5$, the error $R_n(x; 1)$ is no more than 0.0005 in absolute value.

Solution.   We need to find n so that

$$|R_n(x; 1)| = |\ln x - T_n(x; 1)| < 0.0005$$

for all x in the interval $0.5 \leq x \leq 1.5$. In Example 4 of Section 7.1 we found that

$$(2) \qquad T_n(x; 1) = \sum_{k=1}^{n} \frac{(-1)^{k+1}}{k}(x - 1)^k$$

$$= (x - 1) - \frac{(x - 1)^2}{2} + \frac{(x - 1)^3}{3} - \cdots + (-1)^{n+1}\frac{1}{n}(x - 1)^n.$$

We find an acceptable value of n by graphing

$$y = |\ln x - T_n(x; 1)|$$

for different values of n until we find a graph for which the y values stay less than 0.0005 on $0.5 \leq x \leq 1.5$. Verify these graphs with your own graphing calculator or CAS. (Many calculators and most CAS's have routines for computing Taylor polynomials. As you verify the graphs, use these routines to calculate the Taylor polynomials. Although we already have these polynomials in (2), having a machine calculate them will save you some typing.)

Since we want the error to be quite small, we doubt that $T_1(x; 1)$ or $T_2(x; 1)$ will be close enough. We guess $T_5(x; 1)$ and check. Putting $n = 5$ in (2), we get

$$T_5(x; 1) = (x - 1) - \frac{(x - 1)^2}{2} + \frac{(x - 1)^3}{3} - \frac{(x - 1)^4}{4} + \frac{(x - 1)^5}{5}.$$

In Fig. 7.12 we have graphed

$$y = |\ln x - T_5(x; 1)|.$$

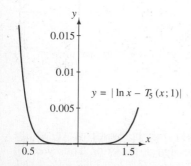

Figure 7.12. Near $x = 0.5$ and $x = 1.5$ the error is larger than 0.0005 in absolute value.

We see that near $x = 0.5$ and $x = 1.5$ the absolute value of the error is larger than 0.0005. We can demonstrate this numerically too. Let $x = 1.4$ and use a calculator to see that

(3) $$|\ln 1.4 - T_5(1.4; 1)| \approx 0.000509 > 0.0005.$$

We expect that a larger n is needed to get a smaller error, so try $n = 8$ next. In Fig. 7.13 we see the graph of

$$y = |\ln x - T_8(x; 1)|.$$

We see from the graph that for $0.5 \leq x \leq 1.5$, the value of

$$|R_8(x; 1)| = |\ln x - T_8(x; 1)|$$

stays well under 0.0005. Thus for $0.5 \leq x \leq 1.5$ we have

$$\ln x \approx T_8(x; 1)$$

and the error is less than 0.0005 in absolute value. As a more concrete check we have

$$\ln 1.4 \approx 0.3364722366 \quad \text{and} \quad T_8(1.4, 1) \approx 0.3364508038$$

and

$$|\ln 1.4 - T_8(1.4; 1)| \approx 2.143 \times 10^{-5} < 0.0005.$$

Note that we never did test $n = 6$ or $n = 7$. We stopped when we saw that $n = 8$ worked. Would $n = 6$ do the job too? How about $n = 7$? What would be the advantages of a value of n smaller than 8? ◼

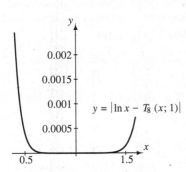

Figure 7.13. The graph shows that $|R_8(x; 1)| < 0.0005$ for $0.5 \leq x \leq 1.5$.

The Remainder Formula

In the previous example, we worked directly with

$$f(x) - T_n(x; 1)$$

to find information about $R_n(x; 1)$. This may work well when n is small and $T_n(x; 1)$ is easy to compute, but it is impractical for large values of n. What if the smallest n satisfying the conditions of Example 1 had been $n = 1000$? In addition, to work with $f(x) - T_n(x; 1)$, we had to know how to calculate f. But if we are trying to approximate f, then we probably do not know how to calculate it. To better answer questions about approximation with Taylor polynomials we need another way of working with and estimating $R_n(x; c)$. We will often use the following result.

The Taylor Inequality

Let f have a continuous $(n + 1)$th derivative on the interval $c - r \leq x \leq c + r$, and let M_{n+1} be the maximum value of $|f^{(n+1)}(x)|$ for x in this interval. Then for $c - r \leq x \leq c + r$ we have

(4) $$|R_n(x; c)| = |f(x) - T_n(x; c)| \leq \frac{M_{n+1}}{(n + 1)!}|x - c|^{n+1}.$$

Investigation

Justify the Taylor inequality result in the special case $n = 3$. (If you thoroughly understand the $n = 3$ case, you will be able to write down a proof for the general case.) For our justification we will need the Mean-value Inequality from Section 4.4. We will use the result in the following, slightly weaker form.

Mean-value Inequality

Let g have a continuous derivative on the interval $I = [a, b]$, and let M be the maximum value of $|g'|$ of I. Then

(5)
$$|g(b) - g(a)| \le M|b - a|.$$

This form follows from the Mean-value Inequality as stated in Chapter 4. See Exercise 41.

We also need more information about R_3 and its derivatives. Consider the function

(6)
$$R_3(t; c) = f(t) - T_3(t; c)$$
$$= f(t) - f(c) - f'(c)(t - c) - \frac{f''(c)}{2!}(t - c)^2 - \frac{f'''(c)}{3!}(t - c)^3$$

for t in the interval $c - r \le t \le c + r$. When we compute the first three derivatives of $R_3(t; c)$ with respect to t, we obtain

(7)
$$R_3'(t; c) = f'(t) - f'(c) - f''(c)(t - c) - \frac{f'''(c)}{2!}(t - c)^2$$
$$R_3''(t; c) = f''(t) - f''(c) - f'''(c)(t - c)$$
$$R_3'''(t; c) = f'''(t) - f'''(c)$$

Substituting $t = c$ in (6) and (7), we find that

$$R_3(c; c) = R_3'(c; c) = R_3''(c; c) = R_3'''(c; c) = 0.$$

Now apply the Mean-value Inequality (5) to the function $g(t) = R_3'''(t; c)$. If M_4 is the maximum value of $|f^{(4)}(t)|$ on the interval $[c - r, c + r]$, then for t in this interval

(8)
$$|R_3'''(t; c)| = |R_3'''(t; c) - 0|$$
$$= |R_3'''(t; c) - R_3'''(c; c)| = |f'''(t) - f'''(c)| \le M_4|t - c|.$$

We now integrate three times. For simplicity we will assume that $x \ge c$. (For the case $x < c$ see Exercise 42.) Integrating $R_3'''(t; c)$ and using (8) gives some information about $R_3''(t; c)$.

$$|R_3''(x; c)| = |R_3''(x; c) - R_3''(c; c)| = \left| \int_c^x R_3'''(t; c) dt \right|.$$

Bring the absolute value signs inside the integral. Note that $c \leq x$ is used here. (How?)

$$|R_3''(x; c)| \leq \int_c^x |R_3'''(t; c)| dt$$

Use (8) to replace the integrand with a larger expression. This gives

$$|R_3''(x; c)| \leq \int_c^x M_4 |t - c| dt.$$

Because $c \leq x$ and the variable t of integration runs between c and x, we also have $c \leq t$. Hence in the last expression, $|t - c| = t - c$. Thus we have

(9) $$|R_3''(x; c)| \leq \int_c^x M_4(t - c) dt = \frac{1}{2} M_4(x - c)^2.$$

Since x is arbitrary in $[c, c + r]$, (9) holds for every t with $c \leq t \leq c + r$. We use this fact when we integrate $R_3''(t; c)$ to find information about $R_3'(t; c)$. Because $c \leq x$, we have

$$|R_3'(x; c)| = |R_3'(x; c) - R_3'(c; c)| = \left| \int_c^x R_3''(t; c) dt \right| \leq \int_c^x |R_3''(t; c)| dt$$

$$\leq \int_c^x \frac{1}{2} M_4 |t - c|^2 dt = \frac{1}{2} M_4 \int_c^x (t - c)^2 dt = \frac{1}{3 \cdot 2} M_4 |x - c|^3.$$

The inequality

$$|R_3'(t; c)| \leq \frac{1}{3 \cdot 2} M_4 |t - c|^3$$

holds for all t satisfying $c \leq t \leq c + r$. We use this to obtain our bound for $|R_3(x; c)|$ by integrating $R_3'(t; c)$. Proceeding as above, we have

$$|R_3(x; c)| = |R_3(x; c) - R_3(c; c)| = \left| \int_c^x R_3'(t; c) dt \right| \leq \int_c^x |R_3'(t; c)| dt$$

$$\leq \frac{1}{3!} M_4 \int_c^x |t - c|^3 dt = \frac{1}{4!} M_4 |x - c|^4$$

This is the bound on the remainder given by the Taylor inequality for $n = 3$.

Finding the exact value of M_{n+1} may be very difficult. Thus when using Taylor's inequality to estimate error, we often use a number larger than M_{n+1} in place of M_{n+1}. This has the disadvantage of making our error estimate larger than necessary but has the advantage of making our calculations easier. Trade-offs of accuracy for calculation time are not unusual in working with approximations. The cost of computations needs to be balanced against the need for a certain level of accuracy.

■ **EXAMPLE 2**   Consider the sine function.

a. Use the Taylor inequality to estimate the maximum error when $T_7(x; 0)$ is used to approximate $\sin x$ on

$$-0.5 \leq x \leq 0.5.$$

Compare this estimate with the graph of

$$y = |\sin x - T_7(x; 0)|$$

on $-0.5 \leq x \leq 0.5$.

b. Suppose we wish to use $T_n(x; 0)$ to approximate $\sin x$ on

$$-\pi \leq x \leq \pi$$

with an error of less than 10^{-6}. How large should n be?

Solution

a. Since

$$\frac{d^8}{dx^8} \sin x = \sin x,$$

we have

$$M_8 = \text{maximum of } |\sin x| \text{ on } [-0.5, 0.5] = \sin 0.5 \leq 1.$$

We choose to use the number 1 instead of $\sin 0.5 \approx 0.479426$ in our error estimate because 1 is easier to work with. By Taylor's inequality, the error when $\sin x$ is approximated by $T_7(x; 0)$ on $-0.5 \leq x \leq 0.5$ is no more than

$$|R_7(x; 0)| \leq \frac{1}{8!} M_8 |x - 0|^8 \leq \frac{|x|^8}{8!}.$$

Since $|x| \leq 0.5$ on the interval in question, we have

$$|R_7(x; 0)| \leq \frac{0.5^8}{8!} \approx 10^{-7}.$$

How does this estimate compare with the actual error? We first compute

$$T_7(x; 0) = x - \frac{1}{3!} x^3 + \frac{1}{5!} x^5 - \frac{1}{7!} x^7.$$

(You should verify this!) Next graph

$$y = |R_7(x; 0)| = |\sin x - T_7(x; 0)|$$

on the interval $[-0.5, 0.5]$. This graph is shown in Fig. 7.14. From this graph we see that the error is less than 6.0×10^{-9}. This is much less than the error estimate obtained using the Taylor inequality.

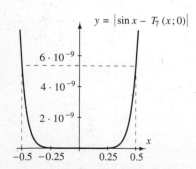

Figure 7.14. The graph shows that $|\sin x - T_7(x; 0)| \leq 6 \times 10^{-9}$ on the interval $-0.5 \leq x \leq 0.5$.

b. If we use $T_n(x; 0)$ to approximate $\sin x$ on $-\pi \le x \le \pi$, the error $R_n(x; 0)$ satisfies

$$(10) \qquad |R_n(x; 0)| \le \frac{1}{(n+1)!} M_{n+1} |x - 0|^{n+1} \le \frac{\pi^{n+1}}{(n+1)!} M_{n+1}.$$

Since

$$(11) \qquad \left| \frac{d^{n+1}}{dx^{n+1}} \sin x \right| = \begin{cases} |\sin x| & n \text{ even} \\ |\cos x| & n \text{ odd} \end{cases}$$

we have $M_{n+1} = 1$. Hence from (10) we have

$$(12) \qquad |R_n(x; 0)| \le \frac{\pi^{n+1}}{(n+1)!}.$$

n	$\dfrac{\pi^{n+1}}{(n+1)!}$
0	3.14
1	4.93
2	5.17
3	4.06
4	2.55
5	1.34
6	0.60
7	0.24
8	0.08
9	0.03
10	$7.37 \cdot 10^{-3}$
11	$1.93 \cdot 10^{-3}$
12	$4.66 \cdot 10^{-4}$
13	$1.05 \cdot 10^{-4}$
14	$2.19 \cdot 10^{-5}$
15	$4.30 \cdot 10^{-6}$
16	$7.95 \cdot 10^{-7}$
17	$1.39 \cdot 10^{-7}$

With a calculator it is not hard to find a value of n for which this last expression is less than 10^{-6}. The values of the right side of (12) for several values of n are shown in the accompanying table. According to the table entries, we should take $n \ge 16$ to guarantee that $T_n(x; 0)$ approximates $\sin x$ on $-\pi \le x \le \pi$ with error of less than 10^{-6}. ∎

■ **EXAMPLE 3**   Consider the square root function $f(x) = \sqrt{x}$.

a. Use the Taylor inequality to estimate the maximum error when $T_6(x; 100)$ is used to approximate $\sqrt{x}$ on

$$90 \le x \le 110.$$

Compare this estimate with the graph of

$$y = |\sqrt{x} - T_6(x; 100)|$$

on this interval.

b. Find n so that $T_n(105, 100)$ approximates $\sqrt{105}$ with error of less than 5.0×10^{-9}.

Solution

a. The seventh derivative of $f(x) = \sqrt{x}$ is

$$f^{(7)}(x) = \frac{-11}{2} \cdot \frac{-9}{2} \cdot \frac{-7}{2} \cdot \frac{-5}{2} \cdot \frac{-3}{2} \cdot \frac{-1}{2} \cdot \frac{1}{2} x^{-13/2}$$

$$= \frac{10{,}395}{128 x^{13/2}}.$$

Hence

$$M_7 = \text{maximum of } |f^{(7)}(x)| \text{ on } [90, 110]$$

$$= \frac{10{,}395}{128(90)^{13/2}} \approx 1.611 \times 10^{-11} < 1.7 \times 10^{-11}.$$

By the Taylor inequality, the error when $\sqrt{x}$ is approximated by $T_6(x; 100)$ on $90 \le x \le 110$ is no more than

$$|R_6(x; 100)| \le \frac{1}{7!} M_7 |x - 100|^7 \le 3.4 \times 10^{-15} |x - 100|^7.$$

Because $|x - 100| \le 10$ on our interval, we have

$$|R_6(x; 100)| \le 3.4 \times 10^{-15} 10^7 = 3.4 \times 10^{-8}.$$

To see how this bound on the error compares with the actual error, we first calculate (by machine—please verify this yourself)

$$T_6(x; 100) = 10 + \frac{(x - 100)}{20} - \frac{(x - 100)^2}{8000} + \frac{(x - 100)^3}{1,600,000}$$
$$- \frac{(x - 100)^4}{256,000,000} + \frac{7(x - 100)^5}{256,000,000,000}$$
$$- \frac{21(x - 100)^6}{102,400,000,000,000}.$$

The graph of

$$y = |R_6(x; 100)| = |\sqrt{x} - T_6(x; 100)|$$

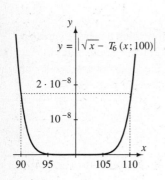

Figure 7.15. The graph shows that $|R_6(x; 100)| < 2 \times 10^{-8}$ for $90 \le x \le 110$.

is shown in Fig. 7.15. From the graph we see that the error is less than 2×10^{-8} on the interval $90 \le x \le 110$. Thus the actual error is a little less than the error estimate that we obtained using the Taylor inequality.

b. By looking for a pattern in the first few derivatives of $f(x) = \sqrt{x}$, it is not hard to guess a formula for the $(n + 1)$th derivative:

$$f'(x) = \frac{1}{2} x^{-1/2}$$

$$f''(x) = -\frac{1}{2^2} x^{-3/2}$$

$$f'''(x) = \frac{(3)(1)}{2^3} x^{-5/2}$$

(13)
$$f^{(4)}(x) = -\frac{(5)(3)(1)}{2^4} x^{-7/2}$$

$$f^{(5)}(x) = \frac{(7)(5)(3)(1)}{2^5} x^{-9/2}$$

$$\vdots$$

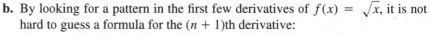

$$f^{(n+1)}(x) = (-1)^n \frac{(2n - 1)(2n - 3) \cdots (5)(3)(1)}{2^{n+1}} x^{(2n+1)/2}.$$

To approximate $\sqrt{105}$ using $T_n(105; 100)$, we need to estimate the error when $T_n(x; 100)$ is used to approximate $\sqrt{x}$ on $100 \le x \le 105$. By the Taylor inequality, we have

$$|R_n(x; 100)| \leq \frac{1}{(n + 1)!} M_{n+1} |x - 100|^{n+1} \leq \frac{5^{n+1}}{(n + 1)!} M_{n+1}.$$

By (13),

$$\left| \frac{d^{n+1}}{dx^{n+1}} \sqrt{x} \right| = \frac{(2n - 1)(2n - 3) \cdots (3)(1)}{2^{n+1}} \frac{1}{|x|^{(2n+1)/2}}$$

$$\leq \frac{(2n - 1)(2n - 3) \cdots (3)(1)}{2^{n+1}} \frac{1}{100^{(2n+1)/2}}$$

on the interval $100 \leq x \leq 105$. The last expression in this equation is M_{n+1}. Hence on $100 \leq x \leq 105$,

n	$\dfrac{5^{n+1}(2n - 1)(2n - 3) \cdots (5)(3)(1)}{(n + 1)!2^{n+1}10^{2n+1}}$
2	$7.81 \cdot 10^{-5}$
3	$2.44 \cdot 10^{-6}$
4	$8.54 \cdot 10^{-8}$
5	$3.20 \cdot 10^{-9}$
6	$1.26 \cdot 10^{-10}$
7	$5.11 \cdot 10^{-12}$
8	$2.13 \cdot 10^{-13}$

$$(14) \qquad |R_n(x; 100)| \leq \frac{5^{n+1}}{(n + 1)!} \cdot \frac{(2n - 1)(2n - 3) \cdots (3)(1)}{2^{n+1}10^{2n+1}}.$$

By experimenting with a calculator or computer, we can find a value of n for which this last expression is less than 5.0×10^{-11}. In the accompanying table we have calculated the value of the right side of (14) for many values of n. From the table we see that we should take $n \geq 7$ if we want to be assured that $T_n(105; 100)$ approximates $\sqrt{105}$ with error of less than 5.0×10^{-11}. ∎

■ **EXAMPLE 4** Let $p(x)$ be a polynomial of degree n and let c be a real number. Show that

$$T_n(x; c) = p(x)$$

for all real x.

Solution. We will have $T_n(x; c) = p(x)$ for all real x precisely when

$$R_n(x; c) = p(x) - T_n(x; c) = 0$$

for all x. Since p is a polynomial of degree n, we have

$$\frac{d^{n+1}}{dx^{n+1}} p(x) = 0$$

for all x. Hence

$$M_{n+1} = \text{maximum of } |p^{n+1}(x)| = 0.$$

By the Taylor inequality, we have

$$|R_n(x; c)| \leq \frac{1}{(n + 1)!} M_{n+1} |x - c|^{n+1} = 0.$$

It follows that $R_n(x; c) = 0$ for all x, so

$$T_n(x; c) = p(x).$$ ∎

Exercises 7.2

Basic

Exercises 1–6:   For the indicated function, find $T_2(x; 0)$. Then use the Taylor inequality to find a bound for $R_2(x; 0)$ on the given interval I. Graph $y = |R_2(x; 0)| = |f(x) - T_2(x; 0)|$ and use this graph to find the *actual* maximum error on I. How does the error compare to the bound found using the Taylor inequality?

1. e^{-2x} on $I = [-0.3, 0.3]$

2. $(3 + x)^{10}$ on $I = [-0.3, 0.3]$

3. $\sin x$ on $I = [-\pi/6, \pi/6]$

4. $\cos x$ on $I = [-\pi/6, \pi/6]$

5. $\dfrac{1}{1 + x}$ on $I = [-1/3, 1/3]$

6. $\arctan x$ on $I = [-0.5, 0.5]$

Exercises 7–12:   For the given c and function f, suppose we wish to use $T_5(x; c)$ to estimate f on $[c - r, c + r]$ with an error of at most 0.01. Use the Taylor inequality to find a value of r for which this is the case. Check your answer by graphing $y = |R_5(x; c)|$ on $[c - r, c + r]$.

7. $f(x) = \sin x$, $c = 0$

8. $f(x) = \dfrac{1}{x}$, $c = 1$

9. $f(x) = x \ln x$, $c = 3$

10. $f(x) = e^x$, $c = \ln 2$

11. $f(x) = 2x^4 - 3x^3 + 6x^2 - 2$, $c = \sqrt{2}$

12. $f(x) = \sqrt{x}$, $c = 1$

13. Verify the expression for $T_6(x; 100)$ given in Example 3.

14. Verify the formula given in (13) for the $(n + 1)$th derivative of the square root function.

15. Let $T_n(x; 100)$ be the nth Taylor polynomial about 100 for the square root function $f(x) = \sqrt{x}$. In Example 3 we showed that $T_7(105; 100)$ is within 5.0×10^{-11} of $\sqrt{105}$. Check this by calculating $T_7(105; 100)$ and $\sqrt{105}$ and then computing the difference between the two results.

16. A function f has nth Taylor polynomial about 1 given by

$$T_n(x; 1) = \sum_{k=0}^{n} \frac{(x - 1)^k}{k! 4^k}.$$

What is the value of $f^{(8)}(1)$?

17. A function g has nth Taylor polynomial about -2 given by

$$T_n(x; -2) = \sum_{k=0}^{n} 3 \frac{(x + 2)^k}{k + 1}.$$

What is the value of $g^{(8)}(-2)$?

18. A function f has nth Taylor polynomial about c given by

$$T_n(x; c) = \sum_{k=0}^{n} (-1)^k \frac{(x - c)^k}{30}.$$

What is $f^{(k)}(c)$?

19. A function g has nth Taylor polynomial about c given by

$$T_n(x; c) = \sum_{k=0}^{n} k^k \frac{(x - c)^k}{(3(k + 1))^{k+1}}.$$

What is $g^{(k)}(c)$?

Growth

20. Let $f(x) = \sin 2x$. Find n so that if $T_n(x; 0)$ is used to approximate f on $[-0.5, 0.5]$, then the error is no more than 0.01.

21. Let $H(x) = \frac{1}{x}$. Find n so that if $T_n(x; 1)$ is used to approximate H on $[2/3, 4/3]$, then the error is no more than 0.01.

22. Let $p(x) = x^4$. Find n so that if $T_n(x; 0)$ is used to approximate p on $[-0.1, 0.1]$, then the error is no more than 0.001.

23. Let $g(t) = \sqrt{t}$. Find n so that if $T_n(t; 4)$ is used to approximate g on $[3, 5]$, then the error is no more than 0.005.

24. Let $r(\theta) = e^{-\theta}$. Find n so that if $T_n(\theta; 0)$ is used to approximate r on $[-3, 3]$, then the error is no more than 0.001.

25. Let $f(x) = \ln(2 - x)$. Find n so that if $T_n(x; 0)$ is used to approximate f on $[-0.5, 0.5]$, then the error is no more than 0.0001.

26. For how many different functions f is $T_2(x; 0) = 1 - x + 2x^2$? Explain your answer.

27. For how many different functions g is

$$T_3(x; -1) = (x + 1) - \frac{(x + 1)}{\pi} + \frac{(x + 1)^2}{e}?$$

Explain your answer.

28. Suppose that we have a Taylor polynomial $T_n(x; c)$ for a function f. Let m be a nonnegative integer with $m < n$. How could we quickly get $T_m(x; c)$ from $T_n(x; c)$?

29. Let $T_9(x; 0)$ be the ninth-degree Taylor polynomial for $\sin x$ about 0 and let $Q_8(x; 0)$ be the eighth-degree Taylor polynomial for $\cos x$ about 0. Show that

$$\frac{d}{dx}T_9(x; 0) = Q_8(x; 0).$$

30. Let $T_6(x; 1)$ be the sixth-degree Taylor polynomial for $\ln x$ about 1 and let $Q_5(x; 1)$ be the fifth-degree Taylor polynomial for $1/x$ about 1. Show that

$$\frac{d}{dx}T_6(x; 1) = Q_5(x; 1).$$

31. Let $T_n(x; c)$ be an nth-degree Taylor polynomial for a function f and let $Q_{n-1}(x; c)$ be an $(n - 1)$th-degree Taylor polynomial for $f'(x)$. Show that

$$\frac{d}{dx}T_n(x; c) = Q_{n-1}(x; c).$$

32. Let $T_6(x; 2)$ be the sixth-degree Taylor polynomial for $1/x$ about 2 and let $Q_5(x; 2)$ be the fifth-degree Taylor polynomial for $-1/x^2$ about 2. Show that

$$T_6(x; 2) = \frac{1}{2} + \int_2^x Q_5(t; 2)\, dt.$$

33. Let $T_9(x; 0)$ be the ninth-degree Taylor polynomial for $\sin x$ about 0 and let $Q_{10}(x; 0)$ be the tenth-degree Taylor polynomial for $\cos x$ about 0. Show that

$$Q_{10}(x; 0) = \cos 0 - \int_0^x T_9(t, 0)\, dt.$$

34. Based on the previous two problems, write an integral formula that relates the Taylor polynomial of a function to the Taylor polynomials of an antideriva-

tive of the function. Give an argument in support of your formula.

35. Let $f(x) = x^{14/3}$ and let $T_n(x; 0)$ be the Taylor polynomial of degree n for f at 0. What is the largest n for which T_n exists?

36. Let $g(x) = |x - 1|^3$ and let $T_n(x; 1)$ be the Taylor polynomial of degree n for g at 1. What is the largest n for which T_n exists?

37. Let $f(x) = (1 - x^2)^{5/2}$. For what values of c does $T_2(x; c)$ exist? For what values of c does $T_5(x; c)$ exist?

38. Verify algebraically that

$$x^3 = ((x - 2) + 2)^3$$
$$= 8 + 12(x - 2) + 6(x - 2)^2 + (x - 2)^3.$$

Show that the expression on the right is $T_3(x; 2)$ for the function $f(x) = x^3$.

39. Verify algebraically that

$$2x^2 - 7x + 1$$
$$= 2((x + 3) - 3)^2 - 7((x + 3) - 3) + 1$$
$$= 40 - 19(x + 3) + 2(x + 3)^2.$$

Show that the expression on the right is $T_2(x; -3)$ for the function $f(x) = 2x^2 - 7x + 1$.

40. Let $p(x)$ be a polynomial of degree n. Based on the previous two exercises, suggest a method for finding $T_n(x; c)$ by expanding the expression $p((x - c) + c)$. How could you get Taylor polynomials $T_m(x; c)$ for $m < n$ by this method?

41. Describe how we can infer the version of the Mean-value Inequality given in this section from the version given in Section 4.4.

42. In Investigation 1 we derived the Taylor inequality for $|R_3(x; c)|$ in the case $x \geq c$. Redo the argument to take care of the case $x < c$.

43. For nonnegative integer n let $T_n(x; 0)$ be the nth-degree Taylor polynomial for the sine function at 0. Is there a value of n such that $T_n(x; 0)$ approximates $\sin x$ to within 0.01 for *all* real numbers x? Justify your answer.

44. For nonnegative integer n let $T_n(x; 1)$ be the nth-degree Taylor polynomial for $\ln x$ at 1. Is there a value of n such that $T_n(x; 1)$ approximates $\ln x$ to within 0.01 for *all* real numbers x in the interval $(0, 2)$? Justify your answer.

45. Show that for positive integer n,

$$(2n - 1)(2n - 3)\cdots(5)(3)(1) = \frac{(2n)!}{2^n n!}.$$

Use this to simplify the bound for $|R_n(x; 100)|$ shown in (14).

Review/Preview _____

Exercises 45–50:   For the given f find $f'(x)$ and $\displaystyle\int_0^x f(t)\,dt$.

46. $f(x) = \displaystyle\sum_{k=1}^{6} \frac{x^k}{k!}$

47. $f(x) = \dfrac{1}{1+x^2}$

48. $f(x) = xe^{x^2}$

49. $f(x) = \sin^2 2x$

50. $f(x) = \displaystyle\sum_{k=0}^{1001} \frac{(x-3)^{2k}}{k^2+1}$

51. $f(x) = \dfrac{x}{x^2 - 4x - 5}$

7.3   SEQUENCES

Set your calculator in radian mode, enter the number 1.86, and repeatedly press the cosine button. What happens? Early in the process the number displayed by the calculator changes each time you hit the cosine button. However, if you continue this process long enough, the number shown in the display stops changing.

For a positive integer n, let x_n be the number that appears on the screen after you press the cosine button for the nth time. If your calculator is set to display six decimal places, you will see

(1)
$$
\begin{aligned}
x_1 &= \cos 1.86 = -0.285189 \\
x_2 &= \cos x_1 = 0.959608 \\
x_3 &= \cos x_2 = 0.573841 \\
x_4 &= \cos x_3 = 0.839822 \\
&\vdots \qquad\qquad \vdots \\
x_{29} &= \cos x_{28} = 0.739080 \\
x_{30} &= \cos x_{29} = 0.739089 \\
&\vdots \qquad\qquad \vdots
\end{aligned}
$$

Although you probably don't want to push the cosine button more than two or three hundred times, the process we have described here actually defines an infinite list of numbers. An infinite list such as the one described here is called a *sequence*. To describe this sequence we use the notation

$$\{x_n\}_{n=1}^{\infty}$$

and provide a description of how x_n can be computed for any positive integer n. As suggested by (1), we can describe this sequence by giving the first number, x_1, and then describing how each subsequent term is produced:

$$x_1 = \cos 1.86 = -0.285189$$

$$x_n = \cos x_{n-1} \quad (n \geq 2)$$

One of the interesting things about this sequence is that it seems to settle toward a fixed number (limit) as n gets large. The value of this fixed number is approximately 0.739085, as you can see by the time you've pressed the cosine button 40 times. We express this fact by writing

$$\lim_{n \to \infty} x_n \approx 0.739085.$$

Many important processes in mathematics can be described using sequences. For example, when we use Newton's method to approximate a solution to an equation, we are actually computing the first few terms of an infinite list (or sequence) that has a limit (we hope!) equal to a solution of the equation. A sequence need not be a list of numbers. The Taylor polynomials for a function f give a sequence

$$\{T_n(x; c)\}_{n=0}^{\infty}.$$

We have seen that when n is large, the Taylor polynomial $T_n(x; c)$ might be a good approximation to f. When this happens we will usually find that the sequence of Taylor polynomials has limit f:

$$\lim_{n \to \infty} T_n(x; c) = f(x)$$

By studying sequences and limits we will better understand Taylor polynomials and their role in approximation of functions.

Sequence Notation

In the preceding discussion we stated that a sequence is an infinite list.

$$x_1, x_2, x_3, x_4, \ldots, x_n, \ldots$$

Each member of the sequence has a definite "position" in the list. The term x_1 is in position 1, term x_2 is in position 2, and so on. Given a position n in the list, there is exactly one term x_n in position n. Thus the term x_n is a *function* of its position n. We use this idea to formally define a sequence.

DEFINITION   sequence

A sequence is a function s with domain equal to the set $\{1, 2, 3, \ldots, n, \ldots\}$ of natural numbers.

If s is a sequence and n is a natural number, we can think of n as referring to a position in a list. The value of the function s at n (i.e., $s(n)$) is simply the nth term in the list. To make notation a bit easier we will usually write s_n for $s(n)$.

Exhibiting a Sequence

Sequences can be described in a number of ways. Sometimes we simply give a formula for the terms of a sequence. For example, the formula

$$s_n = n^2$$

describes the sequence s whose nth term is $s(n) = s_n = n^2$. This sequence might also be denoted by

$$\{n^2\}_{n=1}^{\infty}.$$

The subscript 1 and the superscript ∞ indicate that the domain of the function s described here is the set $\{1, 2, 3, \ldots\}$ of natural numbers.

Sometimes a sequence is described by listing the first few terms of the sequence. When this is done, it is assumed that the terms listed suggest an obvious pattern and that the reader can continue the list and perhaps guess the function $s(n)$ that gives the nth term of the list. For example, it is obvious that the list

$$\frac{1}{2}, \frac{1}{3}, \frac{1}{4}, \cdots$$

is to be continued

$$\cdots, \frac{1}{5}, \frac{1}{6}, \frac{1}{7}, \cdots.$$

The term in position n is

$$s_n = s(n) = \frac{1}{n+1}.$$

Sometimes the value of a term in the sequence may depend on one or more previous terms of the sequence. In these cases it might be very hard to write down a formula for the nth term. However, if we are told how to generate the nth term from the previous terms, we can find the value of any term of the sequence. As an example, consider the sequence s described by

$$s_1 = 1$$

(2)

$$s_n = \frac{1}{2}\left(s_{n-1} + \frac{2}{s_{n-1}}\right) \quad (n \geq 2)$$

We can use (2) to find s_n for any n. For example,

$$s_2 = \frac{1}{2}\left(s_1 + \frac{2}{s_1}\right) = \frac{1}{2}\left(1 + \frac{2}{1}\right) = \frac{3}{2} = 1.5$$

$$s_3 = \frac{1}{2}\left(s_2 + \frac{2}{s_2}\right) = \frac{1}{2}\left((3/2) + \frac{2}{(3/2)}\right) = \frac{17}{12} \approx 1.416667$$

$$s_4 = \frac{1}{2}\left(s_3 + \frac{2}{s_3}\right) = \frac{1}{2}\left((17/12) + \frac{2}{(17/12)}\right) = 577/408 \approx 1.41422$$

$\vdots$

It would be very tedious to find many terms of this sequence by hand. However, with a programmable calculator or CAS it is easy to find several terms of the sequence. What do you think happens to this sequence as n gets large? A sequence for which the value of a term s_n depends on the value of previous terms is called a **recursive sequence.**

Sometimes it is easier to describe a sequence $s(n)$ by letting n vary over a set different from $\{1, 2, 3, \ldots\}$. For example, it seems natural to describe the sequence

$$\frac{1}{4}, \frac{1}{5}, \frac{1}{6}, \ldots$$

by

$$s(n) = \frac{1}{n} \quad \text{for } n = 4, 5, 6, \ldots$$

or by $\{1/n\}_{n=4}^{\infty}$. In practice we will try to give the simplest formula $s(n)$ to describe a sequence and let the variable n come from an appropriate set of integers. Sometimes we will simply write $\{s_n\}$ to denote a sequence when we are not concerned with what the allowable values for n are.

■ **EXAMPLE 1**   Write out the first five terms of the sequence described by each of the following expressions.

a. $\left\{ \dfrac{(-1)^n}{n^2 + 1} \right\}_{n=1}^{\infty}$ **b.** $s_n = \displaystyle\sum_{k=1}^{n} \frac{1}{k}$ **c.** $\begin{aligned} F_1(x) &= x \\ F_{n+1}(x) &= x^{n+1} F_n(x) \quad (n \geq 2) \end{aligned}$

Solution

a. The nth term of the sequence is $s_n = \dfrac{(-1)^n}{n^2 + 1}$. To find the first five terms of the sequence defined by this expression, evaluate it for $n = 1, 2, 3, 4, 5$.

$$s_1 = \frac{(-1)^1}{1^2 + 1} = -\frac{1}{2}$$

$$s_2 = \frac{(-1)^2}{2^2 + 1} = \frac{1}{5}$$

$$s_3 = \frac{(-1)^3}{3^2 + 1} = -\frac{1}{10}$$

$$s_4 = \frac{(-1)^4}{4^2 + 1} = \frac{1}{17}$$

$$s_5 = \frac{(-1)^5}{5^2 + 1} = -\frac{1}{26}$$

b. As in part **a,** evaluate s_n for $n = 1, 2, 3, 4, 5$.

$$s_1 = \sum_{k=1}^{1} \frac{1}{k} = 1$$

$$s_2 = \sum_{k=1}^{2} \frac{1}{k} = 1 + \frac{1}{2} = \frac{3}{2}$$

$$s_3 = \sum_{k=1}^{3} \frac{1}{k} = 1 + \frac{1}{2} + \frac{1}{3} = \frac{11}{6}$$

$$s_4 = \sum_{k=1}^{4} \frac{1}{k} = 1 + \frac{1}{2} + \frac{1}{3} + \frac{1}{4} = \frac{25}{12}$$

$$s_5 = \sum_{k=1}^{5} \frac{1}{k} = 1 + \frac{1}{2} + \frac{1}{3} + \frac{1}{4} + \frac{1}{5} = \frac{137}{60}$$

c. Taking $n = 1, 2, 3, 4, 5$ and using the recursion relation given, we find

$$F_1(x) = x$$
$$F_2(x) = x^2 F_1(x) = x^2 x = x^3$$
$$F_3(x) = x^3 F_2(x) = x^3 x^3 = x^6$$
$$F_4(x) = x^4 F_3(x) = x^4 x^6 = x^{10}$$
$$F_5(x) = x^5 F_4(x) = x^5 x^{10} = x^{15}.$$

The Limit of a Sequence

Let $\{T_n(x; c)\}_{n=0}^{\infty}$ be the sequence of Taylor polynomials associated with a function $f(x)$. We have seen that in some cases, $T_n(x; c)$ is a good approximation to $f(x)$ when n is large, and that the approximation gets better as n gets larger. To better study the behavior of such a sequence for large values of n, we study the limit of a sequence.

DEFINITION **limit of a sequence of real numbers**

Let $\{s_n\}_{n=1}^{\infty}$ be the sequence of real numbers and let L be a real number. We say the sequence has **limit** L if for each $\epsilon > 0$ there is a number $N > 0$ such that

$$|s_n - L| < \epsilon \quad \text{whenever} \quad n > N.$$

When this is the case we write

$$\lim_{n \to \infty} s_n = L.$$

If there is no finite real number L for which the above conditions are satisfied, we say the sequence diverges or that $\lim_{n \to \infty} s_n$ does not exist.

Very roughly, the definition says that if the sequence $\{s_n\}$ has limit L, then when n gets large, s_n gets close to L and stays close to L.

To illustrate the definition graphically, plot the ordered pairs (n, s_n) in the plane and draw the horizontal lines

$$y = L + \epsilon \quad \text{and} \quad y = L - \epsilon.$$

If

$$\lim_{n \to \infty} s_n = L,$$

there must be a positive integer N such that the points

$$(N + 1, s_{N+1}), (N + 2, s_{N+2}), (N + 3, s_{N+3}), \ldots$$

all lie between the two lines. (See Fig. 7.16.) This means that for $n > N$, we have

$$|s_n - L| < \epsilon.$$

Figure 7.16 illustrates another way of picturing a convergent sequence: If $\{s_n\}$ converges to L, then the graph of the points (n, s_n), $n = 1, 2, 3, \ldots$, has horizontal asymptote $y = L$.

The idea of the limit of a sequence is very important in approximations. Suppose that

$$\lim_{n \to \infty} s_n = L$$

and we wish to approximate L with s_n for some n. We know from the above definition that the error $|s_n - L|$ can be made less than a small number ϵ by taking n large enough.

Many times the limit of a sequence can be found by using the same techniques that we used to determine horizontal asymptotes in Chapter 4.

■ **EXAMPLE 2** Let $s_n = \dfrac{3n - 4}{2n + 1}$. Find the limit of the sequence $\{s_n\}_{n=1}^{\infty}$.

Solution. For positive integer n,

$$s_n = \frac{3n - 4}{2n + 1} \cdot \frac{\dfrac{1}{n}}{\dfrac{1}{n}} = \frac{3 - \dfrac{4}{n}}{2 + \dfrac{1}{n}}.$$

Thus

$$\lim_{n \to \infty} s_n = \lim_{n \to \infty} \frac{3 - \dfrac{4}{n}}{2 + \dfrac{1}{n}} = \frac{3}{2}.$$

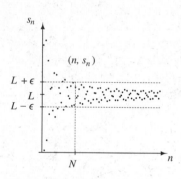

Figure 7.16. When $n > N$, s_n is within ϵ of L. In addition, the line $y = L$ is asymptotic to the graph of the points (n, s_n).

■ **EXAMPLE 3** Let $b_n = 1 + (-1)^n$. Investigate $\lim_{n\to\infty} b_n$.

Solution. A list of the first few elements of the sequence,

$$0, 2, 0, 2, 0, 2, \ldots,$$

shows that the terms of the sequence do not get close to just one number. Hence $\lim_{n\to\infty} b_n$ does not exist. ■

■ **EXAMPLE 4** Let $s_n = \sqrt[n]{n}$. Find the limit of the sequence $\{s_n\}_{n=1}^{\infty}$.

Solution. We use several techniques to investigate the behavior of the sequence. With a calculator or computer we can find the value of s_n for several large values of n. (See the accompanying table.) The data suggest that

$$(3) \qquad \lim_{n\to\infty} \sqrt[n]{n} = 1.$$

We can also use a calculator or CAS to graph the ordered pairs (n, s_n) for many values of n. If the sequence $\{s_n\}_{n=1}^{\infty}$ has limit L, then the points plotted should get close to the line $y = L$. As seen in Fig. 7.17, the points $(n, \sqrt[n]{n})$ quickly get close to the line $y = 1$. This suggests that (3) is correct.

We can also find the limit by analytic means. Note that

$$\ln s_n = \ln\left(n^{1/n}\right) = \frac{\ln n}{n}.$$

In Section 4.4 we saw that

$$\lim_{n\to\infty} \frac{\ln n}{n} = 0.$$

Thus, when n is large, $\ln s_n \approx 0$, so

$$s_n \approx e^0 = 1.$$

This shows that (3) is correct. ■

■ **EXAMPLE 5** Let the sequence $\{h_n\}_{n=1}^{\infty}$ be defined by

$$h_n = \sum_{k=1}^{n} \frac{1}{k} = 1 + \frac{1}{2} + \frac{1}{3} + \cdots + \frac{1}{n}.$$

Investigate $\lim_{n\to\infty} h_n$.

Solution. First use a CAS to find the approximate value of h_n for several large values of n. The data are inconclusive. By the time n reaches 10^6, the sequence still does not appear to be settling down to one number. (See the accompanying table. If a CAS is not available, program your calculator to form the sums for $n = 100$ and $n = 1000$. Otherwise, accept the values listed in the table as correct and go on.)

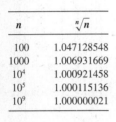

n	$\sqrt[n]{n}$
100	1.047128548
1000	1.006931669
10^4	1.000921458
10^5	1.000115136
10^9	1.000000021

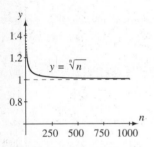

Figure 7.17. When n is large, $n^{1/n}$ is near 1.

n	h_n
100	5.1873775
1000	7.4854709
10^4	9.787606
10^5	12.0901461
10^6	14.3927267

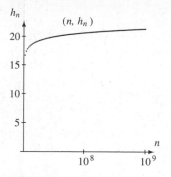

Figure 7.18. The graph of (n, h_n) for $1 \le n \le 10^9$. The graph does not appear to level out, so no limit is suggested.

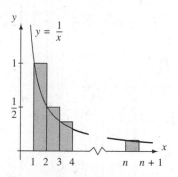

Figure 7.19. The area enclosed by the rectangles is larger than the area under the graph for $1 \le x \le n + 1$.

In Fig. 7.18 we see the graph of the ordered pairs (n, h_n) for $n = 1, 2, 3, \ldots,$ 10^9. The graph does not appear to be leveling out like the graph in the previous example, so no candidate for the limit is suggested by this graph.

To get a better idea about the behavior of h_n, we relate it to something familiar. In the plane construct a rectangle with base the interval $[1, 2]$ on the x-axis and height 1. Next to it put a rectangle with base $[2, 3]$ and height $\frac{1}{2}$, and then a rectangle with base $[3, 4]$ and height $\frac{1}{3}$. Continue this process until you have n rectangles. The last has base of $[n, n + 1]$ and height $1/n$. See Fig. 7.19. The total area enclosed by these rectangles is

$$1 \cdot 1 + 1 \cdot \frac{1}{2} + 1 \cdot \frac{1}{3} + \cdots 1 \cdot \frac{1}{n} = h_n.$$

The graph of $y = 1/x$ passes through the upper left corner of each of these rectangles and lies below the top of each rectangle. Hence the total area enclosed by the n rectangles is greater than the area under the graph of $y = 1/x$ for $1 \le x \le n+1$. Hence we have

$$(4) \qquad h_n \ge \int_1^{n+1} \frac{1}{x}\, dx = \ln(n + 1).$$

From (4) we see that $\{h_n\}$ does not converge to a finite number. In fact, since $\lim_{n\to\infty} \ln(n + 1) = +\infty$, we see that h_n can be made as large as we like by taking n sufficiently large. In this case we say that the sequence $\{h_n\}_{n=1}^{\infty}$ diverges to $+\infty$ and write

$$\lim_{n\to\infty} h_n = +\infty. \qquad \blacksquare$$

The behavior of the sequence in the last example provides an illustration of the following definition.

DEFINITION   infinite limits

Let $\{x_n\}$ be a sequence of real numbers. The sequence diverges to $+\infty$ if for each $M > 0$, there is an $N > 0$ such that $x_n > M$ whenever $n > N$. In this case we write

$$\lim_{n\to\infty} x_n = +\infty.$$

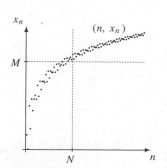

Figure 7.20. When $n > N$, x_n is larger than M.

This definition says that if $\lim_{n\to\infty} s_n = +\infty$, then when n gets large enough, the numbers s_n are also large. To illustrate this graphically, draw the line $y = M$ and plot the points (n, s_n). If $\lim_{n\to\infty} s_n = +\infty$, then there is a positive integer N so that the points

$$(N + 1, s_{N+1}), (N + 2, s_{N+2}), (N + 3, s_{N+3}), \ldots$$

all lie above the line. See Fig. 7.20.

Monotone and Bounded Sequences

Although it is sometimes difficult to decide whether or not a given sequence has a limit, there are some sequences for which this is somewhat routine.

DEFINITION   **monotone sequence, bounded sequence**

Let $\{x_n\}$ be a sequence of real numbers.

- The sequence is **monotone increasing** if

$$x_1 \le x_2 \le x_3 \le \cdots \le x_n \le x_{n+1} \le \ldots.$$

  The sequence is **monotone decreasing** if

$$x_1 \ge x_2 \ge x_3 \ge \cdots \ge x_n \ge x_{n+1} \ge \ldots.$$

- The sequence is **bounded above** if there is a number A such that

$$x_n \le A \quad (n = 1, 2, 3, \ldots).$$

  The sequence is **bounded below** if there is a number B such that

$$B \le x_n \quad (n = 1, 2, 3 \ldots).$$

If a sequence is bounded above and bounded below, we say that the sequence is **bounded**. If a sequence is not bounded, we say that it is **unbounded**.

■ EXAMPLE 6

a. For positive integer n let $h_n = \displaystyle\sum_{k=1}^{n} \frac{1}{k}$. Show that the sequence $\{h_n\}_{n=1}^{\infty}$ is monotone increasing and unbounded.

b. For positive integer n let $x_n = \dfrac{(-1)^n}{n^2}$. Show that the sequence $\{x_n\}_{n=1}^{\infty}$ is bounded but not monotone.

Solution

a. For positive integer n,

$$(5) \qquad h_n = \sum_{k=1}^{n} \frac{1}{k} < \sum_{k=1}^{n} \frac{1}{k} + \frac{1}{n+1} = \sum_{k=1}^{n+1} \frac{1}{k} = h_{n+1}.$$

In (5) substitute $n = 1$, then $n = 2$, then $n = 3$, and so on. From this we have

$$h_1 < h_2 < h_3 < \cdots < h_n < h_{n+1} < \cdots.$$

Therefore, $\{h_n\}_{n=1}^{\infty}$ is monotone increasing. In Example 5 we showed that for any positive integer n,

$$(6) \qquad\qquad h_n \geq \ln(n + 1).$$

From this we conclude that there is no $M > 0$ such that $h_n \leq M$ for all n. This means that the sequence $\{h_n\}$ is unbounded.

b. The first three terms of the sequence $\{x_n\}_{n=1}^{\infty}$ are

$$x_1 = -1, \qquad x_2 = \frac{1}{2}, \qquad x_3 = -\frac{1}{3}.$$

We see that $x_1 < x_2$, but $x_2 > x_3$. Hence the sequence is neither monotone increasing nor monotone decreasing.

The sequence $\{x_n\}_{n=1}^{\infty}$ is bounded because for any n, we have

$$-1 \leq x_n \leq 1.$$

Hence -1 is a *lower bound* for the sequence, and 1 is an *upper bound* for the sequence. ∎

When a sequence is both monotone and bounded, it has a limit. We state this result for the case in which the sequence is monotone increasing. See Exercises 42, 43, and 44 for the case in which the sequence is monotone decreasing.

The Monotone Sequence Theorem

Let $\{x_n\}$ be a monotone increasing sequence of real numbers.

a. If $\{x_n\}$ is bounded above, then $\lim_{n \to \infty} x_n$ exists.

b. If $\{x_n\}$ is not bounded above, then

$$\lim_{n \to \infty} x_n = +\infty.$$

We can get a feel for why the Monotone Sequence Theorem is true by looking at a couple of pictures and giving a few words of explanation. First assume that $\{x_n\}_{n=1}^{\infty}$ is bounded above. Let L be the smallest number such that

$$(7) \qquad\qquad x_n \leq L$$

holds for all positive integers n. Then

$$\lim_{n \to \infty} x_n = L.$$

To see this, suppose ϵ is a small positive number. Then $L - \epsilon < L$. There must be a positive integer N with

$$(8) \qquad\qquad x_N > L - \epsilon.$$

(If this were not the case, then we would have $x_n \leq L - \epsilon$ for all n, contradicting the definition of L.) From (7), (8), and the fact that $\{x_n\}_{n=1}^{\infty}$ is monotone increasing, we have

$$L - \epsilon < x_N \leq x_{N+1} \leq x_{N+2} \leq \cdots \leq L.$$

Hence for all $n > N$, we have $|x_n - L| < \epsilon$. This means that

$$\lim_{n \to \infty} x_n = L.$$

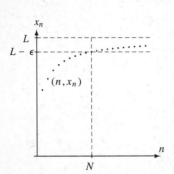

Figure 7.21. For $n > N$ we have $L - \epsilon < x_n < L$.

The argument is illustrated in Fig. 7.21.

Now assume that $\{x_n\}_{n=1}^{\infty}$ is not bounded above. If M is any positive number, then there must be a positive integer N with

$$(9) \qquad\qquad\qquad x_N > M.$$

(Why?) Combining (9) with the fact that $\{x_n\}_{n=1}^{\infty}$ is monotone increasing, we find

$$M < x_N \leq x_{N+1} \leq x_{N+2} \leq \cdots.$$

Hence for all $n > N$, we have $x_n > M$. Referring back to the definition of infinite limits, we see that

$$\lim_{n \to \infty} x_n = +\infty.$$

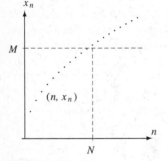

Figure 7.22. For $n > N$ we have $x_n > M$.

See Fig. 7.22 for an illustration of the argument.

The Monotone Sequence Theorem can tell us when a monotone increasing sequence has a limit. However, it does not tell us the value of the limit. Nonetheless, knowing that a sequence has a limit is important information. Once we know that a sequence has a limit, we can attempt to approximate the limit using the terms of the sequence.

■ **EXAMPLE 7**   For positive integer n let

$$p_n = 1 + \frac{1}{2^2} + \frac{1}{3^2} + \cdots + \frac{1}{n^2} = \sum_{k=1}^{n} \frac{1}{k^2}.$$

Show that $\lim_{n \to \infty} p_n$ exists, and approximate the limit to within 0.001.

Solution.   First note that the sequence $\{p_n\}_{n=1}^{\infty}$ is monotone increasing. To show that the sequence is also bounded above we use a technique similar to that introduced in Example 3. For positive integer k consider the rectangle R_k with base the interval $[k-1, k]$ and height $\frac{1}{k^2}$. The total area enclosed by $R_1, R_2, \ldots, R_n$ is

$$1 \cdot 1 + 1 \cdot \frac{1}{2^2} + 1 \cdot \frac{1}{3^2} + \cdots + 1 \cdot \frac{1}{n^2} = p_n.$$

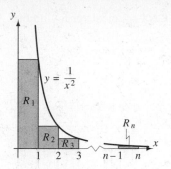

Figure 7.23. Comparing p_n to the area under $y = 1/x^2$.

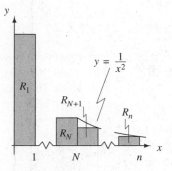

Figure 7.24. $L - p_N$ is less than the area under $y = 1/x^2$ for $N \leq x \leq \infty$.

Now draw the graph of $y = \dfrac{1}{x^2}$. As seen in Fig. 7.23, the graph passes through the upper right corner of each R_k, and each R_k lies under the graph. As suggested by Fig. 7.23, the area under the graph for $1 \leq x \leq n$ is greater than the total area enclosed by $R_2, R_3, \ldots, R_n$. Hence

$$p_n = 1 + \text{area}(R_2) + \text{area}(R_3) + \cdots + \text{area}(R_n)$$

$$\leq 1 + \int_1^n \frac{1}{x^2}\,dx = 1 + \left(1 - \frac{1}{n}\right) < 2.$$

This shows that $\{p_n\}_{n=1}^{\infty}$ is bounded above by 2. Since the sequence $\{p_n\}_{n=1}^{\infty}$ is monotone increasing and bounded above, we know that it has a limit.

Now that we know $\lim_{n\to\infty} p_n$ exists, we can estimate it. To do so we do more work with the rectangles R_k. Let n and N be positive integers with $n > N$. Then

$$p_n = p_N + \frac{1}{(N+1)^2} + \frac{1}{(N+2)^2} + \cdots + \frac{1}{n^2}$$

$$= p_N + \text{area}(R_{N+1}) + \text{area}(R_{N+2}) + \cdots + \text{area}(R_n)$$

$$\leq p_N + \left(\text{area under } y = \frac{1}{x^2} \text{ for } N \leq x \leq n\right)$$

$$= p_N + \int_N^n \frac{1}{x^2}\,dx$$

$$= p_N + \left(\frac{1}{N} - \frac{1}{n}\right)$$

$$< p_N + \frac{1}{N}.$$

It follows that if N is a given positive integer, then

$$p_n \leq p_N + \frac{1}{N}$$

holds for all $n = 1, 2, 3, \ldots$. Now let

$$\lim_{n\to\infty} p_n = L.$$

Because $\{p_n\}_{n=1}^{\infty}$ is monotone increasing and bounded above by $p_N + \dfrac{1}{N}$, we have

$$p_N \leq L \leq p_N + \frac{1}{N}$$

for every positive integer N. Hence L is within $\dfrac{1}{N}$ of p_N. Taking $N = 1000$, we have

$$L \approx p_{1000} = \sum_{k=1}^{1000} \frac{1}{k^2}$$

with an error of no more than 0.001. We can use a CAS or calculator to find that $p_{1000} \approx 1.6439$. Hence

$$\lim_{n \to \infty} p_n \approx 1.6439.$$

As it turns out, the limit of this bounded, monotone sequence is known exactly:

$$\lim_{n \to \infty} p_n = \frac{\pi^2}{6}$$

(For a proof of this fact see page 372 of the Mathematical Association of America publication *Selected Papers in Calculus,* edited by Tom Apostol et al.) ∎

■ **EXAMPLE 8**   Let $a_k = \left(1 + \dfrac{1}{k}\right)^k$. Find $\lim_{k \to \infty} a_k$.

Solution.   To get a feel for how the sequence behaves, calculate a_k for several values of k. The results of some of these calculations are shown in the accompanying table. In Fig. 7.25 we have plotted the points (k, a_k) for $1 \le k \le 1000$. The data in the table and the graph suggest that the sequence $\{a_k\}$ is increasing and bounded above (say, by 3). From the table it appears that

$$\lim_{k \to \infty} a_k = \lim_{k \to \infty} \left(1 + \frac{1}{k}\right)^k \approx 2.71828\ldots.$$

Hence it seems reasonable to guess that $\lim_{k \to \infty} \left(1 + \frac{1}{k}\right)^k = e$.

To verify this guess, we evaluate the limit by using the definition of derivative. Let L be the value of the limit,

(10)
$$L = \lim_{k \to \infty} \left(1 + \frac{1}{k}\right)^k.$$

Take the logarithm of both sides of (10) to get

(11)
$$\ln L = \lim_{k \to \infty} \ln \left(1 + \frac{1}{k}\right)^k = \lim_{k \to \infty} k \ln \left(1 + \frac{1}{k}\right).$$

Now make the substitution $k = 1/h$ in (11). Note that as $k \to +\infty$, we have $h \to 0^+$. Thus

(12)
$$\ln L = \lim_{h \to 0^+} \frac{\ln(1 + h)}{h}.$$

If $f(x) = \ln x$, then by the definition of derivative

$$f'(1) = \lim_{h \to 0} \frac{f(1 + h) - f(1)}{h} = \lim_{h \to 0} \frac{\ln(1 + h) - \ln 1}{h} = \lim_{h \to 0} \frac{\ln(1 + h)}{h}.$$

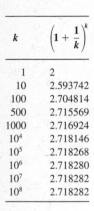

k	$\left(1 + \dfrac{1}{k}\right)^k$
1	2
10	2.593742
100	2.704814
500	2.715569
1000	2.716924
10^4	2.718146
10^5	2.718268
10^6	2.718280
10^7	2.718282
10^8	2.718282

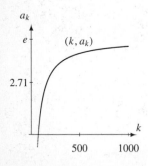

Figure 7.25. The sequence $\{(1 + 1/k)^k\}$ is increasing and bounded above.

Combining with (12), we have

$$\ln L = \lim_{h \to 0} \frac{\ln(1 + h)}{h} = \left. \frac{d}{dx} \ln x \right|_{x=1} = 1,$$

and it follows that $L = e$. Thus

$$\lim_{k \to \infty} \left(1 + \frac{1}{k}\right)^k = e,$$

verifying the guess suggested by the data. ∎

The Arithmetic of Sequences

Sometimes the terms of a sequence $\{c_k\}$ can be written as the sum, product, or quotient of terms of other sequences. If we know the behavior of these other sequences, we can often use the following results to say something about the behavior of $\{c_n\}$. Review some of the results on limits in Section 1.4. You will find the formulas below are very similar to the earlier limit theorems.

Some Limit Theorems

Let $\{a_k\}$ and $\{b_k\}$ be sequences of real numbers with

$$\lim_{k \to \infty} a_k = A \quad \text{and} \quad \lim_{k \to \infty} a_k = B.$$

a. Let r be a real number and let $c_k = ra_k$. Then

$$\lim_{k \to \infty} c_k = \lim_{k \to \infty} (ra_k) = r \lim_{k \to \infty} a_k = rA.$$

b. Let $c_k = a_k + b_k$. Then

$$\lim_{k \to \infty} c_k = \lim_{k \to \infty} (a_k + b_k) = \lim_{k \to \infty} a_k + \lim_{k \to \infty} b_k = A + B.$$

c. Let $c_k = a_k b_k$. Then

$$\lim_{k \to \infty} c_k = \lim_{k \to \infty} (a_k b_k) = \left(\lim_{k \to \infty} a_k\right)\left(\lim_{k \to \infty} b_k\right) = AB.$$

d. Assume that $b_k \neq 0$ for any k and that $B \neq 0$. If $c_k = \dfrac{a_k}{b_k}$, then

$$\lim_{k \to \infty} c_k = \lim_{k \to \infty} \frac{a_k}{b_k} = \frac{\lim_{k \to \infty} a_k}{\lim_{k \to \infty} b_k} = \frac{A}{B}.$$

e. Suppose that the function f is defined for all a_k and that f is continuous at A. If $c_k = f(a_k)$, then

$$\lim_{k \to \infty} c_k = \lim_{k \to \infty} f(a_k) = f\left(\lim_{k \to \infty} a_k\right) = f(A).$$

Sequences of Functions

When we deal with a sequence $\{f_n(x)\}_{n=1}^{\infty}$ of functions we will usually want to know about the convergence of the sequence for all values of x in some interval. When the sequence of functions is a sequence of Taylor polynomials, we will usually be dealing with uniform convergence. Before giving the definition, we need some convenient terminology.

> Let f be a function defined on an interval I and let ϵ be a positive real number. The **ϵ-corridor** about f is the region between the graphs of $y = f(x) + \epsilon$ and $y = f(x) - \epsilon$.

The ϵ-corridor about f is the shaded area in Fig. 7.26. Note that the graph of $y = f(x)$ lies in the ϵ-corridor about f.

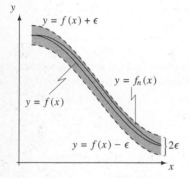

Figure 7.26. When n is large, the graph of $y = f_n(x)$ lies in the ϵ-corridor about f.

DEFINITION   **uniform convergence**

Let $\{f_n\}_{n=1}^{\infty}$ be a sequence of functions each defined on an interval I and let f also be a function defined on I. We say that the sequence $\{f_n\}$ converges uniformly to f on I if for each $\epsilon > 0$ there is a positive integer N so that for all $n > N$, the graph of $y = f_n(x)$ lies in the ϵ-corridor about f.

Actually, the idea introduced in this definition is one we worked with in Section 7.2. First observe that:

> The graph of $y = f_n(x)$ lies in the ϵ-corridor about f exactly when the graph of $y = |f(x) - f_n(x)|$ lies below the line $y = \epsilon$.

In Example 2 of Section 7.2 we showed that if $f(x) = \sin x$ and $T_n(x; 0)$ is the nth Taylor polynomial for f about 0, then

$$|\sin x - T_{16}(x; 0)| \leq 10^{-6} \quad \text{for} \; -\pi \leq x \leq \pi$$

Thus on $-\pi \leq x \leq \pi$, $T_{16}(x; 0)$ lies in the 10^{-6}-corridor about the sine function.

The idea of uniform convergence is very closely related to approximation. When a sequence $\{f_n\}$ converges to a function f uniformly on an interval I, the graph of $y = f_n(x)$ can be made close to the graph of $y = f(x)$ by taking n large enough. See Fig. 7.26. In addition, if the graph of $y = f_n(x)$ lies in the ϵ-corridor about f, then we have

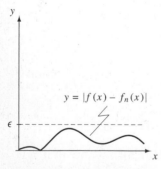

Figure 7.27. f_n is in the ϵ-corridor about f exactly when the graph of $y = |f(x) - f_n(x)|$ lies below the line $y = \epsilon$.

$$|f(x) - f_n(x)| < \epsilon$$

for each $x \in I$. This is illustrated in Fig. 7.27. Hence when ϵ is a small positive number, f_n will be a good approximation to f.

In many cases the sequence of Taylor polynomials $\{T_n(x; c)\}_{n=1}^{\infty}$ for a function f converges uniformly to f on an interval $[c - r, c + r]$ centered at c. This is why Taylor polynomials often provide a good approximation to a function.

■ **EXAMPLE 9**   Let $f(x) = \ln x$ and let $\{T_n(x; 1)\}_{n=1}^{\infty}$ be the sequence of Taylor polynomials for f about $x = 1$.

a. Show that the sequence of Taylor polynomials converges uniformly to $\ln x$ on $\left[\frac{2}{3}, \frac{4}{3}\right]$.

b. Show that the sequence of Taylor polynomials does not converge uniformly to $\ln x$ on $(0, 2)$.

Solution

a. We will show that the absolute error

$$\left| \ln x - T_n(x; 1) \right| = \left| R_n(x; 1) \right|$$

can be made as small as we like on $\left[\frac{2}{3}, \frac{4}{3}\right]$ by taking n large enough. In particular, if we make $\left| R_n(x; 1) \right| < \epsilon$, then $\left| \ln x - T_n(x; 1) \right| < \epsilon$ so the graph of $y = T_n(x; 1)$ will lie in the ϵ-corridor about f. This means that $\{T_n\}$ converges uniformly to f on the desired interval. With $f(x) = \ln x$ we have

$$f'(x) = x^{-1}$$
$$f''(x) = -x^{-2}$$
$$f'''(x) = 2 \cdot 1 x^{-3}$$
$$f^{(4)}(x) = -3 \cdot 2 \cdot 1 x^{-4}$$
$$\vdots$$

Once we see a developing pattern, we jump to a general expression:

$$(13) \qquad \begin{aligned} f^{(n)}(x) &= (-1)^{n-1}(n-1)! x^{-n} \\ f^{(n+1)}(x) &= (-1)^n n! x^{-n-1} \end{aligned}$$

The Taylor estimate for the error $R_n(x; 1)$ on $\left[\frac{2}{3}, \frac{4}{3}\right]$ is

$$(14) \qquad \left| R_n(x; 1) \right| \le \frac{M_{n+1}}{(n+1)!} |x - 1|^{n+1}.$$

where M_{n+1} is the maximum value of $\left| f^{(n+1)}(x) \right|$ on $\left[\frac{2}{3}, \frac{4}{3}\right]$. From (13) we have

$$(15) \qquad \left| f^{(n+1)}(x) \right| = \left| (-1)^n \frac{n!}{x^{n+1}} \right| = \frac{n!}{|x|^{n+1}}.$$

The maximum value for $\dfrac{n!}{|x|^{n+1}}$ on $\left[\frac{2}{3}, \frac{4}{3}\right]$ occurs when the denominator is smallest, that is, when $x = \frac{2}{3}$. Hence

$$M_{n+1} = \frac{n!}{(2/3)^{n+1}} = n! \left(\frac{3}{2}\right)^{n+1}.$$

For $x \in \left[\frac{2}{3}, \frac{4}{3}\right]$ we have $|x - 1| \leq \frac{1}{3}$. Using this bound in (14), we find

$$|R_n(x; 1)| \leq \frac{M_{n+1}}{(n + 1)!}|x - 1|^{n+1}$$

$$= n!\left(\frac{3}{2}\right)^{n+1} \frac{|x - 1|^{n+1}}{(n + 1)!}$$

$$\text{(16)} \qquad = \frac{1}{n + 1}\left(\frac{3}{2}\right)^{n+1}|x - 1|^{n+1}$$

$$\leq \frac{1}{n + 1}\left(\frac{3}{2}\right)^{n+1}\left(\frac{1}{3}\right)^{n+1}$$

$$= \frac{1}{(n + 1)2^n}.$$

Now suppose that we want to make the error

$$\left|\ln x - T_n(x; 1)\right| = |R_n(x; 1)| < \epsilon.$$

As suggested by (16), we can do this by picking n so that

$$\frac{1}{(n + 1)2^n} < \epsilon.$$

For example, we can take n such that $\dfrac{1}{n + 1} < \epsilon.$

b. The sequence $\{T_n(x; 1)\}_{n=1}^{\infty}$ does not converge uniformly to $\ln x$ on $(0, 2)$ because $T_n(x; 1)$ cannot approximate $\ln x$ near 0. Since $T_n(x; 1)$ is a polynomial, it stays bounded when x is close to 0. On the other hand, $\ln x$ tends to $-\infty$ as x goes to 0 in $(0, 2)$. Hence no matter what n is,

$$\left|\ln x - T_n(x; 1)\right|$$

is very large when x is close to 0. This is illustrated in Fig. 7.28. ∎

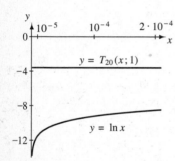

Figure 7.28. For x near 0, we see that $T_n(x; 1)$ is not a good approximation to $\ln x$. Why does the graph of $y = T_{20}(x; 1)$ look like a straight line in this picture?

Exercises 7.3

Basic

Exercises 1–7: Write out the first five terms of each sequence.

1. $\{2^n\}_{n=1}^{\infty}$

2. $\{1 - k + 2k^2\}_{k=0}^{\infty}$

3. $\{3\}_{k=1}^{\infty}$

4. $\left\{\dfrac{j + 1}{j - 2}\right\}_{j=3}^{\infty}$

5. $\left\{\cos\dfrac{k\pi}{4}\right\}_{k=1}^{\infty}$

6. $\left\{\displaystyle\sum_{k=0}^{n} k\right\}_{n=1}^{\infty}$

7. $\left\{\displaystyle\sum_{k=1}^{n} \dfrac{(-1)^{k-1}}{k^k}\right\}_{n=1}^{\infty}$

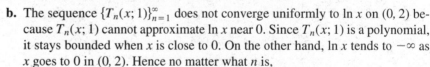

Exercises 8–13: The first four terms of a sequence are shown. Write the most likely next three terms and the tenth term. Then write a formula or description for the nth term of the sequence.

8. $-1, 1, -1, 1, \ldots$

9. $3, 3.1, 3.14, 3.141, \ldots$

10. $1, 3, 6, 10, \ldots$

11. $1, 4, 27, 256, \ldots$

12. $\dfrac{1}{2}, \dfrac{3}{4}, \dfrac{7}{8}, \dfrac{15}{16} \ldots$

13. $1, 1 + x, 1 + x + x^2, 1 + x + x^2 + x^3, \ldots$

Exercises 14–20: A sequence $\{s_n\}$ is defined recursively. Write out the first five terms of the sequence.

14. $s_1 = 1$

$s_n = n + s_{n-1} \quad (n \geq 2)$

15. $s_0 = 1$

$s_n = n s_{n-1} \quad (n \geq 1)$

16. $s_1 = 2$

$s_n = \dfrac{1}{s_{n-1} + 1} \quad (n \geq 2)$

17. $s_1 = 1$

$s_n = \dfrac{n+1}{n+2} s_{n-1} \quad (n \geq 2)$

18. $s_1 = 1$

$s_2 = 1$

$s_n = s_{n-1} + s_{n-2} \quad (n \geq 3)$

19. $s_1 = 1$

$s_2 = 2$

$s_n = \dfrac{1}{2}(s_{n-1} + s_{n-2}) \quad (n \geq 3)$

20. $s_1 = 1$

$s_2 = 2$

$s_n = \sqrt{s_{n-2} s_{n-1}} \quad (n \geq 3)$

Exercises 21–24: The first few terms of a sequence are listed. Write down a recursive formula for the sequence. (Don't forget to define the first term.)

21. $1, 1 + \dfrac{1}{2}, 1 + \dfrac{1}{2} + \dfrac{1}{3}, \ldots$

22. $2, \sqrt{1 + 2}, \sqrt{1 + \sqrt{1 + 2}}, \sqrt{1 + \sqrt{1 + \sqrt{1 + 2}}} \ldots$

23. $2, 2^2, 2^{(2^2)}, 2^{2^{(2^2)}}, \ldots$

24. $1, 1 + x, 1 + x + x^2, 1 + x + x^2 + x^3 \ldots$

Exercises 25–36: Decide if the sequence has a limit. Find or estimate the limit in those cases where it exists. Use analytic, numerical, or graphical means.

25. $\left\{ \dfrac{n-1}{n+1} \right\}$

26. $\left\{ \dfrac{10^k}{k!} \right\}$

27. $\left\{ \dfrac{3j^2 - 2j + 1}{100j + 2000} \right\}$

28. $\left\{ \sqrt{n^2 + n} - n \right\}$

29. $\{\arctan k\}$

30. $\left\{ n \sin \dfrac{1}{n} \right\}$

31. $\left\{ \cos \sqrt{j} \right\}$

32. $\left\{ \displaystyle\sum_{k=1}^{n} \dfrac{1}{2^k} \right\}$

33. $\left\{ \displaystyle\sum_{k=1}^{n} \dfrac{1}{\sqrt{k}} \right\}$

34. $\left\{ \left(1 + \dfrac{1}{k}\right)^{2k} \right\}$

35. $\left\{ \left(1 + \dfrac{2}{k}\right)^{k} \right\}$

36. $\left\{ \left(1 + \dfrac{5}{k}\right)^{k} \right\}$

Growth

37. Let $s_n = \left(1 - \frac{1}{n}\right)^2$. Show that the sequence $\{s_n\}_{n=1}^{\infty}$ is increasing and bounded above. What is the limit of the sequence?

38. Let $s_n = \left(1 - \frac{1}{n}\right)^n$. Provide graphical and/or numerical evidence indicating that the sequence $\{s_n\}_{n=1}^{\infty}$ is increasing and bounded above. What do you think the limit of the sequence is?

39. Let $s_n = (n!)^{1/n}$. Provide graphical and/or numerical evidence indicating that the sequence $\{s_n\}_{n=1}^{\infty}$ is increasing but not bounded above.

40. Let $s_n = (n!)^{1/n}$ as in the preceding problem. Plot several points (n, s_n), including some for large values of n. The points seem to lie on a line. Use two points with large n values to find the slope of the line, and relate this to the graph produced in Exercise 39. (When large n values are used to compute the slope, the answer should be $\approx 1/e$.)

41. Define a sequence recursively by

$$s_1 = \sqrt{2}$$

$$s_n = \sqrt{2 + \sqrt{s_{n-1}}} \quad (n \geq 2).$$

Provide graphical and/or numerical evidence that the sequence is increasing and bounded above. Estimate the limit of the sequence.

42. Define a sequence recursively by

$$s_1 = 4$$

$$s_n = \frac{1}{2}\left(s_{n-1} + \frac{2}{s_{n-1}}\right) \quad (n \geq 2)$$

Provide graphical and/or numerical evidence that the sequence is decreasing and bounded below. Estimate the limit of the sequence.

43. State the decreasing-sequence version of the Monotone Sequence Theorem.

44. Let $h_n = \sum_{k=1}^{n} \frac{1}{k}$.

In Example 3 we showed that $h_n \geq \ln(n + 1)$. Now consider the sequence $\{s_n\}$ where $s_n = h_n - \ln(n + 1)$. Provide graphical and/or numerical evidence that the sequence $\{s_n\}$ is decreasing and bounded below. Estimate the limit of the sequence. (The limit is denoted γ and is called Euler's gamma. Like π and e, it is one of the important constants of mathematics.)

45. For positive integer n let $s_n = \sum_{k=1}^{n} \frac{1}{\sqrt{k}}$.

a. Use rectangles as in Example 3 to show that
$$s_n \geq 2\sqrt{n + 1} - 2.$$

b. Use rectangles as in Example 5 to show that
$$s_n \leq 2\sqrt{n} - 1.$$

46. For positive integer n let $s_n = \sum_{k=1}^{n} \frac{1}{k^3}$.

a. Use rectangles as in Example 3 to show that
$$s_n \geq \frac{1}{2} - \frac{1}{2(n+1)^2}.$$

b. Use rectangles as in Example 5 to show that
$$s_n \leq \frac{3}{2} - \frac{1}{2n^2}.$$

c. Tell how you know that $\lim_{n \to \infty} s_n$ exists, and estimate the limit to within 0.001.

47. For positive integer n let $s_n = \sum_{k=1}^{n} \frac{1}{k^8}$.

a. Use rectangles as in Example 3 to show that
$$s_n \geq \frac{1}{7} - \frac{1}{7(n+1)^7}.$$

b. Use rectangles as in Example 5 to show that
$$s_n \leq \frac{8}{7} - \frac{1}{7n^7}.$$

c. Tell how you know that $\lim_{n \to \infty} s_n$ exists, and estimate the limit to within 10^{-6}.

48. For positive integer n let $s_n = \sum_{k=1}^{n} e^{-k}$.

a. Use rectangles as in Example 3 to show that
$$s_n \geq \frac{1}{e} + \frac{1}{e^2} - \frac{1}{e^{n+1}}.$$

b. Use rectangles as in Example 5 to show that
$$s_n \leq 1 - \frac{1}{e^n}.$$

c. Tell how you know that $\lim_{n \to \infty} s_n$ exists, and estimate the limit to within 10^{-6}.

49. Let c be a positive real number. Define a sequence recursively by

$$s_1 = c$$

$$s_n = c^{s_{n-1}} \quad (n \geq 2).$$

With a calculator or computer algebra system find the first several terms of the sequence for different values of c. For what c does the sequence appear to converge? For what c does it appear to diverge?

50. For integer $n \geq 1$, let

$$s_n = \sum_{k=1}^{n} (-1)^{k-1} \frac{1}{k}$$

$$= 1 - \frac{1}{2} + \frac{1}{3} - \cdots + (-1)^{n-1} \frac{1}{n}.$$

We first study the even indexed terms of the sequence, that is, $s_2, s_4, s_6, \ldots$.

a. Show that for $k \geq 2$,

$$s_{2k} = s_{2k-2} + \left(\frac{1}{2k-1} - \frac{1}{2k} \right).$$

Use this to show that the sequence $s_2, s_4, s_6, \ldots$ is monotone increasing.

b. Show that the sequence $s_2, s_4, s_6, \ldots$ is bounded above by 1.

c. Tell why the sequence $s_2, s_4, s_6 \ldots$ converges.

d. Now consider the whole sequence $s_1, s_2, s_3, s_4,$ .... Give an argument showing that this sequence converges to the same limit as the sequence of even-indexed terms.

e. The limit of this sequence is $\ln 2$. How close is s_{100} to the limit?

51. For integer $n \geq 1$, let

$$s_n = \sum_{k=1}^{n} (-1)^{k-1} \frac{4}{2k-1}$$

$$= 4 - \frac{4}{3} + \frac{4}{5} - \cdots + (-1)^{n-1} \frac{4}{4n-1}.$$

Use the method of the previous problem to show that this sequence converges. What do you guess is the limit of the sequence?

52. Let $a > 0$ be constant. Determine $\lim_{k \to \infty} \left(1 + \frac{a}{k} \right)^k$.

Hint: Use Example 8 and note that $\left(1 + \frac{a}{k} \right)^k = \left(\left(1 + \frac{a}{k} \right)^{k/a} \right)^a$.

53. Use the technique illustrated in Example 8 to show that

$$\lim_{k \to \infty} \left(1 - \frac{1}{k} \right)^k = \frac{1}{e}.$$

Combine this result with the previous exercise to show that for any real a,

$$\lim_{k \to \infty} \left(1 + \frac{a}{k} \right)^k = e^a.$$

Exercises 54–59: For the given function f, real number c, and interval I, show that the sequence $\{T_n(x; c)\}$ of Taylor polynomials converges to f uniformly on I.

54. $f(x) = \cos x$, $c = 0$, $I = [-1, 1]$

55. $f(x) = \sin x$, $c = \frac{\pi}{4}$, $I = \left[0, \frac{\pi}{2} \right]$

56. $f(x) = e^{3x}$, $c = 0$, $I = [-10, 10]$

57. $f(x) = -7x^3 - 3x^2 + 4$, $c = 5$, $I = (-\infty, \infty)$

58. $f(x) = \frac{1}{1-x}$, $c = 0$, $I = [-0.25, 0.25]$

59. $f(x) = \ln x$, $c = 6$, $I = [4, 8]$

Review/Preview

60. Verify that $y = e^{3x}$ is a solution of the differential equation $y'' - 2y' - 3y = 0$.

61. Verify that $y = 2e^{-x} + 4e^{2x}$ is a solution to the differential equation $y'' - y' - 2y = 0$.

62. Write down a simple differential equation that has $y = e^{-7x}$ as one solution.

63. Write down a simple differential equation that has $y = x^2 e^x$ as one solution.

64. Find the values of a_0, a_1, a_2 so that

$$y = a_0 + a_1 x + a_2 x^2$$

is a solution to the differential equation

$$2x^2 y'' - xy' + 3y = x + 1.$$

65. Find the values of a_0, a_1, a_2 so that

$$y = a_0 + a_1(x - 1) + a_2(x - 1)^2$$

is a solution to the differential equation

$$y'' - 3(x - 1)y' + 3y = \frac{3}{4}x^2 - \frac{3}{2}x - 5.$$

66. Estimate the value of $\int_0^3 e^{-x^2} dx$ by interpreting the integral as an area and then using 10 rectangles to approximate the area.

67. Estimate the value of $\int_0^{\sqrt{\pi}} \sin x^2 dx$ by interpreting the integral as an area and then using 10 rectangles to approximate the area.

7.4 INFINITE SERIES

In Example 8 of Section 7.3 we showed that the sequence $\{T_n(x; 1)\}_{n=1}^{\infty}$ of Taylor polynomials for $\ln x$ converges uniformly to $\ln x$ on $I = [2/3, 4/3]$. Hence for $x \in I$ we have

$$(1) \qquad \lim_{n \to \infty} T_n(x; 1) = \lim_{n \to \infty} \sum_{k=1}^{n} \frac{(-1)^{k-1}}{k}(x-1)^k = \ln x.$$

In view of (1) it seems natural to write

$$(2) \qquad \sum_{k=1}^{\infty} \frac{(-1)^{k-1}}{k}(x-1)^k = \ln x.$$

The sum shown in (2) is an example of an infinite series. An infinite series is really just a special kind of sequence, one that can be conveniently described by an expression such as

$$\sum_{k=1}^{\infty} \frac{(-1)^{k-1}}{k}(x-1)^k.$$

Infinite series show up in the solutions of many problems in science, engineering, and mathematics. We will see how infinite series can be used to approximate functions and definite integrals and to solve differential equations. In this section we define infinite series and study some fundamental methods for dealing with series.

Definition of Infinite Series

We will usually think of an infinite series such as (2) as a "sum of infinitely many things." As we've seen before, ideas involving infinity are usually made precise through limits. The definition of infinite series tells us how to use limits to add infinitely many things.

DEFINITION **infinite series**

Let $\{a_k\}_{k=1}^{\infty}$ be a sequence of real numbers or of functions. The **infinite series** with terms $\{a_k\}_{k=1}^{\infty}$ is denoted

$$(3) \qquad \sum_{k=1}^{\infty} a_k$$

or by

$$a_1 + a_2 + a_3 + \cdots.$$

For positive integer n,

$$s_n = \sum_{k=1}^{n} a_k$$

is called the nth **partial sum** of the series (3). The sequence

$$s_1, s_2, s_3, \cdots = \{s_n\}_{n=1}^{\infty}$$

is called the **sequence of partial sums** associated with (3). We say the infinite series (3) has sum S if

$$\lim_{n \to \infty} s_n = S.$$

In this case we write

(4) $$\sum_{k=1}^{\infty} a_k = S.$$

If the sequence of partial sums $\{s_n\}_{n=1}^{\infty}$ has no limit, we say that the series (3) diverges. If $\lim_{n \to \infty} s_n = \infty$, we say that the series (3) diverges to ∞.

___

There are two sequences associated with any infinite series

$$\sum_{k=1}^{\infty} a_k.$$

The first is the sequence $\{a_k\}_{k=1}^{\infty}$ of terms of the sequence. The second is the sequence $\{s_n\}_{n=1}^{\infty}$ of partial sums of the series. Both of these sequences are important in understanding infinite series, but they convey different information about a series. It is important to understand the differences in these two sequences and to understand what each sequence tells us about the behavior of a series.

■ **EXAMPLE 1**   Consider the infinite series

$$\sum_{k=1}^{\infty} \frac{1}{k(k+1)}.$$

Show that this series has sum 1.

Solution.   The sum of an infinite series is defined to be the limit of the sequence of partial sums of the series. For positive integer n, the nth partial sum of the series is

$$s_n = \sum_{k=1}^{n} \frac{1}{k(k+1)}.$$

The first few terms of the sequence of partial sums are

$$s_1 = \sum_{k=1}^{1} \frac{1}{k(k+1)} = \frac{1}{1 \cdot 2} = \frac{1}{2}$$

$$s_2 = \sum_{k=1}^{2} \frac{1}{k(k+1)} = \frac{1}{1 \cdot 2} + \frac{1}{2 \cdot 3} = \frac{2}{3}$$

$$s_3 = \sum_{k=1}^{3} \frac{1}{k(k+1)} = \frac{1}{1 \cdot 2} + \frac{1}{2 \cdot 3} + \frac{1}{3 \cdot 4} = \frac{3}{4}$$

$$s_4 = \sum_{k=1}^{4} \frac{1}{k(k+1)} = \frac{1}{1 \cdot 2} + \frac{1}{2 \cdot 3} + \frac{1}{3 \cdot 4} + \frac{1}{4 \cdot 5} = \frac{4}{5}.$$

From these examples it seems reasonable to guess that

$$(5) \qquad\qquad s_n = \frac{n}{n+1}$$

and hence that the sum of the series is

$$(6) \qquad \sum_{k=1}^{\infty} \frac{1}{k(k+1)} = \lim_{n \to \infty} \sum_{k=1}^{n} \frac{1}{k(k+1)} = \lim_{n \to \infty} s_n = \lim_{n \to \infty} \frac{n}{n+1} = 1.$$

Although guessing and conjecturing are an important part of problem solving, it is a good idea to back your guesses with solid reasoning whenever possible. We can verify the guess given in (5) by first noting

$$\frac{1}{k(k+1)} = \frac{1}{k} - \frac{1}{k+1}.$$

(This is the partial fractions decomposition of $\dfrac{1}{k(k+1)}$.) Hence

$$s_n = \sum_{k=1}^{n} \frac{1}{k(k+1)} = \sum_{k=1}^{n} \left(\frac{1}{k} - \frac{1}{k+1} \right)$$

$$= \left(1 - \frac{1}{2} \right) + \left(\frac{1}{2} - \frac{1}{3} \right) + \left(\frac{1}{3} - \frac{1}{4} \right) + \cdots + \left(\frac{1}{n} - \frac{1}{n+1} \right).$$

In this last expression, all terms but the first and the last cancel with an adjacent term. (Such an expression is called a *telescoping expression*. Can you guess why?) Hence we get

$$s_n = 1 - \frac{1}{n+1} = \frac{n}{n+1}.$$

This verifies (5), so the sum of the series is 1, as shown by (6). ∎

■ **EXAMPLE 2** The harmonic series is the infinite series

$$\sum_{k=1}^{\infty} \frac{1}{k}.$$

Determine whether the harmonic series converges or diverges.

Solution. For positive integer n, the nth partial sum of the harmonic series is

$$s_n = \sum_{k=1}^{n} \frac{1}{k}.$$

In Example 5 of Section 7.3 we showed that the sequence $\{s_n\}_{n=1}^{\infty}$ diverges to $+\infty$. Hence the harmonic series diverges to $+\infty$, and we write

$$\sum_{k=1}^{\infty} \frac{1}{k} = +\infty.$$ ■

■ **EXAMPLE 3** Let r be a real number. Show that the infinite series

$$(7) \qquad 1 + r + r^2 + r^3 + \cdots = \sum_{k=0}^{\infty} r^k$$

converges if $|r| < 1$, but diverges if $|r| \geq 1$.

Solution. To learn about the series, we need to analyze the partial sums. We first consider the special cases $r = 1$ and $r = -1$. When $r = 1$, we have

$$s_n = \sum_{k=0}^{n-1} 1^n = n.$$

Hence when $r = 1$, we have

$$\lim_{n \to \infty} s_n = \lim_{n \to \infty} n = +\infty,$$

so the series diverges to $+\infty$.
 When $r = -1$, we have

$$(8) \qquad s_n = \sum_{k=0}^{n-1} (-1)^k = 1 + (-1) + \cdots + (-1)^{n-1} = \begin{cases} 1 & (n \text{ odd}) \\ 0 & (n \text{ even}). \end{cases}$$

When $r = -1$, the sequence $\{s_n\}_{n=1}^{\infty}$ is alternating 1s and 0s. Hence $\lim_{n \to \infty} s_n$ does not exist, so $\sum_{k=0}^{\infty} r^k$ diverges for $r = -1$.
 When $|r| \neq 1$, we write s_n in a form that is easier to work with. We have

$$(9) \qquad s_n = 1 + r + r^2 + r^3 + \cdots + r^{n-1}.$$

Multiply both sides of (9) by r to obtain

(10) $$rs_n = r + r^2 + r^3 + \cdots + r^{n-1} + r^n.$$

Subtract (10) from (9). After canceling like terms, we have

$$(1 - r)s_n = 1 - r^n,$$

from which we derive

(11) $$s_n = \frac{1 - r^n}{1 - r}.$$

When $|r| < 1$, we have $\lim_{n \to \infty} r^n = 0$. Hence if $|r| < 1$,

$$\lim_{n \to \infty} s_n = \lim_{n \to \infty} \frac{1 - r^n}{1 - r} = \frac{1}{1 - r},$$

so in this case

$$\sum_{k=0}^{\infty} r^k = \frac{1}{1 - r}.$$

When $r > 1$, we have $\lim_{n \to \infty} r^n = \infty$. Thus for $r > 1$,

$$\lim_{n \to \infty} s_n = \lim_{n \to \infty} \frac{1 - r^n}{1 - r} = +\infty,$$

and we see that

$$\sum_{k=0}^{\infty} r^k \text{ diverges to } +\infty.$$

When $r < -1$, the terms of the sequence $\{r^n\}_{n=1}^{\infty}$ alternate in sign and tend toward ∞ in absolute value. Hence when $r < -1$,

$$\lim_{n \to \infty} s_n = \lim_{n \to \infty} \frac{1 - r^n}{1 - r} \text{ does not exist.}$$

Summarizing the results above, we conclude that

$$\sum_{k=0}^{\infty} r^k \begin{cases} = \dfrac{1}{1 - r} & (-1 < r < 1) \\ \text{diverges to } +\infty & (r \geq 1) \\ \text{diverges} & (r \leq -1) \end{cases}$$

The infinite series discussed in the previous example is called a *geometric series*.

> **Geometric Series**
>
> Let a and r be real numbers with $a \neq 0$. The infinite series
>
> $$\sum_{k=0}^{\infty} ar^k = a + ar + ar^2 + ar^3 + \cdots$$
>
> is called a **geometric series**. The number r is called the **common ratio** of the series. If $|r| < 1$, then
>
> (12) $$\sum_{k=0}^{\infty} ar^k = \frac{a}{1-r}.$$
>
> If $|r| \geq 1$, then the series diverges.

■ **EXAMPLE 4** Use geometric series to convert the repeating decimal

$$0.31313131\ldots$$

to a fraction.

Solution. We have

$$0.31313131\ldots = 0.31 + 0.0031 + 0.000031 + 0.00000031 + \cdots$$
$$= \frac{31}{100} + \frac{31}{100^2} + \frac{31}{100^3} + \frac{31}{100^4} + \cdots.$$

This is a geometric series with first term of $\frac{31}{100}$ and common ratio $\frac{1}{100}$. By (12) the sum of the series is

$$\frac{\dfrac{31}{100}}{1 - \dfrac{1}{100}} = \frac{31}{99}.$$

Hence

$$0.31313131\cdots = \frac{31}{99}.$$

How could you check this answer? ■

The Arithmetic of Series

Sometimes we recognize an infinite series $\sum a_k$ as a combination of other series. If we know the behavior (or sum) of these other series, we may be able to determine the behavior (or sum) of $\sum a_k$. Some of the circumstances in which we can do so are summarized below.

Sums and Multiples of Series

Let $\sum a_k$ and $\sum b_k$ be infinite series.

a. Let c be a real number. If $\sum a_k = S$, then

$$\sum ca_k = c \sum a_k = cS.$$

If $c \neq 0$ and $\sum a_k$ diverges, then $\sum ca_k$ also diverges.

b. Suppose that $\sum a_k = S$ and $\sum b_k = T$. Then

$$\sum (a_k + b_k) = \sum a_k + \sum b_k = S + T.$$

These results follow from the results on the arithmetic of sequences given in Section 7.3. See Exercises 52 and 53.

A Test for Divergence

To determine whether an infinite series $\sum a_k$ converges or diverges, we need to determine the behavior of the sequence $\{s_n\}_{n=1}^{\infty}$ of partial sums. In Examples 1 and 3 we found a formula for s_n and then used this formula to determine $\lim_{n \to \infty} s_n$. In Example 2 we did not find a formula for s_n but were able to find estimates from which we could conclude $\lim_{n \to \infty} s_n = +\infty$. Determining whether a series converges or diverges by working with the sequence of partial sums may be effective, but it is usually difficult for all but a few special series. In the next section we will develop several quick tests to determine the behavior of a series. No such test works all of the time, but for most series that we encounter we will be able to find a test that applies. We conclude this section with a simple but useful test that helps us quickly identify some divergent series.

Suppose that the infinite series

$$\sum_{k=1}^{\infty} a_k$$

converges to S. Then we know that

(13)
$$\lim_{n \to \infty} s_n = \lim_{n \to \infty} \sum_{k=1}^{n} a_k = S.$$

Next note that for $n \geq 2$,

(14)
$$a_n = \sum_{k=1}^{n} a_k - \sum_{k=1}^{n-1} a_k = s_n - s_{n-1}.$$

Combining (13) and (14), we have

$$\lim_{n\to\infty} a_n = \lim_{n\to\infty}(s_n - s_{n-1}) = \lim_{n\to\infty} s_n - \lim_{n\to\infty} s_{n-1} = S - S = 0.$$

This shows that if the infinite series $\sum_{k=1}^{\infty} a_k$ converges, then the sequence $\{a_k\}$ of terms of the series has limit 0. Unfortunately, the information $\lim_{k\to\infty} a_k = 0$ does *not* imply that $\sum_{k=1}^{\infty} a_k$ converges. However, when $\lim_{k\to\infty} a_k \neq 0$, we can say the series diverges.

The nth Term Divergence Test

If $\lim_{k\to\infty} a_k$ does not exist or if $\lim_{k\to\infty} a_k$ exists but is not equal to 0, then

$$\sum_{k=1}^{\infty} a_k \text{ diverges.}$$

■ **EXAMPLE 5**   Use the nth term divergence test to show that the geometric series $\sum_{k=0}^{\infty} r^k$ diverges when $|r| \geq 1$.

Solution.   Observe that

(15)
$$\lim_{k\to\infty} r^k \begin{cases} = +\infty & (r > 1) \\ = 1 & (r = 1) \\ \text{does not exist} & (r \leq -1) \end{cases}$$

When $|r| \geq 1$, $\lim_{k\to\infty} r^k \neq 0$. Hence for these values of r, the series $\sum_{k=0}^{\infty} r^k$ diverges.

■ **EXAMPLE 6**   Show that $\lim_{k\to\infty} a_k = 0$ does not imply that $\sum_{k=1}^{\infty} a_k$ converges.

Solution.   Let $a_k = \frac{1}{k}$. The harmonic series $\sum_{k=1}^{\infty} \frac{1}{k}$ diverges, but

$$\lim_{k\to\infty} a_k = \lim_{k\to\infty} \frac{1}{k} = 0.$$

This example demonstrates that $\lim_{k\to\infty} a_k = 0$ does *not* imply that $\sum_{k=1}^{\infty} a_k$ converges.   ■

Exercises 7.4

Basic

Exercises 1–4:   For each infinite series write out the first five terms and the first five partial sums.

1. $\displaystyle\sum_{k=1}^{\infty} k$

2. $\displaystyle\sum_{k=1}^{\infty} \frac{(-1)^{k-1}}{k}$

3. $\displaystyle\sum_{k=1}^{\infty} \frac{(k+1)x^k}{2^k}$

4. $\displaystyle\sum_{k=2}^{\infty} \frac{1}{(\ln k)^2}$

Exercises 5–13:   For each infinite series find a formula for s_n, the nth partial sum of the series. Then find the limit of the sequence of partial sums.

5. $\displaystyle\sum_{k=1}^{\infty} 1$

8. $\displaystyle\sum_{n=1}^{\infty} \left(\frac{1}{\sqrt[3]{n}} - \frac{1}{\sqrt[3]{n+1}} \right)$

11. $\displaystyle\sum_{k=0}^{\infty} e^{-2k}$

6. $\displaystyle\sum_{k=0}^{\infty} \frac{1}{5^k}$

9. $\displaystyle\sum_{k=1}^{\infty} \left(\frac{2}{k+3} - \frac{2}{k+4} \right)$

12. $\displaystyle\sum_{k=1}^{\infty} \left(\frac{1}{k+3} - \frac{1}{k+5} \right)$

7. $\displaystyle\sum_{n=0}^{\infty} (\sqrt{n+1} - \sqrt{n})$

10. $\displaystyle\sum_{k=1}^{\infty} (-4)^{k-1}$

13. $\displaystyle\sum_{k=2}^{\infty} 5(0.8)^k$

Exercises 14–22:   Decide whether the infinite series converges or diverges.

14. $\displaystyle\sum_{k=1}^{\infty} (-2)$

17. $\displaystyle\sum_{k=1}^{\infty} (-1)^k \frac{\ln k + 5}{\ln k + 2}$

20. $\displaystyle\sum_{q=1}^{\infty} \ln \left(\frac{q+2}{q+1} \right)$

15. $\displaystyle\sum_{k=0}^{\infty} \frac{1}{(3.2)^k}$

18. $\displaystyle\sum_{m=1}^{\infty} \frac{2^m}{m^2}$

21. $\displaystyle\sum_{k=1}^{\infty} \left(\frac{1}{2^k} - \frac{5}{3^k} \right)$

16. $\displaystyle\sum_{r=0}^{\infty} (-1)^r \frac{\ln(r+5)}{\ln(r+2)}$

19. $\displaystyle\sum_{n=1}^{\infty} \frac{8}{k}$

22. $\displaystyle\sum_{n=2}^{\infty} \left(\frac{1}{n+2} - \frac{1}{n+5} \right)$

Exercises 23–28:   Convert the decimal to a fraction.

23. $0.15151515\ldots$
24. $0.346346346\ldots$

25. $0.819191919\ldots$
26. $0.00233233233\ldots$

27. $0.45000000\ldots$
28. $0.023444444\ldots$

Exercises 29–31:   Find r.

29. $\displaystyle\sum_{k=0}^{\infty} r^k = 5$

30. $\displaystyle\sum_{k=3}^{\infty} r^k = 10$

31. $\displaystyle\sum_{k=1}^{\infty} r^k = -e$

Exercises 32–35:   A formula for the partial sum s_n of the series $\sum_{k=1}^{\infty} a_k$ is given. Find $a_1, a_2, a_3,$ and a_k.

32. $s_n = n + 2$

34. $s_n = (-1)^n$
35. $s_n = \sin n$

33. $s_n = \dfrac{4+n}{3+n}$

Exercises 36–39:   Find the values of x for which the geometric series converges.

36. $\displaystyle\sum_{k=0}^{\infty} (2x-3)^k$

38. $\displaystyle\sum_{k=0}^{\infty} \left(\frac{1}{x^2-1} \right)^k$

37. $\displaystyle\sum_{k=2}^{\infty} (x^2 - 4x - 5)^k$

39. $\displaystyle\sum_{k=0}^{\infty} (2\sin x)^k$

Growth

40. When a new tennis ball is dropped to a concrete floor from height h, it rebounds to a height of $0.55h$, as illustrated in the figure. Suppose the ball is dropped from a height of 10 feet and after each bounce rebounds to 55 percent of the height it attained on the previous bounce. Find the total up-and-down distance the ball travels.

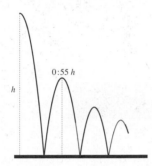

0:55 h

h

41. Let D be the total up-and-down distance traveled by the tennis ball in the previous problem. If we want the ball to travel distance $2D$, from how high should it be dropped?

42. How long (in seconds) does the tennis ball in Exercise 40 keep bouncing? How about the ball in Exercise 41?

43. A tennis ball dropped from a height of 10 feet actually stops bouncing after about 15 bounces. What is the total up-and-down distance traveled by the tennis ball during these bounces? How many seconds do these 15 bounces take?

44. Economists estimate that, on the average, each person spends 65 percent of his or her income. When this money is spent, it becomes income for someone else, who in turn spends 65 percent, and so forth. The effective impact of this money on the economy is the total of the amounts spent as it passes from person to person, each spending 65 percent of what is received. Suppose Marie earns \$1000. What is the effective impact of this sum on the economy?

45. The federal government wishes to stimulate the economy by giving tax rebates to the public so that the effective impact (see Ex. 44) of the rebates on the economy is 50 billion dollars. How much should the government spend on these rebates?

46. Let $p > 0$ and consider the series $\sum_{k=1}^{\infty} \dfrac{1}{k^p}$. Such a series is called a *p-series*. Use the techniques of Examples 5 and 7 in Section 7.3 to show that

$$\int_1^{n+1} \frac{1}{x^p}\,dx \le \sum_{k=1}^{n} \frac{1}{k^p} \le 1 + \int_1^{n} \frac{1}{x^p}\,dx.$$

47. Use the inequality developed in the previous exercise to show that the *p*-series converges if $p > 1$ and diverges if $p \le 1$.

48. Let $p > 1$. Show that

$$\sum_{k=1}^{\infty} \frac{1}{k^p} < 1 + \frac{1}{p-1}.$$

49. Let $p > 1$ and let n be a positive integer. Show that

$$\sum_{k=1}^{\infty} \frac{1}{k^p} - \sum_{k=1}^{n} \frac{1}{k^p}$$

$$< \frac{1}{(n+1)^p} + \frac{1}{p-1} \cdot \frac{1}{(n+1)^{p-1}}.$$

50. Use the estimate given in the previous exercise to estimate

$$\sum_{k=1}^{\infty} \frac{1}{k^5}$$

to within 0.01.

51. Let $\sum_{k=1}^{\infty} a_k$ be an infinite series with $a_k > 0$ for all k. Suppose that $a_{k+1}/a_k > 1$ for every positive integer k. Show that the series $\sum_{k=1}^{\infty} a_k$ diverges.

52. Let $\sum_{k=1}^{\infty} a_k$ be an infinite series and let c be a real number. Use the results on arithmetic of sequences from Section 7.3 to show that

a. If $\sum_{k=1}^{\infty} a_k = S$, then

$$\sum_{k=1}^{\infty} ca_k = c \sum_{k=1}^{\infty} a_k = cS.$$

b. If $\sum_{k=1}^{\infty} a_k$ diverges and $c \ne 0$, then

$$\sum_{k=1}^{\infty} ca_k \text{ also diverges.}$$

53. Let $\sum_{k=1}^{\infty} a_k$ and $\sum_{k=1}^{\infty} b_k$ be convergent infinite series with

$$\sum_{k=1}^{\infty} a_k = S \quad \text{and} \quad \sum_{k=1}^{\infty} b_k = T.$$

Show that the series $\sum_{k=1}^{\infty}(a_k + b_k)$ also converges and that

$$\sum_{k=1}^{\infty}(a_k + b_k) = S + T.$$

54. Let $\sum_{k=1}^{\infty} a_k$ and $\sum_{k=1}^{\infty} b_k$ be convergent infinite series with

$$\sum_{k=1}^{\infty} a_k = S \quad \text{and} \quad \sum_{k=1}^{\infty} b_k = T.$$

Is it true that

$$\sum_{k=1}^{\infty} a_k b_k = ST \ ?$$

If yes, give a supporting argument. If no, give one example showing the assertion is not true.

55. Let $\sum_{k=1}^{\infty} a_k$ and $\sum_{k=1}^{\infty} b_k$ be convergent infinite series with

$$\sum_{k=1}^{\infty} a_k = S \quad \text{and} \quad \sum_{k=1}^{\infty} b_k = T.$$

Suppose that $b_k \neq 0$ for $k = 1, 2, 3, \ldots$ and that $T \neq 0$. Is it true that

$$\sum_{k=1}^{\infty} \frac{a_k}{b_k} = \frac{S}{T} \ ?$$

If yes, give a supporting answer. If no, give one example showing the assertion is not true.

56. *The harmonic bridge.* Imagine that you have access to an unlimited number of identical cards each of length 3 inches. Take n of these cards and stack them on a table in the following way:

The first card is placed so $3 - 1/n$ inches are on the table and $1/n$ inches project beyond the end of the table.

The second card is placed so $3 - 1/(n-1)$ inches of the card are on top of the first card and $1/(n - 1)$ inches project beyond the end of the first card.

The third card is placed so $3-1/(n-2)$ inches of the card are on top of the second card and $1/(n - 2)$ inches project beyond the end of the second card.

$$\vdots$$

The nth card is placed so $3 - 1/1 = 2$ inches of the card are on the $(n - 1)$th card and 1 inch projects beyond the end of the $(n - 1)$th card.

See the accompanying figure.

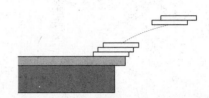

a. Show that the center of gravity of the nth card placed lies over the card below it (that is, over the $(n - 1)$th card). Thus the last card will not "tip" off of the card below it.

b. Show that the center of gravity of the nth and $(n - 1)$th card lies over the $(n - 2)$th card. Hence the last two cards placed will not fall off the $(n - 2)$th card.

c. Show that the center of gravity of the whole set of n cards lies above the table.

d. Based on a, b, and c, give an argument showing that the card structure is stable.

e. How far beyond the edge of the table is the end of the nth card? Tell why with enough cards you could make this end as far beyond the end of the table as you like.

f. Estimate the number of cards required to build a structure so that the end of the last card is 100 feet beyond the end of the table.

Review/Preview _____

57. Find a_0, a_1, and a_2 if

$$3 - 5x + x^2 = a_0 + a_1(x + 2) + a_2(x + 2)^2.$$

58. Find a_0, a_1, a_2, and a_3 if

$$3 + 4x - x^2 + 3x^3$$
$$= a_0 + a_1(x - 5) + a_2(x - 5)^2 + a_3(x - 5)^3.$$

59. Let c be a real number. Find a_0, a_1, and a_2 if

$$7x - 11x^2 = a_0 + a_1(x - c) + a_2(x - c)^2.$$

60. Verify that $y = e^{2x}$ is a solution of the differential equation

$$3y'' - y' - 10y = 0.$$

61. Verify that $y = e^{-x} \cos 2x$ is a solution of the differential equation.

$$y'' + 2y' + 5y = 0.$$

62. Find a and b if $y = e^x$ and $y = 4e^{3x}$ are both solutions of the differential equation

$$y'' + ay' + by = 0.$$

63. Find a and b if $y = e^{x/2} \sin x$ and $y = e^{x/2} \cos x$ are both solutions of the differential equation

$$y'' + ay' + by = 0.$$

64. Find $f^{(n)}(x)$ if $f(x) = \dfrac{2x - 1}{(x - 1)(x - 2)}$.

65. Find $\dfrac{d^n y}{dx^n}$ if $y = \dfrac{1}{2x^2 + x - 10}$.

7.5 TESTS FOR CONVERGENCE

When we have an infinite series $\sum_{k=1}^{\infty} a_k$ there are usually two questions that we are interested in answering:

- Does the series converge?
- If the series converges, what is its sum?

The second of these questions may be difficult or impossible to answer unless the series is of a special type (geometric, or telescoping). Thus we may have to be content with an estimate of the sum of the series. Simply calculating a partial sum s_n for a large value of n is pointless. If we do not know whether the series converges or diverges, we cannot know whether our partial sums are indeed approximating the sum of the series. Thus it is important to answer the first of the questions listed. Only when we know the series converges is it meaningful to try approximating the sum.

In this section we study several tests that can be used to help determine whether or not a series converges.

Series with Nonnegative Terms

Because the sum of an infinite series is defined to be the limit of its sequence of partial sums, many of the things we know about sequences can be applied to infinite series. In Section 7.3 we learned that an increasing sequence either converges to a finite number or diverges to infinity. We use this fact to study series with nonnegative terms.

Let $\sum_{k=1}^{\infty} a_k$ be an infinite series with $a_k \geq 0$ for all k. (We will refer to such a series as a *series of nonnegative terms*.) Then for positive integer n, we have

$$s_n \leq s_n + a_{n+1} = s_{n+1}.$$

Hence

$$s_1 \leq s_2 \leq s_3 \leq \cdots,$$

so the sequence $\{s_n\}_{n=1}^{\infty}$ of partial sums is an increasing sequence. According to the Monotone Sequence Theorem (see Section 7.3),

- If the sequence $\{s_n\}_{n=1}^{\infty}$ is bounded above, then

$$\lim_{n \to \infty} s_n = S$$

  for some a finite real number S. In this case the series has sum S.
- If the sequence $\{s_n\}_{n=1}^{\infty}$ is not bounded above, then

$$\lim_{n \to \infty} s_n = +\infty.$$

In this case the infinite series diverges to (or has sum) $+\infty$.

We have used these ideas in previous work. In Example 2 of Section 7.4 we worked with the harmonic series

$$(1) \qquad\qquad \sum_{k=1}^{\infty} \frac{1}{k}.$$

We saw that the sequence of partial sums

$$s_n = \sum_{k=1}^{n} \frac{1}{k} \quad (n = 1, 2, 3, \ldots)$$

is not bounded above. Hence (1) diverges to $+\infty$.

In Example 7 of Section 7.3 we showed that the sequence $\{s_n\}$ defined by

$$s_n = \sum_{k=1}^{n} \frac{1}{k^2} \quad (n = 1, 2, 3, \ldots)$$

is increasing and is bounded above by 2. Because $\{s_n\}$ is the sequence of partial sums for the series

$$\sum_{k=1}^{\infty} \frac{1}{k^2},$$

we know that this series converges. This information also tells us that the sum of the series is a number less than or equal to 2, but it does not tell us the value of the sum.

The tests for convergence of series with nonnegative terms are simply quick ways to check whether or not the sequence of partial sums is bounded.

The Comparison Tests If two infinite series are similar, knowing the behavior of one of the series might be enough to determine the behavior of the other. The comparison tests indicate one way this can be done.

The Comparison Tests

Let $\sum_{k=1}^{\infty} a_k$ be a series of nonnegative terms.

a. Let $\sum_{k=1}^{\infty} c_k$ be a convergent series of nonnegative terms. If there is a positive integer N such that $a_k \leq c_k$ for all $k > N$, then $\sum_{k=1}^{\infty} a_k$ converges.

b. Let $\sum_{k=1}^{\infty} d_k$ be a divergent series of nonnegative terms. If there is a positive integer N such that $a_k \geq d_k$ for all $k > N$, then $\sum_{k=1}^{\infty} a_k$ diverges.

To see why part a of the comparison test works, let $\sum_{k=1}^{\infty} c_k = C$ and let $s_n = \sum_{k=1}^{n} a_k$. Then the sequence $\{s_n\}_{n=1}^{\infty}$ of partial sums is monotone increasing. Furthermore, if $n > N$, then

$$
\begin{aligned}
s_n &= s_N + (a_{N+1} + a_{N+2} + \cdots + a_n) \\
&\leq s_N + (c_{N+1} + c_{N+2} + \cdots + c_n) \\
&\leq s_N + C.
\end{aligned}
$$

It follows that

$$s_1 \leq s_2 \leq \cdots \leq s_N \leq s_{N+1} \leq \cdots \leq s_n \leq \cdots \leq s_N + C.$$

Since N is a fixed positive integer, $s_N + C$ is a fixed upper bound for $\{s_n\}_{n=1}^{\infty}$. Therefore, by the Monotone Sequence Theorem, $\lim_{n \to \infty} s_n$ is a finite real number S. This means that $\sum_{k=1}^{\infty} a_k$ converges to S.

We leave it to the reader to provide an argument in support of part b. See Exercise 32.

■ **EXAMPLE 1**   Determine whether the infinite series $\sum_{k=1}^{\infty} \dfrac{1}{\sqrt{2k^2 + 3}}$ converges or diverges.

Solution.   When k is large,

$$(2) \qquad\qquad \frac{1}{\sqrt{2k^2 + 3}} \approx \frac{1}{\sqrt{2}\,k}.$$

Thus we expect that the series $\sum_{k=1}^{\infty} \dfrac{1}{\sqrt{2k^2 + 3}}$ will behave like the series $\sum_{k=1}^{\infty} \dfrac{1}{\sqrt{2}\,k}$. The latter series diverges. (Why?) If we wish to use a comparison test to prove divergence, we must show that the terms of $\sum_{k=1}^{\infty} \dfrac{1}{\sqrt{2k^2 + 3}}$ are larger than those of a divergent series of nonnegative terms. Because of (2), it seems reasonable to compare the series with a variation on the harmonic series. For $k \geq 1$ we have

(3) $$\frac{1}{\sqrt{2k^2 + 3}} \geq \frac{1}{\sqrt{2k^2 + 3k^2}} = \frac{1}{\sqrt{5k}}.$$

Since $\displaystyle\sum_{k=1}^{\infty} \frac{1}{\sqrt{5k}}$ diverges, it follows from the Comparison Tests that $\displaystyle\sum_{k=1}^{\infty} \frac{1}{\sqrt{2k^2 + 3}}$ also diverges. ∎

In the preceding example we suspected that $\displaystyle\sum_{k=1}^{\infty} \frac{1}{\sqrt{2k^2 + 3}}$ was divergent as soon as we had estimate (2). The rest of the solution involved the algebra aimed at coming up with the formal comparison in (3). Actually, an estimate such as (2) is enough to determine the behavior of a series. This is more precisely stated in the following result.

The Limit Comparison Test

Let $\sum_{k=1}^{\infty} a_k$ and $\sum_{k=1}^{\infty} b_k$ be series of nonnegative terms. Suppose that

(4) $$\lim_{k \to \infty} \frac{a_k}{b_k} = L \quad \text{where } L \neq 0 \text{ and } L \neq \infty.$$

Then $\sum_{k=1}^{\infty} a_k$ and $\sum_{k=1}^{\infty} b_k$ both converge or both diverge.

Very roughly, (4) says that when k is large, $\dfrac{a_k}{b_k} \approx L$, that is

$$a_k \approx Lb_k.$$

Thus it seems reasonable that for large N,

$$\sum_{k=N}^{\infty} a_k \approx L \sum_{k=N}^{\infty} b_k$$

and hence that both series converge or both diverge. See Exercise 47 for a more formal argument in support of the Limit Comparison Test. See also Exercises 48 and 49 for a discussion of the cases $L = 0$ and $L = +\infty$.

When we use the Limit Comparison Test, we usually know the behavior of one of the two series and wish to determine the behavior of the other.

■ **EXAMPLE 2**   Determine whether the infinite series $\displaystyle\sum_{k=1}^{\infty} \frac{3}{k^2 + 3k + 2}$ converges or diverges.

Solution.   In Example 1 of Section 7.4 we showed that $\displaystyle\sum_{k=1}^{\infty} \frac{1}{k(k + 1)}$ converges.

Since

$$\lim_{k \to \infty} \frac{\dfrac{3}{k^2 + 3k + 2}}{\dfrac{1}{k(k+1)}} = \lim_{k \to \infty} \frac{3k^2 + 3k}{k^2 + 3k + 2} = 3,$$

we conclude that $\displaystyle\sum_{k=1}^{\infty} \frac{3}{k^2 + 3k + 2}$ also converges. ∎

To effectively use the Comparison Tests, we need to have a large stock of series whose behavior we already know. The so-called p-series are used more than any others for this purpose.

The p-Series

Let p be a positive real number. The infinite series

$$\sum_{k=1}^{\infty} \frac{1}{k^p}$$

converges if $p > 1$ and diverges if $p \leq 1$.

For many values of p the behavior of the p-series can be deduced from the comparison tests. For example, if $p \leq 1$, then

$$\frac{1}{k^p} \geq \frac{1}{k} \quad \text{for } k = 1, 2, 3, \ldots.$$

Because the harmonic series $\displaystyle\sum_{k=1}^{\infty} \frac{1}{k}$ diverges, it follows from the Comparison Tests that $\displaystyle\sum_{k=1}^{\infty} \frac{1}{k^p}$ also diverges.

If $p \geq 2$, then

$$\frac{1}{k^p} \leq \frac{1}{k^2} \quad \text{for } k = 1, 2, 3, \ldots.$$

The series $\displaystyle\sum_{k=1}^{\infty} \frac{1}{k^2}$ converges, so by the Comparison Tests, $\displaystyle\sum_{k=1}^{\infty} \frac{1}{k^p}$ also converges.

For $1 < p < 2$ we can prove that the p-series converges by using the rectangle area techniques used in Examples 5 and 7 of Section 7.3. See also Exercises 46 and 47 in Section 7.4 and Exercise 36 in this section.

The Ratio Test In the next section we will work with Taylor series, that is, series whose partial sums are the Taylor polynomials for a function:

$$\sum_{k=0}^{\infty} \frac{f^{(k)}(c)}{k!}(x - c)^k = \lim_{n \to \infty} \sum_{k=0}^{n} \frac{f^{(k)}(c)}{k!}(x - c)^k = \lim_{n \to \infty} T_n(x; c)$$

We will want to know the values of x for which the Taylor series converges. Sometimes this information can be found by working with the remainder $R_n(x; c)$, as in Section 7.2. Sometimes we will use the Ratio Test. The Ratio Test is actually just a quick way to compare a series with a geometric series.

The Ratio Test

Let $\sum_{k=1}^{\infty} a_k$ be an infinite series of nonnegative terms. Suppose that

$$\lim_{k \to \infty} \frac{a_{k+1}}{a_k} = \rho$$

where $0 \le \rho \le \infty$.

a. If $\rho < 1$, then $\sum_{k=1}^{\infty} a_k$ converges.
b. If $\rho > 1$, then $\sum_{k=1}^{\infty} a_k$ diverges.

If $p = 1$, then there is not enough information to decide if the series converges or diverges. Some other method or test must be used.

Since we will use this test often in upcoming sections, we spend some time now showing why the test works.

Suppose

(5)
$$\lim_{k \to \infty} \frac{a_{k+1}}{a_k} = \rho < 1.$$

Let $r = \dfrac{1 + \rho}{2}$. Then $\rho < r < 1$. According to (5), the numbers $\dfrac{a_{k+1}}{a_k}$ get close to ρ as $k \to \infty$. This means that eventually these numbers get closer to ρ than they are to r. Hence there is a positive integer N such that

$$\frac{a_{k+1}}{a_k} < r \quad \text{when} \quad k \ge N.$$

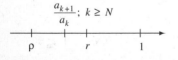

See Fig. 7.29. Since $a_k > 0$ for all k, we can rewrite the last expression as

(6)
$$a_{k+1} < a_k r \quad \text{when} \quad k \ge N.$$

Figure 7.29. When k is large enough, $a_{k+1}/a_k < r$.

Applying (6) with $k = N$, then $k = N + 1$, then $k = N + 2$, and so on, we obtain

$$a_{N+1} < a_N r$$

$$a_{N+2} < a_{N+1} r < a_N r^2$$

$$a_{N+3} < a_{N+2} r < a_N r^3$$

$$\vdots$$

and for $k > 0$

$$a_{N+k} < a_N r^k.$$

Setting $k = j - N$ in this inequality, we get the equivalent statement

$$a_j < a_N r^{j-N} \quad (j > N).$$

Hence for $n > N$ we have

$$s_n = a_1 + a_2 + \cdots + a_{N-1} + a_N + a_{N+1} + \cdots + a_n$$
$$\leq s_{N-1} + a_N + a_N r + a_N r^2 + \cdots + a_N r^{n-N}$$
$$\leq s_{N-1} + \sum_{j=0}^{\infty} a_N r^j.$$

Recall that the sum of a convergent geometric series is the first term over 1 minus the common ratio. Hence the last expression is

$$= s_{N-1} + \frac{a_N}{1 - r}.$$

Thus the sequence $\{s_n\}_{n=1}^{\infty}$ of partial sums of $\sum_{k=1}^{\infty} a_k$ is monotone increasing and bounded above by the number $s_{N-1} + \dfrac{a_N}{1 - r}$. It follows that $\sum_{k=1}^{\infty} a_k$ converges.

Next assume that

$$(7) \qquad\qquad \lim_{k \to \infty} \frac{a_{k+1}}{a_k} = \rho > 1$$

and let $r = \dfrac{1 + \rho}{2}$. Note that now $1 < r < \rho$. Arguing as above, we can find a positive integer N so that

$$(8) \qquad\qquad a_j > a_N r^{j-N} \quad \text{when} \quad j \geq N.$$

Since $r > 1$ and $a_N > 0$, we see that

$$(9) \qquad\qquad \lim_{j \to \infty} a_N r^{j-N} = \infty.$$

Combining (8) and (9), we conclude that

$$\lim_{j \to \infty} a_j \geq \lim_{j \to \infty} a_N r^{j-N} = +\infty.$$

It follows from the nth-term Divergence Test that $\sum_{k=1}^{\infty} a_k$ diverges.

We leave it to the reader to show that $\sum_{k=1}^{\infty} \dfrac{1}{k}$ and $\sum_{k=1}^{\infty} \dfrac{1}{k^2}$ are both series for which $\rho = 1$. Note that the first series diverges but the second series converges. Thus from the information $\rho = 1$ we cannot determine whether a series converges or diverges.

The ratio test is particularly useful for series whose terms a_k involve factorials or powers.

■ **EXAMPLE 3**   For each of the following series tell whether the series converges or diverges.

$$\textbf{a.} \sum_{k=0}^{\infty} \frac{k^2 - k + 2}{2^k} \qquad \textbf{b.} \sum_{k=1}^{\infty} \frac{(k!)^2}{k^{2k+1}}$$

Solution

a. With $a_k = \dfrac{k^2 - k + 2}{2^k}$, we have

$$\frac{a_{k+1}}{a_k} = \frac{\dfrac{(k+1)^2 - (k+1) + 2}{2^{k+1}}}{\dfrac{k^2 - k + 2}{2^k}} = \frac{1}{2} \frac{k^2 + k + 2}{k^2 - k + 2}.$$

Thus

$$\lim_{k \to \infty} \frac{a_k + 1}{a_k} = \lim_{k \to \infty} \frac{1}{2} \frac{k^2 + k + 2}{k^2 - k + 2} = \frac{1}{2} < 1.$$

By the Ratio Test, $\displaystyle\sum_{k=0}^{\infty} \frac{k^2 - k + 2}{2^k}$ converges.

b. Now let $a_k = \dfrac{(k!)^2}{k^{2k+1}}$, so

$$\frac{a_{k+1}}{a_k} = \frac{\dfrac{((k+1)!)^2}{(k+1)^{2k+3}}}{\dfrac{(k!)^2}{k^{2k+1}}}$$

$$= \frac{(k!)^2 (k+1)^2}{(k+1)^{2k+1}(k+1)^2} \cdot \frac{k^{2k+1}}{(k!)^2}$$

$$= \left(\frac{k}{k+1} \right)^{2k+1}.$$

Hence

$$(10) \qquad \lim_{k \to \infty} \frac{a_{k+1}}{a_k} = \lim_{k \to \infty} \left(\frac{k}{k+1} \right)^{2k+1}.$$

To determine the limit in (10), first note that

$$(11) \qquad \left(\frac{k}{k+1} \right)^{2k+1} = \frac{1}{\left(1 + \dfrac{1}{k} \right)^{2k}} \cdot \frac{k}{k+1}$$

for all k. In Example 8 of Section 7.3 we showed that

$$\lim_{k \to \infty} \left(1 + \frac{1}{k} \right)^k = e.$$

Using this when we evaluate the limit of the expression in (11), we find

$$\lim_{k \to \infty} \frac{a_{k+1}}{a_k} = \lim_{k \to \infty} \frac{1}{\left(1 + \dfrac{1}{k}\right)^{2k}} \cdot \frac{k}{k+1} = \frac{1}{e^2} < 1.$$

Hence, by the Ratio Test, the series $\displaystyle\sum_{k=1}^{\infty} \frac{(k!)^2}{k^{2k+1}}$ converges. ■

Absolute and Conditional Convergence

The Comparison Tests and the Ratio Test are tests to determine the convergence or divergence of series of nonnegative terms. What about series that have some positive and some negative terms? For some of these series the Comparison and Ratio Tests can be used to determine whether the series converges or diverges, but in other cases these tests lead to no conclusion. We first look at some series for which these tests can be used successfully.

■ **EXAMPLE 4** Determine whether the series

$$(12) \qquad\qquad \sum_{k=1}^{\infty} (-1)^k \frac{k2^k}{k^2 + 1}$$

converges or diverges.

Solution. We cannot use the Comparison or Ratio Tests directly on (12) because the series is not a series of nonnegative terms. However, we can apply the Ratio Test to the series of absolute values

$$(13) \qquad\qquad \sum_{k=1}^{\infty} \left| (-1)^k \frac{k2^k}{k^2 + 1} \right| = \sum_{k=1}^{\infty} \frac{k2^k}{k^2 + 1}.$$

The kth term of the series of absolute values is $a_k = \dfrac{k2^k}{k^2 + 1}$. Hence

$$\lim_{k \to \infty} \frac{a_{k+1}}{a_k} = \lim_{k \to \infty} \frac{\dfrac{(k+1)2^{k+1}}{(k+1)^2 + 1}}{\dfrac{k2^k}{k^2 + 1}} = \lim_{k \to \infty} 2 \left(\frac{k+1}{k} \right) \left(\frac{k^2 + 1}{k^2 + 2k + 2} \right) = 2.$$

Because the limit of the ratio of terms is greater than 1, the series (13) diverges. But we saw earlier that when a series diverges by the Ratio Test, the terms of the series do not tend to 0. Thus $\lim_{k \to \infty} a_k \neq 0$, and it follows that

$$\lim_{k \to \infty} (-1)^k \frac{k2^k}{k^2 + 1} \neq 0.$$

Hence the series (12) diverges. ■

■ **EXAMPLE 5**   Determine whether the series

$$(14) \qquad \sum_{k=1}^{\infty}(-1)^{k-1}\frac{1}{k^2} = 1 - \frac{1}{2^2} + \frac{1}{3^2} - \frac{1}{4^2} + \cdots$$

converges or diverges.

Solution.   We again start by looking at the series of absolute values,

$$\sum_{k=1}^{\infty}\frac{1}{k^2}.$$

Because this is a *p*-series with $p = 2$, the series converges. We now show that (14) converges. First consider the series

$$(15) \qquad \sum_{k=1}^{\infty}\left((-1)^{k-1}\frac{1}{k^2} + \frac{1}{k^2}\right).$$

The series looks a little funny, but notice that it is a series of nonnegative terms (with many terms equal to 0). Furthermore,

$$\left((-1)^{k-1}\frac{1}{k^2} + \frac{1}{k^2}\right) \le \frac{2}{k^2} \quad (k = 1, 2, 3, \ldots),$$

so the series (15) converges by the Comparison Tests. Now we can write (14) as the difference of two convergent series,

$$\sum_{k=1}^{\infty}(-1)^{k-1}\frac{1}{k^2} = \sum_{k=1}^{\infty}\left((-1)^{k-1}\frac{1}{k^2} + \frac{1}{k^2}\right) - \sum_{k=1}^{\infty}\frac{1}{k^2}.$$

Thus the series (14) converges.   ■

This example illustrates a very important result, called the Absolute Convergence Test.

The Absolute Convergence Test

Let $\sum_{k=1}^{\infty} a_k$ be an infinite series. If the series $\sum_{k=1}^{\infty} |a_k|$ converges, then the series $\sum_{k=1}^{\infty} a_k$ also converges.

To justify this assertion, first observe that for $k \ge 1$,

$$(16) \qquad 0 \le (a_k + |a_k|) \le 2|a_k|.$$

Hence

$$(17) \qquad \sum_{k=1}^{\infty}(a_k + |a_k|)$$

is a series of nonnegative terms. Because $\sum_{k=1}^{\infty} 2|a_k|$ converges, it follows from (16) and the Comparison Tests that (17) also converges. Thus

$$\sum_{k=1}^{\infty} a_k = \sum_{k=1}^{\infty} (a_k + |a_k| - |a_k|) = \sum_{k=1}^{\infty} (a_k + |a_k|) - \sum_{k=1}^{\infty} |a_k|$$

is the difference of two convergent series. It follows that $\sum_{k=1}^{\infty} a_k$ also converges.

The series for which the Absolute Convergence Test applies are called *absolutely convergent series*.

DEFINITION   absolute convergence

If $\sum_{k=1}^{\infty} a_k$ is a series for which $\sum_{k=1}^{\infty} |a_k|$ converges, then we say the series $\sum_{k=1}^{\infty} a_k$ is **absolutely convergent**.

In Example 5 we showed that the series $\sum_{k=1}^{\infty} (-1)^{k-1} \dfrac{1}{k^2}$ is absolutely convergent.

Alternating Series   The absolute convergence test says that if $\sum_{k=1}^{\infty} |a_k|$ converges, then the series $\sum_{k=1}^{\infty} a_k$ also converges. On the other hand, if $\sum_{k=1}^{\infty} |a_k|$ diverges, then the absolute convergence test does not apply. As we shall see, when $\sum_{k=1}^{\infty} |a_k|$ diverges, it may not be possible to conclude anything about the behavior of $\sum_{k=1}^{\infty} a_k$. In some cases this series may converge, and in other cases it may diverge (see Example 4). To decide which is the case may take a careful analysis of the series with the roles of positive and negative terms taken into account. Analysis of this sort can be very difficult. However, if the terms of the series alternate in sign and decrease to 0 in absolute value, we can immediately conclude that the series converges.

The Alternating Series Test

Let $\{b_k\}_{k=1}^{\infty}$ be a sequence of positive real numbers. If

$$b_1 \geq b_2 \geq b_3 \geq \cdots$$

and

$$\lim_{k \to \infty} b_k = 0,$$

then

(18) $$\sum_{k=1}^{\infty} (-1)^{k-1} b_k = b_1 - b_2 + b_3 - b_4 + - \cdots$$

converges.

To justify the Alternating Series Test, we show that the sequence $\{s_n\}$ of partial sums of the series converges. Let $2m$ be an even integer. Then

(19)
$$s_{2m} = (b_1 - b_2) + (b_3 - b_4) + \cdots + (b_{2m-3} + b_{2m-2}) + (b_{2m-1} - b_{2m})$$
$$= s_{2m-2} + (b_{2m-1} - b_{2m}).$$

Because $b_{2m-1} \geq b_{2m}$, it follows from (19) that

$$s_{2m-2} \leq s_{2m}.$$

Thus, the sequence $s_2, s_4, s_6, \ldots, s_{2m}, \ldots$ of even-indexed sums is a monotone increasing sequence. In addition, for any even integer $2m$

$$s_{2m} = (b_1 - b_2) + (b_3 - b_4) + \cdots + (b_{2m-1} - b_{2m})$$
$$= b_1 - (b_2 - b_3) - (b_4 - b_5) - \cdots - (b_{2m-2} - b_{2m-1}) - b_{2m}$$
$$\leq b_1.$$

Thus the sequence of even-indexed partial sums is increasing and bounded above by b_1:

$$s_2 \leq s_4 \leq \cdots \leq s_{2m} \leq \cdots \leq b_1$$

By the Monotone Sequence Theorem, this sequence has a limit which we call S:

(20)
$$\lim_{m \to \infty} s_{2m} = S$$

The sum of an odd number of terms of the series is of the form

$$s_{2m+1} = s_{2m} + b_{2m+1}.$$

Because $\lim_{m \to \infty} b_{2m+1} = 0$, we have

$$\lim_{m \to \infty} s_{2m+1} = \lim_{m \to \infty} s_{2m} + \lim_{m \to \infty} b_{2m+1} = S.$$

Because the sequence of even-indexed partial sums and the sequence of odd-indexed partial sums both converge to S, the sequence $\{s_n\}$ of partial sums also converges to S. That is,

$$\sum_{k=1}^{\infty} (-1)^{k-1} b_k = \lim_{n \to \infty} \sum_{k=1}^{n} (-1)^{k-1} b_k = \lim_{n \to \infty} s_n = S.$$

■ **EXAMPLE 6** Show that the alternating harmonic series

(21)
$$\sum_{k=1}^{\infty} (-1)^{k-1} \frac{1}{k} = 1 - \frac{1}{2} + \frac{1}{3} - \frac{1}{4} + \frac{1}{5} - \cdots$$

converges but is not absolutely convergent.

Solution.   The terms of (21) alternate in sign. Because

$$1 \geq \frac{1}{2} \geq \frac{1}{3} \geq \cdots$$

and

$$\lim_{k \to \infty} \frac{1}{k} = 0,$$

the series converges by the Alternating Series Test.

The series of absolute values is

$$\sum_{k=1}^{\infty} \left| (-1)^{k-1} \frac{1}{k} \right| = \sum_{k=1}^{\infty} \frac{1}{k}.$$

Because the harmonic series diverges, the alternating harmonic series is not absolutely convergent.   ∎

The alternating harmonic series is an example of a *conditionally convergent series.*

DEFINITION   **conditionally convergent series**

If the series $\sum_{k=1}^{\infty} a_k$ converges but the series $\sum_{k=1}^{\infty} |a_k|$ diverges, then we say that the series $\sum_{k=1}^{\infty} a_k$ is **conditionally convergent.**

Estimating the Sum of an Alternating Series   Let $\sum_{k=1}^{\infty} (-1)^{k-1} b_k$ be a series satisfying the conditions of the Alternating Series Test. That is,

$$b_1 \geq b_2 \geq b_3 \geq \cdots$$

and

$$\lim_{k \to \infty} b_k = 0.$$

We showed that under these conditions the sequence of even-indexed partial sums is increasing,

$$(22) \qquad\qquad s_2 \leq s_4 \leq \cdots \leq s_{2m} \leq \ldots,$$

and that these partial sums converge to the sum, S, of the series. On the other hand, the sequence $\{s_{2m+1}\}_{m=0}^{\infty}$ of odd-indexed partial sums is a decreasing sequence that also converges to S:

$$(23) \qquad\quad s_1 \geq s_3 \geq \cdots \geq s_{2m+1} \cdots \quad \text{and} \quad \lim_{m \to \infty} s_{2m+1} = S$$

See Exercise 52. Combining (22) and (23), we have

$$s_2 \leq s_4 \leq \cdots \leq s_{2m} \cdots \leq S \cdots \leq s_{2m+1} \leq \cdots \leq s_3 \leq s_1.$$

Thus any even-indexed partial sum underestimates S while any odd-indexed partial sum overestimates S. Therefore, for any positive integer n we have

(24)
$$s_n \leq S \leq s_{n+1} \quad (n \text{ even})$$
$$s_{n+1} \leq S \leq s_n \quad (n \text{ odd})$$

This is illustrated in Fig. 7.30. Because $|s_{n+1} - s_n| = b_{n+1}$, it follows from (24) that

$$|s_n - S| \leq b_{n+1}$$

This gives us a simple but effective way to estimate the error when we use the partial sum s_n to estimate S.

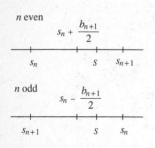

Figure 7.30. Both S and $s_n + (-1)^n b_{n+1}/2$ are between s_n and s_{n+1}.

Estimating the Sum of an Alternating Series

Let $\sum_{k=1}^{\infty}(-1)^{k-1}b_k$ be an alternating series whose terms satisfy the hypotheses of the Alternating Series Test, and let S be the sum of the series. Then for any positive integer n

(25)
$$|s_n - S| = \left| \sum_{k=1}^{n}(-1)^{k-1}b_k - S \right| \leq b_{n+1}.$$

That is, the partial sum s_n differs from S by no more than the absolute value of the next term (i.e., after $(-1)^{n-1}b_n$) in the series.

■ **EXAMPLE 7** Approximate the sum of the alternating harmonic series $\sum_{k=1}^{\infty}(-1)^{k-1}\frac{1}{k}$ with an error of less than 0.01.

Solution. In Example 6 we showed that the alternating harmonic series satisfies the hypotheses of the Alternating Series Test. Let S be the sum of the series. Then for positive integer n

$$|s_n - S| = \left| \sum_{k=1}^{n}(-1)^{k-1}\frac{1}{k} - S \right| \leq \left| (-1)^n \frac{1}{n+1} \right| = \frac{1}{n+1}.$$

If we take $n = 100$, then the difference between s_n and S will be less than $1/100$. Hence (with the aid of a CAS)

$$S \approx s_{100} = \sum_{k=1}^{100}(-1)^{k-1}\frac{1}{k} \approx 0.688172.$$

The actual sum of the alternating harmonic series is $\ln 2$. Is this estimate within 0.01 of $\ln 2$? ■

With a little more work it is possible to do better than (25) in estimating the sum of an alternating series. If we use the midpoint of the interval shown in Fig. 7.30 as our estimate for S, we have

$$S \approx s_n + \frac{(-1)^n b_{n+1}}{2}.$$

In this case the error is

$$\left| S - \left(s_n + \frac{(-1)^n b_{n+1}}{2} \right) \right| \le \frac{b_{n+1}}{2}.$$

In Exercise 52 you are asked to justify this estimate.

Exercises 7.5

Basic

Exercises 1–15: Decide whether or not the infinite series converges or diverges.

1. $\displaystyle\sum_{j=1}^{\infty} \frac{j+3}{j^2 + 2j + 100}$

2. $\displaystyle\sum_{t=1}^{\infty} (-1)^{t-1} \frac{\sqrt{t+1}}{t^2 + 1}$

3. $\displaystyle\sum_{n=0}^{\infty} \frac{n!}{2^n(n+1)}$

4. $\displaystyle\sum_{k=2}^{\infty} \left(\frac{\ln k}{k} \right)^k$

5. $\displaystyle\sum_{l=1}^{\infty} \frac{1}{\sqrt{2l^2 + 3l + 5}}$

6. $\displaystyle\sum_{n=1}^{\infty} \frac{\pi}{\sqrt{(2n^2 + 3n + 5)^3}}$

7. $\displaystyle\sum_{p=1}^{\infty} p e^{-2p}$

8. $\displaystyle\sum_{k=1}^{\infty} (-1)^{k-1} \frac{2}{2k+3}$

9. $\displaystyle\sum_{j=0}^{\infty} \frac{2^j}{j!}$

10. $\displaystyle\sum_{m=2}^{\infty} \frac{1}{(\ln m)^m}$

11. $\displaystyle\sum_{s=2}^{\infty} (-1)^s \frac{1}{\ln s}$

12. $\displaystyle\sum_{i=1}^{\infty} \sin\left(\frac{\pi i}{4} \right)$

13. $\displaystyle\sum_{k=1}^{\infty} \left(\frac{1}{2^k} + (-1)^{k-1} \frac{1}{k} \right)$

14. $\displaystyle\sum_{z=1}^{\infty} \frac{\sin z}{z^2}$

15. $\displaystyle\sum_{n=2}^{\infty} \frac{\ln n}{n^{1.01}}$

Exercises 16–21: Each of the indicated series has some positive and some negative terms. Decide if the series converges absolutely, converges conditionally, or neither.

16. The series in Exercise 2.
17. The series in Exercise 8.
18. The series in Exercise 11.

19. The series in Exercise 12.
20. The series in Exercise 13.
21. The series in Exercise 14.

Exercises 22–27: Verify that the series satisfies the hypotheses of the Alternating Series Test. Then for the given number E, estimate the sum of the series to within E. Use a computer or calculator as needed.

22. $\displaystyle\sum_{k=1}^{\infty} (-1)^{k-1} \frac{1}{k^2 + 1}$, $\quad E = 0.01$

23. $\displaystyle\sum_{r=1}^{\infty} (-1)^{r-1} \frac{1}{\sqrt{r^2 + 1}}$, $\quad E = 0.01$

24. $\displaystyle\sum_{j=0}^{\infty} (-1)^j \frac{1}{4^j}$, $\quad E = 0.0001$

25. $\displaystyle\sum_{k=0}^{\infty} (-1)^{k-1} \frac{1}{2k-1}$, $\quad E = 0.005$

26. $\displaystyle\sum_{\ell=1}^{\infty} (-1)^{\ell-1} \frac{1}{(\ln(\ell+1))^5}$, $\quad E = 0.01$

27. $\displaystyle\sum_{j=0}^{\infty} (-1)^j \frac{1}{j!}$, $\quad E = 10^{-6}$

Exercises 28–31: Let a_k be the kth term of the series. Let $\rho = \displaystyle\lim_{k \to \infty} \left| \frac{a_{k+1}}{a_k} \right|$. For the given value of ρ, find all possible values of x.

28. $\displaystyle\sum_{k=0}^{\infty} \frac{x^k}{3^k}$, $\quad \rho = \frac{1}{2}$

29. $\displaystyle\sum_{k=0}^{\infty} k x^k$, $\quad \rho = 3$

30. $\displaystyle\sum_{k=0}^{\infty} \frac{(x+1)^k}{5^k}$, $\quad \rho = e$

31. $\displaystyle\sum_{k=0}^{\infty} \frac{x^k}{k!}$, $\quad \rho = 0$

Growth

32. Write an argument justifying part b of the Comparison Tests. Your argument can be something like the discussion showing why part a of the test is true.

33. Sometimes students use the following *incorrect* version of the ratio test:

Let $\sum_{k=1}^{\infty} a_k$ be an infinite series with $a_k > 0$ for all k. Suppose that $a_{k+1}/a_k < 1$ for every positive integer k. Then the series $\sum_{k=1}^{\infty} a_k$ converges.

Present an example that shows this statement is not always true.

34. Let $\sum_{k=1}^{\infty} a_k$ be an infinite series with $a_k > 0$ for all k.

Suppose there is a number $\rho < 1$ such that

$$\frac{a_{k+1}}{a_k} < \rho$$

is true for all k. Show that the series $\sum_{k=1}^{\infty} a_k$ converges. How does this test differ from the ratio test? How is it similar?

35. (In this exercise you are asked to establish the Integral Test, another test for convergence of series of nonnegative terms.) Let $f(x)$ be a nonnegative, decreasing function defined for $x \geq 1$.
  a. Using the figure below as a guide, show that

$$\int_1^{n+1} f(x)\,dx \leq \sum_{k=1}^{n} f(k).$$

  b. Based on the inequality established in **a,** show that if $\int_1^{\infty} f(x)\,dx$ diverges, then $\sum_{k=1}^{\infty} f(k)$ diverges.

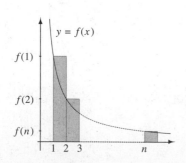

c. Next show that

$$\sum_{k=1}^{n} f(k) \leq f(1) + \int_1^{n} f(x)\,dx.$$

To illustrate your reasoning, draw a figure something like the one used for part **a.**
  d. Use the inequality established in **c** to show that if $\int_1^{\infty} f(x)\,dx$ converges, then $\sum_{k=1}^{\infty} f(k)$ converges.
  e. The Integral Test says that if f is a nonnegative, decreasing function defined for $x \geq 1$, then

$$\sum_{k=1}^{\infty} f(k) \quad \text{and} \quad \int_1^{\infty} f(x)\,dx$$

either both diverge or both converge. Explain why this is true.

36. Use the Integral Test, developed in the previous exercise, to show that the p-series

$$\sum_{k=1}^{\infty} \frac{1}{k^p}$$

converges for $p > 1$ and diverges for $0 < p \leq 1$. (Hint: Set $f(x) = \dfrac{1}{x^p}$. You will have to consider the case $p = 1$ separately.)

37. Use the Integral Test, developed in Exercise 35, to show that the series

$$\sum_{k=2}^{\infty} \frac{1}{k(\ln k)^p}$$

converges for $p > 1$ and diverges for $0 < p \leq 1$.

38. Use the Integral Test, developed in Exercise 35, to show that the geometric series

$$\sum_{k=1}^{\infty} c^k$$

converges for $0 < c < 1$. The Integral Test cannot be used to show that the series diverges for $c > 1$. Why not?

39. (In this exercise we see how the Integral Test can be used to estimate the error when we approximate the sum of a convergent series by a partial sum.) Let f be a nonnegative, decreasing function defined for $x \geq 1$. Suppose that $\int_1^{\infty} f(x)\,dx$ converges. Then we know from the Integral Test that $\sum_{k=1}^{\infty} f(k)$ also converges. Let $\sum_{k=1}^{\infty} f(k) = S$. Show that for positive integer n,

$$\left| S - \sum_{k=1}^{n} f(k) \right| = \left| \sum_{k=n+1}^{\infty} f(k) \right| \leq \int_{n}^{\infty} f(x)\, dx.$$

40. Consider the convergent series $\sum_{k=1}^{\infty} \dfrac{1}{k^{3/2}}$. Let S be the sum of the series.

a. Use the technique developed in the previous problem to show that

$$\left| S - \sum_{k=1}^{100} \frac{1}{k^{3/2}} \right| \leq 0.2.$$

b. What value of n should you take if you want to approximate S by

$$s_n = \sum_{k=1}^{n} \frac{1}{k^{3/2}}$$

with an error of at most 0.005?

41. How many terms of the series $\sum_{k=1}^{\infty} \dfrac{1}{k^{1.01}}$ are needed to approximate the sum with an error of at most 0.001?

42. How many terms of the series $\sum_{k=2}^{\infty} \dfrac{1}{k(\ln k)^2}$ are needed to approximate the sum with an error of at most 0.001?

43. How many terms of the series $\sum_{k=2}^{\infty} \dfrac{1}{k(\ln k)^{1.01}}$ are needed to approximate the sum with an error of at most 0.001?

44. Let $s_n = \sum_{k=1}^{\infty} \dfrac{1}{k}$ be a partial sum of the harmonic series. How large (roughly) must n be to make $s_n > 10$? To make $s_n > 100$?

45. Let $s_n = \sum_{k=1}^{\infty} \dfrac{1}{\sqrt{k}}$ be a partial sum of the p-series with $p = 1/2$. How large (roughly) must n be to make $s_n > 10$? To make $s_n > 1000$?

46. Let $s_n = \sum_{k=1}^{\infty} \dfrac{1}{k \ln k}$. How large (roughly) must n be to make $s_n > 10$? To make $s_n > 100$?

47. Let $\sum_{k=1}^{\infty} a_k$ and $\sum_{k=1}^{\infty} b_k$ be series of positive terms, and suppose that

$$\lim_{k \to \infty} \frac{a_k}{b_k} = L$$

where $0 < L < \infty$.

a. Explain why there is a positive integer K such that for all $k > K$,

$$\frac{1}{2} L < \frac{a_k}{b_k} < \frac{3}{2} L.$$

b. Show that if the assertion in part **a** is true, then for $k > K$,

$$\frac{1}{2} L b_k < a_k < \frac{3}{2} L b_k.$$

c. Explain why it follows from the Comparison Tests that $\sum_{k=1}^{\infty} a_k$ and $\sum_{k=1}^{\infty} b_k$ either both converge or both diverge.

48. Let $\sum_{k=1}^{\infty} a_k$ and $\sum_{k=1}^{\infty} b_k$ be series of positive terms, and suppose that

$$\lim_{k \to \infty} \frac{a_k}{b_k} = 0.$$

a. Show that if $\sum_{k=1}^{\infty} b_k$ converges, then $\sum_{k=1}^{\infty} a_k$ also converges.

b. Show by example that the convergence of $\sum_{k=1}^{\infty} a_k$ does not imply the convergence of $\sum_{k=1}^{\infty} b_k$.

49. Let $\sum_{k=1}^{\infty} a_k$ and $\sum_{k=1}^{\infty} b_k$ be series of positive terms, and suppose that

$$\lim_{k \to \infty} \frac{a_k}{b_k} = \infty.$$

a. Show that if $\sum_{k=1}^{\infty} a_k$ converges, then $\sum_{k=1}^{\infty} b_k$ also converges.

b. Show by example that the convergence of $\sum_{k=1}^{\infty} b_k$ does not imply the convergence of $\sum_{k=1}^{\infty} a_k$.

50. Does the series

$$\sum_{k=1}^{\infty} \frac{1}{k^{1+1/k}}$$

converge or diverge? Justify your answer.

51. Let $\sum_{k=1}^{\infty} a_k$ be a convergent series of positive terms.

Is it always true that $\sum_{k=1}^{\infty} a_k^2$ also converges?

52. Let $\sum_{k=1}^{\infty} (-1)^{k-1} b_k$ satisfy the hypotheses of the Alternating Series Test and let $s_n = \sum_{k=1}^{n} (-1)^{k-1} b_k$.

a. Show that the odd-indexed partial sums form a decreasing sequence. That is, show

$$s_1 \geq s_3 \geq s_5 \geq \cdots \geq s_{2m-1} \geq \cdots.$$

b. Show that if S is the sum of the series $\sum_{k=1}^{\infty} (-1)^{k-1} b_k$ and n is an odd integer, then

$$s_{n+1} \leq S \leq s_n.$$

Establish a similar inequality when n is an even integer.

c. Use the inequalities established in part **b** to show that for positive integer n,

$$\left| S - \left(s_n + \frac{(-1)^n b_{n+1}}{2} \right) \right| \leq \frac{1}{2} b_{n+1}.$$

Review/Preview

53. Find $f(t)$ so that

$$\int_0^x f(t) \, dt = \ln(1 + x).$$

54. Find $h(t)$ such that

$$\int_0^x h(t) \, dt = \arctan x.$$

55. Find $g(t)$ if

$$\int_0^x g(t) \, dt = \arcsin x.$$

56. Use the trapezoid rule with 100 trapezoids to approximate

$$\int_0^{\pi/2} \sin x^2 \, dx.$$

57. Let $P(x) = (x - c)^n$. Find $f''(x)$, $f^{(5)}(x)$, and $f^{(k)}(x)$, $k < n$.

58. Find $T_6(x; \pi/2)$, the sixth Taylor polynomial for $\sin x$ about $\pi/2$.

59. Find $T_6(x; 0)$, the sixth Taylor polynomial for e^{3x} about 0.

60. Find the partial fractions expansion for

$$\frac{3x - 1}{2x^2 - 4x}.$$

61. Find the partial fractions expansion for

$$\frac{1}{(x - 3)(x - 1)(x + 2)}.$$

62. Perform the polynomial multiplication

$$(3x^3 - 2x + 1)(4x^2 - x + 2).$$

63. Find the terms of degree 0, 1, 2, and 3 for the product

$$(1 + x + x^2 + x^3 + x^4)(3 - 4x + x^2 - x^4 - 4x^5).$$

7.6 POWER SERIES AND TAYLOR SERIES

Many of the functions that we work with are easily described by simple formulas. Some examples are

$$(1) \qquad\qquad f(x) = x^2 \quad \text{and} \quad h(t) = \frac{1}{(1 + t)^2}.$$

Given such a formula for a function we can easily compute the value of the function for any real number in the domain of the function. There are other very familiar functions for which we have devised notation, such as

(2) $$s(\theta) = \sin\theta, \quad \ell(t) = \ln t, \quad F(x) = \arctan x.$$

We have done a lot of work with these functions and may feel that we know and understand them. We do not have formulas that tell us how to find the value of such a function for a given input. However, we know that there are algorithms or formulas that we can use to approximate the value of such a function. We may not even know what these algorithms or formulas are, but at least the people that built our calculators and computers do!

There are other important functions that cannot be expressed by simple formulas such as (1). Such functions can show up as solutions to problems in wave phenomena, diffusion, or quantum mechanics, for example. Many of these functions are not as well known as those in (2), so there is no universally known notation to denote them. Many such functions are described by infinite series. For example, the vibration of a piano string can be described by an expression something like

$$d(x, t) = \sum_{k=0}^{\infty} e^{-kt} \sin kt \cos x$$

where at time t, $d(x, t)$ is the height above or below equilibrium of the point of the string x units from one end. Even familiar functions such as those in (2) can be expressed as infinite series. As an example, for any real θ,

(3) $$\sin\theta = \theta - \frac{\theta^3}{3!} + \frac{\theta^5}{5!} - \cdots.$$

In this section we will study infinite series similar to the one shown in (3). Such series are called power series or Taylor series. We sill see how to find the interval of convergence for such a series, how to express a given function as a power series, and how to use the partial sums of these series to approximate the function to a desired accuracy.

Power Series

DEFINITION **power series**

Let c be a real number and $\{a_k\}_{k=0}^{\infty}$ be a sequence of real numbers. An infinite series of the form

(4) $$\sum_{k=0}^{\infty} a_k(x - c)^k = a_0 + a_1(x - c) + a_2(x - c)^2 + \cdots$$

is called a **power series** about c. The numbers $a_0, a_1, a_2, \ldots$ are the coefficients of the power series.

We will usually think of (4) as defining a function, say f, by

$$f(x) = \sum_{k=0}^{\infty} a_k(x - c)^k$$

This function is defined for the values of x for which the series (4) converges. If we know all of the coefficients a_k for the series and if we can evaluate

$$\lim_{k \to \infty} \frac{a_{k+1}}{a_k},$$

then we can use the Ratio Test to find an interval on which f is defined.

■ **EXAMPLE 1**   Show that the power series

$$\sum_{k=1}^{\infty} \frac{kx^k}{3^k}$$

converges when $|x| < 3$ and diverges when $|x| > 3$.

Solution.   The Ratio Test is a test for series with positive terms. Hence we first consider the series

$$\sum_{k=1}^{\infty} \left| \frac{kx^k}{3^k} \right|.$$

To apply the Ratio Test to this series, we look at the ratio of the $(k + 1)$th term to the kth term and take the limit:

(5) $$\rho = \lim_{k \to \infty} \frac{\left| \dfrac{(k + 1)x^{k+1}}{3^{k+1}} \right|}{\left| \dfrac{kx^k}{3^k} \right|} = \lim_{k \to \infty} \frac{1}{3} \left(\frac{k + 1}{k} \right) |x| = \frac{1}{3} |x|.$$

Since

$$\rho = \frac{1}{3} |x| < 1 \quad \text{when} \quad |x| < 3,$$

we see that $\sum_{k=1}^{\infty} \left| \dfrac{kx^k}{3^k} \right|$ converges when $|x| < 3$. By the Absolute Convergence Test of Section 7.5, the series

$$\sum_{k=1}^{\infty} \frac{kx^k}{3^k}$$

also converges when $|x| < 3$.

Next note that

$$\rho = \frac{1}{3} |x| > 1 \quad \text{when} \quad |x| > 3.$$

In our discussion of the Ratio Test in Section 7.5, we showed that when $p > 1$ the terms of the series do not tend to 0. Hence if $|x| > 3$, we have

$$\lim_{k \to \infty} \left| \frac{kx^k}{3^k} \right| \neq 0$$

so that

$$\lim_{k \to \infty} \frac{kx^k}{3^k} \neq 0.$$

Thus, by the nth term divergence test,

$$\sum_{k=1}^{\infty} \frac{kx^k}{3^k}$$

diverges when $|x| > 3$. ∎

When we used the Ratio Test in the previous example, we ignored the special case $x = 0$. The series $\sum_{k=1}^{\infty} \frac{kx^k}{3^k}$ certainly converges when $x = 0$. However, the limit in (5) does not exist when $x = 0$ because for $k \geq 0$ there is a 0 in the denominator of the ratio. In the future we will not mention this special case. We note here that the series $\sum_{k=0}^{\infty} a_k(x - c)^k$ always converges when $x = c$, and we will assume that $x \neq c$ when we apply the Ratio Test to such series.

The process we used in analyzing the series in Example 1 can be streamlined somewhat. Let $\sum_{k=0}^{\infty} a_k(x - c)^k$ be a power series and suppose that

$$\lim_{k \to \infty} \left| \frac{a_{k+1}}{a_k} \right| = \rho$$

where $0 \leq \rho \leq \infty$. Define

$$(6) \qquad R = \begin{cases} \dfrac{1}{\rho} & (0 < \rho < \infty) \\ \infty & (\rho = 0) \\ 0 & (\rho = \infty). \end{cases}$$

Then for $x \neq c$,

$$\lim_{k \to \infty} \frac{|a_{k+1}(x - c)^{k+1}|}{|a_k(x - c)^k|} = \rho|x - c| = \frac{|x - c|}{R} \text{ is } \begin{cases} < 1 & (|x - c| < R) \\ > 1 & (|x - c| > R). \end{cases}$$

Hence $\sum_{k=0}^{\infty} a_k(x - c)^k$ converges for $|x - c| < R$ and diverges for $|x - c| > R$. The number R is called the *radius of convergence* of the power series $\sum_{k=0}^{\infty} a_k(x - c)^k$.

DEFINITION   radius of convergence

Let $\sum_{k=0}^{\infty} a_k(x-c)^k$ be a power series. There is a number $R, 0 \leq R \leq \infty$, such that

$$\sum_{k=0}^{\infty} a_k(x-c)^k \text{ converges when } |x-c| < R.$$

$$\sum_{k=0}^{\infty} a_k(x-c)^k \text{ diverges when } |x-c| > R.$$

The number R is called the **radius of convergence** of $\sum_{k=0}^{\infty} a_k(x-c)^k$. Furthermore, if $\lim_{k \to \infty} \dfrac{a_{k+1}}{a_k} = \rho$ exists, then the radius of convergence is

$$R = \frac{1}{\rho}.$$

(The value of the fraction is interpreted as in (6).)

Let $\sum_{k=0}^{\infty} a_k(x-c)^k$ be a power series and let R be the radius of convergence of the series. Then the series converges for all x satisfying $c - R < x < c + R$. Hence for x in this interval we can define a function f by

(7)
$$f(x) = \sum_{k=0}^{\infty} a_k(x-c)^k.$$

The series in (7) may or may not converge at either of the endpoints of the interval. We shall usually ignore these endpoints, though the behavior of the series at these points can often be determined by using the Comparison Test. See Exercises 19–22. The Ratio Test is ineffective at the endpoints of the interval (why?). However, with the help of the Ratio Test we can approximate $f(x)$ for any x in $(c - R, c + R)$. In fact, we can do more, as stated in the following result.

Uniform Convergence of Power Series

Let $\sum_{k=0}^{\infty} a_k(x-c)^k$ be a power series with radius of convergence R. Define

$$f(x) = \sum_{k=0}^{\infty} a_k(x-c)^k \quad (c - R < x < c + R).$$

If $0 < r < R$, then the series converges uniformly on the interval $[c - r, c + r]$. That is, the sequence

$$\left\{ \sum_{k=0}^{n} a_k(x-c)^k \right\}_{n=0}^{\infty}$$

of partial sums of the power series converges to f uniformly on $[c - r, c + r]$. See Fig. 7.31.

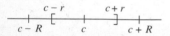

Figure 7.31. A power series converges uniformly on an interval $[c - r, c + r] \subset (c - R, c + R)$.

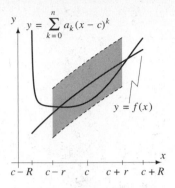

Figure 7.32. When n is large, the graph of the nth partial sum lies in the ϵ-corridor about f for $c - r \leq x \leq c + r$. This need not be the case outside this interval.

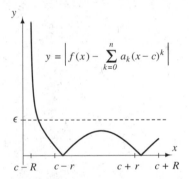

Figure 7.33. On $[c - r, c + r]$, the error $|f(x) - \sum_{k=0}^{n} a_k(x - c)^k|$ is small when n is large.

This result can be interpreted graphically in a couple of ways. Let ϵ be a small positive number and draw the ϵ-corridor around the graph of $y = f(x)$ for $c - r \leq x \leq c + r$. Then if n is large enough, the graph of $y = \sum_{k=0}^{n} a_k(x - c)^k$ lies in the ϵ-corridor about f. This is equivalent to saying that when n is large enough,

$$\left| f(x) - \sum_{k=0}^{n} a_k(x - c)^k \right| < \epsilon$$

for all x satisfying $c - r \leq x \leq c + r$. This means that $\sum_{k=0}^{n} a_k(x - c)^k$ might be a good approximation to $f(x)$ for $c - r \leq x \leq c + r$. See Figs. 7.32 and 7.33.

The following investigation illustrates how the Ratio Test can be used to show that a power series converges uniformly. We also show how information from the Ratio Test can be used to find an approximation to the sum of the series.

Investigation

In Example 1 we showed that the radius of convergence of the power series

$$(8) \qquad \sum_{k=1}^{\infty} \frac{kx^k}{3^k}$$

is 3. For $|x| < 3$ define $f(x)$ to be the sum of this series. Find a partial sum of (8) that approximates $f(x)$ on $[-2, 2]$ with a maximum error of no more than 0.001.

In Section 7.4 we showed that when the ratio test tells us that a series converges, then the terms of the series are eventually dominated by the terms of a convergent geometric series. We will use our knowledge of the geometric series to estimate the error when we use a partial sum of (8) to approximate f.

For convenience let

$$b_k(x) = \frac{kx^k}{3^k}$$

so the series (8) is $\sum_{k=1}^{\infty} b_k(x)$. Let $|x| \leq 2$ and apply the ratio test to (8). As seen in Example 1,

$$\rho = \lim_{k \to \infty} \left| \frac{b_{k+1}(x)}{b_k(x)} \right| = \lim_{k \to \infty} \frac{\left| \frac{(k + 1)x^{k+1}}{3^{k+1}} \right|}{\left| \frac{kx^k}{3^k} \right|} = \lim_{k \to \infty} \frac{1}{3}\left(\frac{k + 1}{k} \right)|x| = \frac{1}{3}|x| \leq \frac{2}{3}.$$

Now pick a number between $\rho = \frac{2}{3}$ and 1, say $\frac{3}{4}$. There is a positive integer K so that for $k \geq K$

$$(9) \qquad \left| \frac{b_{k+1}(x)}{b_k(x)} \right| = \frac{1}{3}\left(\frac{k + 1}{k} \right)|x| \leq \frac{3}{4}.$$

Because $|x| \leq 2$, (9) will be true when

$$\frac{k + 1}{k} \leq \frac{9}{8}.$$

This last inequality is true when $k \geq 8$, so we can take $K = 8$. Hence for $k \geq 8$ and $|x| \leq 2$ we have

(10)
$$|b_{k+1}(x)| \leq \frac{3}{4}|b_k(x)|.$$

Now note that for $|x| \leq 2$

(11)
$$|b_8(x)| = \frac{8|x|^8}{3^8} \leq 8\left(\frac{2}{3}\right)^8 < 0.4.$$

Applying (10) and (11) with $k = 8, 9, 10, \ldots$, we find that when $|x| \leq 2$,

$$|b_9(x)| \leq \left(\frac{3}{4}\right)|b_8(x)| < 0.4\left(\frac{3}{4}\right)$$

$$|b_{10}(x)| \leq \frac{3}{4}|b_9(x)| \leq 0.4\left(\frac{3}{4}\right)^2$$

$$|b_{11}(x)| \leq \frac{3}{4}|b_{10}(x)| \leq 0.4\left(\frac{3}{4}\right)^3$$

$$\vdots$$

and in general, for $k \geq 8$,

(12)
$$|b_k(x)| \leq 0.4\left(\frac{3}{4}\right)^{k-8}.$$

Now let $N > 8$ and estimate the error when we use $\sum_{k=0}^{N} b_k(x)$ to approximate f on $[-2, 2]$. We have

$$\left|f(x) - \sum_{k=0}^{N} b_k(x)\right| = \left|\sum_{k=0}^{\infty} b_k(x) - \sum_{k=0}^{N} b_k(x)\right| = \left|\sum_{k=N+1}^{\infty} b_k(x)\right|$$

$$\leq \sum_{k=N+1}^{\infty} |b_k(x)|.$$

Use (12) on each term of the last sum to obtain

$$\left|f(x) - \sum_{k=0}^{N} b_k(x)\right| \leq \sum_{k=N+1}^{\infty} 0.4\left(\frac{3}{4}\right)^{k-8} = \frac{0.4\left(\frac{3}{4}\right)^{(N+1)-8}}{1 - \frac{3}{4}} = 1.6\left(\frac{3}{4}\right)^{N-7}.$$

We want an error of no more than 0.001, so we need to find N so that the last expression is less than 0.001. We can do this through trial and error with a calculator, or by solving directly. Taking logarithms of both sides, we see that

$$1.6\left(\frac{3}{4}\right)^{N-7} < 0.001$$

is equivalent to

$$(N - 7)\ln\left(\frac{3}{4}\right) < \ln\left(\frac{1}{1600}\right).$$

Dividing by the negative number $\ln\left(\frac{3}{4}\right)$ and then adding 7, we find

$$N > 7 + \frac{\ln(1/1600)}{\ln(3/4)} \approx 32.645.$$

Thus on $[-2, 2]$ we can approximate f to within 0.001 by taking $N = 33$. That is, on $[-2, 2]$

$$f(x) = \sum_{k=1}^{\infty} \frac{kx^k}{3^k} \approx \sum_{k=1}^{33} \frac{kx^k}{3^k}$$

with an error of at most 0.001. In Fig. 7.34 we have graphed $y = f(x)$ and this approximation on the interval $[-3, 3]$. We can detect no difference in the graphs on the interval $[-2, 2]$. In Fig. 7.35 we have graphed

$$y = \left| f(x) - \sum_{k=1}^{33} \frac{kx^k}{3^k} \right|.$$

We see that the error in the approximation is actually much less than 0.001.

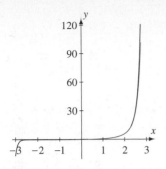

Figure 7.34. Graphs $y = f(x)$ and $y = \sum_{1}^{33} kx^k/3^k$.

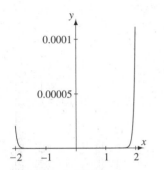

Figure 7.35. When we use $\sum_{1}^{33} kx^k/3^k$ to approximate $f(x)$ on $[-2, 2]$, the error is actually much less than .001.

Taylor Series

In Section 7.1 we introduced the Taylor polynomials. The Taylor polynomials are the partial sums of a special type of power series called a Taylor series.

DEFINITION   **Taylor series**

Let f be a function, let c be a real number in the domain of f, and suppose that f has derivatives of all orders at c. The **Taylor series** for f at the point c is the power series

$$\sum_{k=0}^{\infty} \frac{f^{(k)}(c)}{k!}(x - c)^k.$$

Given a Taylor series for a function, we will usually be interested in answering two questions:

1. What is the radius of convergence, R, of the series?
2. Does the Taylor series converge to f on $c - R < x < c + R$?

A Taylor series is a power series. Hence if we know the coefficients of the Taylor series, we can attempt to answer question 1 by using the Ratio Test. Question 2 is important because it touches on the approximation problem. Suppose that

$$(13) \quad \sum_{k=0}^{\infty} \frac{f^{(k)}(c)}{k!}(x - c)^k = \lim_{n \to \infty} \sum_{k=0}^{n} \frac{f^{(k)}(c)}{k!}(x - c)^k = \lim_{n \to \infty} T_n(x; c) = f(x)$$

for all x in $(c - R, c + R)$. Then for any r between 0 and R the convergence is uniform on the interval $[c - r, c + r]$. This means that for any $\epsilon > 0$ we can find a positive integer n such that

$$|f(x) - T_n(x; c)| < \epsilon \quad \text{for } c - r \le x \le c + r.$$

In other words, on $[c - r, c + r]$ we can approximate f to any desired accuracy with a Taylor polynomial.

In Section 7.2 we showed how to estimate the error

$$|R_n(x; c)| = |f(x) - T_n(x; c)|$$

when the Taylor polynomial $T_n(x; c)$ is used to approximate $f(x)$. If we can show that

$$\lim_{n \to \infty} R_n(x; c) = 0$$

for x in $(c - R, c + R)$, then we can conclude that (13) holds for all such x.

■ **EXAMPLE 2** Find the Taylor series at 0 for $\sin x$. Show that this series converges to $\sin x$ for all real x.

Solution. Let $f(x) = \sin x$. Then

$$f^{(0)}(0) = \sin 0 = 0$$
$$f^{(1)}(0) = \cos 0 = 1$$
$$f^{(2)}(0) = -\sin 0 = 0$$
$$f^{(3)}(0) = -\cos 0 = -1$$
$$f^{(4)}(0) = \sin 0 = 0$$
$$\vdots$$

Since the 0, 1, 0, −1 pattern in the list of values for $f^{(k)}(0)$ repeats, the Taylor series for $\sin x$ is

$$\sum_{k=0}^{\infty} \frac{f^{(k)}(c)}{k!}(x - c)^k = \frac{0}{0!}x^0 + \frac{1}{1!}x! + \frac{0}{2!}x^2 + \frac{(-1)}{3!}x^3 + \frac{0}{4!}x^4 + \frac{1}{5!}x^5 + \cdots$$

$$= x - \frac{x^3}{3!} + \frac{x^5}{5!} - \frac{x^7}{7!} + \frac{x^9}{9!} - \cdots$$

$$= \sum_{j=0}^{\infty} (-1)^j \frac{x^{2j+1}}{(2j + 1)!}.$$

To show this series converges for all real x we use the Ratio Test. For any real number x,

$$\lim_{j \to \infty} \left| \frac{(-1)^{j+1} \dfrac{x^{2j+3}}{(2j+3)!}}{(-1)^{j} \dfrac{x^{2j+1}}{(2j+1)!}} \right| = \lim_{j \to \infty} \frac{|x|^2}{(2j+3)(2j+2)} = 0 < 1.$$

Hence, for any real number x, the series

$$\sum_{j=0}^{\infty} (-1)^{j} \frac{x^{2j+1}}{(2j+1)!}$$

converges. The series has radius of convergence $R = \infty$.

Next we show that for any x the Taylor series has sum $\sin x$. Let $T_n(x; c)$ be the Taylor polynomial of degree n for $\sin x$. Then by the Taylor inequality (Section 7.2),

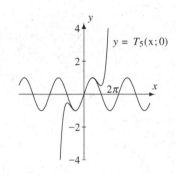

$y = T_5(x; 0)$

(14) $$|\sin x - T_n(x; c)| = |R_n(x; c)| \le \frac{M_{n+1}}{(n+1)!} |x|^{n+1}.$$

Now

$$M_{n+1} = \text{maximum of } \left| \frac{d^{n+1}}{dt^{n+1}} \sin t \right| \text{ for } t \text{ between } 0 \text{ and } x$$

$$= \text{maximum of } |\sin t| \text{ or } |\cos t| \text{ for } t \text{ between } 0 \text{ and } x$$

$$\le 1.$$

Putting this upper bound for M_{n+1} into (14), we find

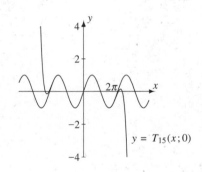

$y = T_{15}(x; 0)$

$$|\sin x - T_n(x; c)| \le \frac{|x|^{n+1}}{(n+1)!}.$$

For fixed real x,

$$\lim_{n \to \infty} \frac{|x|^{n+1}}{(n+1)!} = 0.$$

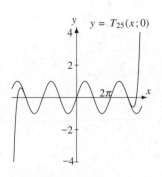

$y = T_{25}(x; 0)$

(Why?) This means that we can make $|\sin x - T_n(x; c)|$ as small as we like by taking n large enough. Hence

$$\sin x = \lim_{n \to \infty} T_n(x; c) = \sum_{k=0}^{\infty} \frac{f^{(k)}(c)}{k!} (x - c)^k = \sum_{j=0}^{\infty} (-1)^j \frac{x^{2j+1}}{(2j+1)!}$$

Figure 7.36. Graphs $y = \sin x$ and $y = T_n(x; 0)$ for $n = 5, 15, 25$.

for all real x. Furthermore, since the radius of convergence of the Taylor (power) series is $R = \infty$, the series converges uniformly to $\sin x$ on intervals of the form $[-r, r]$. In Fig. 7.36 we show the graphs of $y = \sin x$ and some of the Taylor polynomials for $\sin x$. We see that the Taylor polynomials approximate $\sin x$ better and better on larger and larger intervals as n gets bigger.

Sometimes we recognize a series whose behavior we already know about. In such cases it might be possible to show that a Taylor series for a function f converges to f without working with the remainder.

■ **EXAMPLE 3**   Find the Taylor series about 2 for

$$f(x) = \frac{1}{x}.$$

Show that the series converges to $f(x)$ for $0 < x < 4$.

Solution.   We begin by finding the coefficients for the Taylor series. Noting that

$$f^{(0)}(x) = \frac{1}{x}$$

$$f^{(1)}(x) = -\frac{1}{x^2}$$

$$f^{(2)}(x) = \frac{2 \cdot 1}{x^3}$$

$$f^{(3)}(x) = -\frac{3 \cdot 2 \cdot 1}{x^4}$$

$$\vdots$$

we infer that

$$f^{(k)}(x) = (-1)^k \frac{k!}{x^{k+1}}.$$

Hence

$$f^{(k)}(2) = (-1)^k \frac{k!}{2^{k+1}},$$

and the Taylor series is

(15) $$\sum_{k=0}^{\infty} \frac{f^{(k)}(2)}{k!}(x-2)^k = \sum_{k=0}^{\infty}(-1)^k \frac{(x-2)^k}{2^{k+1}}.$$

We can rewrite the last series in (15) as

(16) $$\sum_{k=0}^{\infty} \frac{1}{2}\left(\frac{-(x-2)}{2}\right)^k.$$

This shows that the Taylor series is a geometric series with first term $a = \dfrac{1}{2}$ and common ratio $r = \dfrac{-(x-2)}{2}$. Hence the series converges if and only if

$$|r| = \left|\frac{-(x-2)}{2}\right| < 1, \quad \text{that is, if and only if} \quad 0 < x < 4.$$

For such x the series converges to

$$\frac{a}{1 - r} = \frac{1/2}{1 - \left(\frac{-(x - 2)}{2}\right)} = \frac{1}{x}.$$

Thus for $0 < x < 4$ the Taylor series at 2 for $f(x) = \dfrac{1}{x}$ converges to $\dfrac{1}{x}$. Because the Taylor series is a power series, the convergence is uniform on any closed interval $[a, b]$ contained in $(0, 4)$. ∎

In the next two sections we will look at some applications of power series and Taylor series. For convenience, we list here some of the more important Taylor series and the intervals on which the series converge.

Important Taylor Series

$$\frac{1}{1 - x} = 1 + x + x^2 + x^3 + \cdots = \sum_{k=0}^{\infty} x^k \qquad (-1 < x < 1)$$

$$\sin x = x - \frac{x^3}{3!} + \frac{x^5}{5!} - \frac{x^7}{7!} + \cdots = \sum_{k=0}^{\infty} (-1)^k \frac{x^{2k+1}}{(2k + 1)!}$$
$$(-\infty < x < \infty)$$

$$\cos x = 1 - \frac{x^2}{2!} + \frac{x^4}{4!} - \frac{x^6}{6!} + \cdots = \sum_{k=0}^{\infty} (-1)^k \frac{x^{2k}}{(2k)!}$$
$$(-\infty < x < \infty)$$

$$e^x = 1 + x + \frac{x^2}{2!} + \frac{x^3}{3!} + \cdots = \sum_{k=0}^{\infty} \frac{x^k}{k!} \qquad (-\infty < x < \infty)$$

$$\sqrt{1 + x} = 1 + \frac{1}{2}x - \frac{1}{2^2}\frac{x^2}{2!} + \frac{1 \cdot 3}{2^3}\frac{x^3}{3!} - \frac{1 \cdot 3 \cdot 5}{2^4}\frac{x^4}{4!}x^4 + \cdots$$

$$= 1 + \frac{x}{2} + \sum_{k=2}^{\infty} (-1)^{k-1} \frac{1 \cdot 3 \cdots (2k - 3)}{2^k k!} x^k \qquad (-1 < x \leq 1)$$

Exercises 7.6

Basic

Exercises 1–6: Find the radius of convergence of each power series.

1. $\displaystyle\sum_{k=1}^{\infty} k^2 x^k$

2. $\displaystyle\sum_{k=1}^{\infty} \frac{(-1)^{k-1}}{2^k}(x + 2)^k$

3. $\displaystyle\sum_{j=0}^{\infty} \frac{(j + 1)x^j}{j!}$

4. $\displaystyle\sum_{n=2}^{\infty} \frac{x^{2n}}{(\ln n)^2}$

5. $\displaystyle\sum_{j=1}^{\infty} \frac{j!\left(x - \frac{1}{2}\right)^j}{2^j j^j}$

6. $\displaystyle\sum_{k=0}^{\infty} \frac{k(k + 1)(k + 2)(k + 3)}{5^k}(x + \pi)^k$

Exercises 7–12:   Find the Taylor series for the function f about the point c. Then find the radius of convergence of the Taylor series.

7. $f(x) = 2e^{3x}$, $c = 0$

8. $f(x) = \sqrt{x-2}$, $c = 3$

9. $f(x) = -2x^3 + 4x^2 - 7x + 4$, $c = -2$

10. $f(x) = \dfrac{1}{3+x} + \dfrac{1}{2-x}$, $c = 0$

11. $f(x) = \dfrac{1}{3+x} + \dfrac{1}{2-x}$, $c = 5$

12. $f(x) = \dfrac{1}{3+x} + \dfrac{1}{2-x}$, $c = -5$

Growth

Exercises 13–18:   Show that the power series converges on the interval I. How many terms of the series are needed to approximate the sum to within 0.001 on I?

13. $\displaystyle\sum_{k=1}^{\infty} x^k$, $I = \left[-\dfrac{1}{3}, \dfrac{1}{3}\right]$

14. $\displaystyle\sum_{k=1}^{\infty} \dfrac{x^k}{(2k)!}$, $I = [-1, 1]$

15. $\displaystyle\sum_{j=0}^{\infty} \dfrac{(x-2)^j}{3^{j+1}}$, $I = [1, 3]$

16. $\displaystyle\sum_{m=1}^{\infty} \dfrac{(-1)^{m-1}}{m} x^m$, $I = \left[-\dfrac{1}{2}, \dfrac{1}{2}\right]$

17. $\displaystyle\sum_{j=0}^{\infty} \dfrac{j(j+1)}{j!}(x+3)^j$, $I = [-5, -1]$

18. $\displaystyle\sum_{k=1}^{\infty} \dfrac{x^k}{k^k}$, $I = [-3, 3]$

19. The power series $\displaystyle\sum_{k=1}^{\infty} \dfrac{x^k}{k^2}$ has radius of convergence $R = 1$. Suppose $x = 1$ is substituted into the series. Does the resulting series converge or diverge? What if $x = -1$ is substituted into the series?

20. The power series $\displaystyle\sum_{k=1}^{\infty} \dfrac{x^k}{k}$ has radius of convergence $R = 1$. Suppose $x = 1$ is substituted into the series. Does the resulting series converge or diverge? What if $x = -1$ is substituted into the series?

21. The power series $\displaystyle\sum_{k=1}^{\infty} kx^k$ has radius of convergence $R = 1$. Suppose $x = 1$ is substituted into the series. Does the resulting series converge or diverge? What if $x = -1$ is substituted into the series?

22. Let $\displaystyle\sum_{k=0}^{\infty} a_k(x-c)^k$ be a power series with radius of convergence R, where $0 < R < \infty$. Substitute the value $x = c + R$ into the series. What can you say about the behavior of the resulting series? What if $x = c - R$ is substituted into the series? (See the previous three exercises before answering!)

23. Write down a power series about $c = 5$ with radius of convergence $R = 2$.

24. Write down a power series about $c = -\pi$ with radius of convergence $R = e$.

25. Let $f(x) = \dfrac{1}{1-x}$ and let c be a real number with $-1 < c < 1$. When f is expanded in a Taylor series about c, what is the radius of convergence of the resulting series?

26. Let $f(x) = \dfrac{1}{1-x}$ and let c be a real number with $|c| > 1$. When f is expanded in a Taylor series about c what is the radius of convergence of the resulting series?

27. Let $g(x) = \cos 3x$ and let c be a real number. When g is expanded in a Taylor series about c what is the radius of convergence of the resulting series?

28. For what values of x does the series $\displaystyle\sum_{k=0}^{\infty} \dfrac{k}{x^k}$ converge?

29. For what values of x does the series $\displaystyle\sum_{k=0}^{\infty} \dfrac{(k+1)3^k}{(x-2)^k}$ converge?

30. For what values of x does the series $\displaystyle\sum_{k=0}^{\infty} (2\sin x)^k$ converge?

31. For what values of x does the series $\displaystyle\sum_{k=1}^{\infty} \dfrac{1}{x^2+k^2}$ converge?

Review/Preview

32. Find a_0, a_1, a_2, a_3 so that

$$y = a_0 + a_1 x + a_2 x^2 + a_3 x^3$$

satisfies the differential equation

$$2y'' + 3y' + y = x^3 + 1.$$

33. Find a_0, a_1, a_2, a_3 so that

$$y = a_0 + a_1 x + a_2 x^2 + a_3 x^3$$

satisfies the differential equation

$$y'' - 4y' + 4y = 2x^2.$$

34. Find a and b so that

$$y = a \cos 2x + b \sin 2x$$

is a solution of the differential equation

$$y'' + y' + y = -\sin 2x.$$

35. Find a and b so that

$$y = a \cos 4x + b \sin 4x$$

is a solution of the differential equation

$$-y'' + 4y' + y = 2 \cos 4x - 2 \sin 4x.$$

36. Show that

$$\sum_{k=1}^{10} \frac{k}{k^2 + 1} = \sum_{j=0}^{9} \frac{j+1}{(j+1)^2 + 1}.$$

37. Show that

$$\sum_{k=0}^{\infty} k(k-1)x^{k-2} = \sum_{j=-2}^{\infty} (j+2)(j+1)x^j.$$

38. Find m and n so that

$$\sum_{k=3}^{140} \frac{2^{k-8}}{(k+1)(k+2)^3} = \sum_{j=m}^{n} \frac{2^j}{(j+9)(j+10)^3}.$$

39. Find m so that

$$\sum_{k=0}^{\infty} \frac{3^k x^{k+3}}{(k+1)(k+2)(k+3)} = \sum_{j=m}^{\infty} \frac{3^{j-3} x^j}{(j-2)(j-1)j}.$$

40. Combine into one sum with terms involving only x^k:

$$\sum_{k=0}^{30} k x^{k-1} + \sum_{k=0}^{29} (k+1)x^k$$

41. Combine into one sum with terms involving only x^k:

$$\sum_{k=0}^{\infty} k(k-1)x^{k-2} - 2 \sum_{k=0}^{\infty} x^k$$

7.7 COMPUTING POWER SERIES

When we have calculated the Taylor series

$$\sum_{k=0}^{\infty} \frac{f^{(k)}(c)}{k!}(x-c)^k$$

for a function f at given c, much of our effort has gone into finding the values of $f^{(k)}(c)$ for $k = 0, 1, 2, 3, \ldots$. Sometimes there is no other way to proceed. However, there are often short cuts available. In this section we shall see that we can quickly generate many Taylor series by applying differentiation, integration, and algebra to the five series listed at the end of Section 7.6.

These techniques can help us solve another problem too. In Section 7.8 we will see how to express the solutions of some differential equations as Taylor series. Such series solutions can be easier to work with if we can write them in terms of familiar functions. This can be done if we recognize how our series solution can be obtained by manipulating known Taylor series.

Algebraic Manipulation of Taylor Series

New Taylor series can be obtained from known series by adding or multiplying series, or by substitution in a known series.

Substitution Many Taylor series can be found by starting with a known series and substituting a new expression for the variable in the series.

■ **EXAMPLE 1** Find the Taylor series about $\frac{\pi}{2}$ for $\sin x$.

Solution. First note that

$$\sin x = \cos\left(x - \frac{\pi}{2}\right).$$

Hence a series for $\sin x$ can be found by substituting $x - \frac{\pi}{2}$ for t in the series

$$(1) \qquad \cos t = \sum_{k=0}^{\infty} (-1)^k \frac{t^{2k}}{(2k)!} \quad (-\infty < t < \infty).$$

Making this substitution, we obtain

$$(2) \qquad \sin x = \cos\left(x - \frac{\pi}{2}\right) = \sum_{k=0}^{\infty} (-1)^k \frac{\left(x - \frac{\pi}{2}\right)^{2k}}{(2k)!}.$$

Since (1) is valid for all real t, equation (2) is true for all real x. ■

■ **EXAMPLE 2** Find the Taylor series about -2 for the function

$$f(x) = \frac{9 - 2x}{x^2 - 9}.$$

What is the radius of convergence of the series?

Solution. We will obtain the series for f by substitution in the geometric series. First use partial fractions to break $f(x)$ into a sum of simpler fractions:

$$\frac{9 - 2x}{x^2 - 9} = \frac{1}{x - 3} - \frac{2}{x + 3}.$$

We will find a Taylor series for each of these fractions and then add the two series to get a series for f.

Take $\frac{1}{x - 3}$ first. Because we are looking for a series about -2, the series we obtain will be in powers of $(x - (-2)) = (x + 2)$. Hence we rewrite $\frac{1}{x - 3}$ with the realization that the final answer will be given in terms of $x + 2$. This gives

$$(3) \qquad \frac{1}{x - 3} = \frac{1}{(x + 2) - 5} = -\frac{1}{5}\left(\frac{1}{1 - \frac{x + 2}{5}}\right).$$

The last expression in (3) was written to prepare for substitution into the geometric series

$$(4) \qquad \frac{1}{1-t} = \sum_{k=0}^{\infty} t^k \quad (-1 < t < 1).$$

Substitute $\dfrac{x+2}{5}$ for t in (4) to obtain

$$(5) \qquad \frac{1}{1 - \dfrac{x+2}{5}} = \sum_{k=0}^{\infty} \left(\frac{x+2}{5} \right)^k = \sum_{k=0}^{\infty} \frac{1}{5^k}(x+2)^k.$$

Since (4) is valid for $|t| < 1$, it follows that (5) is true when $\left| \dfrac{x+2}{5} \right| < 1$. Multiplying (5) by $-\frac{1}{5}$ and using (3), we see that

$$(6) \qquad \frac{1}{x-3} = -\frac{1}{5} \sum_{k=0}^{\infty} \frac{1}{5^k}(x+2)^k = -\sum_{k=0}^{\infty} \frac{1}{5^{k+1}}(x+2)^k.$$

Now apply a similar procedure to the fraction $\dfrac{2}{x+3}$. We have

$$(7) \qquad \begin{aligned} \frac{2}{x+3} &= \frac{2}{(x+2)+1} = 2 \frac{1}{1-(-(x+2))} \\ &= 2 \sum_{k=0}^{\infty} (-(x+2))^k = 2 \sum_{k=0}^{\infty} (-1)^k (x+2)^k. \end{aligned}$$

for $|x+2| < 1$. Combining (6) and (7), we see that

$$(8) \qquad \begin{aligned} \frac{9-2x}{x^2-9} &= \frac{1}{x-3} - \frac{2}{x+3} \\ &= -\sum_{k=0}^{\infty} \frac{1}{5^{k+1}}(x+2)^k + 2\sum_{k=0}^{\infty} (-1)^k(x+2)^k \\ &= \sum_{k=0}^{\infty} \left((-1)^k 2 - \frac{1}{5^{k+1}} \right)(x+2)^k. \end{aligned}$$

Because (6) is valid for $|x+2| < 5$ and (7) is valid for $|x+2| < 1$, we conclude that (8) is true for $|x+2| < 1$. Hence the series has radius of convergence 1. ∎

Multiplication of Series   Power or Taylor series about the same point c can be multiplied in much the same way as we multiply polynomials. Although a general formula for the product of two series can be given (see Exercise 26), we will usually be satisfied with using multiplication to find the first few terms of a series.

■ **EXAMPLE 3**   Let $f(x) = e^x \sin x$. Find the terms through degree 3 of the series for f about 0.

Solution. Start by writing out the first few terms of the series for e^x and $\sin x$.

$$e^x = 1 + x + \frac{x^2}{2} + \frac{x^3}{6} + \frac{x^4}{24} + \frac{x^5}{120} + \cdots$$

$$\sin x = x - \frac{x^3}{6} + \frac{x^5}{120} - \cdots$$

To find the product of these two series, think of them as "infinitely long polynomials" and proceed as you would if multiplying two polynomials. First write one series above the other, keeping terms with like powers of x aligned in columns. Then multiply every term in the series on the bottom by every term in the series on the top. Keep the results of these multiplications aligned in columns by powers of x. When we have all products that will give a term of degree 3 or less, stop and add the columns. The result is the beginning of the series for $e^x \sin x$.

e^x series: $\qquad 1 + x + \dfrac{x^2}{2} + \dfrac{x^3}{6} + \dfrac{x^4}{24} + \dfrac{x^5}{120} + \cdots$

$\sin x$ series: $\qquad x \qquad\quad - \dfrac{x^3}{6} \qquad\quad + \dfrac{x^5}{120} - \cdots$

Multiply top row by x: $\qquad x + x^2 + \dfrac{x^3}{2} + \dfrac{x^4}{6} + \dfrac{x^5}{24} + \cdots$

Multiply top row by $-\dfrac{x^3}{6}$: $\qquad\qquad\qquad - \dfrac{x^3}{6} - \dfrac{x^4}{6} - \dfrac{x^5}{12} - \cdots$

Multiply top row by $\dfrac{x^5}{120}$: $\qquad\qquad\qquad\qquad\qquad + \dfrac{x^5}{120} + \cdots$

Add the columns: $\qquad x + x^2 + \dfrac{x^3}{3} \qquad\quad - \dfrac{x^5}{30} \pm \cdots$

In the multiplication demonstrated above we actually produced the series for f out to the x^5 term. Hence the first few terms of the Taylor series about 0 for $e^x \sin x$ are

$$x + x^2 + \frac{x^3}{3} - \frac{x^5}{30} \cdots$$

How well does the fifth-degree Taylor polynomial for f approximate f on $[-1, 1]$? ∎

Integration and Differentiation

Integration Integration is a powerful tool for developing new power series from old ones. Suppose that we have

(9) $$f(x) = \sum_{k=0}^{\infty} a_k(x - c)^k \quad \text{for} \quad c - R < x < c + R.$$

Suppose also that for x in this interval we have $F'(x) = f(x)$. It seems reasonable that we can obtain a series for F by integrating the series for f. That is, for $c - R < x < c + R$,

$$F(x) - F(c) = \int_c^x f(t)\,dt = \int_c^x \sum_{k=0}^{\infty} a_k(t - c)^k dt$$

$$= \sum_{k=0}^{\infty} \left(\int_c^x a_k(t - c)^k dt \right) = \sum_{k=0}^{\infty} \frac{a_k}{k+1}(x - c)^{k+1}.$$

Summarizing the result of this work, we have

Integration of Power Series

Suppose that $F'(x) = f(x)$ for $(c - R < x < c + R)$ and that (9) is true. Then

$$F(x) = F(c) + \sum_{k=0}^{\infty} \frac{a_k}{k+1}(x - c)^{k+1} \quad \text{for } (c - R < x < c + R).$$

In particular, the series for F about c has the same radius of convergence as the series for f about c.

■ **EXAMPLE 4** Find a power series about 0 that is equal to $\arctan x$ in an interval centered at 0.

Solution. Because

$$(10) \qquad \arctan x = \arctan x - \arctan 0 = \int_0^x \frac{1}{1 + t^2}\,dt,$$

we start by finding a series about 0 for $\dfrac{1}{1 + t^2}$. We do this by substituting $-t^2$ for r in the geometric series

$$\frac{1}{1 - r} = \sum_{k=0}^{\infty} r^k.$$

This equality holds for $|r| < 1$. Hence when $|t| < 1$, we have

$$\frac{1}{1 + t^2} = \frac{1}{1 - (-t^2)} = \sum_{k=0}^{\infty} (-1)^k t^{2k}.$$

Integrating this last series and using (10), we obtain

$$\arctan x = \int_0^x \frac{1}{1 + t^2}\,dt = \int_0^x \left(\sum_{k=0}^{\infty} (-1)^k t^{2k} \right) dt$$

$$= \sum_{k=0}^{\infty} \left(\int_0^x (-1)^k t^{2k}\,dt \right) = \sum_{k=0}^{\infty} \frac{(-1)^k}{2k+1} x^{2k+1}.$$

This series converges to $\arctan x$ for $-1 < x < 1$. (Why?)   ■

We can also use these ideas to estimate the values of definite integrals.

■ **EXAMPLE 5**   Estimate the value of

(11)
$$\int_0^3 e^{-x^2}\,dx$$

to within 0.001.

Solution.   We will solve this problem in two ways. First recall that

$$e^t = \sum_{k=0}^{\infty} \frac{t^k}{k!} \quad \text{for } (-\infty < t < \infty).$$

Hence for all real x, we have

(12)
$$e^{-x^2} = \sum_{k=0}^{\infty} \frac{(-x^2)^k}{k!} = \sum_{k=0}^{\infty} \frac{(-1)^k x^{2k}}{k!}.$$

Thus for all real w,

$$\int_0^w e^{-x^2}\,dx = \sum_{k=0}^{\infty} \left(\int_0^w (-1)^k \frac{x^{2k}}{k!}\,dx \right) = \sum_{k=0}^{\infty} \frac{(-1)^k}{(2k+1)k!} w^{2k+1}.$$

Setting $w = 3$, we have

(13)
$$\int_0^3 e^{-x^2}\,dx = \sum_{k=0}^{\infty} (-1)^k \frac{3^{2k+1}}{(2k+1)k!}.$$

We know that the series in (13) converges, and we would like to approximate the sum with an error of at most 0.001. We need to find out how many terms of the series in (13) are needed to approximate the sum to the desired accuracy. We do this by using the error estimate for the alternating series test. Let

$$b_k = \frac{3^{2k+1}}{(2k+1)k!}.$$

Then the series in (13) is $\sum_{k=0}^{\infty}(-1)^k b_k$. We know that $\lim_{k\to\infty} b_k = 0$ because the series converges. In addition, we have

$$b_0 < b_1 < b_2 < \cdots < b_7 > b_8 > b_9 > b_{10} > \cdots.$$

Thus the sequence $\{b_k\}$ is decreasing once we get past the first six terms. Hence, except for the first six terms, (13) satisfies the hypotheses of the alternating series test. Therefore, for $n \geq 7$,

$$\left| \int_0^3 e^{-x^2}\,dx - \sum_{k=0}^{n} (-1)^k \frac{3^{2k+1}}{(2k+1)k!} \right| \leq b_{n+1} = \frac{3^{2n+3}}{(2n+3)(n+1)!}.$$

The difference between the integral and the partial sum will be less than 0.001 if we choose $n \geq 7$ with $b_{n+1} < 0.001$. We can find such n by trial and error. In the accompanying table we show some values for b_{n+1}. From the table we see that

n	$b_{n+1} = \dfrac{3^{2n+3}}{(2n+3)(n+1)!}$
6	189.800
10	102.543
15	8.05133
20	0.149418
24	0.0027225
25	0.000906842
26	0.000291289

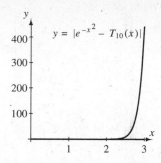

Figure 7.37. $T_{10}(x)$ is not a good approximation to e^{-x^2} on $[0, 3]$.

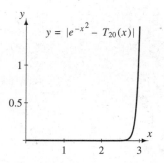

Figure 7.38. $|e^{-x^2} - T_{20}(x)|$ is not small enough for all x in the interval $[0, 3]$.

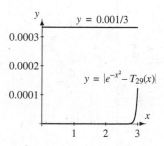

Figure 7.39. The graph shows $|e^{-x^2} - T_{29}(x)| < 0.001/3$ for $0 \le x \le 3$.

$b_{n+1} < 0.001$ for $n = 25$. Hence

$$\int_0^3 e^{-x^2}\, dx \approx \sum_{k=0}^{25} (-1)^k \frac{3^{2k+1}}{(2k+1)k!} \approx 0.885522,$$

and the error in this approximation is no more than 0.001.

For a second solution to this problem, we rely more on trial and error and graphs. Our first objective is to find a partial sum of the series (12) that differs from e^{-x^2} by less than $0.001/3$ on the interval $0 \le x \le 3$. Let $T_n(x)$ be such a partial sum. Then we have

$$\left| \int_0^3 e^{-x^2}\, dx - \int_0^3 T_n(x)\, dx \right| \le \int_0^3 \left| e^{-x^2} - T_n(x) \right| dx < \int_0^3 \frac{0.001}{3}\, dx = 0.001.$$

This means that

$$\int_0^3 e^{-x^2}\, dx \approx \int_0^3 T_n(x)\, dx$$

with an error of less than 0.001. We find $T_n(x)$ by graphing

$$y = \left| e^{-x^2} - T_n(x) \right|$$

for $0 \le x \le 3$ and just trying values of n until this graph stays below the line $y = 0.001/3$. In Figs. 7.37, 7.38, and 7.39 we show these graphs for $n = 10$, 20, 29. We used a CAS to find the partial sums $T_n(x)$ then graphed the desired difference. By trial and error we found $n = 29$ was the smallest integer for which $\left| e^{-x^2} - T_n(x) \right| < 0.001/3$ on all of $[0, 3]$. Hence we have

$$\int_0^3 e^{-x^2}\, dx \approx \int_0^3 \left(\sum_{k=0}^{29} (-1)^k \frac{x^{2k}}{k!} \right) dx = \sum_{k=0}^{29} (-1)^k \frac{3^{2k+1}}{(2k+1)k!} \approx 0.886201$$

with an error of no more than 0.001.   ■

Differentiation   Power series can also be differentiated to obtain power series for new functions.

Differentiation of Power Series

Suppose that

$$f(x) = \sum_{k=0}^{\infty} a_k(x - c)^k \quad \text{for } c - R < x < c + R.$$

Then

$$f'(x) = \sum_{k=0}^{\infty} ka_k(x - c)^{k-1} \quad \text{for } c - R < x < c + R.$$

In particular, the power series for f' has the same radius of convergence as the power series for f.

■ **EXAMPLE 6**   Find a power series about 0 for $\dfrac{1}{(1-x)^2}$.

Solution.   Because

$$\frac{1}{(1-x)^2} = \frac{d}{dx}\left(\frac{1}{1-x}\right),$$

we can find a series for $\dfrac{1}{(1-x)^2}$ by differentiating the geometric series. Recall that

$$\frac{1}{1-x} = \sum_{k=0}^{\infty} x^k \quad \text{for } (-1 < x < 1).$$

Thus

$$\frac{1}{(1-x)^2} = \left(\frac{1}{1-x}\right)' = \left(\sum_{k=0}^{\infty} x^k\right)' = \sum_{k=0}^{\infty} (x^k)' = \sum_{k=0}^{\infty} kx^{k-1} = \sum_{j=0}^{\infty} (j+1)x^j$$

for $-1 < x < 1$. How could we have used series multiplication to get this result? ■

Exercises 7.7

Basic

Exercises 1–12:   Use the techniques illustrated in this section to find the Taylor series for f about the point $x = c$. Also give the interval of convergence for each series. (Use the five Taylor series given at the end of the previous section.)

1. $f(x) = e^{-3x}, \quad c = 0$

2. $f(x) = \dfrac{3}{x+4}, \quad c = 0$

3. $f(x) = 2\cos x - 3\sin x, \quad c = 0$

4. $f(x) = \sqrt{1 - x^2}, \quad c = 0$

5. $f(x) = \cos^2 x, \quad c = 0$

6. $f(x) = (x^2 + 1)\sin 2x, \quad c = 0$

7. $f(x) = \cos x, \quad c = \dfrac{\pi}{4}$

8. $f(x) = \dfrac{1}{\sqrt{1+x}}, \quad c = 0$

9. $f(x) = \sqrt{3 + 2x}, \quad c = -1$

10. $f(x) = \dfrac{x+2}{x^2 - 3x}, \quad c = 2$

11. $f(x) = \ln(x+2), \quad c = 0$

12. $f(x) = x^2 \sin x, \quad c = 1$

Exercises 13–16:   Find the first four nonzero terms of the Taylor series for g about $x = c$.

13. $g(x) = \dfrac{e^x}{1-x}, \quad c = 0$

14. $g(x) = (\sin x)\sqrt{1+x}, \quad c = 0$

15. $g(x) = \arcsin x, \quad c = 0$

16. $g(x) = \sqrt{\dfrac{1+x}{1-x}}, \quad c = 0$

Growth

17. Let $\displaystyle\sum_{k=0}^{\infty} a_k(x-c)^k$ be a power series with

$$\lim_{k\to\infty} \frac{a_{k+1}}{a_k} = R.$$

Show that the series $\displaystyle\sum_{k=0}^{\infty} ka_k(x-c)^{k-1}$ also has radius of convergence R.

18. Let $\displaystyle\sum_{k=0}^{\infty} a_k(x-c)^k$ be a power series with

$$\lim_{k\to\infty} \frac{a_{k+1}}{a_k} = R.$$

Show that the series $\displaystyle\sum_{k=0}^{\infty} \frac{a_k}{k+1}(x-c)^{k+1}$ also has radius of convergence R.

19. Let $\displaystyle f(x) = \sum_{k=1}^{\infty} \frac{\sqrt{k}}{3^k}x^k$ for $-3 < x < 3$. Approxi-

mate $\displaystyle\int_0^1 f(x)\,dx$ with an error of less than 0.001.

20. Let $\displaystyle f(x) = \sum_{k=0}^{\infty}(-1)^k \frac{x^k}{(k!)^2}$ for $-\infty < x < \infty$.

Approximate $\displaystyle\int_{k=0}^{\infty} f(x)\,dx$ with an error of less than 0.001.

21. Approximate $\displaystyle\int_0^{\pi/2} \sin(x^2)\,dx$ with an error of less than 0.001. Compare with your answer to Exercise 56 in Section 7.5.

Exercises 22–25: Find the sum of the series. Each series was obtained by applying substitution, integration, and/or differentiation to one of the series listed at the end of Section 7.6.

22. $\displaystyle\sum_{k=0}^{\infty}\left(\frac{x-3}{5}\right)^k$

23. $\displaystyle\sum_{k=0}^{\infty}(-1)^{k-1}\frac{(x+2)^{2k}}{k!}$

24. $\displaystyle\sum_{k=0}^{\infty}(k+1)(k+2)x^k$

25. $\displaystyle\sum_{k=1}^{\infty}\frac{1}{k}\left(\frac{x-2}{2}\right)^k$

26. Suppose the product of the two power series $\displaystyle\sum_{k=0}^{\infty} a_k x^k$ and $\displaystyle\sum_{k=0}^{\infty} b_k x^k$ is

$$\left(\sum_{k=0}^{\infty} a_k x^k\right)\left(\sum_{k=0}^{\infty} b_k x^k\right) = \sum_{c=0}^{\infty} c_n x^n.$$

Show that

$$c_0 = a_0 b_0$$
$$c_1 = a_0 b_1 + a_1 b_0$$
$$c_2 = a_0 b_2 + a_1 b_1 + a_2 b_0.$$

and that in general

$$c_n = \sum_{k=0}^{n} a_k b_{n-k}$$
$$= a_0 b_n + a_1 b_{n-1} + \cdots + a_k b_{n-k}$$
$$+ \cdots + a_n b_0.$$

27. By equating the coefficients of x^n on each side of

$$\sum_{k=0}^{\infty} \frac{2^k}{k!}x^k = e^{2x}$$
$$= e^x e^x = \left(\sum_{k=0}^{\infty}\frac{1}{k!}x^k\right)\left(\sum_{k=0}^{\infty}\frac{1}{k!}x^k\right),$$

show that

$$\frac{2^n}{n!} = \frac{1}{0!}\frac{1}{n!} + \frac{1}{1!}\frac{1}{(n-1)!} + \cdots$$
$$+ \frac{1}{k!}\frac{1}{(n-k)!} + \cdots + \frac{1}{n!}\frac{1}{0!}.$$

28. Assume that the Taylor series for $\sec x$ at $c = 0$ is $\displaystyle\sum_{k=0}^{\infty} b_k x^k$ and that this series converges to $\sec x$ for x in some interval $-R < x < R$. Use the identity

$$1 = (\cos x)(\sec x)$$
$$= \left(\sum_{k=0}^{\infty}(-1)^{k-1}\frac{x^{2k}}{(2k)!}\right)\left(\sum_{k=0}^{\infty} b_k x^k\right)$$

to find the first three nonzero terms of the series for $\sec x$.

29. With the aid of the identity

$$(\cos x)(\tan x) = \sin x,$$

find the first four nonzero terms of the series for $\tan x$ about $c = 0$.

30. Let $f(x) = \sum_{k=0}^{\infty} a_k(x - c)^k$ for $c - R < x < c + R$.

Show that the power series is also the Taylor series for f at $x = c$. That is, show that $a_k = \dfrac{f^{(k)}(c)}{k!}$.

31. Suppose that $\sum_{k=0}^{\infty} a_k(x - c)^k = 0$ for $c - R < x <$

$c + R$. Show that $a_k = 0$ for $k = 0, 1, 2, \ldots$.

32. Suppose that

$$\sum_{k=0}^{\infty} a_k(x - c)^k = \sum_{k=0}^{\infty} b_k(x - c)^k$$

for $c - R < x < c + R$. Show that $a_k = b_k$ for $k = 0, 1, 2, \ldots$. (Hint: Do Exercise 31 first.)

33. Suppose that

$$f(c) = f'(c) = f''(c) = \cdots = f^{(m-1)}(c) = 0$$

and that

$$f(x) = \sum_{k=0}^{\infty} a_k(x - c)^k$$

for $c - R < x < c + R$. Show that for x in this interval

$$f(x) = (x - c)^m \sum_{k=0}^{\infty} a_{k+m}(x - c)^k.$$

With the help of this result, evaluate

$$\lim_{x \to c} \frac{f(x)}{(x - c)^m}.$$

Exercises 34–39: *Using power series to evaluate limits. If $\lim_{x \to c} f(x) = \lim_{x \to c} g(x) = 0$ and both f and g are equal to their power series about c, then we can use the power series to determine $\lim_{x \to c} \dfrac{f(x)}{g(x)}$. Use the idea developed in the previous exercise to replace each of $f(x)$ and $g(x)$ by power series multiplied by a factor of the form $(x - c)^{m-1}$. Cancel common factors and then determine the limit.*

34. $\lim_{x \to 0} \dfrac{\sin x}{2x}$

35. $\lim_{x \to 0} \dfrac{\cos x - 1}{x \sin x - x}$

36. $\lim_{x \to 0} \dfrac{e^{x^2} - 1 - x^2}{x^2 \cos x}$

37. $\lim_{x \to 1} \dfrac{(\ln x)^2}{2(x - 1)^2}$

38. $\lim_{x \to 0} \dfrac{\ln(1 + ax)}{x}$, $\quad a > 0$

39. $\lim_{x \to \pi/2} \dfrac{\cos x - (x - \pi/2)}{(x - \pi/2)^3}$

Review/Preview

40. Given that $a_0 = 1$ and that

$$a_k = ka_{k-1}, \quad (k \ge 1)$$

find $a_1, a_2, a_3,$ and a_4. Give a formula for a_n.

41. Given that $a_0 = 5$ and that

$$a_k = \frac{a_{k-1}}{2k}, \quad (k \ge 1)$$

find $a_1, a_2, a_3,$ and a_4. Give a formula for a_n.

42. Given that $a_0 = 1$, $a_1 = -1$ and that

$$a_k = \frac{a_{k-2}}{2}, \quad (k \ge 1)$$

find $a_2, a_3, a_4, a_5,$ and a_6. Give a formula for a_n. (You will actually need two formulas, one for even n and one for odd n.)

43. Given that $a_0 = \pi$, $a_1 = 0$, and that

$$a_k = \frac{ka_{k-2}}{k + 2}, \quad (k \ge 1)$$

find $a_2, a_3, a_4, a_5,$ and a_6. Give a formula for a_n. (You will actually need two formulas, one for even n and one for odd n.)

44. Let $\mathbf{v} = (-1, 2)$ and $\mathbf{w} = (3, 4)$. Find the length of $\mathbf{v}$ and the length of $\mathbf{w}$. Find the angle determined by the two vectors.

45. Let $\mathbf{v} = (2, -3)$ and $\mathbf{w} = (0, 8)$. Find the length of $\mathbf{v}$ and the length of $\mathbf{w}$. Find the angle determined by the two vectors.

46. Let $\mathbf{v} = (-1, 3)$. Find a vector perpendicular to $\mathbf{v}$.

47. Let $\mathbf{w} = (1, -\pi)$. Find a vector perpendicular to $\mathbf{w}$.

48. A particle moves in the plane in such a way that its position at time t is given by $\mathbf{r}(t) = (2t + 1, -3t)$, $t \geq 0$. Sketch the path of the particle. Also find the speed and the velocity of the particle.

49. A particle moves in the plane in such a way that its position at time t is given by $\mathbf{r}(t) = (5 \cos 2t, 5 \sin 2t)$, $t \geq 0$. Sketch the path of the particle. Also find the speed and the velocity of the particle.

7.8   SOLVING DIFFERENTIAL EQUATIONS

Some differential equations have solutions that cannot be expressed as a finite combination of elementary functions (e.g., polynomials, trigonometric functions, exponentials, and logarithms). However, in many cases a solution to such an equation can be expressed as a power series. Even if we are unable to find a closed-form expression for the series, the partial sums of the series can be used to approximate the solution to the differential equation.

Consider an initial value problem

$$(1) \qquad y' = F(t, y), \qquad y(t_0) = y_0.$$

Assume (1) has a solution y that can be expressed as a power series

$$(2) \qquad y = y(t) = \sum_{k=0}^{\infty} a_k(t - t_0)^k.$$

Note that we have taken the point about which the series is expanded to be t_0, the point at which the initial value data are given. We may determine a_0 by using the initial condition in (1). Setting $t = t_0$ in (2), we find

$$y_0 = y(t_0) = \sum_{k=0}^{\infty} a_k(t_0 - t_0)^k = a_0.$$

To determine $a_1, a_2, \ldots$ we substitute (2) into (1) and, after some algebra, obtain an equation involving one or more power series. We then match coefficients $(t - t_0)^n$ for each positive integer n and solve the resulting equations for the remaining coefficients. We illustrate this process with a simple example.

■ **EXAMPLE 1**   Use series to solve the initial value problem

$$(3) \qquad y' = 2y, \qquad y(0) = 1.$$

Solution.   We assume that (3) has a solution that can be expressed as a power series about 0. Hence

$$(4) \qquad y = y(t) = \sum_{k=0}^{\infty} a_k t^k.$$

Substituting $t = 0$ into this last expression, we find

$$(5) \qquad 1 = y(0) = a_0.$$

In preparation for the substitution of (4) into (3), first note that

$$y' = \sum_{k=0}^{\infty} k a_k t^{k-1}.$$

Hence (3) becomes

(6)
$$\sum_{k=0}^{\infty} k a_k t^{k-1} = 2 \sum_{k=0}^{\infty} a_k t^k.$$

We want to work with coefficients of like powers of t. This is easiest when all series involved display the same power of t. Thus we adjust the sum on the left of (6) so it involves t^k instead of t^{k-1}. We do this by making the substitution $j = k - 1$ (or $k = j + 1$) in the sum on the left. We have

$$\sum_{k=0}^{\infty} k a_k t^{k-1} = \sum_{j+1=0}^{\infty} (j + 1)a_{j+1} t^j = \sum_{j=-1}^{\infty} (j + 1)a_{j+1} t^j.$$

Renaming the index j as k in this last sum and substituting into (6), we obtain

$$\sum_{k=-1}^{\infty} (k + 1)a_{k+1} t^k = 2 \sum_{k=0}^{\infty} a_k t^k.$$

Subtracting the left side of this equation from both sides of the expression, we find

(7)
$$\sum_{k=-1}^{\infty} (k + 1)a_{k+1} t^k - 2 \sum_{k=0}^{\infty} a_k t^k$$

$$= (-1 + 1)a_0 t^{-1} + \sum_{k=0}^{\infty} ((k + 1)a_{k+1} - 2a_k)t^k = 0.$$

(The expression $(-1 + 1)a_0 t^{-1} = 0$ is the $k = -1$ term from the first sum. This term had to be handled separately because the other sum did not have such a term. Two summations can be combined only if they sum over the same set of indices, in this case $k = 0$ to ∞.) In (7) we have a power series that is always equal to 0. This can happen only if the coefficient of each t^k is 0. (Why?) Setting each coefficient to 0 gives us a collection of equations involving the coefficients a_k.

coefficient of t^0: $(0 + 1)a_{0+1} - 2a_0 = 0$

coefficient of t^1: $(1 + 1)a_{1+1} - 2a_1 = 0$

coefficient of t^2: $(2 + 1)a_{2+1} - 2a_2 = 0$

$\qquad \vdots \qquad\qquad\qquad \vdots \qquad \vdots$

coefficient of t^k: $(k + 1)a_{k+1} - 2a_k = 0$

We can solve each of these equations to express a coefficient a_{k+1} in terms of the previous coefficient a_k. Recalling that $a_0 = 1$, we find

$$a_1 = \frac{2}{1}a_0 = \frac{2}{1}$$

$$a_2 = \frac{2}{2}a_1 = \frac{2^2}{2 \cdot 1}$$

$$a_3 = \frac{2}{3}a_2 = \frac{2^3}{3!}$$

$$\vdots \qquad \vdots$$

$$a_k = \frac{2}{k}a_{k-1} = \frac{2^k}{k!}$$

Hence we see that for $k = 0, 1, 2, \ldots$ we have $a_k = \frac{2^k}{k!}$. Substituting these values for the coefficients into (4), we have

$$y = \sum_{k=0}^{\infty} \frac{2^k}{k!}t^k.$$

We recognize this as the Taylor series about 0 for e^{2t}. Check that $y = e^{2t}$ works in (3). ∎

The technique illustrated in the previous example can also be used to find solutions of second-order differential equations, that is, differential equations that involve second derivatives.

■ **EXAMPLE 2**   Solve the initial value problem

(8) $y'' - ty' - y = 0, \qquad y(0) = 0, \quad y'(0) = 1.$

Check that the solution is correct by verifying that it works in (8).

Solution.   Assume that the solution to (8) can be expressed in the form

(9) $y = y(t) = \sum_{k=0}^{\infty} a_k t^k.$

Differentiating (9), we find

$$y' = \sum_{k=0}^{\infty} k a_k t^{k-1}$$

$$y'' = \sum_{k=0}^{\infty} k(k-1) a_k t^{k-2}.$$

Substituting into (8), we find

$$0 = y'' - ty' - y$$

(10)
$$= \sum_{k=0}^{\infty} k(k-1)a_k t^{k-2} - t \sum_{k=0}^{\infty} k a_k t^{k-1} - \sum_{k=0}^{\infty} a_k t^k$$

$$= \sum_{k=2}^{\infty} k(k-1)a_k t^{k-2} - \sum_{k=0}^{\infty} k a_k t^k - \sum_{k=0}^{\infty} a_k t^k.$$

We need to combine these three summations into one, so rewrite the first of the sums in the last line of (10) so it involves t^k instead of t^{k-2}. Making the substitution $k - 2 = j$ (or $k = j + 2$) in this sum, we have

$$\sum_{k=2}^{\infty} k(k-1)a_k t^{k-2} = \sum_{j+2=2}^{\infty} (j+2)(j+1)a_{j+2} t^j = \sum_{j=0}^{\infty} (j+2)(j+1)a_{j+2} t^j.$$

Rename j as k in the last sum and substitute into (10). We now find

$$\sum_{k=0}^{\infty} (k+2)(k+1)a_{k+2} t^k - \sum_{k=0}^{\infty} k a_k t^k - \sum_{k=0}^{\infty} a_k t^k$$

$$= \sum_{k=0}^{\infty} ((k+2)(k+1)a_{k+2} - (k+1)a_k) t^k = 0.$$

For $k = 0, 1, 2, \ldots$, the coefficient of t^k is 0. This leads to the following equations.

coefficient of t^0:	$(0+2)(0+1)a_{0+2} - (0+1)a_0 = 0$
coefficient of t^1:	$(1+2)(1+1)a_{1+2} - (1+1)a_1 = 0$
coefficient of t^2:	$(2+2)(2+1)a_{2+2} - (2+1)a_2 = 0$
$\vdots$	$\vdots$ $\vdots$
coefficient of t^k:	$(k+2)(k+1)a_{k+2} - (k+1)a_k = 0$

Before solving these equations, use the initial conditions in (8) to see that $a_0 = y(0) = 0$ and $a_1 = y'(0) = 1$. We then find that

$$a_2 = \frac{1}{2 \cdot 1} a_0 = 0$$

$$a_3 = \frac{2}{3 \cdot 2} a_1 = \frac{1}{1 \cdot 3}$$

$$a_4 = \frac{3}{4 \cdot 3} a_2 = 0$$

$$a_5 = \frac{4}{5 \cdot 4} a_3 = \frac{1}{1 \cdot 3 \cdot 5}$$

$$a_6 = \frac{5}{6 \cdot 5} a_4 = 0$$

$$\vdots \qquad \qquad \vdots$$

$$a_k = \frac{k-1}{k(k-1)} a_{k-2} = \frac{1}{k} a_{k-2}.$$

Based on the pattern that emerges on solving for the a_k's and on the recursion relation $a_k = \dfrac{1}{k} a_{k-2}$, we see that

$$a_k = \begin{cases} 0 & (k \text{ even}) \\ \dfrac{1}{1 \cdot 3 \cdot 5 \cdots k} & (k \text{ odd}). \end{cases}$$

Hence the solution to (8) is

$$y = a_1 t + a_3 t^3 + a_5 t^5 + \cdots + a_{2j+1} t^{2j+1} + \cdots$$

(11)
$$= \sum_{j=0}^{\infty} a_{2j+1} t^{2j+1}$$

$$= \sum_{j=0}^{\infty} \frac{t^{2j+1}}{1 \cdot 3 \cdot 5 \cdots (2j+1)}.$$

What is the radius of convergence of the series?

We can check the solution to almost any differential equation by making sure that the solution works in the differential equation. We do this informally by seeing that the first few terms of the series solution "work" in (8). Since

(12)
$$y = y(t) = \sum_{j=0}^{\infty} \frac{t^{2j+1}}{1 \cdot 3 \cdot 5 \cdots (2j+1)}$$

$$= t + \frac{t^3}{1 \cdot 3} + \frac{t^5}{1 \cdot 3 \cdot 5} + \frac{t^7}{1 \cdot 3 \cdot 5 \cdot 7} + \frac{t^9}{1 \cdot 3 \cdot 5 \cdot 7 \cdot 9} + \cdots,$$

we have

(13)
$$y' = 1 + t^2 + \frac{t^4}{1 \cdot 3} + \frac{t^6}{1 \cdot 3 \cdot 5} + \frac{t^8}{1 \cdot 3 \cdot 5 \cdot 7} + \cdots$$

and

(14)
$$y'' = 2t + \frac{4t^3}{1 \cdot 3} + \frac{6t^5}{1 \cdot 3 \cdot 5} + \frac{8t^7}{1 \cdot 3 \cdot 5 \cdot 7} + \frac{10t^9}{1 \cdot 3 \cdot 5 \cdot 7 \cdot 9} \cdots.$$

Substituting (12), (13), and (14) into (8), we have

$$y'' - ty' - y$$

$$= \left(2t + \frac{4t^3}{1 \cdot 3} + \frac{6t^5}{1 \cdot 3 \cdot 5} + \frac{8t^7}{1 \cdot 3 \cdot 5 \cdot 7} + \frac{10t^9}{1 \cdot 3 \cdot 5 \cdot 7 \cdot 9} + \cdots\right)$$

$$- t\left(1 + t^2 + \frac{t^4}{1 \cdot 3} + \frac{t^6}{1 \cdot 3 \cdot 5} + \frac{t^8}{1 \cdot 3 \cdot 5 \cdot 7} + \cdots\right)$$

$$- \left(t + \frac{t^3}{1 \cdot 3} + \frac{t^5}{1 \cdot 3 \cdot 5} + \frac{t^7}{1 \cdot 3 \cdot 5 \cdot 7} + \frac{t^9}{1 \cdot 3 \cdot 5 \cdot 7 \cdot 9} + \cdots\right)$$

$$= (2 - 1 - 1)t + \left(\frac{4}{3} - 1 - \frac{1}{3}\right)t^3 + \left(\frac{6}{5 \cdot 3} - \frac{1}{3} - \frac{1}{5 \cdot 3}\right)t^5$$

$$+ \left(\frac{8}{7 \cdot 5 \cdot 3} - \frac{1}{5 \cdot 3} - \frac{1}{7 \cdot 5 \cdot 3}\right)t^7$$

$$+ \left(\frac{10}{9 \cdot 7 \cdot 5 \cdot 3} - \frac{1}{7 \cdot 5 \cdot 3} - \frac{1}{9 \cdot 7 \cdot 5 \cdot 3}\right)t^9 + \cdots$$

$$= 0t + 0t^3 + 0t^5 + 0t^7 + 0t^9 + \cdots$$

$$= 0.$$

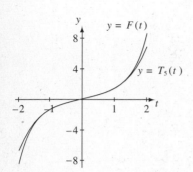

Figure 7.40. The graphs of $y = F(t)$ and $y = T_5(t)$ on the interval $-2 \le t \le 2$.

This shows that the y defined by (11) is a solution to the differential equation. Because this $y = y(t)$ also satisfies the conditions $y(0) = 0$ and $y'(0) = 1$, it is a solution to (8).

The partial sums of the series solution can be used to give a good approximation to the solution near $t = 0$. Let

$$T_5(t) = t + \frac{t^3}{3} + \frac{t^5}{15}$$

and

$$T_9(t) = t + \frac{t^3}{3} + \frac{t^5}{15} + \frac{t^7}{105} + \frac{t^9}{945}.$$

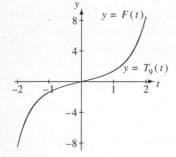

Figure 7.41. The graphs of $Y = F(t)$ and $y = T_9(t)$ on the interval $-2 \le t \le 2$.

In Figs. 7.40 and 7.41 we have graphed these partial sums along with $y = F(t)$, where $F(t)$ is an approximation to the solution found using Euler's method. We can see quite a difference between $F(t)$ and $T_5(t)$ near $t = 2$ and $t = -2$. However, the graphs of $y = F(t)$ and $y = T_9(t)$ are indistinguishable for $-2 \le t \le 2$. ∎

Sometimes it is difficult or impossible to find a formula for the coefficients a_k in a series solution. However, we can sometimes use the differential equation itself to generate the first few coefficients of the solution.

■ **EXAMPLE 3** Assume that the initial value problem

$$(15) \qquad y'' - t^2y' + 3y^2 = t, \qquad y(0) = -1, \quad y'(0) = 1$$

has a solution of the form

$$(16) \qquad\qquad y = \sum_{k=0}^{\infty} a_k t^k.$$

Find the first five nonzero terms in the series.

Solution. From the initial conditions given in (15), we find $a_0 = -1$ and $a_1 = 1$. To find a_2, start by finding the second derivatives of the series in (16). The result is

$$(17) \qquad\qquad y''(t) = \sum_{k=2}^{\infty} k(k-1) a_k t^{k-2}.$$

Substituting $t = 0$, we find $y''(0) = 2a_2$, from which we derive

$$(18) \qquad\qquad a_2 = \frac{y''(0)}{2}.$$

Next solve (15) for y'' to find that

$$(19) \qquad\qquad y'' = y''(t) = t^2 y'(t) - 3y(t)^2 + t.$$

Setting $t = 0$ in this expression, we find

$$y''(0) = 0^2 y'(0) - 3y(0)^2 + 0 = 0^2 a_1 - 3a_0^2 + 0 = -3.$$

Substituting into (18), we obtain

$$a_2 = -\frac{3}{2}.$$

To find a_3, differentiate (17) to get

$$(20) \qquad\qquad y'''(t) = \sum_{k=3}^{\infty} k(k-1)(k-2) a_k t^{k-3}.$$

Set $t = 0$ and solve for a_3 to obtain

$$(21) \qquad\qquad a_3 = \frac{y'''(0)}{3!}.$$

Next differentiate both sides of (19). This gives

$$(22) \qquad \begin{aligned} y'''(t) &= (t^2 y'(t) - 3y(t)^2 + t)' \\ &= 2ty'(t) + t^2 y''(t) - 6y(t)y'(t) + 1. \end{aligned}$$

Set $t = 0$ in this expression to find

$$\begin{aligned} y'''(0) &= 2 \cdot 0 y'(0) + 0^2 y''(0) - 6y(0)y'(0) + 1 \\ &= 2 \cdot 0 a_1 + 0^2 (2a_2) - 6a_0 a_1 + 1 \\ &= -6(-1)1 + 1 \\ &= 7. \end{aligned}$$

Combining with (21), we find

$$a_3 = \frac{7}{6}.$$

To find a_4, differentiate (20), set $t = 0$, and solve for a_4. This leads to

$$(23) \qquad\qquad a_4 = \frac{y^{(4)}(0)}{4!}.$$

Next, differentiate (22), substitute $t = 0$, and use the values for the previously found coefficients. The result is

$$y^{(4)}(0) = 2y'(t) + 4ty''(t) + t^2 y'''(t) - 6y'(t)^2 - 6y(t)y''(t)\big|_{t=0} = -22.$$

Hence

$$a_4 = \frac{-22}{4!} = -\frac{11}{12}.$$

This gives us five nonzero coefficients for the series in (16). Hence the solution to (15) is

$$y = y(t) = \sum_{k=0}^{\infty} a_k t^k = -1 + t - \frac{3}{2}t^2 + \frac{7}{6}t^3 - \frac{11}{12}t^4 + \cdots.$$

Is

$$T_4(t) = -1 + t - \frac{3}{2}t^2 + \frac{7}{6}t^3 - \frac{11}{12}t^4$$

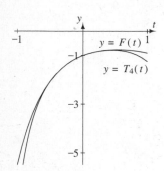

Figure 7.42. The graphs of $y = F(t)$ and $y = T_4(t)$ on the interval $-1 \le t \le 1$.

a good approximation to the actual solution of the differential equation? In Fig. 7.42 we have graphed $y = T_4(t)$ and $y = F(t)$ on the same set of axes. The function $F(t)$ is an approximation to the solution of the differential equation and was found using an Euler method. The two approximations agree pretty well for $-1/2 \le t \le 1/2$, but they are not very close outside of this interval. ∎

Exercises 7.8

Basic

Exercises 1–4:   By equating coefficients of like powers of x on both sides of the equation, find the value of a_k for $k = 0, 1, 2, \dots$.

1. $\displaystyle\sum_{k=0}^{\infty}(a_k - \sqrt{k})t^k = t - 3t^2$

2. $\displaystyle a_0 + \sum_{k=1}^{\infty}(a_k - k^2 a_{k-1})t^k = 1$

3. $\displaystyle a_0 + (2a_1 - a_0)t + \sum_{k=2}^{\infty}(a_k - a_{k-1}a_{k-2})t^k = 0$

4. $\displaystyle a_0 + \sum_{k=1}^{\infty} 2a_k a_{k-1} t^k = \frac{2}{1-t}$

Exercises 5–8:   Combine into one sum involving x^k.

5. $\displaystyle\sum_{k=1}^{\infty}(2k+1)x^{k-1} - \sum_{k=1}^{\infty} x^k$

7. $\displaystyle\sum_{k=0}^{\infty} x^{k-2} + \sum_{k=0}^{\infty} x^{k-1} + \sum_{k=0}^{\infty} x^k$

6. $\displaystyle\sum_{k=0}^{\infty}\frac{1}{k^2+1}x^{k+1} - \sum_{k=0}^{\infty} k(k-1)x^{k-2}$

8. $\displaystyle\sum_{k=0}^{\infty} 2^{k+1}x^{k-2} - \sum_{k=0}^{\infty} 2^{k-1}x^{k-1} + \sum_{k=0}^{\infty} 3\cdot 2^k x^k$

Exercises 9–12:   Find a series solution for the initial value problem. Verify your answer by checking that your solution satisfies the differential equation.

9. $y' + 3y = 0$, $y(0) = 2$

11. $(1+t)y' + 3y = 0$, $y(0) = \pi$

10. $y' - t^2 y = 2$, $y(0) = -1$

12. $(1 - t^2)y' + 3ty = t$, $y(0) = 0$

Growth

Exercises 13–20:   Assume the initial value problem has a solution of the form $y = \sum_{k=0}^{\infty} a_k t^k$. Find the first five nonzero terms of the series.

13. $y' - t^2 y = 2t$, $y(0) = 1$

14. $y' - y = \sin t$, $y(0) = 1$

15. $y'' + y' + y = 0$, $y(0) = 1$, $y'(0) = -1$

16. $y'' - (t^2+1)y = 0$, $y(0) = 1$, $y'(0) = 2$

17. $y'' - (t^2+1)y' = 0$, $y(0) = 1$, $y'(0) = 2$

18. $y'' + (\sin t)y = e^t$, $y(0) = 1$, $y'(0) = 0$

19. $y'' - (y')^2 = 1$, $y(0) = 1$, $y'(0) = 1$

20. $y'' + (t^2+1)y^3 = 0$, $y(0) = 1$, $y'(0) = 0$

Exercises 21–26:   Find a series solution to the initial value problem. Verify your answer by checking that your solution satisfies the differential equation.

21. $y'' + 4y = 0$, $y(0) = 1$, $y'(0) = 0$

24. $y'' - ty = 0$, $y(0) = 0$, $y'(0) = 1$

22. $y'' + 4y = 0$, $y(0) = 2$, $y'(0) = -1$

25. $(1-t)y'' + y = 0$, $y(0) = 1$, $y'(0) = 0$

23. $y'' - ty = 0$, $y(0) = 1$, $y'(0) = 0$

26. $(2+t^2)y'' - ty' + 4y = 0$, $y(0) = 0$, $y'(0) = 1$

27. Let $y_1(t)$ be a solution to the initial value problem

$$y'' + A(t)y' + B(t)y = 0, \quad y(0) = 1, \quad y'(0) = 0,$$

and let $y_2(t)$ be a solution to the initial value problem

$$y'' + A(t)y' + B(t)y = 0, \quad y(0) = 0, \quad y'(0) = 1.$$

Show that $ay_1(t) + by_2(t)$ is a solution to

$$y'' + A(t)y' + B(t)y = 0, \quad y(0) = a, \quad y'(0) = b.$$

28. Let $y_0(t)$ be a solution to the differential equation

$$y'' + A(t)y' + B(t)y = 0,$$

and let $y_p(t)$ be a solution to the differential equation

$$y'' + A(t)y' + B(t)y = C(t).$$

Show that for any real number c, $y_p(t) + cy_0(t)$ is also a solution to

$$y'' + A(t)y' + B(t)y = C(t).$$

Review/Preview

29. Let $\mathbf{r}(t) = (3t, 5t)$ denote the position of a particle at time t. How far does the particle travel from $t = 1$ to $t = 6$?

30. Let $\mathbf{r}(t) = (-2t, \pi t)$ denote the position of a particle at time t. How far does the particle travel from $t = 2$ to $t = 7$?

31. Let $\mathbf{r}(t) = (at, bt)$ denote the position of a particle at time t. How far does the particle travel from $t = c$ to $t = d$? Assume $d > c$.

32. Find a vector of length 1 perpendicular to $(-1, 2)$.

33. Find a vector of length 1 perpendicular to $(3, 2)$.

34. Let $\mathbf{r}(t) = (3t^2, 2\sqrt{t} - 5)$, $t \geq 0$, be the position of a particle at time t. Find a vector of length 1 tangent to the path of the particle at position $\mathbf{r}(1)$.

35. Let $\mathbf{r}(t) = (3\sin 4t, 2e^t \tan t)$, $0 \leq t \leq \pi/2$, be the position of a particle at time t. Find a vector of length 1 tangent to the path of the particle at position $\mathbf{r}(\pi/4)$.

REVIEW OF KEY CONCEPTS

Approximation is important in every area of science, engineering, and mathematics. In this chapter we improved on the tangent line approximation by using Taylor polynomials. The Taylor polynomial $T_n(x; c)$ for a function f takes the value $f(c)$ at $x = c$. In addition, the first n derivatives of T_n and f are the same at $x = c$. Because of this, the Taylor polynomial $T_n(x; c)$ is a good approximation to $f(x)$ when x is near c. When $n = 1$, the Taylor polynomial is the tangent line approximation. For $n \geq 1$, the Taylor polynomial is usually a better approximation to f than the tangent line expression.

To use an approximation effectively, we need to know something about the error in the approximation. The error in the Taylor polynomial approximation is

$$R_n(x; c) = f(x) - T_n(x; c).$$

In Section 7.2 we found an upper bound for $|R_n(x; c)|$ and used this bound to estimate the error in an approximation by Taylor polynomials. We showed how to use this estimate to find a Taylor polynomial for which $|R_n(x; c)| < E$, where E is a given upper bound for the error.

To better understand Taylor polynomials, we next studied sequences and infinite series. We observed that the Taylor polynomials for a function f can be used to give a good approximation to f if the sequence $\{T_n(x; c)\}$ of Taylor polynomials converges to f. Because the sequence of Taylor polynomials for f is the sequence of partial sums of the Taylor series for f, some questions about the behavior of Taylor polynomials were answered by using results from our study of infinite series.

Next we studied power series and Taylor series. Using results developed in earlier sections, we were able to find the interval of convergence for such series. In addition, we saw that if the Taylor series for a function f converges to f on an interval $c - R < x < c + R$, then the corresponding sequence $\{T_n(x; c)\}$ of Taylor polynomials for f converges uniformly to f on any closed interval $[c - r, c + r]$ contained in $(c - R, c + R)$. This means that the Taylor polynomials for f can be used to approximate f as closely as we like on $[c - r, c + r]$.

If Taylor series and Taylor polynomials are to be a useful tool, then it is important that we be able to generate Taylor series and polynomials quickly and accurately. We saw that with knowledge of just five Taylor series (see Section 7.6) we can use substitution, algebra, differentiation, and integration to find Taylor series for many other functions. We concluded the chapter by showing how power series can be used to solve some differential equations.

Chapter Summary

The Comparison Tests

a. Let $\sum_{k=1}^{\infty} b_k$ be a series of nonnegative terms and let $\sum_{k=1}^{\infty} c_k$ be a convergent series, also with nonnegative terms. If there is a positive integer N such that

$$b_k \leq c_k \quad \text{for all } k \geq N,$$

then

$$\sum_{k=1}^{\infty} b_k \quad \text{also converges.}$$

b. Let $\sum_{k=1}^{\infty} b_k$ be a series of nonnegative terms and let $\sum_{k=1}^{\infty} d_k$ be a divergent series of nonnegative terms. If there is a positive integer N such that

$$d_k \leq b_k \quad \text{for all } k \geq N,$$

then

$$\sum_{k=1}^{\infty} b_k \quad \text{also diverges.}$$

c. Let $\sum_{k=1}^{\infty} a_k$ and $\sum_{k=1}^{\infty} b_k$ both be series of positive terms. If

$$0 < \lim_{k \to \infty} \frac{a_k}{b_k} < \infty,$$

then the two series both converge or both diverge.

The Ratio Test

Let $\sum_{k=1}^{\infty} a_k$ be a series of positive terms. Suppose that

$$\lim_{k \to \infty} \frac{a_{k+1}}{a_k} = \rho, \quad \text{where } 0 \leq \rho \leq \infty.$$

If

$p < 1$	the series converges
$p > 1$	the series diverges
$p = 1$	the series could either converge or diverge.

Absolute Convergence Test

Let $\sum_{k=1}^{\infty} a_k$ be an infinite series. If

$$\sum_{k=1}^{\infty} |a_k|$$

converges, then

$$\sum_{k=1}^{\infty} a_k$$

also converges.

Alternating Series Test

Let $\{b_k\}_{k=1}^{\infty}$ be a sequence of nonnegative real numbers. If

$$b_1 \geq b_2 \geq b_3 \geq \cdots \geq b_k \geq b_{k+1} \geq \cdots$$

and

$$\lim_{k \to \infty} b_k = 0,$$

then

$$\sum_{k=1}^{\infty} (-1)^{k-1} b_k$$

converges.

Taylor Polynomials

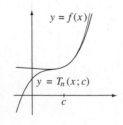

Let f be a function. If $f(c)$, $f'(c)$, $f''(c), \ldots f^{(n)}(c)$ are all defined, then the nth Taylor polynomial for f at c is

$$T_n(x; c) = \sum_{k=0}^{n} \frac{f^{(k)}(c)}{k!}(x - c)^k.$$

When n is large, $T_n(x; c)$ is often a good approximation to $f(x)$ for x values near c.

If $f(x) = \ln x$ and $c = 3$, then

$$f(c) = \ln 3$$

$$f'(c) = 1/3$$

$$f''(c) = -1/3^2 = -1/9.$$

Hence the second Taylor polynomial about $c = 3$ for f is

$$T_2(x; 3) = \ln 3 + \frac{1}{3}(x - 3)$$
$$- \frac{1}{18}(x - 3)^2.$$

Error in a Taylor Approximation

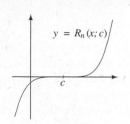

$y = R_n(x; c)$

Let f be a function and let $T_n(x; c)$ be the nth Taylor polynomial for f at c. The difference between $f(x)$ and $T_n(x; c)$ is

$$R_n(x; c) = f(x) - T_n(x; c)$$

and is called the error in the approximation by the Taylor polynomial of degree n. If $|R_n(x; c)|$ is small, then $T_n(x; c)$ may be a good approximation to $f(x)$.

Let $f(x) = \ln x$ and $c = 3$. If we use the approximation

$$\ln x \approx T_2(x; 3),$$

then the error is

$$R_n(x; 3)$$
$$= \ln x - T_2(x; 3)$$
$$= \ln x - \left(\ln 3 + \frac{(x - 3)}{3}\right.$$
$$\left. - \frac{(x - 3)^2}{18}\right).$$

Estimating the Error

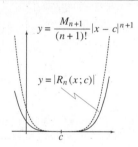

$y = \dfrac{M_{n+1}}{(n+1)!}|x - c|^{n+1}$

$y = |R_n(x; c)|$

Let f be a function and $T_n(x; c)$ the nth Taylor polynomial for f at c. Suppose that $|f^{(n+1)}(x)| \le M_{n+1}$ on the interval $c - r \le x \le c + r$. Then for x in this interval,

$$|R_n(x; c)| \le \frac{M_{n+1}}{(n + 1)!}|x - c|^{n+1}.$$

This estimate for the error is useful in estimating the error in the approximation $f(x) \approx T_n(x; c)$.

Let $f(x) = \ln x$, $c = 3$, and $n = 2$. Then

$$f^{(n+1)}(x) = \frac{d^3}{dx^3}\ln x = \frac{6}{x^3}.$$

For $2 \le x \le 4$, we have

$$\left|\frac{6}{x^3}\right| \le \frac{3}{4}.$$

Hence on this interval

$$|R_2(x; 3)| \le \frac{3/4}{3!}|x - 3|^3 \le \frac{1}{8}.$$

The Limit of a Sequence

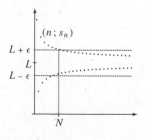

$(n; s_n)$

$L + \epsilon$

L

$L - \epsilon$

N

Let $\{s_n\}$ be a sequence of real numbers. The sequence converges to L if for any number $\epsilon > 0$ we can find a number $N > 0$ such that

$$|s_n - L| < \epsilon$$

whenever $n > N$. Roughly, this means that the numbers s_n get close to L as n gets large.

Let $s_n = 1 + (-1)^n \dfrac{1}{\sqrt{n}}$. When n is very large, $\dfrac{1}{\sqrt{n}}$ is close to 0. Hence for large n,

$$s_n = 1 + (-1)^n \frac{1}{\sqrt{n}} \approx 1.$$

Thus

$$\lim_{n \to \infty} s_n = \lim_{n \to \infty}\left(1 + (-1)^n\frac{1}{\sqrt{n}}\right) = 1.$$

Increasing, Bounded Above

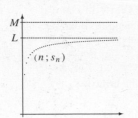

Let $\{s_n\}$ be a sequence of real numbers. The sequence is increasing if

$$s_1 \leq s_2 \leq \cdots \leq s_n \leq s_{n+1} \leq \cdots.$$

The sequence is bounded above if there is a number M such that

$$s_n \leq M$$

for all n. If a sequence of real numbers is increasing and bounded above, then it converges. That is,

$$\lim_{n \to \infty} s_n = L$$

for some number L.

Let $s_n = \left(1 + \dfrac{1}{n}\right)^n$. Experimentation with a calculator or computer suggests that the sequence $\{s_n\}$ is increasing and that $s_n \leq 3$ for all n. Hence the sequence has a limit. In Section 7.3 we saw that

$$\lim_{n \to \infty} s_n = \lim_{n \to \infty} \left(1 + \frac{1}{n}\right)^n = e.$$

Uniform Convergence

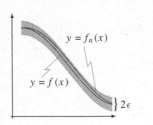

Let $\{f_n\}$ be a sequence of functions defined on an interval I, and let f be another function defined on I. We say that the sequence $\{f_n(x)\}$ converges to $f(x)$ uniformly on I if for each number $\epsilon > 0$ there is a number N so that whenever $n > N$, we have

$$|f(x) - f_n(x)| \leq \epsilon$$

for all x in the interval I. This means that for $n > N$, the graph of $y = f_n(x)$ lies in the ϵ-corridor about the graph of $y = f(x)$.

Let

$$T_n(x) = \sum_{k=0}^{n} \frac{x^k}{k!}$$

be the nth Taylor polynomial for f about 0. Then the sequence $\{T_n(x)\}$ converges uniformly to e^x on any interval of the form $[-r, r]$.

Estimating an Alternating Series

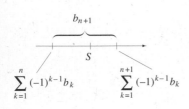

Suppose that $\sum_{k=1}^{\infty} (-1)^{k-1} b_k$ converges by the alternating series test, and let S be the sum of the series. For positive integer n,

$$\left| S - \sum_{k=0}^{n} (-1)^{k-1} b_k \right| \leq |b_{n+1}|.$$

Hence if we use a partial sum of the series to estimate the sum of the series, the error is no more than the absolute value of the first omitted term.

It can be shown that

$$\sum_{k=1}^{\infty} \frac{(-1)^{k-1}}{2k-1} = 1 - \frac{1}{3} + \frac{1}{5} - \cdots$$

$$= \frac{\pi}{4}.$$

Hence

$$\left| \frac{\pi}{4} - \sum_{k=1}^{1000} \frac{(-1)^{k-1}}{2k-1} \right| \leq \frac{1}{2001}.$$

Radius of Convergence

	Consider a power series	Consider the power series

div $\qquad$ conv $\qquad$ div

$c - R \quad c \quad c + R$

Consider a power series

$$\sum_{k=0}^{\infty} a_k(x - c)^k.$$

There is a number R, $0 \leq R \leq \infty$, such that

The series converges when
$$|x - c| < R.$$

The series diverges when
$$|x - c| > R.$$

The number R is called the radius of convergence of the power series. If

$$\lim_{k \to \infty} \left| \frac{a_{k+1}}{a_k} \right| = \rho$$

where $0 \leq \rho \leq \infty$, then

$$R = 1/\rho.$$

Consider the power series

$$\sum_{k=0}^{\infty} \frac{x^k}{k!}.$$

We have

$$\lim_{k \to \infty} \left| \frac{a_{k+1}}{a_k} \right| = \lim_{k \to \infty} \frac{1/(k + 1)!}{1/k!} = 0.$$

The radius of convergence of the power series is

$$R = \frac{1}{\rho} = \frac{1}{0} = \infty.$$

This means that the series converges for every real number x.

CHAPTER REVIEW EXERCISES

Exercises 1–9:   Find the limit of the sequence.

1. $\left\{ \dfrac{10^{2n}}{n!} \right\}$

2. $\{ \sqrt[n]{100} - 1 \}$

3. $\{\tan j\}$

4. $\left\{ \displaystyle\sum_{j=0}^{n} \dfrac{1}{3^j} \right\}$

5. $\left\{ k^{(-1)^k} \right\}$

6. $\left\{ \left(1 + \dfrac{2}{k}\right)^k \right\}$

7. $\{ \sqrt{2n^2 + n + 1} - \sqrt{2}n \}$

8. $\{ n^{3/2} - (n + 1)^{3/2} \}$

9. $\left\{ \displaystyle\sum_{k=0}^{n} \dfrac{1}{k!} \right\}$

Exercises 10–18:   Decide whether the series converges or diverges.

10. $\displaystyle\sum_{k=1}^{\infty} \dfrac{2k}{\sqrt{k^2 + 1}}$

11. $\displaystyle\sum_{k=1}^{\infty} (\sqrt[k]{k} - 1)^k$

12. $\displaystyle\sum_{j=0}^{\infty} 6(0.123)^{2j}$

13. $\displaystyle\sum_{m=0}^{\infty} \dfrac{10^m}{m!}$

14. $\displaystyle\sum_{k=1}^{\infty} (-1)^{k-1} \dfrac{k + 1}{2k}$

15. $\displaystyle\sum_{\ell=1}^{\infty} \dfrac{\ell!}{\ell^\ell}$

16. $\displaystyle\sum_{q=1}^{\infty} (e - 1)^q$

17. $\displaystyle\sum_{r=10}^{\infty} (-1)^r \dfrac{1}{\ln(\ln r)}$

18. $\displaystyle\sum_{j=2}^{\infty} \dfrac{1}{j^{3/2} - (j - 2)^{3/2}}$

Exercises 19–22: Find the radius of convergence of the power series.

19. $\displaystyle\sum_{k=0}^{\infty} k^{10} x^k$

20. $\displaystyle\sum_{j=0}^{\infty} \frac{2.7^j}{j!}(x-2)^{2j}$

21. $\displaystyle\sum_{m=1}^{\infty} \frac{1 \cdot 3 \cdot 5 \cdots (2m-1)}{3 \cdot 6 \cdot 9 \cdots (3m)} x^m$

22. $\displaystyle\sum_{r=1}^{\infty} \frac{2 \cdot 6 \cdot 10 \cdots (4r-2)}{3 \cdot 7 \cdot 11 \cdots (4r-1)}(x+3)^r$

Exercises 23–28: Find the Taylor series for f about the point c. State the radius of convergence of the series.

23. $f(x) = \cos 2x, \quad c = 0$

24. $f(x) = e^{2x}, \quad c = -\ln 2$

25. $f(x) = \dfrac{x + 10}{(x+1)(x+4)}, \quad c = 1$

26. $f(x) = x^{10}, \quad c = -2$

27. $f(x) = \displaystyle\int_0^x \sin(t^2)\, dt, \quad c = 0$

28. $f(x) = \sin^2 4x, \quad c = \pi/8$

Exercises 29–34: Find the sum of the series. (Some of these series were obtained by substituting a number into a known Taylor series.)

29. $\displaystyle\sum_{k=0}^{\infty} 0.9^k$

30. $\displaystyle\sum_{j=0}^{\infty} \frac{2}{(j+1)(j+3)}$

31. $\displaystyle\sum_{k=1}^{\infty} \left(\frac{e-1}{e}\right)^k$

32. $\displaystyle\sum_{r=0}^{\infty} \frac{2^r}{r!}$ (Hint: Recall the series for e^x.)

33. $\displaystyle\sum_{k=1}^{\infty} (-1)^{k-1} \frac{1}{k2^k}$ (Hint: Recall the series for $\ln(1-x)$.)

34. $\displaystyle\sum_{k=0}^{\infty} (-1)^k \frac{\pi^{2k+1}}{(2k+1)!}$

35. Find a value of n so that $\displaystyle\sum_{k=1}^{n} 1/\sqrt[3]{k} > 1000$.

36. Find n so that the nth partial sum of the series $\displaystyle\sum_{k=2}^{\infty} \frac{(-1)^{k-1}}{\ln k}$ is within 0.001 of the sum of the series.

37. Find n so that the nth partial sum of the series $\displaystyle\sum_{k=1}^{\infty} 1/k^4$ is within 0.0001 of the sum of the series.

38. Find n so that the nth partial sum of the series $\displaystyle\sum_{k=2}^{\infty} \frac{1}{k(\ln k)^4}$ is within 0.001 of the sum of the series.

39. Find n so that the nth partial sum of the series $\displaystyle\sum_{k=2}^{\infty} \frac{(-1)^{k-1}}{\sqrt{k}}$ is within 10^{-9} of the sum of the series.

40. Let $T_n(x;0)$ be the nth Taylor polynomial for e^{2x} about the point $x = 0$. Find n so that $T_n(x;0)$ is within 0.01 of e^{2x} for all x in the interval $-2 \le x \le 2$.

41. Let $T_n(x;\pi/4)$ be the nth Taylor polynomial for $\cos x$ about the point $x = \pi/4$. Find n so that $T_n(x;\pi/4)$ is within 0.001 of $\cos x$ for all x in the interval $\pi/4 - 1 \le x \le \pi/4 + 1$.

42. Let $T_n(x;0)$ be the nth Taylor polynomial for $\arctan x$ about the point $x = 0$. Find n so that $T_n(x;0)$ is within 0.01 of $\arctan x$ for all x in the interval $-3/4 \le x \le 3/4$.

43. Let $T_n(x;-1)$ be the nth Taylor polynomial for $\dfrac{1}{1-x}$ about the point $x = -1$. Find n so that $T_n(x;-1)$ is within 0.001 of $\dfrac{1}{1-x}$ for all x in the interval $-2 \le x \le 0$.

44. Use series to solve the differential equation

$$y'(t) - 2ty(t) = 0, \quad y(0) = 1.$$

The solution is the power series of an elementary function. What is the function?

45. Use series to solve the differential equation

$$(1-t)y'(t) - y(t) = 0, \quad y(0) = 1.$$

The solution is the power series of an elementary function. What is the function?

46. Let $i = \sqrt{-1}$, so $i^2 = -1$, $i^3 = -i$, and so on. Substituting $i\theta$ for x in the series identity

$$e^x = \sum_{k=0}^{\infty} \frac{x^k}{k!},$$

do some algebraic manipulations to arrive at the identity

$$e^{i\theta} = \cos\theta + i\sin\theta.$$

Use this identity to evaluate $e^{i\pi}$. Also evaluate $e^{i\pi/4}$.

47. Let $i = \sqrt{-1}$. Use series to prove the identities

$$\cos x = \frac{e^{ix} + e^{-ix}}{2}$$

and

$$\sin x = \frac{e^{ix} - e^{-ix}}{2i}.$$

48. The hyperbolic cosine function, denoted cosh, is defined by

$$\cosh x = \frac{e^x + e^{-x}}{2}.$$

Find the Taylor series about $x = 0$ for the hyperbolic cosine function.

49. The hyperbolic sine function, denoted sinh, is defined by

$$\sinh x = \frac{e^x - e^{-x}}{2}.$$

Find the Taylor series about $x = 0$ for the hyperbolic sine function.

50. By combining the results of the previous three problems, show that

$$\cos(ix) = \cosh x$$

and

$$-i\sin(ix) = \sinh x.$$

Student Project

Signal Processing and Series

Suppose that the power series $\sum_{k=0}^{\infty} a_k x^k$ has radius of convergence 1. Then the series converges for each $x \in (-1, 1)$, so we can define a function f on this interval by $f(x) = \sum_{k=0}^{\infty} a_k x^k$. Now suppose that a friend across the country needs to approximate $f(x)$ for several values of x and asks us to tell her what f is. We could just write out the first several terms of the series and send that, but there's a shorter way. We could instead just send the first few series coefficients, $\{a_0, a_1, a_2, \ldots, a_n\}$, and tell our friend that these are the coefficients of the first $n + 1$ terms of a power series about 0. With this information our friend could reconstruct a partial sum of the series and obtain her approximations. Why does this work? It works because a function that can be expressed as a power series about 0 is completely determined by its power series coefficients.

This idea is used several million times every day in the communications industry, though the series involved are Fourier series rather than power series. This application is based on the following fact.

Fourier Series and Convergence

Let f be a function defined on the interval $0 \le t \le 2$. If f is "nice" enough, then there are real numbers $a_0, a_1, a_2, \ldots$ and $b_1, b_2, b_3, \ldots$ such that

(1) $$f(t) = a_0 + \sum_{k=1}^{\infty} (a_k \cos(k\pi t) + b_k \sin(k\pi t))$$

for all $0 < t < 2$. Furthermore, the series (1) converges uniformly to $f(t)$ on each interval of the form $[\delta, 2 - \delta]$, where $0 < \delta < 1$.

We will not attempt to define "nice" here. However it can be shown that any function with a continuous first derivative on [0, 2] is "nice." This means that most of the functions we use on a day-to-day basis are "nice" functions.

A series like that in (1) is called a **Fourier series.** The coefficients $\{a_k\}$ and $\{b_k\}$ are called the **Fourier coefficients** for f. In practice, $f(t)$ might be the amplitude or strength of a signal (e.g., a sound) at time t. To "transmit" the signal, several of the Fourier coefficients are computed and the values transmitted, often in binary form. At the receiving end the Fourier coefficients are used to obtain a partial sum of the Fourier series for f and hence an approximation for f. With this information an approximation to the signal can be constructed, perhaps resulting in a voice or a picture from space. In this project we look at some examples to see how the partial sums of a Fourier series approximate a function, and we then indicate how the approximation process can be implemented in practice.

Problem 1: Computing Fourier Coefficients

a. Let k and m be positive integers with $k \neq m$. Verify that

$$\int_0^2 \cos k\pi t \, dt = 0, \qquad \int_0^2 \sin k\pi t \, dt = 0$$

and that

$$\int_0^2 \cos k\pi t \sin m\pi t \, dt = 0.$$

Show also that

$$\int_0^2 \cos^2 k\pi t \, dt = 1 = \int_0^2 \sin^2 k\pi t \, dt.$$

b. Assume that f is continuous and that (1) is true on [0, 2]. By integrating both sides of (1), show that

$$(2) \qquad\qquad a_0 = \frac{1}{2} \int_0^2 f(t) \, dt.$$

c. Assume that f is continuous and (1) is true on [0, 2]. Let m be a positive integer. Multiply both sides of (1) by $\cos m\pi t$ then integrate from 0 to 2. Tell why this leads to

$$(3) \qquad\qquad a_m = \int_0^2 f(t) \cos m\pi t \, dt.$$

Use a similar argument to show that

$$(4) \qquad\qquad b_m = \int_0^2 f(t) \sin m\pi t \, dt.$$

Problem 2: An Example of Fourier Approximation.   Let $f(t) = t$ for all t in $[0, 2]$.

a. Use (2), (3), and (4) to find the Fourier coefficients for f.

b. With the values for the a_k's and b_k's found in a, graph

$$y = a_0 + \sum_{k=1}^{5} (a_k \cos k\pi t + b_k \sin k\pi t).$$

How well does this partial sum of the Fourier series approximate f on $[0, 2]$?

c. Next graph

$$y = a_0 + \sum_{k=1}^{20} (a_k \cos k\pi t + b_k \sin k\pi t).$$

How well does this partial sum of the Fourier series approximate f on $[0, 2]$. Where does the partial sum appear to be a good approximation to $f(t) = t$? Where is it a poor approximation?

Problem 3: An Example of Fourier Approximation.   Repeat the previous problem with $f(t) = t^2$.

Problem 4: An Example of Fourier Approximation.   Repeat Problem 2 with the "sawtooth" function

$$f(t) = \begin{cases} t & (0 \le t \le 1) \\ 2 - t & (1 < t \le 2) \end{cases}$$

The graph of this function is shown in Fig. 7.43.

In practice, a formula for the function f that gives the strength of a signal at time t is not known. In such cases the signal is sampled at several different times t, and its strength $f(t)$ at each of these times is recorded. With these data the Fourier coefficients (2), (3), and (4) are then approximated using numerical integration.

As an example, suppose a signal is sampled at the 10 times $0.2, 0.4, 0.6, \ldots,$ 2.0 to give the data shown in the accompanying table. The 10 sample times t are equally spaced and divide the interval $[0, 2]$ into 10 equal pieces each of length 0.2. Using 0.2 as the width of a rectangle and the $f(t)$ values as the "heights" of rectangles, we can approximate the Fourier coefficients using Riemann sums. For example,

$$b_1 = \int_0^2 f(t) \sin(1 \cdot \pi t)\,dt \approx \sum_{j=1}^{10} f(0.2j)(\sin(1 \cdot \pi 0.2 j))(0.2)$$

$$\approx (2.412)(\sin(0.2\pi))0.2 + (5.138)(\sin(0.4\pi))0.2$$
$$+ (-1.334)(\sin(0.6\pi))0.2 + (-0.0605)(\sin(0.8\pi))0.2$$
$$+ (4.0)(\sin(1.0\pi))0.2 + (-2.412)(\sin(1.2\pi))0.2$$
$$+ (-5.138)(\sin(1.4\pi))0.2 + (1.334)(\sin(1.6\pi))0.2$$
$$+ (0.0605)(\sin(1.8\pi))0.2 + (-4.0)(\sin(2.0\pi))0.2$$

$$\approx 2.0.$$

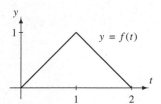

Figure 7.43. The "sawtooth" function.

t	$f(t)$
0.2	2.412
0.4	5.138
0.6	−1.334
0.8	−0.0605
1.0	4.0
1.2	−2.412
1.4	−5.138
1.6	1.334
1.8	0.0605
2.0	−4.0

Hence in the Fourier series for the signal f, the coefficient of $\sin(\pi t)$ is approximately 2. The constant term a_0 of the series is approximated by

$$a_0 = \frac{1}{2}\int_0^2 f(t)\,dt \approx \sum_{j=1}^{10} f(0.2\,j)0.2 \approx 0.$$

Problem 5: Approximation of Fourier Coefficients.   For the data given in the table, compute approximations to a_1, a_2, a_3, b_2, and b_3. Plot the data and the graph of

$$y = a_0 + \sum_{k=1}^{3}(a_k \cos(k\pi t) + b_k \sin(k\pi t))$$

on the same set of axes. How well does this partial sum of the Fourier series fit the data?

Problem 6: Approximation of Fourier Coefficients.   Suppose that the 100 ordered pairs

$$(0.02\,j, (0.02\,j)\sin(0.02\,j + 1)), \quad j = 1, 2, 3, \ldots, 100$$

are obtained as data points for a signal f defined for $0 \le t \le 2$. Approximate the Fourier coefficients $a_0, a_1, \ldots, a_{10}$ and $b_1, b_2, \ldots, b_{10}$ for f. Plot the data points and the graph of

$$y = a_0 + \sum_{k=1}^{10}(a_k \cos(k\pi t) + b_k \sin(k\pi t))$$

on the same set of axes. How well does this partial sum fit the data?

Problem 7: Approximation of Fourier Coefficients.   Your instructor will provide 100 ordered pairs of data points for a signal f defined for $0 \le t \le 2$. Use these data to approximate the Fourier coefficients $a_0, a_1, \ldots, a_{10}$ and $b_1, b_2, \ldots, b_{10}$ for f. Plot the data points and the graph of

$$y = a_0 + \sum_{k=1}^{10}(a_k \cos(k\pi t) + b_k \sin(k\pi t))$$

on the same set of axes. How well does this partial sum fit the data?

In practice, the important process of using numerical integration to approximate the Fourier coefficients for a signal is now accomplished by means of the fast Fourier transform, or FFT. Mathematicians and engineers have studied the numerical integration process as it is used to approximate Fourier coefficients. They found many inefficiencies in these computations; for example, many multiplications performed in approximating one Fourier coefficient are repeated in approximating another Fourier coefficient. The FFT eliminates many of these redundancies and provides a fast algorithm for approximating Fourier coefficients. With today's high-speed computers the FFT can be implemented several times every second, resulting in rapid, efficient, and accurate communications.

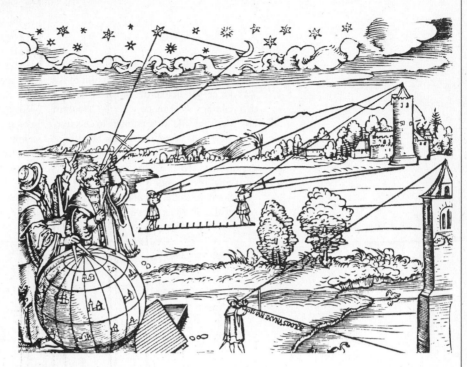

Figure 7.44. The baculum.

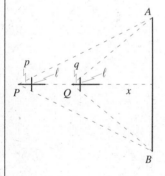

Figure 7.45. Using the baculum.

The baculum, also known as the arbalete, geometric cross, cross-staff, or Jacob's staff, was an ingenious device for measuring distances. The baculum consisted of a rod about four feet long, and a perpendicular crosspiece fitted in such a way that it could slide back and forth along the rod. Equally spaced marks were drawn on the rod so that one could gauge the position of the crosspiece. (See Fig. 7.44.)

To use the baculum to measure a horizontal distance AB, a person stood at a point P from which his line of sight to the midpoint of the distance to be measured was perpendicular to $\overline{AB}$. The baculum was then held at eye level and pointed toward the midpoint of $\overline{AB}$. The crosspiece was then slid along the rod until its endpoints coincided with A and B. At this time the position of the crosspiece on the rod was noted. The person then moved to another point Q along the line of sight to the midpoint, again adjusted the crosspiece to just cover $\overline{AB}$, and then noted the new position of the crosspiece on the rod. Using the distance PQ between the two observation positions and the difference between the two positions of the crosspiece, the distance AB could be computed.

The distance computation is based on similar triangles and is diagramed in Fig. 7.45. Assume Q is closer to the midpoint of $\overline{AB}$ than P and let x be the distance from Q to the midpoint of $\overline{AB}$. In addition, let l be the length of the crosspiece and let p and q be the distances of the crosspiece from the end of the rod in the two positions. Working first with $\triangle AQB$ and the similar smaller triangle determined by the crosspiece, we find

$$(5) \qquad \frac{x}{q} = \frac{AB}{\ell}.$$

Next, from $\triangle APB$ we have

$$(6) \qquad \frac{x + PQ}{p} = \frac{AB}{\ell}.$$

Solving (5) and (6) simultaneously, we find that

(7)
$$AB = \frac{(PQ)\ell}{p - q}.$$

The baculum could also be used when the line from the user to the midpoint of $\overline{AB}$ was not perpendicular to $\overline{AB}$. Just as above, two readings were made with the baculum, but when the rod was pointed to the midpoint of $\overline{AB}$ the crosspiece had to be tilted so its endpoints covered A and B. In this case the distance AB is again given by (7). Try verifying this on your own.

Sometimes the baculum was designed so that the distance between marks on the rod was equal to the length of the crosspiece. The crosspiece was set at one mark and the user positioned himself so that the endpoint of the crosspiece covered A and B. Then the crosspiece was moved to the next mark on the rod and the user again positioned himself so the crosspiece just covered $\overline{AB}$. Why did this method make the calculation of AB easier? Suppose the length of AB was known. How could the baculum be used to find your distance from the midpoint of AB? The baculum was also used to measure the angular distance between two stars. How do you think this was done?

Try making your own baculum and use it to measure the width of a large building or the distance traveled by a car.

Answers to Selected Exercises

If you need further help with algebra, you may want to obtain a copy of the *Student's Solution Manual* for this book. It contains solutions to all the answers in this answer section. Your college bookstore has this manual (ISBN 0-06-502359-5) or can order it for you.

In this section we provide the answers that we think most students will obtain when they work the exercises using the methods explained in the text. If your answer does not look exactly like the one given here, it is not necessarily wrong. In many cases there are other equivalent forms of the answer. In general, if your answer does not agree with the one given in the text, see whether it can be transformed into an equivalent form. If it can, then it is the correct answer. If you still have doubts, talk with your instructor.

Section 1.1

1. -9
2. 10.205
3. $-12 - 7\sqrt{3}$
4. $t^2 + (2h + 1)t + (h^2 + h - 3)$
5. $8x^3 + 12x^2(2h + 1) + 6x(4h^2 + 4h + 1) +$
   $(8h^3 + 12h^2 + 6h + 1)$
6. $\dfrac{2x + y}{5x - 3y}$
7. $\{x : x \text{ a real number }\}$
8. $\{t : t \text{ a real number}, t \neq -4\}$
9. $\{x : x \leq -1 \text{ or } x \geq 0\}$
10. $\{r : r < a \text{ or } r > b\}$
11. $\{r : r > b\}$
12. $\{t : t \neq 0, \pm\pi, \pm2\pi, \pm3\pi, \ldots\}$
13. Domain: all real numbers; Range: $\{4\}$
14. Domain $=$ Range $=$ all real numbers
15. Domain: all real numbers; Range: $\{y : 0 < y < 1\}$
16. Domain: $\{x : x \neq 2\}$; Range: $\{y : y \neq 1\}$
17. Domain: $\{x : x \neq 0\}$; Range: $\{-1, 1\}$

18.

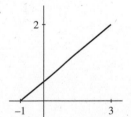

19. $f(x) = \sqrt{-(x + 1)(x - 3)}$
    (other answers are possible)

20. $g(x) = \dfrac{1}{\sqrt{x^2 - 1}}$ (other answers are possible)

21.

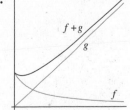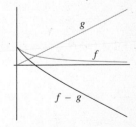

22.

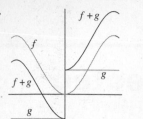

23. $(f \circ g)(x) = \dfrac{1}{x^2 - 6x + 8}$;

Domain: $\{x : x \neq 2, x \neq 4\}$

24. $(r \circ r)(u) = \sqrt{1 - \sqrt{1 - u}}$;

Domain: $\{u : 0 \leq u \leq 1\}$

25. $f(g(t)) = \dfrac{|\sin t|}{\sin t}$;

Domain: $\{t : t \neq 0, \pm\pi, \pm 2\pi, \ldots\}$

26. $h(k(x)) = \sqrt{(x - 1)(x - 2)(x - 3)^2(x - 4)}$;

Domain: $\{x : 1 \leq x \leq 2 \text{ or } x \geq 4\}$

27.

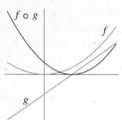

28.

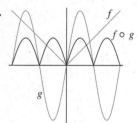

29. $g(x) = \dfrac{2}{x}$, $h(x) = x^3 - 3x^2 + 1$; $f = g \circ h$

30. $f(x) = \sin x$, $g(x) = \sqrt{x}$, $h(x) = x + 5$;

$H = f \circ g \circ h$

31. $F(x) = f(x)$, $H(x) = \dfrac{1}{x}$, $G(x) = g(x)$;

$h = F \circ H \circ G$

32. $f(x) = x^{1/2}$, $g(x) = \dfrac{1 - 2x}{1 + 2x}$, $h(x) = \sin x$;

$r = f \circ g \circ h$

51. slope $= \dfrac{4}{7}$, y-intercept $= \dfrac{5}{7}$

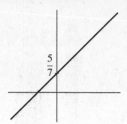

53. $x = 8$

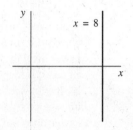

55. Height $= h(t) = 5000 - 12t$;

hits ground at $t = 416\frac{2}{3}$ seconds

57. 11 kph

59. $t + x + 2$

61. $y = -3x + 1$

Section 1.2

1. slope $= -\dfrac{3}{7}$, y-intercept $= -\dfrac{5}{7}$

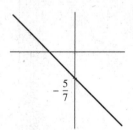

2. 6 ft/s^2, 15 ft/s; rate of change of v with respect to t is 6 ft/s^2; acceleration

3. $\dfrac{2}{15}$ in./min; $d = \dfrac{2}{15}t + 136$

4. ≈ -0.16 atm/liter

5. **a.** $h(t) = \dfrac{5}{32\pi}t$ **b.** $\dfrac{5}{32\pi}$ cm/s

    c. 128π seconds

6. a. $C = \dfrac{5}{9}F - \dfrac{160}{9}$ **c.** $\dfrac{5}{9}°C/°F$

   d. $F = \dfrac{9}{5}C + 32$ **f.** $\dfrac{9}{5}°F/°C$

b. **e.**

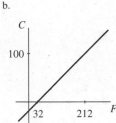

23. a. $-\dfrac{11}{4}$ **b.** $2\sqrt{5} + 3$ or $-2\sqrt{5} + 3$

   c. $-\dfrac{11}{20}$

25. $y = \sqrt{3}x + (3 + \sqrt{3})$

27. $x^4 + 8x^2 + 12$

29. $\left(\dfrac{7 + \sqrt{89}}{4}, 10 + \sqrt{89}\right), \left(\dfrac{7 - \sqrt{89}}{4}, 10 - \sqrt{89}\right)$

31. $\dfrac{48}{5}$

Section 1.3

1. 4

2. 4

3. 5

4. -2

5. 12

6. $3a^2$

7. $-\dfrac{1}{2}$

8. -6

9. $\dfrac{1}{6}$

10. $\dfrac{1}{2\sqrt{a}}$

12. at $x = 0, \approx 1$; at $x = \dfrac{2\pi}{3}, \approx -\dfrac{1}{2}$

13. at $x = 0, \approx 0.693147$
   at $x = 1, \approx 1.386294$
   at $x = 2, \approx 2.772589$
   at $x = 4, \approx 11.090355$

14. undefined for $t = \dfrac{1}{2}$;

   rate of change is 2 for $x > \dfrac{1}{2}$;

   rate of change is -2 for $x < \dfrac{1}{2}$

15. undefined at $x = -3$ and $x = \dfrac{4}{3}$;

   r.o.c. $= \begin{cases} 5 & x < -3 \\ 7 & -3 < x < 4/3 \\ -5 & x > 4/3 \end{cases}$

16. a. $0 \le t \le \dfrac{\sqrt{125}}{2} \approx 5.59$ s

   b. ft/s, tells how fast ball is falling

   c. $-32t$ ft/s

   d. $t = \dfrac{\sqrt{125}}{2}$ s, $-16\sqrt{125} \approx -178.885$ ft/s

17. $-\dfrac{3}{2}$ millibars/h; pressure is falling; most rapid rise

   is where graph rises most steeply; most rapid drop
   is where graph falls most steeply

18.

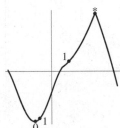

27. a. 3 **b.** $x + a + 2$

   c. $\dfrac{3}{(x + 2)(a + 2)}$ **d.** $\dfrac{-4ax + 12}{(x^2 + 3)(a^2 + 3)}$

29. two solutions for $|b| > \sqrt{28}$
   one solution for $b = \pm\sqrt{28}$
   no solutions for $|b| < \sqrt{28}$

31. a. $0 < \theta < \dfrac{\pi}{2}$

   b. $\dfrac{\pi}{2} < \theta < \pi$ and $\dfrac{3\pi}{2} < \theta < 2\pi$

   c. $\dfrac{\pi}{4} < \theta < \dfrac{5\pi}{4}$

33. a. $a = 0.003, b = 0.001$
   b. $a = -0.0005, b = 0.000001$
   c. $a = 10^{-2}, b = 10^{-11}$
   d. $a = 10^{-11}, b = -10^{-2}$

35. a. 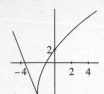   b. 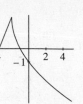   c.

d.   e.   f.

Section 1.4

1. $-\dfrac{2}{3}$

2. -1

3. 1

4. -9

5. Does not exist

6. 0

7. 45

8. $-\dfrac{2}{49}$

9. $\dfrac{1}{2}$

10. 0

11. Does not exist

12. Does not exist

13. $\dfrac{1}{2}$

14. 1

15. Does not exist

16. 2

17. **a.** $-1 \le c \le 0$ and $c = 2$ **b.** $c = 1$
   c. 1 **d.** Does not exist   **e.** Does not exist

18. **a.** $c = -4, c = 2$ **b.** Does not exist
   c. 9 **d.** Does not exist

39. $\dfrac{10}{3}(x + 1)\left(x - \dfrac{1}{2}\right)(x - 3)$

41. $\dfrac{6}{25}\left(x + \dfrac{3}{2}\right)^2 (x - 5)$

43. $t = 6\pi$

45. $\sin \theta = \dfrac{12}{13}, \cos \theta = -\dfrac{5}{13}$

47. To start: by the distance formula,
$$c^2 = (a \cos \theta - b)^2 + (a \sin \theta)^2$$

Section 1.5

1. **a.** 6 **b.** Any positive $\delta < 0.01$ will work.

2. **a.** $\pi - 2$
   b. Any positive $\delta < 0.005$ will work.

3. **a.** 6 **b.** Any $\delta > 0$ will work.

4. **a.** 2 **b.** Any positive $\delta < 0.03$ will work.

5. **a.** -3 **b.** Any positive $\delta < 0.1$ will work.

6. **a.** $-\dfrac{1}{4}$ **b.** Any positive $\delta < 0.4$ will work.

7. **a.** -13 **b.** $1.9995 < x < 2.0005$

8. **a.** $\dfrac{1}{\sqrt{2}}$ **b.** $0.95 < x < 1.05$

9. **a.** 1 **b.** $-0.06 < x < 0.06$

10. **a.** $\ln 10 \approx 2.302585$
   b. $-0.003 < x < 0.003$

11. **a.** $2\sqrt{5}$ **b.** $4.96 < x < 5.04$

12. **a.** 2 **b.** $0.782 < x < 0.790$

13. **a.** $-\dfrac{1}{4}$ **b.** $0.93 < x < 1.07$

14. **a.** 10 **b.** $3x^2 - 2x - 8 \approx 10(x - 2)$
   c. $1.985 < x < 2.015$
   d. Error of about 0.0012

15. **a.** $-\dfrac{4}{3}$ **b.** $\sqrt{2r^2 + 1} \approx 3 - \dfrac{4}{3}(r + 2)$
   c. $-2.15 < r < -1.85$
   d. Error of about 1.495×10^{-5}

16. a. $9 \ln 3 \approx 9.888$ **b.** $3^t \approx 9 + 9.888(t - 2)$
c. $1.97 < x < 2.03$
d. Error of about 0.00123

17. a. 2   **b.** $\tan x \approx 1 + 2\left(x - \dfrac{\pi}{4}\right)$
c. $0.77 < x < 0.80$
d. Error of about 0.00239

18. As seen in Fig. 1.46,

$$\sin h = PR < QS = h.$$

Therefore, $\sin x < x$ for $x > 0$. Thus, in the first quadrant, the graph of $y = x$ is the upper graph. If $x < 0$, then $-x > 0$, so

$$-\sin x = \sin(-x) < -x.$$

Therefore,

$$x < \sin x.$$

Thus, in the third quadrant, the graph of $y = x$ is the lower graph.

19. a. $\lim_{t \to 0^+} h(t) = 1$, $\lim_{t \to 0^-} h(t) = -1$
b. Does not exist   **c.** No, because $h(0)$ is undefined and because $\lim_{t \to 0} h(t)$ does not exist.

20. a.

b. $\lim_{w \to 2^-} P(w) = 0.55$, $\lim_{w \to 2^+} P(w) = 0.78$
c. $\lim_{w \to a} P(w)$ does not exist when a is a positive integer.

21. a. 1, 1   **b.** 1, 1   **c.** 2, 0   **d.** -1 and 1
22. a. $-4, 2$ **b.** 2, 2   **c.** 3, -3
d. $-4, -2$, and 2

43. a. $x^4 - 4x^3 + 3x^2 + 2x - 1$
b. $x^4 - 6x^3 + 9x^2$ **c.** $x^4 - 4x^3 + 4x^2$
d. $x^8 - 8x^7 + 24x^6 - 32x^5 + 14x^4 + 8x^3 - 8x^2 + 1$

45. a. $x = \dfrac{1}{2}$, $x = 3$, or $x = \dfrac{9}{2}$
b. $x = -5$, $x = 3$, or $x = 6$

47. a. 3   **b.** -3 **c.** $\dfrac{5}{2}$ **d.** $\dfrac{2}{3}$

49. a. 2   **b.** 0   **c.** 6   **d.** 4

51. $48\sqrt{143} \approx 573.997$ cm^3

53. $\dfrac{4 \tan 13° \tan 8°}{\tan 13° - \tan 8°} \approx 1.437$ miles ≈ 7586.5 ft

Section 1.6

1. $2x$
2. $4x$
3. $-6t + 4$
4. $-\dfrac{2}{(u + 2)^2}$
5. $\dfrac{4}{(2 - x)^2}$
6. $1 - \dfrac{1}{\sqrt{t}}$
7. $2 + \dfrac{3}{2\sqrt{t + 9}}$
8. $2as + b$
9. $-\dfrac{a}{(ax + b)^2}$
10. $\dfrac{a}{2\sqrt{at + b}}$
11. $y = -(x - 1) - 2$
12. $y = 10(t + 2) - 7$
13. $y = 4(r - 1) + 6$
14. $y = -\dfrac{1}{a^2}(x - a) + \dfrac{1}{a}$
15. $y = -\dfrac{8}{9}(u - 2) + \dfrac{2}{3}$
16. $y = \dfrac{1}{\sqrt{2b - 1}}(x - b) + \sqrt{2b - 1}$
17. $y = 2(x - 2) + 4$
18. $4.14^2 \approx 8(4.14 - 4) + 16 = 17.12$, error ≈ 0.0196
$3.91^2 \approx 8(3.91 - 4) + 16 = 15.28$, error ≈ 0.0081
19. $\dfrac{1}{3.15} \approx -\dfrac{1}{9}(3.15 - 3) + \dfrac{1}{3} \approx 0.316667$, error ≈ 0.000794
$\dfrac{1}{2.85} \approx -\dfrac{1}{9}(2.85 - 3) + \dfrac{1}{3} = 0.35$, error ≈ 0.000877
20. a. $y = \dfrac{1}{2}\left(x - \dfrac{\pi}{3}\right) + \dfrac{\sqrt{3}}{2}$
b. $\sin\left(\dfrac{\pi}{3} + 0.1\right) \approx \dfrac{1}{2}(0.1) + \dfrac{\sqrt{3}}{2} \approx 0.916025$

21. a. $y = 2\left(x - \dfrac{\pi}{4}\right) + 1$

b. $\tan\left(\dfrac{\pi}{4} - 0.05\right) \approx 2(-0.05) + 1 = 0.9$,

error ≈ 0.004686

22. -4.988×10^{-5} N/m

23. -6.051×10^{-7} N/m

24. a. $C'(N) = \dfrac{500}{\sqrt{N}} + 0.1N - 0.1$

b. Extra cost $\approx C'(10) \approx \$159.01$

39. $x^{n-1} + ax^{n-2} + a^2x^{n-3} + \cdots + a^{n-2}x + a^{n-1}$

41. $(3x - 1)(2x + 5)(x + 7)$

43. $(x - 3)(x^2 + 5x + 12)$

45. Because $x = a$ is a solution to $p(x) - p(a) = 0$

47. 32

49. a. 2 **b.** -2 **c.** $\dfrac{3}{2}$ **d.** 0

Section 2.1

1. 4

2. $9t^2 - 2$

3. $x^3 + x^2 + x + 1$

4. $3ar^2 + 2br + c$

5. 0

6. $(n + 3)t^{n+2}$

7. $4x + 3$

8. $48t^2 + 192t + 192$

9. $x^2, \quad \dfrac{1}{3}x^3 + 10$

10. $-3, \quad -3x - 44\sqrt{2}$

11. $4x^3, \quad \dfrac{1}{12}x^4$

12. $(n + 1)x^n, \quad \dfrac{1}{n+1}x^{n+1}$ and $\dfrac{1}{n+1}x^{n+1} - 36$

13. $-\dfrac{51}{2}$

14. $\dfrac{20}{3}$

15. -4

16. $y = 5(x - 2) + 6$

17. $y = 2\sqrt{2}$

18. $y = (2ad + b)(x - d) + (ad^2 + bd + c)$

19. $y = na^{n-1}(x - a) + a^n$

31. $-\dfrac{25}{9}$

33. There are no such a.

35. If $\log_{10} x = b$, then $10^b = x$. For any real number b, 10^b is positive. Hence x must be positive.

37. $\tan\theta; \sec\theta; 1 + \tan^2\theta = \sec^2\theta$

39. -1

Section 2.2

1. $(2x - 3)(2x + 4) + (x^2 - 3x + 1)2$

2. $2(3x^3 - 3x + \pi) + 2x(9x^2 - 3)$

3. $2(-9t^5 + 6t^4 - 21t^2 + 8t - 1)(-45t^4 + 24t^3 - 42t + 8)$

4. $\dfrac{2(2t + 1) - 2(2t - 1)}{(2t + 1)^2}$

5. $\dfrac{(4x - 3)(x^2 - 7) - 2x(2x^2 - 3x + 10)}{(x^2 - 7)^2}$

6. $-6x + 2 - \dfrac{4}{x^2} - \dfrac{441}{x^{22}}$

7. $2\left(1 - \dfrac{1}{r} + \dfrac{2}{r^2}\right)\left(\dfrac{1}{r^2} - \dfrac{4}{r^3}\right)$

8. $\dfrac{[(2u - 2)(4u^2 - u + 1) + (u^2 - 2u + 3)(8u - 1)](u^3 + 8) - 3u^2(u^2 - 2u + 3)(4u^2 - u + 1)}{(u^3 + 8)^2}$

9. $2\theta(\theta^3 - 3)(\theta^4 - 4) + (\theta^2 - 2)(3\theta^2)(\theta^4 - 4) + (\theta^2 - 2)(\theta^3 - 3)(4\theta^3)$

10. $(z - z^{-1})(2z^2 - 4 + 2z^{-2}) + z(1 + z^{-2})(2z^2 - 4 + 2z^{-2}) + z(z - z^{-1})(4z - 4z^{-3})$

11. $\dfrac{1}{2}\dfrac{2r + 2}{\sqrt{r^2 + 2r + 7}}$

12. $\dfrac{1}{2}\left(\dfrac{v^2 - 3v + 1}{2v + 5}\right)^{-\frac{1}{2}}\left(\dfrac{(2v - 3)(2v + 5) - 2(v^2 - 3v + 1)}{(2v + 5)^2}\right)$

13. $\dfrac{1}{2}\dfrac{2}{\sqrt{2s - 1}}\sqrt{6s + 7} + \sqrt{2s - 1}\left(\dfrac{1}{2}\dfrac{6}{\sqrt{6s + 7}}\right)$

14. $\dfrac{1}{2}\dfrac{1}{\sqrt{\sqrt{x + 1}}}\dfrac{1}{2\sqrt{x + 1}}$

15. 13

16. 3

17. 0

18. 10

19. $\left(-\dfrac{7}{9} + \dfrac{5\sqrt{10}}{18}, \dfrac{238}{243} - \dfrac{625\sqrt{10}}{243}\right) \approx (-1.65619, \ 9.11285)$

$\left(-\dfrac{7}{9} - \dfrac{5\sqrt{10}}{18}, \dfrac{238}{243} + \dfrac{625\sqrt{10}}{243}\right) \approx (0.100633, \ -7.15401)$

20. $\left(-\dfrac{1}{3}, 2\sqrt{\dfrac{5}{3}}\right) \approx (-0.333333, 2.58199)$

21. $y = \dfrac{113}{225}(x - 2) + \dfrac{1}{15}$

22. $y = -(x + 2)$

33. $F(x) = \dfrac{1}{x}$, $G(x) = \sin x$, $H(x) = \sqrt{x}$,

$J(x) = x - 3$, $f = F \circ G \circ H \circ J$

35. $x^4 + x$ and $x^4 + x - 42$. The graphs are "parallel." The graph of $y = x^4 + x - 42$ is 42 units below the graph of $y = x^4 + x$.

37. $\dfrac{ac^2}{2a^2 - c^2}$

39. $(2^a)^b = 2^{ab}$ for all real numbers a, b. 2^{a^b} means $2^{(a^b)}$. Note that the exponent a^b is usually different from the exponent ab.

41. $+\infty$

Section 2.3

1. $\dfrac{26}{3}x^{10/3}$

2. $\dfrac{1}{5}(u - 7)^{-4/5}$

3. $23(2x^2 - 3x + 1)^{22}(4x - 3)$

4. $\dfrac{1}{2}\dfrac{2}{\sqrt{2t - 7}}$

5. $\dfrac{5}{4}\left(x^3 - 5x + \dfrac{1}{2}\right)^{1/4}(3x^2 - 5)$

6. $12(2\theta - 1)^5(4\theta + \pi)^8 + 32(2\theta - 1)^6(4\theta + \pi)^7$

7. $\dfrac{1}{3}\left(\dfrac{4u - 1}{3u + 7}\right)^{-2/3}\left(\dfrac{4(3u + 7) - 3(4u - 1)}{(3u + 7)^2}\right)$

8. $\dfrac{1}{8}\dfrac{1}{\sqrt{1 + \sqrt{1 + \sqrt{x + 1}}}}\dfrac{1}{\sqrt{1 + \sqrt{x + 1}}}\dfrac{1}{\sqrt{x + 1}}$

9. $24z(3z^2 - 1)^3(4z^{-3} + 2z^{-2} + 4)\sqrt{1 - \dfrac{1}{z}}$

$+ (3z^2 - 1)^4(-12z^{-4} - 4z^{-3})\sqrt{1 - \dfrac{1}{z}}$

$+ (3z^2 - 1)^4(4z^{-3} + 2z^{-2} + 4)\left(1 - \dfrac{1}{z}\right)^{-1/2}\dfrac{1}{2z^2}$

10. $-10\dfrac{1 + 2x}{(1 + x + x^2)^{11}}$

11. $(2(x^2 - 3x + 1) - 3)(2x - 3)$

12. $\dfrac{-8}{(1 + (1 + t^{-2})^{-2})^3(1 + t^{-2})^3 t^3}$

13. $-3x^{-5/2}(x^{-3/2} + 1)$

14. $-\dfrac{1}{(2x - 1)^{3/2}}$

15. $2(-6x^2 + 1)\sin(-2x^3 + x + 1)$

16. $\dfrac{6(3x - 1)}{(3x - 1)^2 + 1}$

17. $\dfrac{-3x^2 + 1}{2 + 3y^2}$

18. $\dfrac{3y - 4x}{2y - 3x}$

19. $\dfrac{y^2 + 2y - 2x}{2(4y - xy - x)}$

20. $\dfrac{2xy^3\sqrt{x^2 + y^2} + \sqrt{x^2 + y^2} - x}{y - \sqrt{x^2 + y^2} - 3x^2y^2\sqrt{x^2 + y^2}}$

21. $-\dfrac{1}{3}$

22. -1

37. $(x + 2)^{-3}$ and $(x + 2)^{-3} - 6\sqrt{3}$

39. $\cos\theta$ decreases from 1 to -1 as θ grows from 0 to π. $\cos\theta$ increases from -1 to 1 as θ grows from π to 2π.

41. period $= \pi$, amplitude $= 5$

43. The graphs cross twice.

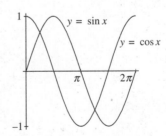

45. $\tan(\pi + 1) \approx 1.557408$

$\tan(2\pi - 1) = \tan(\pi - 1) \approx -1.557408$

Section 2.4

1. $\theta = \dfrac{\pi}{6}, \dfrac{\pi}{6} \pm 2\pi, \dfrac{\pi}{6} \pm 4\pi, \dfrac{\pi}{6} \pm 6\pi, \ldots$ or

$\theta = \dfrac{5\pi}{6}, \dfrac{5\pi}{6} \pm 2\pi, \dfrac{5\pi}{6} \pm 4\pi, \dfrac{5\pi}{6} \pm 6\pi, \ldots$

2. $x = \dfrac{\pi}{4}, \dfrac{\pi}{4} \pm \pi, \dfrac{\pi}{4} \pm 2\pi, \dfrac{\pi}{4} \pm 3\pi, \ldots$

3. $t \approx 0.927295, 0.927295 \pm 2\pi,$
$0.927295 \pm 4\pi, 0.927295 \pm 6\pi, \ldots$ or
$t \approx -0.927295, -0.927295 \pm 2\pi,$
$-0.927295 \pm 4\pi, -0.927295 \pm 6\pi, \ldots$

4. $\theta = \dfrac{4\pi}{9} + \dfrac{2\pi}{3}k, \quad k = 0, \pm 1, \pm 2, \pm 3, \ldots$ or

$\theta = \dfrac{5\pi}{9} + \dfrac{2\pi}{3}k, \quad k = 0, \pm 1, \pm 2, \pm 3, \ldots$

5. $x = \pm\pi, \pm 3\pi, \pm 5\pi, \pm 7\pi, \ldots$

6. $x \approx 1.271088 + k\pi, \quad k = 0, \pm 1, \pm 2, \pm 3, \ldots$ or
$x \approx -0.890581 + k\pi, \quad k = 0, \pm 1, \pm 2, \pm 3, \ldots$

7. $2\cos(2x + 3)$

8. $\sec t \tan t$

9. $\dfrac{x\sin x + \cos x}{x^2}$

10. $\dfrac{-4\cos 2u}{(1 + \sin 2u)^2}$

11. $\dfrac{2}{(1 - \theta)^2} \sec^2\left(\dfrac{1 + \theta}{1 - \theta}\right)$

12. $-3\csc^2 3t$

13. $-4\csc 4v \cot 4v$

14. $\cos^2 r - \sin^2 r \quad (= \cos 2r)$

15. $(\cos x - x\sin x)\cos(x\cos x)$

16. $5(2 - \sqrt{\sin w})^4 \dfrac{\cos w}{2\sqrt{\sin w}}$

17. $4\tan(2x)\sec^2(2x)$

18. $(12\sin^3 t - 6\sin^2 t + 1)\cos t$

19. $-\sqrt{2}$

20. 1

21. 1

22. -1

23. $\dfrac{\theta}{2}$ must be a third-quadrant angle.

24. $\ldots, -\dfrac{17\pi}{2} < \theta < -\dfrac{15\pi}{2}$, or $-\dfrac{9\pi}{2} < \theta < -\dfrac{7\pi}{2}$,

or $-\dfrac{\pi}{2} < \theta < \dfrac{\pi}{2}$, or $\dfrac{7\pi}{2} < \theta < \dfrac{9\pi}{2}$, or

$\dfrac{15\pi}{2} < \theta < \dfrac{17\pi}{2}, \ldots$

This is equivalent to saying that $\dfrac{\theta}{2}$ is coterminal with

an angle α where $-\dfrac{\pi}{4} < \alpha < \dfrac{\pi}{4}$. That is, the ter-

minal side of $\dfrac{\theta}{2}$ must lie in the shaded region of the

following diagram.

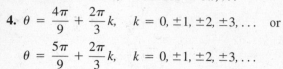

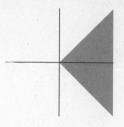

25. $\dfrac{\theta}{3}$ must be coterminal with an angle α

where $\dfrac{\pi}{2} < \alpha < \dfrac{2\pi}{3}$ or $\dfrac{5\pi}{6} < \alpha < \pi$ or

$\dfrac{3\pi}{2} < \alpha < \dfrac{5\pi}{3}$ or $\dfrac{11\pi}{6} < \alpha < 2\pi$.

That is, the terminal side of $\dfrac{\theta}{3}$ must lie in the shaded

region of the following diagram.

26. $\theta = \ldots, -9\pi, -5\pi, -\pi, 3\pi, 7\pi, 11\pi, \ldots$

39. $x = \dfrac{7y + 2}{2y - 3}$

41. $(5, 27), \left(-1, \dfrac{27}{8}\right)$

43. $199, 16, 7a^2$

45.

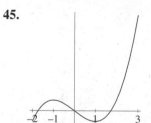

a. $-\sqrt{3}, 0, \sqrt{3}$

b. On $-2 \le x \le -1$ and $1 \le x \le 3$

c. On $-1 \le x \le 1$

d. At $x = -1$ the value of f is greater than the value at any nearby point. At $x = 1$ the value of f is less than the value at any nearby point.

Section 2.5

1. $-\dfrac{4}{3}$

2. $t = \log_5 2 \approx 0.430677$

3. 1

4. $r = \ln(1/2) \approx -0.693147$

5. $x = 0$ or $x = \ln 2 \approx 0.693147$

6. No solution

7. $u = \ln(5 \pm 2\sqrt{6}) = \pm 2.29243$

8. $2e^{2x}$

9. $4e^{2x}\cos(2e^{2x})$

10. $C_4 4^t \approx 1.38629 \cdot 4^t$

11. $e^t e^{e^t}$

12. $\dfrac{2e^{2x} - e^x}{2\sqrt{e^{2x} - e^x + 3}}$

13. $4r^3 e^{4r} + 4r^4 e^{4r}$

14. $C_2 2^s e^{4s} + 4(2^s e^{4s}) \approx 0.693147(2^s e^{4s}) + 4(2^s e^{4s})$

15. $\sec^2\!\left(\dfrac{1 + e^t}{1 - e^t}\right)\dfrac{2e^t}{(1 - e^t)^2}$

16. $2\tau e^{\tau^2} + 2\tau^3 e^{\tau^2}$

17. $\dfrac{2e^u(e^u + 2)\sin 3u - 3(e^u + 2)^2 \cos 3u}{\sin^2 3u}$

18. $\dfrac{e^t - e^{-t}}{2}$

19. $\dfrac{e^t + e^{-t}}{2}$

20. $\dfrac{2x + 2y - ye^x}{e^x - 2x - 2y}$

21. $-\dfrac{y}{y + 1}$

22. $\dfrac{1 - e^x \tan(xy^2) - y^2 e^x \sec^2(xy^2)}{2xye^x \sec^2(xy^2) - 3}$

23. $\dfrac{y}{2ye^{y^2} - x}$

25. $e^2 \approx 7.389056$

26. a. $C_0 = 150,\ k = \ln(2/3) \approx -0.405465$
    b. ≈ 19.75 ppm   **c.** After ≈ 6.68 years

28. a. Suppose that a magnitude x star is k times as bright as a magnitude $x + 1$ star. Since a magnitude 1 star is 100 times as bright as a magnitude 6 star, $k^5 = 100$. This gives $k = \sqrt[5]{100}$.
    b. $b_1/b_2 = (\sqrt[5]{100})^{5.3 - 1.2} \approx 43.652$

45.

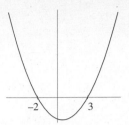

$x < -2$ or $x > 3$

47. $-9x^2 + 2 + \dfrac{12}{x^5}$

49. 4

51. The x-intercept of the line is $(a, 0)$, and the y-intercept is $(0, b)$.

53. $(x - 2)^2 + (y - 3)^2 < 9$

Section 2.6

1. 4

2. $-\dfrac{2}{3}$

3. 3

4. 1995

5. 5

6. 0

7. $\dfrac{1}{3}\log_2 5 = \dfrac{1}{3}\dfrac{\ln 5}{\ln 2} \approx 0.773976$

8. $\ln(1/2) \approx -0.693147$

9. No solution

10. -0.0403458

11. $\dfrac{32}{3}$

12. $\sqrt{8}$

13. 1

14. $\exp\!\left(\dfrac{1}{1/\ln 2 + 1/\ln 5 - 1/\ln 10}\right) \approx 1.84706$

15. $\dfrac{1}{x + 2}$

16. $1 + \ln t$

17. $\dfrac{\cos \theta}{\sin \theta}$ $(= \cot \theta)$

18. $\dfrac{1}{u \ln u}$

19. $\dfrac{10}{t}(1 + \ln(2t))^9$

20. $\dfrac{3}{x(2 - \ln x)^2}$

21. $\dfrac{-2 + 3z^2}{1 - 2z + z^3} - \dfrac{\sin z + z \cos z}{z \sin z}$

22. $-\dfrac{1}{2x(\ln x)^{3/2}}$

23. $\dfrac{2 \ln \theta}{\theta} e^{(\ln \theta)^2}$

24. $(1 + \ln x)e^{x \ln x}$

25. $\dfrac{x^2}{2x + 1}\left(\dfrac{2}{x} - \dfrac{2}{2x + 1}\right)$

26. $x^{x+1}\left(1 + \dfrac{1}{x} + \ln x\right)$

27. $\sqrt{x \tan x}\left(\dfrac{1}{2x} + \dfrac{\sec^2 x}{2 \tan x}\right)$

28. $\left(\dfrac{1 + e^x}{1 - e^x}\right)^{-10}\left(\dfrac{-10e^x}{1 + e^x} - \dfrac{10e^x}{1 - e^x}\right)$

29. $b^{\log_b xy} = xy = b^{\log_b x}b^{\log_b y} = b^{\log_b x + \log_b y}$

30. $b^{\log_b x^r} = x^r = (b^{\log_b x})^r = b^{r \log_b x}$

31. $\log_1 x = y$ would mean that $1^y = x$. If $x \neq 1$, there is no possible y value.

32. $y = (9.02501)^x \approx e^{2.2x}$

33. $y = \left(\dfrac{1}{20.0855}\right)^x \approx e^{-3x}$

34. $\ln x$ is not defined for $x \le 0$.

55. $x = \dfrac{4 - y}{3}$

57. $x = \dfrac{2 \pm \sqrt{-3y - 20}}{3}$

59. $(4, 4), (4, 20)$

61. $5\sqrt{2}$

63. $\arctan 2 \approx 1.107149$ radians $\approx 63.435°$

Section 2.7

1. $\dfrac{\pi}{3}$

2. $-\dfrac{\pi}{4}$

3. $-\dfrac{\pi}{2}$

4. $6 - 2\pi$

5. $-\dfrac{\pi}{4}$

6. $\dfrac{3}{5}$

7. $\dfrac{2}{\sqrt{1 - 4x^2}}$

8. $\arctan t + \dfrac{t}{1 + t^2}$

9. $-\dfrac{1}{1 + \theta^2}$

10. $\dfrac{2}{(1 - x)^2}\dfrac{1}{\sqrt{1 - \left(\dfrac{1 + x}{1 - x}\right)^2}}$

11. $\dfrac{1}{\sqrt{1 - t^2}} - \dfrac{t}{|t|\sqrt{1 - t^2}}$

12. $\dfrac{2x}{1 + x^4}e^{\arctan x^2}$

13. Domain of f is $\{x : x \neq 3/2\}$.
Domain of f^{-1} is $\{x : x \neq -1/2\}$.
$$f^{-1}(x) = \dfrac{2 - 3x}{1 + 2x}$$

14. Domain of g^{-1} is the set of real numbers.

15. Show that $G(F(t)) = t$.

16. $l^{-1}(x) = \dfrac{x - b}{m}$

18. a. The graph of such a function is symmetric about the line $y = x$.

b.

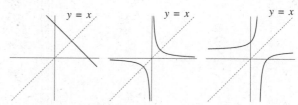

c. Examples are $f(x) = \dfrac{1}{x}$ and $y = -x + 2001$.

19. $\approx 18.3°$ on January 31, $\approx 21.4°$ on October 31.

20. For $20°$ the date is around February 7 or November 5.
For $40°$ the date is around April 11 or September 3.

37. Center $= (2, -4)$, radius $= \sqrt{22}$

39. We'd see two "line segments," one of slope 6 and one of slope -6.

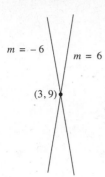

$m = -6$ $m = 6$

$(3, 9)$

41. a. $a = \dfrac{2}{3}, b = -\dfrac{1}{3}, c = -1$

b. $d = 2, f = 1, g = -1$

c. $(-1, 0), (0, -1), (2, 1)$

43. $\dfrac{8s}{4s^2 + 3}$

45. $-\dfrac{e^\theta}{(1 + e^\theta)^2}$

Section 2.8

1. a. $C = 4.0 \cdot 10^6, k \approx 0.0280637,$
$P(t) \approx 4.0 \cdot 10^6 e^{0.0280637t}$

b.

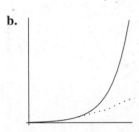

c. $C \approx 4.54 \cdot 10^6, k \approx 0.0263031,$
$P(t) \approx 4.54 \cdot 10^6 e^{0.0263031t}$

d.

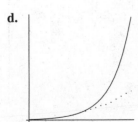

2. a. $\dfrac{dA}{dt} = kA$

b. $A(t) = 10,000 e^{(t/2)\ln(1.25)} \approx 10,000 e^{0.111572t}$

c. $A(t) = (16,000/\sqrt[3]{11}) e^{(t/3)\ln(11/8)}$
$\approx 7194 e^{0.106151t}$

3. a. Increasing when $P = 250$ million, decreasing when $P = 400$ million.

b. $P = k/c \approx 3.00 \cdot 10^8$. If the population ever reaches k/c, then it stays at this figure.

4. a. The term $1.496P$ may be the term that represents growth under uncrowded conditions. The term $0.121P^{1.35}$ becomes more important as P gets larger and so could represent the effect of overcrowding.

b. Let $c = (1.496/0.121)^{1/0.35} \approx 1319.5$. Population is increasing when $P < c$, decreasing when $P > c$, and unchanging when $P = c$.

5. b. $dA/dt = (Ce^{-kt})' = C(-ke^{-kt})$
$= -k(Ce^{-kt}) = -kA$

c. $C = 10, k = \ln 2/5568 \approx -0.000124488$

6. $\dfrac{dC}{dt} = -kC, k$ a positive constant.

7. $\dfrac{dP}{dt} = -kP, k$ a positive constant.

8. $\dfrac{dT}{dt} = -k(T - T_0), k$ a positive constant.

15. a. $8(4x - 3)$ **b.** $(2 + x)/(x + 1)^{3/2}$

c. $2e^{2x}\sin 3x + 3e^{2x}\cos 3x$

17. $\sqrt{10}/3$ cm/s

19. $t = 1$ second; $(33, 44)$

Section 3.1

1. $v(1.5) = -14.7$ m/s, $x(4.5) \approx 0$ m

t_1	t_2	$v(t_1, t_2)$
1.5	1.6	-15.19
1.5	1.52	-14.798
1.5	1.51	-14.749
1.5	1.501	-14.7049

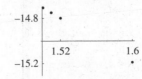

-14.8

1.52 1.6

-15.2

2. $v(1.25) = -12.25$ m/s, $x(3.9) \approx 0$ m

t_1	t_2	$v(t_1, t_2)$
1.25	1.35	−12.74
1.25	1.27	−12.348
1.25	1.26	−12.299
1.25	1.251	−12.2549

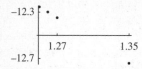

3. $A = 3.8, B = 1, C = 1.8$;
$x(1.3) \approx 0$ m; $x'(2.9) \approx 0$ m/s

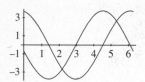

4. $A = 4.9, B = 1, C = 2.3$;
$x(0.85) \approx 0$ m; $x'(2.4) \approx 0$ m/s

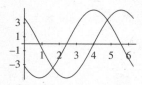

5. $v(2.74721) \approx -72.8$ m/s
6. $v(11.1111) = -18$ m/s
7. $b \approx 128.575$ s; $t_{max} \approx 64.2857$ s;
$x_{max} = 2.0 \times 10^4$ m
8. $b \approx 122.45$ s; $t_{max} \approx 61.2245$ s;
$x_{max} = 1.8 \times 10^4$ m
9. $x(3) = 271,986$ miles; $v_{max} = 71.3021$ mph
10. $x(2) = 48,228.9$ miles; $v_{max} = 65.928$ mph
11. $x(2.1) = 36.591$
12. $x(3.9) = 114.771$
13. $x(1.2) = -0.149432$
14. $x(3.6) = 1.07885$
15. $x_{max} = 737.194$
16. $x_{max} = 858.745$
53. $(-1.60229, 1.19694)$; (x, y) is on circle of radius 2
55. $r = 26.926, \theta = 2.7611$
57. Points on a circle of radius 3, centered at $(0, 0)$
59. $\theta = 0.28379$
61. 2.24899 or $128.857°$

Section 3.2

1. $\|\mathbf{r}\| = \sqrt{2}, \theta = 0.78540$
2. $\|\mathbf{r}\| = \sqrt{5}, \theta = 0.463648$
3. $\|\mathbf{r}\| = 5, \theta = 2.21430$
4. $\|\mathbf{r}\| = 5\sqrt{2}, \theta = 1.71269$
5. $\|\mathbf{a}\| = 6.59469, \theta = 4.00212$
6. $\|\mathbf{v}\| = \sqrt{101}, \theta = 4.61272$
7. $\|\mathbf{v}\| = 1, \theta = 5.81954$
8. $\|\mathbf{r}\| = 1, \theta = 5.90268$

9. $y = \dfrac{3}{5}x + 2$

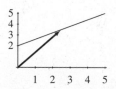

10. $y = -\dfrac{3}{4}x - \dfrac{7}{4}$

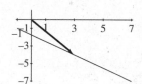

11. $x^2 + y^2 = 3^2$

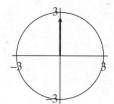

12. $x^2 + y^2 = 2^2$

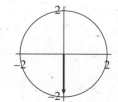

13. $\sqrt{34}$ m/s
14. 5 m/s
15. 3 m/s

16. 2 m/s

33. $1/\sqrt{2x+1}$

35. $6x\cos(3x^2)$

37. $2e^{2x-3}$

39. $1/(2\sqrt{x}\sqrt{1-x})$

41. $(1-3x^2)/(2\sqrt{x}(x^2+1)^2)$

43. $2x/(x^2+9)$

Section 3.3

1. $\overrightarrow{PQ} = \overrightarrow{ST} = (3, -2)$

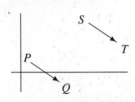

2. $\overrightarrow{PQ} = \overrightarrow{ST} = (-3, 1)$

3. From $(4, -4)$ to $(4, -4) + (2, 3) = (6, -1)$

4. From $(-2, -5)$ to $(-2, -5) + (-5, 2) = (-7, -3)$

5. $\|\overrightarrow{PQ}\| = 8, \theta = 0$

6. $\|\overrightarrow{PQ}\| = 2\sqrt{5}, \theta = 5.81954$

7. $\|\overrightarrow{PQ}\| = 2\sqrt{17}, \theta = 4.46741$

8. $\|\overrightarrow{PQ}\| = 6\sqrt{5}, \theta = 4.24874$

9. $(-0.260166, 17.7471)$

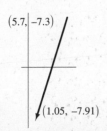

10. $(1.05129, -7.90525)$

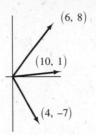

11. $-(\mathbf{F}_1 + \mathbf{F}_2) = -(10, 1)$

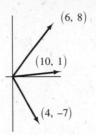

12. $-(\mathbf{F}_1 + \mathbf{F}_2) = -(1150, 1115)$

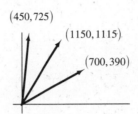

13. Magnitude ≈ 62.9621; direction $\approx 332.860°$

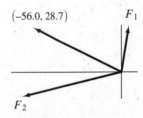

14. Magnitude ≈ 36.3646; direction $\approx 71.7532°$

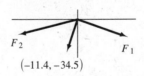

15. $\mathbf{a} = (5, 2), \mathbf{b} = (2, 3), \mathbf{c} = (7, 5); T = (10, 9)$

16. $\mathbf{a} = (-2, -2), \mathbf{b} = (3, 1), \mathbf{c} = (1, -1);$
    $T = (2, -4)$

17. Single displacement is $(-10, 18); T = (-10, 22)$

18. Single displacement is $(-3, -4); T = (-7, 5)$

19. $\mathbf{a} - \mathbf{b} = (1, 2)$

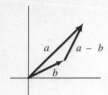

20. $\mathbf{a} - \mathbf{b} = (3, 3)$

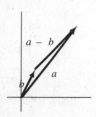

21. $m = b/a, \mathbf{r} = (1, m)$
22. $(-8, 18)$
23. $(11, -13)$

37. $r = \sqrt{2}, \theta = \pi/4$
39. $r = \sqrt{17}, \theta = 6.03821$
41. $r = \sqrt{34}, \theta = 2.60117$
43. $x = 1, y = 1$
45. $x = -2.91287, y = 0.717748$
47. $x = -0.412302, y = 3.17333$

Section 3.4

1. $\mathbf{v}(1) = (2, 6)$ m/s, $\|\mathbf{v}(1)\| = 2\sqrt{10}$ m/s

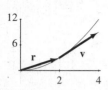

2. $\mathbf{v}(1) = (4, 2)$ m/s, $\|\mathbf{v}(1)\| = 2\sqrt{5}$ m/s

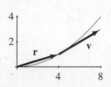

3. $\mathbf{v}(1) = (-1, 4)$ m/s, $\|\mathbf{v}(1)\| = \sqrt{17}$ m/s

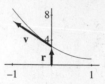

4. $\mathbf{v}(2) = (-3, 12)$ m/s, $\|\mathbf{v}(2)\| = 3\sqrt{17}$ m/s

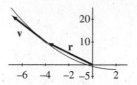

5. $\mathbf{v}(\pi/6) = 10(-\sqrt{3}/2, 1/2)$ m/s,
$\|\mathbf{v}(\pi/6)\| = 10$ m/s

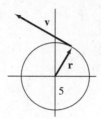

6. $\mathbf{v}(\pi/2) = (0, 6)$ m/s, $\|\mathbf{v}(\pi/2)\| = 6$ m/s

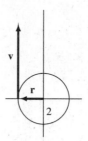

7. $\mathbf{v}(2) = (1, 1/2)$ m/s, $\|\mathbf{v}(2)\| = \sqrt{5}/2$ m/s

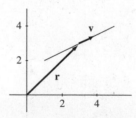

8. $\mathbf{v}(1) = \langle -2, 0 \rangle$ m/s, $\|\mathbf{v}(1)\| = 2$ m/s

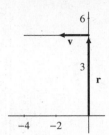

9. $\langle 1, 0.839604 \rangle$ m/s, $\langle 1, 0.863511 \rangle$ m/s,
$\langle 1, 0.865775 \rangle$ m/s; $\mathbf{v}(\pi/6) = \langle 1, 0.866025 \rangle$ m/s

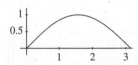

10. $\langle 1, -0.889562 \rangle$ m/s, $\langle 1, -0.868511 \rangle$ m/s,
$\langle 1, -0.866275 \rangle$ m/s; $\mathbf{v}(\pi/3) = \langle 1, -0.866025 \rangle$ m/s

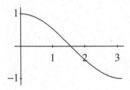

11. $\langle 1, 2.85884 \rangle$ m/s, $\langle 1, 2.73192 \rangle$ m/s, $\langle 1, 2.71964 \rangle$ m/s;
$\mathbf{v}(1.0) = \langle 1, 2.71828 \rangle$ m/s

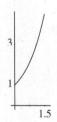

12. $\langle 1, 0.487902 \rangle$ m/s, $\langle 1, 0.498754 \rangle$ m/s,
$\langle 1, 0.499875 \rangle$ m/s; $\mathbf{v}(1.0) = \langle 1, 0.5 \rangle$ m/s

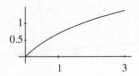

19.

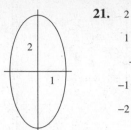

21.

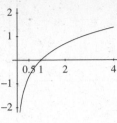

23.

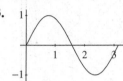

25.

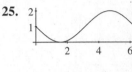

Section 3.5

1. $\mathbf{r}(-2) = \langle 9, 0 \rangle$, $\mathbf{r}(4) = \langle 19, 5 \rangle$, $x = 1 + 2(y - 2)^2$

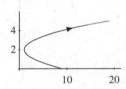

2. $\mathbf{r}(-3) = \langle 24, -2 \rangle$, $\mathbf{r}(3) = \langle 0, 4 \rangle$, $x = y^2 - 6y + 8$

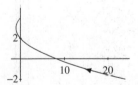

3. $\mathbf{r}(-2) = \langle 0, 2 \rangle$, $\mathbf{r}(2) = \langle 4, 6 \rangle$,
$4x^2 + y^2 - 4xy + 2x - 2y = 0$

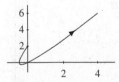

4. $\mathbf{r}(-3) = \langle 12, 3 \rangle$, $\mathbf{r}(2) = \langle 2, 8 \rangle$,
$x^2 + y^2 - 2xy - 6x - 3y = 0$

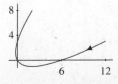

5. $\mathbf{r}(-2) = (-3, -3), \mathbf{r}(3) = (7, 2), y = \dfrac{1}{2}x - \dfrac{3}{2}$

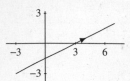

6. $\mathbf{r}(0) = (3, 5), y = 2x - 1$

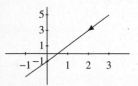

7. $y = 3x - 5$

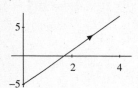

8. $y = -\dfrac{5}{2}x - \dfrac{13}{2}$

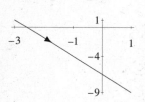

9. $y = x$

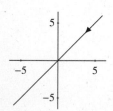

10. $y = x$

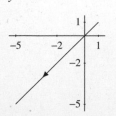

11. $\mathbf{r}'(1) = (1, 2)$

12. $\mathbf{r}'(1) = (1, 3)$

13. $\mathbf{r}'(\pi/4) = (1, 2)$

14. $\mathbf{r}'(\pi/6) = (1, 2/3)$

15. $\mathbf{r}'(4) = (1, 2)$

16. $\mathbf{r}'(-1) = (1, 2)$

17. $\mathbf{r}'(1) = (1, 2e^2)$

18. $\mathbf{r}'(1) = (1, e/2)$

19. $\mathbf{r}'(1) = (1, 0)$

20. $\mathbf{r}'(1) = (1, 1/2)$

21.

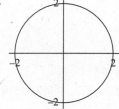

22.

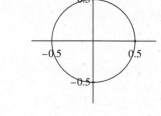

23.

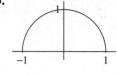

24.

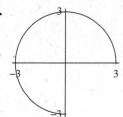

25.

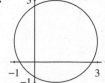

26.

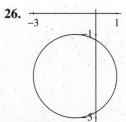

27.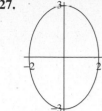

28.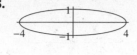

29.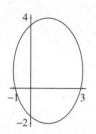

30.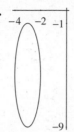

31.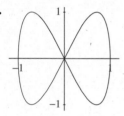

32.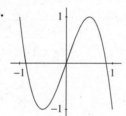

33.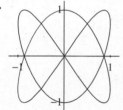

34.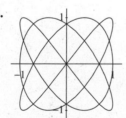

51.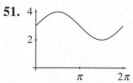

53.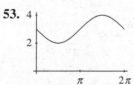

55.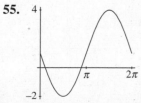

57.

59.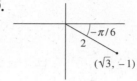

61.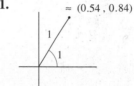

63. Polar form is $\approx (5, 5.64)$.

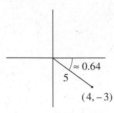

65. Polar form is $\approx (5, 4.07)$.

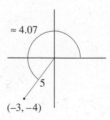

Section 3.6

1.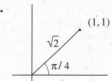

2.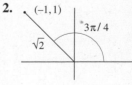

3. 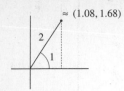 ≈ (1.08, 1.68)

4. (0.21, 2.99)

5.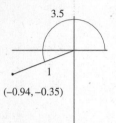

6.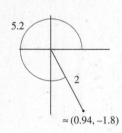

7.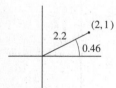

8.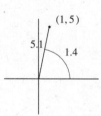

9.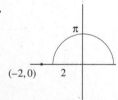

10.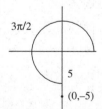

11.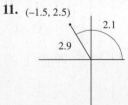

12.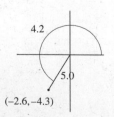

13.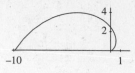

14.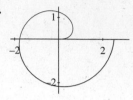

15.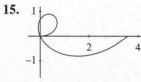

16.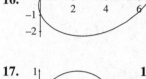

17.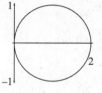

18.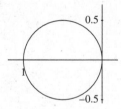

19.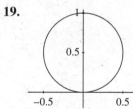

20.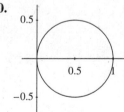

21.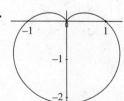

22.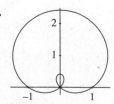

23.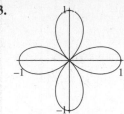

24.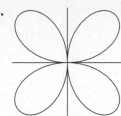

31. $y - 4 = -\dfrac{1}{2}(x - 3)$

33. $y - 6 = \dfrac{5}{2}(x + 1/2)$

Section 3.7

1. -12
2. -6
3. $3 - 2t$
4. -12
5. 0
6. 0
7. $5/\sqrt{2}$
8. $7\sqrt{3}/2 + 5/2$
9. $10/\sqrt{26}$
10. $41/\sqrt{109}$
11. $3/2$
12. 0.760713
13. $(-13/50, 91/50)$
14. $(11/2, 11/2)$
15. 0.432408
16. 1.24905
17. $135°$
18. $149.165°$
19. $75°$ (by inspection)
20. $223°$ (by inspection)
21. Yes
22. Yes
23. No
24. No
25. $\mathbf{t} = 28(0, \sin 315°)$, $\mathbf{c} = 28(\cos 315°, 0)$
26. $\mathbf{t} \approx (3.5, -19.8)$, $\mathbf{c} \approx (-28.2, -5.0)$
27. $\mathbf{h} \approx (-10.8, 10.8)$, $\mathbf{c} \approx (-9.09, -9.09)$
28. $\mathbf{h} \approx (6.8, 1.2)$, $\mathbf{c} \approx (-6.8, 39)$
29. $\mathbf{F}_2$
30. $\mathbf{F}_2$
31. 5303.30 J
32. 5177.84 J
59. $\mathbf{v} = (1/\sqrt{2}, 2/5)$, $\|v\| = 0.812404$

61. $(12, -8), (p_1, p_2) + (12, -8)$
63. $2/3$
65. $-3/(2\sqrt{6})$
67. 1
69. $\frac{1}{3}\ln 3$
71. $\mathbf{r} = (t, \frac{1}{2}t^2 + 1)$
73. $\mathbf{r} = \frac{1}{2}(-\cos 2t, 1 + \sin 2t)$

Section 3.8

1. Decreases it by approximately 4 m.
2. Increases it by approximately 192 m.
3. Approximately 20,000 m.
4. Approximately 390 m.
5. $100\pi/3$ m/s
6. 24 radians/s
7. 28,101 km/h; 87.7 minutes
8. 28,058 km/h; 88.1 minutes
9. 229 km
10. 225 km
11. $\mathbf{r} = \left(2t + 1, \frac{1}{2}t^2 + 1\right)$
12. $\mathbf{r} = \left(\frac{1}{2}t^2 - t, \frac{1}{3}t^3\right)$
13. $\mathbf{r} = \left(t^2 + 5t + 2, \frac{2}{3}t^{3/2} + 2\right)$
14. $\mathbf{r} = \left(3t^3 - 3t^2 + t + 1, \frac{2}{9}(2 + 3t)^{3/2} + 2 - \frac{4}{9}\sqrt{2}\right)$
15. $\mathbf{r} = \left(3t^3 - 3t^2 + t - 2, \frac{2}{9}(2 + 3t)^{3/2} + 3 - \frac{4}{9}\sqrt{2}\right)$
16. $\mathbf{r} = \left(\tan^{-1} t, 3\ln(t + 1)\right)$
17. $\mathbf{r} = \left(-\frac{1}{2}\cos(2t + 1) + 2 + \frac{1}{2}\cos(1), \frac{1}{2}\sin(2t + 1) + 2 - \frac{1}{2}\sin(1)\right)$
18. $\mathbf{r} = (2 + \tan t, 1 + \sec t)$

Section 4.1

1.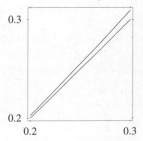

2.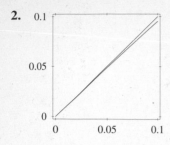

3. Plotting $x - \sin x$ and $1 - \cos x$ for $-0.2 \le x \le 0.2$ shows that the first of these is a much closer approximation. See figure. The table shows the same thing. Values for $x < 0$ can be obtained by symmetry.

x	$x - \sin x$	$1 - \cos x$
0.05	0.000021	0.0012498
0.1	0.000167	0.004996
0.15	0.0005619	0.011229

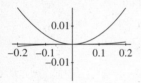

4. Use the tangent line approximation
$$f(a + h) \approx f(a) + h f'(a)$$
in the form (let $a + h = x$)
$$f(x) \approx f(a) + f'(a)(x - a).$$
With $a = \pi/2$,
$$\sin x \approx \sin(\pi/2) + \cos(\pi/2)(x - \pi/2) = 1$$
$$\cos x \approx \cos(\pi/2) - \sin(\pi/2)(x - \pi/2)$$
$$= \pi/2 - x.$$

Thus, $0.98\ldots = \sin 1.4 \approx 1$ and $0.16\ldots = \cos 1.4 \approx \pi/2 - 1.4 \approx 0.17$. The following figure shows $\sin x - 1$ and $\cos x - (\pi/2 - x)$. Both numerical and graphical data show that the tangent line approximation to $\cos(1.4)$ is better than that to $\sin(1.4)$.

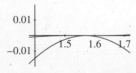

5. Use the tangent line approximation
$$f(a + h) \approx f(a) + h f'(a)$$

in the form (let $a + h = x$)
$$f(x) \approx f(a) + f'(a)(x - a).$$
With $a = 0$, $e^x \approx 1 + x$. From the figure,
$$|e^x - (1 + x)| < 0.8 \quad \text{on } [-0.5, 1.0].$$

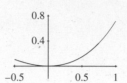

6. With $a = 1$, $2^{-x} \approx 2^{-1} - 2^{-1} \ln 2(x - 1)$. From the figure,
$$|2^{-x} - (2^{-1} - 2^{-1} \ln 2(x - 1))| < 0.16 \quad \text{on } [0, 2].$$

7. With $a = 4$, $\sqrt{x} \approx 2 + (1/4)(x - 4)$. From the figure,
$$|\sqrt{x} - (2 + (1/4)(x - 4))|$$
$$< |\sqrt{6} - (2 + (1/4)(6 - 4))| = 0.05\ldots$$

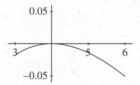

8. $\ln 2/\ln 1.08 \approx 9.006$ years; rule of 72 result is 9, showing that the rule of 72 is very accurate for r near 8 percent.

9. Within 0.00050 m.

10. Within 0.00074 m.

11. $\mathbf{r} = \left(-21.1, 9.4\right)$

12. $\mathbf{r} = \left(8.1, -13.4\right)$

13. 4.87 m

14. 0.03257 units

15. $\left(-0.062035, 3.605017\right)$

16. $52h$

17. $-\frac{4}{9}h$

18. eh

19. $-3e^{-3}h$

20. $0 \cdot h = 0$

21. $-11h/100$

22. $2h/\sqrt{3}$

23. $h/2$

38. e^{2t}, $e^{2t} + 2$

41.

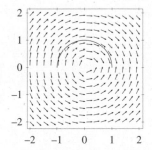

Section 4.2

1. Approximately 44.4 h. The discrete model gives 41.1 h.

2.

n	nh = t	Euler	Exact	n	nh = t	Euler	Exact
0	0.0	0.500	0.500	10	0.5	1.344	1.414
1	0.05	0.552	0.555	11	0.55	1.484	1.569
2	0.1	0.609	0.616	12	0.6	1.639	1.741
3	0.15	0.673	0.683	13	0.65	1.809	1.932
4	0.2	0.743	0.758	14	0.7	1.997	2.144
5	0.25	0.820	0.841	15	0.75	2.205	2.378
6	0.3	0.905	0.933	16	0.8	2.434	2.639
7	0.35	0.999	1.035	17	0.85	2.687	2.928
8	0.4	1.103	1.149	18	0.9	2.966	3.249
9	0.45	1.218	1.275	19	0.95	3.275	3.605
				20	1	3.615	4.000

3. $a = 1$. Yes.

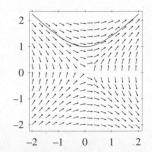

4. $a = 1$. Yes.

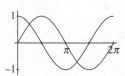

5. $T(1) \approx 97.270$, $T(23) \approx 80.176$, $T(24) \approx 79.767$. So the time of death was approximately 11 A.M. or noon of the previous day.

6. From the example, after 1 second the trooper's velocity is approximately 5.5 m/s. During the first second his velocity went from 54 m/s to 5.5 m/s. So he fell $(54 + 5.5)/2 = 29.75$ m, approximately. This leaves $610 - 30$ m, which at 5.5 m/s takes 105 s. So, about 106 seconds.

19. $w(p) = 0$; $(1.5, 2.25)$; $w(x) = x^2 - x^3 + 1$

21. For $\sqrt{5}$, $x_1 = 2$, $x_2 = 2.250000$,
$x_3 = 2.236111$, $x_4 = 2.236068$;
this is close to $\sqrt{5}$, which is $2.236067977\ldots$;
$x_2 = (a/x_1 + x_1)/2$; $x_{n+1} = (a/x_n + x_n)/2$

Section 4.3

1. Between 2.0 and 2.1.

2. Between 0.3 and 0.4; between 4.0 and 4.1.

3. Between 0.9 and 1.0.

4. Between 0.5 and 0.6.

5. Between 0.8 and 0.9.

6. Between 0.8 and 0.9; two others.

7. Between 0.1 and 0.2; one other.

8. Between 1.4 and 1.5.

9. 1.7100

10. 0.7391

11. 4.4934

12. 2.2356

13. -0.5785

14. 0.3099 and 4.000

15. 0.0007, 0.0047, 0.0131

16. 2.0946

37. The graph of the sine function increases where cosine is positive and decreases where cosine is negative.

39. $[0.19, 0.95]$

Section 4.4

1. Increasing since $0 < u < v$ implies
$v^4 - u^4 = (v - u)(v + u)(v^2 + u^2) > 0$.

2. Increasing since $u < v$ implies
$v^3 - u^3 = (v - u)(v^2 + uv + u^2) > 0$
$(v^2 + u^2 \geq uv$ for all v and $u)$.

3. Decreasing since $0 < u < v$ implies $1/v > 1/u$.

4. Decreasing since $0 < u < v$ implies
$0 < u^2 < v^2$ implies $0 < u^2 + 1 < v^2 + 1$
implies $1/(v^2 + 1) < 1/(u^2 + 1)$.

5. Decreasing since $u < v < -1/2$ implies
$v^2 + v - u^2 - u = (v - u)(u + v + 1) < 0$,
since $u + v + 1 < -1/2 - 1/2 + 1 = 0$.

6. Increasing since $1/2 < u < v$ implies
$v^2 - v - u^2 + u = (v - u)(u + v - 1) > 0$,
since $u + v - 1 > 1/2 + 1/2 - 1 = 0$.

7. Decreasing since $0 < u < v$ implies
$(v + 1)/v - (u + 1)/u = (u - v)/(uv) < 0$.

8. Decreasing since $1 < u < v$ implies
$v/(v-1) - u/(u-1) = (u-v)/((v-1)(u-1)) < 0$.

9. Decreasing on $(-\infty, 0]$ and increasing on $[0, \infty)$.

10. Increasing on $(-\infty, \infty)$.

11. Decreasing on $(-\infty, 0)$ and $(0, \infty)$.

12. Increasing on $(-\infty, 0]$ and decreasing on $[0, \infty)$.

13. Decreasing on $(-\infty, -1/2]$ and increasing
on $[-1/2, \infty)$.

14. Decreasing on $(-\infty, 1/2]$ and increasing
on $[1/2, \infty)$.

15. Decreasing on $(-\infty, 0)$ and decreasing on $(0, \infty)$.

16. Decreasing on $(-\infty, 1)$ and decreasing on $(1, \infty)$.

17. Increasing on $[0, 1.2]$, decreasing on $[1.2, 1.7]$, and
increasing on $[1.7, 2]$.

18. Decreasing on $[0, 1.46]$, increasing on $[1.46, 1.85]$,
and decreasing on $[1.85, 2]$.

19. Increasing on $[0, \pi/2]$ and decreasing on $[\pi/2, \pi]$.

20. Decreasing on $(-\infty, -1]$, increasing on $[-1, 1]$, and
decreasing on $[1, \infty)$.

21. Increasing on $(0, 1]$ and decreasing on $[1, \infty)$.

22. Increasing on $(0, \infty)$.

23. Decreasing on $(-\infty, -1]$ and increasing on $[-1, \infty)$.

24. Increasing on $(-\infty, 1]$ and decreasing on $[1, \infty)$.

25. Concave up on $(0, \pi/4)$ and $(3\pi/4, \pi)$, concave
down on $(\pi/4, 3\pi/4)$, i.p. at $\pi/4$ and $3\pi/4$.

26. Concave down on $(0, \pi/4)$ and $(3\pi/4, \pi)$, concave
up on $(\pi/4, 3\pi/4)$, i.p. at $\pi/4$ and $3\pi/4$.

27. Concave up on $(-\infty, -0.58)$ and $(0.58, \infty)$, concave
down on $(-0.58, 0.58)$, i.p. at ± 0.58.

28. Concave down on $(-\infty, -1)$ and $(0, 0.79)$, concave
up on $(-1, 0)$ and $(0.79, \infty)$, i.p. at 0 and 0.79.

29. Concave down on $(-\infty, -2)$, concave up on
$(-2, \infty)$, i.p. at -2.

30. Concave down on $(-\infty, 2)$, concave up on $(2, \infty)$,
i.p. at 2.

47. $-\sqrt{7}/5$

49. -0.2733

51. 1.9165

Section 4.5

1. $d\theta/dt \le 0.0436$ (radians/second)

2. Yes (motorist's speed > 55.7 mph).

3. 0.110959; graph is evidence;
also, function $y^3 + 9y - 1$ has a positive derivative,
is therefore increasing, and, hence, can have at most
one zero.

4. 0.85 m/s

5. 1.7 m/s

6. More rapidly; 1.4 m/s

7. The formula gives 0, $\pi a^2/2$, and πa^2, respectively,
which are correct, by considering the segment relative to a circle; $A = \pi/3 - \sqrt{3}/4 \approx 0.61418\ldots$.

9. 0.077 m/min

10. 0.028 m/min

11. $15/4$ m/s

12. $dw/dt = (du/dt)(\sqrt{w^2 + 1})/w$

13. $dw/dt = (du/dt)w^2/(w \cos w - \sin w)$

14. -0.0041 cm/s

15. 0.78 qt/h

16. 9.4×10^{-6} m/s

17. 4.7×10^{-22} m/year decrease at half-volume;
5.9×10^{22} centuries

18. 1.7π in.^{3}/s

35.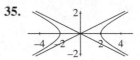

37. $h = 1$ is the inflection point.

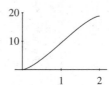

Section 4.6

1. Vertical asymptotes: $x = \pi/2$, $x = 3\pi/2$

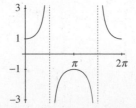

2. Horizontal asymptote: $y = 0$

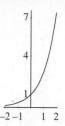

3. Horizontal asymptotes: $y = -\pi/2$, $y = \pi/2$

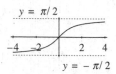

4. Asymptotes: $y = \pm x$

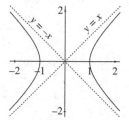

7. The perpendicular distance from a point on the curve to the line $y = x$ appears to approach zero as $|x|$ becomes large. The perpendicular distance from a point on the curve to the curve $y = 1/x$ appears to approach zero as $|x|$ approaches 0.

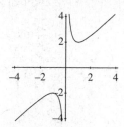

8. The perpendicular distance from a point on the curve to the parabola with equation $y = x^2$ appears to approach zero as $|x|$ becomes large. The perpendicular distance from a point on the curve to the

curve $y = 1/x$ appears to approach zero as $|x|$ approaches 0.

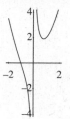

9. $y = \pm \sqrt{2}/3$
10. $y = 1$
11. $x \approx 6.1072$. Graphical argument: graph crosses x-axis once. Analytical argument: since derivative is $3x(x - 4)$, graph increases on $(-\infty, 0)$, rising to the point $(0, -4)$; decreases on $(0, 4)$; and thereafter increases; so only one zero is possible.
12. $x = -1$ is v.a.; $y = 2$ is h.a.

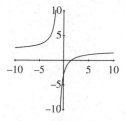

13. $x = 1$ is v.a.; $y = 3$ is h.a.

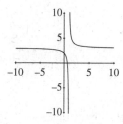

14. $x = 0$ is v.a.; $y = 1$ is h.a.

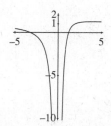

15. $x = 0$ is v.a.; $y = 0$ is h.a.

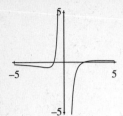

16. $x = 1$ is v.a.; $y = 1$ is h.a.

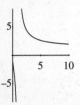

17. $y = \pm 2$ is h.a.

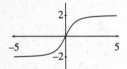

18. $x = -1$ and $x = 2$ are v.a.s; $y = 1$ is h.a.

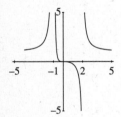

19. $x = -4$ and $x = 1/2$ are v.a.s; $y = 1/2$ is h.a.

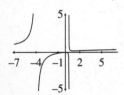

20. $x = 0$ is v.a.

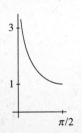

21. $x = 0$ and $x = \pi/2$ are v.a.s

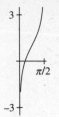

22. $x = 1$ is v.a.

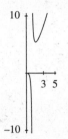

23. $x = 1$ is v.a.

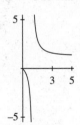

24. $y = 0$ is h.a.

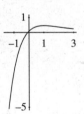

25. $y = 0$ is h.a.

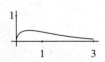

26. $y = 0$ is h.a.

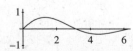

27. $y = 0$ is h.a.

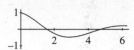

28. $x = -1$ is v.a.; $y = 0$ is h.a.

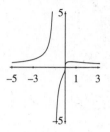

29. $x = -1$ is v.a.; $y = 0$ is h.a.

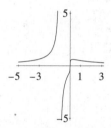

45. No. It could be downhill all the way from one town center to the next.

47. $f(2) = -0.7568\ldots$, $f(0) = 0$, $f(\sqrt{\pi/2}) = 1$; for $0 \le x \le 2$, f takes on its largest value at $x = \sqrt{\pi/2}$ (a zero of f') and its smallest value at $x = 2$ (an end point of the domain of f).

Section 4.7

1.

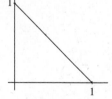

2.

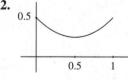

3.

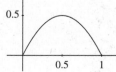

4.

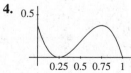

5.

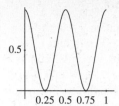

6.

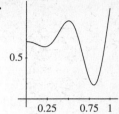

7. Since $\sqrt{2} \approx 1.41$ is relatively close to 1.5, it may be close enough to use 1.5.

8. For $-2 \le x \le -1/3$, $2x - 3 < 0$ and $3x + 1 < 0$. Recall that $|Q| = Q$ when $Q \ge 0$ and $|Q| = -Q$ when $Q < 0$.

9. $-7x + 9$ and $5x - 9$

10. Concave up on $[-3, 5/2]$ and concave down on $[5/2, 5]$.

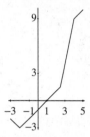

11. Local maximum of 0.641832 at $x = 0.653271$.

13. Extrema at -2 (loc. min.), -1 (glo. max.), 2 (glo. min.), 3 (loc. min.); glo. max. is 8 and glo. min. is -19.

14. Extrema at -1 (loc. min.), $-1/2$ (loc. max.), 1 (glo. min.), 2 (glo. max.); glo. max. is 6 and glo. min. is -7.

15. Extrema at -3 (loc. max.), -1 (glo. min.), 1 (glo. max.), 3 (loc. min.); glo. max. is $1/2$ and glo. min. is $-1/2$.

16. Extrema at -3 (loc. max.), $1 - \sqrt{2}$ (glo. min.), 1 (glo. max.); glo. max. is 0 and glo. min. is -1.20711.

17. Extrema at $1/2$ (glo. max.), 2 (glo. min.), 3 (loc. max.); glo. max. is 16.125 and glo. min. is 6.

18. Extrema at 1 (loc. max.), 1.35721 (glo. min.), 2 (glo. max.); glo. max. is $15/2$ and glo. min. is 6.52605.

19. Extrema at 0 (loc. max.), $1/4$ (glo. min.), 5 (glo. max.); glo. max. is 1.22365 and glo. min. is -0.244979.

20. Extrema at 0 (loc. max.), $1/2$ (glo. min.), 4 (glo. max.); glo. max. is 1.48766 and glo. min. is -0.244979.

21. Extrema at -1 (glo. min.), 1 (glo. max.);
    glo. max. is 24 and glo. min. is -4.
22. Extrema at -1 (loc. max.), 1 (glo. min.),
    2 (glo. max.); glo. max. is 13 and glo. min. is 2.
23. Extrema at 0 (glo. min.), $\pi/2$ (glo. max.),
    π (glo. min.), $3\pi/2$ (glo. max), 2π (glo. min.);
    glo. max. is 1 and glo. min. is 0.
24. Extrema at 0 (glo. max.), 1 (glo. min.), 2 (loc. min.);
    glo. max. is 0 and glo. min. is -4.
45. 1180 m
47. -4.65539 m
49. 1485.07 m

Section 4.8

1. $(2, 2)$
2. $(0, 0)$
3. $(1/5, 7/5)$
4. $(4/13, -20/13)$
5. Closest: $(1.72189, 1.19858)$;
   furthest: $(0.00466216, 0.167056)$
6. Closest: $(0, 0)$; furthest: $(2, 1.58740)$
7. $\beta = 4.118°$
8. $\beta = 14.87°$
9. Base $= 0.873580$ ft; height $= 1.31037$ ft
10. Height $=$ width $= 1/4$ m
11. $a/2$ by $\sqrt{3}a/4$
12. $\theta = \pi/3$
13. $5/2$ by $3/2$
14. $x = 0.117712$
39. 0.666667 m
41. -1.5 m
43. -4.96338 m
45. 2.47654 m
47. $A(b) = \frac{3}{8}b^2$; $A'(b) = \frac{3}{4}b$,

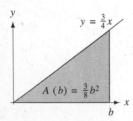

which has the form of $y = \dfrac{3}{4}x$.

49. 0.0223607 m²/min

Section 5.1

1. $x(1/6) \approx (1^2 + \cdots + 9^2)h^3 \approx 1.32 \times 10^{-3}$;
   $x(1/3) \approx (1^2 + \cdots + 19^2)h^3 \approx 1.14 \times 10^{-2}$

Given a function f defined on an interval $[a, b]$, to relate the ideas of an antiderivative of f and the area beneath the graph of f.

2. $x(t) = \frac{1}{4}t^4$
3. $L_{15000} = 1.1248875025$, $U_{15000} = 1.1251125025$

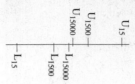

not to scale

Upper sums decrease and lower sums increase as the number of subdivisions increases.

4. $L_{15000} = 1.26393806$, $U_{15000} = 1.26731306$
5. $1^2 + 2^2 + 3^2 + 4^2 + 5^2 = 55$
6. $1^4 + 2^4 + 3^4 = 98$
7. $(1^2 + 1) + (2^2 + 1) + (3^2 + 1) + (4^2 + 1) + (5^2 + 1) + (6^2 + 1) = 97$
8. $(2 \cdot 1^2 - 5 \cdot 1 + 1) + (2 \cdot 2^2 - 5 \cdot 2 + 1) + (2 \cdot 3^2 - 5 \cdot 3 + 1) = 1$
9. $\sqrt{2} + \sqrt{3} \approx 3.14626$
10. $\sqrt{1} + \sqrt{4} + \sqrt{7} + \sqrt{10} \approx 8.80803$
11. $\sin(0/10) + \sin(1/10) + \sin(2/10) + \sin(3/10) + \sin(4/10) + \sin(5/10) \approx 1.46287$
12. $\cos(0/10) + \cos(1/10) + \cos(2/10) + \cos(3/10) + \cos(4/10) + \cos(5/10) + \cos(6/10) + \cos(7/10) \approx 7.31923$
13. $1 + 1/3 + 1/5 + 1/7 + 1/9 + 1/11 = 6508/3465 \approx 1.87821$
14. $-5/3 - 2/3 + 1/3 + 4/3 + 7/3 + 10/3 = 5$
15. $1 + 1 + 1 = 3$
16. $1 + 1 + 1 = 3$
17. $h^3 \sum_{j=1}^{10} j^2$
18. $h^2 \sum_{j=1}^{7} j^2$
19. $2 \sum_{j=3}^{35} 1/j$
20. $(1/13) \sum_{j=7}^{27} j$
21. $\left(|h|/\sqrt{19}\right) \sum_{j=1}^{21} \sqrt{j}$
22. $(1/|h|) \sum_{j=3}^{37} \sqrt{j}$
23. $e \sum_{j=1}^{21} e^j$

24. $0.01 \sum_{j=2}^{50} \ln(j)$

25. $\sum_{j=1}^{15} j/(j+1)$

26. $\sum_{j=1}^{n} x^j$

39. $\frac{1}{6}x^6$

41. $\frac{1}{6}(2x+1)^3$

43. $\frac{2}{3}x^{3/2}$

45. $-\cos(x-1)$

47. $\tan x$

49. $2^x/(\ln 2)$

51. $\frac{1}{2}e^{2x}$

53. $\arctan x$

55. $\frac{1}{3}x^3 + \frac{2}{3}x^{3/2}$

57. $\ln x$ if $x > 0$

Section 5.2

1. $\int_0^1 x^2\,dx = 1/3, \int_1^2 x^2\,dx = 7/3, \int_0^2 x^2\,dx = 8/3$

2. $\int_0^1 x^3\,dx = 1/4, \int_1^2 x^3\,dx = 15/4, \int_0^2 x^3\,dx = 4$

3. $m(b-a) = 12, \int_a^b f(x)\,dx = 21.333\ldots,$
$M(b-a) = 48$

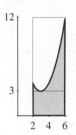

4. $m(b-a) \approx 2.618, \int_a^b f(x)\,dx = 4.122,$
$M(b-a) = 7.854$

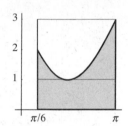

5. $A = 1$

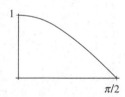

6. $A = \sqrt{3}/2$

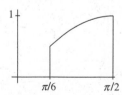

7. $A = 4$

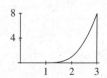

8. $A = 33/15$

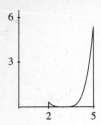

9. $A = e - 1$

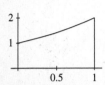

10. $A = 1/\ln 2$

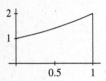

11. $A = 1$

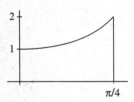

12. $A = 1$

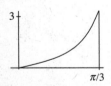

13. $A = \arctan(\pi/4)$

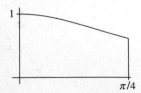

14. $A = \pi/6$

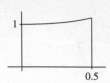

15. $A = \ln 2$

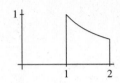

16. $A = 2$

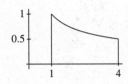

17. $\ln x$
18. $\sin x^2$
19. $\sqrt{1 + x^2}$
20. $\sqrt{1 - x^2}$
21. $2x \ln x^2 = 4x \ln x$
22. $3x^2 \cos x^{3/2}$
23. $\sqrt{1 + x}/(2\sqrt{x})$
24. $-(\tan x^{-2})/x^2$
25. $\left(\arctan(e^x) - e^{2x}\right)e^x$
26. $2x \cos x^2 \sin\left(\sin^2 x^2\right)$
51. $f'(x) = -(\sin\sqrt{x})/(2\sqrt{x})$
53. $f'(x) = 3^{2x} + 2x3^{2x} \ln 3$
55. $f'(x) = \tan x^2$
57. $f'(x) = -1/(x^2 + 1)$
59. $2e^{2x} \sin(1 - 3x) - 3e^{2x} \cos(1 - 3x)$
61. $2\left(x - \frac{5}{4}\right)^2 - \frac{17}{8} = 0; x = (5 \pm \sqrt{17})/4$

Section 5.3

1. $\frac{1}{3}(2\sqrt{2} - 1)$

2. $343/9$

3. $1/4$

4. $1/4$

5. $2e(e-1)$

6. $4/\ln 2$

7. $3\sqrt{3}/5$

8. $16(2+\sqrt{2})/15$

9. $\ln 4 - \ln 2 \ (=\ln 2)$

10. $\frac{1}{2}\ln 9$

11. $u = 2x - 1$, (A)

12. $u = 5 + 2x$, (A)

13. $u = 5 - x$, (A)

14. $u = 3x - 8$, (A)

15. $u = 2x + 1$, (B)

16. $u = 3x + 2$, (B)

17. $u = 1 - 2x$, (C)

18. $u = \frac{1}{2}x + 3$, (C)

19. $u = \frac{1}{3} + \frac{2}{7}x$, (D)

20. $u = 1 - x$, (D)

21. $u = 2x + 3$, (E)

22. $u = -0.1x + 0.6$, (E)

23. $u = -x + 1$, (F)

24. $u = 7x + 3$, (F)

25. $\sqrt{2}x = u$, (G); or rewrite as $x^2 + \left(1/\sqrt{2}\right)^2$ and use (G)

26. $\sqrt{3}x = u$, (G); or rewrite as $x^2 + \left(\sqrt{2/3}\right)^2$ and use (G)

27. $u = 2x + 1$, (G)

28. $u = 5x - 2$, (G)

29. $\sqrt{2}x = u$, (H); or rewrite as $\sqrt{\left(1/\sqrt{2}\right)^2 - x^2}$ and use (H)

30. $\sqrt{3}x = u$, (H); or rewrite as $\sqrt{\left(1/\sqrt{3}\right)^2 - x^2}$ and use (H)

31. $\sqrt{2}x = u$, (H); or rewrite as $\sqrt{\left(\sqrt{7/2}\right)^2 - x^2}$ and use (H)

32. $\sqrt{3}x = u$, (H); or rewrite as $\sqrt{\left(\sqrt{5/3}\right)^2 - x^2}$ and use (H)

33. $u = 2x + 1$, (H)

34. $u = 7 + 2x$, (H)

71. $(4, 5)$ and $(8, 21)$

73. $(-0.203058, 0.979167)$ and $(0.203058, 0.979167)$, approximately

Section 5.4

1. 36 square units

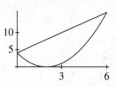

2. $4/3$ square units

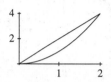

3. $1/12$ square units

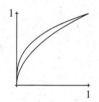

4. $1/20$ square units

5. ≈ 0.322188 square units

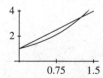

6. ≈ 0.427724 square units

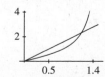

7. 16/3 square units

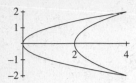

8. ≈0.387102 square units

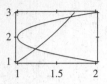

9. ≈0.777437 square units

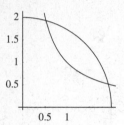

10. ≈0.846188 square units

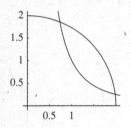

11. 1/3 square units

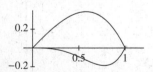

12. 2/3 square units

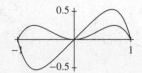

13. ≈1.00857 square units

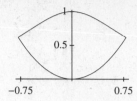

14. ≈1.23262 square units

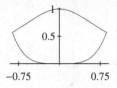

15. ≈0.256692 square units

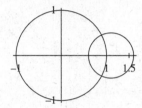

16. ≈1.60451 square units

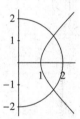

17. $4\sqrt{2} \approx 5.65685$ square units

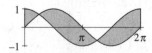

18. ≈3.63974 square units

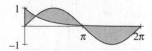

22. No; $t \approx 0.72$

Section 5.5

1. $(e^2 + 1)/4$

2. $(2e - 3)/e^2$

3. $x(\ln x - 1) + C$

4. $\frac{1}{9}x^3(3\ln 2x - 1) + C$

5. $(2\ln 2 - 1)/(\ln 2)^2$

6. $\frac{3}{10}(-333 + 667\ln 10)/(\ln 10)^2$

7. $\frac{2}{9}x^{3/2}(3\ln 2 - 2) + C$

8. $\frac{3}{16}x^{4/3}(4\ln x - 3) + C$

9. $\sin x - x\cos x + C$

10. $\cos x + x\sin x + C$

11. $\sqrt{1 - x^2} + x\arcsin x + C$

12. $x\arctan x - \frac{1}{2}\ln(1 + x^2) + C$

13. $e^{ax}(a\cos bx + b\sin bx)/(a^2 + b^2) + C$

14. $2e^{\sqrt{x}}(\sqrt{x} - 1) + C$

37. $a = 2, b = 1, c = -3$

Section 5.6

1. $Q(x) = 3x + 2, r(x) = -4x - 3$

2. $Q(x) = \frac{1}{2}x + \frac{1}{4}, r(x) = \frac{11}{4}x - \frac{3}{4}$

3. $Q(x) = 1, r(x) = -4x^2 + 7x + 13$

4. $Q(x) = 1, r(x) = 2$

5. $Q(x) = 0, r(x) = 2x - 1$

6. $Q(x) = 0, r(x) = x^3$

7. $(x - 1)(x + 2)(x^2 + 1)$

8. $(x - 4)(x + 1)(2x - 1)(x^2 + 1)$

9. $(x - 1)(x + 1)(x^2 + 1)$

10. $(x - 1)(x + 1)(x^2 - x + 1)(x^2 + x + 1)$

11. $(x - 1)(x + 1)(x^2 + x + 1)$

12. $(x - 2)(x - 3)(x^2 + 2x + 3)$

13. $\approx (x + 2.8662)(x - 0.210756)(x - 1.65544)$

14. $\approx (x - 1.91229)(x - 3.05164)(x^2 + 1.96392x + 2.91315)$

15. $\approx (x + 2.92870)(x + 0.609909)(x^2 - 0.538605x + 1.11967)$

16. $\approx (x - 0.322548)(x - 1.74576)(x - 4.53662)(x - 9.39507)$

17. $\approx 16(x - 0.524648)(x + 0.524648)(x - 1.65068)(x + 1.65068)$

18. $\approx 192x(x - 0.798214)(x + 0.798214)(x - 0.442930)(x + 0.442930)$

19. $3/(x + 1) - 5/(x + 4)$

20. $2/(x - 3) - 1/(x + 5)$

21. $x + 1/(x + 2) - 2/(x - 3)$

22. $2x - 1 + 3/(x - 1) - 2/(x + 1)$

23. $1/(x + 3) + 1/(x + 3)^2$

24. $3/(x + 1) + 5/(x + 1)^2$

25. $1/(x - 2) + (2x - 3)/(x^2 + 1)$

26. $1/(2x + 1) + 3/(x^2 + x + 1)$

27. $2/(x - 2) - 1/(x + 3) - 1/(2x + 3)$

28. $(2/3)/(x - 1) - (1/2)/(x + 1) - 1/(x + 2)$

29. $1/(x + 1) + (x + 2)/(x^2 + 3x + 1)^2 - (x + 2)/(x^2 + 3x + 1)$

30. $\dfrac{1}{x + 1} + \dfrac{4 - x}{(x^2 - 3x - 3)^2} + \dfrac{4 - x}{x^2 - 3x - 3}$

31. $\ln(5/2)$

32. $\ln(25\sqrt{7}/4)$

33. $-1/(x + 1) - (3/2)\arctan(x/2) + C$

34. $(-5/2)/(x + 2)^2 + \frac{1}{5}\arctan(x/5) + C$

35. $\ln(x + 1) + (1/\sqrt{2})\arctan((x + 1)/\sqrt{2}) + C$

36. $-2\ln(x + 9) + (1/2)\arctan((x + 3)/2) + C$

37. $(-1/2)/(x^2 + 4) + (1/2)\arctan(x/2) + C$

38. $(x - 2)/(4(x^2 + 2)) + (1/(4\sqrt{2}))\arctan(x/\sqrt{2}) + C$

Section 5.7

1. $T_5 = 1.110268; \ln 3 - T_5 = -0.011655;$
$T_{10} = 1.101562; \ln 3 - T_{10} = -0.002950$

2. $M_5 = 1.092857; \ln 3 - M_5 = 0.005755;$
$M_{10} = 1.097142; \ln 3 - M_{10} = 0.001470$

3. $S_5 = 1.098661; \ln 3 - S_5 = -0.000048;$
$S_{10} = 1.098616; \ln 3 - S_{10} = -0.000003$

4. $T_5 = 0.340537; T_{10} = 0.341143;$
$T_{15} = 0.341255$

5. $M_5 = 0.341749; M_{10} = 0.341446;$
$M_{15} = 0.341390$

6. $S_5 = 0.341345; S_{10} = 0.341345;$
$S_{15} = 0.341345$

7. $n = 2$ is sufficient; $S_2 = 0.523616;$
$|S_2 - \arcsin 0.5| = 0.000018$

8. With $f(x) = 4a\sqrt{1 - e^2\sin^2 x}$ and using astronomical units, $L \approx 245.8669$ AU.

19. 0.25π cm^3

21. $e^x(x - 1) + C$

23. $27/2$

Section 6.1

1. $V = \frac{4}{3}\pi a^3$

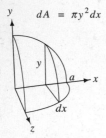

2. $V = \frac{4}{3}\pi ab^2$

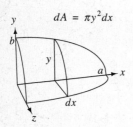

3. $V = 32\pi/5$

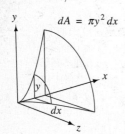

4. $V = 2187\pi/896 \approx 7.67$

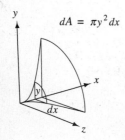

5. $V = 2\pi$

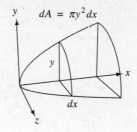

6. $V = 6\sqrt[3]{4}\pi/5 \approx 5.98$

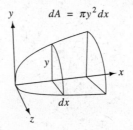

7. $V = \pi^2/2$

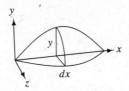

8. $V = \pi^2/4$

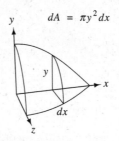

9. $V = \pi$

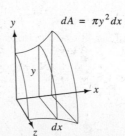

10. $V = \pi - \pi^2/4 \approx 0.67$

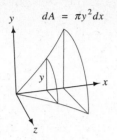

11. $V = 2\pi/35$

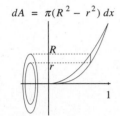

12. $V = \pi/10$

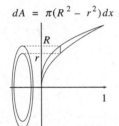

13. $V = 384\pi/5$

14. $V = 72\pi/5$

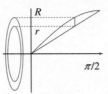

15. $V = \pi^2/12$

16. $V = \frac{2}{9}\pi\left(\left(4 + \sqrt{3}\right)\pi - 6\sqrt{3}\right) \approx 5.32$

17. Intersections at $x = 1, 2.37018$; $V \approx 1.43$

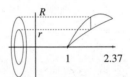

18. Intersections at $x = 0.0735591, 2.18584$; $V \approx 432.52$

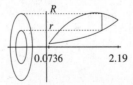

19. 2.59761×10^6 m^3

20. Start with a pyramid of height H and base a. Lop a smaller pyramid off its top, with base b and height $H-h$. Show that $H = ah/(a-b)$ and then subtract two pyramid formulas.

21. Referring to the figure, show that $R = r - (r/h)z$ and $dV = \pi R^2 dz$.

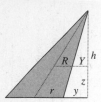

22. Observe that $dV = \pi r^2 dz$.

23. Simply note that
$(R - r)^2 = R^2 - 2Rr + r^2 \neq R^2 - r^2$.

24. Referring to the figure and notation of Example 4,
$z = y \tan\theta$, $dV = \left(\frac{1}{2}\tan\theta\right)y^2 dx$,
and $V = \frac{2}{3}a^3 \tan\theta$.

49. $\|\mathbf{r}'(t)\| = \sqrt{10}$; hence object traveled $3\sqrt{10}$.

51. $\frac{1}{2}\sqrt{5} + \frac{1}{4}\ln(\sqrt{5} + 2)$

Section 6.2

1. $25\sqrt{61}$ m/s

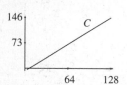

2. $75\sqrt{61}$ m/s

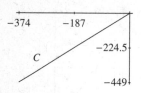

3. $18,000\pi$ m/s

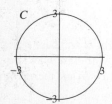

4. $27,000\pi$ m/s

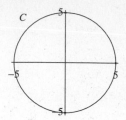

5. 24 m/s

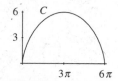

6. 32 m/s

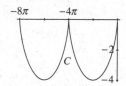

7. $\sqrt{17} + \frac{1}{4}\ln\left(4 + \sqrt{17}\right)$ m

8. $5\sqrt{26}/2 + \frac{1}{2}\ln\left(5 + \sqrt{26}\right)$ m

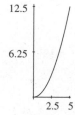

9. $335/27$ m

10. $61/54$ m

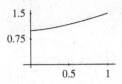

11. 4 m

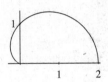

12. 4 m

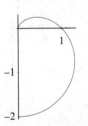

13. $\sqrt{2}(e^{\pi} - 1)$ m

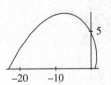

14. $\dfrac{(2^{\pi} - 1)\sqrt{1 + (\ln 2)^2}}{\ln 2} \approx 13.74$ m

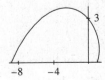

15. $\pi^2/2$ m

16. $\dfrac{1}{2}\pi\sqrt{\pi^2 + 4} + 2\ln((\pi + \sqrt{\pi^2 + 4})/2) \approx 8.32$ m

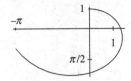

17. $\dfrac{1}{2}\sqrt{2} + \dfrac{1}{2}\ln(1 + \sqrt{2}) \approx 1.15$ m

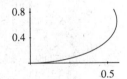

18. $\dfrac{1}{2}\sqrt{2} + \dfrac{1}{2}\ln(1 + \sqrt{2}) \approx 1.15$ m

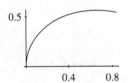

19. 3.97 m

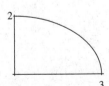

20. 2.42 m

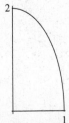

21. 1.55 m

22. 1.13 m

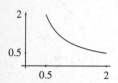

23. $\mathbf{T}(\pi/4) = (1/\sqrt{13})(-3, 2)$

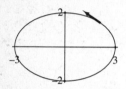

24. $\mathbf{T}(\pi/6) = (1/\sqrt{13})(-1, 2)$

25. $\mathbf{T}(1) = (1/\sqrt{10})(1, 3)$

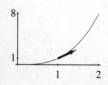

26. $\mathbf{T}(1) = (1/\sqrt{2})(1, -1)$

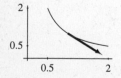

27. The arc length is approximately 2.27322. Too big. Approximation would improve on this flatter portion of curve.

28. Both integrals evaluate to 2π.

35. Polar coordinates: the pole, $(1/\sqrt{2}, \pi/4)$; rectangular coordinates: $(0, 0)$, $(1/2, 1/2)$.

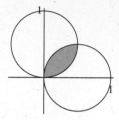

37. Polar coordinates:
the pole, $(0.73908\ldots, 0.73908\ldots)$;
rectangular coordinates:
$(0, 0)$, $(0.54624\ldots, 0.49785\ldots)$.

Section 6.3

1. 6 square units
2. 210 square units
3. 24 square units
4. 4 square units
5. $A = 4\pi$ square units

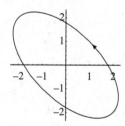

6. $A = 6\pi$ square units

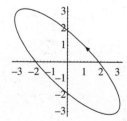

7. $A = 1/60$ square units

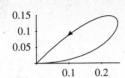

8. $A = 1/240$ square units

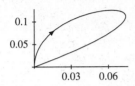

9. $A = (e^{\pi} + 1)/2$ square units

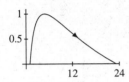

10. $A = 2\pi$ square units

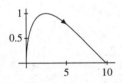

11. $A = \pi^3/3$ square units

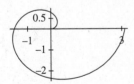

12. $A = (2^{4\pi} - 1)/(4\ln 2)$ square units

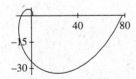

13. $A = 3\pi/2$ square units

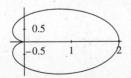

14. $A = \pi/4$ square units

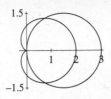

15. $A = (\pi - 2)/8$ square units

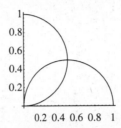

16. $A = \frac{1}{6}a^2(4\pi - 3\sqrt{3})$ square units

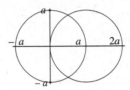

17. $A = \pi - 3\sqrt{3}/2$ square units

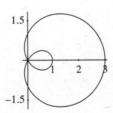

18. $A = \pi - 3\sqrt{3}/2$ square units

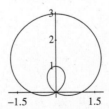

19. $A = 8\pi - 3\sqrt{7} - 16\arcsin(3/4) \approx 3.63$ square units

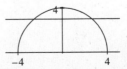

20. $A = 4\pi/3 - \sqrt{3}$ square units

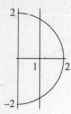

45. $\mathbf{P} = 10\cos 20°(\cos 210°, \sin 210°)$; $\mathbf{N} = \mathbf{F} - \mathbf{P}$

Section 6.4

1. $3mg$ J
2. $3mg$ J
3. $18g$ J
4. $6g$ J
5. -22.5 J
6. -12 J
7. $2k$ J
8. $2k$ J
21. No.

Section 6.5

1. $R = 7/9$
2. $R = 64/11$, from "start"
3. $\mathbf{R} = (7/26, -9/26)$
4. $\mathbf{R} \approx (0.654148, 0.747598)$
5. $m = 1.675$ kg; $R = 0.258706$
6. $m \approx 1.925$ kg; $R \approx 0.356364$
7. $\mathbf{R} = (13a/10, 9a/10)$
8. $\mathbf{R} = (-2a/9, 4a/9)$
9. With center at origin, $\mathbf{R} = (4a/(3\pi), 4a/(3\pi))$
10. $m = \sigma(e - 1)$, $\mathbf{R} = (1/(e - 1), (e + 1)/4)$
11. $m = 2\sigma$, $\mathbf{R} = (\pi/2, \pi/8)$
29. $(1/\sqrt{37}, 6/\sqrt{37})$

Section 6.6

1. $\mathbf{T}(0) = (1, 0)$, $\mathbf{N}(0) = (0, 1)$, $\kappa(0) = 1/2$;
curvature is greatest at $\mathbf{r}(0)$ and least at $\mathbf{r}(2)$.

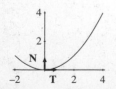

2. $\mathbf{T}(0) = (0, 1)$, $\mathbf{N}(0) = (1, 0)$, $\kappa(0) = 1$;
curvature is greatest at $\mathbf{r}(0)$ and least at $\mathbf{r}(2)$.

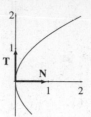

3. $\mathbf{T}(\pi/4) = (-3/\sqrt{13}, 2/\sqrt{13})$,
$\mathbf{N}(\pi/4) = (-2/\sqrt{13}, -3/\sqrt{13})$,
$\kappa(\pi/4) = 12\sqrt{2}/(13\sqrt{13})$;
curvature is greatest at $\mathbf{r}(0)$ and $\mathbf{r}(\pi)$ and least at
$\mathbf{r}(\pi/2)$ and $\mathbf{r}(3\pi/2)$.

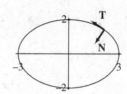

4. $\mathbf{T}(\pi/6) = (-1/\sqrt{13}, 2\sqrt{3/13})$,
$\mathbf{N}(\pi/6) = (-2\sqrt{3/13}, -1/\sqrt{13})$,
$\kappa(\pi/6) = 16/(13\sqrt{13})$;
curvature is greatest at $\mathbf{r}(\pi/2)$ and least at $\mathbf{r}(0)$.

5. At $(1, 1)$, $\mathbf{T} = (1/\sqrt{10}, 3/\sqrt{10})$,
$\mathbf{N} = (-3/\sqrt{10}, 1/\sqrt{10})$, and $\kappa = 3/(5\sqrt{10})$;
curvature is least at $\mathbf{r}(0)$ and greatest at
$\mathbf{r}(45^{-1/4})$.

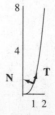

6. At $(1, 1)$, $\mathbf{T} = (1/\sqrt{2}, -1/\sqrt{2})$,
$\mathbf{N} = (1/\sqrt{2}, 1/\sqrt{2})$, and $\kappa = 1/\sqrt{2}$;
curvature is greatest at $\mathbf{r}(1)$ and least at $\mathbf{r}(2)$.

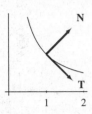

7. At the point with polar coordinates $(2\sqrt{2}, \pi/4)$,
$\mathbf{T} = (-1, 0)$, $\mathbf{N} = (0, -1)$, and $\kappa = 1/2$; curvature is constant.

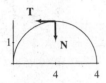

8. At the point with polar coordinates $(3, \ln 3)$,

$$\mathbf{T} = ((\cos \ln 3 - \sin \ln 3)/\sqrt{2}, (\cos \ln 3 + \sin \ln 3)/\sqrt{2})$$

$$\mathbf{N} = (-(\cos \ln 3 + \sin \ln 3)/\sqrt{2}, (\cos \ln 3 - \sin \ln 3)/\sqrt{2})$$

and $\kappa = 1/(3\sqrt{2})$; curvature is greatest at the point with polar coordinates $(1, 0)$ and least at $(e^{\pi/2}, \pi/2)$.

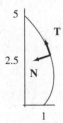

11. $\mathbf{r}(\pi/6) = (0, -18)$, $\mathbf{r}''(\pi/6) = (0, -18)$

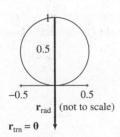

12. $\mathbf{r}(\pi/2) = (3e^{\pi/2}, -4e^{\pi/2})$,
$\mathbf{r}''(\pi/2) = (3e^{\pi/2}, -4e^{\pi/2})$

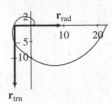

27. $A(10) = 0.9$, $A(100) = 0.99$, $A(1000) = 0.999$. It appears that the area of the region C_∞ is 1 square unit.

Section 6.7

1. Divergent
2. Divergent
3. Converges to 2
4. Converges to $2/3$
5. Divergent
6. Converges to 5
7. Divergent
8. Divergent
9. Converges to $\pi/2$
10. Converges to $\ln(2 + \sqrt{3})$
11. Converges to $\pi/3$
12. Divergent
13. Divergent
14. Divergent
15. Converges to $\ln 2$
16. Converges to $\ln 2$
17. Converges to -1
18. Divergent
19. Converges to π
20. Divergent
23. Area is 1
24. Area is $\frac{1}{2}\ln 2$
25. Convergent
26. Convergent
27. Convergent
28. Divergent
29. Converges to $3/13$ (use formula (23) from the table of integrals).

51. First, $f''(1) = \pm 6\sqrt{2}$. Assume $f''(1) = 6\sqrt{2}$ (other case similar).

$$P_0(x) = 2$$
$$P_1(x) = 2 + 1 \cdot (x - 1)$$
$$P_2(x) = 2 + 1 \cdot (x - 1) + (1/2) \cdot 6\sqrt{2} \cdot (x - 1)^2$$

At $(1, 2)$, these polynomials have, respectively, the right height, the right height and slope, and the right height, slope, and curvature.

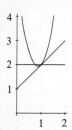

Section 7.1

1. $T_0(x; 0) = 1, T_1(x; 0) = 1 - x$,
$T_2(x; 0) = 1 - x + x^2$,
$T_3(x; 0) = 1 - x + x^2 - x^3$,
$f(0.1) = 1/1.1 \approx 0.909091, T_3(0.1; 0) = 0.909$

2. $T_0(x; 0) = 3^{10}, T_1(x; 0) = 3^{10} + 10 \cdot 3^9 x$,
$T_2(x; 0) = 3^{10} + 10 \cdot 3^9 x + 45 \cdot 3^8 x^2$,
$T_3(x; 0) = 3^{10} + 10 \cdot 3^9 x + 45 \cdot 3^8 x^2 + 120 \cdot 3^7 x^3$,
$f(0.1) = (3 + 0.1)^{10} = 81,962.8286981$,
$T_3(0.1; 0) = 81,946.89$

3. $T_0(x; 0) = 0, T_1(x; 0) = T_2(x; 0) = x$,
$T_3(x; 0) = x - x^3/6$,
$f(0.1) = \sin(0.1) \approx 0.099834$,
$T_3(0.1; 0) \approx 0.099833$

4. $T_0(x; 0) = T_1(x; 0) = 1$,
$T_2(x; 0) = T_3(x; 0) = 1 - x^2/2$,
$f(0.1) = \cos(0.1) \approx 0.995004$,
$T_3(0.1; 0) = 0.995$

5. $T_0(x; 0) = 0, T_1(x; 0) = T_2(x; 0) = x$,
$T_3(x; 0) = x - x^3/3$,
$f(0.1) = \arctan(0.1) \approx 0.0996687$,
$T_3(0.1; 0) \approx 0.099667$

6. $T_0(x; 0) = 1, T_1(x; 0) = 1 + 2x$,
$T_2(x; 0) = 1 + 2x + 2x^2$,
$T_3(x; 0) = 1 + 2x + 2x^2 + 4x^3/3$,
$f(0.1) = e^{0.2} \approx 1.221403, T_3(0.1; 0) = 1.221333$

7. $T_0(x; -2) = -1, T_1(x; -2) = -1 - (x + 2)$,
$T_2(x; -2) = -1 - (x + 2) - (x + 2)^2$,
$T_3(x; -2) = -1 - (x + 2) - (x + 2)^2 - (x + 2)^3$,
$f(0.1) = 1/(-0.95) \approx -1.05263$,
$T_3(0.1; -2) = -1.052625$

8. $T_0(x; \pi/4) = 1, T_1(x; \pi/4) = 1 + 2(x - \pi/4)$,
$T_2(x; \pi/4) = 1 + 2(x - \pi/4) + 2(x - \pi/4)^2$,
$T_3(x; \pi/4) = 1 + 2(x - \pi/4) + 2(x - \pi/4)^2$
$\qquad + 8(x - \pi/4)^3/3$,
$f(\pi/4 - 0.03) = \tan(\pi/4 - 0.03) \approx 0.941731$,
$T_3(\pi/4 - 0.03; \pi/4) = 0.941728$

9. $T_0(x; 3) = 9 \ln 3$,
$T_1(x; 3) = 9 \ln 3 + (3 + 6 \ln 3)(x - 3)$,
$T_2(x; 3) = 9 \ln 3 + (3 + 6 \ln 3)(x - 3)$
$\qquad + (3/2 + \ln 3)(x - 3)^2$,
$T_3(x; 3) = 9 \ln 3 + (3 + 6 \ln 3)(x - 3)$
$\qquad + (3/2 + \ln 3)(x - 3)^2$
$\qquad + (x - 3)^3/9$,
$f(2.95) = 2.95^2 \ln 2.95 \approx 9.414409$,
$T_3(2.95; 3) \approx 9.414410$

10. $T_0(x; 3\pi/4) = 1/\sqrt{2}$,
$T_1(x; 3\pi/4) = 1/\sqrt{2} - (x - 3\pi/4)/\sqrt{2}$,
$T_2(x; 3\pi/4) = 1/\sqrt{2} - (x - 3\pi/4)/\sqrt{2}$
$\qquad - (x - 3\pi/4)^2/2\sqrt{2}$,
$T_3(x; 3\pi/4) = 1/\sqrt{2} - (x - 3\pi/4)/\sqrt{2}$
$\qquad - (x - 3\pi/4)^2/2\sqrt{2}$
$\qquad + (x - 3\pi/4)^3/6\sqrt{2}$,
$f(3\pi/4 + 0.1) = \sin(3\pi/4 + 0/1) \approx 0.632981$,
$T_3(3\pi/4 + 0.1; 3\pi/4) \approx 0.632978$

11. $T_0(x; 1) = \pi/4, T_1(x; 1) = \pi/4 + (x - 1)/2$,
$T_2(x; 1) = \pi/4 + (x - 1)/2 - (x - 1)^2/4$,
$T_3(x; 1) = \pi/4 + (x - 1)/2 - (x - 1)^2/4$
$\qquad + (x - 1)^3/12$,
$f(1.05) = \arctan(1.05) \approx 0.809784$,
$T_3(1.05; 1) \approx 0.809784$

12. $T_0(x; \ln 3) = 3, T_1(x; \ln 3) = 3 + 3(x - \ln 3)$,
$T_2(x; \ln 3) = 3 + 3(x - \ln 3) + 3(x - \ln 3)^2/2$,
$T_3(x; \ln 3) = 3 + 3(x - \ln 3) + 3(x - \ln 3)^2/2$
$\qquad + (x - \ln 3)^3/2$,
$f(1) = e \approx 2.7182818, T_3(1; \ln 3) \approx 2.718270$

13. The top graph.

14. The graph of $y = \tan x$ is the top graph in the first quadrant and is the bottom graph in the third quadrant.

17. $(0, 1)$ is the only point on the graph of $y = e^x$ for which both the x and y coordinates are rational numbers.

18. a. $T_6(x;0) = x - x^3/6 + x^5/120$

b.

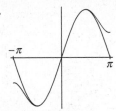

c.

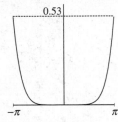

d. ≈ 0.53

e. $\sin(-0.2) \approx -0.198669$,
$T_6(-0.2;0) = -0.198669$,
$\sin 0 = 0$, $T_6(0;0) = 0$, $\sin(0.5) \approx 0.479426$,
$T_6(0.5;0) \approx 0.479427$, $\sin 1 \approx 0.841471$,
$T_6(1;0) \approx 0.841667$

19. a. $T_{17}(x;2) = -3 + 4(x-2) + 3(x-2)^2 + (x-2)^3$

b. The two graphs are the same.

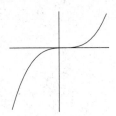

c. The graph is the line $y = 0$.

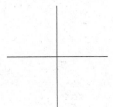

20. a. $T_6(x;0) = 1 + x + x^2 + x^3 + x^4 + x^5 + x^6$

b.

c.

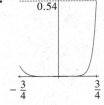

d. 0.54

e. $1/(1 - (-0.9)) \approx 0.526316$,
$T_6(-0.9;0) = 0.778051$,
$1/(1 - (-0.05)) \approx 0.952381$,
$T_6(-0.05;0) = 0.952381$,
$1/(1 - (0.5)) = 2$, $T_6(0.5;0) = 1.984375$,
$1/(1 - (-0.7)) \approx 3.333333$,
$T_6(0.7;0) = 3.058819$

21. a. $T_6(x;0) = 1 + (x-1)/2 - (x-1)^2/8$
$+ (x-1)^3/16$
$- 5(x-1)^4/128$

b.

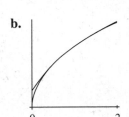

c.

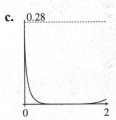

d. 0.28

e. $\sqrt{0.4} \approx 0.632456$, $T_4(0.4;1) = 0.6364375$,
$\sqrt{0.8} \approx 0.894427$, $T_4(0.8;1) = 0.8944375$,
$\sqrt{1} = 1$, $T_4(1;1) = 1$, $\sqrt{1.4} \approx 1.183216$,
$T_4(1.4;1) = 1.183$, $\sqrt{1.8} \approx 1.341641$,
$T_4(1.8;1) = 1.336$

41. $1, 4, 9, 16$

43. $3/2, 17/12, 577/408, 665,857/470,832$

47. $2^{21}\cos 2\theta$

49. $a^n e^a$

51. $b = -\pi/720$, $c = 0$

Section 7.2

1. $T_2(x;0) = 1 - 2x + 2x^2$,
$|R_2(x;0)| \le 8e^{0.6}(0.3)^3/3! \approx 0.0655963$

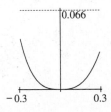

Actual maximum error ≈ 0.042.

2. $T_2(x; 0) = 3^{10} + 10 \cdot 3^9 x + 45 \cdot 3^8 x^2$,
$|R_2(x; 0)| \le 720(3.3)^7(0.3)^3/3! \approx 13809$

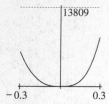

Actual maximum error ≈ 8488.

3. $T_2(x; 0) = x$,
$|R_2(x; 0)| \le 1(\pi/6)^3/3! \approx 0.024$

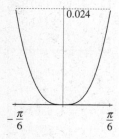

Actual maximum error ≈ 0.024.

4. $T_2(x; 0) = 1 - x^2/2$,
$|R_2(x; 0)| \le (\sin \pi/6)(\pi/6)^3/3! \approx 0.012$

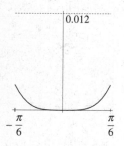

Actual maximum error ≈ 0.0031.

5. $T_2(x; 0) = 1 - x + x^2$,
$|R_2(x; 0)| \le 30.375(1/3)^3/3! = 0.1875$

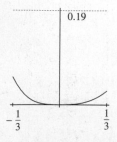

Actual maximum error ≈ 0.056.

6. $T_2(x; 0) = x$,
$|R_2(x; 0)| \le 2(.5)^3/3! \approx 0.0417$

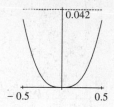

Actual maximum error ≈ 0.036.

7. $r \le 1.38$

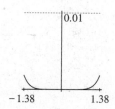

8. $r \le 0.3$

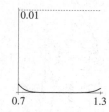

9. $r \le 1.28$

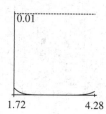

10. $r \le 1.04$

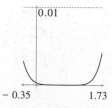

11. $f(x) = T_5(x; \sqrt{2})$ for all real x. Hence the error in the estimate $f(x) \approx T_5(x; \sqrt{2})$ is 0. r can be any positive number.

12. $r \leq 0.48$

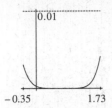

15. $\sqrt{105} \approx 10.2469507659595983832$
$T_7(105; 100) \approx 10.2469507659645080566$
$T_7(105; 100) - \sqrt{105} \approx 4.9096734 \times 10^{-12}$

16. $1/4^8$

17. $3 \cdot 8!/9$

18. $(-1)^k k!/30$

19. $k^k k!/(3k + 3)^{k+1}$

47. $-2x/(1 + x^2)^2$, arctan x

49. $4 \sin 2x \cos 2x$, $(4x - \sin 4x)/8$

51. $(-5 - x^2)/(x^2 - 4x - 5)^2$,
$\dfrac{5}{6} \ln|x - 5| + \dfrac{1}{6} \ln|x + 1| - \dfrac{5}{6} \ln 5$

Section 7.3

1. $2, 4, 8, 16, 32$

2. $1, 2, 7, 16, 29$

3. $3, 3, 3, 3, 3$

4. $4, 5/2, 2, 7/4, 8/5$

5. $\sqrt{2}/2, 0, -\sqrt{2}/2, -1, -\sqrt{2}/2$

6. $1, 3, 6, 10, 15$

7. $-1, -3/4, -85/108, -5413/6912,$
$-16,922,537/21,600,000$

8. $-1, 1, -1$
10th term $= 1$, nth term $= (-1)^n$

9. $3.1415, 3.14159, 3.141592$
10th term $= 3.141592653$,
nth term $= \pi$ to n digits

10. $15, 21, 28$
10th term $= 55$,
nth term $= 1 + 2 + \cdots + n = n(n + 1)/2$

11. $3125, 46,656, 823,543$
10th term $= 10^{10}$, nth term $= n^n$

12. $31/32, 63/64, 127/128$
10th term $= 1023/1024$, nth term $= (2^n - 1)/2^n$

13. $1 + x + x^2 + x^3 + x^4, 1 + x + x^2 + x^3 + x^4 + x^5,$
$1 + x + x^2 + x^3 + x^4 + x^5 + x^6$
10th term $= 1 + x + x^2 + \cdots + x^9 = \displaystyle\sum_{k=0}^{9} x^k$
nth term $= 1 + x + x^2 + \cdots + x^{n-1} = \displaystyle\sum_{k=0}^{n-1} x^k$

14. $1, 3, 6, 10, 15$

15. $1, 1, 2, 6, 24$

16. $2, 1/3, 3/4, 4/7, 7/11$

17. $1, 3/4, 3/5, 1/2, 3/7$

18. $1, 1, 2, 3, 5$

19. $1, 2, 3/2, 7/4, 13/8$

20. $1, 2, \sqrt{2}, \sqrt{2\sqrt{2}}, \sqrt{\sqrt{2}\sqrt{2\sqrt{2}}}$

21. $s_1 = 1$
$s_n = s_{n-1} + \dfrac{1}{n}, \quad n \geq 2$

22. $s_1 = 2$
$s_n = \sqrt{1 + s_{n-1}}, \quad n \geq 2$

23. $s_1 = 2$
$s_n = 2^{s_{n-1}}, \quad n \geq 2$

24. $s_1 = 1$
$s_n = s_{n-1} + x^{n-1}, \quad n \geq 2$

25. Converges, limit is 1

26. Converges, limit is 0

27. Diverges

28. Converges, limit is $1/2$

29. Converges, limit is $\pi/2$

30. Converges, limit is 1

31. Diverges

32. Converges, limit is 1

33. Diverges

34. Converges, limit is e^2

35. Converges, limit is e^2

36. Converges, limit is e^5

63. $y' = 2xe^x + x^2e^x$
$y(0) = 0$

65. $a_0 = -7/4, a_2 = -1/4,$
a_1 can be any real number.

67. ≈ 0.89

Section 7.4

1. terms: $1, 2, 3, 4, 5$
partial sums: $1, 3, 6, 10, 15$

2. terms: $1, -1/2, 1/3, -1/4, 1/5$
partial sums: $1, 1/2, 5/6, 7/12, 47/60$

3. terms: $1, x, 3x^2/4, x^3/2, 5x^4/16$
partial sums: $1, 1 + x, 1 + x + 3x^2/4,$
$1 + x + 3x^2/4 + x^3/2,$
$1 + x + 3x^2/4 + x^3/2 + 5x^4/16$

4. terms: $1/(\ln 2)^2, 1/(\ln 3)^2, 1/(\ln 4)^2,$
$1/(\ln 5)^2, 1/(\ln 6)^2, 5$
partial sums (decimal approx.): 2.08137, 2.9099,
3.43025, 3.8163, 4.12779

5. $s_n = n, \lim\limits_{n\to\infty} s_n = \infty$

6. $s_n = \dfrac{5}{4}\left(1 - \dfrac{1}{5^n}\right), \lim\limits_{n\to\infty} s_n = 5/4$

7. $s_n = \sqrt{n+1} - 1, \lim\limits_{n\to\infty} s_n = \infty$

8. $s_n = 1 - \dfrac{1}{\sqrt[3]{n+1}}, \lim\limits_{n\to\infty} s_n = 1$

9. $s_n = \dfrac{1}{2} - \dfrac{2}{n+4}, \lim\limits_{n\to\infty} s_n = 1/2$

10. $s_n = (1 - (-4)^n)/5, \lim\limits_{n\to\infty} s_n$ does not exist

11. $s_n = (1 - e^{-2n})/(1 - e^{-2}), \lim\limits_{n\to\infty} s_n = e^2/(e^2 - 1)$

12. $s_n = \dfrac{9}{20} - \dfrac{1}{n+4} - \dfrac{1}{n+5}, \lim\limits_{n\to\infty} s_n = \dfrac{9}{20}$

13. $s_n = \sum\limits_{k=2}^{n+1} 5(0.8)^k = 16(1 - 0.8^n), \lim\limits_{n\to\infty} s_n = 16$

14. Diverges
15. Converges
16. Diverges
17. Diverges
18. Diverges
19. Diverges
20. Diverges
21. Converges
22. Converges
23. $5/33$
24. $346/999$
25. $811/990$
26. $233/99900$
27. $9/20$
28. $211/9000$
29. $4/5$
30. ≈ 0.9217
31. No solution
32. $a_1 = 3, a_2 = 1, a_3 = 1, a_k = 1$
33. $a_1 = 5/4, a_2 = -1/20, a_3 = -1/30,$
$a_k = \dfrac{-1}{(k+2)(k+3)}$
34. $a_1 = -1, a_2 = 2, a_3 = -2, a_k = (-1)^k 2$

35. $a_1 = \sin 1, a_2 = \sin 2 - \sin 1, a_3 = \sin 3 - \sin 2,$
$a_k = \sin k - \sin(k-1)$
36. $1 < x < 2$
37. $2 - \sqrt{10} < x < 2 - 2\sqrt{2}$ and
$2 + 2\sqrt{2} < x < 2 + \sqrt{10}$
38. $|x| > \sqrt{2}$
39. $-\dfrac{\pi}{6} + \pi l \le x \le \dfrac{\pi}{6} + \pi l, \quad l = 0, \pm 1, \pm 2, \dots$
57. $a_0 = 17, a_1 = -9, a_2 = 1$
59. $a_0 = 7c - 11c^2, a_1 = 7 - 22c, a_2 = -11$
63. $a = -1, b = 5/4$
65. $\dfrac{(-1)^n n!}{9(x-2)^{n+1}} - \dfrac{(-1)^n n! 2^{n+1}}{9(2x+5)^{n+1}}$

Section 7.5

1. Diverges
2. Converges
3. Diverges
4. Converges
5. Diverges
6. Converges
7. Converges
8. Converges
9. Converges
10. Converges
11. Converges
12. Diverges
13. Converges
14. Converges
15. Converges
16. Absolutely convergent
17. Conditionally convergent
18. Conditionally convergent
19. Neither conditionally convergent nor absolutely convergent
20. Conditionally convergent
21. Absolutely convergent

22. 9 terms, $\sum\limits_{k=1}^{9} (-1)^{k-1} \dfrac{1}{k^2 + 1} \approx 0.369$

23. 99 terms, $\sum\limits_{r=1}^{99} (-1)^{r-1} \dfrac{1}{\sqrt{r^2 + 1}} \approx 0.446$

24. 7 terms, $\sum\limits_{j=0}^{6} (-1)^j \dfrac{1}{4^j} \approx 0.800$

25. 100 terms, $\sum\limits_{k=1}^{100} (-1)^{k-1} \dfrac{1}{2k - 1} \approx 0.783$

26. 11 terms, $\displaystyle\sum_{l=1}^{11}(-1)^{l-1}\frac{1}{(\ln(l+1))^5}\approx 5.766$

27. 10 terms, $\displaystyle\sum_{j=0}^{9}(-1)^j\frac{1}{j!}\approx 0.368$

28. $x=\pm 3/2$

29. $x=\pm 3$

30. $x=-1\pm 5e$

31. All real x

53. $1/(1+t)$

55. $1/\sqrt{1-t^2}$

57. $f''(x)=n(n-1)(x-c)^{n-2}$
$f^{(5)}(x)=n(n-1)(n-2)(n-3)(n-4)(x-c)^{n-5}$
$f^{(k)}(x)=n(n-1)\cdots(n-k+1)(x-c)^{n-k}$

59. $1+3x+\dfrac{9x^2}{2}+\dfrac{27x^3}{6}+\dfrac{81x^4}{24}+\dfrac{243x^5}{120}+\dfrac{729x^6}{720}+\cdots$

61. $\dfrac{1}{10(x-3)}-\dfrac{1}{6(x-1)}+\dfrac{1}{15(x+2)}$

63. $3-x+0x^2+0x^3-x^4-\cdots$

Section 7.6

1. 1

2. 2

3. ∞

4. 1

5. $2e$

6. 5

7. $\displaystyle\sum_{k=0}^{\infty}\frac{2\cdot 3^k x^k}{k!},\quad R=\infty$

8. $1+\dfrac{x-3}{2}$
$\qquad+\displaystyle\sum_{k=2}^{\infty}(-1)^{k-1}\frac{1\cdot 3\cdot 5\cdots(2k-3)}{2^k k!}(x-3)^k,$
$\qquad R=1$

9. $50-47(x+2)+16(x+2)^2-2(x+2)^3,\quad R=\infty$

10. $\displaystyle\sum_{k=0}^{\infty}\left(\frac{(-1)^k}{3^{k+1}}+\frac{1}{2^{k+1}}\right)x^k,\quad R=2$

11. $\displaystyle\sum_{k=0}^{\infty}(-1)^k\left(\frac{1}{8^{k+1}}-\frac{1}{3^{k+1}}\right)(x-5)^k,\quad R=3$

12. $\displaystyle\sum_{k=0}^{\infty}\left(\frac{1}{7^{k+1}}-\frac{1}{2^{k+1}}\right)(x+5)^k,\quad R=2$

33. $a_0=3/4,\ a_1=1,\ a_2=1/2,\ a_3=0$

35. $a=66/545,\ b=-2/545$

39. 3

41. $\displaystyle\sum_{k=0}^{\infty}(k^2+3k)x^k$

Section 7.7

1. $\displaystyle\sum_{k=0}^{\infty}(-1)^k\frac{3^k x^k}{k!},\quad -\infty<x<\infty$

2. $\dfrac{3}{4}\displaystyle\sum_{k=0}^{\infty}(-1)^k\frac{x^k}{4^k},\quad -4<x<4$

3. $\displaystyle\sum_{k=0}^{\infty}(-1)^k\left(\frac{2x^{2k}}{(2k)!}-\frac{3x^{2k+1}}{(2k+1)!}\right),\quad -\infty<x<\infty$

4. $1-\dfrac{x^2}{2}-\displaystyle\sum_{k=2}^{\infty}\frac{1\cdot 3\cdot 5\cdots(2k-3)}{2^k k!}x^{2k},$
$\qquad -1\le x\le 1$

5. $1+\displaystyle\sum_{k=1}^{\infty}(-1)^k\frac{2^{2k-1}x^{2k}}{(2k)!},\quad -\infty<x<\infty$

6. $2x+\displaystyle\sum_{k=1}^{\infty}(-1)^k\left(\frac{2^{2k+1}}{(2k+1)!}-\frac{2^{2k-1}}{(2k-1)!}\right)x^{2k+1},$
$\qquad -\infty<x<\infty$

7. $\dfrac{\sqrt{2}}{2}\displaystyle\sum_{k=0}^{\infty}(-1)^k\left(\frac{(x-\pi/4)^{2k}}{(2k)!}-\frac{(x-\pi/4)^{2k+1}}{(2k+1)!}\right),$
$\qquad -\infty<x<\infty$

8. $1+2\displaystyle\sum_{k=1}^{\infty}(-1)^k\frac{1\cdot 3\cdot 5\cdots(2k-1)}{2^{k+1}k!}x^k,$
$\qquad -1<x\le 1$

9. $1+(x+1)$
$\qquad+\displaystyle\sum_{k=2}^{\infty}(-1)^{k-1}\frac{1\cdot 3\cdot 5\cdots(2k-3)}{k!}(x+1)^k,$
$\qquad -\dfrac{3}{2}\le x\le -\dfrac{1}{2}$

10. $\displaystyle\sum_{k=0}^{\infty}\left(\frac{(-1)^{k-1}}{3\cdot 2^k}-\frac{5}{3}\right)(x-2)^k,\quad 1<x<3$

11. $\ln 2+\displaystyle\sum_{k=1}^{\infty}(-1)^{k-1}\frac{x^k}{k\cdot 2^k},\quad -2<x\le 2$

12. $\sin 1 + (2\sin 1 + \cos 1)(x - 1)$

$$+ \sum_{k=0}^{\infty}(-1)^k\left(\frac{\sin 1}{(2k)!} + \frac{2\cos 1}{(2k+1)!} - \frac{\sin 1}{(2k+2)!}\right)(x-1)^{2k+2}$$

$$+ \sum_{k=0}^{\infty}(-1)^k\left(\frac{\cos 1}{(2k+1)!} - \frac{2\sin 1}{(2k+2)!} - \frac{\cos 1}{(2k+3)!}\right)(x-1)^{2k+3},$$

$-\infty < x < \infty$

13. $1 + 2x + 5x^2/2 + 8x^3/3 + \cdots$

14. $x + x^2/2 - 7x^3/24 - x^4/48 - \cdots$

15. $x + x^3/6 + 3x^5/40 + 5x^7/112 + \cdots$

16. $1 + x + x^2/2 + x^3/2 + \cdots$

41. $a_1 = \dfrac{5}{2}, a_2 = \dfrac{5}{8}, a_3 = \dfrac{5}{48},$

$a_4 = \dfrac{5}{384}, a_n = \dfrac{5}{2^n n!}$

43. $a_2 = \pi/2, a_3 = 0, a_4 = \pi/3,$
$a_5 = 0, a_6 = \pi/4;$
$a_n = 0$ if n is odd,

$a_n = \dfrac{\pi}{n/2+1}$ if n is even

45. $\|\mathbf{v}\| = \sqrt{13}, \|\mathbf{w}\| = 8,$
$\theta = \arccos(-3/\sqrt{13}) \approx 2.554$

47. $(\pi, 1)$

49. $\mathbf{v}(t) = (-10\sin 2t, 10\cos 2t), \|\mathbf{v}(t)\| = 10$

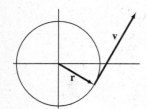

Section 7.8

1. $a_0 = 0, a_1 = 2, a_2 = -3 + \sqrt{2}, a_k = \sqrt{k}, k \geq 3$

2. $a_0 = 1, a_1 = (1!)^2 = 1, a_2 = (2!)^2 = 4,$
$a_k = (k!)^2$

3. $a_k = 0$

4. $a_0 = 2, a_1 = 1/2, a_2 = 2; a_k = 1/2, k$ odd;
$a_k = 2, k$ even

5. $3 + \displaystyle\sum_{k=1}^{\infty}(2k+2)x^k$

6. $-2 + \displaystyle\sum_{k=1}^{\infty}\left(\frac{1}{(k-1)^2+1} - (k+1)(k+2)\right)x^k$

7. $\dfrac{1}{x^2} + \dfrac{2}{x} + \displaystyle\sum_{k=0}^{\infty}3x^k$

8. $\dfrac{2}{x^2} + \dfrac{7}{2x} + \displaystyle\sum_{k=0}^{\infty}5\cdot 2^{k+1}x^k$

9. $y = 2\displaystyle\sum_{k=0}^{\infty}(-1)^k\frac{3^k}{k!}t^k = 2e^{-3t}$

10. $y = -\displaystyle\sum_{k=0}^{\infty}\frac{t^{3k}}{3^k k!} + 2\sum_{k=0}^{\infty}\frac{t^{3k+1}}{1\cdot 4\cdot 7\cdots(3k+1)}$

11. $y = \dfrac{\pi}{2}\displaystyle\sum_{k=0}^{\infty}(-1)^k(k+2)(k+1)t^k$

12. $y = \dfrac{t^2}{2} - \dfrac{t^4}{8} - \displaystyle\sum_{k=3}^{\infty}\frac{1\cdot 3\cdot 5\cdots(2k-5)}{2^k k!}t^{2k}$

29. $5\sqrt{34}$

31. $(d-c)\sqrt{a^2+b^2}$

33. $(2/\sqrt{13}, -3/\sqrt{13})$

35. $\left(\dfrac{-2}{\sqrt{e^{\pi/2}+4}}, \dfrac{e^{\pi/4}}{\sqrt{e^{\pi/2}+4}}\right) \approx (-0.673799, 0.738915)$

Index

Exponents and Radicals

For real numbers r and s:

$$x^r x^s = x^{r+s}$$

$$\frac{x^r}{x^s} = x^{r-s} \quad x \neq 0$$

$$(x^r)^s = x^{rs}$$

$$x^{-r} = \frac{1}{x^r} \quad x \neq 0$$

$$(xy)^m = x^m y^m$$

$$\left(\frac{x}{y}\right)^m = \frac{x^m}{y^m} \quad y \neq 0$$

For positive integers m and n:

$$\sqrt[m]{\sqrt[n]{x}} = \sqrt[mn]{x}$$

$$\sqrt[m]{x}\,\sqrt[m]{y} = \sqrt[m]{xy}$$

$$\frac{\sqrt[m]{x}}{\sqrt[m]{y}} = \sqrt[m]{\frac{x}{y}} \quad y \neq 0$$

The Quadratic Equation

If $ax^2 + bx + c = 0$ and $a \neq 0$

then $\quad x = \dfrac{-b \pm \sqrt{b^2 - 4ac}}{2a}$.

Plane Geometry

Circle

Circumference $= 2\pi r$

Area $= \pi r^2$

Sector

Arc length $= r\theta$

Area $= \frac{1}{2}\theta r^2$

Triangle

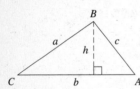

Perimeter $= a + b + c$

Area $= \frac{1}{2}bh = \frac{1}{2}ab \sin C$

Equilateral Triangle

Area $= \dfrac{\sqrt{3}}{4}s^2$

Parallelogram

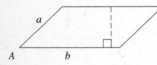

Area $= bh = ab \sin A$

Trapezoid

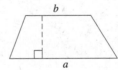

Area $= \frac{1}{2}(a + b)h$

Pythagorean Theorem

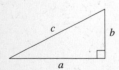

$c^2 = a^2 + b^2$

The Law of Cosines

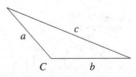

$c^2 = a^2 + b^2 - 2ab \cos C$

Solid Geometry

Sphere

Volume $= \frac{4}{3}\pi r^3$

Surface area $= 4\pi r^2$

Rectangular Parallelepiped

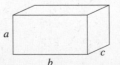

Volume $= abc$

Surface area $= 2ab + 2ac + 2bc$

Cylinder

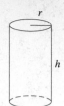

Volume $= \pi r^2 h$
Lateral surface area $= 2\pi rh$

Cone

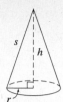

Volume $= \frac{1}{3}\pi r^2 h$
Lateral surface area $= \pi rs$

Trigonometry

Right Triangle

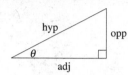

$$\sin\theta = \frac{\text{opp}}{\text{hyp}}, \qquad \cos\theta = \frac{\text{adj}}{\text{hyp}}$$

Unit Circle

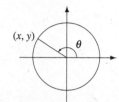

$$\sin\theta = y, \qquad \cos\theta = x$$

Other Trigonometric Functions

$$\tan\theta = \frac{\sin\theta}{\cos\theta}, \qquad \cot\theta = \frac{\cos\theta}{\sin\theta}, \qquad \sec\theta = \frac{1}{\cos\theta}, \qquad \csc\theta = \frac{1}{\sin\theta}$$

30-60-90 Triangle

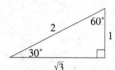

$$\sin 30° = \cos 60° = \frac{1}{2}$$

$$\cos 30° = \sin 60° = \frac{\sqrt{3}}{2}$$

45-45-90 Triangle

$$\sin 45° = \cos 45° = \frac{\sqrt{2}}{2}$$

Radian Measure

$$360° = 2\pi \text{ radians}$$

The Law of Sines

$$\frac{\sin A}{a} = \frac{\sin B}{b} = \frac{\sin C}{c}$$

Identities

$$\cos^2\theta + \sin^2\theta = 1$$
$$\sin(-\theta) = -\sin\theta$$

$$\cos(-\theta) = \cos\theta$$

$$1 + \tan^2\theta = \sec^2\theta$$
$$\tan(-\theta) = -\tan\theta$$

$$\sin(\alpha + \beta) = \sin\alpha\cos\beta + \cos\alpha\sin\beta \qquad \cos(\alpha + \beta) = \cos\alpha\cos\beta - \sin\alpha\sin\beta$$

$$\sin 2\theta = 2\sin\theta\cos\theta \qquad\qquad \cos 2\theta = \cos^2\theta - \sin^2\theta \qquad\qquad \tan 2\theta = \frac{2\tan\theta}{1 - \tan^2\theta}$$

$$\sin\left(\frac{\theta}{2}\right) = \pm\sqrt{\frac{1 - \cos\theta}{2}} \qquad\qquad\qquad\qquad \cos\left(\frac{\theta}{2}\right) = \pm\sqrt{\frac{1 + \cos\theta}{2}}$$

Table of Integrals

(A) $\displaystyle\int x^n\,dx = \frac{x^{n+1}}{n+1} + C, \qquad n \neq -1$

(B) $\displaystyle\int \frac{1}{x}\,dx = \ln|x| + C$

(C) $\displaystyle\int \sin x\,dx = -\cos x + C$

(D) $\displaystyle\int \cos x\,dx = \sin x + C$

(E) $\displaystyle\int e^x\,dx = e^x + C$

(F) $\displaystyle\int a^x\,dx = \frac{a^x}{\ln a} + C$

(G) $\displaystyle\int \frac{dx}{x^2 + a^2} = \frac{1}{a}\arctan\frac{x}{a} + C$

(H) $\displaystyle\int \frac{dx}{\sqrt{a^2 - x^2}} = \arcsin\frac{x}{a} + C$

(1) $\displaystyle\int \sec^2 x\,dx = \tan x + C$

(2) $\displaystyle\int \sec x\tan x\,dx = \sec x + C$

(3) $\displaystyle\int \tan x\,dx = -\ln|\cos x| + C$

(4) $\displaystyle\int \sec x\,dx = \ln|\sec x + \tan x| + C$

(5) $\displaystyle\int \ln x\,dx = x(\ln x - 1) + C$

(6) $\displaystyle\int \frac{dx}{\sqrt{x^2 \pm a^2}} = \ln\left|x + \sqrt{x^2 \pm a^2}\right| + C$

(7) $\displaystyle\int \frac{dx}{x^2 - a^2} = \frac{1}{2a}\ln\left|\frac{x-a}{x+a}\right| + C$

(8) $\displaystyle\int \frac{dx}{x\sqrt{a^2 \pm x^2}} = -\frac{1}{a}\ln\left|\frac{a + \sqrt{a^2 \pm x^2}}{x}\right| + C$

(9) $\displaystyle\int \sqrt{x^2 \pm a^2}\,dx = \frac{x}{2}\sqrt{x^2 \pm a^2} \pm \frac{a^2}{2}\ln\left|x + \sqrt{x^2 \pm a^2}\right| + C$

(10) $\displaystyle\int \frac{\sqrt{x^2 \pm a^2}}{x^2}\,dx = -\frac{\sqrt{x^2 \pm a^2}}{x} + \ln\left|x + \sqrt{x^2 \pm a^2}\right| + C$

(11) $\displaystyle\int \frac{dx}{x\sqrt{x^2 - a^2}} = \frac{1}{a}\operatorname{arcsec}\frac{x}{a} + C$

(12) $\displaystyle\int \sqrt{a^2 - x^2}\,dx = \frac{x}{2}\sqrt{a^2 - x^2} + \frac{a^2}{2}\arcsin\frac{x}{a} + C$

(13) $\displaystyle\int x^2\sqrt{a^2 - x^2}\,dx = \frac{x}{8}(2x^2 - a^2)\sqrt{a^2 - x^2} + \frac{a^4}{8}\arcsin\frac{x}{a} + C$

(14) $\displaystyle\int \frac{\sqrt{a^2 - x^2}}{x^2}\,dx = -\frac{1}{x}\sqrt{a^2 - x^2} - \arcsin\frac{x}{a} + C$

(15) $\displaystyle\int \frac{x^2}{\sqrt{a^2 - x^2}}\,dx = -\frac{x}{2}\sqrt{a^2 - x^2} + \frac{a^2}{2}\arcsin\frac{x}{a} + C$

(16) $\displaystyle\int \frac{\sqrt{a^2 \pm x^2}}{x}\,dx = \sqrt{a^2 \pm x^2} - a\ln\left|\frac{a + \sqrt{a^2 \pm x^2}}{x}\right| + C$